Algorithmic
Foundations
of Robotics

Algorithmic
Foundations
of Robotics

WITHDRAWN

edited by

Ken Goldberg
University of Southern California
Los Angeles, California

Dan Halperin
Stanford University
Stanford, California

Jean-Claude Latombe
Stanford University
Stanford, California

Randall Wilson
Sandia National Laboratories
Alburquerque, New Mexico

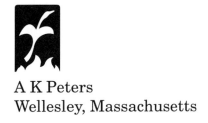

A K Peters
Wellesley, Massachusetts

 94
The Workshop on the Algorithmic
Foundations of Robotics

Editorial, Sales, and Customer Service Office

A K Peters, Ltd.
289 Linden Street
Wellesley, MA 02181

Library of Congress Cataloging-in-Publication Data

Algorithmic foundations of robotics / edited by Ken Goldberg
     p. cm.
    Includes bibliographical references.
    1-56881-045-8
    1. Robotics. 2. Algorithms. I. Goldberg, Ken.
  TJ211.A443 1995
  629.8'92--dc20   95-7181
                                     CIP

Cover illustration: Mechanical drawing of Leonardo da Vinci. From *Leonardo da Vinci*, Cod. Atl., fol. 357 r-a.Reynal and Company, New York, p. 498. Copyright in Italy by the Istituto Geografico De Agostini - Novara - 1956.

Printed in the United States of America
99 98 97 96 95 10 9 8 7 6 5 4 3 2 1

Contents

Foreword

Robotics involves the combined application of several disciplines, such as kinematics, dynamics, control, and programming. Robot *algorithms* are abstractions describing motion and perception acts that, when executed in the physical world, achieve goal arrangements of physical objects. Such algorithms may be implemented in various ways, e.g., as software modules in a programmable robot controller or as pure mechanisms. Their specification, design, and analysis raise a unique combination of basic questions having their roots in computer science and combinatorial geometry, as well as in control theory.

Although interest in robot algorithms has increased in recent years, most of the research on this topic has been presented at conferences and journals that cover a broad range of topics in either robotics or computational geometry. To better understand how algorithms might serve as a coherent and unifying subfield within robotics, we proposed a small workshop to bring together researchers who have been active in this area for several years. We also invited graduate students and representatives from industry.

We did not anticipate that the response would be so enthusiastic. Although the conference was solely funded by participants, fifty-two invited researchers travelled to the historic Queen Anne Hotel in San Francisco, 17-19 February 1994, to attend the Workshop on the Algorithmic Foundations of Robotics (WAFR).

Thirty-seven papers were presented in a single track over three days. Although the topics ranged widely, computational issues such as discretization and complexity served as unifying themes. There was also an expected dichotomy between those researchers who work largely in the area of computational geometry and those from robotics: the age-old tension between mathematical elegance and practical usefulness. The two are rarely achieved simultaneously. Fortunately, the WAFR participants shared an appreciation for both objectives and were eager to learn from the other camp.

Since our aim was to have an impact on the broader community, we sought to publish the papers in book form; A. K. Peters has kindly agreed to serve as publisher. Besides the editors of this volume, the program committee included R. Brost, J. Canny, B. Donald, M. Erdmann, E. Grimson, L. Guibas, S. Hutchinson, L. Joskowicz, J.P. Laumond, M. Mason, B. Mishra, and J. Reif. It was a pleasure to have such a talented and accomplished group in the same room, and we hope to organize another workshop with this theme in the near future. We are grateful to all of the participants listed below for their contributions and good spirits.

Ken Goldberg
Dan Halperin
Jean-Claude Latombe
Randy Wilson

July 1994

Participants

Rachid Alami	LAAS–CNRS
Ruzena Bajcsy	University of Pennsylvania
Saugata Basu	New York University
Pierre Bessière	LIFIA
Jean-Daniel Boissonnat	INRIA
Randy C. Brost	Sandia National Laboratories
John Canny	University of California, Berkeley
Raja Chatila	LAAS–CNRS
Alan Christiansen	Tulane University
Thomas Dean	Brown University
Bruce Donald	Cornell University
Michael Erdmann	Carnegie Mellon University
Ayman O. Farahat	Texas A&M University
Ken Goldberg	University of Southern California
Leonidas J. Guibas	Stanford University
Dan Halperin	Stanford University
Seth Hutchinson	University of Illinois, Urbana-Champaign
Daniel P. Huttenlocher	Cornell University
Leo Joskowicz	IBM T.J. Watson Research Center
Takeo Kanade	Carnegie Mellon University
Lydia Kavraki	Stanford University
Oussama Khatib	Stanford University
Pradeep K. Khosla	Carnegie Mellon University
Daniel E. Koditschek	University of Michigan
Yotto Koga	Stanford University
Krasimir Kolarov	Interval Research
David J. Kriegman	Yale University
Jean-Claude Latombe	Stanford University
Christian Laugier	LIFIA
Jean-Paul Laumond	LAAS–CNRS
Tsai-Yen Li	Stanford University
Kevin M. Lynch	Carnegie Mellon University
Matthew T. Mason	Carnegie Mellon University
Brian Mirtich	University of California, Berkeley
Rajeev Motwani	Stanford University
Balas K. Natarajan	Hewlett Packard Labs
Mark H. Overmars	Utrecht University
Christos Papadimitriou	University of California, San Diego

Christiaan J. J. Paredis   Carnegie Mellon University
Richard Pollack   New York University
Jean Ponce   University of Illinois, Urbana-Champaign
Anil Rao   Utrecht University
John H. Reif   Duke University
Ari Requicha   University of Southern California
Daniela Rus   Cornell University
Micha Sharir   Tel Aviv University
Camillo J. Taylor   Yale University
Marek Teichmann   New York University
Bob Tilove   General Motors
Jeffrey C. Trinkle   Texas A&M University
Hongyan Wang   Duke University
Randall H. Wilson   Sandia National Laboratories

Robot Algorithms

Jean-Claude Latombe, *Stanford University, Stanford, CA, USA*

Robotics is often viewed as the combined application of component disciplines, such as mechanisms, dynamics, control, and programming, to create programmable mechanical systems capable of performing a great variety of tasks in the physical space. This paper takes the stand that, like computer science, robotics is fundamentally about algorithms. Like computer algorithms, robot algorithms are abstract descriptions of processes whose design and analysis are, to some extent, independent of any particular implementation technology. But robot algorithms also differ in significant ways from computer algorithms. While the latter have full control over the data to which they apply, robot algorithms deal with physical objects, which they attempt to control despite, or thanks to, the laws of nature. This leads robot algorithms to blend basic control issues (controllability and observability) and computational issues (calculability and complexity) in a unique fashion. Investigating robot algorithms casts new light on the true nature of robotics: As our understanding of these algorithms matures, robots will more and more appear as machines executing robot algorithms, just as computers are machines executing computation algorithms, independent of any ephemeral technology.

1   What is Robotics?

Robots are versatile mechanical devices (e.g., articulated arms, wheeled vehicles) equipped with actuators and sensors under the control of a computing system. They perform tasks by executing motions in the physical space. This space is populated by various objects and is subject to the laws of nature. A typical task consists of achieving a goal spatial arrangement of physical objects from a given initial arrangement, e.g., assembling a product.

Robots are intended to be programmable. That is, they may perform a variety of tasks by simply changing the program controlling them. This view has led to the development of robot programming languages, such as AL, VAL, LM, RCC, AML, and KAREL to only cite a few (e.g., see [35]). All these languages are essentially classical programming languages (such as C) extended by robot-specific constructs, e.g., geometric data types and MOVE statements. Unfortunately, they only embed a very superficial understanding of how robots execute tasks. In particular, programs in these languages only describe the motions to be performed by the robots. For the most part, the effects of these motions on the physical space are left implicit. Designing such programs takes a long time and debugging them is even more painful. Resulting programs are often ad hoc and unreliable. In industrial settings, reliability is achieved through massive engineering of the environment, rather than by clever programming.

This is only the tip of the iceberg. More generally, robotics has been mainly viewed so far as the combined application of various component disciplines, such as mechanisms, dynamics, control, and programming [17]. This view still prevails at the educational level, where courses and textbooks often introduce robotics by first presenting its technological components, taking the classical manipulator arm as an example: mechanical structure, actuators, sensors, controller; from there, a typical course or textbook moves to analyzing the specifics of the component disciplines in this context; finally, it discusses some application tasks to illustrate the concept of robot programming. [By the way, the first paragraph of this introduction isn't much different!] Most research conference proceedings adopt a similar organization.

Some years ago, an analogous approach was taken to introduce computer science. The technological components (e.g., transistors, core ferrite, digital electronics) of a computer were presented prior to any programming concepts, not to speak about more abstract al-

Algorithmic Foundations of Robotics
1-56881-045-8

1

gorithms. It is now widely recognized that algorithms, along with calculability and complexity issues, form the core of computer science. Algorithms are abstractions that can be described and analyzed independently of any specific computer technology. Analyzing their complexity has enormously contributed to the progress of computer software and hardware technology.

In this paper I take the stand that robotics is also fundamentally about algorithms. Robot algorithms are abstract descriptions of processes consisting of motion and perception actions executed in the physical space. Like their computer counterparts, they can be designed and analyzed independently of their particular implementation. However, they also differ in significant ways from computer algorithms. Indeed, the latter have full control over, and perfect access to the data they use, letting aside a few problems related to finite-precision arithmetics or memory sharing by parallel processes. In contrast, robot algorithms apply to physical objects in the real world, which they attempt to control despite (or thanks to) the fact that these objects are subject to the independent and imperfectly modelled laws of nature. Data acquisition through sensor reading is also subject to errors. Robot algorithms hence raise controllability and observability issues that are classical in control theory, but not present in traditional computer algorithms. The complexity of the physical space and its possible variations within the same robot task also raise combinatorial questions, such as finding a collision-free path among obstacles and matching sensory data against a prior model. Therefore, algorithmic complexity issues that are central to computer science must be addressed in robot algorithms. Such issues are not critical in classical control theory because it deals with processes with no or small combinatorial variations. I believe that this unique blend of control and computational issues is at the core of robotics science. I also think that progress in understanding, designing, and analyzing robot algorithms may have an impact on robotics similar in magnitude to the impact of studying computer algorithms on computing systems.

Below I present what I see as the main characteristics of robot algorithms. I also show how the formal investigation of robot algorithms can yield more reliable and more efficient robot systems, and how it can ultimately encourage the development of new, more appropriate robot technology. This paper does not present any new results. My main goal is to propose a new perspective relating a variety of old and recent results, and to motivate researchers to develop further the core notion of robot algorithm.

By no means does this paper attempt to demonstrate that developing new robotic technology and building robotic systems are not important. Indeed, robot algorithms, technology, and systems will considerably benefit from each other, as their computer counterparts have made progress concurrently and synergistically. The interaction is likely to be even greater in robotics. As I will show in this paper, the design of a robot algorithm is not limited to "programming" per se. For instance, engineering a robot and its environment are an integral part of this design. Moreover, some robot algorithms can be implemented as pure mechanisms. Conversely, many man-made mechanisms can be interpreted as implementations of algorithms. As I said above, robot algorithms propose a unifying perspective over a variety of results. The analytical concepts and tools that derive from this perspective (e.g., correctness and complexity of robot algorithms) should help in motivating and facilitating the development of new robotic technology and systems.

2   Robot Algorithms Control

The primary goal of a robot algorithm is to describe a procedure for controling a subset of the real world – the environment – in order to achieve a given goal (e.g., a spatial arrangement of several objects). The real world is subject to the laws of nature, e.g., gravity, inertia, friction. For that reason, we can regard it as an independent agent performing its own actions (e.g., applying forces). These actions are not under the direct control of the robot algorithm, but they are not arbitrary either. To some extent, they can be modelled, hence predicted. Ignoring them may prevent robot algorithms from achieving their goals. On the other hand, taking advantage of them may sometimes yield simpler algorithms.

The robot is only the *medium* for controlling the physical environment. The robot algorithm specifies the operations to be executed by the robot to control the environment. To perform these operations, the robot needs in turn to be controlled. Most of the research on robot control has addressed this second problem. Instead, a robot algorithm is basically about the former problem, i.e., controlling the environment.

However, the robot is itself an important object of the environment. For example, we don't want it to collide with obstacles. The robot algorithm should therefore control the *relation* between the robot and the rest of the environment, but it does not specify how the robot's actuators are internally controlled. If the robot is a wheeled vehicle whose goal is to dock against a machine, the algorithm should give both a path allowing the robot to roll toward the machine without colliding with other objects and a (possibly sensor-based) motion strategy to achieve the docking relation with enough precision. How the robot wheels are controlled to generate the robot motions is not part of the algorithm. In fact, the robot controller, along with the various sensor modules (e.g., force, vision) available to acquire and preprocess data about the environment, determines a set of primitive commands, the *instruction set* of the robot, with which one can build a robot algorithm. Different robots may have very different instruction sets.

The design of a robot algorithm requires identifying a possibly infinite set of relevant environment states (one being the goal) and selecting robot primitive commands in the available instruction set whose execution will take the environment through a sequence of states ending at the goal.[1] The robot may not be able to perfectly control the environment so that an operation may transform a state into one among several possible states. Incomplete knowledge of the initial state may also yield ambiguity in achieved states. Then the algorithm must contain state-recognition functions. But, because environment sensing is incomplete (e.g., local) and noisy, states are often not directly recognizable (i.e., no sensor or combination of sensors returns a state's identity). Thus, state-recognition functions may be quite complicated. In particular, they may depend on the previous state, the command just executed, and sensing history during the execution of this command [19].

As a first illustration, consider the task of orienting a part P on a horizontal table. This is a classical subtask in industrial part feeding. If a vision system (for instance, a classical 2D binary vision system) is available with the camera looking from above, the robot can use this sensor to measure P's orientation. This yields the following sort of robot algorithm:

1. Determine P's initial orientation using the vision system.
2. Move the gripper to the grasp position of P.
3. Close the gripper (i.e., grasp P).
4. Rotate the gripper by the difference between P's initial and goal orientations.
5. Open the gripper (i.e., ungrasp P).
6. Move the gripper to a resting position.

As expressed above, the algorithm is still incomplete (e.g., it does not specify the grasp position of P), but its underlying principle is nevertheless clear. The states of interest are defined by the orientations of P, though only the initial and goal states are explicitly considered by the algorithm. Step 1 acquires the initial state. Step 4 achieves the goal state. Steps 3, 5, and 6 do not affect the state of the environment (as far as P's orientation is concerned). Step 3 is a prerequisite to executing step 4. Strictly speaking, steps 5 and 6 are not needed. However, if the goal is that P be at some orientation *and* the robot be away from P, then they achieve the second subgoal. The algorithm assumes that the vision system gets the exact initial orientation of P and that the robot's gripper rotates by the exact amount requested. These assumptions cannot be met by any real robot, but, on the other hand, no part orienting task does require that the goal be exactly achieved. Therefore, the goal state (as well as any other state) is a small interval of orientations of P centered at a nominal one. If the vision system and the robot are precise enough, the above algorithm achieves the goal.

Let now P be a $2\frac{1}{2}$D n-sided convex polygon. Another, very different algorithm for orienting P consists of squeezing P between two parallel jaws a few times, with different orientations of the jaws at each time (see Fig. 1). This algorithm does not require any environment sensing. It is based on the following principle developed in [25, 26]: Let P be at an arbitrary initial orientation. If it is squeezed between two translating parallel jaws, it will achieve a new orientation, which necessarily belongs to a set of $2n$ possible orientations determined by the geometry of the polygonal cross-section of the part and the orientation of the jaws.[2]

---

[1] The goal may also be a permanent one.

[2] Actually, this is true only if the frictional forces exerted on P by the table are large relative to the inertial force

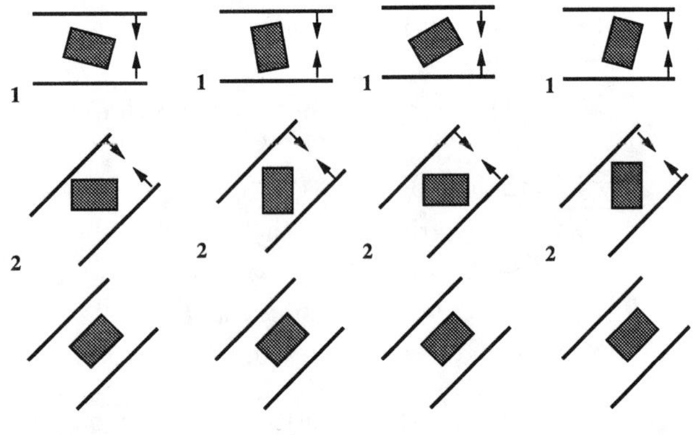

Figure 1: Orienting a convex polygon [25]

Thus, the squeezing operation has reduced the original infinite set of potential orientations of P to a set of $2n$ orientations. If the robot releases the part on the table and squeezes it again with another orientation of the jaws, this finite set of orientations may be reduced further. In fact, it can be shown that, for any n-sided convex polygon P, there exists a sequence of squeezes of length $2n-1$ achieving a single, well-defined orientation of P [25]. From there, a final rotation of the gripper achieves the desired goal orientation of P. The states considered by this algorithm are individual orientations of P and sets of orientations. The state achieved by each squeeze is determined by the jaws' orientation and the previous state. Its prediction is made possible by the fact that we have enough prior understanding of the mechanics of the operation. The result that any convex polygon admits a finite sequence of squeezes ending at a unique orientation guarantees that any goal is reachable from any state, including the one where the orientation can be anyone in $[0, 2\pi)$. For many other tasks, the question of whether the goal state is reachable, or not, is more problematic.

This discussion shows that *the same task can be performed by running very different robot algorithms which may use different instruction sets*. They also illustrate two ways of recognizing a state: the first algorithm uses sensing, the second model-based prediction. But in both cases, state recognition is fairly easy. In many tasks, it is much harder. Consider, for example, a mobile robot navigating in an office environment. A

typical mobile robot controls its motion using dead-reckoning techniques, e.g., by counting the number of turns of the wheels. Not only these techniques yield important cumulative errors, but they also do not apply well when the model of the environment (e.g., object locations) is imprecise. Hence, many robot algorithms sense environment features (e.g., a wall, the corner made by two walls, a door) to localize the robot more precisely. However, some features may be confused with similar features, when they are observed locally (which is all what a sensor can do, not to speak about noise). This may cause major localization mistakes if instantaneous sensing is used exclusively. The designer of the robot algorithm must therefore make sure that, as the robot moves from the initial to the goal state, enough environment features will be sensed along the way to allow each successive state to be recognizable by a state-recognition function taking sensing history into account. The algorithm may then become quite involved.

To illustrate, consider the environment of Fig. 2(a). Obstacles are shown black. The robot is a point. The goal state is defined as the edge G contained in the wall W. The robot is currently located anywhere in the disk I (initial state). A possible algorithm to attain G is to first move perpendicularly to W until the robot touches it. Even in the presence of control errors upper-bounded by the angle θ, the robot is guaranteed to hit W in the region denoted by H. From there it can slide along the wall toward G. The robot is guaranteed to eventually reach G. But can this achievement be reliably recognized? If W contains no marks that can be sensed during sliding, the answer depends on the length of G relative to that of H and on the size of the dead-reckoning localization error along the wall. If H and/or localization errors are too large, it may very well happen that the algorithm stops the robot before it has attained G (wrongly believing that the robot is in G) or that it lets the robot traverse the goal without stopping it (believing that G has not been attained yet). Hence, a reliable robot algorithm must guarantee that the goal be eventually reached *and* recognized.

Assume now that the environment is that of Fig. 2(b). The additional wall W' makes it possible to propose another algorithm: The robot first moves toward W' until it touches it and then slides along it toward W; at the extremity of W', it heads toward W. Despite directional uncertainty θ, the robot is guaran-

---

applying to P, and if the contact between P and at least one jaw is almost frictionless. As shown in [25, 26], these assumptions can easily be achieved.

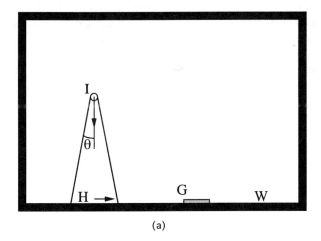

(a)

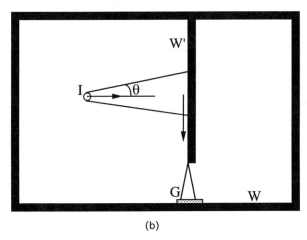

(b)

Figure 2: Goal recognition in mobile robot navigation

teed to be in G when it senses that it has touched W. Thus, this algorithm is guaranteed to both bring the robot to the goal and stop it there.

In summary, robot algorithms control (the environment). As such, they raise fundamental controllability and observability issues, which are classical in control theory [28]: Given a set of states and a set of primitive commands, can the goal state be reached from all possible initial states? If executing a command may yield several possible states, can the attained states be recognized?

The *controllability issue* has recently attracted a lot of attention in connection to nonholonomic robots, e.g., car-like robots, tractor-trailers, flying vehicles, submarines, fingers with spherical tips [4, 38, 41]. For such

robots, the state space (configuration space) is a manifold whose dimension is greater than the one of the control space (a subset of the tangent space). For example, a car-like robot modelled as a two-dimensional object in the plane has a configuration space of dimension 3 (two parameters define the car's position, one defines its orientation). However, if we assume pure rolling contact between the wheels and the ground, the robot's linear velocity is supported by a line whose orientation relative to the robot is fixed. Hence, at any configuration, the control space (the velocity space) is a two-dimensional subspace of the tangent space at this configuration. Bounds on the steering angle further restrict the control space to a subset (still of dimension 2) of this subspace [36]. Controllability questions are: Can the robot span its configuration space? If yes, does this remains true, if that space is punctuated by obstacles? Such questions have been studied for relatively simple robot systems (e.g., see [4]). But they may also arise in a more complicated fashion when physical laws are taken into account. In general, the state space of a robot and its environment can be modelled as a high-dimensional manifold, but the dimension of the set of robot controls available in each state is relatively very small. Is it possible to move in the environment manifold from one given state to another by applying an appropriate sequence of controls? For example, such a question is investigated in [44] in the context of pushing a sliding object, in the presence of frictional forces between the pushing agent and the sliding object and between the sliding object and the ground. But, in general, the question has barely been touched so far.

The *observability issue* is ubiquitous in such tasks as mobile robot navigation and mechanical assembly (part mating), where control and sensing errors can be much larger than the tolerances allowed by the task. In robotics, the two problems of building a controller (i.e., selecting a command to achieve a state) and building an observer (i.e., selecting a function to recognize a state) are generally not independent. The subtle relation between state reachability and state recognizability was first pointed out in [43] and analyzed in greater detail in [19, 55]. As suggested in [19], one way to simplify the problem is to consider state recognition and state achievement in sequence. Let a kernel of a state s be any subset k_s of s such that, if k_s is reached, the achievement of s can be recognized using instantaneous sensing only, hence independently of the way it has

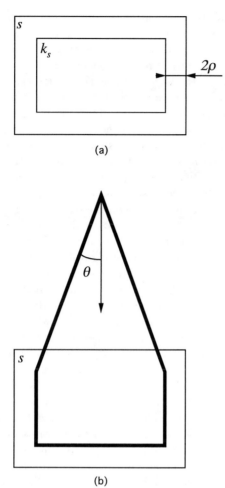

(a)

(b)

Figure 3: Kernel of a Rectangular Goal

been reached. For example, if the goal of a point robot in the plane is to move and stop into a rectangle s and the robot's localization error is bounded by a constant $\rho > 0$, the maximal kernel k_s is obtained by shrinking s by 2ρ as shown in Fig. 3(a). If the command to achieve s from the previous state is guaranteed to reach k_s, it is also guaranteed that the achievement of s will be recognizable (though it will usually be recognized after the robot has penetrated s by some amount). For example, in Fig. 3, if the command to achieve s is to move downward and if the maximal possible directional error in executing this command is θ, s is recognizably achievable from any state contained in the region shown with a bold contour in Fig. 3(b). However, by treating the recognizability and reachability issues sequentially, one

will not be able to build a robot algorithm if the task can only be reliably accomplished by considering the more intricate relation between reachability and recognizability. For instance, in the example of Fig. 3, the kernel vanishes if any side of s is shorter than 4ρ. Nevertheless, it is shown in [55] that rectangles having sides shorter than 4ρ can be reached recognizably from other states by using more involved state-recognition functions.

3 Robot Algorithms Plan

The robot algorithms sketched in the previous section for orienting a part are very simple. They explicitly consider a small number of states of the environment. Selecting commands to go from one state into another, until the goal state is ultimately reached, is rather straightfoward (though the second algorithm, performing successive squeezes, requires non-trivial insight of the task mechanics). Most tasks are not so simple.

Consider the following variant of the part-orienting task. Parts are successively fed with arbitrary orientations on the table by a machine that is not under the robot's control. These parts can have different and arbitrary $2\frac{1}{2}$D convex polygonal geometry, but whenever a part arrives, the feeding machine provides its geometric model to the robot, along with the goal orientation for the part. Assume that the robot has no vision sensor. The iterative squeeze technique can still be used, but now the robot algorithm must embed a planner that computes automatically the sequence of jaws' orientations to attain a unique orientation of the incoming part.

As another example, consider the classical pick-and-place task where a manipulator arm must transport an object from some initial location to a goal one. It may be the case that the obstacles to the arm and the moved object are not known in advance, i.e., when the robot algorithm is designed. The algorithm must then use sensing to detect and localize the obstacles. It also needs a planning component to generate a collision-free path among these obstacles. If the sensors only provide a local (hence, incomplete) map of the obstacles, but move with the robot, the algorithm may have an iterative sense-plan-act loop structure.

Other examples, e.g., mobile robot navigation in indoor or outdoor environments, would make the same point: *For most tasks the set of states that may have to be considered at execution time is too large to be explicitly anticipated.* Hence, the robot algorithm must incorporate a planner. In first approximation, the planner may be regarded as a separate algorithm that automatically generates a control algorithm (i.e., a robot algorithm as presented in the previous section) for achieving a given task. In fact, in many circumstances, the planner must be invoked on-line several times, so that it is an actual component of the robot algorithm. The case where the planner is called only once corresponds to one particular structure of a robot algorithm. A more general structure is one which blends the planning algorithm and the control algorithms (the plans) generated by the planning algorithm.

The role of planning in a robot algorithm is to automatically synthesize control algorithms. Hence, whenever it is invoked, planning must automatically (1) select a set of relevant states, (2) synthesize commands to transform these states into one another until the goal is achieved, and (3) synthesize the needed state-recognition functions. Commands and functions synthesized in (2) and (3) must be constructed using the robot's instruction set. *The effect of planning is to dynamically change the control part of the robot algorithm, by dynamically changing the set of states of the environment that are explicitly considered.*

In general, planning requires exploring combinatorial spaces and thus introducing planning in robot algorithms raises deep computational issues. This is not to say that such issues were totally absent in the previous section (indeed, control algorithms must perform under strict real-time constraints), but the combinatorial issues raised by planning are much more formidable. One basic set of questions is the following: At any one time, can a planner generate a control algorithm whose execution is guaranteed to achieve the goal, if one such algorithm exists given the currently available knowledge? If no such algorithm exists, can a planner generate an algorithm that either succeed or fail, while, if it fails, providing new information pertinent to achieving the goal? If enough information has been acquired to safely conclude that the task is not feasible, can a planner indicate failure? How? At what computational cost?

Over the past years, a variety of motion planning methods have been proposed which address such ques-

tions under more or less restrictive assumptions [36]. Several of these methods are sound and complete under well-defined hypotheses (e.g., robot control is perfect, errors are bounded), i.e., they return a correct control algorithm (one which is guaranteed to succeed if the mentioned assumptions are actually satisfied) if one exists, and failure otherwise. Other methods have weaker formal properties. For example, some planners generate correct control algorithms that either succeed or fail *recognizably* [14]. Others generate randomized control algorithms that only converge toward the goal in probability, with no bound of time [20, 21].

For example, for the polygon-orienting task, there exists a complete planner to generate a sequence of squeezes achieving a single orientation for a given convex polygonal part. This planner takes quadratic time in the number of edges of the part [25]. Unfortunately, most planning problems have higher computational complexity. For example, if perfect robot control and perfect knowledge of the environment geometry are assumed, planning the motion of a manipulator robot between two configurations reduces to computing a collision-free path connecting these configurations. This problem, though simple in apparence, is known to have exponential worst-case time complexity in the number of degrees of freedom of the manipulator [52, 54], yielding complete algorithms with daunting running times even for a classical six-degree-of-freedom arm (and tasks requiring the coordination of several robots are likely to involve many more degrees of freedom). There also exist complete and fairly general planning methods to generate sensor-based motion strategies that are guaranteed to succeed despite bounded errors in robot control and in sensing, but these planners may take double-exponential time in the complexity of the environment (e.g., the number of edges of the objects) [8].

The study of the inherent complexity of motion planning has led to establishing correspondences between robotic and computational problems. By showing that robots deal with physical objects very much like computers deal with data, these correspondences emphasize the profound algorithmic nature of robotics.[3] For example, it has been proven that path planning for a

---

[3] Actually, the calculator of Blaise Pascal and the mechanical computer of Charles Babbage are previous illustrations of this relation.

robot with an arbitrary number of degrees of freedom is PSPACE-hard [52], which is a strong indication that it requires exponential time in the number of degrees of freedom. The proof uses the robot's degrees of freedom to both encode the configuration of a polynomial space bounded Turing machine and design obstacles which force the robot's motions to simulate the computation of this machine. As another example, various versions of the assembly sequencing problem – Given the description of an assembly product, in which order can the constituent parts be assembled? – have been shown NP-complete by reducing in polynomial time any instance of the 3-SAT problem (a well-known NP-complete computational problem [24]) to the geometric description of a mechanical assembly such that the separation of this assembly into two subassemblies gives the solution of the 3-SAT problem instance [29].

Despite the fact that many important motion planning problems have been shown to have high computational complexity, remarkable progress has been made in practical planning over the past few years. Fast path planners have been developed for robots with few degrees of freedom. Probabilistically complete path planners deal with many degrees of freedom, often in reasonable time [3, 30]. Efficient planners for simple nonholonomic robots (e.g., single-body car-like robots) have been developed [2, 37, 39]. Promising methods have been proposed and implemented for manipulation planning [1, 32], assembly planning [56], and motion planning with bounded uncertainty [15, 19, 40]. Furthermore, from a theoretical point of view, the very existence of complete planning methods is an extremely positive achievement, since these methods provide us with a means to decide whether some tasks are feasible, or not.

Computational complexity of robot algorithms has also some interesting relation with robot controllability. For example, consider a classical car-like robot (nonholonomic robot) modelled as a 2D polygon P moving in the plane among stationary polygonal obstacles made of n edges. Such a car is fully controllable [38], i.e.: If there exists a collision-free path for P between any two given configurations, assuming no kinematic constraints, then there also exists a collision-free path for the nonholonomic P. Furthermore, a nonholonomic path can be obtained by introducing a finite number of backup maneuvers in a holonomic path. Hence, deciding if a path exists for the car-like robot has the

same complexity as planning a collision-free path for a polygon translating and rotating freely, i.e. is nearquadratic in n [27]. However, the number of maneuvers needed to transform a holonomic path into a nonholonomic one, hence the computational cost of actually generating a nonholonomic path, is not bounded by any function of n. If the linear velocity of the car has fixed sign (i.e., no backup maneuver is possible) and the turning radius is lower-bounded by a positive constant, the existence of a holonomic path does not imply that a nonholonomic one exists (the robot is not fully controllable). Then deciding whether there exists a collision-free nonholonomic path between two given configurations is significantly harder. See [23], which describes a decision algorithm that takes doubly exponential time in n.

Algorithmic complexity and state observability also interact. There is strong evidence that planning a motion strategy for a robot with a fixed (possibly small) number of degrees of freedom to reliably achieve a goal region in the presence of bounded errors in both control and sensing is intractable [7]. The planner described in [8] takes at least double-exponential time in the complexity of the environment. On the other hand, if sensing is perfect, planning is relatively easy and takes polynomial time [22]. If sensing uncertainty is null in some regions and infinite elsewhere, then planning is also polynomial [40]. Another form of state recognition (sticking) yields a single-exponential planner [15].

The potentially high cost of planning and the fact that it may have to be done on-line raise an additional issue: *A robot algorithm must carefully allocate time between computations aimed at planning and computations aimed at controlling and sensing the environment.* For example, if the environment is changing dynamically (say, under the influence of other agents), spending too much time planning may result in control algorithms that are obsolete at the time they are to be executed; on the other hand, not enough planning may yield irreversible failures (e.g., destroying a resource that will be needed later) that could have been avoided. The problem of allocating time between planning and control is still poorly understood, but it currently attracts considerable interest and several promising ideas have been proposed. For example, it has been suggested to develop planners which return a plan whatever amount of time is allocated to them (*anytime planning*). If more time becomes available later, they can be called back to *incrementally* improve the previous plan within the new interval of time that is allocated to them [5]. Effective incremental methods have been proposed, which can update a plan at marginal (though perhaps not constant) computational cost when new information becomes available about the environment. Opportunistic planners that avoid large precomputations before actually searching for a motion plan (e.g., [3, 9]) are closely related. Deliberative techniques have also been developed to decide which amount of time should be given to each type of computation (planning and control) and update this decision as more information is collected. This trend results in robot algorithms where planning and control are increasingly intermingled [11]. Techniques are also being developed to evaluate the *competitiveness* of online algorithms, by comparing their efficiency to that of algorithms having access to all relevant data ahead of time [13, 51].

In summary, most tasks require robot algorithms to incorporate a planning component capable of dynamically changing the portion of algorithm controlling the environment. This need raises two sorts of computational problems. One issue is related to planning itself: How computationally complex is it? How can we design efficient planners with good formal properties (forms of soundness and completeness)? The other issue is: How to balance time between planning and control? According to which criterion?

4 Robot Algorithms Reason About Geometry

Imagine a robot whose task is to autonomously maintain a botanic garden. To set and update its goal agenda, this robot needs knowledge in domains like botany, climatology, pest control, and fertilization. However, the algorithms using this knowledge can barely be considered parts of a robot algorithm. Otherwise, almost anything would become robotics. On the other hand, all robots, including gardener robots, accomplish tasks by eventually moving objects (including themselves) in the real world. Hence, at some point, all robots must reason about the geometry of their environment, i.e., the shape of the physical objects and their spatial relationships. Actually, geometry is not enough, since objects have mass inducing gravitational

and inertial forces, while contacts between objects generate frictional forces. All robots must therefore reason with classical mechanics. However, usual concepts of mechanics translate into geometric constructs (e.g., forces are represented as vectors in some appropriate space), so that most of the reasoning of a robot eventually involves dealing with geometry.

Computing with continuous geometric models raises very basic discretization issues. To illustrate, consider the problem of planning a collision-free path for a robot. A planner must first discretize the continuous free space (the set of collision-free configurations of the robot) into a graph to which well-known search algorithms can be applied. Various discretization approaches are possible. For example, one can blindly place a regular grid across configuration space and search that grid for a sequence of adjacent points in free space. The grid is just a computational tool and has no physical meaning. Its resolution is arbitrarily chosen despite its critical role in the computation: if it is too coarse, planning is likely to fail; if it is too fine, planning will take too much time.

Instead, *criticality-driven discretizations* have been proposed, whose underlying principle is widely applicable. They consists of partitioning the continuous space of interest into regions, called cells, such that some pertinent property remains invariant over each cell. This property changes when the boundary (criticality) separating two cells is crossed. For example, the algorithm for orienting a convex polygon is based on such a discretization. The set of all possible orientations of the part is represented as the unit circle S^1 (i.e., the cyclic interval $[0, 2\pi)$). For a given orientation of the gripper, this circle can be partitioned into cells (open arcs and points) such that, for all initial orientations of the part in the same cell, the part's final orientation will be the same after the gripper has closed its jaws. The final orientation is the invariant associated with the cell. From this decomposition, it is a relatively simple matter to plan a squeezing sequence.

Consider path planning. Several criticality-driven discretizations of the robot's free space have been proposed. Here, a cell is typically a connected region in free space such that a path connecting any two configurations in this region can be computed in constant time [53, 54]. The partitioning of the free space yields a connectivity graph representing the adjacency rela-

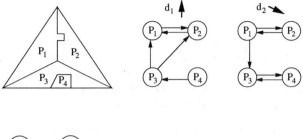

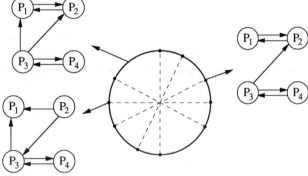

Figure 4: Criticality-based decomposition induced by the blocking structure of an assembly

tion between the cells. Path planning is thus reduced to searching this graph for a path between the node corresponding to the cell containing the robot's current configuration and the node corresponding to the cell containing the goal configuration. The boundary between two adjacent cells is a critical surface (a curve in two dimensions, a hypersurface in dimensions greater than three) where the constraints on robot motion change discontinuously, e.g., contact with a certain edge is possible of one side, but not on the other.

As another example consider assembly planning. A basic computation step for an assembly planner is to partition a given assembly A into two proper subassemblies S and $A \backslash S$ such that S can be moved away from $A \backslash S$ as a rigid body without disturbing $A \backslash S$. (The reverse operation, assembling $A \backslash S$ and S into A is then feasible.) The blocking graph G of A for some direction d is defined as follows [56]: The nodes of G are the parts composing A; an arc connects part P to part P' iff an arbitrarily small translation of P leads the interiors of P and P' to overlap. A necessary condition for being able to partition the assembly A into two subassemblies by a single translation is that G be not strongly connected. While the set of directions d is continuous

(the unit circle S^1 in 2D and the unit sphere S^2 in 3D), it can be decomposed into cells such that the blocking graph remains constant in each cell [56, 57]. The blocking graph changes only when the boundary between two cells is crossed. Hence, for the purpose of computing an assembly sequence, it is sufficient to consider only one direction per cell. Fig. 4 shows the decomposition of S^1 for a 4-part planar polygonal assembly and some of the associated blocking graphs. This decomposition can be extended to infinite translations as well as to more complicated motions combining translation and rotation [58]. It can be computed in polynomial time and exploited to produce assembly sequences also in polynomial time [56].

Here, it is interesting to notice the difference between motion planning as used in robot algorithms and planning methods developed in Artificial Intelligence (and often illustrated with robot examples) [49]. The latter take the implicit description of a discrete space as input. This description is defined by both an initial state and operators to generate new states. Most of AI planning thus focuses on finding smart ways to search this discrete space. In contrast, the key issue in motion planning is the discretization of some continuous geometric space into the smallest possible graph.

Criticality-based discretization techniques are the basis of a variety of sound and complete motion planning methods. Other such discretization techniques are used in connection to sensing. The aspect graph of an object is one of them [33]: When the viewpoint of an object changes continuously, the topology of the visible contours of this object remains constant, except at critical viewpoints partitioning the manifold of all viewpoints into cells. Similarly, the set of physical features that can be detected by the sensors of a mobile robot partition the configuration space of this robot into cells such that over any given cell, the same features are visible to the sensors.

When there are errors in control and in sensing, criticality-based discretization techniques can become quite involved. Then a state-recognition function F partitions the free space of a robot into cells of three types: those where F will certainly evaluate to true, despite various possible errors (e.g., errors in sensing); those where it will certainly evaluate to false; and those where it may evaluate to true or false depending on the actual errors. The robot algorithm must associate motion commands and state-recognition functions such that no motion command is ever executed if it may lead the robot to traverse a cell of the third type. This yields the notion of a preimage [43]: The preimage of a goal for a given motion command and its associated state-recognition function is the set of all robot's configurations such that, if the robot starts executing the motion command at any of these points, it is guaranteed to reach the goal and the state-recognition function will reliably recognize goal achievement. While the shape of a preimage changes continuously when one slightly varies parameters in the motion command and the state-recognition function (e.g., the commanded direction of motion), the topology of its boundary remains constant, except at critical values of the parameters. This idea is the basis for several planners dealing with uncertainty [15, 40]. Other criticalities are used in [16] to partition the robot's configuration space into perceptual equivalence classes, yielding the automatic construction of recognizable states.

Criticality-based discretization originated many algorithmic techniques in robotics. Several of them are based on previous work in computational geometry [18], e.g.: plane-sweep, topological sweep, computing over arrangements, visibility analysis, etc. Although robotics and computational geometry have several interests in common, there are some differences between them. Robot algorithms tend to operate in non-Euclidean large-dimensional spaces (manifolds and their tangent spaces). They are more likely to involve time and uncertainty. As new information may become available while they are being executed, robot algorithms tend to operate incrementally, in order to avoid recomputing everything whenever a new relevant piece of information is obtained. But these differences are mostly nuances that will vanish as time passes.

Robot algorithms often require dealing with high-dimensional spaces. Although criticality-based discretization methods apply to such spaces in theory (for instance, see [54]), their computational complexity is then overwhelming. This has led to the development of Monte-Carlo-like techniques that efficiently approximate the topology and geometry of such spaces. See [3, 30, 50]. These techniques are still empirical, but the approach they suggest might well be the only practical one to deal with high-dimensional spaces.

In summary, all robots end up moving physical objects and sensing their physical environment. Hence, robot algorithms must reason about the continuous geometry of the robots' environment (as well as about mechanical laws). The key step in such reasoning is to discretize a continuous geometrical space into cells over which some interesting property remains invariant.

5 Robot Algorithms Are Abstractions

A robot algorithm is eventually expressed in a formal language. Thus, it can be submitted to some mathematical analysis. The following question then arises: Is it correct? In other words: Will it guarantee the robot to achieve the goal? Sometimes this question may be too demanding, that is, its answer be a swift 'no'. Then one may consider a weaker, but still useful, notion of correctness, e.g.: What is the probability of success of the algorithm? In any case, if nothing, or too little, can be said about the chances of success or failure of an algorithm, this algorithm is of little use, if any. Of course, one can also experimentally evaluate an algorithm by actually running it on sample tasks, but the real world is full of surprises. For many tasks, pure experimental analysis turns into a long and painful experience with no clearcut outcome.

On the other hand, many people (mostly researchers) often expect too much from robot algorithms. They typically want the robot to face any sort of situation in a reasonable manner. Such a goal is hopeless. Indeed, robot algorithms cannot deal with all aspects of the real world. For one reason, even the most complex models of the physical world cannot be perfectly accurate. For another, increasing model complexity tends to increase the complexity of the algorithms even faster, especially the complexity of their planning components. Robot algorithms are necessarily based on some abstraction of the behavior of the robot and the physical environment.

Abstraction is the result of making assumptions. These assumptions should be made as explicit as possible when one designs a robot algorithm. For example, one may assume that robot control is perfect, or that control errors are bounded, or that they have Gaussian distribution. A robot algorithm must be correct (for some formal definition of what 'correct' means, ideally the strongest one) under the assumptions made. For example, if a robot arm has to move between two configurations, a collision-free path computed using some geometric model of the environment is a correct algorithm under the rather stringent assumptions that robot control is perfect and the environment model is complete and accurate. Perhaps the same algorithm may remain correct under weaker assumptions, e.g.: if control errors are small, a collision-free path achieving sufficient clearance with obstacles is still correct. However, relaxing assumptions further will eventually make the algorithm incorrect. For example, if the model of the environment used by the planner is incomplete, there may be unexpected obstacles in the environment; then the algorithm must be changed to include sensor readings to detect them. If some obstacles obstruct the path, the algorithm must amend the path possibly by calling its path planner again, with an updated model of the environment.

Research on autonomous robots is aimed at making assumptions on the robot and its environment as mild as possible. But this does not mean that robot tasks should be too under-specified. As assumptions become less severe, robot algorithms achieving some form of correctness tend to become more complicated. For example, allowing errors in robot control and environment sensing yields robot algorithms with sophisticated state-recognition functions. Furthermore, the embedded planner must usually anticipate several courses of events to generate control programs robust enough to avoid too frequent replanning. At some point, the conceptual and computational complexities of the robot algorithm may become too high, requiring that either assumptions be made more stringent, or that the definition of correctness be made weaker. In both cases, something is lost. On the other hand, the literature is filled with robot systems whose design does not make clear assumptions. Such systems may work beautifully on some tasks, but fail miserably on some others. Who knows in advance? The issue is not: Can we avoid making assumptions?, but: What assumptions are the most appropriate considering the task domain of the robot?

Making assumptions requires enforcing them, if they are not naturally satisfied. This is done by *engineering* the environment and/or the robot accordingly. *Choosing a robot arm able to position its end-effectors with a given accuracy, deciding to equip a mobile robot with a certain type of range sensor, and installing infrared beacons in the environment are all engineering decisions aimed at making sure that some assumptions* made in a robot algorithm are satisfied. When the robot's workspace cannot be engineered (e.g., in some hostile or distant environments), engineering focuses on equipping the robot with appropriate sensors and effectors. *There is a strong dependence between the complexity of an algorithm and the amount (hence, the cost) of the engineering work this algorithm requires to operate reliably.* This dependence has been noticed for a long time (for example, we know that guiding mobile robots by tracking wires buried in the ground makes navigation algorithms simpler), but it is still poorly understood. As we will see in the next section, recent formal results cast new light on this relation.

Notice that the need for experimenting with a robot algorithm does not disappear after it has been proven correct under well-stated assumptions. However, *the purpose of the experimentation shifts from demonstrating that the algorithm works as expected on a sample of tasks, to verifying that the amount of engineering induced by the assumptions is acceptable.* Hopefully, in the future, we will understand better than we do today how assumptions translate into engineering, and conversely. Some sets of assumptions will be known in advance to require reduced engineering costs, while others will be known to be impractical. New technologies in domains like sensing will make new assumptions practical and other assumptions less interesting.

Since robot algorithms are abstractions, they can be designed hierarchically, that is, at successive levels of abstraction. At high levels of abstraction, one may use a very simple and general instruction set (specifying a 'virtual' robot), which can be implemented using different robot technologies. Comparing algorithms at such a level of abstraction may help not only design algorithms at lower levels of abstraction, but also choose the most appropriate robot technologies for the task at hand. This is particularly important if the task (or variants of the task) will be performed many times. For example, consider the task of assembling a mechanical product (e.g., a CD player). At a high level of abstraction, one may totally ignore the robot by assuming that the constituent parts of the product are free-flying, massless geometric objects with accurate dimensions. An algorithm at this level is an assembly sequence defining the order in which the parts can be put together without collisions. At a lower level, one may introduce non-geometric properties of the parts (e.g., mass) and dimensional uncertainties. At an even lower level, the algorithm may deal with the fact that parts need to be grasped before they are moved, requiring that some faces be accessible by the grippers. The actual structure of the manipulator(s) can be introduced only at the lowest level of abstraction. In fact, the same assembly sequence can be expanded in various robot algorithms executable by very different robots, e.g., a six-degree-of-freedom revolute-joint arm or a combination of simpler robots having few degrees of freedom each.

While the robot algorithm is being refined, decisions can be made about the robotic system that will execute it. If many instances of the same product have to be manufactured, it may be more efficient to end up with a hardwired automation system. In fact, just as some computer algorithms are more usefully implemented in the form of a VLSI chip rather than in software, *it may be advantageous to implement certain robot algorithms as hardwired automation.* For example, the iterative squeeze algorithm to orient a polygon does not require a manipulator arm; it can be easily translated into specific mechanical hardware. Similarly, vibratory bowls used in industry to feed parts at a given position and orientation are specific mechanical implementations of randomized robot algorithms for orienting parts (but a new bowl must be designed for each new part). The RCC wrist (remote compliance center) used to insert pegs into holes despite uncertainty [12] is another example of a pure mechanical implemention of a robot algorithm. Again, some flexibility is lost, since changes in the geometry of the peg may require designing a different wrist. At another extreme, robot algorithms may also end up being executed by humans.

In summary, a robot algorithm is necessarily based on some abstract view of the physical environment. This has two major implications. Firstly, robot algorithms can be made formally correct by making explicit assumptions about the robot and the environment, and by enforcing these assumptions through appropriate engineering. Secondly, robot algorithms can be designed at successive levels of detail, allowing robot systems to be designed concurrently.

6 Robot Algorithms Have Physical Complexity

We saw that robot algorithms raise important computational complexity issues. They also have physi-

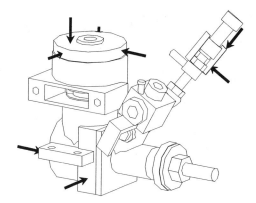

Figure 5: Grasping an assembly

cal complexity. Just as we define the complexity of a computation by the amount of time and memory it requires, we can define the physical complexity of a robot algorithm by the amount of physical resources it takes, e.g., the number of hands, motions, beacons. Some resources, like the number of motions, relate to the time spent executing the algorithm. Others, like the number of hands or beacons, relate to the engineering cost induced by the algorithm.

For example, one complexity measure of the algorithm orienting a convex polygon by squeezing it at successive orientations is the maximal number of squeeze operations this algorithm must perform. As another example, consider an assembly operation merging several parts into a subassembly. The number of subsets of parts moving relative to each other (all parts in the same subset keep a constant relative position) measures the number of grippers necessary to hold these subsets during the operation. All grippers, except one, must move during the operation. The number of grippers (or hands) required by an assembly sequence is the maximal number of hands needed for its various operations. It is a lower-bound on the number of grippers that a robotic system must have to execute this assembly sequence. The number of fingers to safely grasp or fixture an object (e.g., to generate form-closure grasps) is another complexity measure. There has been several studies on this topic for a rigid object [34, 47, 45]. The problem is not as well-understood when the object is an assembly of parts not rigidly attached to each other. For example, Fig. 5 shows positions of frictionless fingers on an assembly (a model-aircraft engine) such that no single subassembly can translate away [57].

Though there is a strong conceptual analogy between computational and physical complexities, there are also major differences between the two notions:

• Physical complexity must be measured along many more dimensions than computational complexity. The number of hands, fingers, and motions are only three possible dimensions. Others include the number of degrees of freedom, the type and precision of sensors, the number of environmental features to be used by sensors, etc.

• Computational complexity typically measures an asymptotic trend, that is, the order of magnitude of the time or space needed by an algorithm when the number of data to which this algorithm applies grows very large. A tighter evaluation is needed for physical complexity since robot tasks involve relatively few objects.

For example, estimating the number of infra-red beacons needed by an indoor navigation algorithm as being proportional to, say, the logarithm of the number of rooms in the environment when this number grows very large is probably not very useful. Indeed, this number may not be greater than a few dozens, so that proportionality constants are likely to be more important than the asymptotic trend.

Evaluating the physical complexity of robot algorithms is important to compare them. It also yields the notion of the *inherent physical complexity of a task*, a notion analogous to the inherent complexity of a computational problem. For example, to orient a convex polygon with n edges, $2n - 1$ squeezes are needed in the worst-case. This means that no correct algorithm can perform better in all cases; but, of course, there may exist non-optimal correct algorithms that execute more squeezes in all cases. Consider the task of assembling a product. At a high level of abstraction, one may assume that all components of the product are free-flying geometric objects. By generating all feasible assembly sequences, one could determine the number of hands needed by each sequence and return the smallest number. This number is a measure of the inherent complexity of assembling the product. It is a lower bound on the complexity of any robot algorithm for assembling the product, i.e., no particular assembly algorithm can require less hands.

Evaluating the inherent physical complexity of a task may lead to redefining the task, if it turns out to be too complex. (In a similar way, knowing that a computational problem is inherently intractable often leads to considering more specific versions of this problem.) For example, it has been shown that a product made of n parts may need up to n hands for its assembly [48], hence requiring the delicate coordination of $n - 1$ motions. Perhaps a product whose assembly requires several hands could be redesigned so that two hands are sufficient, as is the case for most industrial products. Actually, designers of industrial products try to reduce physical complexity along various dimensions. For instance, many mass-produced devices are designed so that they can be assembled with translations only, along very few directions (possibly a single one). A tool that would automatically evaluate the inherent physical complexity of a product while it is being designed would be very useful to help designers create products that are easier to manufacture, service, and recycle.

Determining the complexity of a computational problem is known to be a delicate problem in computer science. Much less is known about how to evaluate the physical complexity of a robot task. Generating all possible algorithms and comparing their respective complexities is clearly not feasible in general. Nevertheless, several practical complexity measures have been defined and are in use in industry [6]. Moreover, limited progress has been made recently to automatically estimate the complexity of assembly tasks using the blocking relation illustrated in Fig. 4. The cell decomposition (of the set of motion directions) and the corresponding blocking graphs form an implicit representation of all assembly sequences (for some family of motions). As shown in [57], it can be efficiently used, not only to generate particular assembly sequences, but also to estimate complexity measures, despite the fact that in the worst case the number of valid assembly sequences is exponential in the number of parts composing an assembly.

The inherent physical complexity of a robot task is certainly not a completely new concept, but its formal application to task analysis is recent and still little developed. Nevertheless, its potential impact is considerable. Computational complexity theory has enormously contributed to the progress of computer software and hardware technology. It has been instrumental to assess the performance of particular algorithms, leading to both very fast specialized hardware modules and efficient software packages capable of solving problems orders of magnitude larger than those previously solved. It has yielded the fundamental distinction between tractable and intractable problems, which is now key to the development of many computer systems. It has also led to identifying basic computational problems such that if a better algorithm is ever produced for one of them, a broad range of applications immediately benefits from this improvement [24]. In areas such as computer graphics, some algorithms have been turned into new technology to achieve levels of performance that could not be attained by pure software. Physical complexity measures could have a similar impact on robot algorithms, bringing improvements that would outpace any advances in the underlying technologies themselves. It may also contribute to the emergence of new robot technologies where a large number of very simple, massively produced actuators and sensors are combined together in highly modular robotic systems. The RISC approach to robotics is an illustration of this trend [10]. Modular robots are another illustration [31, 59]. Such systems will be significantly more reliable and more easily reconfigurable to deal with changes in the tasks. As this trend develops, future robots may look very different from those of today.

An interesting issue is how the computational and the physical complexities of a robot algorithm relate to each other. For example, planning for mobile robot navigation with bounded errors in control and sensing requires exponential time in the complexity of the environment. On the other hand, burying wires in the ground or placing enough infra-red beacons allows robots with appropriate sensors to navigate reliably at small computational cost. In fact, such engineering eliminates the source of the intractability of robot navigation. In other words, assumptions under which robot navigation becomes a relatively simple computational problems have been enforced. But isn't it too much? The intractability of planning with errors in control and sensing can perhaps be eliminated with less stringent assumptions requiring less costly engineering. Along this line, it has been recently possible to define a formal type of landmark (environment feature) such that, if several such landmarks are distributed in the environment, planning for mobile robot navigation is reduced to a polynomial-time computation [40]. Creating such landmarks requires much less engineering than burying wires, while providing more navigation flexibility.

In summary, robot algorithms have both computational and physical complexity. Evaluating physical complexity allows comparison of several alternative algorithms. It also yields the important concept of the inherent physical complexity of a task. Understanding the relation between computational and physical complexities is key to designing computationally efficient robot algorithms while minimizing engineering costs.

7 Conclusion

Robots carry out tasks by executing algorithms. The efficiency and reliability of robot systems critically depend on the properties of these algorithms. In contrast, robot technology appears a secondary (but still important) issue. Indeed, robot algorithms can be ultimately implemented in different ways, for instance as software modules for very versatile mechanical devices massively equipped with sensors, or as sensorless hardwired automation. Robot algorithms may even help select appropriate robot technology. *They also have the potential to considerably influence the instruction sets of future robots*, helping defining minimal instruction sets for classical task domains and benefiting from such sets in return. They will probably influence other domains (e.g., graphics).

The interest in robot algorithms is not new, but it mainly arised during the past decade. As mentioned previously, it was already present in the machines of Pascal and Babbage. However, the current interest derives from more contemporary work, in particular the work of Reif [52], who established the first mapping between a Turing machine and a geometric robot, Lozano-Pérez [42], who popularized the concept of a configuration space in Robotics, Schwartz and Sharir [53, 54], who introduced criticality-based decomposition techniques for path planning, and Mason [46], who emphasized the role of task mechanics in robot algorithms. Originally, this trend was often seen as very narrow, since it mainly focused on geometric path planning with complete prior models. Over the years, new issues related to on-line execution, controllability, incompleteness of knowledge, uncertainty in control and sensing, multi-robot coordination, etc, have been addressed.

As our understanding of robot algorithms matures further, their impact on robotics will grow. Robotics will more and more appear as the discipline whose core is the design, analysis, and implementation of robot algorithms. And robots will simply become machines that execute robot algorithms, just as computers are machines executing computation algorithms, independent of any ephemeral technology.

8 Acknowledgments

The author is supported in part by ARPA grant N00014-92-J-1809, NSF grant IRI-9306544-001, and a grant from the Stanford Integrated Manufacturing Association (SIMA). This paper was motivated by Bruce Donald's review of the author's book "Robot Motion Planning" [17]. The ideas presented in this paper have benefited from old and recent discussions with many people, including Rachid Alami, Jérôme Barraquand, Bruce Donald, Mike Erdmann, Malik Ghallab, Ken Goldberg, Jose-Luis Gordillo, Leo Guibas, Danny Halperin, Lydia Kavraki, Oussama Khatib, Yoshihito Koga, Christian Laugier, Jean-Paul Laumond, Anthony Lazanas, Tsai-Yen Li, Emmanuel Mazer, Rajeev Motwani, Shashank Shekhar, Brian Subirana, Jocelyne Troccaz, Randy Wilson, and Mark Yim.

References

[1] R. Alami, T. Siméon, and J.P. Laumond. A Geometrical Approach to Planning Manipulation Tasks. The Case of Discrete Placements and Grasps, *Robotics Research 5*, H. Miura and S. Arimoto (eds.), MIT Press, Cambridge, MA, 1990, pp. 453-463.

[2] J. Barraquand and J.C. Latombe. On Nonholonomic Mobile Robots and Optimal Maneuvering, *Revue d'Intelligence Artificielle*, 3(2), Hermes, Paris, 1989, pp. 77-103.

[3] J. Barraquand and J.C. Latombe. Robot Motion Planning: A Distributed Representation Approach, *Int. J. of Robotics Research*, 10(6), 1991, pp. 628-649.

[4] J. Barraquand and J.C. Latombe. Nonholonomic Multibody Mobile Robots: Controllability and Motion Planning in the Presence of Obstacles, *Algorithmica*, 10(2-3-4), Aug.-Sept.-Oct. 1993, pp. 121-155.

[5] M. Boddy and T.L. Dean. Solving Time-Dependent Planning Problems, *Proc. 11th Int. Joint Conf. on Artificial Intelligence*, 1989, pp. 979-984.

[6] G. Boothroyd. *Assembly Automation and Product Design*, Marcel Dekker, Inc., New York, NY, 1991.

[7] J.F. Canny and J. Reif. New Lower Bound Techniques for Robot Motion Planning Problems, *Proc. Ann. Symp. on Foundations of Comp. Sci.*, Los Angeles, CA, 1987, pp. 49-60.

[8] J.F. Canny. On Computability of Fine Motion Plans, *Proc. IEEE Int. Conf. on Robotics and Automation*, Scottsdale, AZ, 1989, pp. 177-182.

[9] J.F. Canny and M.C. Lin. An Opportunistic Global Path Planner, *Proc. IEEE Int. Conf. of Robotics and Automation*, Cincinnati, OH, 1990, pp. 1554-1559.

[10] J.F. Canny and K.Y. Goldberg. *A "RISC" Paradigm for Industrial Robotics*, Tech. Rep. No. ESRC 93-4, Dept. of Electrical Engineering, U.C. Berkeley, 1993.

[11] T.L. Dean and M.P. Wellman. *Planning and Control*, Morgan Kaufmann Publishers, San Mateo, CA, 1991.

[12] T.L. De Fazio, D.S. Seltzer, and D.E. Whitney. The Instrumented Remote Center Compliance, *The Industrial Robot*, 11(4), 1984, pp. 238-242.

[13] X. Deng, T. Kameda, and C.H. Papadimitriou. How to Learn an Unknown Environment, *Proc. 32nd Ann. Symp. on Foundations of Comp. Sci.*, Oct. 1991. pp. 298-303,

[14] B.R. Donald. A Geometric Approach to Error Detection and Recovery for Robot Motion Planning with Uncertainty, *Artificial Intelligence J.*, 37(1-3), 1988, pp. 223-271.

[15] B.R. Donald. The Complexity of Planar Compliant Motion Planning Under Uncertainty, *Algorithmica*, 5, 1990, pp. 353-382.

[16] B.R. Donald and J. Jennings. Sensor Interpretation and Task-Directed Planning Using Perceptual Equivalence Classes, *Proc. IEEE Int. Conf. on Robotics and Automation*, Sacramento, CA, 1991, pp. 190-197.

[17] B.R. Donald. Review of [36], *IEEE Tr. on Robotics and Automation*, 8(2), 1992, pp. 290-292.

[18] H. Edelsbrunner. *Algorithms in Combinatorial Geometry*, Springer-Verlag, New York, NY, 1987.

[19] M. Erdmann. *On Motion Planning with Uncertainty*, Tech. Rep. 810, AI Lab., MIT, Cambridge, MA, 1984.

[20] M. Erdmann. *On Probabilistic Strategies for Robot Tasks*, Ph.D. Dissertation, Tech. Rep. 1155, AI Lab., MIT, Cambridge, MA, 1990.

[21] M. Erdmann. Randomization in Robot Tasks. *Int. J. of Robotics Research*, 11(5), 1992, pp. 399-436.

[22] M. Erdmann. Action Subservient Sensing and Design, *Proc. IEEE Int. Conf. Robotics and Automation*, Atlanta, GA, 1993, Vol. 2, pp. 592-598.

[23] S. Fortune and G. Wilfong. Planning Constrained Motions, *Proc. STOCS*, Chicago, 1988, pp. 445-459.

[24] M.R. Garey and D.S. Johnson. *Computers and Intractability. A Guide to the Theory of NP-Completeness*, W.H. Freeman and Co., New York, NY, 1979.

[25] K.Y. Goldberg. Orienting Polygonal Parts without Sensors, *Algorithmica*, 10(2-3-4), Aug.-Sept.-Oct. 1993.

[26] K.Y. Goldberg. Feeding and Sorting Algorithms for the Parallel-Jaw Gripper, *Preprints of the 6th Int. Symp. on Robotics Research*, Hidden Valley, PA, Oct. 1993.

[27] D. Halperin and M. Sharir. Near-Quadratic Bounds for the Motion Planning Problem for a Polygon in a Polygonal Environment, *Proc. FOCS*, 1993.

[28] A. Isodori. *Nonlinear Control Systems*, 2nd edition, Springer-Verlag, New York, NY, 1989.

[29] L. Kavraki, J.C. Latombe, and R.H. Wilson. On the Complexity of Assembly Partitioning, *Information Processing Letters*, 48, 1993, pp. 229–235.

[30] L. Kavraki and J.C. Latombe. Randomized Preprocessing of Configuration Space for Fast Path Planning, *Proc. IEEE Int. Conf. on Robotics and Automation*, San Diego, CA, May 1994, pp. 2138–2145.

[31] P. Khosla. Reconfigurable Robotic Systems, *Preprints of the 6th Int. Symp. on Robotics Research*, Hidden Valley, PA, Oct. 1993.

[32] Y. Koga and J.C. Latombe. Experiments in Dual-Arm Manipulation Planning, *Proc. IEEE Int. Conf. Robotics and Automation*, Nice, May 1992, pp. 2238–2245.

[33] D.J. Kriegman and J. Ponce. Computing Exact Aspect Graphs of Curved Objects: Solids of Revolution, *Int. J. of Computer Vision*, 5(2), 1990, pp. 119-135.

[34] K. Lakshminarayana. The Mechanics of Form Closure, *ASME*, Paper No. 78-DET-32, 1978.

[35] J.C. Latombe, C. Laugier, J.M. Lefebvre, E. Mazer, and J.F. Miribel. The LM Robot Programming System, *Robotics Research 2*, H. Hanafusa and H. Inoue (eds.), MIT Press, 1985, pp. 377–391.

[36] J.C. Latombe. *Robot Motion Planning*, Kluwer Academic Publishers, Boston, MA, 1991.

[37] J.C. Latombe. A Fast Path Planner for a Car-Like Indoor Mobile Robot, *Proc. 9th Nat. Conf. Artificial Intelligence*, AAAI, Anaheim, CA, July 1991, pp. 659–665.

[38] J.P. Laumond. Feasible Trajectories for Mobile Robots with Kinematic and Environment Constraints, *Proc. Int. Conf. on Intelligent Autonomous Systems*, Elsevier Science Publishers B.V., Amsterdam, The Netherlands, 1986. pp. 346-354.

[39] J.P. Laumond, P. Jacobs, M. Taix, and R. Murray. A Motion Planner for Nonholonomic Mobile Robots, to appear in *IEEE Tr. on Robotics and Automation*.

[40] A. Lazanas and J.C. Latombe. *Landmark-Based Robot Navigation*, Tech. Rep. STAN-CS-92-1428, Dept. of Computer Science, Stanford, CA, 1992. (To appear in *Algorithmica*.)

[41] Z. Li and J.F.Canny. Motion of Two Rigid Bodies with Rolling Constraint, *Proc. IEEE Tr. on Robotics and Automation*, 6(1), 1990, pp. 62–72.

[42] T. Lozano-Pérez. Spatial Planning: A Configuration Space Approach, *IEEE Tr. on Computers*, 32(2), 1983, pp. 108-120.

[43] T. Lozano-Pérez, M.T. Mason, and R.H. Taylor. Automatic Synthesis of Fine-Motion Strategies for Robots, *Int. J. of Robotics Research*, 3(1), 1984, pp. 3-24.

[44] K.M. Lynch. Planning Pushing Paths, *Proc. JSME Int. Conf. on Advanced Mechatronics*, Tokyo, Aug. 1993, pp. 451–456.

[45] X. Markenscoff, L. Ni, and C.H. Papdimitriou. The Geometry of Grasping, *Int. J. of Robotics Research*, 9(1), 1990, pp. 61–74.

[46] M.T. Mason. Mechanics and Planning of Manipulator Pushing Operations, *Int. J. of Robotics Research*, 5(3), 1986, pp. 53-71.

[47] B. Mishra, J.T. Schwartz, and M. Sharir. On the Existence and Synthesis of Multifinger Positive Grips, *Algorithmica*, 2, 1987, pp. 541–558.

[48] B.K. Natarajan. On Planning Assemblies, *Proc. ACM Symp. on Computational Geometry*, 1988, pp. 299-308.

[49] N.J. Nilsson. *Principles of Artificial Intelligence*, Morgan Kaufmann Publishers, San Mateo, CA, 1981.

[50] M. Overmars. A Random Approach to Path Planning, *Tech. Rep. 32*, Utrecht Univ., The Netherlands, 1992.

[51] C.H. Papadimitriou and M. Yanakakis. Shortest Paths Without a Map, *Theoretical Computer Science*, 84, 1991, pp. 127-150.

[52] J.H. Reif. Complexity of the Mover's Problem and Generalizations, *Proc. FOCS*, 1979, pp. 421-427.

[53] J.T. Schwartz and M. Sharir. On the Piano Movers' Problem: I. The Case if a Two-Dimensional Rigid Polygonal Body Moving Amidst Polygonal Barriers,

Comm. on Pure and Applied Mathematics, 36, 1983, pp. 345-398.

[54] J.T. Schwartz and M. Sharir. On the 'Piano Movers' Problem: II. General Techniques for Computing Topological Properties of Real Algebraic Manifolds, *Advances in Applied Mathematics*, Academic Press, 4, 1983, pp. 298-351.

[55] S. Shekhar and J.C. Latombe. On Goal Recognizability in Motion Planning with Uncertainty, *Proc. IEEE Int. Conf. on Robotics and Automation*, Sacramento, CA, 1991, pp. 1728–1733.

[56] R.H. Wilson. *On Geometric Assembly Planning*, PhD Thesis, Rep. STAN-CS-1416, Dept. of Computer Science, Stanford Univ., March 1992.

[57] R.H. Wilson and J.C. Latombe. Geometric Reasoning About Mechanical Assembly, *Artificial Intelligence J.*, 71(1), 1995.

[58] R.H. Wilson, L. Kavraki, T. Lozano-Pérez, and J.C. Latombe. *Two-Handed Assembly Sequencing*, Rep. No. STAN-CS-93-1478, Dept. of Computer Science, Stanford Univ., June 1993. To appear in *Int. J. of Robotics Research*.

[59] M. Yim. A Reconfigurable Modular Robot with Many Modes of Locomotion, *Proc. JSME Int. Conf. on Advanced Mechatronics*, Tokyo, Aug. 1993, pp. 283–288.

A Probabilistic Learning Approach to Motion Planning

Mark H. Overmars, Petr Švestka
Department of Computer Science, Utrecht University
P.O.Box 80.089, 3508 TB Utrecht, the Netherlands
e-mail: markov@cs.ruu.nl, petr@cs.ruu.nl

In this paper a new paradigm for robot motion planning is proposed. We split the motion planning process into two phases: the learning phase and the query phase. In the learning phase we construct a probabilistic roadmap in configuration space. This roadmap is a graph where nodes correspond to randomly chosen configurations in free space and edges correspond to simple collision-free motions between the nodes. These simple motions are computed using a fast local method. The longer we learn, the denser the roadmap becomes and the better it is for motion planning. In the query phase we can use this roadmap to find paths between different pairs of configurations. If a possible path is not found one can always extend the roadmap by learning further. This gives a very flexible scheme in which learning time and success for queries can be balanced.

We will demonstrate the power of the paradigm by applying it to various instances of motion planning : free flying planar robots, planar articulated robots and car-like robots (with non-holonomic constraints). We expect it to be applicable in many other instances as well.

1 Introduction

The *motion planning problem* is well-known in the field of robotics. It asks for computing feasible paths for a given robot $\mathcal{A}$ in a workspace containing some obstacles. Two versions of the problem can be formulated. In one version, a start configuration s and a goal configuration g are given before hand, and the objective is to compute a feasible path for $\mathcal{A}$ from s to g. In the second version, no start and goal configurations are specified, and the objective is to compute a data-structure, which can later be used for queries with arbitrary start and goal configurations. We refer to the former case as

a *single shot* problem, and to the latter as a *learning* problem.

The 'classical' approaches to motion planning can roughly be divided in the following three classes: *roadmap methods*, *cell decomposition methods*, and *potential field methods*. For a thorough discussion of these approaches see e.g. [10] and [7].

Let $\mathcal{C}$ denote the space of all configurations for the robot, and let $\mathcal{C}_f$ be the robots free configuration space, i.e., the subset of $\mathcal{C}$ where the robot does not intersect any obstacles. The roadmap approach (or *skeleton approach*) consists of capturing the connectivity of $\mathcal{C}_f$ in the form of a network of one dimensional curves - the *roadmap* - lying in $\mathcal{C}_f$. After a roadmap ρ has been constructed, the path planning is reduced to connecting the start and goal configurations to ρ, and searching ρ for a path.

The principle of the cell decomposition approach is to decompose the robots free configuration space $\mathcal{C}_f$ into a collection of non-overlapping regions (cells), whose union is (exactly or approximately) $\mathcal{C}_f$. This *cell decomposition* is then used for constructing the *connectivity graph* G which represents the adjacency relation among the constructed cells. Every node in G corresponds to a cell, and two nodes are connected by an edge if and only if their corresponding cells are adjacent. The path planning is then performed by finding a path in G from the node corresponding to the start cell (= the cell containing the start configuration) to the node corresponding to the goal cell (= the cell containing the goal configuration).

We see that both the roadmap approach as well as the cell decomposition approach consist of constructing a global data structure that can later be used for solving one or more motion planning problems. This means that both approaches are suitable for learning problems (as well as single shot problems). Another strong point is that cell-decomposition and roadmap algorithms are typically complete, i.e., whenever a path exists a path will be found. Drawbacks though are that (1) the computations of the data structures tend to be very expensive in both time and memory, and (2) they do not seem to be suitable for robots with non-holonomic constraints, like for example car-like robots or multi-body mobile robots.

In the potential field approach no data structure is built. Globally the idea is that the robot (represented by a configuration in configuration space) is treated as a particle under the influence of an *artificial potential field* whose variations are expected to reflect the 'structure' of the free configuration space C_f. The potential field is typically defined by a function $C \rightarrow R$ which is a weighed sum of an *attractive* potential pulling the robot towards the goal configuration, and a number of *repulsive* potentials pushing the robot away from the obstacles. The motion planning is performed by repeatedly computing the most promising direction of motion, and moving in this direction by some step size.

A typical problem with potential field methods is that the robot can get stuck in a local minimum of the potential field. I.e., the robot gets to a configuration m where the (weighed) sum over all the potentials is equal to the null-vector. Recently though much progress has been made in defining good potential functions with few local minima, and efficient techniques have been developed for escaping from local minima. Currently there exist practical potential field planners for robots with many degrees of freedom, as well as for some types of non-holonomic robots (see for example [2]). So it seems that the potential field approach does not have the disadvantages of the two former approaches. A major drawback of the potential field approach though is that the whole concept is unsuitable for learning problems, due to the fact that every goal configuration defines a distinct potential field.

In this paper we describe a new paradigm to the learning motion planning problem, by combining a global roadmap approach with a local planner. In the learning phase, a probabilistic roadmap is built up by repeatedly generating random free configurations and trying to connect these to other (earlier added) configurations by some simple motion planning algorithm. The network thus formed is stored in a graph G. The configurations are stored as nodes in G, and the links,

which are paths in free configuration space, are stored as edges in G. After the learning is done, a query consists of trying to connect the given start and goal configurations to some nodes (in the same connected component) of G, with paths which are feasible for the robot. Next, the path between these nodes in G can be transformed into a feasible path in configuration space.

We claim that our paradigm is a very powerful one, and to support this claim, we apply it to a number of different robots : free flying robots, articulated robots, and car-like robots. For each robot type and for a number of different scenes, we test how much learning time is required to be able to solve a certain percentage of 'all' queries in the scene. It turns out that often only very little time (in the order of seconds) is required to solve a percentage of nearly 100%.

The learning phase of this approach is closely related to earlier work that we have done on single shot planners. In [13] a probabilistic single shot planner for free flying planar robots is described in detail, and [14] deals with two types of car-like robots, i.e., normal ones, and cars which can only move forwards. In both papers we gave a lot of experimental results, some of which have guided certain choices made in the learning approach described in this paper. Other related work has independently been done by L.Kavraki and J.-C.Latombe. In [8] they describe a probabilistic method for configuration space preprocessing, which also builds up a probabilistic road map that, in a query phase, can be used for motion planning. They deal with articulated robots of high degree of freedom, with links that are line segments. Their method though seems to be more restricted than ours.

This paper is organized in the following way: In section 2 the learning algorithm is described in general terms. In section 3 the query phase is described. In sections 4, 5 and 6 we apply the paradigm to, respectively, free flying planar robots, car-like robots, and (planar) articulated robots. For each of these robot types we fill in the details and give experimental results. Finally, in section 7, we draw some conclusions.

2   The learning paradigm

The learning phase can be described in general terms, without focussing on any specific robot type. The idea is that the data structure built up during the learning phase later (in the query phase) is used to solve individual motion planning problems.

The learning algorithm is a *two-level* approach, consisting of a *global method* and a *local method*. The local method is a (primitive) motion planner, which tries to compute (simple) feasible[1] paths connecting two given free configurations. It is allowed to fail now and then, but it is essential that it always terminates and that it is deterministic. The global method uses the local method to build the mentioned probabilistic roadmap. The basic algorithm is extremely simple. It builds up an undirected graph $G = (V, E)$, by repeatedly generating a random free configuration c, adding c to V, computing a set $N(c) \subset V$ (c's *neighbors*), and adding an edge (c, n) to E for every $n \in N(c)$ to which the local method can connect from c.

Assume now that we are dealing with a robot $\mathcal{A}$, and that L is a local method for $\mathcal{A}$. To describe the global method formally, we need the following :

- A *symmetrical* function $L_d \in \mathcal{C} \times \mathcal{C} \to boolean$, that returns whether the local method can compute a feasible path for $\mathcal{A}$ between its two argument-configurations. We refer to this function as L's decision function.

- A function $D \in \mathcal{C} \times \mathcal{C} \to \mathcal{R}^+$. It defines the metric[2] used, and should give a suitable notion of distance for arbitrary pairs of configurations, taking the properties of the robot $\mathcal{A}$ into account. We assume that D is symmetrical.

The algorithm can now be described as follows:

The global method:

$V = \emptyset$
$E = \emptyset$
loop
    $c = $ A randomly chosen free configuration.
    $V = V \cup \{c\}$
    $N(c) = $ A set of neighbors of c, chosen from V.
    $E = E \cup \{(c, a) \mid a \in N(c) \wedge L_d(c, a)\}$

---

[1] A path P is feasible for a robot $\mathcal{A}$ iff P lies in free configuration space, and the motion described by P is achievable by $\mathcal{A}$, i.e. it respects $\mathcal{A}$'s constraints.

[2] By metric we simply mean a function of type $\mathcal{C} \times \mathcal{C} \to \mathcal{R}^+$, without any restrictions.

When the learning is done, queries can be performed. Given a start configuration s and goal configuration g, we try to connect s and g to nodes $\tilde{s}$ and $\tilde{g}$ in the (same connected component of) the graph. If this succeeds, we compute the shortest path in the graph from $\tilde{s}$ to $\tilde{g}$. For each edge on this path we (re)construct a feasible path in configuration space using the local planner. In this way we obtain a feasible path connecting s to g. If $\tilde{s}$ and $\tilde{g}$ cannot be found, the query fails. We will go into some more details about the query phase in section 3.

The paradigm described above leaves a number of choices to be made. I.e., a local method must be chosen, a metric must be defined, and it must be defined what the neighbors of a node are. Some choices must be left open as long as we do not focus on a particular robot type, but certain global remarks can be made that apply to all robot types.

Local method

One of the crucial ingredients in the learning phase is the local method. As mentioned before, the local method must compute paths which are *feasible* for $\mathcal{A}$. The reason for this is that the global paths computed in the query phase are basically just concatenations of local paths (i.e., paths computed by the local method), so clearly in order for the global paths to be feasible, the local paths must be as well. Furthermore, the local method should be deterministic. Otherwise the existence of a path in G between two nodes a and b does not guarantee that a feasible path in configuration space connecting a and b can be reconstructed in the query phase. A final requirement is that the local method always terminates (some potential field methods do not have this property).

There are still many possible choices for such an algorithm. On one hand one could take a very powerful method. Such a method would very often succeed in finding a feasible path when one exists, and, hence, relatively few nodes would be required in order to obtain a graph which captures the connectivity of the free configuration space well. Such a local method would (probably) be slow, but one could hope that this is compensated by the fact that only a few executions of the method need to be performed. On the other hand, one could choose a very simple and fast algorithm that is much less successful. In this case many more nodes will have to be added in order to obtain a reasonable graph, which means that many more executions of the local method will be required. But this might be compensated by the fact that each execution is very cheap. So it is clear that there is a trade off, and it is not trivial to make a smart choice here.

Clearly, the choice of the local method should be aimed at maximizing the amount of 'knowledge' acquired by the global method per time unit. E.g., if we define the amount of 'knowledge' stored in a graph(V, E) as $|E|$, then we should search for a local method with $\frac{\text{chance of success}}{\text{average running time}}$ as high as possible. This is though a rather vague criterion, and we have guided the choice of our local methods by experiments. These clearly indicated that very fast (and, hence, not very powerful) local methods lead to the best performance of the global method.

Neighbors and edge adding methods

Another important choice to be made is that of the neighbors $N(c)$ of a (new) node c. As is the case for the choice of the local method, the choice of $N(c)$ has large impact on the performance of the global method. Reasons for this are that the choice of the node neighbors strongly influences the overall structure of the graph, and that, regardless of how the local method is exactly chosen, the executions of the local method are by far the most time-consuming operations of the global method. This is caused by the fact that the local method must perform intersection-tests (of the robot with the obstacles) for each path that it, successfully or not, computes.

So it is clear that 'useless' executions of the local method should be avoided as much as possible. To start with, an execution of the local method which fails is useless, in the sense that it does not extend the 'knowledge' stored in the graph. To prevent too many failures of the local method, we only consider nearby nodes (with respect to the metric D), i.e. nodes within distance *maxdist* of the new node (where *maxdist* is some well chosen real constant). Thus

$$N(c) \subset \{\tilde{c} \in V \,|\, D(c, \tilde{c}) \leq maxdist\} \qquad (1)$$

Now what is the value of successful executions of the local method? This depends on what kind of paths we want to get during the query phase.

One possibility is that we do not care about what these paths look like, as long as many queries succeed. In this case adding any edge (a, b) which creates a cycle in the graph is worthless, because a and b were already connected, and, hence, no query can ever succeed *thanks to (a, b)*. This suggests that we try to connect c at most once to each connected component in the graph, which, in combination with (1), leads to the following definition :

Definition 1

$$N(c) = \{n \in \tilde{V} \,|\, D(c, n) \leq maxdist$$
$$\wedge$$
$$\forall m \in \tilde{V} - \{n\} : \; connected(m, n)$$
$$\Rightarrow$$
$$D(c, n) < D(c, m)\}$$

where $\tilde{V} = V - \{c\}$

So in every connected component of G, the nearest node to c is a neighbor of c, under the condition that $D(c, n) \leq maxdist$. We refer to the edge adding method which results from this definition as the *forest* method, because it leads to a graph that is a forest.

Now suppose that we prefer short paths above arbitrary ones in the query phase. The forest method is not suitable for this purpose. An edge in a cycle is no longer necessarily worthless, because removing it may, in the query phase, result in much longer paths. So we should allow cycles now. One possibility is to just pick the k nearest nodes as the neighbors. We refer to the edge adding method resulting from this choice as the *nearest-k* method. Using this method (with say $k \geq 4$) one obtains a very dense graph and near-optimal paths in the query phase, but unfortunately it is quite expensive.

A way to obtain reasonable paths at relatively low expense is what we refer to as the *heuristic loops* method. The idea is that, when a new node c has been added, first edges are added using the forest definition. After this is done, one or more nodes in c's connected component are picked as extra neighbors, in order to obtain some loops containing c. These extra neighbor(s) should be chosen using some heuristics, aimed at reducing (some) path lengths in c's connected component. We have decided on picking as (only) extra neighbor the node $n \in V$ which minimizes the ratio between $D(c, n)$ and the length of the current shortest path in G from node c to node n.

See figures 1, 2, and 3 for examples of graphs obtained with, respectively, the forest method, the nearest-4 method, and the heuristic loops method, all in the same scene with a rectangular free flying robot. As can be seen, the heuristic loops method captures the topology of the free space in the best way, which results in short paths. (Note that in the figures some edges go through obstacles. This is not an error though. The reason is that we display an edge between two nodes as a straight line segment, while the actual path corresponding to the edge might be different.)

Distance

We have seen that the distance function D is used for choosing the neighbors $N(c)$ of a new node c. For example in the forest method, from every connected component in the neighborhood of the new node c, that node is picked which is nearest to c with respect to the 'metric' D. By picking the nearest node from a component C, we want to maximize the chance of successfully connecting from c to C. But this means that D should be defined in a way, such that $D(a, b)$ (for arbitrary a and b) somehow reflects the chance that the local method will *fail* to compute a feasible path from a to b. For this we define the distance between two configurations a and b as the *length*[3] of the path from a to b computed by the local method in absence of obstacles. In this way any local method induces its own metric.

Adaptive node adding

In the algorithm described above the nodes are added in a fully random way. Experiments indicate that this random node adding strategy performs quite well, e.g. better than adding nodes in some regular pattern. It is though possible to use some heuristics during the node generation, such that more nodes are added in certain 'interesting' areas of free configuration space than in others.

For example, areas where where two or more connected component of G 'overlap' are interesting, be-

---

[3]For car-like robots we define the length of a path $P \in [0, 1] \to \mathcal{C}$ to be the length of P's projection in $\mathcal{R}^2$. For free flying robots and articulated robots we take the actual length of the path in configuration space, but scaled appropriately along the angular axis, taking into account the robots geometry.

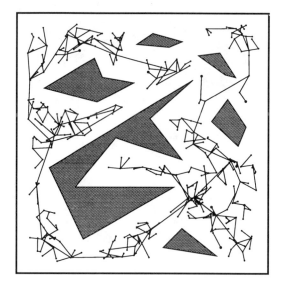

Figure 1: *A graph obtained with the forest method.*

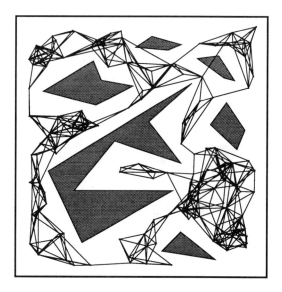

Figure 2: *A graph obtained with the nearest-4 method.*

cause adding nodes in those areas is likely to result in the 'merging' of those components. The *adaptive adding heuristic* is aimed at this. When a new configuration c has been (randomly) generated, it is not added to the graph immediately, but instead the *adding*

Figure 3: *A graph obtained with the heuristic loops method.*

chance of the configuration is computed. If there are no nodes very near to c, or if there are two or more different connected components very near to c, then c's adding chance is high (e.g. 1). Otherwise it is low (e.g. 0.05). After this adding chance has been computed, c is added with this probability, and otherwise it is discarded.

The underlying idea is that if two or more different connected components lie very near to c, then adding c is likely to connect these components. Hence c should be added. If there are no nodes very near to c, then clearly not enough learning has yet been done in c's near environment, and hence c should be also added. In the case that c has exactly one connected component very near (which typically is the case most often), then c's near environment is already explored (to some extent), and c is not very likely to connect some different connected components, so probably c is not a very valuable node. Hence its adding chance is low. So adaptive adding adds nodes mainly in 'empty' areas of $\mathcal{C}_f$, and in 'overlapping' areas of different connected components.

3 Queries

During the query phase, paths are to be found between arbitrary start and goal configurations, using the graph

computed in the learning phase. The idea is that, given a start configuration s and goal configuration g, we try to connect s and g to nodes $\tilde{s}$ and $\tilde{g}$ of the same connected component in G, with some feasible paths P_s and P_g. If this succeeds, then we compute the shortest path P_G in G connecting $\tilde{s}$ to $\tilde{g}$, and a feasible path in configuration space from s to g is constructed by concatenating P_s, the subpaths computed by the local method when applied to pairs of consecutive nodes in P_G, and P_g. Otherwise the query fails.

The main question is how to compute the paths P_s and P_g. The queries should preferably terminate 'instantaneously', so no expensive algorithm is allowed here. One possibility is to use the local method. We have decided on a simple randomized planner, which performs a Brownian-like motion in configuration space for a short period of time (e.g. 0.25 seconds), and during this motion at regular intervals tries to connect to the graph using the local method. For free flying robots and articulated robots, the Brownian-like motion consists of repeatedly picking a random direction in configuration space, and moving in this direction until an obstacle is collided with, or the time runs out. When a collision occurs, a new random direction is chosen. So the paths computed by this planner are sequences of path seqments which are straight in configuration space, followed by a path segment computed by the local method. For car-like robots the Brownian-like motion is generated by repeatedly picking random values in control space[4]. (i.e., random values for the steering angle and the velocity of the robot), and performing the motion thus defined until a collision occures. Experiments show that this Brownian-like method works fine, e.g., better than just using the local method.

3.1 Smoothing

Paths computed in the query phase can, especially when the graph has been built using the forest edge adding method, be quite ugly and unnecessarily long.

To improve this, one can apply some path smoothing techniques on these 'ugly' paths. The smoothing routine that we implemented is very simple. It just repeatedly picks a pair of random configurations (c_1, c_2)

---

[4]Instead of the actual control space $[-\phi_{max}, \phi_{max}] \times [-v_{max}, v_{max}]$ we use the discretisation $\{-\phi_{max}, \phi_{max}\} \times \{-1, 1\}$, where ϕ_{max} is the robots maximal steering angle, and v_{max} its maximal velocity

on the 'to be smoothed' path P_C, tries to connect these with a feasible path Q_{new} using the local method, and if this is successfully accomplished and Q_{new} is shorter than the path segment Q_{old} in P_C from c_1 to c_2, then it replaces Q_{old} by Q_{new} (in P_C). So basically it just tries to replace randomly picked segments of the path by shorter ones, using the local method. The longer this is done, the shorter (and nicer) the path gets. Typically, this method smoothes a path very well in just a few seconds.

Still one can argue that a few seconds is too much for a query. In that case one should either accept the ugly paths, or use a more expensive edge adding method, like for example the nearest-k method or the heuristic loops method. The gain obtained by these edge adding methods (in terms of shorter paths in the query phase) is hard to measure accurately, and is very dependent of the scenes that we are dealing with. Experiments though show that for many scenes it is significant. In some cases the path lengths obtained with e.g. the nearest-4 edge adding method are, on the average, only about half of those obtained with the forest method.

4 Free flying planar robots

As a first example of the use of the general paradigm introduced in section 2 we apply it to free flying planar robots, i.e., the robot is a polygon that can rotate and translate freely in the plane among a set of polygonal obstacles. This setting of the motion planning problem is rather simple because the robot has only three degrees of freedom and no non-holonomic constraints. It has been studied in great detail in the past and many methods have been proposed to deal with it. Still the problem is far from trivial. We first fill in a few details of the method, and then give some experimental results.

4.1 Details of the method

To use the global paradigm we need a suitable local method, which we describe below. For the distance between configurations the induced metric will be used. Also we will introduce a non-random node adding strategy, which makes use of the geometry of the workspace, and can be used in combination with the random adding strategy. This strategy is only applicable to planar solid robots, such as free flying and car-like ones.

Local method

For free flying robots we have done a large number of experiments with different types of local methods, and, based on the results, we have decided on the following method that can be regarded as a very simple potential field method: Let ϵ be some small step size. The method will let the robot take steps of this size. To avoid collisions during a step we blow up the robot with a factor ϵ. (In fact we use different step sizes for the translation and rotation of the robot. The rotation step size is chosen automatically such that during a rotation step no part of the robot moves more than a distance of ϵ.) In pseudo C-code the method looks as follows:

ALGORITHM LOCAL METHOD

```
1.  config = source;
2.  loop
3.      if (config == goal) return goal reached;
4.          newconf = config plus step of size ε towards goal;
5.      if newconf in free space
6.          config = newconf;
7.      else
8.          for each of the 26 direct neighbors of config
9.              determine distance to goal;
10.         Sort neighbors by distance in a list neighbor[];
11.         for (i=0; i<13; i++)
12.             if neighbor[i] in free space
13.                 config = neighbor[i]; break;
14.         if (i == 13) return goal not reached;
```

The algorithm simply loops until the goal is reached (line 3) or no progress can be made (line 14). In each loop first we try to make a step in the direction of the goal (line 4-6). If we fail all 26 direct neighbors of the current configuration (by taking an ϵ-step in the x- y- and θ-direction) are considered. We treat them in order of their progress towards the goal (i.e. their distance to the goal) by computing these distances (line 8-9) and sorting them (line 10). Note that only the first 13 of these configurations are better than the current configuration. Hence, we only look at these 13 neighbors and take the first one that is in free space as the new configuration (line 11-13). When none of the neighbors is possible we conclude that no path could be found (line 14). (In the actual implementation the order of the steps is changed to obtain some slight speed-up.)

One can view this method as a potential field method. Here the attracting potential created by the goal is simply inversely proportional to the distance. The repulsive potential is infinite when the robot gets nearer than ϵ to an obstacle and is 0 otherwise. The motion computed this way can turn around corners but often leads to a local minimum where no progress can be made.

The big advantage of the approach is that it is very fast and general. The only basic test we need is whether the (slightly blown-up) robot intersects any of the obstacles. Such a test can be performed very fast after some preprocessing of robot and obstacles; much faster than the distance computations that are normally required for other potential field methods. Note that, when the robot is not near to an obstacle, the step towards the goal will succeed and, hence, only one of the basic tests is performed. This makes things even faster.

Geometric node adding

Many practical scenes have the form of corridors and rooms. In such cases one can boost the planners performance by adding configurations at important positions, in particular along edges of obstacles (i.e., along walls) and next to vertices of obstacles (to facilitate the robot to move around corners). This is implemented in the following way: For the robot we determine the geometric axis, i.e., the axis such that the width of the robot becomes minimal (with respect to the axis). For each edge and vertex of the obstacles (in random order) we determine the outer normal. Next we place the robot with its geometric axis perpendicular to the normal along the edge or convex vertex at such a distance that the robot has some small clearance with respect to the edge or vertex. See figure 4 for an example. If this is a free configuration it is added to the graph, otherwise it is discarded. Often geometric adding captures most of the connectivity of the free configuration space. See figure 5 for an example of a graph obtained after geometric adding (and the nearest-4 edge adding method). Once all edges and convex vertices have been treated, the geometric adding stops. In order for the learning process to continue at this point, the geometric adding should be used in combination with random adding. This means that after all the geometrically obtained nodes have been used, we continue by adding random nodes.

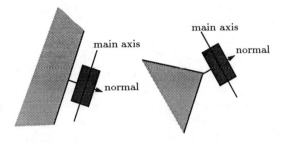

Figure 4: *Adding a rectangular robot along an edge or vertex.*

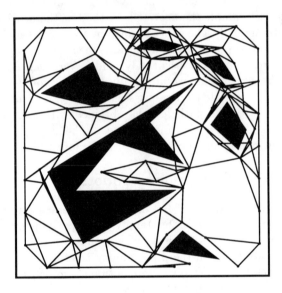

Figure 5: *A graph obtained after geometric adding, for a rectangular robot.*

Experimental results

We now present a number of experimental results for free flying planar robots, obtained by applying our learning approach to a number of different scenes. The method is implemented in C++ and the tests were performed on a Silicon Graphics indigo workstation with an R3000 processor running at 33 MHZ. This machine is rated on the SPECMARKS benchmark with 24.2 SPECfp92 and 22.4 SPECint92.

In the test scenes that will be used, the coordinates of all workspace obstacles lie in the unit square, and all node configurations are chosen such that they project on this unit square. Paths between configurations are allowed to run outside this square, but we prevented this in the test scenes by adding a small obstacle boundary around the unit square.

We say the programs 'knowledge' of a scene S is $k\%$, if $k\%$ of 'all' solvable queries in scene S are solved successfully (and $(100-k)\%$ are not). Given a particular scene, we want to measure the 'knowledge' that the program acquired about the scene after having learned for a certain amount of time. It is not possible to measure this 'knowledge' exactly, but we approximate it in the following probabilistic way : First we generate a (large) set $V_c \subset \mathcal{C}_f$ of randomly picked free configurations. Then we generate a (large) set $V_p \subset V_c \times V_c$ of free random configuration pairs, but we discard all pairs (a, b) where a lies in a different connected component of $\mathcal{C}_f$ than b. Finally, we count for how many pairs $(a, b) \in V_p$ the query with arguments a and b succeeds. If s is the number of successful queries, than $\frac{s}{|V_p|} \cdot 100\%$ is our approximation of the programs knowledge. One can expect this approximation to be a good one, if V_p is taken sufficiently large (we use 1000). Instead of giving arrays of numbers, we show some figures where the programs knowledge is set against the learning time. Clearly, the knowledge as defined above not only depends on the graph, but also on the time we allow for answering a query. Remember that during a query we perform a Brownian-like motion for some given amount of time, to connect the start and goal configurations to the graph. The more time we allow for this Brownian-like motion, the higher is the chance that the query will succeed. In our tests we kept the maximal query time very low, i.e., at 0.3 seconds.

In section 2 a number of different edge adding methods have been described, as well as two node adding heuristics. Furthermore, in section 4.1 a geometrically based non-random node adding method has been introduced. So there are many different settings possible. We give results for the tree edge adding method (with $maxdist=0.25$) combined with random node adding, adaptive random node adding, and adaptive geometric node adding. With adaptive random node adding we mean the random node adding method combined with the adaptive adding heuristic, and with geometric node adding we mean the geometric node adding method, followed by adaptive random node adding (when all geometric nodes have been added). So we do not give results for any of the described edge adding methods which allow loops in the graph. These typically slow

down the learning process a bit. Experiments that we have performed indicated that the learning process slows down by a factor of approximately 2 when the heuristic loops method is applied, and by a factor of about 3 when the nearest-4 method is used. In return though, as mentioned before, these edge adding methods build graphs which give shorter paths in the query phase than the graphs obtained with the forest method, reducing the average path lengths by up to 50%.

The test scenes

We tested the method on the four scenes shown in figures 6 to 9. In each scene a (smoothed) path computed by our planner is indicated by a number of steps. Scene 1 is a relatively easy scene, consisting of one long corridor which requires the long robot to make some sharp curves. In scene 2 we have a triangular robot amongst a large number of small triangular obstacles. Most motion planning problems in this scene are easy, but there also exist many narrow areas in free configuration space, corresponding to areas in workspace where the robot is tightly surrounded by three or more obstacles. If either the start or the goal configuration of a query is positioned in such an area, then this query is far from trivial. The difficulty in scene 3 is the long and thin non-convex robot. Scene 4 is the most difficult of the four scenes. It contains many narrow passages, it gives the robot little freedom of movement, and most queries can only be solved by relatively long paths.

The test results

Figures 10 to 13 show the amount of knowledge $K(t)$ that the planner has acquired after having learned for a certain time t. The different curves in each figure correspond to different node adding strategies. The dashed ones correspond to random adding, the dotted ones to adaptive random adding, and the solid ones to adaptive geometric adding. The edges are added with the tree method.

Figure 10 shows that the learning problem for scene 1 is solved for 100% within a few seconds. When random adding is used, this takes about 8 seconds. Adaptive random adding gives a significant improvement if the learning time exceeds about 3 seconds, giving 100% knowledge in just over 5 seconds. The relatively bad performance for short learning times is caused by the fact that adaptive adding only begins to pay off when a

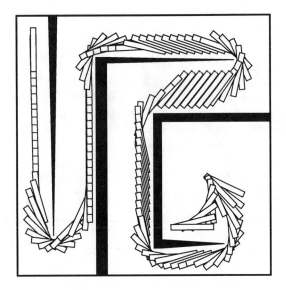

Figure 6: *A free flying robot in scene 1.*

Figure 8: *A free flying robot in scene 3.*

Figure 7: *A free flying robot in scene 2.*

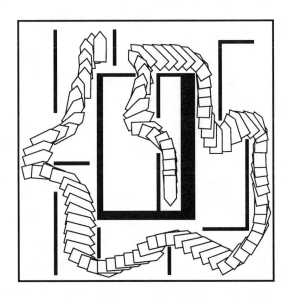

Figure 9: *A free flying robot in scene 4.*

number of large components are present, which, in their 'overlapping' areas, define (not too many) 'promising' areas in configuration space. Initially the graph has no structure, and adaptive adding only slows down the node generation. We see that adaptive geometric adding gives by far the best performance for all learning times. In a scene with many long and straight corri-

dors, like scene 1, the graph obtained by adding the geometric nodes typically captures most of $\mathcal{C}_f$'s connectivity by itself, not requiring many random nodes to be added. In scene 1 the geometric learning phase (the phase where the geometric nodes are added) takes about $1\frac{1}{2}$ seconds, and the knowledge acquired during this phase is about 75%. Note that the initial knowledge, i.e. the knowledge present after 0 seconds of learning, is already about 22%. This is caused by the fact that our Brownian-like planner, which is used for computing connections between the query configurations and the graph, can solve some easy problems in scene 1 by itself.

The plot in figure 11 shows a similar structure as the one in figure 10. Adaptive geometric adding is the best adding method, giving about 95% knowledge in 10 seconds, and the adaptive random method is better than the normal random method if the learning time exceeds some bound, which is about 8 seconds for scene 2. The difference in performance between the geometric adding method and the two random ones is quite small in comparison to scene 1. This is due to the rather chaotic and unstructured character of the scene. Another difference with scene 1 is that the knowledge converges much slower towards 100%. The presence of some very narrow areas in $\mathcal{C}_f$ causes this.

The learning problem in scene 3 is solved for almost 100% in about 12 seconds. We see that geometric adding does not help here. Apparently, the non-convex shape of the robot causes the required motions along obstacle edges to intersect the obstacles too often. Again adaptive adding helps considerably.

The last scene is again one with many straight corridors, and, hence, geometric adding is useful. The adaptive geometric adding strategy solves the learning problem in about 10 seconds. We see that in this relatively difficult scene, which, hence, requires a large graph, adaptive adding really helps a lot. With normal random node adding, it takes more than a minute to solve the problem (for 100%).

The graph in which the computed motions are stored is typically quite small, and, hence, cheap to store. In scene 1 the graph grows with about 80 nodes per second, in scenes 3 and 4 with about 40 nodes per second, and in scene 2 (where the large number of small obstacles causes the intersection tests to be relatively expensive) with about 30 nodes per second. With each of the edge adding methods described in section 2, the number of edges in the graph is linear in the number of nodes. So, if the program has learned for a time t in one of the test scenes, the graph size is something between $30 \cdot t$ and $80 \cdot t$.

As stated before, we kept the query time under 0.3 seconds. If we increase this time, the performance of

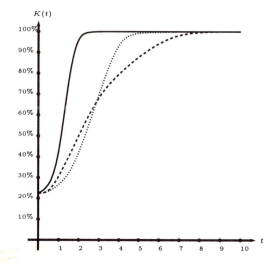

Figure 10: *Learning in scene 1.*

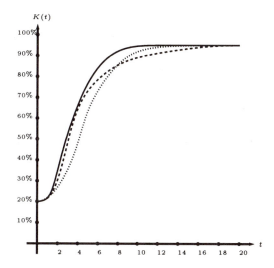

Figure 11: *Learning in scene 2.*

the method improves. E.g., in scene 2, when we allow 2.0 seconds per query, the knowledge acquired after having learned for time t is about equal to the knowledge acquired after having learned for time $2t$ with 0.3 as maximal query time. So the learning goes twice as fast.

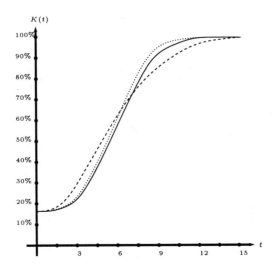

Figure 12: *Learning in scene 3.*

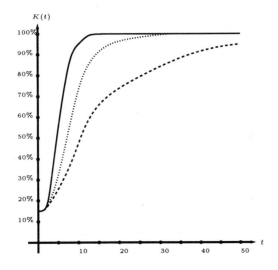

Figure 13: *Learning in scene 4.*

5   Car-like robots

In this section we apply the learning paradigm to car-like robots. As free flying planar robots, car-like robots have three degrees of freedom, but their motions are subject to certain *non-holonomic constraints*, i.e. constraints on their achievable velocities, which makes motion planning considerably more difficult.

Although the motion planning problem for car-like robots has received considerable attention in the past years, there still exist only a few practical (single shot) planners for these robots (see [11], [3]). It turns out that the learning paradigm can easily deal with non-holonomic constraints. We will briefly indicate the results here. For a more extensive description we refer to [14].

We model a car-like robot $\mathcal{A}$ as a polygon that moves in the plane. It has fixed to itself a *reference point* $R_{\mathcal{A}}$ and line $A_{\mathcal{A}}$, referred to as $\mathcal{A}$'s *main axis*, and also it has defined a *minimal turning radius* $r_{\mathcal{A}} \in \mathcal{R}^+$. Furthermore, we refer to the line which is perpendicular to $A_{\mathcal{A}}$ and goes through $R_{\mathcal{A}}$, as $\mathcal{A}$'s *minor axis*. $\mathcal{A}$ can perform certain translations and rotations in the plane, which can be defined in terms of the above. The only translations that $\mathcal{A}$ can perform are translations along $A_{\mathcal{A}}$, and the only rotations that it can perform are such where the center rotation lies on $\mathcal{A}$'s minor axis, at a distance of at least $r_{\mathcal{A}}$ of $R_{\mathcal{A}}$. See also figure 14.

In real life, car-like robots can perform more complex motions also, but it has been shown that if a feasible motion between two configurations exists for a car-like robot, then there also exists one which is a sequence of the described rotations and translations.

For applying our paradigm to car-like robots, we need a (good) local method that computes feasible paths for car-like robots.

Local method

As for free flying planar robots, we have done many experiments with different types of car-like local methods. In particular, we have experimented with some simple potential field methods, conceptually similar to the local method for free flying robots. A problem with car-like robots though is, that locally they can only move approximately along their main axis, which makes it hard to 'slide' them along obstacle boundaries. So potential fields should be used which keep them at some distance from the obstacles, but such potential fields turn out to be much too expensive to evaluate for our purposes. As a result, we have decided on a local method which is even simpler than the one for free flying robots.

We refer to a path describing a rotational motion of $\mathcal{A}$ as a *rotational path*, and one describing a translational motion as a *translational path*. If a rotational path describes a rotation of radius $r_{\mathcal{A}}$, then we refer to

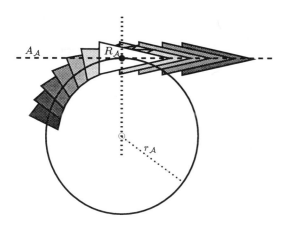

Figure 14: *Some simple car-like motions.*

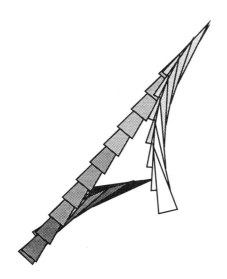

Figure 15: *A RTR-path.*

it as a *maximally curved* rotational path. A *RTR-path* is now defined as the concatenation of a maximally curved rotational path, a translational path, and another maximally curved rotational path (see also figure 15). Analog, a *TRT-path* is defined as being the concatenation of a translational path, a maximally curved rotational path, and another translational path. With these two path constructs we define our local method: Given two argument configurations a and b, if either the shortest RTR-path connecting a to b or the shortest TRT-path connecting a to b intersects[5] no obsta-

---

[5]We have implemented efficient routines which test

cles, one of these two paths is returned (depending on which one is free), and otherwise failure is returned. The RTR-path is always tested first (it is most of the time a bit shorter, and, hence, has less chance of intersecting obstacles), which means that the induced metric is the 'RTR-metric', i.e., the distance between a and b is defined as the length of the shortest RTR-path connecting a to b. See [14] for more details on the exact shapes and computation of TRT- and RTR-paths.

Experimental results

We give experimental results for scene 1, scene 2, and scene 4 (see section 4). Figure 16 shows a (smoothed) path for a car-like robot, computed by our planner in scene 4. So we skip scene 3. This scene is not very 'realistic' if the robot has car-like constraints.

The geometric node adding strategy is applicable to car-like robots, but instead of choosing the geometric axis such that it minimizes the robots width, one should take $A_\mathcal{A}$ as geometric axis. In this way the geometric nodes correspond to configurations of the robot where it can move 'parallel' along the obstacle boundaries.

See figures 17 to 19 for plots that again show the amount of knowledge $K(t)$ about the corresponding scenes acquired by our planner after having learning for some time t. Again the edges are added with the tree method. The dashed curves correspond to random adding, the dotted ones to adaptive random adding, and the solid ones to adaptive geometric adding. The experiments have been performed with $maxdist=0.5$ and $r_\mathcal{A}=0.1$.

The plot in figure 17 shows the same global structure as the plot for free flying robots (see figure 10). So also for car-like robots does adaptive geometric adding give the best performance in scene 1, followed by adaptive random adding. There are two main differences with the results for free flying robots. Firstly, the learning process is about three times slower, and secondly, although the knowledge is almost complete in about 10 seconds (approximately 95% with adaptive geometric adding), it then converges only very slowly towards 100%. E.g., it takes more than a minute to reach 98%.

---

whether the sweep volume of a polygon during a rotational or a translational motion intersects any obstacles. These tests do not require the robot to be 'blown up'.

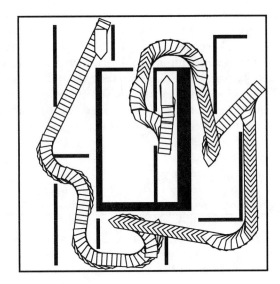

Figure 16: *A car-like path in scene 4.*

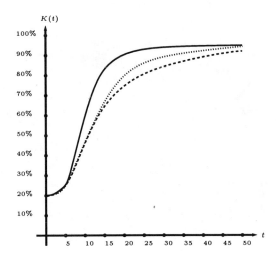

Figure 18: *Learning in scene 2.*

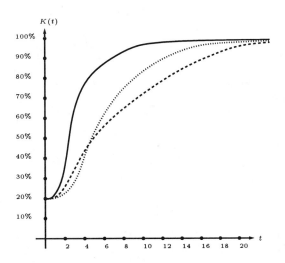

Figure 17: *Learning in scene 1.*

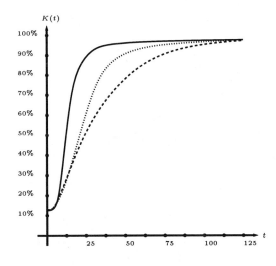

Figure 19: *Learning in scene 4.*

The cause for this slow convergence is that, although all queries performed in the tests are solvable (only configuration pairs in same connected components of $\mathcal{C}_f$ are generated), some queries can be very hard to solve for a robot with car-like constraints. This is the typically the case when either the start configuration or the goal configuration of the query is such, that the robot positioned at that configuration faces obstacles very near in front of itself *and* behind itself, thus requiring the

robot to perform a large number of reversals to either leave or reach the configuration. Such configurations though can be regarded as unlikely in practice. Hence, basically all 'reasonable' paths are found, even though the knowledge is not 100%.

We see that in scene 2 the knowledge reaches 90% in about 20 seconds with the adaptive geometric adding method, which is again about three times slower than for free flying robots. It takes about three minutes to reach 98%.

Finally, the plot for scene 4 shows learning 90% knowledge in scene 4 takes about 25 seconds with adaptive geometric adding.

The graph grows with about 25 nodes per second in scene 1, 15 nodes per second in scene 2, and about 20 nodes per second in scene 4.

6 Articulated robots

As a final example of the use of our learning paradigm, we apply it to planar articulated robots. Many solutions to the motion planning problem for articulated robots have been proposed, but still only a few planners exist that can deal with high degree of freedom articulated robots, and often only in restricted cases (see [5], [4], [9], [1]).

A planar articulated robot $\mathcal{A}$ consists of n *links* $L_1, \ldots, L_n$, which are some solid planar bodies (we use polygons), connected to each other by $n-1$ *joints* $J_2, \ldots, J_n$. Furthermore, the first link L_1 is connected to some *base point* in the workspace by a joint J_1. Each joint can be either a *prismatic joint*, or a *revolute joint*. If J_i is a prismatic joint, then link L_i can translate along some vector, which is fixed to link L_{i-1} (or to the workspace, if $i = 1$), and if J_i is a revolute joint, then link L_i can rotate round some point which is fixed to link L_{i-1} (or to the workspace, if $i = 1$). The range of the possible translations or rotations of each link L_i is constrained by J_i's *joint bounds*, consisting of a lower bound low_i and an upper bound up_i. The configuration space of a n-linked planar articulated robot can, hence, be represented by $[low_1, up_1] \times [low_2, up_2] \times \ldots \times [low_n, up_n]$. In the scenes we show, prismatic joints are denoted by straight arrows, and revolute ones by curved arrows.

Local method

We have implemented and tested different local methods for planar articulated robots. Firstly, we implemented a potential field method, which is analog to the local method for free flying robots. In configuration space, we perform a walk by repeatedly computing all direct neighbors of the 'current' configuration c, and going to the best one (with respect to its distance to the goal configuration) which is free, if it is better than c. Of course, we must define what the direct neighbors of a configuration are. For free flying

robots we took the set $c + (\{-\epsilon, 0, \epsilon\}^3 - \{(0, 0, 0)\})$. The analog definition for planar articulated robots would be $c + (\{-\epsilon, 0, \epsilon\}^n - \{(0, 0, 0)\})$. For simple robots, with not more than say three degrees of freedom, the definition works fine. The problem though is that, if the number of degrees of freedom gets high, a configuration will have too many direct neighbors. E.g., if the robot has eight links, a configuration will have 6561 direct neighbors. A possibility is to redefine the direct neighbors. For example, we can take the set $\{(\epsilon, 0, \ldots, 0), (-\epsilon, 0, \ldots, 0), (0, \epsilon, 0, \ldots, 0), (0, -\epsilon, 0, \ldots, 0), \ldots, (0, \ldots, 0, \epsilon), (0, \ldots, 0, -\epsilon)\}$. The number of direct neighbors is then only linear in the number of degrees of freedom. The price of such a 'cheaper' definition though is, that the local method will more often get stuck in a local minimum. Which definition works best very much depends on the scene, and in particular on the number of degrees of freedom of the robot.

Experiments show that for robots with many degrees of freedom, the best results are obtained by very simple local methods, e.g., the local method which just constructs the straight path (in configuration space) connecting its two argument configurations, and succeeds iff this path intersects no obstacles. Because we are mainly interested in robots with many degrees of freedom, we will give experimental results using this last local method.

Experimental results

We demonstrate the performance of our learning paradigm applied to planar articulated robots with four example scenes, where the robots have, respectively, three, five, four, and nine degrees of freedom. We give results for two different node adding strategies, i.e., the random strategy and the adaptive random strategy, combined with the tree edge adding method. Geometric node adding, as described in section 4.1, cannot be applied to articulated robots. The distance between two configurations c_1 and c_2 is defined as Euclidean distance between c_1 and c_2 in configuration space, but scaled in a way that this distance reflects the sweep volume (in work space) of the straight path (in configuration space) from c_1 to c_2. In the test scenes used, the coordinates of all workspace obstacles again will lie in the unit square.

The test scenes

See figures 20 to 23 for the test scenes. In each scene a (smoothed) path for the corresponding robot, computed by our planner, is indicated by a few steps. The darkest step corresponds to the start configuration, and the white step to the end configuration of the path. Furthermore, for every joint it is indicated whether it is revolute (a curved arrow) or prismatic (a straight arrow).

Scene 1 is a relatively easy scene. We see a three degrees of freedom robot, with three revolute joints, in a workspace containing two obstacles. Scene 2 is more difficult, due to the robots five degrees of freedom (it has five revolute joints), and the presence of some narrow areas in free configuration space. Furthermore there is an obstacle boundary. In scene 3 we have a four degrees of freedom robot, with three revolute joints, and one prismatic joint. The final scene is the most difficult one, with a nine degrees of freedom robot.

The configuration spaces of the four robots are, respectively, $[-120, 120]^3$, $[-120, 120]^5$, $[-90, 90] \times [0, 0.2] \times [-90, 90] \times [-120, 120]$, and $[-120, 120]^9$, so the links can rotate over at most 240°. The robot is not allowed to intersect itself, restricting the possible motions even further.

The test results

See figures 24 to 27 for plots of $K(t)$, the knowledge acquired by the planner after having learned for time t. The dashed plots correspond to random node adding, and the dotted plots to adaptive random node adding. We see that the learning problem in scene 1 is solved for practically 100% in about four seconds with random node adding, and in about two seconds when the adaptive adding strategy is applied. In scene 2 it takes almost ten seconds to acquire 95% knowledge with adaptive random adding, which is again a clearly better performance than that achieved by normal random adding. In scene 3 the problem is solved for 90% in about fifteen seconds, with both adding strategies. Scene 4 takes much longer, i.e., more than two minutes are required for 90% knowledge.

The growth of the graphs varies from about 8 nodes per second in scene 4 to about 100 nodes per second in scene 1.

Figure 20: *A 3DOF articulated robot in scene 1.*

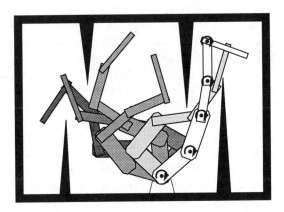

Figure 21: *A 5DOF articulated robot in scene 2.*

7 Conclusions and related work

In this paper we have presented a probabilistic technique for the learning motion planning problem. This technique, which is conceptually is simple, proves to be very fast for simple robots (e.g., free flying planar robots, easy articulated robots), and powerful enough to deal reasonably well with non-holonomicaly constrained robots (car-like robots) and robots with high degrees of freedom.

A nice property furthermore is the great flexibility of the method. In order to apply the method to some particular robot type, all that is needed is a local method which computes feasible paths for this robot type, and some (induced) metric. Experimental results (see [12], [14]) indicate that very primitive local methods achieve

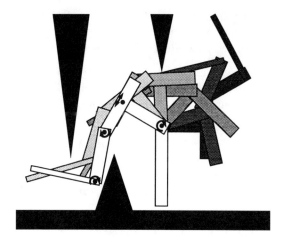

Figure 22: *A 4DOF articulated robot in scene 3.*

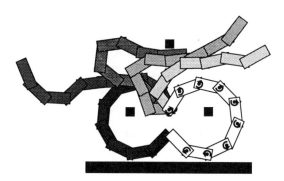

Figure 23: *A 9DOF articulated robot in scene 4.*

the best results. Hence, it should normally be no problem to find a good local method for some given robot type.

Currently we are working on the application of the method to tractor-trailer robots, and furthermore we are planning extensions of the method in various directions, aimed at solving more difficult motion planning problems, such as motion planning in scenes with multiple robots, moving obstacles, or perhaps some amount of uncertainty. Also we are planning to apply our method to 3-dimensional workspaces. Finally, we are also considering some possibilities to do the node adding in a non-random manner, which guarantees completeness in the query phase.

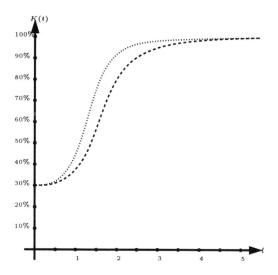

Figure 24: *Learning in scene 1.*

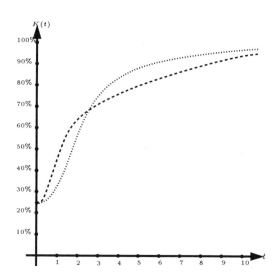

Figure 25: *Learning in scene 2.*

Acknowledgments

This research was partially supported by the ESPRIT III BRA Project 6546 (PROMotion) and by the Dutch Organisation for Scientific Research (N.W.O.).

We would like to thank Geert-Jan Giezeman for the implementation of many crucial 'geometric' routines (some of which are contained in the *Plageo* library, see [6]), Erik Vermeer, who has written most of the code

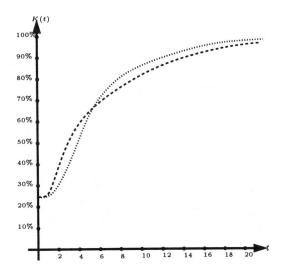

Figure 26: *Learning in scene 3.*

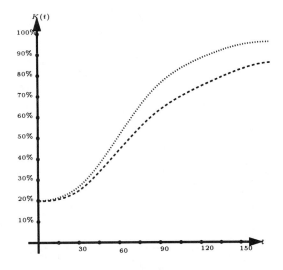

Figure 27: *Learning in scene 4.*

for articulated robots, Jules Vleugels, Erik van Wessel, and Otfried Schwarzkopf for *Ipe*, with which all figures in this paper where either created or edited.

References

[1] J. Barraquand and J.-C. Latombe. A monte-carlo algorithm for path planning with many degrees of freedom. In *Proc. IEEE Intern. Conf. on Robotics and Automation*, pages 1712–1717, 1990.

[2] J. Barraquand and J.-C. Latombe. Robot motion planning: A distributed representation approach. *Internat. J. Robot. Res.*, 10:628–649, 1991.

[3] J. Barraquand and J.-C. Latombe. Nonholonomic multibody mobile robots: Controllability and motion planning in the presence of obstacles. *Algorithmica*, 10:121–155, 1993.

[4] B. Faverjon and P. Tournassoud. A local approach for path planning of manipulators with a high number of degrees of freedom. In *Proc. IEEE Intern. Conf. on Robotics and Automation*, pages 1152–1159, 1987.

[5] B. Faverjon and P. Tournassoud. A practical approach to motion planning for manipulators with many degrees of freedom. In *Proc. 5th Intern. Symp. on Robotics Research*, pages 65–73, 1989.

[6] G.-J. Giezeman. *PlaGeo—A Library for Planar Geometry*. Dept. Comput. Sci., Utrecht Univ., Utrecht, the Netherlands, August 1993.

[7] Y. Hwang and N. Ahuja. Gross motion planning— a survey. *ACM Comput. Surv.*, 24(3):219–291, 1992.

[8] L. Kavraki and J.-C. Latombe. Randomized preprocessing of configuration space for fast path planning. In *Proc. IEEE Internat. Conf. on Robotics and Automation*, San Diego, CA, 1994.

[9] K. Kondo. Motion planning with six degrees of freedom by multistrategic bidirectional heuristic free-space enumeration. *IEEE Transactions on Robotics and Automation*, 7(3):267–277, 1991.

[10] J.-C. Latombe. *Robot Motion Planning*. Kluwer Academic Publishers, Boston, 1991.

[11] J.-P. Laumond, M. Taïx, and P. Jacobs. A motion planner for car-like robots based on a global/local approach. In *Proc. IEEE Internat. Workshop Intell. Robots Syst.*, pages 765–773, 1990.

[12] J. Mastwijk. Motion planning using potential field methods. master thesis, Utrecht, the Netherlands, August 1992.

[13] M. Overmars. A random approach to motion planning. Technical Report RUU-CS-92-32, Dept. Comput. Sci., Utrecht Univ., Utrecht, the Netherlands, October 1992.

[14] P. Švestka. A probabilistic approach to motion planning for car-like robots. Technical Report RUU-CS-93-18, Dept. Comput. Sci., Utrecht Univ., Utrecht, the Netherlands, April 1993.

The "Ariadne's clew"[1] algorithm:

Global planning with local methods[2]

Pierre Bessière[3], *Centre National de la Recherche Scientifique, Laboratoire LIFIA, Grenoble, FRANCE*

Juan-Manuel Ahuactzin, *Laboratoire IMAG-LIFIA, Grenoble, FRANCE*

El-Ghazali Talbi, *Laboratoire IMAG-LGI, Grenoble, FRANCE*

Emmanuel Mazer, *Centre National de la Recherche Scientifique, Laboratoire LIFIA, FRANCE*

The goal of the work described in this paper is to build a path planner able to drive a robot in a dynamic environment where the obstacles are moving.

In order to do so, we propose a method, called "ARIADNE'S CLEW algorithm", to build a global path planner based on the combination of two local planning algorithms : an EXPLORE algorithm and a SEARCH algorithm. The purpose of the EXPLORE algorithm is to collect information about the environment with an increasingly fine resolution by placing landmarks in the searched space. The goal of the SEARCH algorithm is to opportunistically check if the target can be easily reached from any given placed landmark.

The ARIADNE'S CLEW algorithm is shown to be very fast in most cases allowing planning in dynamic environments. Hence, it is shown to be complete, which means that it is sure to find a path when one exists. Finally, we describe a massively parallel implementation of this algorithm.

1   Introduction

The goal of this work is to build a path planner able to drive a robot in a dynamic environment where the obstacles are moving.

Designing a path planner is a central question in robotics research. A review of the existing approaches can be found in Latombe's book [1]. There are two main ways to deal with this problem : the global and the local approaches. The global approaches suppose that a complete representation of the configuration space has been computed before looking for a path. The global approaches are complete in the sense that if a path exists it will be found. Unfortunately, computing the complete configuration space is very time consuming, worst, the complexity of this task grows exponentially as the number of degrees of freedom increases. Consequently, today most of the robot path planners are used off-line : the planner is invoked with a model of the environment, it produces a plan which is passed to the robot controller which, in turn, executes it. In general, the time necessary to achieve this, is not short enough to allow the robot to move in a dynamic environment. The local approaches need only partial knowledge of the configuration space. The decisions to move the robot are taken using local criteria and heuristics to choose the most promising directions. Consequently, the local methods are much faster. Unfortunately, they are not complete, it may happen that a solution exists and is not found. The local approaches consider planning as an optimisation problem, where finding a path to the target corresponds to the optimisation of some given function. As any optimisation technique, the local approaches are subject to get trapped in some local optima, where a path to the goal has not been found and from which it is impossible or, at least, very difficult to escape.

The ultimate goal of a planner is to find a path in the configuration space from the initial position to the target. However, while searching for this path, an interesting sub-goal to consider may be to try to collect information about the free space and about the possible paths to go about that space. The ARIADNE'S CLEW algorithm tries to do both at the same time. An EXPLORE algorithm collects information about the free space with an increasingly fine resolution, while, in parallel, a SEARCH algorithm opportunistically checks if the target can be reached. The EXPLORE algorithm works by placing landmarks in the searched space in such a way that a path from the initial position to any landmark is known. In order to learn as much as possible about the free space the EXPLORE algorithm tries to spread the landmarks all over the space. To do so, it tries to put the landmarks as far as possible from one another.

For each new landmark produced by the EXPLORE algorithm the SEARCH algorithm checks with a local method if the target may be reached from that landmark. The ARIADNE'S CLEW algorithm is very fast, however, we will show that it is a complete planner which will find a path if one exits. The resolution at which the space is scanned and the time spent to do so, automatically adapts to the difficulty of the problem.

Both the EXPLORE and the SEARCH algorithms may be seen as solving optimisation problems. We first introduce the optimisation technique we are using, namely, genetic algorithms. We then describe successively in some details, the SEARCH algorithm, the EXPLORE algorithm and the concatenation of both. We finally explain a massively parallel implementation of our method and present some results proving that using this method we are able to drive a robot in a dynamic environment. We conclude with a discussion and some perspectives for future work.

2   Principle of genetic algorithms

Genetic algorithms are programs used to deal with optimisation problems. They have first been introduced by Holland [2]. Their goal is to find optimum of a given function F on a given search space S. For instance, the search space S may be 2^N, a point of S is then described by a vector of N bits and F is a function able to compute a real value for each of the 2^N vectors.

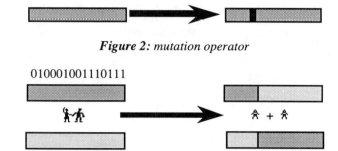

Figure 2: *mutation operator*

010001001110111

Figure 3: *cross-over operator*

In an initialisation step a set of points of the search space S (called a "population" of "individuals"), is drawn at random (the "genotype" of each individual is a vector of N bits). Then, the genetic algorithm iterates over the following 4 steps until a satisfying optimum is reached (see figure 1) :

1 - **Evaluation**: The function F is computed for each individual, ordering the population from the worst to the best.

2 - **Selection**: Pairs of individuals are selected, best individuals having more chance to be selected than poor ones (one individual may appear in different pairs).

3 - **Reproduction**: New individuals (called "offspring") are produced from these pairs.

4 - **Replacement**: A new population is generated by replacing some of the individuals of the old population by the new ones.

Reproduction is done using some "genetic operators". Number of them may be used but the two most common are mutation and crossing-over. The mutation operator picks at random some mutation locations among the N possible sites in the vector and flip the value of the bits at these locations as represented in figure 2. The cross-over operator selects at random a cut point among the N possible sites in the binary genotype and exchanges the last parts of the two parents vectors as shown in figure 3.

Genetic algorithms have many

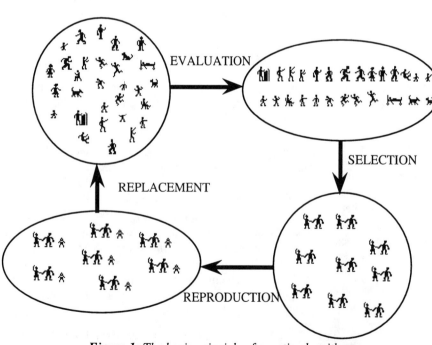

Figure 1: *The basic principle of genetic algorithms*

applications and exhibit very impressive optimisation capabilities compare to other optimisation techniques especially when the search space is big ($\approx 2^{300}$) and F quite irregular (see [3] for a recent survey).

Beside their scientific interest as a model of biological evolution, genetic algorithms have two main technological interests:

1 - They are very robust techniques able to deal with a very large class of optimisation problems.

2 - They are very easy to program in parallel and the acceleration obtained by doing so is considerable (see [4]).

We proposed a parallel genetic algorithm and developed an implementation on a massively parallel machine based on Transputers (see [5]). This algorithm and the performances obtained by the parallel implementation have been an essential achievement for the success of the work described in this paper.

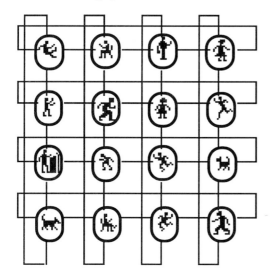

Figure 4: *the principle of the parallel genetic algorithm*

The principle of this parallel genetic algorithm is described by figure 4. It consists in one parallel process running for each individual in the population. The processes are organised in a torus structure where each process has 4 neighbours. At each generation all the individuals, in parallel, choose among their 4 neighbours with whom they want to breed and reproduce with the chosen bride. The parallel genetic algorithm iterates over the following 4 steps until a satisfying optimum is reached :

1 - **Evaluation**: Evaluate *in parallel* all the individuals.

2 - **Selection**: Select *in parallel*, among the four neighbours, the bride with the best evaluation.

3 - **Reproduction**: Reproduce *in parallel* with the chosen bride.

4 - **Replacement**: Replace *in parallel* the parents by the offspring.

3   The search algorithm

The purpose of the SEARCH algorithm is to determine if the target τ may be reached "simply" from a given point π. In order to do so, it looks for fixed length Manhattan motions in the configuration space starting at π and ending at τ.

Given a system with N degrees of freedom $\{\theta_1, \theta_2, ..., \theta_N\}$, a Manhattan motion of length 1 consists in moving each degree of freedom θ_i successively once by $\Delta\theta_i$. A Manhattan motion of length L is a succession of L Manhattan motions of length 1 or of LxN elementary motions of a single degree of freedom. Such a Manhattan motion M is denoted :

$$M = \left(\Delta\theta_1^1, \Delta\theta_2^1, ..., \Delta\theta_i^1, ..., \Delta\theta_N^1, \Delta\theta_1^2, \Delta\theta_2^2, ..., \Delta\theta_N^L\right)$$

Let us call τ_i^j the point reached in the configuration space after ixj elementary motions. Let us call τ_a^b the furthest point reached along M before a collision occurred. We are looking for a collision free Manhattan motion such that $\tau_a^b = \tau_N^L = \tau$.

The SEARCH algorithm may be expressed as an optimisation problem for the parallel genetic algorithm where:

- The search space S_S is the set of all Manhattan motions of length L starting at π.
- The evaluation function F_S applied to a Manhattan motion M given a target τ is defined as follow :
  $F_S(M,\tau)=0$ if any τ_i^j of M preceding τ_a^b is in the BACKPROJECTION$_S$ of τ. (The BACKPRO-JECTION$_S$ of τ is the set of all points of the searched space from which τ may be reached by a Manhattan motion of length 1).
  Otherwise, $F_S(M,\tau)=\|\tau - \tau_a^b\|$.

The SEARCH algorithm tries to minimise the evaluation function $F_S(M,\tau)$ over the search space S_S.

Manhattan motions have been chosen because for the NxL elementary motions of M, it is possible to compute simply in parallel, both the corresponding τ_i^j and the collision-free test on the path from π to τ_i^j (see [6]).

Furthermore, in a 3 dimension physical space, the collision-free test itself consists in three processes running in parallel checking respectively that there is no vertex-to-plan collisions, plan-to-vertex collisions and edge-to-edge collisions. Finally, each of these three processes may be expressed as the parallel evaluation of AxB processes where A is the number of elements in the first set of the test (A is the number of vertices, plans or edges) and where B is the number of elements in the second set (B is the number of plan, vertices or edges).

The SEARCH algorithm may be used as a planner by itself. It has been used as such for several applications. Let us describe briefly two of them (a more detailed presentation may be found in [7]).

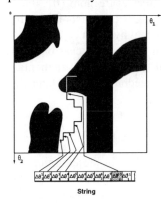

Figure 5.a

The first application is a planner for a planar arm with two degrees of freedom. By restricting ourselves to two dimensions we can graphically represent the configuration space and give the reader a better feeling of the method. However, the proposed method does not make any hypothesis about the number of degrees of freedom and can be used without modification for arms with a much larger number of degrees of freedom. Figure 5.a shows a Manhattan motion in the configuration space and the associated "individual" of the genetic algorithm. Figure 5.b shows the initial and final configuration of the arm in the operational space. Figure 5.c shows the path found in the operational

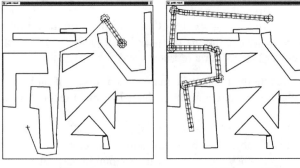

Figure 6.a   **Figure 6.b**

space and Figure 5.d the path found in the configuration space. Finally, Figure 5.e shows the portion of the configuration space which has been evaluated. It should be noticed that only a very restricted part of the configuration is really computed, this is one of the main explanation of the efficacy of the algorithm and this is why this algorithm is able to handle planning in dynamic environments.

The second application is a planner for a holonomic mobile robot. Figure 6 shows how the planner behaves in a dynamic environment. Figure 6.a shows the initial found path. Figure 6.b shows the path re-planned after the closing of the door. The used version of SEARCH has been implemented on a massively parallel Transputers machine. The planning time for a given path was less than 1 second on a machine of 64 Transputers.

As shown by the two previous examples, the used of SEARCH has a planner is very interesting. However, it may happen that the genetic algorithm gets trapped in some local minima. In that case the planner does not find a solution even if one exists. The SEARCH algorithm is not complete, this is its main drawback.

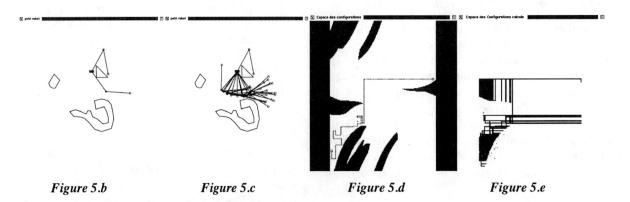

Figure 5.b   **Figure 5.c**   **Figure 5.d**   **Figure 5.e**

4 The explore algorithm

The purpose of the EXPLORE algorithm is to collect information about the free space. The EXPLORE algorithm works by placing landmarks in the searched space in such a way that a collision free Manhattan path from the initial position π to any landmark λ_k is known. In order to learn as much as possible about the free space the EXPLORE algorithm tries to spread the landmarks all over the space. To do so, it tries to put the landmarks as far as possible from one another. Let us call $\Lambda = \{\lambda_1, \lambda_2, ...\lambda_k, ...\}$ the set of already placed landmarks at a given step of the program. It is possible to define the distance between a point α of the searched space and the set Λ by $D(\alpha, \Lambda) = Min \|\lambda_k - \alpha\|$ on all landmarks $\lambda_k \in \Lambda$.

The EXPLORE algorithm may be expressed as an optimisation problem for the parallel genetic algorithm where:

- The search space S_e is the set of all Manhattan motions of length L starting from any landmark λ_k of Λ.
- The evaluation function F_e applied to a Manhattan motion M of S_e is defined as follows :
$F_e(M) = D(\tau_a^b, \Lambda)$ where τ_a^b is still the furthest point reached along M before a collision occurred.

The EXPLORE algorithm tries to maximise over the search space S_e the evaluation function $F_e(M)$. Figure 7 shows how the landmarks spread in the environment.

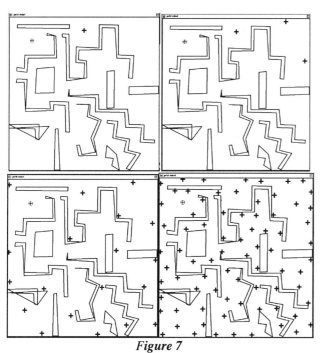

Figure 7

Figure 8 shows "the ARIADNE'S CLEW" : a tree of landmarks allowing to go about the free space.

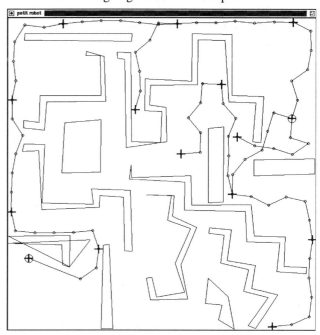

Figure 8: *The ARIADNE'S CLEW*

5 The Ariadne's clew algorithm

The purpose of the ARIADNE'S CLEW algorithm is to find a path from a given point π to a target τ.

The ARIADNE'S CLEW algorithm is the following:

1 - Use the SEARCH algorithm to find if a "simple" path exist between π and τ.
2 - If no "simple" path found by step 1, then do until a path is found
    2.1   *   Use EXPLORE to generate a new landmark λ.
    2.2   *   Use SEARCH to look for a "simple" path from λ to τ.

It is interesting to notice that SEARCH may be seen as a backprojection function for EXPLORE. SEARCH could be called BACKPROJECTION$_e$ because it plays relatively to EXPLORE the exact same role than BACKPROJECTION$_s$ relatively to SEARCH. A quite complicated backprojection function indeed, which usually produces very big backprojection allowing EXPLORE to stop after placing just a few landmarks.

The ARIADNE'S CLEW algorithm has three very important qualities:

- It reduces to the very fast SEARCH algorithm for most of the cases.

- It is complete, in the sense that if a path exists it will be found (see proposition 2 below).

- It automatically adapts the resolution at which it scans the space to the complexity of the problem (see proposition 3 below).

Figure 9 shows two complex paths found by the ARIADNE'S CLEW algorithm.

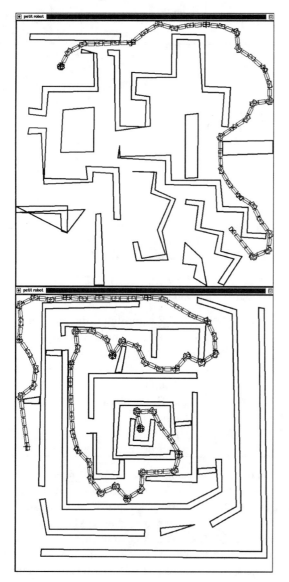

Figure 9

In the sequel of this section, three important propositions concerning the ARIADNE'S CLEW algorithm

will be established. However, given the restricted length of this paper, only sketches of proofs are proposed.

Definition 1: a PATH P from an initial point π to a target τ in an N dimensions metric space is defined as an N-uplet (F1(t), F2(t), ..., FN(t)) of N continuous functions from [0,1] -> $\Re$ such that (F1(0), F2(0), ..., FN(0)) are the co-ordinates of π and (F1(1), F2(1), ..., FN(1)) are the co-ordinates of τ.

Definition 2: a MANHATTAN MOTION OF LENGTH 1 in an N dimensions space is defined as an N-uplet $M_1 = \left(\Delta\theta_1^1, \Delta\theta_2^1, ..., \Delta\theta_i^1, ..., \Delta\theta_N^1 \right)$ where each $\Delta\theta_i^1$ is an integer corresponding to the length of the move along dimension i expressed in some given elementary length unit υ.

Definition 3: a MANHATTAN MOTION OF LENGTH L in an N dimension space is defined as an NxL-uplet: $M_L = \left(\Delta\theta_1^1, \Delta\theta_2^1, ..., \Delta\theta_i^1, ..., \Delta\theta_N^1, \Delta\theta_1^2, \Delta\theta_2^2, ..., \Delta\theta_N^L \right)$ where each $\Delta\theta_i^j$ is an integer corresponding to the length of the jth move along dimension i expressed in some given elementary length unit υ.

Proposition 1: for any $\varepsilon>0$, for any path P, it is possible to find υ, L and a Manhattan motion M_L of length L, such that the path P is approximated by M_L with an error less than ε.

Sketch of proof:
- Direct application of the Stone-Weirstrass theorem.

Proposition 2: the ARIADNE'S CLEW algorithm is complete, which means that, for any given $\varepsilon>0$, if a path exists from the initial point π to the target τ it will find (in a finite time) L and a Manhattan motion of length L M_L starting at π and ending at τ with an error less than ε.

Sketch of proof:
- Proposition 1 insures that such a Manhattan motion M_L exists.
- The ARIADNE'S CLEW algorithm searches a discrete finite space.
- The ARIADNE'S CLEW algorithm insures that all the produced Manhattan motions are different.
Consequently, M_L will be produced after a finite amount of time.

Remark: In fact Proposition 2 proves that any algorithm producing Manhattan motions without producing twice the same is complete. This is true either for an algorithm enumerating the Manhattan motions or for an algorithm drawing randomly the Manhattan motions (without drawing twice the same). Of course the ARIADNE'S CLEW algorithm is doing much, much, better than those two.

Definition 4: for a given ε, let us call the COMPLEXITY OF THE PROBLEM the minimum number C of identical tiles necessary to do a paving of the space, the biggest dimension of a tile being equal to ε.

Definition 5: let us call RESOLUTION R the number of landmarks generated by the ARIADNE'S CLEW algorithm to find a solution.

Proposition 3: resolution R is always inferior or equal to complexity C.

Sketch of proof:
- as long as R<C, two different landmarks may not be in the same tile given that the ARIADNE'S CLEW algorithm maximises the distance between the landmarks.
- for R=C, there is exactly one landmark in each tile.
- in that case, it exists a Manhattan motion starting at a distance of π less than ε (starting at the landmark in the same tile than π) and ending at a distance of τ less than ε (ending at the landmark in the same tile than τ).

Remark: In practice, experiences prove that R<<C. There are two main reasons for this. First, most of the time, SEARCH stops EXPLORE after the generation of just a few landmarks. Second, the ARIADNE'S CLEW algorithm adapts locally its resolution to the surrounding free space, generating a lot of landmarks where narrow doors or corridors have to be found and generating just a few of them when in an open free space.

6 Parallel implementation

It is possible to design a massively parallel implementation of the ARIADNE'S CLEW algorithm with 5 embedded levels of parallelism (see figure 10) :

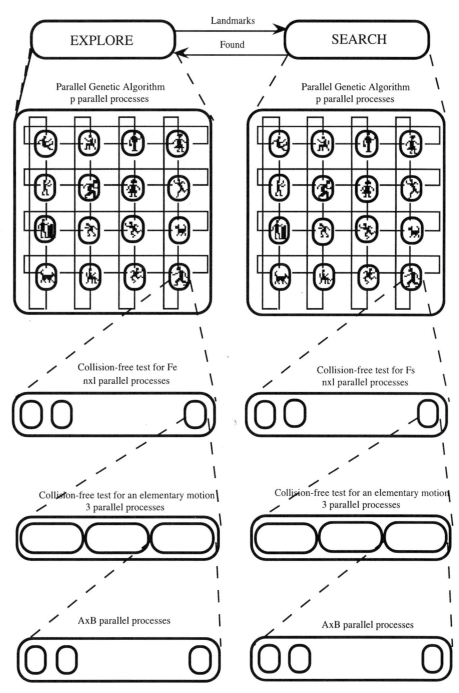

Figure 10

1 - At top level of parallelism, SEARCH and EXPLORE may run in parallel. EXPLORE produces a landmark while SEARCH exploits the previous one. SEARCH stops EXPLORE as soon as it reaches the target.

2 - Both SEARCH and EXPLORE need to run a genetic algorithm which can be implemented in parallel as described in section 2.

3 - Fs and Fe, the evaluation functions of SEARCH and EXPLORE, consist mainly in testing collision on paths. This may be done by NxL parallel processes, where N is the number of degrees of freedom and L the length of the considered Manhattan motions.

4 - Each of this NxL processes may be further decomposed as three parallel processes testing respectively vertex-to-plan, plan-to-vertex and edge-to-edge collisions.

5 - Finally, each of this test needs AxB parallel processes where A is the number of elements in the first set of the test (A is the number of vertices, plans or edges) and where B is the number of elements in the second set (B is the number of plans, vertices or edges).

The methodology used to implement this on a parallel machine consists in writing the application fully in parallel as if there were as many processors available as the number of processes. Of course, in practice, this is not the case. However, we use languages and tools (particularly the PAROS parallel operating system and the PARX communication kernel developed in the SUPERNODE II project, see [8]) which allows to conceive a parallel program independently of the architecture of the target machine.

We implement this on SUPER-NODE machine made of 128 Transputers.

7   Conclusion, results and perspectives

We have presented a general method to search a continuous configuration space. This method is implemented using a minimisation technique based on parallel genetic algorithms. We have demonstrated the validity of the method on a set of complex path planning problems. Finally, we proposed a massively parallel implementation of the method which permits on-line re-planning.

Our experimental set-up used to test our algorithm for an actual six degrees of freedom arm is represented on figure 11.

RobotI is under the control of the MegaNode (128 T800 Transputers parallel machine) running a parallel implementation of the ARIADNE'S CLEW algorithm. RobotII is used as a dynamically obstacle : it is manually controlled via KALI (KALI is a robot control software initially developed at Mc Gill University). First we use a CAD system called ACT[4] which permits precise geometrical description of the arms and obstacles and which is able to present 3D simulations. We compile this representation into a special format which is downloaded to the MegaNode. A final position is then specified to RobotI, the MegaNode quickly (**2 seconds**) produces a plan which assume RobotII is standing still, should the position of RobotII change under manual control, RobotI stops and the MegaNode (re)computes another path.

We plan in the next months to work on the problems of grasping, re-grasping and fine motion synthesis.

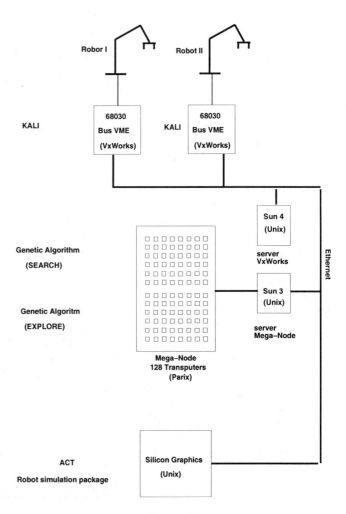

Figure 11

8 Bibliography

[1]   J-C. Latombe; "*Robot motion planning*", Ed. Kluwer Academic Publisher, 1991.

[2]   J.H. Holland; "*Adaptation in natural and artificial systems*"; Ann Arbor: Univ. of Michigan Press, 1975.

[3]   D.E. Goldberg; "*Genetic algorithms in search, optimization, and machine learning*"; Addison-Wesley, 1989.

[4]   E-G. Talbi & P.Bessière; "*A parallel genetic algorithm for the graph partitioning problem*"; ACM Int. Conf. on Supercomputing, Cologne, Germany, June 1991.

[5]   E-G. Talbi & T.Muntean; "*A parallel genetic algorithm for process-processors mapping*", Int. Conf. on High Speed Computing II, Montpellier, M.Durand and F.El Dabaghi (Editors), Elsevier Science Pub., North-Holland, pp.71-82, Oct 1991.

[6]   T. Lozano-Pérez, J.L. Jones, E. Mazer & P.A. O'Donnell; "*HANDEY A robot task planner*"; the MIT Press, 1992

[7]   J.M. Ahuactzin, A-G. Talbi, P. Bessière & E. Mazer; "*Using genetic algorithms for robot motion planning*"; ECAI92, Vienna, Austria, 1992

[8]   T. Muntean, N. Gonzalez & Y. Langue; "*PARX Kernel for the PAROS parallel operating system*"; ESPRIT '91; Kluewer Academic Publishers, 1991.

---

[1]Once upon a time, Ariadne, daughter of Minos, King of Crete, helped Theseus to kill the Minotaur who lived in the Labyrinth, a huge maze build by Daedalus. The main difficulty for Theseus was to find his way through out the Labyrinth. Ariadne imagined to give him a thread to unwind in order to find his path back.

[2]This work has been made possible by two EEC ESPRIT project: SUPERNODE II (n°2528) & PAPAGENA (n°6857) and CONACYT Mexico

[3]Telephone: (33)76.57.46.73; E-mail: bessiere@imag.fr; Fax: (33)76.57.46.02.

[4]Kali andACT are comercial products of Aleph Technologies

Motion Planning of Legged Robots

Jean-Daniel Boissonnat, Olivier Devillers, Sylvain Lazard
INRIA, Centre de Sophia-Antipolis, BP 93 06902 Sophia-Antipolis, France
Phone : +33 93 65 77 77 – Fax : +33 93 65 76 43 – E-mail: lazard@sophia.inria.fr

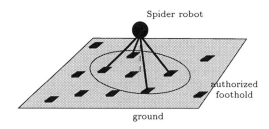

Figure 1: *The spider robot*

We study the problem of computing the set $\mathcal{F}$ of accessible and stable placements of a simple legged robot called the spider robot. The body of this robot is a single point and the legs are line segments attached to the body. The robot can only put its feet on some regions, called the foothold regions. Moreover, the robot is subject to two constraints: Each leg has a maximal extension R (accessibility constraint) and the body of the robot must lie above the convex hull of its feet (stability constraint). We present an efficient algorithm to compute $\mathcal{F}$. If the foothold regions are polygons with n edges in total, our algorithm computes $\mathcal{F}$ in $O(n^2 \alpha_8(n) \log n)$ time and $O(n^2 \alpha_8(n))$ space where $n\alpha_8(n) = \lambda_8(n)$ is the maximum length of the Davenport-Schinzel sequence of order 8 on n symbols. $\Omega(n^2)$ is a lower bound for the size of $\mathcal{F}$.

1 Introduction

Although legged robots have already been studied in robotics [8, 9], only a very few papers consider the motion planning problem amidst obstacles [6, 5, 1, 2]. In [6, 5] some heuristic approaches are described while, in [1, 2] efficient and provably correct geometric algorithms are described for a restricted type of legged robot, the so-called spider robots to be defined precisely below, and for finite sets of point footholds.

Compared to the classic piano movers problem, legged robots introduce new types of constraints. We assume that the environment consists of regions in the plane, called *foothold regions*, where the robot can safely put its legs. Then the legged robot must satisfy two different constraints: the accessibility and the stability constraints. A *placement* (position of the body of the robot) is called *accessible* if the legs of the robot can reach the footholds, and is called *stable* if the center

of mass of the robot lies above the interior of the convex hull of its feet. A placement that is both feasible and stable is called *admissible*, and the set of admissible placements is clearly the relevant information for a legged robot : we call this set *the free space* of the legged robot.

A first simple instance of a legged robot is the spider robot. The spider robot has been inspired by the Ambler, developed at Carnegie Mellon University [7]. The body of the spider robot is a single point, all its legs are attached to the body and can reach any foothold at distance less than R from the body (see Figure 1). The problem of the spider robot moving in an environment of point footholds has already been studied in [2] but the method used cannot be generalized to more complex environments. This paper proposes a new method to compute the free space of the spider robot based on a transformation between this problem and the problem of moving a half-disk amidst obstacles. The algorithm is simpler than the one described in [2] and the method can be extended to the case of polygonal foothold regions.

Once the free space has been computed, it can be used to find trajectories and sequences of legs assignments as described in [1].

The paper is organized as follows: Some notations and results of [2] are recalled in the next section. Sec-

Algorithmic Foundations of Robotics
1-56881-045-8

49

tion 3 shows the transformation between the spider robot problem and the half-disk problem. We present in Section 4 an algorithm that computes the free space of a spider robot for point footholds. The last section shows how the algorithm can be extended to polygonal foothold regions.

2 Notations and previous results

Let us introduce some notations: $\mathcal{S}$ is the set of discrete footholds $\{s_1, \ldots, s_n\}$. $\mathcal{F}$ and $\delta(\mathcal{F})$ denote respectively the free space of the spider robot and its boundary. C_i denotes the circle of radius R centered at s_i. $\mathcal{A}$ is the arrangement of the circles C_i for $1 \leq i \leq n$, i.e. the subdivision of the plane induced by the circles. This arrangement has an important geometric meaning in our problem and we will express the complexity results in term of $|\mathcal{A}|$, the size of $\mathcal{A}$. In the worst-case, $|\mathcal{A}| = \Gamma(n^2)$ but if k denotes the maximum number of disk $D(s_i, R)$ that can cover a point of the plane, it can be shown that $|\mathcal{A}| = O(kn)$ [10]. Clearly k is not larger than n and in case of sparse footholds $|\mathcal{A}|$ is linearly related to the number of footholds. $\mathrm{CH}(\mathcal{E})$ denotes the interior of the convex hull of a set $\mathcal{E}$, $\mathcal{C}(\mathcal{E})$ the complementary set of $\mathcal{E}$ and $\overline{\mathcal{E}}$ the closure of $\mathcal{E}$.

The algorithm described in [2] uses extensively the arrangement $\mathcal{A}$. In a cell Γ of $\mathcal{A}$, the set of footholds that can be reached by the robot is fixed, thus the part of Γ that belongs to $\mathcal{F}$ is exactly the intersection of Γ with the convex hull of the reachable footholds. Therefore the edges of $\delta(\mathcal{F})$ are either circular arcs belonging to $\mathcal{A}$ or portions of line segments joining two footholds; moreover a vertex of $\delta(\mathcal{F})$ which is the intersection of two straight line edges is a foothold (see Figure 2). The complexity of $\mathcal{F}$ has been proved to be $|\mathcal{F}| = \Theta(|\mathcal{A}|)$ [2].

An algorithm based on the same basic idea as above is also presented in [2]. It uses sophisticated data structures allowing the offline maintenance of convex hulls. Its time complexity is $O(|\mathcal{A}| \log n)$.

The algorithm described in this paper has the same time complexity, only uses simple data structures and can be extended to the case where the foothold regions are polygons (see Section 5).

3 From spider robots to half-disk robots

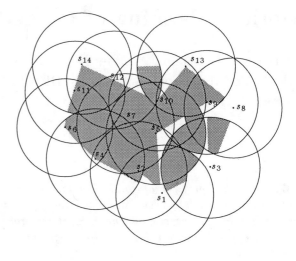

Figure 2: *An example of the free space of the spider robot.*

Theorem 1 *The spider robot does not admit a stable and accessible placement at point P if and only if there exists a half-disk (of radius R) centered at P that does not contain any foothold of $\mathcal{S}$ (Figure 3).*

Proof: Let G denotes the body of the robot. The spider robot is not stable if and only if the interior CH of the convex hull of all the reachable footholds does not contain G. $G \notin CH$ if and only if there exists a straight line L separating G and CH, which is equivalent to: there exists an open half-disk centered at G which does not contain any foothold (Figure 3). $\square$

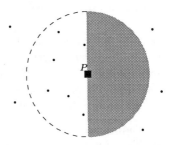

Figure 3: *A placement which is not stable.*

The next theorem will establish the connection between the free space of the spider robot and the free space of a half-disk robot moving by translation and rotation amidst n point obstacles.

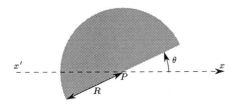

Figure 4: $HD(P,\theta)$

Definition 2 *Let $HD(P,\theta)$ be the open half-disk of radius R centered at P (see Figure 4) defined by :*

$$\begin{cases} (x - x_P)^2 + (y - y_P)^2 < R^2 \\ (x - x_P)\sin\theta - (y - y_P)\cos\theta < 0 \end{cases}$$

We define the free space $\mathcal{L}$ of the half-disk robot moving by translation and rotation amidst the obstacles of $\mathcal{S}$ as the set of (P,θ) such that the robot $HD(P, \theta + \pi)$ does not collide with any obstacle. (P,θ) denotes the position of the robot $HD(P, \theta + \pi)$.

Though this definition may seem unnatural, it will be convenient in the sequel (see Proposition 5).

Notice that $\mathcal{L}$ is defined in $\mathbb{R}^2 \times S^1$ (where $S^1 = \mathbb{R}/2\pi\mathbb{Z}$). For convenience, we will often identify S^1 and the interval $[0, 2\pi]$ of $\mathbb{R}$ and speak of the θ-axis. This identification allows us to define $p_{//\theta}$ the orthogonal projection onto $\mathbb{R}^2$.

With our definition of a half-disk robot, we can rewrite Theorem 1 as:

Theorem 3 $\mathcal{F} = \mathcal{C}(p_{//\theta}(\mathcal{L}))$, *where $\mathcal{L}$ is the free space of the half-disk robot moving amidst the footholds $s_1, \ldots, s_n$ considered as point obstacles.*

Definition 4 $\forall s_i \in \mathcal{S}$ $(1 \le i \le n)$ *let us define:*

$$\mathcal{H}_i = \{(P,\theta) \in \mathbb{R}^2 \times S^1 \ / \ P \in HD(s_i, \theta)\}$$

$$\mathcal{H} = \bigcup_{i=1}^n \mathcal{H}_i$$

$$\mathcal{C}_i = C_i \times S^1$$

$\mathcal{H}_i$ *will be called the helicoidal volume centered at s_i(Figure 5). $\mathcal{C}_i$ is a torus in $\mathbb{R}^2 \times S^1$.*

Notice the typographical difference between the circle C_i defined in $\mathbb{R}^2$ and $\mathcal{C}_i$ defined in $\mathbb{R}^2 \times S^1$. As a general rule, calligraphic or greek letters will denote

objects in $\mathbb{R}^2 \times S^1$ and angles, while roman letters will denote objects in $\mathbb{R}^2$.

The boundary of the half-disk $HD(s_i, \theta)$ consists of a half circle $HC_i(\theta)$ and a diameter $D_i(\theta)$ of C_i. The boundary of $\mathcal{H}_i$ is composed of a cylindrical surface $\mathcal{S}_i$ contained in $\mathcal{C}_i$ and a helicoidal surface $\mathcal{D}_i$:

$$\begin{aligned} \mathcal{S}_i &= \{HC_i(\theta), \theta \in S^1\} \\ \mathcal{D}_i &= \{D_i(\theta), \theta \in S^1\} \end{aligned}$$

We will distinguish further the two radii $r_i(\theta)$ and $r_i(\theta + \pi)$ of $D_i(\theta)$, and consider the boundary of $\mathcal{H}_i$ as composed of the three patches

$$\begin{aligned} \mathcal{S}_i &= \{HC_i(\theta), \theta \in S^1\} \\ \mathcal{R}_i^+ &= \{r_i(\theta), \theta \in S^1\} \\ \mathcal{R}_i^- &= \{r_i(\theta + \pi), \theta \in S^1\} \end{aligned}$$

Let Π_{θ_0} denotes the plane $\theta = \theta_0$ in $\mathbb{R}^2 \times S^1$.

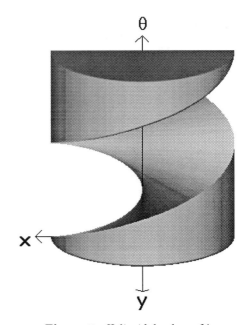

Figure 5: *Helicoidal volume $\mathcal{H}_i$*

Proposition 5 *The free space of the half-disk robot moving by translation and rotation amidst the obstacles $s_1, \ldots, s_n$ is the complementary set, in $\mathbb{R}^2 \times S^1$, of the union of the n helicoidal volumes centered at the s_i $(1 \le i \le n)$:*

$$\mathcal{L} = \mathcal{C}(\mathcal{H})$$

Proof: The free space $\mathcal{L} \cap \Pi_\theta$ of the half-disk robot moving by translation keeping a fixed orientation θ amidst the obstacle $s_1, \ldots, s_n$ is the complementary set of the union of the Minkowski's sum $s_i - HD(O, \theta + \pi) = HD(s_i, \theta)$ (Definition 2). Thus we have:

$$\forall \theta \in S^1 \quad \mathcal{L} \cap \Pi_\theta = \mathcal{C}(\bigcup_{1 \le i \le n} HD(s_i, \theta))$$

and Definition 4 yields the result.   □

Theorem 3 and Proposition 5 give $\mathcal{F}$ in terms of $\mathcal{H}$:

Theorem 6 $\mathcal{F} = \mathcal{C}(p_{//\theta}(\mathcal{C}(\mathcal{H})))$ *where* $\mathcal{C}(\mathcal{E})$ *denotes the complementary set of* $\mathcal{E}$ *in* $\mathbb{R}^2$ *or* $\mathbb{R}^2 \times S^1$.

Remark 7 $\mathcal{C}(p_{//\theta}(\mathcal{C}(\mathcal{H})))$ *is the vertical projection of the largest cylinder included in* $\mathcal{H}$, *whose axis is parallel to the* θ-*axis (see Figure 6). The basis of this cylinder is in fact* $\mathcal{F}$.

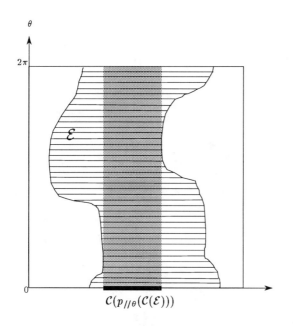

Figure 6: $\mathcal{C}(p_{//\theta}(\mathcal{C}(\mathcal{E})))$

4   Computation of $\mathcal{F}$

We know that each arc of the boundary $\delta(\mathcal{F})$ of $\mathcal{F}$ is either a straight line segment belonging to a line joining two footholds or an arc of a circle C_i (Section 2). The circular arcs $\delta(\mathcal{F}) \cap C_i$ are computed first (Sections 4.1, 4.2, and 4.3) and linked together with the line segments in a second step (Section 4.4 and 4.5).

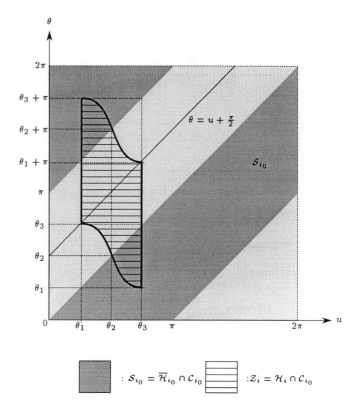

Figure 7: *Intersection of* $\mathcal{H}_i$ *with* $\mathcal{C}_{i_0}$ *for* $|s_{i_0}, s_i| = \sqrt{2}\,R$, $\theta_1 = \pi/4$, $\theta_2 = \pi/2$, $\theta_3 = 3\pi/4$.

4.1   Computation of $\delta(\mathcal{F}) \cap C_i$

We compute the contribution of each circle C_i to $\delta(\mathcal{F})$ in turn. Let $\mathcal{C}_{i_0}$ be the torus $C_{i_0} \times S^1$ for some $i_0 \in \{1, \ldots, n\}$. The natural parameters of $\mathcal{C}_{i_0}$ will be denoted u and θ. The portion $\mathcal{S}_{i_0}$ of the boundary of the helicoidal volume $\mathcal{H}_{i_0}$ which is included in $\mathcal{C}_{i_0}$ is shown in dark grey in Figure 7.

Let $\mathcal{Z}_i$ denotes the intersection $\mathcal{H}_i \cap \mathcal{C}_{i_0}$. Figure 7 shows an example of such a region $\mathcal{Z}_i$ through the natural parametrization.

According to Theorem 6 and Remark 7, the largest vertical strip Σ_{i_0} included in $\cup_{i \ne i_0} \overline{\mathcal{Z}_i} \cup \mathcal{S}_{i_0}$ (*i.e.* $\cup_i \overline{\mathcal{H}_i} \cap \mathcal{C}_{i_0}$) projects, along a direction parallel to the θ-axis, onto the portion of $\mathcal{C}_{i_0}$ which contributes to the closure of $\mathcal{F}$. The largest vertical strip Σ'_{i_0} included in $\cup_{i \ne i_0} \mathcal{Z}_i$ (*i.e.* $\cup_i \mathcal{H}_i \cap \mathcal{C}_{i_0}$) projects, along a direction parallel to the θ-axis, onto the portion of $\mathcal{C}_{i_0}$ which contributes

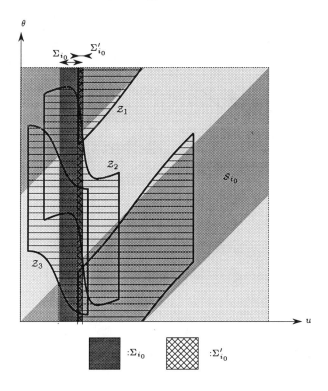

Figure 8: *Contribution of C_{i_0} to $\delta(\mathcal{F})$ $(0 < |s_1, s_{i_0}| < R$, $R \leq |s_2, s_{i_0}| < \sqrt{2}\,R$, $\sqrt{2}\,R \leq |s_3, s_{i_0}| < 2R)$.*

to the interior of $\mathcal{F}$. It follows that the contribution of C_{i_0} to $\delta(\mathcal{F})$ is the vertical projection onto C_{i_0} of the vertical strip $\Sigma_{i_0} \setminus \Sigma'_{i_0}$ (see Figure 8).

The main problem is to compute the union of the regions $\mathcal{Z}_i$ traced on C_{i_0} by the $\mathcal{H}_i$ in $O(k_{i_0} \log k_{i_0})$ time where k_{i_0} is the number of helicoidal volumes $\mathcal{H}_i$ intersecting C_{i_0}.

This is possible because the $\mathcal{Z}_i$ have a special shape that allows to reduce the computation of the union of the $\mathcal{Z}_i$ to the computation of a small number of lower envelopes of curves drawn on C_{i_0} with the property that two of them intersect at most once. The geometric properties of the $\mathcal{Z}_i$ are discussed in Section 4.2 and, in Section 4.3, we present and analyze the algorithm for constructing $\delta(\mathcal{F}) \cap C_{i_0}$.

4.2 Properties of the $\mathcal{Z}_i$

For convenience, we will use the vocabulary of the plane when describing objects on the torus. For instance, the

curve drawn on the torus C_{i_0} with equation

$$a\,\theta + b\,u + c = 0$$

will be called a line. The line $u = u_0$ will be called vertical and oriented according to increasing θ. Lower and upper will refer to this orientation.

Proposition 8 *For $i \neq i_0$, $\mathcal{Z}_i$ is a connected region bounded by two vertical line segments of length π, and two curved edges ρ_i^+ and ρ_i^- which are translated copies of one another. Specifically $\rho_i^+ = \rho_i^- + (0, 0, \pi)$.*

Proof: Consider the two points of intersection between the circles C_{i_0} and C_i $(i \neq i_0)$. Let $u = \theta_1$ and $u = \theta_3$ be the parameters of these two points through the parametrisation of C_{i_0}. The half circle $HC_i(\theta)$ intersects C_{i_0} at $u = \theta_3$ $(u = \theta_1)$ for $\theta \in [\theta_1, \theta_1 + \pi]$ $(\theta \in [\theta_3, \theta_3 + \pi])$ (Figure 9). Hence $\mathcal{S}_i \cap C_{i_0}$ consists of the two vertical line segments of length π $\{(u, \theta) \in \{\theta_3\} \times [\theta_1, \theta_1 + \pi]\}$ and $\{(u, \theta) \in \{\theta_1\} \times [\theta_3, \theta_3 + \pi]\}$ (Figure 7).

It is clear from their definitions that

$$\mathcal{R}_i^+ = \mathcal{R}_i^- + (0, 0, \pi).$$

Therefore the two curved edges $\rho_i^- = \mathcal{R}_i^- \cap C_{i_0}$ and $\rho_i^+ = \mathcal{R}_i^+ \cap C_{i_0}$ are translated copies of one another (Figure 7). □

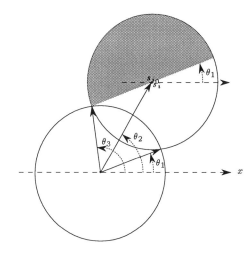

Figure 9: *Definition of θ_1, θ_2 and θ_3 $(0 < |s_{i_0}, s_i| < 2R)$.*

When not necessary, we will not specify which one of ρ_i^+ or ρ_i^- (resp. $\mathcal{R}_i$ or $\mathcal{R}_i^+$) is considered, and we will simply use the notation ρ_i (resp. $\mathcal{R}_i$). ρ_i' or ρ_i^j will denote a portion of ρ_i.

Proposition 9 *If $\sqrt{2}\,R \leq |s_{i_0}, s_i| < 2R$, $\mathcal{Z}_i$ is intersected by the line $\theta = u + \frac{\pi}{2}$ and ρ_i^+ and ρ_i^- are separated by that line.*

Proof: To show that ρ_i^+ and ρ_i^- are separated by the line $\theta = u + \frac{\pi}{2}$, we prove that these two curved edges do not intersect the line, except possibly at their end points.

Let (u_P, θ_P) be a point of ρ_i and P the point of C_{i_0} with parameter u_P. By definition, we have $u_P = \angle(\vec{x}, \overrightarrow{s_{i_0}P})$ $[2\pi]$ and $\theta_P = \angle(\vec{x}, \overrightarrow{s_i P})$ $[\pi]$ (Figure 10). Let $\gamma = \angle(\overrightarrow{Ps_{i_0}}, \overrightarrow{Ps_i})$ $[2\pi]$. If $\sqrt{2}\,R \leq |s_{i_0}, s_i| < 2R$, then $\gamma \in [\frac{\pi}{2}, \frac{3\pi}{2}]$ and $\gamma = \frac{\pi}{2}$ $[\pi]$ only when $|s_{i_0}, s_i| = \sqrt{2}\,R$ and (u_P, θ_P) is an end point of ρ_i. As $\gamma = \theta_P - u_P [\pi]$, ρ_i^+ and ρ_i^- are separated by the line $\theta = u + \frac{\pi}{2}$.

It remains to show that the line $\theta = u + \frac{\pi}{2}$ intersects $\mathcal{Z}_i$. This follows from the fact that the point of C_{i_0} $(\theta_2, \theta_2 + \frac{\pi}{2})$, where $\theta_2 = \angle(\vec{x}, \overrightarrow{s_{i_0}s_i})$ $[2\pi]$, belongs to the line $\theta = u + \frac{\pi}{2}$ and also to $\mathcal{Z}_i$ if $R < |s_{i_0}, s_i| < 2R$ (Figure 11). $\qquad\square$

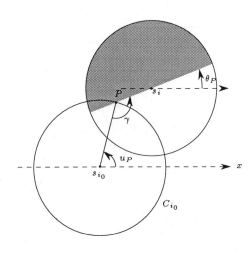

Figure 10: *For the definition of P, u_P, θ_P, γ.*

Proposition 10 *Let ρ_i' and ρ_j' be portions of ρ_i and ρ_j respectively which are both graphs of functions of θ defined over θ-intervals smaller than π. Then ρ_i' and ρ_j' intersect at most once.*

Proof: Let (u_I, θ_I) be an intersection point between ρ_i' and ρ_j' and I be the point of the circle C_{i_0} with parameter u_I. Because, by definition, ρ_i' is included in the intersection of $\mathcal{C}_{i_0}$ and $\mathcal{R}_i$, I is an intersection point

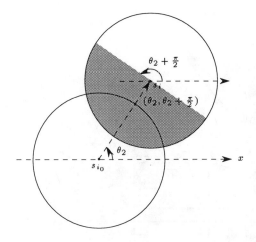

Figure 11: *Section of $\mathcal{H}_i \cap \mathcal{C}_{i_0}$ by the plane $\Pi_{\theta_2 + \frac{\pi}{2}}$*

of C_{i_0} and the diameter of $HD(s_i, \theta_I)$. Therefore the line passing through s_i and I has slope θ_I.

By applying the same argument to ρ_j', we obtain that s_i and s_j belong to the same straight line of slope θ_I. Therefore, if ρ_i' and ρ_j' intersect twice, at (u_I, θ_I) and (u_J, θ_J), then $\theta_I = \theta_J [\pi]$. Since the curves are the graphs of functions of θ defined on θ-intervals of length less than π, they intersect at most once. $\qquad\square$

Proposition 11 *If $\sqrt{2}\,R \leq |s_{i_0}, s_i| < 2R$, ρ_i is the graph of a function of θ defined over a θ-interval smaller than π.*

Proof: If $\sqrt{2}\,R \leq |s_{i_0}, s_i| < 2R$, $r_i(\theta)$ (and also $r_i(\theta + \pi)$) intersects C_{i_0} in at most one point, which proves that ρ_i is the graph of a function of θ. Furthermore, the θ-interval where ρ_i is defined is smaller than π. $\square$

As a consequence of Propositions 10 and 11, if the distances $|s_{i_0}, s_i|$ and $|s_{i_0}, s_j|$ belong to $[\sqrt{2}\,R, 2R[$, ρ_i and ρ_j intersect at most once.

Proposition 12 *If $0 < |s_{i_0}, s_i| < R$ $(R \leq |s_{i_0}, s_i| < \sqrt{2}\,R)$, ρ_i can be subdivided into two (three) sub-curves denoted ρ_i^k $k \leq 2$ (3) such that each piece is the graph of a function of θ defined over a θ-interval smaller than π.*

Proof: • **Case 1** : $0 < |s_{i_0}, s_i| < R$.

Any radius of C_i intersects C_{i_0} at most once. Hence ρ_i is the graph of a function of θ.

ρ_i is defined over a θ-interval greater than π but smaller than 2π. By splitting this interval in two equal parts, we split ρ_i in two sub-curves ρ_i^1 and ρ_i^2 which are defined over a θ-interval smaller than π.

• **Case 2 :** $R \le |s_{i_0}, s_i| < \sqrt{2}\,R$.

In that case, the θ-interval where $r_i(\theta)$ $(r_i(\theta + \pi))$ intersects C_{i_0} is smaller than π. This implies that ρ_i is defined over a θ-interval smaller than π.

There can be two intersection points between $r_i(\theta)$ and C_{i_0}. In order to overcome this difficulty, we split $r_i(\theta)$ into two segments (of fixed lengths, independent of θ) as follows (Figure 12). Let $r_i(\theta_T)$ be one of the two radii of C_i which are tangent to C_{i_0}. Let T be the point where $r_i(\theta_T)$ and C_{i_0} are tangent and $T(\theta)$ the point of $r_i(\theta)$ which is identical to T when $\theta = \theta_T$. Cutting $r_i(\theta)$ at $T(\theta)$ define two sub-radii $r_i'(\theta)$ and $r_i''(\theta)$ that intersect C_{i_0} in at most one point each; $r_i'(\theta)$ denotes the sub-radius not connected to s_i. We define

$$\mathcal{R}_i'^{+} = \{r_i'(\theta), \theta\} \qquad \mathcal{R}_i'^{-} = \{r_i'(\theta + \pi), \theta\}$$
$$\mathcal{R}_i''^{+} = \{r_i''(\theta), \theta\} \qquad \mathcal{R}_i''^{-} = \{r_i''(\theta + \pi), \theta\}$$

The intersection of $\mathcal{R}_i'$ and C_{i_0} consists of one continuous curve ρ_i^2, which is the graph of a decreasing function of θ. The intersection of $\mathcal{R}_i''$ and C_{i_0} consists of two continuous curves ρ_i^1 and ρ_i^3, which are both graphs of increasing functions of θ (see $\mathcal{R}_2$ on Figure 8). □

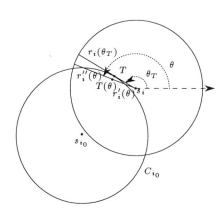

Figure 12: *For the definition of T, $T(\theta)$, $r_i'(\theta)$, $r_i''(\theta)$.*

As $\mathcal{R}_i'^{+} = \mathcal{R}_i'^{-} + (0, 0, \pi)$ and $\mathcal{R}_i''^{+} = \mathcal{R}_i''^{-} + (0, 0, \pi)$, ρ_i^{k+} and ρ_i^{k-} are translated copy of one another by vector $(0, 0, \pi)$. We denote $\mathcal{Z}_i^k$ the sub-region of $\mathcal{Z}_i$ below ρ_i^{k+} and above ρ_i^{k-}.

Proposition 13 *If $0 < |s_{i_0}, s_i| < R$, the two sub-curves ρ_i^{1+} and ρ_i^{1-} are separated by the line $\theta = u - \frac{\pi}{2}$ and the same holds for the two sub-curves ρ_i^{2+} and ρ_i^{2-}.*

If $R \le |s_{i_0}, s_i| < \sqrt{2}\,R$, the two sub-curves ρ_i^{2+} and ρ_i^{2-} are separated by the line $\theta = u + \frac{\pi}{2}$. The two sub-curves ρ_i^{1+} and ρ_i^{1-} are separated by the line $\theta = u - \frac{\pi}{2}$ and the same holds for the two sub-curves ρ_i^{3+} and ρ_i^{3-}.

Moreover, in each case, the region above ρ_i^{k-} and below ρ_i^{k+} is intersected by its separating line.

Proof: The proof is similar to the proof of Proposition 9. Let (u_P, θ_P) denote a point of a curve ρ_i^k. P is the point of C_{i_0} with parameter u_P and $\gamma = \angle(\overrightarrow{Ps_{i_0}}, \overrightarrow{Ps_i})\,[2\pi]$. We show that $\gamma \ne \frac{\pi}{2}\,[\pi]$, except when (u_P, θ_P) is an end point of ρ_i^k. As $\gamma = \theta_P - u_P$, ρ_i^{k+} and ρ_i^{k-} are separated by the lines $\theta = u \pm \frac{\pi}{2}$,

Case 1 : $0 < |s_{i_0}, s_i| < R$.

$\gamma \in [-\frac{\pi}{3}, \frac{\pi}{3}]$ for any P. Therefore the two curves ρ_i^+ and ρ_i^- are separated by the line $\theta = u - \frac{\pi}{2}$, and the same holds for the sub-curves of ρ_i^+ and ρ_i^-. Moreover, the point of C_{i_0} $(\theta_2, \theta_2 - \frac{\pi}{2})$ belongs to the line and to $\mathcal{Z}_i$, which proves that the line $\theta = u - \frac{\pi}{2}$ intersects $\mathcal{Z}_i$.

Case 2 : $R \le |s_{i_0}, s_i| < \sqrt{2}\,R$.

Let (u_{P_1}, θ_{P_1}) be the point connecting ρ_i^1 and ρ_i^2, and (u_{P_2}, θ_{P_2}) be the point connecting ρ_i^2 and ρ_i^3. P_1 and P_2 are the points where the tangent to C_{i_0} passes through s_i. Thus P_1 and P_2 are the only points of C_{i_0} where $\gamma = \frac{\pi}{2}\,[\pi]$. Therefore the curves ρ_i^{k-} and ρ_i^{k+} $(k = 1, 2, 3)$ are separated by the lines $\theta = u \pm \frac{\pi}{2}$.

If $R \le |s_{i_0}, s_i| < \sqrt{2}R$, it is easily shown that $(\theta_2, \theta_2 + \frac{\pi}{2})$ belongs to the region above ρ_i^{2-} and below ρ_i^{2+}, $(\theta_1, \theta_1 - \frac{\pi}{2})$ belongs to the region above ρ_i^{1-} and below ρ_i^{1+}, $(\theta_3, \theta_3 - \frac{\pi}{2})$ belongs to the region above ρ_i^{3-} and below ρ_i^{3+}. □

4.3 Construction of $\delta(\mathcal{F}) \cap C_{i_0}$

We have seen in section 4.1 that the contribution of a circle C_{i_0} to $\delta(\mathcal{F})$ is given by the computation of $\cup_{i \ne i_0} \mathcal{Z}_i$ and $\cup_{i \ne i_0} \overline{\mathcal{Z}_i} \cup \mathcal{S}_{i_0}$. Propositions 9 and 13 show that the set of regions $\mathcal{Z}_i$ can be split into two subsets Ω_1 and Ω_2 as follows. A region $\mathcal{Z}_i$ (or $\mathcal{Z}_i^k$) of Ω_1 or Ω_2 is the region below the curve ρ_i^+ (or ρ_i^{k+}) and above ρ_i^- (or ρ_i^{k-}). For convenience, if not specified, ρ_i will denote either ρ_i or ρ_i^k. Each region of Ω_1 (Ω_2) is intersected by the line $\theta = u + \frac{\pi}{2}$ ($\theta = u - \frac{\pi}{2}$) and ρ_i^+, ρ_i^-

are separated by this line. Furthermore, the curves ρ_i do not intersect the lines $\theta = u \pm \frac{\pi}{2}$ (proofs of Propositions 9, 13). Thus, for $\mathcal{Z}_i \in \Omega_1$, we can consider the curves ρ_i^+ in the domain $\{(u, \theta) \in S^1 \times [u + \frac{\pi}{2}, u + \frac{3\pi}{2}]\}$ and the curves ρ_i^- in $\{(u, \theta) \in S^1 \times [u - \frac{\pi}{2}, u + \frac{\pi}{2}]\}$. In these domains, the region below the set of ρ_i^+ and above the set of ρ_i^- is given by the upper envelope of the ρ_i^+ and the lower envelope of the ρ_i^-. That yields the union of the regions of Ω_1 in $S^1 \times S^1$ without extra computation. We compute similarly the union of the regions of Ω_2.

Let us estimate the complexity of the above construction. The k_{i_0} helicoidal volumes $\mathcal{H}_i$ that intersect $\mathcal{C}_{i_0}$ can be found in $O(k_{i_0})$ time once the Delaunay triangulation of the footholds has been computed which can be done in $O(n \log n)$ time [3]. Since two curves ρ_i and ρ_j intersect each other at most once (Propositions 10, 11, 12) the upper and lower envelopes can be computed in $O(k_{i_0} \log k_{i_0})$ time and $O(k_{i_0} \alpha(k_{i_0}))$ space where α is the pseudo inverse of the Ackerman's function [4]. The union of Ω_1 and Ω_2 can be done in $O(k_{i_0} \alpha(k_{i_0}))$ time because each envelope is the graph of a function of u and two arcs of the two envelopes intersect each other at most once.

The computation of $\cup_{i \neq i_0} \overline{\mathcal{Z}_i} \cup \mathcal{S}_{i_0}$ can be done in the same time as follows. Let $\overline{\rho_{i_0}^+}$ and $\overline{\rho_{i_0}^-}$ denote the upper and lower edges of $\mathcal{S}_{i_0}$. $\overline{\rho_{i_0}^+}$ and $\overline{\rho_{i_0}^-}$ are the line $\theta = u$ and $\theta = u - \pi$. Proposition 10 is still true if only one of the two considered curved edges is a graph of a function of θ defined over a θ-interval smaller than π. Thus we can add to set Ω_2 the region $\mathcal{S}_{i_0}$, which satisfies the other conditions. Then the computation $\cup_{i \neq i_0} \overline{\mathcal{Z}_i} \cup \mathcal{S}_{i_0}$ follows as described above. Its time and space complexity is the same as above.

According to Theorem 6 and Remark 7, the contribution of C_{i_0} to the interior of $\mathcal{F}$ is $\mathcal{C}(p_{//\theta}(\mathcal{C}(\cup_{i \neq i_0} \mathcal{Z}_i)))$, the projection onto C_{i_0} of the largest vertical strip Σ'_{i_0} included in $\cup_{i \neq i_0} \mathcal{Z}_i$. This projection is easily computed by projecting the curved edges of $\cup_{i \neq i_0} \mathcal{Z}_i$ (Figures 6, 8). The computation of the contribution of C_{i_0} to the closure of $\mathcal{F}$, $\mathcal{C}(p_{//\theta}(\mathcal{C}(\cup_{i \neq i_0} \overline{\mathcal{Z}_i} \cup \mathcal{S}_{i_0})))$, is done similarly by projecting the strip Σ_{i_0}. These computations yield the contribution of C_{i_0} to $\delta(\mathcal{F})$ (Theorem 6). This step can plainly be done in $O(k_{i_0} \log k_{i_0})$ time.

Moreover we label an arc of $\delta(\mathcal{F})$ either by i if the arc belongs to the circle C_i or by (i, j) if the arc belongs to the straight line segment $[s_i, s_j]$. The labels

of the edges of $\delta(\mathcal{F})$ incident to C_{i_0} can be found as follows, without increasing the complexity. An arc of $\delta(\mathcal{F}) \cap C_{i_0}$ corresponds to a vertical strip $\Sigma_{i_0} \setminus \Sigma'_{i_0}$. An end point P of an arc is either the projection of a vertical edge of the strip or the projection of an intersection point between two curved edges. In the first case, P is the intersection of C_{i_0} with some C_i and in the second case, P is the intersection of C_{i_0} with some line segment $[s_i, s_j]$. Hence the labels of the edges of $\delta(\mathcal{F})$ incident to C_{i_0} can be found with no overcost during the construction.

Since $\mathcal{A}$ is the arrangement of the circles of radius R centered at the footholds, $|\mathcal{A}| = \sum_{i=1}^{n} k_i$. The above considerations yield the following theorem:

Theorem 14 *We can compute $\delta(\mathcal{F}) \cap \mathcal{A}$ and the labels of the edges of $\delta(\mathcal{F})$ incident to the arcs of $\delta(\mathcal{F}) \cap \mathcal{A}$ in $O(|\mathcal{A}| \log n)$ time and $O(|\mathcal{A}| \alpha(n))$ space.*

4.4 Computation of the arcs of $\delta(\mathcal{F})$ issued from a foothold

The above section has shown how to compute all the vertices of $\mathcal{F}$ that are incident to at least one arc of circle. It remains to find the vertices of $\mathcal{F}$ incident to two straight edges. Such vertices are known to be footholds of $\mathcal{S}$ (see Section 2). Consider a foothold s_{i_0}, and let CH be the interior of the convex hull of the footholds contained in the disk $D(s_{i_0}, R)$. We assume CH to be non empty, otherwise $s_{i_0} \notin \delta(\mathcal{F})$.

- If s_{i_0} belongs to CH, then s_{i_0} belongs to the interior of $\mathcal{F}$ because $s_{i_0} \in \mathcal{F}$ and $\mathcal{F}$ is an open set. Hence in this case $s_{i_0} \notin \delta(\mathcal{F})$.

- If s_{i_0} is a vertex of CH, then, in a neighborhood of s_{i_0}, CH and $\mathcal{F}$ are identical. It follows that two edges of $\delta(\mathcal{F})$ are incident to s_{i_0}. These two edges are contained in the two edges of CH incident to s_{i_0}.

The k'_{i_0} footholds contained in the disk $D(s_{i_0}, R)$ can be found in linear time because we have already computed the Delaunay triangulation of the footholds [3]. Thus we can decide if s_{i_0} belongs to the interior of CH, or else, find the two vertices of CH adjacent to s_{i_0} in $O(k'_{i_0})$ time and space. As the sum of the k'_i for $\{1, \ldots, n\}$ is bounded by the size of $\mathcal{A}$, we have the following theorem:

Theorem 15 *The footholds belonging to $\delta(\mathcal{F})$ and the labels of the arcs of $\delta(\mathcal{F})$ issued from these footholds can be found in $O(|\mathcal{A}|)$ time and space.*

4.5 Construction of $\mathcal{F}$

Theorem 16 *The free space of the spider robot can be computed in $O(|\mathcal{A}|\log n)$ time and $O(|\mathcal{A}|\alpha(n))$ space.*

Proof: By Theorem 14, we have computed all the circular arcs of $\delta(\mathcal{F})$ and the labels of the edges of $\delta(\mathcal{F})$ incident to them. By Theorem 15, we have computed all the vertices of $\delta(\mathcal{F})$ that are incident to two straight edges of $\delta(\mathcal{F})$ and the label of these two edges. It remains to sort, along each line (s_i, s_j) (circle C_i), the vertices of $\delta(\mathcal{F})$ belonging to that line (circle). This can easily be done in $O(|\mathcal{A}|\log n)$ time since $|\delta(\mathcal{F})| = \Theta(|\mathcal{A}|)$ [2]. It is then an easy task to deduce a complete description of $\delta(\mathcal{F})$. $\qquad\square$

5 Generalization to polygonal foothold regions

We consider now the case where the set of footholds is no longer a set of points but a set $\mathcal{S}$ of polygonal regions bounded by a set of n line segments $e_1, \ldots, e_n$. Clearly $\mathcal{S}$ is a subset of the free space $\mathcal{F}$. Let $\mathcal{F}'$ denotes the free space of the spider robot using as foothold regions only the edges of $\mathcal{S}$. Suppose that the spider robot admits an admissible placement outside $\mathcal{S}$ with its legs inside some polygonal footholds ; then the placement remains admissible if it retracts its legs on the boundary of these polygonal regions. Hence $\mathcal{F} = \mathcal{F}_s \cup \mathcal{S}$. We first show how to compute $\mathcal{F}'$.

All we have done in Section 3 remains true if the foothold regions are line segments provided that C_i, $\mathcal{H}_i$ and $\mathcal{A}$ are replaced by C_{e_i}, $\mathcal{H}_{e_i}$ and $\mathcal{E}\mathcal{A}$, where the generalized circles C_{e_i} and the generalized helicoidal volumes $\mathcal{H}_{e_i}$ are defined by (see Figures 13 and 14):

$$C_{e_i} = \{P \in I\!\!R^2 \,/\, |P, s| = R, \ s \in e_i\}$$

$$\mathcal{H}_{e_i} = \{(P, \theta) \in I\!\!R^2 \times S^1 \,/\, P \in HD(s, \theta), \ s \in e_i\}.$$

In order to avoid confusion, the objects C_i, $\mathcal{C}_i$, $\mathcal{H}_i$, $\mathcal{Z}_i$, $\mathcal{S}_{i_0}$, ρ_i, associated to a site s_i will be, up to the end, denoted by C_{s_i}, $\mathcal{C}_{s_i}$, $\mathcal{H}_{s_i}$, $\mathcal{Z}_{s_i}$, $\mathcal{S}_{s_{i_0}}$, ρ_{s_i}.

Let $\mathcal{E}\mathcal{A}$ denote the arrangement of the n generalized circles C_{e_i}. Notice that $|\mathcal{E}\mathcal{A}| = \Theta(n^2)$.

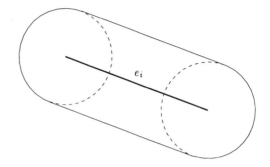

Figure 13: C_{e_i}

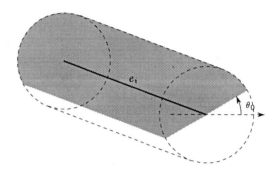

Figure 14: $\mathcal{H}_{e_i} \cap \Pi_{\theta_0}$

Each arc of the boundary $\delta(\mathcal{F}')$ of $\mathcal{F}'$ is either an arc of C_{e_i}, corresponding to a maximal extension of one leg, or an arc corresponding to a subset of unstable equilibriums of the spider robot. As we have done in Section 4 the contribution of each C_{e_i} to $\delta(\mathcal{F}')$ is computed first in turn.

C_{e_i} is composed of two half circles and two straight line segments. The contribution of C_{e_i} to $\delta(\mathcal{F}')$ will be computed in two steps, the contribution of the half circles and the contribution of the straight line segments. For convenience, we will not compute the contribution of each half circle to $\delta(\mathcal{F}')$ but the contribution of the whole circle to $\delta(\mathcal{F}')$. The way we compute these contributions does not depend of which circle is considered. Let $C_{s_{i_0}}$ denote one of the two circles, s_{i_0} its center and $\mathcal{C}_{s_{i_0}} = C_{s_{i_0}} \times S^1$. The computation $\delta(\mathcal{F}') \cap \mathcal{C}_{s_{i_0}}$ is done as explained in Section 4.1 but the properties of the new regions $\mathcal{H}_{e_i} \cap \mathcal{C}_{s_{i_0}}$ are not the same as properties of $\mathcal{H}_{s_i} \cap \mathcal{C}_{s_{i_0}}$ described in Section 4.2. The regions $\mathcal{Z}_{e_i} = \mathcal{H}_{e_i} \cap \mathcal{C}_{s_{i_0}}$ and $\mathcal{S}_{e_i} = \overline{\mathcal{H}_{e_i}} \cap \mathcal{C}_{s_{i_0}}$ will be studied in Sections 5.1 and 5.2. Section 5.3 generalizes

Section 4.3.

The contribution of the straight line segments of C_{e_i} to $\delta(\mathcal{F}')$ will be studied in Section 5.4.

Finally we will study the arcs of $\delta(\mathcal{F}')$ corresponding to the placements where the spider robot is in an unstable equilibrium (Section 5.5) and conclude in Section 5.6.

5.1 Properties of the $\mathcal{Z}_{e_i}$

We study here the region $\mathcal{Z}_{e_i}$ which is the intersection of $\mathcal{H}_{e_i}$ and of the torus $\mathcal{C}_{s_{i_0}}$. We use for the torus $\mathcal{C}_{s_{i_0}}$ the vocabulary introduced in Section 4.2. We will study first the region $\mathcal{Z}_{D_i}$ drawn on $\mathcal{C}_{s_{i_0}}$ by the helicoidal volume $\mathcal{H}_{D_i}$ induced by the straight line D_i, of slope γ_i, supporting e_i.

Proposition 17 *If D_i is a straight line of slope γ_i such that $s_{i_0} \notin D_i$, $\mathcal{Z}_{D_i}$ is a connected region bounded by two vertical line segments of length π, and two curved edges $\rho_{D_i}^+$ and $\rho_{D_i}^-$ which are symmetrical with respect to the line $\theta = \gamma_i$ (Figure 15, 16).*

Remark 18 *Since γ_i is defined modulo π, $\rho_{D_i}^+$ and $\rho_{D_i}^-$ are also symmetrical with respect to the line $\theta = \gamma_i + \pi$.*

Proof: Since $s_{i_0} \notin D_i$, $\mathcal{Z}_{D_i}$ is bounded by two vertical line $u = u_1$ and $u = u_2$ where u_1 and u_2 are the parameters on the circle $C_{s_{i_0}}$ of the two points I_1 and I_2 at distance R from D_i (Figure 17). Let s be a generic point of D_i and s_1 (s_2) be the point at distance R from I_1 (I_2). $\mathcal{Z}_s$ denotes the region traced on $\mathcal{C}_{s_{i_0}}$ by the helicoidal volume $\mathcal{H}_s$ associated to s.

$\mathcal{Z}_{D_i}$ is the union of the regions $\mathcal{Z}_s$, $s \in D_i$. As $\mathcal{Z}_s$ is connected (Proposition 8) and continuous with respect to s, $\mathcal{Z}_{D_i}$ is connected.

For any $s \neq s_1$ ($s \neq s_2$) the circle of radius R centered at s does not contain I_1 (I_2), then $\mathcal{Z}_{s_1}$ ($\mathcal{Z}_{s_2}$) coincides with $\mathcal{Z}_{D_i}$ on the line $u = u_1$ ($u = u_2$). Therefore $\mathcal{Z}_{D_i}$ is a region bounded by two vertical line segments of length π (Proposition 8).

The region $\mathcal{Z}_{D_i}$ is symmetrical with respect to the line $\theta = \gamma$ because the Minkowski's sum of $HD(P, \gamma + \theta)$ and D_i is equal to the Minkowski's sum of $HD(P, \gamma - \theta)$ and D_i.   □

Proposition 19 *If D_i is a straight line not intersecting $C_{s_{i_0}}$, the line $\theta = u + \frac{\pi}{2}$ intersects the region $\mathcal{Z}_{D_i}$ but not its upper and lower edges $\rho_{D_i}^+$ and $\rho_{D_i}^-$, except possibly at their end points (Figure 15).*

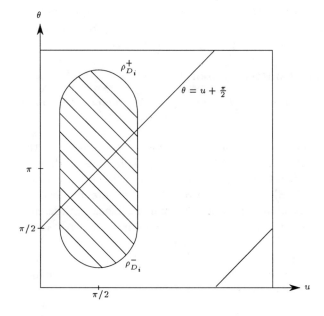

Figure 15: *$\mathcal{Z}_{D_i}$ for C_{i_0} of radius 4 centered at the origin and for $D_i : y = 6$*

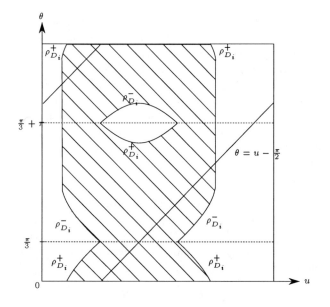

Figure 16: *$\mathcal{Z}_{D_i}$ for C_{i_0} of radius 4 centered at the origin and for $D_i : -\frac{\sqrt{3}}{2}x + \frac{1}{2}y = 2$ the line of slope $\gamma_i = \frac{\pi}{3}$ at distance 2 from s_{i_0}*

Proof: We use here the notations of the previous proof.

Let I'_1 (I'_2) denotes the intersection point distinct from I_1 (I_2) between $C_{s_{i_0}}$ and the circle of radius R centered at s_1 (s_2). Let u'_1 (u'_2) be its parameter on $C_{s_{i_0}}$ (Figure 17). By the definition of s_1 and I_1, the lines ($s_1 I_1$) and D_i are orthogonal; therefore ($s_{i_0} I'_1$) and D_i are also orthogonal. The same argument shows that ($s_{i_0} I'_2$) and D_i are orthogonal. Therefore $I'_1 = I'_2$ or $[I'_1 I'_2]$ is a diameter of $C_{s_{i_0}}$. By the definitions of I'_1 and I'_2, the distance from I'_1 and I'_2 to D_i is smaller than R. As ($s_{i_0} I'_1$) $\perp D_i$, ($s_{i_0} I'_2$) $\perp D_i$ and $s_{i_0} \notin D_i$, $[I'_1 I'_2]$ can not be a diameter of $C_{s_{i_0}}$. Then, I'_1 is equal to I'_2 and $u'_1 = u'_2$.

On the other hand, since D_i does not intersect $C_{s_{i_0}}$, the angle $\angle(\overrightarrow{I_1 s_{i_0}}, \overrightarrow{I_1 s_1})$ is greater than $\pi/2$ and therefore $|s_{i_0}, s_1| \geq \sqrt{2}\,R$ (Figure 17). $\mathcal{Z}_{D_i}$ contains $\mathcal{Z}_{s_1}$ ($\mathcal{Z}_{s_2}$) which contains on the u-interval $[u_1, u'_1]$ ($[u'_2, u_2]$) the line $\theta = u + \frac{\pi}{2}$ (Proposition 9). Since $u'_1 = u'_2$ and, as we have seen in the previous proof, $\mathcal{Z}_{D_i}$ is defined for u in $[u_1, u_2]$, $\mathcal{Z}_{D_i}$ contains the line $\theta = u + \frac{\pi}{2}$ for $u \in [u_1, u_2]$. Furthermore for $u = u_1$ ($u = u_2$) this line intersects the vertical edge of $\mathcal{Z}_{s_1}$ ($\mathcal{Z}_{s_2}$) (Proposition 9) which coincides with the vertical edge of $\mathcal{Z}_{D_i}$ (see previous proof). □

Since D_i intersects $C_{s_{i_0}}$, the angle $\angle(\overrightarrow{I_1 s_{i_0}}, \overrightarrow{I_1 s_1})$ is smaller than $\pi/2$; therefore $|s_{i_0}, s_1| < \sqrt{2}\,R$ and the same holds for $|s_{i_0}, s_2|$. Let s be a point of D_i between s_1 and s_2, I and I' the two intersection points between $C_{s_{i_0}}$ and the circle of radius R centered at s, u_I and $u_{I'}$ their parameters on $C_{s_{i_0}}$. When s moves along $[s_1 s_2]$, u_I and $u_{I'}$ describe $[u_1, u'_2]$ and $[u'_1, u_2]$. The arguments in the last proof that yield $I'_1 = I'_2$ still apply. Thus, for any $U \in [u_1, u_2]$, there exists s in D_i such that one of the two vertical edges of $\mathcal{Z}_s$ lies on the line $u = U$. Furthermore, since the distance between s_{i_0} and s is smaller than $\sqrt{2}\,R$ (Proposition 13), that vertical edge of $\mathcal{Z}_s$, contained in $\mathcal{Z}_{D_i}$, is intersected by the line $\theta = u - \frac{\pi}{2}$. That holds for any U in $[u_1, u_2]$. As $\mathcal{Z}_{D_i}$ is bounded by the lines $u = u_1$ and $u = u_2$, $\mathcal{Z}_{D_i}$ is intersected by the line $\theta = u - \frac{\pi}{2}$ and $\rho_{D_i}^+$ and $\rho_{D_i}^-$ are not intersected by that line. □

Proposition 21 *If D_i is a straight line of slope γ_i passing through s_{i_0}, $\mathcal{Z}_{D_i}$ is the region shown in Figure 18. The curves $\rho_{D_i}^+$ and $\rho_{D_i}^-$ are defined on $S^1 \setminus \{\gamma_i + \frac{\pi}{2}, \gamma_i + \frac{3\pi}{2}\}$ and belongs to the lines $\theta = u$, $\theta = u + \pi$, $\theta = -u + 2\gamma_i$, $\theta = -u + 2\gamma_i + \pi$.*

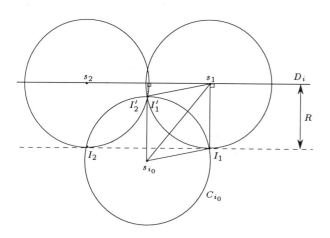

Figure 17: *For the proof of Propositions 17 and 19*

Proposition 20 *If D_i is a straight line not passing through s_{i_0} and intersecting $C_{s_{i_0}}$, the line $\theta = u - \frac{\pi}{2}$ intersects the region $\mathcal{Z}_{D_i}$ but not its upper and lower edges $\rho_{D_i}^+$ and $\rho_{D_i}^-$, except possibly at their end points (Figure 16).*

Proof: We use the notations of the two last proofs.

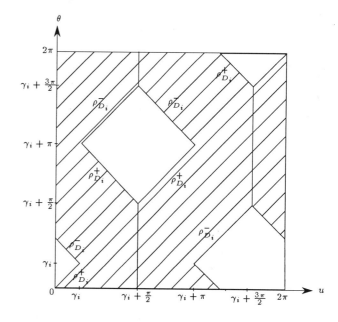

Figure 18: $\mathcal{Z}_{D_i}$ *for D_i of slope γ_i passing through s_{i_0}*

Proof: For any $\theta_0 \in [\gamma_i, \gamma_i + \frac{\pi}{2}]$, $\mathcal{H}_{D_i} \cap \Pi_{\theta_0}$ intersects $C_{s_{i_0}}$ for any u such that $u \notin [\theta_0 + \pi, 2\gamma_i - \theta_0]$ (Figure 19)

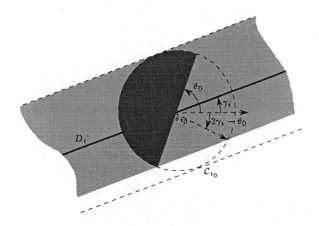

Figure 19: $\mathcal{H}_{D_i} \cap \Pi_{\theta_0}$, $s_{i_0} \in D_i$

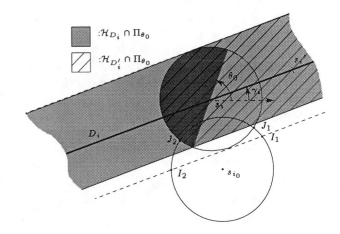

Figure 20: *Illustration of the case where $\theta_0 \in [\gamma_i, \gamma_i + \pi]$: $\mathcal{H}_{D_i'}$ and $\mathcal{H}_{D_i}$ coincide in $\mathcal{D}_R(s_i)$. For any $\theta_0 \in S^1$, $\mathcal{H}_{D_i'} \cap \Pi_{\theta_0}$ does not intersect the arc of circle J_2I_2 and coincides with $\mathcal{H}_{D_i} \cap \Pi_{\theta_0}$ on the arc I_1J_1.*

and $u \neq \gamma_i + \frac{\pi}{2}$ ($\mathcal{H}_{D_i}$ is an open set).

$\mathcal{Z}_{D_i}$ is symmetrical with respect to the line $\theta = \gamma_i$ (see the proof of Proposition 17). Moreover γ_i is defined modulo π, which gives the result. $\qquad\square$

Proposition 22 *Let s_i and s_i' be the two end points of the straight line segment e_i and D_i the supporting line of e_i. The region $\mathcal{Z}_{e_i}$ can be split into a bounded number of vertical strips such that each strip is bounded from above and from below by an arc of one of the curves ρ_{D_i}, ρ_{s_i} or $\rho_{s_i'}$. Here ρ_{D_i}, ρ_{s_i} and $\rho_{s_i'}$ denote the upper or lower edges of $\mathcal{Z}_{D_i}$ (Proposition 17), $\mathcal{Z}_{s_i}$ and $\mathcal{Z}_{s_i'}$ respectively (Proposition 8). Moreover, if $s_{i_0} \notin D_i$, on each strip, the upper and lower edges of $\mathcal{Z}_{e_i}$ are separated by the line $\theta = u + \pi/2$ or $\theta = u - \pi/2$.*

Proof: Let D_i' be the half straight line $[s_i, s_i')$ and $\gamma_i = \angle(\vec{x}, \overrightarrow{s_i s_i'})$ $[2\pi]$. Let $D_R(s_i)$ be the open disk of radius R centered at s_i and $\mathcal{D}_R(s_i)$ be $D_R(s_i) \times S^1$. Let u_1, u_2, v_1 and v_2 be the parameters of the points I_1, I_2, J_1, J_2 on the circle $C_{s_{i_0}}$ (Figure 20); I_1 and I_2 are the two points at distance R from D_i, J_1 and J_2 are the two intersection points between $C_{s_{i_0}}$ and C_i. $\mathcal{Z}_{s_i}$ is defined on $]v_1, v_2[$ and if $s_{i_0} \notin D_i$, $\mathcal{Z}_{D_i}$ is defined on $]u_1, u_2[$. Notice that $]v_1, v_2[$ is included in $]u_1, u_2[$ since $s_i \in D_i$. Let $\mathcal{H}_{D_i'}$ be the helicoidal volume induced by D_i' and $\mathcal{Z}_{D_i'}$ its intersection with $C_{s_{i_0}}$.

- **Case 1:** $s_i \neq s_{i_0}$.

  For $\theta \in [\gamma_i, \gamma_i + \pi]$, $\mathcal{H}_{D_i'}$ coincides with $\mathcal{H}_{D_i}$ in $\mathcal{D}_R(s_i)$ (Figure 20). Therefore $\mathcal{Z}_{D_i'}$ coincides with $\mathcal{Z}_{D_i}$ for $(u, \theta) \in]u_1, u_2[\times[\gamma_i, \gamma_i + \pi]$.

  For $\theta \in [\gamma_i - \pi, \gamma_i]$, $\mathcal{H}_{D_i'}$ coincides with $\mathcal{H}_{s_i}$ in $\mathcal{D}_R(s_i)$ (Figure 21). Therefore $\mathcal{Z}_{D_i'}$ coincides with $\mathcal{Z}_{s_i}$ for $(u, \theta) \in]u_1, u_2[\times[\gamma_i - \pi, \gamma_i]$.

  If $s_{i_0} \notin D_i$, then depending of the respective position of D_i' and s_{i_0}, $\mathcal{Z}_{D_i'}$ coincides, if not empty, with $\mathcal{Z}_{D_i}$ in the strips $]u_1, v_1] \times S^1$ and $[v_2, u_2[\times S^1$ (Figure 20). If $s_{i_0} \in D_i$, then $\mathcal{Z}_{D_i'}$ coincides with $\mathcal{Z}_{D_i}$ $\forall u \notin]v_1, v_2[$.

- **Case 2:** $s_i = s_{i_0}$.

  $\mathcal{Z}_{D_i'}$ coincides with $\mathcal{Z}_{D_i}$ for $u \in [\gamma_i - \frac{\pi}{2}, \gamma_i + \frac{\pi}{2}]$ and is empty for $u \in [\gamma_i + \frac{\pi}{2}, \gamma_i + \frac{3\pi}{2}]$ (Figure 20).

Moreover, since $s_i \in D_i'$, $\mathcal{Z}_{s_i}$ is defined for $(u, \theta) \in]v_1, v_2[\times S^1$ and included in $\mathcal{Z}_{D_i'}$. Then the upper and lower edges of $\mathcal{Z}_{D_i'}$ in the strip $]v_1, v_2[\times S^1$ are separated by the line $\theta = u \pm \pi/2$ (Propositions 9 and 13). On the other hand, if $s_{i_0} \notin D_i$, on a strip $U \times S^1$ where $\mathcal{Z}_{D_i'}$ coincides with $\mathcal{Z}_{D_i}$ the upper and lower edges of $\mathcal{Z}_{D_i'}$ are separated by the line $\theta = u \pm \pi/2$ (Propositions 19 and 20).

Hence the computation of $\mathcal{Z}_{s_i}$ and $\mathcal{Z}_{D_i}$ provides in constant time $\mathcal{Z}_{D'_i}$ which can be subdivided into a bounded number of vertical strips bounded from above and from below by an arc of the curves ρ_{D_i} or ρ_{s_i}. If $s_{i_0} \notin D_i$, these arcs are separated by the line $\theta = u + \pi/2$ or $\theta = u - \pi/2$.

Since $s_i{}' \in D_i{}'$, $\mathcal{Z}_{s'_i}$ is included in $\mathcal{Z}_{D'_i}$. Therefore $\mathcal{Z}_{e_i}$ can be deduced from $\mathcal{Z}_{D'_i}$ in the same way. Then $\mathcal{Z}_{e_i}$ can be deduced from $\mathcal{Z}_{D_i}$, $\mathcal{Z}_{s_i}$ and $\mathcal{Z}_{s'_i}$ in constant time and $\mathcal{Z}_{e_i}$ can be subdivided into a bounded number of vertical strips bounded from above and from below by the arc, included in the strip, of one of the curves ρ_{D_i}, ρ_{s_i} or $\rho_{s'_i}$. $\quad\square$

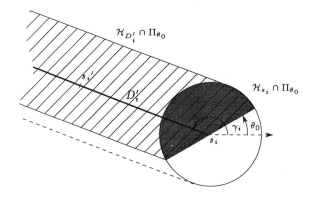

Figure 21: *When $\theta_0 \in [\gamma_i - \pi, \gamma_i]$, $\mathcal{H}_{D'_i}$ and $\mathcal{H}_{s_i}$ coincide in $\mathcal{D}_R(s_i)$.*

Lemma 23 *Let (u_I, θ_I) be a point of $\rho_{D_i}^-$ $(\rho_{D_i}^+)$ and I be the point of the circle $C_{s_{i_0}}$ with parameter u_I. The point A_I (B_I) at distance R from I in the direction θ_I $(\theta_I + \pi)$ belongs to D_i (Figure 22).*

Proof: We suppose that D_i is not the line of slope θ_I passing through I. The other case is easy.

If $(u_I, \theta_I) \in \rho_{D_i}^-$, then $\exists s \in D_i$ such that $(u_I, \theta_I) \in \rho_s^-$. That means that I belongs to the radius $r_{s_i}(\theta_I + \pi)$ of $HD(s, \theta_I)$ because $\rho_s^- = \mathcal{R}_s^- \cap \mathcal{C}_{s_{i_0}}$ (see the proof of Proposition 8). Hence s belongs to the straight line segment $[IA_I]$. If $s \neq A_I$ the half-disk robot at position (I, θ_I) that is $HD(I, \theta_I + \pi)$ (Definition 2) collides with the straight line D_i and so (u_I, θ_I) belongs to the interior of $\mathcal{Z}_{D_i}$. Hence $A_I = s$ and belongs to D_i.

By applying the same argument if (u_I, θ_I) is a point of $\rho_{D_i}^+$, we obtain that B_I belongs to D_i. $\quad\square$

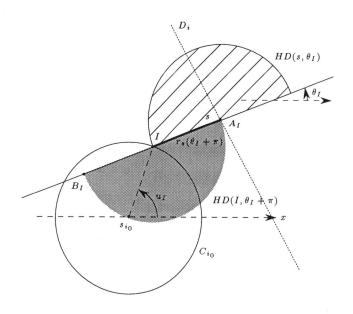

Figure 22: *Case where $s \in [IA[$*

Proposition 24 *Two curves $\rho_{D_i}^+$ and $\rho_{D_j}^+$ $(\rho_{D_i}^-$ and $\rho_{D_j}^-)$ intersect at most twice.*

Proof: A point on $\rho_{D_i}^+$ $(\rho_{D_i}^-)$ corresponds to a point on D_i (Lemma 23). As only two points on $\rho_{D_i}^+$ $(\rho_{D_i}^-)$ correspond to the same point on D_i, three points on $\rho_{D_i}^+$ $(\rho_{D_i}^-)$ define D_i. Therefore, if there exists three intersection points between $\rho_{D_i}^+$ and $\rho_{D_j}^+$ $(\rho_{D_i}^-$ and $\rho_{D_j}^-)$, then $D_i = D_j$. $\quad\square$

Proposition 25 *Two curves $\rho_{D_i}^+$ and $\rho_{D_j}^-$ intersect at most four times.*

Proof: Let (u_I, θ_I) be an intersection point between $\rho_{D_i}^+$ and $\rho_{D_j}^-$, and I be the point of the circle $C_{s_{i_0}}$ with parameter u_I. We can consider I point of $C_{s_{i_0}}$ as the middle of a ladder of length $2R$ and slope θ_I with one end point on D_i and the other on D_j (Lemma 23). The set of the middle points of such a ladder moving along D_i and D_j is an ellipse. There is at most four intersection points between an ellipse and a circle thus the curves $\rho_{D_i}^+$ and $\rho_{D_j}^-$ intersect at most four times. $\quad\square$

Proposition 26 *Two curves $\rho_{D_i}^{\pm}$ and $\rho_{s_j}^{\pm}$ intersect at most six times.*

Proof: Let (u_I, θ_I) be an intersection point between ρ_{D_i} and ρ_{s_j}, and I be the point of the circle $C_{s_{i_0}}$ with parameter u_I. The point at distance R from I in the direction θ_I (or $\theta_I + \pi$) belongs to D_i (Lemma 23) and s_j belongs to the line of slope θ_I passing through I (see the proof of Proposition 10).

We can consider the set of the end points P of a ladder of length R (or $-R$) where the other end point moves along the circle $C_{s_{i_0}}$ and the supporting line of the ladder contains s_j. This set is a circular conchoid and we show in the Appendix B that it intersects a straight line at most six times. □

5.2 Properties of the $\mathcal{S}_{e_i}$

We study here the regions $\mathcal{S}_{e_i}$ intersection of $\overline{\mathcal{H}_{e_i}}$ and $\mathcal{C}_{s_{i_0}}$. We use for the torus $\mathcal{C}_{s_{i_0}}$ the vocabulary introduced in Section 4.2.

Plainly, if s_{i_0} is not an end point of e_i, $\mathcal{S}_{e_i}$ is the closure of $\mathcal{Z}_{e_i}$ (see for example Figure 14).

Let $s_i = s_{i_0}$ and s_i' be the end points of e_i and $\gamma_i = \angle(\vec{x}, \overrightarrow{s_i s_i'})\ [2\pi]$.

Restricted to the portion of $\mathcal{C}_{s_{i_0}}$ parametrized by $(u, \theta) \in [\gamma_i - \frac{\pi}{2}, \gamma_i + \frac{\pi}{2}] \times S^1$, $\overline{\mathcal{H}_{e_i}} \cap \mathcal{C}_{s_{i_0}}$ is equal to $\overline{\mathcal{H}_{e_i} \cap \mathcal{C}_{s_{i_0}}}$ (Figure 20). Therefore $\mathcal{S}_{e_i}$ coincides with $\overline{\mathcal{Z}_{e_i}}$ for $(u, \theta) \in [\gamma_i - \frac{\pi}{2}, \gamma_i + \frac{\pi}{2}] \times S^1$.

Restricted to the portion of $\mathcal{C}_{s_{i_0}}$ parametrized by $(u, \theta) \in [\gamma_i + \frac{\pi}{2}, \gamma_i + \frac{3\pi}{2}] \times S^1$, $\overline{\mathcal{H}_{e_i}} \cap \mathcal{C}_{s_{i_0}}$ is equal to $\overline{\mathcal{H}_{s_{i_0}}} \cap \mathcal{C}_{s_{i_0}}$ (Figure 21). Therefore $\mathcal{S}_{e_i}$ coincides with $\mathcal{S}_{s_{i_0}}$ for $(u, \theta) \in [\gamma_i + \frac{\pi}{2}, \gamma_i + \frac{\pi}{2}] \times S^1$.

As $\mathcal{Z}_{e_i}$ is not defined in the strip $[\gamma_i + \frac{\pi}{2}, \gamma_i + \frac{3\pi}{2}] \times S^1$ (proof of Proposition 22), $\mathcal{S}_{e_i}$ is the union of $\overline{\mathcal{Z}_{e_i}}$ and the portion of $\mathcal{S}_{s_{i_0}}$ included in that strip.

We sum up the above results in the following proposition:

Proposition 27 *Let s_i and s_i' be the end points of e_i and $\gamma_i = \angle(\vec{x}, \overrightarrow{s_i s_i'})\ [2\pi]$.*

- *If s_{i_0} is not an end point of e_i then $\mathcal{S}_{e_i} = \overline{\mathcal{Z}_{e_i}}$.*

- *If $s_i = s_{i_0}$ then
  $\mathcal{S}_{e_i} = \overline{\mathcal{Z}_{e_i}} \cup (\mathcal{S}_{s_{i_0}} \cap [\gamma_i + \frac{\pi}{2}, \gamma_i + \frac{3\pi}{2}] \times S^1)$.*

5.3 Construction of $\delta(\mathcal{F}') \cap C_{s_{i_0}}$

Let $\lambda_k(n)$ denote the maximum length of the Davenport-Schinzel sequence of order k on n symbols and $\alpha_k(n) = \lambda_k(n)/n$.

Sections 5.1 and 5.2 allow us to compute the contribution of the circle $C_{s_{i_0}}$ to $\delta(\mathcal{F}')$ in the same way we have proceeded in Section 4.3.

The k_{i_0} helicoidal volumes $\mathcal{H}_{e_i}$ that intersect $C_{e_{i_0}}$ can be found in $O(k_{i_0})$ time once the arrangement $\mathcal{E}\mathcal{A}$ of the C_{e_i} have been computed in $O(|\mathcal{E}\mathcal{A}| \log n)$ Thanks to Propositions 22, 24, 25 and 26, we can compute the union of the set of $\mathcal{Z}_{e_i}$ such that the supporting line of e_i does not contain s_{i_0}. Since any two curves involved in the upper and lower envelopes intersect at most six time, the computation can be done in $O(\lambda_7(k_{i_0}) \log k_{i_0})$ time and $O(\lambda_8(k_{i_0}))$ space [4].

Since the e_i are the edges of the polygons of $\mathcal{S}$, assumed to be in general position, there are exactly two straight line segments ending at a vertex s_{i_0} of $\mathcal{S}$ and the straight lines supporting the other straight line segments do not contain s_{i_0}. Therefore, by Propositions 24, 25 and 26, we can compute the union of all the $\mathcal{Z}_{e_i}$ in $O(\lambda_7(k_{i_0}) \log k_{i_0})$ time and $O(\lambda_8(k_{i_0}))$ space.

Proposition 27 implies that we can compute the union of all the $\mathcal{S}_{e_i}$ within the same time and space bounds.

Since $\mathcal{E}\mathcal{A}$ is the arrangement of the set of the C_{e_i} and $\mathcal{A}$ is the set of circles of radius R centered at the end points of each straight line segment e_i, the generalization of Section 4.3 yields the following theorem:

Theorem 28 *We can compute $\delta(\mathcal{F}') \cap \mathcal{A}$ and the labels of the edges of $\delta(\mathcal{F}')$ incident to the arcs of $\delta(\mathcal{F}') \cap \mathcal{A}$ in $O(|\mathcal{E}\mathcal{A}| \alpha_7(n) \log n)$ time and $O(|\mathcal{E}\mathcal{A}| \alpha_8(n))$ space.*

5.4 Contribution of the straight line segments of C_{e_i} to $\delta(\mathcal{F}')$

We want to compute the contribution of the straight line segments of C_{e_i} to $\delta(\mathcal{F}')$. This can be done in a way similar to what we have done in Section 4.3.

Let e_0 be one of the two straight line segments of an $C_{e_{i_0}}$, D_0 its supporting line of slope γ_0 and $\Pi = D_0 \times S^1$. Π is a section of a cylinder but for convenience we will use the vocabulary of the plane when describing objects on Π. We want to compute the contribution of

D_0 to $\delta(\mathcal{F}')$ by computing the union of the $\mathcal{H}_{e_i} \cap \Pi$ and the union of the $\overline{\mathcal{H}_{e_i}} \cap \Pi$. We study first the region $\mathcal{Y}_s = \mathcal{H}_s \cap \Pi$ where s is a point after what we study the region $\mathcal{Y}_{D_i} = \mathcal{H}_{D_i} \cap \Pi$ where D_i is the line supporting e_i.

Proposition 29 *The region $\mathcal{Y}_s = \mathcal{H}_s \cap \Pi$ is a connected region bounded by two vertical line segments of length π, and two curved edges ϱ_s^+ and ϱ_s^- which are translated copies of one another by vector $(0, 0, \pi)$.*

Proof: The proof is an easy generalization of Proposition 8. □

Proposition 30 *For any s, the two curved edges ϱ_s^+ and ϱ_s^- are not intersected by the lines $\theta = \gamma_0$ and $\theta = \gamma_0 + \pi$. Furthermore $\mathcal{Y}_s$ is intersected by one of these lines.*

Proof: ϱ_s is the intersection between Π and the helicoidal portion $\mathcal{R}_s$ of the boundary of $\mathcal{H}_s$. Since $s \notin D_0$, the diameter of the half-disk $HD(s, \gamma_0)$ or $HD(s, \gamma_0 + \pi)$ does not intersect the line D_0 of slope γ_0 while one of these two half-disks intersects D_0. □

Proposition 31 *If D_i and D_0 are not parallel, the region $\mathcal{Y}_{D_i} = \mathcal{H}_{D_i} \cap \Pi$ is a connected region bounded by two vertical line segments of length π, and two curved edges $\varrho_{D_i}^+$ and $\varrho_{D_i}^-$ which are symmetrical with respect to the line $\theta = \gamma_i$ (Figure 15, 16).*

Remark 32 *Since γ_i is defined modulo π, $\varrho_{D_i}^+$ and $\varrho_{D_i}^-$ are also symmetrical with respect to the line $\theta = \gamma_i + \pi$.*

Proof: The proof is an easy generalization of Proposition 8 where I_1 and I_2 are the two points of D_0 at distance R from D_i. □

Proposition 33 *The curves ϱ_{D_i} can be subdivided into two sub-curves denoted $\varrho_{D_i}^k$ ($k \in \{1, 2\}$) such that:*

- *The line $\theta = \gamma_0$ intersects the region bounded from above by $\varrho_{D_i}^{1+}$ and from below by $\varrho_{D_i}^{1-}$ but does not intersect the two curves.*

- *The line $\theta = \gamma_0 + \pi$ intersects the region bounded from above by $\varrho_{D_i}^{2+}$ and from below by $\varrho_{D_i}^{2-}$ but does not intersect the two curves.*

Proof: Plainly, if D_i and D_0 are parallel, one of the lines $\theta = \gamma_0$ and $\theta = \gamma_0 + \pi$ intersects $\mathcal{Y}_{D_i}$ and does not intersect ϱ_{D_i}.

Let I be the intersection between the two lines D_0 and D_i. Let A and B be the two points of D_0 at distance R from D_i. Let μ_I, μ_A and μ_B their parameters on D_0.

Assume $\gamma_i \in]\gamma_0, \gamma_0 + \pi/2]$. For all $\theta \in [\gamma_i - \frac{\pi}{2}, \gamma_i + \frac{\pi}{2}]$, the Minkowski's sum of $HD(I, \theta)$ and D_i contains $]IA[$. Since $\gamma_0 \in \bigcap_{\gamma_i \in]\gamma_0, \gamma_0 + \pi/2]} [\gamma_i - \frac{\pi}{2}, \gamma_i + \frac{\pi}{2}]$, $\mathcal{H}_{D_i} \cap \Pi$ contains $]\mu_I, \mu_A[\times \{\gamma_0\}$.

In the same way, $\mathcal{H}_{D_i} \cap \Pi$ contains $]\mu_I, \mu_B[\times \{\gamma_0 + \pi\}$.

If $\gamma_i \in]\gamma_0 + \pi/2, \gamma_0 + 3\pi/2]$, we have the same result with μ_A and μ_B exchanged.

As the curves ϱ_{D_i} are defined in $]\mu_A, \mu_B[\times S^1$, the curves can be subdivided into the sub-curves defined in $]\mu_A, \mu_I] \times S^1$ and $[\mu_I, \mu_B[\times S^1$. □

Proposition 34 *Let s_i and s_i' be the two end points of the straight line segment e_i and D_i the supporting line of e_i. The region $\mathcal{Y}_{e_i}$ can be split into a bounded number of vertical strips such that each strip is bounded from above and from below by the arc of the curves ϱ_{D_i}, ϱ_{s_i} or $\varrho_{s_i'}$ which is included in the strip. Here ϱ_{D_i}, ϱ_{s_i} and $\varrho_{s_i'}$ denote the upper or lower edges of $\mathcal{Y}_{D_i}$ (Proposition 31), $\mathcal{Y}_{s_i}$ and $\mathcal{Y}_{s_i'}$ respectively (Proposition 29). Moreover, on each strip $U \times S^1$, the upper and the lower edges of $\mathcal{Y}_{e_i}$ are separated by the line $\theta = \gamma_0$ or $\theta = \gamma_0 + \pi$.*

Proof: The proof is an easy generalization of Proposition 22. □

Proposition 35 *Two curves ϱ_{s_i} and ϱ_{s_j} intersect at most once.*

Proof: The proof of Proposition 10 does not require that $\mathcal{H}_{s_i}$ is intersected by a torus. Hence, Proposition 10 remains valid for the curves ϱ_{s_i}.

$r_i(\theta)$ and $r_i(\theta + \pi)$ denote the two radii of the half-disk $HD(s_i, \theta)$.

For all θ in S^1, except γ_0 and $\gamma_0 + \pi$ if $s_i \in D_0$, the two radii $r_i(\theta)$ and $r_i(\theta + \pi)$ intersect D_0 at most once and if $s_i \in D_0$, the half-disks $HD(s_i, \gamma_0)$ and $HD(s_i, \gamma_0 + \pi)$ do not intersect D_0 because the half-disk is defined as an open set. Furthermore, $r_i(\theta)$ and $r_i(\theta + \pi)$ must intersect D_0 on a θ-interval smaller than π. Therefore ϱ_{s_i} is the graph of a function of θ defined over a θ-interval smaller than π. □

If we replace the curves $\rho_{D_i}^{\pm}$ by $\varrho_{D_i}^{\pm}$, Lemma 23 still holds, which yields the two following propositions:

Proposition 36 *Two curves $\varrho_{D_i}^{\pm}$ and $\varrho_{D_j}^{\pm}$ intersect at most twice.*

Proof: The proof of Proposition 24 still holds.

Since a straight line intersects an ellipse at most twice, the proof of Proposition 25 yields the result. $\square$

Proposition 37 *Two curves ϱ_{D_i} and ϱ_{s_j} intersect at most four times.*

Proof: Each intersection point between ϱ_{D_i} and ϱ_{s_j} corresponds to an intersection point between a conchoid and a straight line (see the proof of Proposition 26). That two curves intersect at most four times (Appendix A). $\square$

Plainly, if D_i and D_0 are not parallel lines at distance R, $\overline{\mathcal{Y}_{D_i}} \cap \Pi$ is equal to $\overline{\mathcal{Y}_{D_i} \cap \Pi}$. Otherwise $\mathcal{Y}_{D_i} \cap \Pi = \emptyset$ and $\overline{\mathcal{Y}_{D_i}} \cap \Pi$ is equal to $D_0 \times [\gamma_0 - \frac{\pi}{2}, \gamma_0 + \frac{\pi}{2}]$ or $D_0 \times [\gamma_0 + \frac{\pi}{2}, \gamma_0 + \frac{3\pi}{2}]$, depending of the respective positions of D_i and D_0. Furthermore, in general position, there exists exactly one line D_i at distance R from D_0 and the others are not parallel to D_0.

Therefore the computation of $\delta(\mathcal{F}') \cap D_0$ can be done in the same way as in Section 4.3. Here, instead of $\theta = u \pm \pi/2$, the two lines $\theta = \gamma_0$ and $\theta = \gamma_0 + \pi$ are the two lines which induce the two sets Ω_1 and Ω_2 introduced in Section 4.3.

Since the k_{i_0} $\mathcal{H}_{e_i}$ that intersect $\mathcal{C}_{e_{i_0}}$ can be found in linear time since the arrangement $\mathcal{EA}$ have been already computed, the analysis of the algorithm follows from Propositions 35, 36 and 37. Therefore the contribution of $\delta(\mathcal{F}')$ to D_0 and so to e_0, can be computed in $O(\lambda_5(k_0) \log(k_0))$ time and $O(\lambda_6(k_0))$ space and if $\mathcal{B}$ denotes the set of straight line segments of $\mathcal{EA}$, we have the following theorem:

Theorem 38 *We can compute $\delta(\mathcal{F}') \cap \mathcal{B}$ and the labels of the edges of $\delta(\mathcal{F}')$ incident to the arcs of $\delta(\mathcal{F}') \cap \mathcal{B}$ in $O(|\mathcal{EA}|\alpha_5(n) \log n)$ time and $O(|\mathcal{EA}|\alpha_6(n))$ space.*

5.5 Arcs of $\delta(\mathcal{F}')$ corresponding to the placements where the spider robot is in an unstable equilibrium

An edge of $\mathcal{F}'$ corresponding to an accessibility limit of the spider robot is an arc of some C_{e_i} $(1 \leq i \leq n)$.

The arcs of $\delta(\mathcal{F}')$ corresponding to an unstable equilibrium of the spider robot belong the 2-contact curves drawn by the midpoint of a ladder of length $2R$, moving amidst the foothold regions considered as obstacles and maintaining two points in contact with the boundary of the foothold regions. In the case where the footholds are points, the 2-contact curves are straight line segments. When the set of foothold regions is the set of straight line segments $\{e_i \ / \ 1 \leq i \leq n\}$, the 2-contact curves of the ladder are either straight line segments, arcs of ellipses or arcs of conchoid (see Figure 23). The 2-contact straight lines can be either part of the straight line joinning an e_i and an e_j or part of an e_i. We can compute the 2-contact arcs of a ladder moving by translation and rotation amidst straight line barriers in $O(|\mathcal{EA}| \log n)$ time and $O(|\mathcal{EA}|)$ space [11].

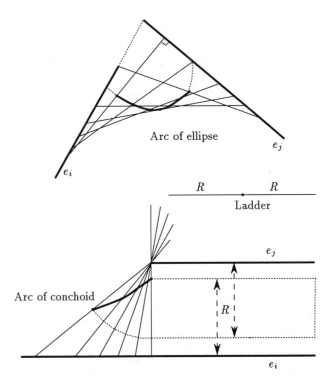

Figure 23: *Arc of an ellipse (conchoid) drawn by the midpoint of a ladder keeping contact with two edges (an edge and a point). The arcs are incident to the contribution of C_{e_i} and C_{e_j} to $\delta(\mathcal{F}')$ (in dotted line).*

Notice that the portions included in $\delta(\mathcal{F}')$ of a 2-contact arc of a ladder moving by translation and rotation in the presence of two barrier e_i and e_j must end in e_i, e_j, C_{e_i} or in C_{e_j}. If the arc ends on C_{e_i} then it must be connected to an arc of $\delta(\mathcal{F}') \cap C_{e_i}$ and if it ends on e_i then it must be connected to the 2-contact arc of the ladder slipping along e_i.

5.6 Construction of $\mathcal{F}'$ and $\mathcal{F}$

Let $\lambda_k(n)$ denote the maximum length of the Davenport-Schinzel sequence of order k on n symbols and $\alpha_k(n) = \lambda_k(n)/n$.

Theorem 39 *Given as foothold regions a set of n straight line segments, we can compute the free space $\mathcal{F}'$ of the spider robot in $O(|\mathcal{EA}|\alpha_8(n)\log n)$ time and $O(|\mathcal{EA}|\alpha_8(n))$ space.*

Proof: By Theorems 28 and 38, we have computed the contribution off all the C_{e_i} to $\delta(\mathcal{F}')$ and the label of the edges of $\delta(\mathcal{F}')$ incident to them. By Section 5.5, we have computed a set of arcs containing the arcs of $\delta(\mathcal{F}')$ corresponding to the placements where the spider robot is in an unstable equilibrium. It remains to sort along these arcs the vertices of $\delta(\mathcal{F}')$. This can be done in $O(|\mathcal{EA}|\alpha_8(n)\log n)$ time since Theorems 28 and 38 and Section 5.5 give a bound for $|\delta(\mathcal{F}')|$ in $O(|\mathcal{EA}|\alpha_8(n))$. $\square$

As we have seen above, if the set of straight line segments is the boundary of a set of polygonal foothold regions, adding the interior of the polygons to $\mathcal{F}'$ gives the whole free space $\mathcal{F}$ of the spider robot moving on these polygons. This does not increase the geometric complexity of the free space nor the complexity of the computation. Thus we have the following theorem:

Theorem 40 *Given a set of polygonal foothold regions with n edges in total, we can compute the free space $\mathcal{F}$ of the spider robot in $O(|\mathcal{EA}|\alpha_8(n)\log n)$ time and $O(|\mathcal{EA}|\alpha_8(n))$ space.*

The function $\alpha_8(n)$ growing extremely slowly and can be considered as a small constant in practical situations. This result is almost optimal since, as shown in [2], $\Omega(|\mathcal{EA}|)$ is a lower bound for the size of $\mathcal{F}$.

6 Conclusion

We have seen in Theorem 16 that, when the foothold regions are n points in the plane, the free space of the spider robot can be computed in $O(|\mathcal{A}|\log n)$ time and

$O(|\mathcal{A}|\alpha(n))$ space where $\alpha(n)$ is the pseudo inverse of the Ackerman's function and $\mathcal{A}$ the arrangement of the n circles of radius R centered at the footholds. By [2] the size of $\mathcal{F}$ is known to be $\Theta(|\mathcal{A}|)$. The size of $\mathcal{A}$ is $O(n^2)$ but, if k denotes the maximum number of disks of radius R centered at the footholds that can cover a point of the plane, it can be shown that $|\mathcal{A}| = O(kn)$ [10]. Thus, in case of sparse footholds, the sizes of $\mathcal{A}$ and $\mathcal{F}$ are linearly related to the number of footholds. Moreover $|\mathcal{F}|$ is usually much smaller that $|\mathcal{A}|$, even when $\mathcal{A}$ has quadratic size.

When the footholds regions are n straight line segments or polygons with n edges in total, the free space of the spider robot can be computed in $O(|\mathcal{EA}|\alpha_8(n)\log n)$ time and $O(|\mathcal{EA}|\alpha_8(n))$ space. $n\alpha_k(n) = \lambda_k(n)$ is the maximum length of the Davenport-Schinzel sequence of order k on n symbols; $\mathcal{EA}$ is the arrangement of the n curves consisting of the points lying at distance R from the straight line edges. Notice that $\alpha(n) = \alpha_3(n)$. The size of $\mathcal{EA}$ is $O(n^2)$. The term $\alpha_8(n)$ in the previous bounds comes from Proposition 26; It could be improved to $\alpha_6(n)$ by a careful analysis (too long to be reported here).

It should be observed that, in the case of point footholds, our algorithm implies that $O(|\mathcal{A}|\alpha(n))$ is an upper bound for $|\mathcal{F}|$. However, this bound is not tight since $|\mathcal{F}| = \Theta(|\mathcal{A}|)$ [2]. We let as an open question whether $\Omega(|\mathcal{EA}|)$ is an upper bound for the size of $\mathcal{F}$.

Once the free space $\mathcal{F}$ is known, several questions can be answered. In particular, given two points in the same connected component of $\mathcal{F}$, one can compute a motion, *i.e.* a sequence of leg assignments that join the two points [1].

Appendix A: Conchoid

A conchoid is the curve traced by the end point P of a straight line segment $[PQ]$ of length L when Q moves along a fixed straight line D and when the line (PQ) contains a fixed point O (Figure 24).

Let H be the orthogonal projection of O onto D and h the euclidean distance OH. We give a parametric equation of the conchoid in the orthonormal reference frame (H, D, HO).

Let X and Y denote the coordinates of P in this reference frame. Let α denote the angle $\angle(D, (PQ))$. We have:

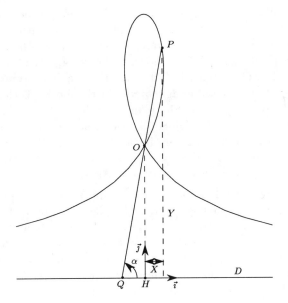

Figure 24: *Conchoid*

$$\left\{ \begin{array}{l} X = L\cos(\alpha) - \frac{h}{\tan(\alpha)} \\ Y = L\sin(\alpha) \end{array} \right.$$

Considering $t = \tan(\alpha/2)$, we have:

$$\left\{ \begin{array}{l} X = \frac{ht^4 - 2Lt^3 + 2Lt - h}{2t(1+t^2)} \\ Y = \frac{2Lt}{1+t^2} \end{array} \right.$$

Equation $aX + bY + c = 0$ is a polynomial of degree four in t.

Appendix B: Circular conchoid

A circular conchoid is the curve traced by the end point P of a straight line segment $[PQ]$ of length L when Q moves along a fixed circle C of radius R centered at O_1 and when the line (PQ) contains a fixed point O (Figure 25).

We give a parametric equation of the circular conchoid in the orthonormal reference frame with O_1 as its origin and O_1O as its Y-axis.

Let X and Y denote the coordinates of P in this reference frame. Let α denote the angle $\angle(\vec{i}, \overrightarrow{QP})$, ϕ the angle $\angle(\vec{i}, \overrightarrow{O_1Q})$ and h the euclidean distance OO_1. We have:

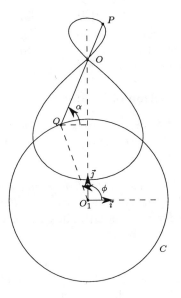

Figure 25: *Circular conchoid*

$$\left\{ \begin{array}{l} X = R\cos(\phi) + L\cos(\alpha) \\ Y = R\sin(\phi) + L\sin(\alpha) \end{array} \right.$$

where

$$\begin{array}{rcl} \cos(\alpha) & = & \frac{-R\cos(\phi)}{\sqrt{R^2\cos^2(\phi) + (h - R\sin(\phi))^2}} \\ & = & \frac{-R\cos(\phi)}{\sqrt{R^2 + h^2 - 2hR\sin(\phi)}} \\ \sin(\alpha) & = & \frac{h - R\sin(\phi)}{\sqrt{R^2\cos^2(\phi) + (h - R\sin(\phi))^2}} \\ & = & \frac{h - R\sin(\phi)}{\sqrt{R^2 + h^2 - 2hR\sin(\phi)}} \end{array}$$

With $t = \tan(\alpha/2)$, we set:

$$\left\{ \begin{array}{l} X = \frac{1}{1+t^2}\left(R(1 - t^2) - \frac{LR(1 - t^2)}{\sqrt{f(t)}} \right) \\ Y = \frac{1}{1+t^2}\left(2Rt + \frac{L(h(1+t^2) - 2Rt)}{\sqrt{f(t)}} \right) \end{array} \right.$$

with:

$$f(t) = \frac{(R^2 + h^2)(1 + t^2) - 4hRt}{1 + t^2}$$

Equation $aX + bY + c = 0$ is a polynomial of degree six in t.

References

[1] J.-D. Boissonnat, O. Devillers, L. Donati, and F. Preparata. Motion planning for a spider robot. In *Proc. 9th IEEE Internat. Conf. Robot. Autom.*, pages 2321–2326, 1992.

[2] J.-D. Boissonnat, O. Devillers, L. Donati, and F. P. Preparata. Stable placements of spider robots. In *Proc. 8th Annu. ACM Sympos. Comput. Geom.*, pages 242–250, 1992.

[3] M. Dickerson and R. L. Drysdale. Fixed radius search problems for points and segments. *Inform. Process. Lett.*, 35:269–273, 1990.

[4] J. Hershberger. Finding the upper envelope of n line segments in $O(n \log n)$ time. *Inform. Process. Lett.*, 33:169–174, 1989.

[5] S. Hirose and O. Kunieda. Generalized standard foot trajectory for a quadruped walking vehicle. *The International Journal of Robotics Research*, 10(1), February 1991.

[6] S. Hirose, M. Nose, H. Kikuchi, and Y. Umetani. Adaptive gait control of a quadruped walking vehicule. In *Int. Symp. on Robotics Research*, pages 253–277. MIT Press, 1984.

[7] J.Bares and W.L.Whittaker. Configuration of an autonomous robot for mars exploration. In *World Conference on Robotics Research*, volume 1, pages 37–52, May 1989.

[8] Special issue on legged locomotion. *International Journal of Robotics Research*, 3(2), 1984.

[9] Special issue on legged locomotion. *International Journal of Robotics Research*, 9(2), April 1990.

[10] M. Sharir. On k-sets in arrangements of curves and surfaces. *Discrete Comput. Geom.*, 6:593–613, 1991.

[11] S. Sifrony and M. Sharir. A new efficient motion planning algorithm for a rod in two-dimensional polygonal space. *Algorithmica*, 2:367–402, 1987.

Vision-Based Motion Planning and Exploration Algorithms for Mobile Robots

Camillo J. Taylor *Electrical Engineering, Yale University, New Haven, CT 06520-8267*
David J. Kriegman *Electrical Engineering, Yale University, New Haven, CT 06520-8267*

This paper focuses on the problem of designing algorithms that would enable a mobile robot equipped with a visual recognition system to carry out a systematic exploration of an unfamiliar environment in search of one or more recognizable targets. As a by-product of this exploration process, the algorithm constructs a representation of the environment that the robot can use in future navigation tasks. One of the main advantages of the proposed algorithm is that the robot is not required to accurately determine its absolute position in a global frame of reference. This algorithm has been implemented on our mobile robot platform RJ, and results from these experiments are presented.

1 Introduction

Many of the tasks we would like to have mobile robots perform are posed in a relational manner - get the copies from the copying machine, fetch the screwdriver from the tool bench, load these boxes into the truck - where the robot's final destination is specified relative to some recognizable target. In many of these cases, the user may not know the actual coordinates of the destination in an absolute frame of reference and must rely on the robot's ability to recognize these targets from sensor data. This paper focuses on the problem of designing algorithms that would enable a mobile robot equipped with a visual recognition system to carry out a systematic exploration of an unfamiliar environment in search of one or more recognizable targets. As a by-product of this exploration process, these algorithms construct a representation of the environment that the robot can use in future navigation tasks.

Some of the earliest work on employing vision-based recognition systems for robotic tasks was done at SRI by Nilsson et. al. [16]. In this project, the information from a video camera mounted on a mobile robot was used to recognize specific objects which the robot was supposed to manipulate. Since then, several researchers have developed mobile robot navigation algorithms that rely on various types of recognition systems. Kuipers and Byun [8, 9] proposed a navigation scheme based on the idea of a place graph. In their work, a place was defined as a specific point in the world that the robot could recognize from sonar readings. The edges in their place graph represented navigation operations, like wall following, that would take the robot from one place to another. Mataric later designed and implemented a similar algorithm that constructed a place map of portions of an office environment from sonar data [14]. The basic idea behind both of these approaches is to recast the navigation problem in terms of finding a route through the place graph from one region to another. Neither of these researchers addressed the problem of conducting a systematic exploration an environment containing multiple obstacles.

The Achilles heel of these place graph algorithms is that they rely on poorly defined heuristics to partition the world into recognizable places. It is not clear whether a robot employing these heuristics would be able to successfully subdivide an arbitrary environment into places nor is it clear that the selected places would correspond to meaningful regions of the workspace. In this paper, we will explain how the obstacle boundaries and the visibility of specific landmarks induce a well-defined partitioning of the freespace which the robot can learn online.

Levitt et. al. [11] have proposed a method for partitioning an outdoor area into regions based on the visibility of pairs of recognizable landmarks. They assume that the robot can always determine its position with respect to a set of *Landmark Pair Boundaries* (LPBs) which are virtual lines drawn between pairs of landmarks in the environment. They suggest that the robot could plan a path from one place to another by deducing the sequence of LPBs that it would have to cross to get from one region to another. However, the analysis provided in this paper does not take into consideration the problems that would occur when landmarks become occluded from the robot's field of view or when the robot encounters unexpected obstacles.

Lazanas and Latombe [10] have investigated the problem of developing provably correct navigation algorithms that rely on the robot's ability to recognize and localize a specific set of landmarks at known positions in the environment. They assume that the positions of the landmarks and the obstacles are known a'priori and that their task is to construct a navigation plan that will pilot the robot to the target location even in the presence of significant control uncertainty. The problem considered in this paper is quite different since the robot will not be provided with any prior information about the structure of its environment. Instead, it will be expected to discover the landmarks and obstacles that are present in its workspace during the course of its exploration.

In recent years, the problem of exploring an unknown environment has received considerable attention from researchers in the computer science theory community. Chin and Ntafos [18] considered the problem of finding a closed route through the interior of a simple polygon such that every point on the interior of the polygon would be visible from some point on the path.[1] They were able to show that the general problem is NP-hard, however, for the special case of a rectilinear polygon they described an $O(n \log n)$ for computing such a tour. Deng and Papadimitriou [3] investigated the problem of exploring an unknown polygonal room with a bounded number of polygonal obstacles. They were interested in comparing the length of the path taken by a robot that was trying to learn the environment for the first time with the length of the shortest watchman's tour that the robot could have taken to see all the points on all the obstacles boundaries. Kalyanasundaram and Pruhs [7] considered the problem of conducting a systematic exploration of an unknown environment containing a number of convex polygonal obstacles.

All of these algorithms assume that the environment is populated with polygonal obstacles and that the robot can accurately determine its position with re-

---

[1]This problem is often referred to as the nightwatchman's problem.

spect to a global frame of reference. In practice, it is quite difficult to accurately estimate the position of a mobile robot with respect to an arbitrary frame of reference. Most of the mobile robot systems currently in use rely on some form of odometry to determine their global position. Every odometric system suffers from the problem of cumulative error; as the robot moves further and further away from its starting point, errors in the estimates of the robot's position will grow monotonically. There are several robot localization systems that require the user to go through the trouble and expense of installing a set of beacons at known locations in the robot's workspace. Other global positioning techniques like DGPSS are not accurate enough to be used to construct metric maps of indoor environments where the obstacles are generally less than a metre apart [6].[2] One of the main advantages of the algorithm presented in this paper is that it does not rely on an accurate global positioning system, which makes it much easier to implement.

2   World Model

This research was motivated by a desire to produce algorithms that would enable mobile robot systems to navigate successfully in unstructured office environments like the one shown in Figure 1. In order to tackle this problem, we needed to develop a model of the environment which was simple enough to be tractable, but realistic enough to allow us to implement the resulting algorithms on our mobile robot platform.

Figure 2 shows the major aspects of the world model which will be described in more detail in this section. The robot is modelled as a holonomic vehicle with a circular cross section travelling through a planar workspace. These assumptions allow us to represent the robot as a point moving in a two-dimensional configuration space.

The robot is equipped with a vision-based recognition system which allows it to recognize and localize some (but not all) of the objects in the environment. More specifically, we assume that there is a set of distinct, recognizable objects in the world which will be

---

[2]The best accuracies achieved so far with DGPSS were on the order of one metre, this might be suitable for navigating on a harbor, but it is insufficient for navigating in tight corridors and cluttered office environments.

Figure 1: *An actual office environment.*

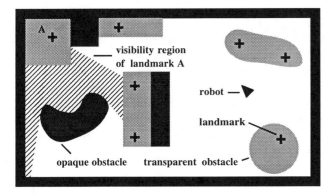

Figure 2: *Major features of the world model.*

termed *landmarks*. These landmarks are modeled as points in the workspace, and the robot should be able to determine the range and bearing to every landmark that is visible from its current position. We also assume that the robot is capable of looking in all directions; this capability can be realized by mounting the camera system on a pan-tilt head.[3] The consequences of using a slightly stronger set of assumptions about the capabilities of the vision system are discussed in [21].

The robot must also contend with a set of obstacles which are modeled as simple closed curves in the pla-

---

[3]Note that we do not assume that the robot can determine the orientation of the landmark; consequently, it cannot determine its position with respect to a fixed coordinate frame centered around the landmark.

nar configuration space. There are a finite number of distinct obstacles in the configuration space, and each of them has a finite length perimeter. Since the robot has a finite diameter, a single configuration space obstacle may actually represent several disjoint obstacles in the workspace. For the sake of simplicity, we assume that the freespace of the robot is bounded by one of the configuration space obstacles.

The obstacles in our model come in two flavors, opaque and transparent. Opaque obstacles will occlude landmarks from the view of the robot, while transparent ones do not. In the real world, opaque obstacles might include walls or bookshelves; transparent obstacles will block the robot's progress, but the robot can either see through or over them (e.g. tables, chairs and waste paper baskets). Note that this notion of transparency has more to do with the fact that we have chosen to represent a 3-d world with a 2-d configuration space than with the optical properties of the obstacles themselves. Every landmark in the environment must be contained within the boundary of an obstacle, however, a single obstacle may contain more than one landmark.

The robot should also be equipped with a proximity sensor which can be used to detect imminent collisions and to perform boundary following. While this could be implemented using vision, a number of other sensors are more commonly used: sonar, infrared sensors, laser range finders or light striping systems. Obstacle avoidance and boundary following are two of the most basic behaviors that one could imagine implementing on a mobile robot, and a number of researchers have demonstrated systems that perform these tasks quite well.

The fact that a landmark can only be observed if it is not occluded is sometimes referred to as the *visibility constraint*. This constraint induces a natural partitioning of the robot's workspace into regions where particular landmarks are visible. Let us define the *visibility region* of a landmark to be the area in space where that landmark is visible. As Figure 3 shows, the visibility region of any landmark will be a simply connected, closed, star-shaped set. The boundary curve of the visibility region can be described in a coordinate system centered at the landmark by:

$$\begin{bmatrix} x \\ y \end{bmatrix} = r(\theta) \begin{bmatrix} \cos\theta \\ \sin\theta \end{bmatrix}$$

where the angle θ is measured in a coordinate system centered around the landmark, and $r(\theta)$ is a piecewise continuous function which represents the extent of the visibility region at each angle. At the discontinuities, the region is delimited by radial line segments whose endpoints are defined by the left and right limits of $r(\theta)$. Due to sensor resolution and measurement noise, a landmark will only be visible within some finite radius r_{max} which serves as an upper bound for $r(\theta)$.

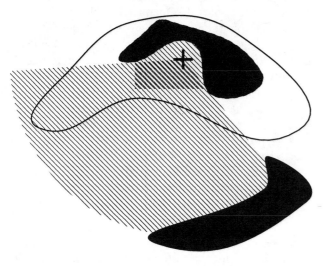

Figure 3: *The textured region represents the visibility region of the landmark.*

Note that in order for a landmark to be visible in the robot's freespace, it must be visible along some portion of the enclosing obstacle boundary. This proposition can be proved quite simply by constructing a straight line between the landmark and the point in freespace where the landmark is visible as shown in Figure 4. Since light travels in straight lines, the landmark must also be visible from every point along that line and since the landmark lies in the interior of an obstacle, the line must cross the boundary of that obstacle at least once by the Jordan curve theorem.

This observation implies the following conclusion:

Theorem 1 *If the robot circumnavigates all the obstacles that contain landmarks, it will eventually discover all the visible landmarks in the workspace.*

Proof: Since every landmark must be visible from the boundary of the obstacle that encloses it, a robot that

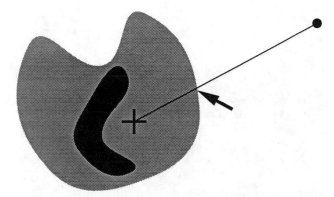

Figure 4: *If a landmark is visible in the robot's freespace then it must be visible from some point along the boundary of the obstacle that encloses it.*

circumnavigates all the obstacle boundaries will eventually find all the landmarks. $\square$

This theorem is particularly relevant to our task since it means that the robot does not have to investigate every point in the 2-dimensional configuration space in order to find all the landmarks; it can accomplish its goal simply by following a finite length path around the boundaries of the obstacles. The exploration algorithm described in Section 4 is based on this observation.

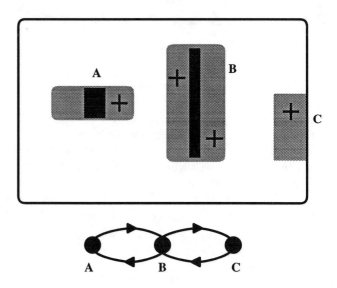

Figure 5: *Boundary place graph representation for a simple environment.*

A *boundary place* is defined as the boundary of a configuration space obstacle as shown in Figure 5. Two boundary places A and B will termed *connected* if a landmark contained inside boundary B is visible from some point on boundary A. These definitions allow us to define a *place graph* as shown in Figure 5 where the nodes represent boundary places, and the directed arcs represent connections between these places.

3   Navigation Using the Place Graph

Given two places, A and B, that are connected in the boundary place graph of the environment, we can define a simple, sensor-based strategy that would take the robot from boundary place A to boundary place B. Since the two places are connected, we know that at least one landmark located inside boundary place B is visible from boundary place A. The robot could simply circumnavigate boundary place A until one of these landmarks became visible and then employ the approach algorithm outlined below to move to boundary place B. This approach algorithm is actually a slightly modified version of the Bug1 algorithm proposed by Lumelsky and Stepanov [12, 13].

Approach Algorithm:
1) Head towards landmark until obstacle encountered
2) Circumnavigate the obstacle boundary and let L
   denote the point on the section of the boundary
   from which the landmark is visible where the robot
   comes closest to the target
3) **If** the target landmark appears to lie inside the
   obstacle boundary at the point L
   then
      terminate
   else
      follow the boundary back to L
4) goto step 1

Figure 6 shows an example of the execution of this approach algorithm. Note that the robot may leave the visibility region of the landmark during the execution of this algorithm.

It is possible to prove that this algorithm will cause the robot to converge to the boundary of the obstacle that contains the landmark in a finite amount of time. The proof given below has been divided into two parts:

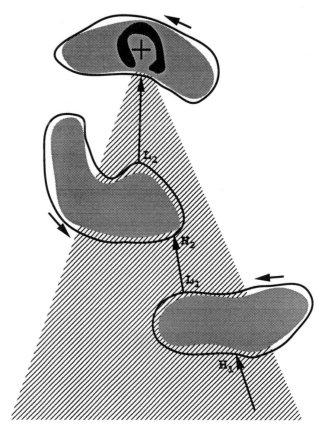

Figure 6: *Execution of the approach algorithm.*

the first part discusses the procedure that the robot uses to decide whether the target landmark lies inside or outside of a configuration space obstacle; the second part demonstrates that the robot will, in fact, converge to the correct obstacle boundary and gives an upper bound on the total distance that the robot would have to travel before termination. This proof is based on the one provided in [13].

Consider the portion of the obstacle boundary from which the landmark is visible, and let L denote the point in this section that is closest to the target as shown in Figure 7. In the sequel, this point will be referred to as the *closest observable point*.[4] The robot can locate the closest observable point as it circumnavigates the obstacle boundary by keeping track of

---

[4] Note that this point may be different from the point on the boundary that is closest to the target since the landmark may not be visible from every point on the obstacle boundary.

the range to the target whenever the landmark is visible. Note that the robot does not have to record the actual coordinates of the closest observable point, it will suffice to record the distance to the target at this point. Whenever the robot needs to return to the closest observable point, it can simply track the obstacle boundary until the landmark is visible and the range to the target is equal to the range recorded at the closest observable point.

If the landmark appears to be inside the boundary at the closest observable point, then it must lie inside the obstacle boundary, otherwise it must lie outside. This implies that the termination condition given in step 3 of the algorithm will only succeed when the obstacle has circumnavigated the obstacle that contains the target.

Interior Landmark

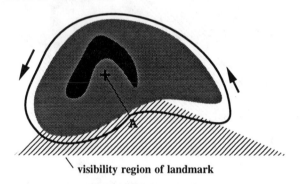

visibility region of landmark

Exterior Landmark

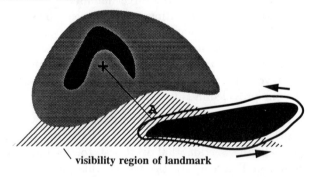

visibility region of landmark

Figure 7: *Determining whether a landmark is inside or outside an obstacle boundary.*

Lemma 1 *The line segment between the closest observable point and the target landmark will only touch the obstacle boundary at the closest observable point.*

Proof: If there were another point on the obstacle boundary that lay on the line segment between the closest observable point and the landmark, then that point would also lie in the visibility region of the landmark and it would be closer to the target than the supposed closest observable point. □

Theorem 2 *If the line segment between the closest observable point and the landmark is directed into the obstacle at the closest observable point, then the landmark must lie within the obstacle boundary otherwise it must lie outside.*

Proof: Lemma 1 states that the line segment between the closest observable point and the landmark will never pass through the obstacle boundary again so if the line segment is directed into the obstacle boundary at the closest observable point, we can infer that the target must lie inside the obstacle boundary by the Jordan Curve Theory. Similarly, if the line segment is directed away from the obstacle boundary at that point, we can conclude that it must lie outside. □

The following Lemmas prove that the robot will converge to the boundary of the obstacle that contains the target landmark in a finite amount of time.

Lemma 2 *When the robot leaves an obstacle boundary in order to head towards the target landmark, it never returns to any point on that obstacle.*

Proof: Let O be an obstacle that the robot encounters along its path that does *not* contain the target landmark. Let H_i denote the point where the robot first encounters the obstacle and L_i denote the point where it leaves that obstacle boundary to head towards the target as shown in Figure 6. The integer i indicates the order in which the obstacles are encountered.

Since the robot encounters H_i while it is moving in a straight line towards a visible target, we can deduce that H_i must lie in the visibility region of the landmark. The algorithm will choose the leave point, L_i, to be the closest observable point on the obstacle boundary.[5] From the definition of the closest observable point, we

---

[5] If there are many points on the obstacle boundary that could serve as the closest observable point, the robot can choose any one of these without affecting the correctness of the algorithm.

can infer that $d(H_i) > d(L_i)$ where $d(P)$ denotes the distance between the point P and the target landmark. We can also deduce that $d(L_i) > d(H_{i+1})$ since the robot travels directly towards the target when it leaves the obstacle boundary.

Taken together, these observations imply that if we consider the sequence of hit and leave points that the robot encounters along its journey, the distance between the robot and the target landmark decreases monotonically until the robot reaches the boundary of the obstacle that encloses the landmark.

The only way that the robot could return to a previously visited obstacle is if it encounters the obstacle at a new hit point H'. This hit point H' would have to be closer to the target than the previous leave point associated with that obstacle, L_i, which would imply that L_i was not the closest observable point on the boundary after all. □

Lemma 3 *If d_i denotes the perimeter of an obstacle that the robot encounters during its journey, then the robot will travel a distance no more than $2d_i$ along the boundary of that obstacle.*

Proof: The robot will circumnavigate every obstacle it encounters during its journey which means it will travel a distance of at least d_i along the obstacle boundary. In addition, it may have to travel a maximum distance of d_i along the boundary to get back to the leave point associated with that obstacle. □

If the robot had some means of measuring the distance it has travelled along the obstacle boundary, it could choose the shortest path along the boundary back to the leave point which would reduce this upper bound to $1.5d_i$.

Lemma 4 *The robot can only encounter obstacles that intersect the portion of the landmarks visibility region that lies within a disc of radius D centered around the target landmark where D denotes the distance between the robot's start point and the landmark.*

Proof: In Lemma 2 we proved that the distance between the hit points and the target landmark decreases monotonically over time which implies that every hit point must be less than D units away from the target landmark. We also noted that every hit point must lie within the landmarks visibility region. Taken together, these observations imply that every obstacle that the landmark encounters must have some section of its boundary intersect the portion of the landmarks visibility region that lies within a disc of radius D centered around the target landmark. □

Theorem 3 *The maximum distance that the robot will have to travel before it converges to the boundary of the obstacle that contains the target landmark is given by $D + 2\sum d_i$ where $\sum d_i$ represents the sum of the perimeters of the obstacles that intersect the portion of the landmarks visibility region that lies within a disc of radius D centered around the target landmark.*

Proof: If there were no extraneous obstacles, the robot would have to travel a distance of at most D before it encountered the boundary of the obstacle that enclosed the target landmark.

Lemma 4 states that the robot will only encounter those obstacles that intersect the portion of the landmarks visibility region that lies within a disc of radius D centered around the target landmark. Lemmas 2 and 3 imply that the robot will travel no more than $2\sum d_i$ along the boundaries of those obstacles. □

In order to execute this approach algorithm, the robot must use its vision system to determine the range and bearing to the target landmark from its current position whenever the landmark is visible. It must also be able to detect when it has completely circumnavigated an obstacle boundary. Note, however, that the robot does not have to be able to determine its absolute position with respect to a fixed frame of reference.

4 Exploration Algorithm

Theorem 1 states that if the robot were able to circumnavigate all of the obstacles that contained landmarks, it would eventually discover all the visible landmarks. The algorithm presented in this section was based on this result. Effectively, the algorithm will cause the robot to perform a tour of the boundary place graph of the environment where visiting a node in the place graph corresponds to circumnavigating the boundary of that obstacle. We will term a particular landmark in the environment *explored* iff the robot has circumnavigated the configuration space obstacle that encloses that landmark. By exploring each of the landmarks that it sees, the robot can incrementally learn the entire boundary place graph of the environment. Once the entire graph has been explored, the robot can use the representation it has constructed during its travels for further navigation tasks.

The entire exploration algorithm is outlined below in pseudo code.

Exploration Algorithm:
Find first landmark
While ⟨ unexplored landmarks ⟩
      Select unexplored landmark, λ
      Plan path through explored part of the
          boundary place graph to region where
          λ is visible
      Approach λ
      Circumnavigate boundary that contains λ, and
          record any observed landmarks
      Update the place graph

The robot maintains two data structures during this exploration procedure: $\mathcal{L}$ the list of landmarks it has seen and $\mathcal{B}$ the list of boundary places it has circumnavigated. For each boundary place, the robot records which landmarks are visible from that obstacle boundary. These landmarks are divided into two categories: interior landmarks which lie inside the obstacle boundary and exterior landmarks which lie outside. These lists, $\mathcal{L}$ and $\mathcal{B}$, can also be thought of as a representation for the portion of the boundary place graph that the robot has explored so far.

Figure 8 shows a typical environment that a robot might encounter along with its boundary place graph representation. Figure 9 shows the progress of the exploration algorithm at various stages. In stage 1, the robot circumnavigates obstacle A and determines that there are two unexplored landmarks visible from the boundary of A which need to be explored. These unexplored landmarks correspond to unexplored edges in the robot's representation of the boundary place graph. In stage 2 the robot traverses one of the unexplored edges and ends up circumnavigating obstacle B; after it has finished its circuit around this obstacle, it concludes that there is still one unexplored edge in the graph, and so it plans a path back through the graph to boundary A and traverses the unexplored edge to

boundary place C. In the final stage it visits obstacle D and concludes that there are no more unexplored landmarks in the environment. By this stage, the robot has built up a complete representation for the boundary place graph of the environment which it can use in future navigation tasks. Notice that whenever the robot visits a node in the boundary place graph, it discovers all the edges emanating from that node. It can determine which of these edges lead to previously visited places and which lead to unexplored nodes.

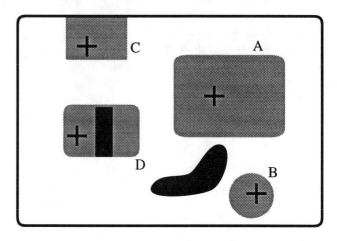

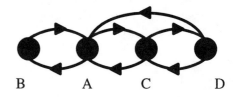

Figure 8: *An example environment and its place graph.*

Since the exploration problem has been cast in terms of a graph traversal, it should not be surprising that the necessary and sufficient condition required to ensure success should be framed in the following manner.

Necessary and sufficient condition: The exploration algorithm presented above will discover all the landmarks that are visible in the robot's freespace iff the boundary place graph of the environment is *strongly connected*. A directed graph is termed strongly connected iff it is possible to construct a path from any node in the graph to any other node.

Proof: Proving that the condition is necessary follows from the definition of a strongly connected graph. If the graph is not strongly connected, then we can always

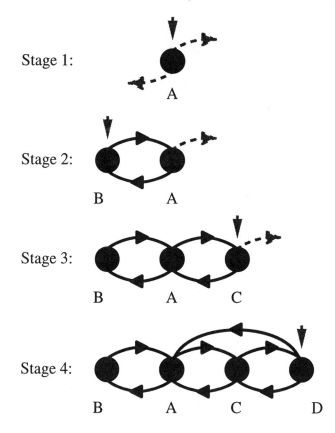

Stage 1:

Stage 2:

B   A

Stage 3:

B   A   C

Stage 4:

B   A   C   D

Figure 9: *Progress of the exploration algorithm. The dashed lines denote unexplored edges while the grey arrow indicates the current position of the robot in the graph.*

find two nodes A and B such that there is no path from A to B through the graph. This means that if the exploration procedure were started out on node A it would never be able to be able to carry out a complete tour of the graph since it would never be able to get to node B.

In order to prove that this condition is in fact sufficient, we first divide the place graph into two parts, the explored section E and the unexplored section U. If there is no unexplored section, the exploration algorithm will terminate normally. Otherwise, we can select any node T in the unexplored section of the graph and consider a path from the robot's current location, S, to that node; since the place graph is strongly connected, we know that such a path must exist. Since the robot's current position, S must lie in the explored part of the graph, we can conclude that at some point

along this path there must be an unexplored edge that connects a node in the explored part of the graph to a node in the unexplored part in the graph. This implies that the robot can always plan a path through the explored part of the graph to an unexplored node. □

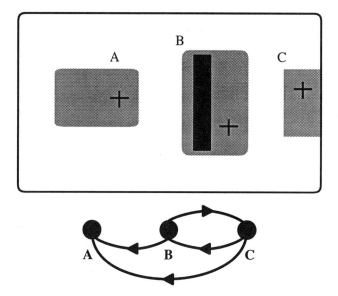

Figure 10: *The exploration algorithm will find all the visible landmarks iff the boundary place graph of the environment is strongly connected. This environment could cause problems for the exploration algorithm.*

Figure 10 shows an example of an environment that could cause problems for the exploration algorithm. Notice that boundary place A is not connected to any of the other boundary places in the environment. If the robot were to visit this obstacle during the course of an exploration procedure, it would not be able to use the approach algorithm described in Section 3 to navigate to another boundary place so the exploration algorithm would terminate with an error.

One of the most important features of this exploration algorithm is the fact that it only requires the robot to perform two types of navigation operations: boundary following and navigation with respect to a visible landmark. Both of these operations can be carried out using simple, closed loop control strategies that only rely on local sensor data.

It is also interesting to note that the robot only

records a simple topological representation of its environment instead of a metric map. The robot's ability to navigate successfully through this environment does *not* depend on its ability to locate itself with respect to an absolute metric map, it relies instead on its ability to recognize and localize salient landmarks in its immediate neighborhood. This approach to the exploration problem is fundamentally different from that taken by a number of other researchers who have worked in this area [1, 2, 5, 17, 22, 19].

Since the algorithm does not rely on having an accurate metric map of the environment, it avoids the usual problems associated with trying to construct such a representation from sensor data. The algorithm can be used to navigate over an arbitrarily large area, and the size of the robot's internal representation will only grow linearly with the number of landmarks encountered and the number of obstacles in the environment.

4.1 Complexity Results

This section discusses some of the complexity issues related to the online graph exploration algorithm described earlier in this section. It is quite easy to prove a lower bound on the number of edges that the robot has to traverse in order to carry out a complete exploration of a particular graph. Given a graph with n nodes the robot will have to traverse at least one edge for each new node it visits, this implies that the robot will have to traverse at least $(n-1)$ edges in order to visit all n nodes.

Proving an upper bound is only slightly harder. At any stage in the exploration process, the worst that could happen is that the robot might have to travel through all the previously visited nodes in order to get to an unexplored edge. That is, at stage n it might have to traverse n edges which implies that the total number of edges required to explore the entire graph would be $\sum_{i=1}^{n} i = n(n+1)/2$.

It is quite simple to construct graphs where the number of edges that the robot has to traverse in order to visit all of the nodes is $O(n^2)$. Consider the graph shown in Figure 11 where the n nodes are divided evenly between two sets: the trunk nodes and the leaf nodes. To visit a leaf node in this graph the robot has to travel through all of the trunk nodes. This means that the total number of edges that the algorithm will traverse is given by $(n/2) \times (n/2) = (n^2/4)$.

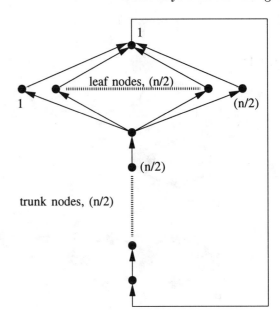

Figure 11: *It will take $O(n^2)$ edge traversals to completely explore this graph.*

By a similar construction, we can show that there is no competitive deterministic online algorithm for exploring an arbitrary strongly connected graph. An online algorithm is termed competitive iff the number of steps required by that algorithm is no more than a constant times the minimum number of steps required by an offline algorithm with complete information [3, 20].

Consider the graph shown in Figure 12. This graph is similar to the one shown in Figure 11, and the only difference being that in this case the leaf nodes are connected together by a sequence of edges. By adding these edges we have made it possible to completely explore the graph with the minimum number of edge traversals $(n-1)$. The online algorithm, however, does not have the benefit of complete knowledge of the graph, so in the worst case it would have to carry out the same exploration sequence that it would have used if the extra edges weren't there. That is the leaf nodes would get visited in the order indicated on Figure 12, 1 through $n/2$. The ratio between the cost of the online algorithm and the offline algorithm is $O(n)$ so the online algorithm is *not* competitive.

Deng and Papadimitriou [4] have considered the problem of exploring all the edges on a directed graph. They were able to construct competitive online algo-

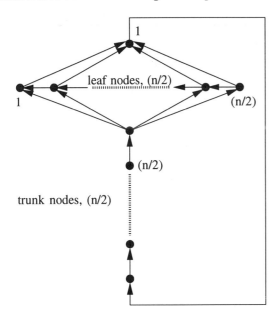

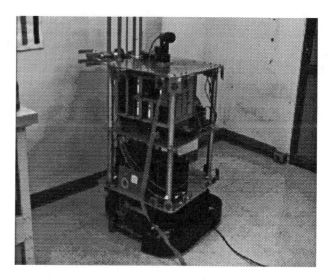

Figure 13: *RJ, an experimental mobile robot platform.*

Figure 12: *The ratio between the number of edge traversals taken by the best online algorithm and the worst offline algorithm is $(n-1) : (n/2+1)(n/2)$.*

rithms for exploring graphs with bounded deficiency. The deficiency of a directed graph was defined as the minimum number of edges which would have to be added in order to make it Eulerian. Our problem differs from theirs since we are only interested in visiting every node in the graph rather than traversing every edge.

5 Implementation

The algorithm described in the previous section has been implemented on our experimental mobile robot platform, RJ. This section describes the architecture of the actual mobile robot system and discusses the experimental trials that were carried out. Our mobile robot system, was built on top of a Labmate base from Transitions Research Corporation which we equipped with an onboard power system, an image processing subsystem, a sonar subsystem and a multiprocessor control system composed of a network of Inmos transputers.

Each transputer is a single chip RISC microprocessor (20 Mips, 2 MFlops) which is equipped with 4 bidirectional serial lines (20 Mbits/s each) that it can use to communicate with other transputers. These transputers were designed to be interconnected in networks

so that demanding computational applications could be distributed over multiple processors. This model is particularly appropriate for our mobile robot control system which has to process the information from several disparate sensor systems and control multiple servo mechanisms simultaneously in real-time, an extremely challenging task for a single processor computer. This organization also offers a significant advantage over more traditional bus-based architectures since the designer can apply dedicated microprocessors to each sensor subsystem without having to worry about conflicts for bandwidth on a shared bus.

There are three major subsystems on our mobile robot platform: the vision subsystem which controls the image capture hardware, the digital signal processing subsystem and the pan-tilt head; the motion control subsystem which controls the mobile robot base and the sonar module; and the main control subsystem which is responsible for coordinating the robot's overall behavior and performing higher-level planning operations. A block diagram of the system is shown in Figure 14.

A list of the robot's onboard equipment:

- Labmate mobile base.
- Image processing subsystem : CCD camera, frame grabber, pan tilt head, DSP system.
- Sonar subsystem.

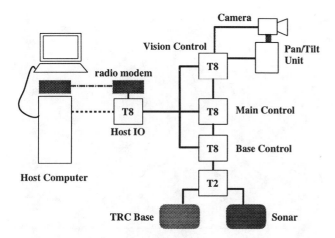

Figure 14: *Architecture of the mobile robot system.*

- LIDAR range finder.
- Multiprocessor control network.
- Radio modem

To execute the exploration algorithm, the robot has to be able to perform two basic operations: boundary following and landmark recognition. The robot combines the information from its sonar subsystem with its odometry to form a range map of a 2 metre square area centered around its current position. It uses this local map to avoid obstacles and track boundaries. Figure 15 shows an example of the kind of map the robot would build up from several sonar scans.

The simple bar-code targets shown in Figure 16 were employed as landmarks in our experiments. These targets were recognized by making using of a projective invariant known as the cross-ratio [15]. Let $X_1 - X_4$ be the coordinates of 4 consecutive points along a straight line and let $x_1 - x_4$ be the projections of those points onto an imaging line as shown in Figure 17.

It is quite simple to show that the following equation will be true irrespective of the camera position or focal length.

$$\frac{(X_3 - X_1)(X_4 - X_2)}{(X_3 - X_2)(X_4 - X_1)} = \frac{(x_3 - x_1)(x_4 - x_2)}{(x_3 - x_2)(x_4 - x_1)} \quad (1)$$

This means that the cross-ratio function $\left(\frac{(x_3-x_1)(x_4-x_2)}{(x_3-x_2)(x_4-x_1)}\right)$ could be used to recognize this configuration of points in a given image. Note however,

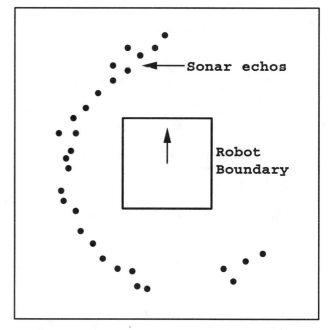

Figure 15: *Typical local sonar map built up by a mobile robot as it moves through its environment. The dots in the figure correspond to sonar returns detected by the sonar subsystem. The map covers a 2 metre square area centered around the robot's current position.*

Figure 16: *Bar code landmarks used in our experiments. Note that the recognition algorithm must find these targets even in cluttered indoor environments with many vertical edges.*

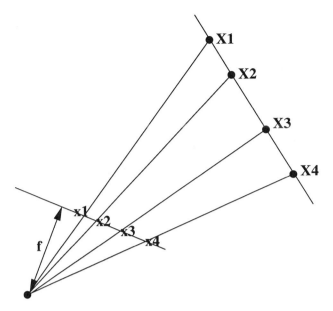

Figure 17: *Projection of points onto a 1-dimensional camera.*

that this cross-ratio test provides a necessary but not sufficient condition for target recognition since it is not possible to verify the actual 2-dimensional configuration of the observed features from a single image.

The bar-code targets were recognized by capturing images of the scene with the onboard video camera, extracting the vertical edges in these images and locating groups of six consecutive edges with the appropriate cross-ratios. Every target consisted of 5 vertical stripes which could take on one of three colors: black, white and gray. Once a target had been detected, the robot could use the gray-level values in the image to discern the color of each stripe and thereby distinguish between different landmarks.

This recognition system offered several advantages over other schemes that were considered: it was non-heuristic, it could be implemented in real-time, and it worked well in cluttered office environments with many extraneous vertical edges. The recognition algorithm was implemented on the robot's image processing subsystem which allowed it to process an entire frame of image data in under 20 milliseconds. The edge extraction was performed by the onboard DSP system while the cross-ratio computations were carried out on the transputer. The computational complexity of the recognition algorithm varies linearly with the number of vertical edges found in the scene.

Since the exploration algorithm does not depend upon the details of the recognition system, it would be quite easy, in principle, to replace this target recognition scheme with more sophisticated algorithms that were capable of recognizing naturally occurring structures like doors, chairs and tables.

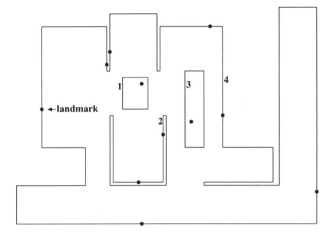

Figure 18: *During the course of the exploration, the robot circumnavigated each of the c-space obstacles in this office complex in the order indicated by their labels (1 thru 4).*

Figure 18 shows an approximate floor plan of the office complex (not to scale) that was used to test the exploration algorithm. This environment consisted of 4 distinct configuration space obstacles and 11 landmarks distributed over 3700 sq. ft. of floor space. The robot circumnavigated each of the configuration space obstacles in the order indicated and eventually discovered all the landmarks. The robot travelled several hundred metres during the course of this exploration procedure which took a little over half an hour.

6 Conclusion

The main contribution of this research has been to provide a new algorithm that enables a mobile robot equipped with a vision-based recognition system to carry out a systematic exploration of its environment in search of one or more recognizable objects. A detailed analysis of this algorithm was presented in Section 4 along with a necessary and sufficient condition for the robot to be able to accomplish its task. This research demonstrates that it is possible to design provably correct navigation algorithms that rely on a robot's ability to recognize certain landmarks in the environment.

The algorithm presented in this paper offers several advantages over other approaches that have been proposed. Unlike the exploration algorithms developed by Ntafos [18], Papadimitriou[3], Kalyanasundaram [7] and Lumelsky [22], this approach does *not* require an accurate global positioning system; this makes our algorithm much easier to implement. We have been able to demonstrate that a robot can carry out complex navigation tasks without the benefit of an accurate global map. Our experience with this algorithm leads us to conclude that it is much more important for the robot to record the relationship between the obstacles in its environment than their absolute positions in a global frame of reference.

In the course of this paper, we have explained how the visibility regions of the landmarks and the boundaries of the configuration space obstacles induce a well-defined partitioning of the freespace of the robot into places. One of the most important features of this partitioning is that the robot can learn it online from sensor data and can build up a place graph representation for its environment. This means that the robot does not have to rely on ad-hoc heuristics to subdivide the world into recognizable places.

Another important features of this exploration algorithm is the fact that it only requires the robot to carry out two types of navigation operations: boundary following and navigation with respect to a visible landmark. Both of which can be carried out using simple, closed loop control strategies that only rely on local sensor data. Rodney Brooks and other researchers have mounted a strong attack against deliberative robots that spend most of their time building up complicated internal models of the environment before performing any action. He pointed out that by coupling actions directly to sensing through a feedback loop, you can get robots to exhibit more robust low-level behaviors. However, if you intend to design a robot that can navigate from one end of campus to another, reactive behaviors alone will not be sufficient. The robot will have to record some representation of its environment which it can use to plan an appropriate path. This navigation algorithm combines some of the best features of both approaches; the lower level navigation operations are performed by sensor-based control strategies while the robot constructs a global map of the environment that it can use to plan paths from one end of its workspace to another.

Acknowledgements

Thanks to Jefferey Westbrook for providing the examples shown in figures 11 and 12. Support for this work has been provided in part by NSF Young Investigator Award, IRI-9257990 and DDM-9112458.

References

[1] J.F. Canny. *The Complexity of Robot Motion Planning*. MIT Press, 1988.

[2] R. Chatila and J.P. Laumond. Position referencing and consistent world modeling for mobile robots. In *IEEE Int. Conf. on Robotics and Automation*, 1985.

[3] Xiaotie Deng, Tiko Kameda, and Christos Papadimitriou. How to learn an unknown environment. In *Proceedings of the 32nd Annual Symposium on Foundations of Computer Science*, page 298, October 1991.

[4] Xiaotie Deng and Christos H. Papadimitriou. Exploring an unknown graph. In *Proceedings of the 31st Annual Symposium on Foundations of Computer Science*, page 355, October 1990.

[5] Alberto Elfes. Sonar-based real-world mapping and navigation. *IEEE Journal of Robotics and Automation*, 3(3):249–265, June 1987.

[6] Ivan A. Getting. The global positioning system. *IEEE Spectrum*, 30(12):36, December 1993.

[7] Bala Kalyanasundaram and Kirk Pruhs. A competitive analysis of algorithms for searching unknown scenes. *Computational Geometry: Theory and Applications*, 3:139–155, 1993.

[8] B. Kuipers and T. Levitt. Navigation and mapping in large-scale space. *AI Magazine*, 9(2):25–43, 1988.

[9] Benjamin Kuipers and Yung-Tai Byun. A robot exploration and mapping strategy based on a semantic hierarchy of spatial representations. *Robotics and Autonomous Systems*, 8:47–63, 1991.

[10] Anthony Lazanas and Jean-Claude Latombe. Landmark-based robot navigation. In *Proc. Am. Assoc. Art. Intell.*, July 1992.

[11] Tod. S. Levitt, Daryl T. Lawton, David M. Chelberg, and Philip C. Nelson. Qualitative navigation. In *Proc. Image Understanding Workshop*, pages 447–465, February 1987.

[12] V.J. Lumelsky and A.A. Stepanov. Dynamic path planning for a mobile automaton with limited information on the environment. *IEEE Trans. on Automatic Control*, 31(11):1058–63, 1986.

[13] Vladimir J. Lumelsky and Alexander A. Stepanov. Path-planning strategies for a point mobile automaton moving amidst unknown obstacles of arbitrary shape. *Algorithmica*, 2:403–430, 1987.

[14] M.J. Mataric. Integration of representation into goal directed behavior. *IEEE Trans. on Robotics and Automation*, 8(3):304–312, June 1992.

[15] J. Mundy and A. Zisserman. *Geometric Invariance in Computer Vision.* MIT Press, Cambridge, Mass., 1992.

[16] Nils J. Nilsson. Shakey, the robot. Technical Note 323, SRI, April 1984.

[17] B.J. Oomen, S.S. Iyengar, N.S.V. Rao, and R.L. Kashyap. Robot navigation in unknown terrains using learned visibility graphs. part i: The disjoint convex obstacle case. *IEEE Journal of Robotics and Automation*, RA-3(6):672–681, December 1987.

[18] Wei pang Chin and Simeon Ntafos. Optimum watchman routes. *Information Processing Letters*, 28:39–44, May 1988.

[19] N.S.V. Rao and S.S. Iyengar. Autonomous robot navigation in unknown terrains: Incidental learning and environmental exploration. *IEEE Systems, Man, and Cybernetics*, 20(6), November/December 1990.

[20] Daniel D. Sleator and Robert E. Tarjan. Amortized efficiency of list update and paging rules. *Communications of the ACM*, 28(2):202, February 1985.

[21] Camillo J. Taylor and David J. Kriegman. Exploration strategies for mobile robots. In *IEEE Int. Conf. on Robotics and Automation*, May 1993.

[22] Snehasis Mukhopadhyay Vladimir J. Lumelsky and Kang Sun. Dynamic path planning in sensor based terrain acquisition. *IEEE Journal of Robotics and Automation*, 6(4):462–472, August 1990.

Visually-Guided Navigation by Comparing Edge Images

Daniel P. Huttenlocher, Michael E. Leventon and William J. Rucklidge, *Computer Science Department, Cornell University, Ithaca NY USA*

We present a method for navigating a robot from an initial position to a specified landmark in its visual field, using a sequence of monocular images. The location of the landmark with respect to the robot is determined based on the change in size and location of the landmark in the image, as a function of the motion of the robot. The change in size and location of the landmark in the image are determined by matching intensity edges in successive frames. The method does not require prior calibration of the camera. We show some examples of the operation of the system.

1 Introduction

In this paper we describe a method for using two-dimensional shape information to determine the location of a mobile robot with respect to some visual landmark in the world. The task is for the robot to navigate to a specified target or landmark in its visual field, possibly in the presence of obstacles. The landmark is initially specified either by marking some portion of an image (containing the landmark) or by providing a prior model. The location of the landmark with respect to the robot is recovered from the change in apparent size and position of the landmark in the image, as a function of the motion of the robot. The size and position of the landmark are determined by comparing two-dimensional shapes from successive images taken as the robot moves.

The key aspects of the method are,

1. The position of the landmark is expressed in terms of the robot-centered quantities range (distance) and bearing (orientation), rather than world coordinates.

2. The range and bearing are calculated using the change in size and location of the object in the image, as a function of the translation and rotation of the robot. The methods do not require camera calibration.

3. The identification of the landmark in the image data is performed by comparing two-dimensional shapes that can translate and scale, without the use of three-dimensional shape information. The range and bearing to the landmark are used to predict its size and location in the image, in order to speed up the comparison process.

Our approach differs from much of the previous work in visually guided robot navigation since it uses recognition of a specific object in the world instead of extracting some more global image properties; see, for example, the use of stereo in [2] and image deformations in [6]. While this only gives us essentially a single depth reading, it allows us to concentrate on the landmark, and so obtain a high degree of overall accuracy in motion, as we continually confirm that we are on track towards the specified target. In contrast with work such as [7], we do not construct any explicit three-dimensional models. Our work also complements the system described in [8] for finding distinctive landmarks along a route, by providing an effective means of navigating from one landmark to the next.

The key observation underlying the approach is that when a camera moves directly towards an object, the range to that object is given by $m/(s-1)$ where m is the distance that the camera moved and s is the change in the apparent size of the object in the image. Thus a straightforward method for navigating to a landmark is to determine the bearing to that landmark, rotate so that the robot (and camera) is heading in that direction, move forward some distance, use the change in apparent size of the landmark to compute the range to the landmark, move to the landmark. As the robot moves towards the landmark the range and bearing estimates can be updated and the path corrected based on these measurements. By correcting the path as the robot moves, the method can compensate for the camera being misaligned (not pointing in the same direction as the "forward" motion of the robot) and for the robot not moving in exactly the commanded direction.

The configuration of our system is a camera mounted on a wheeled-robot that moves across a relatively flat surface. The camera is mounted at a fixed position on the robot, with its focal point at approximately the center of rotation of the robot, its optic axis approximately in the direction of forward motion of the robot, and its image plane approximately perpendicular to the ground plane. There are thus two degrees of freedom for the camera: a translation in a plane parallel to the ground plane (and approximately perpendicular to the image plane of the camera), and a rotation about the focal point. The translational degree of freedom has the primary effect of changing the imaged size of an object, and the rotational degree of freedom has the primary effect of changing the location of the image x-coordinate of an object.

The overall operation of the navigation method consists of the following steps: (1) grab an image of the current visual field of the robot, (2) use a two-dimensional model to localize the landmark in the current image, (3) compute the range and bearing using the localized landmark and the robot motion since the previous image was taken, (4) construct a new two-dimensional model to use in localizing the landmark in the next image, and (5) command the robot to make the next motion (a forward motion or a rotation), whereupon return to step 1. Initially we will describe the method assuming that the robot stops moving during steps 1-4, but in practice these operations actually happen concurrently with the robot motion.

We will first discuss the computation of the range from the change in size of the landmark in the image, and then consider the computation of the bearing from the location (x-position) of the landmark in the image. In section 4 we then describe the shape comparison method that is used to determine the scale and location of the landmark in the image, and discuss the use of the range and bearing estimates to speed up the shape comparison by predicting the location and size of the landmark in the image. Sections 5 and 6 discuss the navigation method, including the obstacle avoidance, and present some examples. The navigation

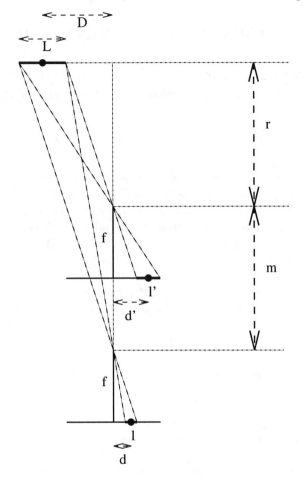

Figure 1: *Determining the range to the landmark. A top-down view of the camera as it it moves (translates) distance m in the direction of the camera optic axis. The landmark is of length L with a reference point at its center that is distance D from the optic axis.*

method has been implemented on a mobile robot platform (developed by [5]); the shape comparison computation is done off-board on a Sun SPARCstation. We have experimented extensively with the navigation system, and it generally brings the robot into contact with a target that is approximately half a meter in each dimension.

2 Determining the Range to the Landmark

In this section we describe the method used to compute the range, r, to a landmark, and discuss the assumptions underlying the method. The only quantities that are measured are the change (increase) in the apparent size of the landmark in the image, s, and the distance that the robot moved, m. Consider a pinhole camera with focal length f, and a landmark of length L whose center lies D away from the optical axis, as illustrated in Figure 1. The camera moves a distance m in the direction of the optic axis (thus towards the landmark, as the landmark is visible). Before the motion, the image of the landmark is of length l, and after the motion is of length l'. The offset of the center of the landmark from the center of the image is d in the first image and d' in the second. From the projection equations,

$$\frac{f}{d+l/2} = \frac{r+m}{D+L/2}$$

and

$$\frac{f}{d'+l'/2} = \frac{r}{D+L/2}$$

and thus

$$(r+m)(d+l/2) = r(d'+l'/2).$$

Similarly $f/d = (r+m)/D$ and $f/d' = r/D$, and thus $(r+m)d = rd'$. From this we obtain $(r+m)l/2 = rl'/2$ and rewriting in terms of r yields

$$r = \frac{m}{\frac{l'}{l} - 1} = \frac{m}{s-1} \tag{1}$$

where $l'/l = s$ is the change in size of the landmark in the image.

The quantity $r = m/(s-1)$ depends only on the size change s and the distance moved m, but not the camera parameters. Implicitly, however, the accuracy of the range measurement depends on the motion being nearly directly towards the landmark. This is because r only measures the actual range to the landmark when the motion is directly towards the object (i.e., the quantity D is zero). There are two sources of inaccuracy when $D \neq 0$. The first is that r only measures the component of the range in the direction of the motion. That is, the true range is $\sqrt{r^2 + D^2}$ whereas the computed distance is r.

The second source of inaccuracy when $D \neq 0$ comes from that fact that the derivation of $r = m(s-1)$ assumes that the landmark is in a plane perpendicular to the direction of motion of the camera (see Figure 1). When $D = 0$ the landmark projects directly onto the optical center, and this assumption is not necessary.

Thus when D is nonzero and the landmark is in a plane that is not perpendicular to the direction of motion (or for instance is not planar) there will be error in computing the range. These two sources of inaccuracy thus lead to a navigation strategy of heading directly towards the landmark, within some tolerance, so that the resulting errors are small. Experimentally, we have determined that the robot generally navigates successfully to the target when the landmark is allowed to be no more than ± 20 pixels of the camera center in an image 360 pixels wide (until the robot is close enough that the object fills much of the field of view, at which point the landmark is kept within a pixel of the center because small rotation errors at such a close range can cause the object to be lost).

Another possible source of inaccuracy in the computation of the range arises from the fact that the image of the landmark may change shape in ways other than translation and scaling. In particular, perspective effects and the fact that different parts of the object will be visible in successive images will result in changes in the image shape. This may influence the computation of the scale change that is used in computing the range. We use a shape comparison method that operates on dense sets of features (intensity edges), and that does not compute a correspondence between features of two shapes being compared (see Section 4). Thus, the method is less sensitive to errors in the locations of individual features than are methods based on correspondences between a small number of features (where errors in a few features may cause a significant error in the shape comparison). Using this shape comparison method, we have found the effects of perspective and correspondence to be negligible until the camera is quite close to the object (within a meter or so using a camera with a 16 mm lens and an image width of about $\pm 15°$), as will be seen in the experimental results section below.

3 Centering the Landmark in the Image

The above method for computing the range to a landmark requires centering the landmark in the image, so that the robot can head directly towards the landmark in order to estimate the range. Centering the landmark in the image is straightforward given an estimate of the

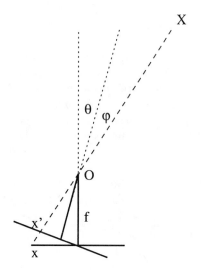

Figure 2: *Computing focal length.*

focal length of the camera, and assuming that the camera is mounted as described in the introduction, such that the x-axis of the image is approximately parallel to the ground plane and the focal point of the camera is near the center of rotation of the robot. Small inaccuracies in these assumptions are not an issue (e.g., we do not do any careful alignment or calibration of the camera with respect to its mount on the robot).

Given these assumptions (the assumption that the robot moves in a straight line can be relaxed, as discussed below), a rotation of the robot simply changes the x-coordinate of an object in the image. Thus if the landmark is centered at some image location $c = (c_x, c_y)$, a rotation by

$$\theta = \tan^{-1}\left(\frac{c_x}{f}\right) \qquad (2)$$

will place the landmark at the image center ($x = 0$), where f is the focal length of the camera. We do not assume that f is known, but rather estimate it from the motion of the landmark in the image as a function of the rotation of the camera. The initial estimate of f is obtained by rotating the camera a fixed, known, amount, and is then refined during subsequent motion towards the landmark, until the robot is close enough that perspective effects and other sources of error become significant (when partial matching starts; see Section 5).

Figure 2 illustrates the manner in which we compute an approximation to f by rotating the camera

some given amount about the focal point. Consider a point X in the world that projects to a point x in the image. If the camera is rotated about the focal point by some given amount θ, then X will project to some new location x' in the resulting image. Let φ be the angle between the optic axis of the rotated camera and the line $\overline{XO}$ (where O is the focal point of the camera). Let $\psi = \theta + \varphi$ be the angle between the optic axis of the original camera and the line $\overline{XO}$.

We know that $\tan(\psi) = x/f$ and $\tan(\varphi) = x'/f$, and hence

$$\tan(\psi) - \tan(\varphi) = \frac{x - x'}{f}.$$

This can be used to approximate f by noting that since $\theta = \psi - \varphi$, when $\tan(\psi) - \tan(\varphi) \approx \tan(\psi - \varphi)$ then

$$f \approx \frac{x - x'}{\tan(\theta)}. \qquad (3)$$

This approximation is close when $\cos(\psi - \varphi) \gg \sin(\psi)\sin(\varphi)$ which clearly holds when ψ and φ are both small, but which also holds relatively well over the range of rotations that occur with our camera (which has a field of view of about $\pm 15°$). We choose the rotation of the camera to be $\theta = \pm 8°$ (where the sign of the rotation is in the direction that will move the object towards the image center). Thus if the initial angle is, for example, $\psi = 12°$, then $\varphi = 4°$ and $\cos(\psi - \varphi) \approx .99 \gg .01 \approx \sin(\psi)\sin(\varphi)$. In an image with a visual angle of $\pm 15°$ the initial angle between the optic axis and the image of the landmark is rarely more than $12°$ because the landmark has some width, and thus its center (which is the reference point on the landmark) is generally not all the way on the edge of the image.

If the robot does not move in a straight line (which we discovered was the case for our robot — it actually moves on a large circle), then when the robot is traveling large distances to the landmark, the landmark will drift away from the center of the image as the robot moves. Thus this effect is most pronounced when the robot begins at considerable distance from the landmark. We have observed it most when starting 8-10 meters away from the landmark. This source of error can be corrected for using the amount that the center of the landmark translates in the image, assuming that the landmark is stationary. We initially center the landmark ($x = 0$), then move forward a distance of m

and measure the new position x' of the landmark. The quantity

$$R = \frac{\tan^{-1}(x'/f)}{m}$$

gives the degrees/meter of compensating rotation needed to keep the robot moving in a straight line.

We estimate the drift, R, only when the robot is initially 8 or more meters from the landmark, and then use it to correct the robot's motion locally throughout the path (i.e., local differential rotations are performed during a translation to keep the robot headed in the correct direction). When the robot is initially closer than 8 meters, we have experimentally found that the drift computation is not very accurate. Note that when the drift is not computed, the robot will generally still find the target; however some additional error and expense is introduced because the robot drifts away from the target direction, re-centers, and so forth. As the error in estimating range is proportional to the degree to which the heading is off, it is better to estimate and correct for the drift directly when possible.

4 Locating the Landmark in the Image

As the robot moves, the landmark changes size and position in the camera image. After every motion, the robot recomputes the range and bearing to the landmark so that it may adjust its course, in order to ensure that it continues to move directly towards the landmark.

The robot's time frame is divided into time steps; each time step corresponds to the robot completing a motion (either a forward translation or a rotation). At the beginning of time step t, the robot has a model of the landmark object M_t, an estimate r_t of the range to the landmark, and the position x_t of the landmark in the camera image (i.e. the bearing to the landmark). It performs some motion, then acquires a new frame from the camera and detects the intensity edges (using a detector similar to [1]). This produces a binary edge image I_{t+1}, in which the robot must locate the landmark, using its previous model M_t.

Since the robot has moved, the landmark will have moved (translated) in the image. It will also have enlarged, as the robot is getting closer to it. Also, if the robot rotates about a vertical axis, or does not move directly towards the landmark, then the landmark may appear to scale slightly in x. We therefore

search for a transformation $T = (p_x, p_y, s_x, s_y)$ of the model, where each point $w = (w_x, w_y)$ of M_t is mapped to $T(w) = (s_x w_x + p_x, s_y w_y + p_y)$. The difference between s_x and s_y will, however, be small, as the primary influence on these scale values is the overall enlarging of the landmark as the robot approaches.

The stages in the processing of I_{t+1} are as follows. First, the robot uses the information that it knew from the previous time step (the size and position of the landmark) together with its knowledge of the motion it performed to estimate a new size and position, and searches around that location for a section of I_{t+1} which closely resembles M_t. It then finds the scaling and translation of M_t which brings it into the best possible alignment with I_{t+1}. Next, the p_x component of this transformation is converted into the new bearing to the landmark (as described in Equation 2) and the two scale values s_x and s_y are averaged together to give s, the overall scale, which is converted into the new estimate of the range to the landmark (as described in Equation 1); it has now computed r_{t+1} and x_{t+1}. The robot then builds a new model of the landmark, M_{t+1}, determines what motion should be carried out next, and proceeds.

The following subsections explain in detail how each of these steps is carried out.

4.1 The Hausdorff Distance

In order to precisely locate M_t in I_{t+1}, we must have a measure of similarity. Here, we use the minimum Hausdorff distance under translation and scaling of the model M_t.

M_t and I_{t+1} will be finite point sets, where each point represents a pixel where an edge was detected. The Hausdorff distance between the image I_{t+1} and a transformation of the model $T(M_t)$ is defined as

$$H(I_{t+1}, T(M_t)) = \max(h(I_{t+1}, T(M_t)), \ h(T(M_t), I_{t+1}))$$
(4)

where

$$h(A, B) = \max_{a \in A} \min_{b \in B} \|a - b\|, \qquad (5)$$

and $\|\cdot\|$ is some underlying norm. Here, we use the L_2 or Euclidean norm.

The function $h(A, B)$ is called the *directed* Hausdorff distance from A to B. In effect, $h(A, B)$ ranks each point of A based on its distance to the nearest point of B, and then the largest ranked such point (the most mismatched point of A) specifies the value of the distance. Intuitively, if $h(A, B) = \epsilon$, then each point of A must be within distance ϵ of some point of B, and there also is some point of A that is exactly distance ϵ from the nearest point of B (the most mismatched point). Note that in general $h(A, B)$ and $h(B, A)$ can attain very different values (the directed distances are not symmetric).

The computation of $H(I_{t+1}, T(M_t))$ does not involve determining an explicit pairing (or correspondence) of points of I_{t+1} with points of $T(M_t)$ (for example many points of I_{t+1} may be close to the *same* point of $T(M_t)$). This contrasts with most model-based recognition methods which determine a correspondence between points of the model and the image.

The directed Hausdorff measure can be represented in terms of set containment. Let $I_{t+1}^{\epsilon} = I_{t+1} \oplus C_{\epsilon}$, where C_{ϵ} is a disk of radius ϵ and $\oplus$ is the Minkowski sum (for two point sets P and Q, $P \oplus Q = \{p + q | p \in P, q \in Q\}$). Intuitively, I_{t+1}^{ϵ} is the set obtained by replacing each point of I_{t+1} with a disk of radius ϵ, and taking the union of all of these disks. By definition $h(T(M_t), I_{t+1}) \leq \epsilon$ if and only if $T(M_t) \subseteq I_{t+1}^{\epsilon}$, because in order for every point of $T(M_t)$ to be within distance ϵ of I_{t+1} it must be contained in I_{t+1}^{ϵ}. We therefore extend the Hausdorff distance by defining, for some ϵ,

$$f_{\epsilon}(T(M_t), I_{t+1}) = \frac{\#(T(M_t) \cap I_{t+1}^{\epsilon})}{\#(T(M_t))}$$

where $\#(S)$ is the number of points in the set S. This is the *fraction* of the points of $T(M_t)$ which lie inside I_{t+1}^{ϵ}, or (equivalently) the fraction of the points of $T(M_t)$ that lie within ϵ of some point of I_{t+1}

Each model, M_t is enclosed by some box, defining its extent. This box also defines a transformed box bounding $T(M_t)$. When calculating the fraction of image points that lie within ϵ of some transformed model point, we are only really interested in the image points that lie within this transformed box. Image points that lie outside this transformed box are likely to be parts of other objects also present in the image. We therefore define $f_{\epsilon}'(I_{t+1}, T(M_t))$ to be the fraction of the points of I_{t+1} lying inside the bounding box of $T(M_t)$, that fall within ϵ of some point of $T(M_t)$.

4.2   Rasterizing Transformation Space

We impose a raster grid on the four-dimensional space of translations and scales, and search on this grid for the new location and scale of the landmark. We do not search the entire space of possible translations and scales, but instead bound the range of allowable transformations; we describe this in the following subsection.

If T and T' are two transformations which are adjacent in the rasterized transformation space (i.e. neighbors on the grid of transformations, considered here to be 8-connected: each location is considered to be connected to its immediate neighbors only), we want the points of $T(M_t)$ to be adjacent (on the image grid) to the points of $T'(M_t)$. This leads us to the following rules:

1. The translational component of the transformation should be rasterized to an accuracy of one pixel: translations should have integer components.

2. If for each point $w = (w_x, w_y) \in M_t$, we have $0 \leq w_x \leq x_{\max}$ and $0 \leq w_y \leq y_{\max}$, then we should rasterize the x-scale component to an accuracy of $1/x_{\max}$ and the y-scale component to an accuracy of $1/y_{\max}$. Thus, if $T = (p_x, p_y, s_x, s_y)$ and $T' = (p_x, p_y, s_x, s_y')$ are adjacent in the grid of transformations, $s_y' = s_y \pm 1/y_{\max}$ and so for any $w \in M_t$, the y-coordinate of $T'(w)$ will be at most $w_y/y_{\max} \leq 1$ away from the y-coordinate of $T(w)$. Our rule is therefore that the x-scale component should be an integer multiple of $1/x_{\max}$, and the y-scale component should likewise be an integer multiple of $1/y_{\max}$. Note that this rasterization depends on the model M_t.

Transformations can therefore be represented by four integers; the quadruple (i_x, i_y, j_x, j_y) would represent the transformation $(i_x, i_y, j_x/x_{\max}, j_y/y_{\max})$.

4.3   Predicting the Scale and Location of the Match

We can bound the range of the search on the grid of transformations by predicting the position and scale of the landmark, based on the previous values for its size, range and bearing. Initially, suppose that our estimate of the range is r_t, the size (of the image of the landmark) is l_t, and the x-position of the center of the landmark is x_t. We wish to compute $\tilde{r}_{t+1}$, $\tilde{l}_{t+1}$ and $\tilde{x}_{t+1}$ which are predictions of r_{t+1}, l_{t+1} and x_{t+1}, the range, size and x-position of the landmark after a forward motion of magnitude m. The motion will be close to directly towards the landmark.

$\tilde{r}_{t+1}$ is simply $r_t - m$, as we are moving towards the landmark. The predicted value of the size of the landmark and thus the next scale factor can then be calculated as a function of $\tilde{r}_{t+1}$ and m: $\tilde{r}_{t+1} = m/(\tilde{l}_{t+1}/l_t - 1)$, and so $\tilde{l}_{t+1} = l_t(m/\tilde{r}_{t+1} + 1)$; the scale is predicted to be $\tilde{l}_{t+1}/l_t$.

The predicted position of the landmark can also be determined from its previous position. Ideally, the landmark would have been centered in the image, and the robot would have moved directly towards it; in the next image, it would therefore continue to be centered. In practice, though, it will tend to move outward from the image center as the robot moves, as it will not have been exactly centered, and the robot may not have moved directly forwards. Our prediction of its new position is simply $\tilde{x}_{t+1} = r_t x_t/\tilde{r}_{t+1}$, as the drift is proportional to the distance from the camera center. We then use the predicted location of the landmark in order to restrict the search space: we search for a match only in a small portion of the transformation space, surrounding the anticipated position and scale. Since the x and y scales of the landmark will typically be close, we also only consider transformations where they are within 1% of each other. This allows us to correct for small unanticipated effects due to errors in the scale and position estimates, or errors in motion (including possible motion of the landmark).

Given two fractions f_1 and f_2, and some threshold ϵ, we consider all transformations (lying on the grid) inside the current transformation bounds for which $f_\epsilon(T(M_t), I_{t+1}) \geq f_1$ and $f'_\epsilon(I_{t+1}, T(M_t)) \geq f_2$. For each such transformation T, we also compute

1. ϵ_0, the minimum value of ϵ for which $f_\epsilon(T(M_t), I_{t+1}) \geq f_1$

2. $f_{\epsilon_0}(T(M_t), I_{t+1})$

3. ϵ'_0, the minimum value of ϵ for which $f'_\epsilon(I_{t+1}, T(M_t)) \geq f_2$

4. $n_{\epsilon'_0}(I_{t+1}, T(M_t))$, the actual number of points of I_{t+1} which lie inside $M_t^{\epsilon'_0}$.

Such a search can be performed very efficiently using methods described in [3]. These transformations are then ranked based on these four items, in lexicographic order (i.e. ϵ_0 is most important, followed by $f_{\epsilon_0}(T(M_t), I_{t+1})$, etc). All but the top 10% are discarded. The remaining transformations are grouped into connected components in transformation space (using 80-connectedness adjacency on the four-dimensional grid: each transformation is considered to be connected to all those transformations whose parameters differ by no more than one in each dimension) and the centroid of each component is found. Within each component, each of these values is averaged over all the matches in that component, giving four average values. The components are then ranked (lexicographically) by these four values; the best component is then chosen as the one most likely to contain the image of the landmark, and its centroid determines the new values for the size and position of the landmark, and thus its range and bearing.

If the landmark is not found in this search (i.e. no transformations satisfying these criteria can be found inside the search range), then the range of the search is enlarged, f_1 and f_2 are lowered, and the search is repeated until the landmark is located.

Once the landmark has been located in I_{t+1}, we build M_{t+1}, which will be used to find the landmark in the next time step. This is done by simply cutting out a rectangular box around the location in I_{t+1} where M_t was found.

Initially, ϵ, f_1 and f_2 are set to fixed values (currently 2 pixels, 90% and 85% respectively). After each time step, the average values of ϵ_0 etc. generated by the best component are used to determine the values of ϵ etc. to be used in the next time step.

5 The Overall Navigation Method

Initially the robot grabs an image of the visual field which is presumed to contain the landmark. An initial model is given, which consists of a set of two-dimensional intensity edges (as discussed above this model can be extracted from an image of the scene, or can be obtained in some other manner such as from a modeling system). The navigation process is divided into an initialization stage and an approach stage. In the initialization stage a crude approximation to the range is obtained and the focal length of the camera is estimated. In the approach stage, the robot moves in a direction that is the current best estimate of the heading to the landmark, and updates the range and focal length values, as well as computing some other quantities discussed below. The approach stage is divided into steps, where at each step the robot grabs an image of the visual field and matches a model from the previous step to this image, in order to update the range and bearing estimates.

The initialization stage consists of two preprogrammed movements that are used to calculate the range to the landmark and the focal length of the camera. Before the robot executes either of these movements, an image is grabbed and the initial model is matched to this image, in order to determine the initial model size. The robot then moves forward a known distance, currently 50cm, and grabs another image. The initial model is matched to this image; this yields the scale change s of the landmark over a 50cm motion. The range to the landmark is calculated from s and the motion using Equation 1. A new model is then generated by selecting the rectangular portion of the image where the landmark was found.

The robot then executes the second of the preprogrammed initialization movements, in order to estimate the focal length of the camera. The robot rotates a known amount, currently 8 degrees, in the direction that would result in the landmark being closer to the center of the image frame and grabs an image. After matching the model to the image, the robot uses the amount of rotation and the landmark's change in x-position in the image to calculate an estimate of the camera's focal length, as described by Equation 3.

In general the object will not be centered in the image frame after the two initial movements. Therefore, before beginning the approach, the robot calculates the angle needed to rotate in order to center the landmark, using the focal length of the camera and the number of pixels that the object is off center by Equation 2. After rotating this amount, the robot grabs an image and searches for the landmark to be sure that the centering worked and to recalculate the focal length of the camera.

At this point, the landmark is centered in the frame, and the robot enters the approach phase in which it proceeds towards the object. Ideally, moving forward

when the landmark is centered in the image frame should always keep the robot on a direct course, until the robot eventually makes contact with the landmark. However, factors such as the object not being completely centered and the robot not moving exactly as commanded may cause the landmark to drift off center in the image. During the approach, the landmark may drift in the frame, signifying that the robot is slightly off target. Before each forward motion, the robot therefore calculates how much rotation is required to center the landmark in the frame, and performs this rotation simultaneously with the forward motion.

During each step of the approach process, the scale change between the current image and the previous image is computed. Rather than simply computing the range from this scale change, the scale changes for the last several images are combined in order to produce a more accurate estimate of the range. That is, the overall scale change across the sequence of images and the total motion during that sequence are used to compute the range from Equation 1. All of the images taken since the last rotation of the robot are used in this process.

5.1   Concurrent Moving and Matching

In practice, the amount of time required to grab an image, extract its edges, and perform a matching step is usually about 3-4 seconds using a Sun 670 (for a 360×240 image). At the robot's usual translational velocity, this corresponds to about 20cm of forward motion. We can therefore improve the overall speed of the process by performing motion and matching in parallel: the robot grabs a frame, and begins searching for the landmark. It simultaneously begins moving forward. Usually, the search locates the landmark within a few seconds. The robot knows the range and bearing to the landmark that it had when it grabbed that frame. It can therefore predict the current range and bearing, and compute a rotation which will keep the landmark centered (and the robot on course). It can then update its model, execute this rotation (while continuing to move forward), grab another image, and begin another matching step.

In order to ensure that the robot does not go too far off course by continuing to move far away from the location where the last matched image was grabbed, we set a "safe distance" limit (currently 20cm) on the distance that it is allowed to travel while processing a match. In almost all cases, the robot completes the matching step before traveling this far; it therefore rarely has to stop moving.

5.2 Obstacle Avoidance

The obstacle avoidance algorithm was added to handle the case when there are obstacles in the robot's path to the landmark. Since our navigation is completely visually guided, we assume that the obstacles do not interfere with the camera's line of sight to the landmark, but just impede the robot's path. We also assume that the surfaces of the obstacles are flat.

As the robot is moving forward, it will stop and register an impediment if one of its contact bumpers has been pressed. The robot uses its estimate of range to determine whether the impediment is an obstacle or the landmark. If the calculated range is greater than some value, currently one meter, then the robot assumes the impediment is an obstacle and begins its obstacle avoidance routine. If the range from the target is less then one meter, the robot assumes that it has arrived, and returns a successful hit.

The goal of the obstacle avoidance is to successfully move around the obstacle, while keeping the landmark in the image frame. The first step in the obstacle avoidance routine is for the robot to align itself with the obstacle. The robot we used has eight contact bumpers attached around the front portion of the cylindrical body. The robot is considered aligned when the robot's heading is normal to the face of the obstacle. This active alignment can be achieved by rotating while remaining in contact with the object until only the middle two contact bumpers are pressed [4].

At this point, the robot backs up a small amount and rotates 90 degrees in the direction that would still be making progress towards the landmark. The robot begins repeatedly pinging the sonar on the side where the obstacle lies, as it translates forward. By the flat surface assumption, the sonar on the side of the robot facing the obstacle will register a large change in distance once the robot has cleared the obstacle.

If another obstacle is encountered as the robot is translating to clear the first obstacle, then the robot rotates 180 degrees, and attempts to move around the obstacle in the opposite direction in a similar way. If the robot encounters yet another obstacle as it attempts to

clear the box in this direction, then the robot is boxed in (in that it must actually move away from the target to get to it), and gives up.

During this process, the robot maintains the angle and distance traveled since first contact with the obstacle. From these parameters and the previous range to the landmark, the robot calculates the new range and bearing to the target, and rotates to that bearing. The robot then grabs an image, locates the landmark, recenters, and resumes the approach.

5.3 Partial Matching

As the robot approaches the object, the landmark gets larger in the image frame, and at some range portions of the landmark may begin to fall outside the image. When this happens, the model from a given frame will not match the image at the next frame very well, because parts of the object will be missing. At this point, the matching process is changed to allow for partial matches of the model (by lowering the fraction of model points that must be near points of the image, recall this fraction f was discussed in Section 4).

The navigation system goes into the partial matching phase once the robot has gotten close enough to the object that part of it lies near the boundary of the image (within 15 pixels of any border). In this phase, the matching method only requires an initial fraction of $f = .8$ when looking for the model in the image. In addition, the robot moves smaller intervals between successive images (the safe distance is reduced), since the shape change in the landmark is relatively great when the robot is close.

6 Some Experimental Results

Here we describe the results of some experiments we have performed using this technique. In each case, the robot was started some distance away from the landmark, with the landmark in view. The user then outlined a rectangular area which enclosed the landmark; this was used as the first model.

Figures 3 and 4 show examples of the robot navigating to its target. Each figure consists of ten rows representing ten of the time steps from a navigation sequence. Each row of the figures contains three images: the edge image acquired by the robot at that time step (I_t), the model from the previous time step overlaid on

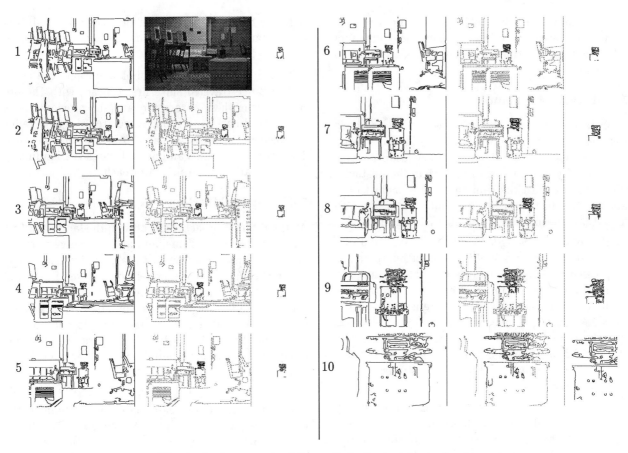

Figure 3: *An example of the method. See text for explanation.*

the image at the location where it was found (M_{t-1} overlaid on I_t), and the model extracted from the image (M_t). In the first example, the target is one of the lab's other mobile robots. Only the top portion of the robot is initially visible, so this is used as the initial model. The target in the second example is the person sitting on the couch. Note that the images are quite cluttered.

In both figures, rows 1, 2 and 3 show the initial calibration movements described in Section 5. Row 1 contains the edge image generated before any movement is performed. It does not have a previous model, so no overlay is shown; the original image is shown instead. Row 2 is after a forward movement of 50cm to determine the initial range, and Row 3 is after the 8° rotation to determine the camera focal length. Row 4 is at a position intermediate between the initial position and the first obstacle. Row 5 is just before the

robot makes contact with the first obstacle; Row 6 is just after it has cleared that obstacle. Rows 7 and 8 bracket the avoidance of the second obstacle. Row 9 is intermediate between Row 8 and the location of the landmark. Row 10 was taken just before the robot made contact with the target.

In the first example, the total distance traveled was 9m; 43 images in total were processed. The total distance traveled in the second example was 10m and 46 images were processed. The time between images acquisitions in these examples was typically around 5 seconds, and the forward movement between acquisitions was usually slightly less than 20cm, so the robot was usually able to begin the next movement without actually coming to a stop. If the robot encountered an obstacle and had to navigate around it, this of course took more time.

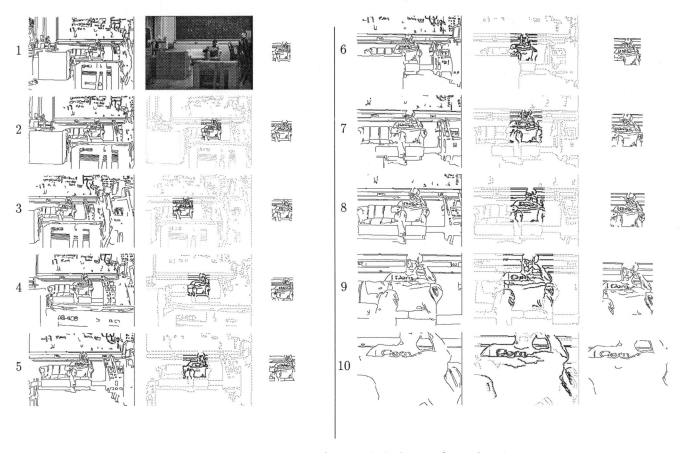

Figure 4: *A second example of the method. See text for explanation.*

7 Summary

We have presented a method for using visual information to navigate a mobile robot to a landmark in its visual field. The method operates by comparing two-dimensional edge images in order to recover the change in location and size of the landmark in the image as the robot moves. The change in image location is used to keep the landmark centered in the robot's visual field (and thereby keep the robot moving straight towards the landmark). The change in image scale is used to estimate the range to the landmark (given that the robot moves straight towards the landmark).

The method does not make use of absolute world coordinates, or of any three-dimensional information. The robot always maintains an estimate of the range and bearing to the landmark from the last place that a picture was taken, and updates that estimate after each successive image is obtained. The method has been used in our laboratory to control a mobile robot, and some examples of its operation were presented. The matching technique is fast enough (on a SPARCstation) that it can run concurrently with them motion of the robot. Overall, the method is quite simple, not requiring complex representations of objects, or accurate calibration of the camera system.

8 Acknowledgments

This work was supported in part by National Science Foundation PYI grant IRI-9057928 and matching funds from Xerox Corp., and in part by Air Force contract AFOSR-91-0328.

References

[1] J.F. Canny. A computational approach to edge de-

tection. *IEEE Trans. Pat. Anal. and Mach. Intel.*, 8(6):34–43, 1986.

[2] Enrico Grosso, Missimo Tistarelli, and Giulio Sandini. Active/dynamic stereo for navigation. In *Proc. 2nd European Conf. on Computer Vision*, pages 516–525, Santa Margherita Ligure, Italy, May 1992.

[3] D.P. Huttenlocher and W.J. Rucklidge. A multiresolution technique for comparing images using the Hausdorff distance. In *Proc. Computer Vision and Pattern Recognition*, pages 705–706, New York, NY, 1993.

[4] J. Jennings and D. Rus. Active model acquisition for near-sensorless manipulation with mobile robots. In *Proc. IASTED International Conference on Robotics and Manufacturing*, Oxford, England, September 1993.

[5] Jonathan Rees and Bruce Donald. Program mobile robots in Scheme. In *Proc. IEEE International Conference on Robotics and Automation*, Nice, 1992.

[6] Carlo Tomasi and Jianbo Shi. Direction of heading from image deformations. In *Proc. Computer Vision and Pattern Recognition*, pages 422–427, New York, NY, 1993.

[7] Z. Zhang and O.D. Faugeras. Building a 3d world model with a mobile robot. In *Proc. 10th Intl. Conf. Pattern. Recoc.*, 1990.

[8] J.Y. Zheng, M. Barth, and S. Tsuji. Qualitative route scene description using autonomous landmark detection. In *Proc. Third International Conference on Computer Vision*, pages 558–562, Osaka, Japan, 1990.

Mobile Robot Navigation Algorithms

Raja Chatila, *LAAS-CNRS, 7 Ave. du Colonel Roche, 31077 Toulouse, France*

Navigation is an instance of the general robotic task "perceive-decide-act". We discuss how this task is translated into algorithms such that its execution context is taken into account in order to produce the more appropriate robot behavior. To illustrate the discussion, we consider two navigation cases and experiments, one indoors and the other outdoors. We present the processes we developed and discuss the environment representations necessary to perform navigation in both cases. Different motion planning algorithms are used, according to the nature of the terrain (flat, uneven) and the constraints of the environment (free, cluttered). Navigation strategies (actually the algorithms for performing navigation itself) for selecting perception tasks, subgoals for motion planning, and the motion modes are discussed.

1 Introduction

Navigation is the basic task that has to be solved by an autonomous mobile robot. By navigation, we understand the task of reaching an arbitrarily distant goal, the path being not predefined or given to the robot. Navigation is in general an *incremental process* that can be summarized in four main steps:

1. Environment perception and modelling: any motion requires a representation of the local environment at least, and often a more global knowledge. The navigation process has to decide where, when and what to perceive.

2. Localization: the robot needs to know where it is with respect to its environment and goal.

3. Motion decision and planning: the robot has to decide where or which way to go, locally or at the longer term, and possibly a trajectory;

4. Motion execution: the commands corresponding to the motion decisions are executed by control processes - possibly sensor-based - either for following a geometrical trajectory, or a direction, etc.

Navigation is an instance of the general robotic task paradigm "perceive - decide - act". It may embed this paradigm at several levels (e.g., the "act" part may itself include this loop). In implementing the actual navigation task, its four main general steps may be more or less simple or complex. Their complexity depends on the general context in which this task is to be executed:

- the *environment:* may be initially known or unknown (or partially known), static or with moving objects, with a variable nature of terrain and obstacle density, etc.,

- the *goal:* may be specified by landmarks or by coordinates,

- the *navigation task* itself: may have constraints, e.g., time bounds, best path criteria, completeness, etc.,

- the robot's *abilities:* computing power, sensors and their uncertainties, robot size and kinematics, etc. determine how the task will be executed.

The solution to the navigation problem, the actual execution of the task, will strongly depend on all these constraints.

For example, if the robot moves in a very dynamic environment - with respect to its own velocity and computing power - then there is no point in planning a trajectory, as it would be obsolete before it is even started. The algorithms to solve navigation in this case should be very fast and rely on local data. The motion planning and perception steps can then be made very simple, for example an attraction by the goal with local avoidance, and a collision test respectively.

On the other hand, if much is known in advance on the environment and if perturbations are transient, then the solution would be to plan ahead - especially if there is an efficiency constraint on the path (and there always is, at least because of energy or time limitations) - and adapt execution to local perturbations.

An environment model is needed for navigation in order to plan a collision-free path to reach the goal with respect to a criterion (distance, time, security, etc.). It is also needed for the robot to self-locate with respect to it. Otherwise, reaching a given goal (that is not in sight) will be impossible in general, considering the uncertainties of odometry or inertial sensors.

In addition, sensors are always imperfect: their data are uncomplete (lack of information concerning existing features) and inaccurate. They generate artefacts (information on non-existing features) and errors (wrong information concerning existing features). The same area when perceived again can therefore be differently represented. Hence environment representations must take into account uncertainties and tolerate variations. Models are also different in indoor or outdoor environments: natural environments are not intrinsically structured, as compared to indoor environments where simple geometric primitives (e.g., planar surfaces) are close to match reality. Finally, the representations must be at several levels of abstraction, i.e., adapted to the level of the processes using them.

Path finding, which is based on geometrical and topological representations, cannot - or at least very inefficiently - be solved in general by local methods driving the robot directly (which is equivalent to performing the search physically) without mapping. A variety of motion planning algorithms [14] exist. According to their complexity and the representations they manipulate, they are adapted to be used in different situations that the robot as a system must recognize: constrained space, uneven terrain, etc. Several motion planners should be embedded in the system; selecting the more appropriate according to the situation is part of the navigation algorithm.

Finally, to cope with perception and motion uncertainties, it is more robust to plan the motion in terms of closed-loop sensor-based processes directly [8, 15, 1], rather than executing a geometrical trajectory relying on odometry or inertial data. Another approach is to compute a trajectory that takes into account possible error reduction by sensing [16, 22].

The question we want to address is how to instanciate the general navigation task we just overviewed, to take into account the variations due to the execution context. In other words, what is the general **navigation algorithm**, to be run by the robot control system, that solves the problem while adapting its complexity to the actual external constraints and the available time and resources? In order to try to answer this question, we first present two examples of navigation that mainly differ by the nature of the environment: indoors (§2) , and outdoors (§3). In both cases, the task is to reach a goal (a known landmark) in an initially unknown environment.

2 Indoor Navigation

The environment being unknown, reaching a goal efficiently requires building an environment model incrementally as the robot moves, and maintaining a consistent representation of it: otherwise, the robot would not be able to plan its motion. Furthermore, motion planning should have the property of completeness: otherwise, the navigation algorithm would not be able to decide that the task has failed (i.e., that no trajectory is possible).

2.1 Perception and Modelling

In indoor navigation, objects have mostly planar surfaces. The ground is considered as being flat. It is therefore adequate to use planar surfacic representations (or edges in 2D). In order to cope with sensor uncertainties, and because of the incremental modelling process, filtering techniques are used [17, 18].

The process incrementally builds object surfaces - or edges in 2D - also considered as features for robot localization, and their uncertainties. Free space is also incrementally updated.

The environment model is built in our case from data acquired by a horizontally scanning laser range finder. Knowing the variance on each depth point measurement (from the sensor uncertainty model), points may be grouped into line segments with their variances (using a Kalman filter). The result of this step is a local surfacic model shown in figure 1.a. Matching this model with the global model built up to this step (figure 1.b and c.) by geometrical correspondance, and fusing them by stochastic fusion techniques provides for position correction of the robot, and a better estimate of the global model itself, that integrates new parts (figure 1.d). Robot localization is done at each perception when the newly perceived parts are matched and fused with the current global model.

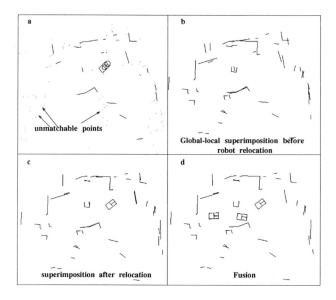

Figure 1: *A complete step of surfacic modelling.*

Path planning needs a representation of free space. The local free space is built by interpolating successive surface measurements (edges) ordered by sensor acquisition. A geometric reasoning on visibility is made in order to build a consistent representation that takes into account sensor uncertainties and missing data.

The local and global (previously built) free spaces are then superimposed, and the global free space model updated after processing possible inconsistencies (figure 2).

The result of this modelling step of the navigation algorithm is a planar surfacic model of objects and a polygonal representation of free space.

2.2 Motion planning

Motion planning techniques in polygonal environments are well known [14] and will not be overviewed here. Rather, we want to emphasize that their use should be adapted to the task context. For example, in a large free space area with few obstacles, isotropic obstacle growing is sufficient. In a very constrained space, an

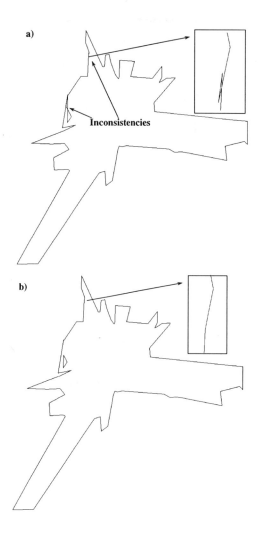

Figure 2: *Global free space model updating with inconsistency correction.*

(x, y, θ) planner is necessary, taking into account kinematic and non-holonomic constraints if necessary, according to the robot's kinematics. In a corridor-like region, a "wall-following" closed-loop process efficiently implements the task, etc.

The motion planner used in the experimental example here is a complete planner (because of the exploratory nature of the task); the robot, Hilare 1.5, is approximated by a circle, the environment being not constrained. Figure 3 shows the execution of incremental indoor navigation with environment modelling.

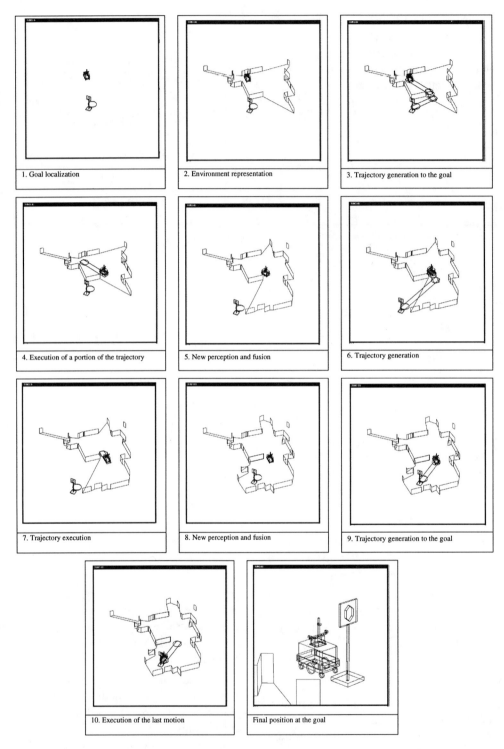

Figure 3: *Experimental execution of an indoor navigation task.*

3   Outdoor Navigation

The complexity of outdoor navigation stems essentially from the diversity and lack of structure of natural environments: some areas can be totally flat (possibly scattered with easily detectable obstacles, e.g., large rocks), whereas others can be much more cluttered, such as uneven rocky area. This variety induces different robot behaviors, and constrains both the perception and motion planning processes.

Several systems for outdoor navigation have already been developed in the ALV [6] and UGV projects [13], for the Navlab [23] and Ambler [11]. Some approaches integrate planning and control [20, 11]. However, these systems make use of a single mode for navigation, which limits their use to rather flat terrain in general, except maybe for Ambler, a legged machine, but then the system is rather inefficient since it runs the same process of footfall search even if the environment is simple and does not require it.

A navigation algorithm, as we present it here, must adapt the perception, planning and execution functions to the nature of the environment. This implies different environment representations and the integration of various motion planning algorithms, to be used according to the nature of the area to cross [5]. By adapting the robot behavior to the nature of the terrain, it can execute its navigation task more efficiently.

We consider three navigation modes:

• A **"2D" planned** mode: Applied when the terrain is mostly flat - or with an admissible slope for the robot -, it relies on the planning and execution of a 2D trajectory, using a binary description of the environment in terms of Crossable/Non-Crossable areas.

• A **"3D" planned** mode: Applied when an uneven area has to be crossed, it requires a precise model of the terrain, on which a trajectory taking inot account robot stability is planned and executed[1].

• A **reflex** mode: The robot locomotion commands are determined on the basis of *(i)* a goal (heading or position) and *(ii)* the information provided by "obstacle detection" sensors. This navigation mode does not require by itself a modeling of the terrain, that has simply been labeled as open space, with few possible

---

[1]Since the robot is always in contact with the ground, this mode could also be called "2.5 D"

obstacles. However, obstacle avoidance needs to have some representation of the obstacles.

The adequate navigation mode is selected after a quick analysis of the raw 3D data produced in our experiments either by a laser range finder (LRF) or by a stereovision algorithm. This quick analysis provides a classification of the terrain regions in terms of {flat (with slope value), uneven, obstacle, unknown}, that is incrementally udated to maintain a global description of the environment. All "strategic" navigation decisions are taken on the basis of this global representation. They concern the determination of the intermediate positions, the choice of the navigation mode to apply to reach them, as well as the definition of the next perception task to execute (Which sensor to use? With what operating modalities? How should the data be processed?). Such an approach involves the development of various perception and motion planning processes, and emphasizes the importance of the navigation algorithm, which is in charge of the strategic decisions.

3.1   Perception and Modelling

Representing the terrain becomes especially difficult in the case of outdoor environments, where the lack of geometric structure and the sensors imperfections complicate the building and the management of the representations. Moreover, various kind of informations are required during navigation: a localisation procedure requires a description of landmarks, whereas a trajectory planning procedure requires a continuous description of the zone to cross for instance. Building a "universal" terrain model that contains all the necessary informations is extremely difficult and not efficient, and we therefore chose to build a *multi-layered heterogeneous* terrain model wherein a particular representation is built or updated only when required by a given task. This involves the development of various perception processes, each of them being dedicated to the extraction of specific representations (*multi-purpose perception*). The various representations we use are required for:

• Trajectory planning. The 2D planner requires in our case a **binary bitmap description**, and the 3D planner builds its own data structure on the basis of an **elevation map** on a regular Cartesian grid.

• Localization. Several representations are useful

for this task: **a set of 3D points** in the case of a correlation-based localization (iconic matching), or a global **map of detected landmarks** in the case of a object-based localization. This last representation requires geometrical models. Currently we are developing an approach using superquadric primitives [4].

• Strategic decisions. To decide for the navigation strategies (sub-goal selection, perception decision, etc.), we build a **labelled topological connection graph**, on which a heuristic search is performed.

The multi-level environment model embeds the different representations (figure 3). The relationships between the various representations make explicit their building rules, and defines a constructive dependency graph between them. On the figure, the arrows labeled "S" represent updatings systematically performed, arrow labeled "C" represent updatings performed, *i.e.*, only when required.

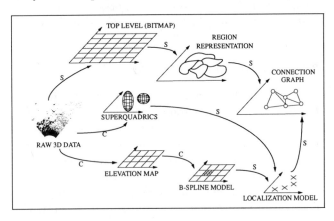

Figure 3: *Terrain representations used in the system and their constructive relations*

3.1.1 Global Representations

These representations used by the navigation algorithm are systematically built when data are acquired.

Fast Terrain Analysis

This process produces a description of the perceived terrain in term in *navigation classes*[12], using a discretization of the perceived area that respects the sensor resolution. This discretization defines "cells", on which different global characteristics are determined: density (number of points in a cell compared with a nominal density defined by the discretization rates), mean altitude, variance on the altitude, mean normal

vector and corresponding variances. These informations allow to heuristically label each cell as one of { *Flat (with slope value), Uneven, Obstacle, Unknown*}. An important point is the ability to estimate a confidence value on the label of each cell: the quantitative estimations of this confidence value are statistically determined (this value obviously decreases with the distance of the cell to the sensor).

This classification procedure takes less than half a second (on a Sparc-10) to process a 10 000-points depth image. It has proved its robustness on a large number of different images, produced either by the LRF or a stereovision correlation algorithm, and is especially weakly affected by sensor noise (uncertainties, artefacts and errors).

Global Model Construction

The global terrain model is a bitmap structure, in which the main informations provided by the classification are encoded (label and corresponding confidence, elevation). This structure is flexible: other informations can easily be encoded in a pixel attribute without modifying the entire description.

Fusion of the classifier output (figure 4) is a simple and fast procedure: each cell is written in the bitmap using a polygon filling algorithm. When a pixel has already been perceived, the possible conflict with the new perception is solved by comparing the label confidence values. Many experiments have proved the robustness of the method.

Connection Graph

Once the global representation is updated, it is structured with classical region merging and contour following algorithms. Regions are areas of uniform label, elevation and confidence (figure 5). The graph is deduced from the adjacency relations between the regions: each node corresponds to the middle point of a border, and each arc corresponds to the crossing of a region. Note that the constrained areas (Uneven, Obstacle and Unknown) are first grown by the robot radius: the robot is considered to be in a constrained area when any point of its body is. This procedure is executed within a few seconds (essentially depending on the ratio of constrained areas to grow); as an example, the fusion of four acquisitions covering an area of approximately 200 square meters produces 400 regions, and the graph contains 1500 nodes.

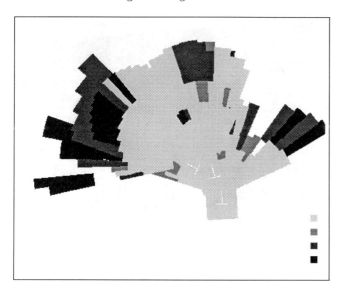

Figure 4: *Navigation in natural terrain (3 steps; robot positions are marked by a cross). Terrain classification after fusion of 3 perceptions is shown (from clear to dark: unknown, horizontal, flat with slope, uneven, obstacle).*

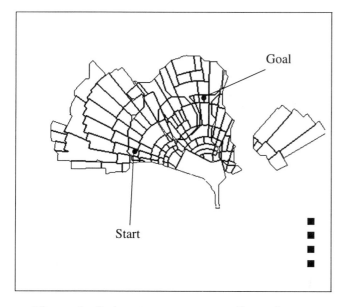

Figure 5: *Region structure corresponding to figure 4*

3.1.2 Elevation Map

This terrain model is necessary for motion planning on uneven terrain, and for localization with iconic techniques. It is constructed by using a generic interpolation method [19] that builds a discrete representa-

tion $z = f(x, y)$ on a regular Cartesian grid from a 3D spherical image.

A fusion of different local elevation maps (i.e., built from one perception) into a global map is needed for trajectory planning if the uneven area cannot be entirely perceived from a single viewpoint. Once the estimation of the new robot's position is done (section 3.1.3), we combine the new local map and the former global one into a new global map. The combination is based on a weighted average that takes into account uncertainties.

3.1.3 Localization Model

The localization process relies on matching terrain peaks, considered as salient features [9]. More complex object-based localization using a global model of objects and their topological relationships is also under development [4].

The specific terrain representation used to extract the elevation peaks is a B-Spline surface based model, built upon an elevation map. Such a model is very compact, and can be made hierarchical: a coarse level B-Spline representation is first computed on a uniform mesh, and a test based on the least-square errors points out the areas where some refinement is needed. A new mesh with smaller size patches is then defined, and a new B-Spline representation is computed, which ultimately leads to a tree model, in which each node corresponds to a B-Spline surface.

This analytic model allows to extract features such as high curvature points, valleys or ridges. We implemented a peak extraction procedure based on a quick analysis of the matrix expression of the B-Spline surfaces. Once the peaks are extracted, we apply a feature matching localization method, cooperating with an iconic one: the iconic method is only performed in the neighborhood of the detected features (peaks). Hence, using small correlation windows, we avoid the long processing time usually encountered with such methods.

3.2 Motion Planning

The reflex navigation mode skips the global modeling and motion planning steps: when the terrain to be crossed is flat and almost obstacle free (which is detected by a characterization of the image), the robot heads directly to its goal. If an obstacle is detected by a proximity sensor (the surveillance mode of the LRF), the robot stops[2], and the nominal procedure of terrain modeling (*i.e.*, fusion of the classification results and structuration) is started.

When motion planning is necessary, depending on the label of the region to cross, the 2D or the 3D planner is selected:

A. Flat Terrain

The trajectory is searched with a simplified and fast method, based on bitmap and potential fields techniques. In a natural environment, and given the uncertainties of motion, perception and modelling, we consider it sufficient to approximate the robot by a circle, and its configuration space is hence two-dimensional, corresponding to the robot's position in the horizontal plane. Path planning is done according the following procedure: first, a binary bitmap *free/obstacle* is extracted from the global bitmap model over the region to be crossed; a distance propagation method (similar to [3]) then produces a distance map and a discrete voronoï diagram, from which a path reaching the sub-goal is extracted. This trajectory is then transformed into a sequence of line segments and rotations.

Search time depends only on the bitmap discretization, and not on the complexity of the environment. The final trajectory is obtained in at most 2.5 seconds on a 256×256 bitmap, on a Sparc 10 workstation.

B. Uneven Terrain

On uneven terrain, irregularities are important enough and stability of the robot must also be taken into account. The path planner therefore requires a 3D description of the terrain, based on the elevation map, and a precise model of the robot geometry in order to produce collision-free trajectories that guarantee vehicle stability and take into account its kinematic constraints. This planner, described in [21], computes a motion verifying these constraints by exloring a graph of configurations obtained by applying sequences of constant controls (rotations and straight line segments).

In the case of incremental exploration of the environment, an additional constraint must be taken into account: the existence of unknown areas on the terrain elevation map. Indeed, any terrain irregularity may

---

[2]Local obstacle avoidance is this reflex mode is under developement.

Figure 6: *A 3D trajectory planned on a real elevation map*

hide part of the ground. When it is possible (this caution constraint can be more or less relaxed), the path planner must avoid such unknown areas. If not, it must search the best way through unknown areas, and provide the best perception point of view on the way to the goal. In order to improve the heuristic guidance of the search, a new heuristic distance to the goal is built by bitmap techniques: a cost bitmap is computed, including the difficulty of the terrain, and the proportion of unknown areas around the patch currently explored. The procedure is as follows:

1. The unknown areas of the elevation map are filled by an interpolation operation, which provides a continuous elevation map.

2. The planner then searches a path through unknown parts of the map. Unknown areas have a higher cost.

3. A potential propagation from the goal generates a distance information corresponding to the best way to the goal through terrain relief and unknown areas.

Finally, once a trajectory reaching the goal is obtained, it is truncated at the last point that enables to perceive the first unknown area to cross.

Search time strongly depends on the difficulty of the terrain. The whole procedure takes between tens of seconds to a few minutes, on an Indigo R4000 Silicon Graphics workstation. Figure 6 shows a trajectory computed on a real terrain, where darker areas correspond to interpolated unknown terrain.

Figure 7: *The mobile robot Adam.*

3.3 Experiment

The systems and algorithms presented above have been actually implemented and used in the "EDEN" experiment [5] carried out with the mobile robot ADAM[3] (figure 7). ADAM is equipped with a 3D scanning laser range finder, two orientable color cameras, and a 3-axis inertial platform. On-board computing equipment is composed of two VME racks running under the real time operating system VxWorks. The robot has six non orientable wheels, with passive suspensions. Its maximum speed is 28 cm/s.

The navigation procedures are written in C-PRS [10]. The task is executed in an initially unknown environment that is gradually discovered by the robot. A landmark to reach is an object known by its model, which can be recognized and localized on a video image. The landmark position is given approximately, and is not necessarily in the field of view of the initial position of the robot. From the approximate given position of the landmark, goal region coordinates are determined, and navigation is started.

The environment is a 20 by 50 meters area, composed of flat, sloppy, uneven rocky areas, and of big obstacle rocks.

---

[3]ADAM: "Advanced Demonstrator for Autonomy and Mobility" is property of FRAMATOME and MATRA MARCONI Space, currently lent to LAAS.

4 Navigation Algorithms

The navigation algorithms perform the decisions concerning perception and motion strategies, which imply the definition of intermediate goals and the choice of navigation modes to apply to reach them. They actually control a variety of algorithms for perception and modelling, localization and motion planning and execution.

The general approach that we have implemented is an incremental navigation that proceeds until the final goal is reached. The navigation algorithm basically plans a route in the environment using an A^*-like search in a labelled connection graph built by several perception procedures. This search selects a path, *i.e.* a succession of connected regions, that defines the intermediate goal and the motion mode to activate. One main difference between the indoor and outdoor cases are the navigation modes adapted to the terrain.

The algorithms have *modalities* associated with them that enable to better control the decisions between several choices (e.g., the reflex mode activation, the extra cost of crossing uneven terrain, etc.).

We present here the outdoor algorithm. It is embedded in a more general task that achieves also target recognition and localization. The algorithm uses a graph of labelled regions, and the valuation of the arcs (that connect the region borders) is obviously enables to implement different strategies. Currently an arc value is a heuristic function of two criteria:

• **Arc label**: to plan a route that minimizes the execution time, the region labels are taken into account. Uneven areas are more costly and are therefore avoided when possible.

• **Arc confidence**: considering only the former constraint, the artefacts produced by the classification procedure (essentially badly labeled "obstacle" cells) would mislead the navigation. The arc label criterion is therefore weighted by its confidence, which allows the planner to cross some obstacle areas for instance, which actually triggers the execution of a new perception task when facing such areas.

Other criteria can also be taken into account during the search, such as altitude variation (for saving energy), or a *localization ability value*, derived using the localization model, to select paths along which a localization is possible.

Once the optimal path is found in a graph, it is analyzed to determine the next *perceptual task*:

• A localization task is planned when the position estimation uncertainty exceeds a threshold (this uncertainty estimation is simply derived from the path length and crossed region labels);

• A "discover" task (fast classification) is planned when reaching unknown areas: it is determined in order to perceive the unknown areas that corresponds to the path returned by the search;

• A model refinement task is planned when reaching constrained areas labeled with a low confidence, or when a an uneven area has to be crossed (fine modeling).

The general navigation algorithm can be summarized by the (simplified) procedure NAVIGATE-TO given below. It selects between the reflex and the planned modes according to the terrain analysis. Goal position is assumed to be given here.

NAVIGATE-TO(robot-position, goal, modalities)
begin
  while ¬ (robot-position **in** goal-region)
    if (local data) not available
      Perceive-environment(parameters);
    endif
    if open-area /* (fast test) */
      start **Reflex-mode**
    else
    start **Planned-navigation**(robot-position, goal, modalities)
    endif

  endwhile

end

The "Planned-mode" actually performs the selection between the cases of falt and uneven terrain, and calls the adequate perception and motion planning and execution functions.

The planned-navigation procedure first builds the navigation map (step 1.), then finds a path (step 2.) using a cost function that can be possibly modified by the modalities associated with the task, refines the path (step 3.) by selecting the appriopriate planners corresponding to the nature of crossed regions. This process plans only a part of the path, by taking into account the region label quality (other criteria are possible of course). When the trajectories have been computed,

and before execution, landmarks can be selected for localization during motion for reducing robot position uncertainty. Finally, the trajectories are executed (step 4.). Only part of the motion may be actually executed, again with respect to a criterion (here a maximum distance threshold for example).

PLANNED-NAVIGATION(robot-position, goal, modalities)
begin
  1. **Build navigation-map** /* region structure */
  2. **Find-path**(region-graph; cost function
    $f = \sum_{i=0}^{goal} \alpha_i \text{dist}(r_i, r_{i+1})$,
    α_i(node-label, localization-features, other)

  3. **Refine-path** (according to modalities):
    forall nodes
      if node-type = uneven:
        Build-and-fuse elevation map;
      endif
      while computed crossed distance <
          threshold(node label quality, other)
        Select-planner(2D/3D);
        Select-localization-features

      endwhile

    endforall

  4. **Execute-motion**(modalities: dist-max, speed-max, other)

end

Note that these algorithms are controllable at each decision. This enables to take into account the constraints attached to the task at each step.

5 Conclusion

Autonomous navigation requires complex decisional and modelling processes, especially in a natural environment which presents variable terrain and features. In autonomous navigation, the only knowledge the robot possesses about its environment relies on the interpretation of its uncertain sensory data. A navigation algorithm controls a number of processing algorithms for intermediate goal selection, planning of perception actions, trajectory generation and robot localization, each of which needs adequate representations of the environment. These representations must correctly express the actual environment features, e.g., planar indoors, unstructured outdoors.

We have presented examples of navigation tasks and a navigation algorithm. The algorithm must select the basic processes it manipulates (motion planner, perception processes, etc.) according to the actual situation of the robot. It should therefore be placed at the level of the general decisional processes of the robot, as part of other task accomplishing algorithms.

Acknowledgements

The algorithms, systems and experiments were developed in collaboration with R. Alami, B. Dacre-Wright, B. Degallaix, P. Fillatreau, M. Devy, S. Fleury, M. Herrb, S. Lacroix, P. Moutarlier, F. Nashashibi, C. Proust, and T. Siméon.

References

[1] R. Alami and T. Siméon. Planning robust motion stategies for a mobile robot. *Technical Report LAAS-93005*, October 93.

[2] N. Ayache and O. Faugeras. Maintaining representations of the environment of a mobile robot. In *Robotics Research : The Fourth International Symposium, Santa Cruz (USA)*, Août 1987.

[3] J. Barraquand and J.-C. Latombe, "Robot motion planning: A distributed representation approach", *in International Journal of Robotics Research*, 1991.

[4] S. Betge-Brezetz, R. Chatila, M. Devy. Natural Scene Understanding for Mobile Robot Navigation. '94 IEEE Conf. on R & A. San Diego, May 1994.

[5] R. Chatila, S. Fleury, M. Herrb, S. Lacroix, C. Proust. Autonomous Navigation in Natural Environment. 3rd ISER, Kyoto, Japan, October 1993.

[6] M. Daily, J. Harris, D. Kreiskey, K. Olion, D. Payton, K. Reseir, J. Rosenblatt, D. Tseng, and V. Wong, "Autonomous cross-country navigation with the ALV", *in IEEE International Conference on Robotics and Automation, Philadelphia (USA)*, 1988.

[7] H.F. Durrant-Whyte. *Integration, Coordination and Control of Multi-Sensor Robot Systems*. Kluwer Academic Publ., Boston, MA, 1987.

[8] M. Erdmann. Using backprojections for fine motion planning with uncertainty. *Int. Journal of Robotics Research*, 5(1):19–45, Spring 1986.

[9] P. Fillatreau, M. Devy, and R. Prajoux, "Modelling of Unstructured Terrain and Feature Extraction using B-spline Surfaces", *in '93 International Conference on Advanced Robotics (ICAR), Tokyo (Japan)*, Nov. 1993.

[10] Ingrand, F. F., Georgeff, M. P., and Rao, A. S. An Architecture for Real-Time Reasoning and System Control. *IEEE Expert, Knowledge-Based Diagnosis in Process Engineering*, 7(6):34–44, December 1992.

[11] E. Krotkov, J. Bares, T. Kanade, T. Mitchell, R. Simmons, and R. Whittaker. Ambler: A six-legged planetary rover. In *'91 ICAR, Pisa (Italy)*, pages 717–722, June 1991.

[12] S. Lacroix, P. Fillatreau, F. Nashashibi, R. Chatila, and M. Devy, "Perception for Autonomous Navigation in a Natural Environment", *in Workshop on Computer Vision for Space Applications, Antibes, France*, Sept. 1993.

[13] D. Langer, J.K. Rosenblatt, and M. Hebert. An Integrated System for Autonomous off-Road Navigation. *in IEEE International Conference on Robotics and Automation, San Diego, CA*, May 1994.

[14] J-C. Latombe. *Robot Motion Planning*, Kluwer Academic Publishers, 1991

[15] J-C. Latombe, A. Lazanas, S. Shekhar. Robot motion planning with uncertainty in control and sensing. *Artificial Intelligence*, 52(1):1–47, Nov 1991.

[16] A. Lazanas, J-C. Latombe. Landmark-based robot navigation. Technical Report STAN-CS-92-1428, Stanford University, May 1992.

[17] P. Moutarlier and R. Chatila. Stochastic Multisensory Data Fusion for Mobile Robot Location and Environment Modelling. In 5th ISRR, Miura and Arimoto Eds, MIT Press, 1990.

[18] P. Moutarlier and R. Chatila, Incremental free-space modelling from uncertain data by an utonomous mobile robot, IROS'91, Osaka (Japan, Nov. 1991.

[19] F. Nashashibi and M. Devy, "3D Incremental Modeling and Robot Localization in a Structured

Environment using a Laser Range Finder", *in IEEE Transactions on Robotics and Automation, Atlanta (USA)*, May 1993.

[20] D. Payton, J. Rosenblatt, and D. Keirsey, "Plan Guided Reaction", *IEEE Transactions on Systems, Man, and Cybernetics*, vol. 20, pp. 1370–1382, Nov./Dec. 1990.

[21] T. Siméon and B. Dacre Wright, "A Practical Motion Planner for All-terrain Mobile Robots", *in IEEE International Workshop on Intelligent Robots and Systems (IROS '93) Japan*, July 1993.

[22] H. Takeda, J-C. Latombe. Sensory uncertainty field for robot navigation. In *IEEE ICRA '92*, Nice.

[23] Charles Thorpe, Martial H. Hebert, Takeo Kanade, and Steven A. Shafer. Vision and navigation for the Carnegie-Mellon NAVLAB. *IEEE Transaction on Pattern Analysis and Machine Intelligence*, 10(3), May 1988.

Two manipulation planning algorithms

R. Alami, J.P. Laumond, and T Siméon

LAAS / CNRS, 7, Avenue du Colonel Roche, 31077 Toulouse, France

This paper addresses the motion planning problem for a robot in presence of movable objects. Motion planning in this context appears as a constrained instance of the coordinated motion planning problem for multiple movable bodies.

Indeed, a solution path (in the configuration space of the robot and all movable objects) is a sequence of transit-paths, where the robot moves alone, and transfer-paths where a movable object "follows" the robot. A major problem is to find the set of configurations where the robot has to "grasp" or "release" objects.

Based on [1, 5], the paper gives an overview of a general approach which consists in building a manipulation graph *whose connected components characterize the existence of solutions. Two planners developed at LAAS/CNRS illustrate how the general formulation can be instantiated in specific cases.*

1 The manipulation planning problem

Robot motion planning usually consists in planning collision-free paths for robots moving amidst fixed obstacles. Nevertheless a robot may have to perform tasks which are more difficult than planning motions only for itself. In some situations, a robot may be able to move objects and to change the structure of its environment. In such a context, the robot moves amidst obstacles but also movable objects. A movable object cannot move by itself; it can move only if it is grasped by the robot. According to the standard terminology, considering movable objects appears as a constrained instance of the *coordinated motion planning problem*, that we call the *manipulation planning problem* [1].

---

[1] Note that this problem is related to Pick&Place and re-Grasping tasks and not to the dextrous manipulation of an object by a multi-fingered robot hand.

As stated in [1] (see also Latombe's book [10]), a general geometric formulation of the problem can be defined as follows.

1.1 Manipulation task and configuration space(s)

The environment is a 3D (resp. 2D) workspace which consists of three types of bodies: (1) static obstacles, (2) movable objects and (3) a robot.

For the robot and for each object we consider its associated configuration space. Object configuration spaces are 6D (resp. 3D); the robot configuration space is n-dimensional, where n is its number of degrees of freedom. Let CS denote the cartesian product of all objects and robot configuration spaces.

In the following, we will say that c is an incompletely specified configuration when some parameters in c are left unspecified; besides, we will say that a configuration c' "verifies" c when it is included in the subspace defined by c. Such a terminology will be used in order to denote partially specified goals.

Furthermore, we introduce a function Free which gives, for each domain of CS, the set of its free configurations (i.e. configurations where the bodies do not overlap).

A manipulation task is clearly a particular path in $\mathsf{Free}(CS)$. The converse does not hold: all paths in $\mathsf{Free}(CS)$ do not necessarily correspond to a manipulation task. Indeed a manipulation path is a constrained path in $\mathsf{Free}(CS)$. We have now to define geometrically these constraints. There are two types of constraints:

- constraints on the placements of objects; these constraints model the physics of the manipulation context (any object must be in a stable position in the environment),

- constraints on object motions; any object motion is a motion induced by a robot motion.

1.2   Placement constraints

All configurations in $\mathsf{Free}(CS)$ do not necessarily correspond to a physically valid environment configuration. For example an object can not "levitate", and must be in a stable position. Geometrically speaking, we have to reduce the space of free configurations to a subspace which contains all valid configurations. These constraints concern only the objects. For example, if we constrain a polyhedron to be placed only on top of horizontal faces of polyhedral obstacles or of other objects (which are already in a stable position); its placement constraints will then define a finite number of 3-dimensional manifolds in its configuration space,

We call $PLACEMENT$ the subspace of $\mathsf{Free}(CS)$ containing all valid placements for all objects, i.e. placements which respect the physical constraints of the manipulation context. With this definition, all the objects have a fixed and known geometrical relations with the obstacles or with other objects.

$PLACEMENT$ is not more precisely defined; the definition depends on the context, and appears clearly for each context. For a mobile robot in a 2-dimensional euclidean space, amidst movable objects, $PLACEMENT = \mathsf{Free}(CS)$.

For the planner presented in Section 2, we assume that each object has a finite number of placements in the environment; then $PLACEMENT$ appears as a finite union of n-dimensional manifolds, where n is the number of degrees of freedom of the robot.

For the planner presented in Section 3, the movable object can be placed anywhere in the environment.

1.3   Motion constraints

We define a grasp mapping G_O^{T} as a mapping from the configuration space of the robot (noted CSR) into the configuration space of a given object O (noted CSO), which verifies $G_O^{\mathsf{T}}(cr) = co$, where $cr \in CSR$ and $co \in CSO$. This mapping models the geometrical relation which is defined by a grasping operation (T denotes a homogeneous transform between the robot gripper frame and the object reference frame). Such mappings define geometrically the semantics of grasping for a particular manipulation context. They can be

in finite or infinite number, and can be given explicitly (as in Section 2) or implicitly (they are defined for example by a contact relation between the robot or the object as in Section 3).

1.4   Problem statement

Definition A *transfer-path* is a path in $\mathsf{Free}(CS)$ such that there is one object O and one grasp mapping G_O^{T} verifying:

- the configuration parameters of any object $O' \neq O$ are constant along the path

- for any configuration of the path, $G_O^{\mathsf{T}}(cr) = co$, where cr and co designate respectively the configuration parameters of the robot and O.

Two configurations of $\mathsf{Free}(CS)$ connected by a transfer-path are said to be *g-connected*.

We call $GRASP$ the subspace of $\mathsf{Free}(CS)$ containing the configurations which are *g-connected* with a configuration of $PLACEMENT$.

Definition: A *transit-path* is a path in $\mathsf{Free}(CS)$ such that the configuration parameters of the objects are constant along the path. Two configurations in $\mathsf{Free}(CS)$ connected by a transit-path are said to be *t-connected*.

Remark: a transit-path is included in $PLACEMENT$ (but every path in $PLACEMENT$ is not necessary a transit-path).

We are now in position for defining any manipulation task as a manipulation path in $\mathsf{Free}(CS)$:

Definition: A *manipulation-path* is a path in $\mathsf{Free}(CS)$ which is a finite sequence of transit-paths and transfer-paths. Two configurations in $\mathsf{Free}(CS)$ connected by a manipulation-path are said to be *m-connected*.

A manipulation planning problem can then be defined as:

Manipulation planning problem: An initial configuration i and a final (completely or incompletely specified) configuration f being given, does there exists a configuration verifying f which is *m-connected* with i ? If the answer is yes, give a *manipulation path* between i and some configuration verifying f.

1.5 Manipulation graph

The previous definitions lead to a property which models the structure of the solution space:

Lemma: A transit-path and a transfer-path are connected iff both have a common extremity in $PLACEMENT \cap GRASP$.

The manipulation planning problem then appears as a constrained path finding problem inside the various connected components of $PLACEMENT \cap GRASP$ and between them.

In the case of a discrete number of placements and grasps, $PLACEMENT \cap GRASP$ consists of a finite set of configurations.

When the environment contains only one movable object, even if there is an infinite number of placements and grasps, we can prove that two configurations which are in a same connected component of $GRASP \cap PLACEMENT$ are *m-connected* (see Appendix). This property leads to reduce the problem.

In both cases, it is sufficient to study the connectivity of the various connected components of $GRASP \cap PLACEMENT$ by transit-paths and transfer-paths.

We then define a graph whose nodes are the connected components of $GRASP \cap PLACEMENT$. There are two types of edges. A *transit* (resp. *transfer*) edge between two nodes indicates that there exists a transit-path (resp. transfer-path) path linking two configurations of the associated connected components.

This graph is called a *manipulation graph (MG)*. It verifies the fundamental property:

Property: An initial configuration i and a goal (completely or incompletely specified) configuration g being given, there exists a configuration f verifying g and m-connected with i iff:

- there exist a node N_i in MG and a configuration c_i in the associated connected component of $GRASP \cap PLACEMENT$, such that i and c_i are t-connected or g-connected;
- there exist a node N_f in MG and a configuration c_f in the associated connected component of $GRASP \cap PLACEMENT$ such that:
  - c_f and f are t-connected or g-connected;
  - N_i and N_f are in the same connected component of MG

In order to use this method for particular instances of the problem, one needs:

1. to compute the connected components of $GRASP \cap PLACEMENT$;
2. to determine the connectivity of these connected components using transit-paths and transfer-paths;
3. and to provide a method for planning a path in a given connected component of $GRASP \cap PLACEMENT$.

We present, in the sequel, two manipulation planners working respectively when $PLACEMENT \cap GRASP$ is reduced to a finite set of points (Section 2) and when the environment contains only one movable object (Section 3).

2 The case of discrete placements and grasps for several movable objects

In this section, we present a description of a manipulation task planner for the case of discrete placements and grasps for objects[2]. It is directly derived from the general scheme above. It is based on the fact that the connected components of $GRASP \cap PLACEMENT$ are given *a priori* by some discretization and that the construction of transit-paths and transfer-paths can be obtained using a collision-free path planner for a robot amidst stationary obstacles.

It leads to an effective construction of the manipulation graph.

For simplicity reasons, we give a presentation considering only two objects. The extension to a finite number of objects is straightforward. The presentation will be illustrated using the example of Figure 1, i.e. a 2D world where all bodies are polygonal and where the robot is allowed to move only in translation. However, the solution we propose is general.

2.1 Notations

We designate the robot by R and the objects by A and B. Let cr, ca and cb be the configuration pa-

---

[2] This planner has been first introduced in [1]. In this current presentation we have added new experimental results.

Figure 1: *A manipulation task generated automatically*

rameter vectors and CSR, CSA and CSB the associated configuration spaces. Let n be the dimension of CSR. The configuration space of all movable bodies is: $CS = CSR \times CSA \times CSB$.

Object placements: We assume that each object has a finite number of possible placements in the environment that will physically correspond to stable positions when the robot does not hold the object.

We designate by $p_A^1, \ldots p_A^i \ldots \in CSA$ and by $p_B^1, \ldots p_B^j \ldots \in CSB$ the authorized placements for A and B respectively (Figure 2).

Remark: we may also give explicitly - or give means to compute - all authorized placements combinations $(p_A^i, p_B^i) \in CSA \times CSB$, in order to take into account, for example, the possibility of stacking an object on another object.

Object grasps: We assume that each object has a finite number of possible grasps. A given grasp for

object A is specified by providing a mapping $G_A^i : CSR \longrightarrow CSA$.

Let G_A^1, G_A^2 G_A^3 and G_B^1, G_B^2 be the authorized grasps for A and B (Figure 2).

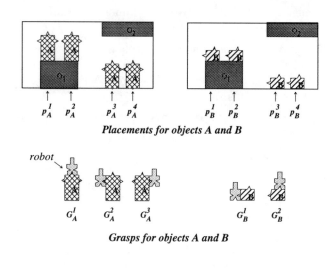

Placements for objects A and B

Grasps for objects A and B

Figure 2: *Placements and grasps for objects* A *and* B

2.2 Building the manipulation graph

2.2.1 Building the nodes

Intuitively, $GRASP \cap PLACEMENT$ is simply the set of all configurations where the robot is authorized to grasp or un-grasp an object in a given placement, taken into account all possible placement combinations for the other objects.

Let I(*grasp* ; *placementA* ; *placementB*) designate the set of all free configurations where the robot can grasp one object using *grasp* while objects are placed in *placementA* and *placementB*.

Thus, $GRASP \cap PLACEMENT$ is the union of all I(G_A^i ; p_A^j ; p_B^k) and I(G_B^i ; p_A^j ; p_B^k).

By definition, I(G_A^i ; p_A^j ; p_B^k) equals

$$\mathsf{Free}(\{cr \in CSR \mid G_A^i(cr) = p_A^j\} \times \{p_A^j\} \times \{p_B^k\})$$

The computation of $\{cr \in CSR \mid G_A^i(cr) = p_A^i\}$ is done using the robot inverse geometric model. If we assume that the robot is not redundant and that the solution does not include a robot singular configuration, then I(G_A^i ; p_A^j ; p_B^k) consists of a finite (and small) number of configurations.

Each node of MG corresponds to a configuration which can be computed easily; it represents a connected component of $GRASP \cap PLACEMENT$ reduced to a single configuration.

2.2.2 Building the edges

The second step in building MG consists of establishing edges between nodes. Two nodes N_1 and N_2 are linked by a transit (resp. transfer) edge if there exists a transit-path (resp. transfer-path) between two configurations of their associated connected components.

All paths involving a node have an extremity in common: the single configuration represented by the node. For a node in $I(G_A^i \; ; \; p_A^j \; ; \; p_B^k)$, all transit-paths will correspond to paths where the robot moves alone while A and B are in p_A^j and in p_B^k, and all transfer-paths will correspond to paths where the robot holds A using grasp G_A^i while B is in p_B^k. This is why we define the concept of *task state* which denotes the fact that the robot is in a situation where it moves alone, or it is holding a given object.

Task states and CS slices: We define a *task state* by giving for each object its current placement and, for at most one object, its current grasp: $(grasp \; ; \; placement A \; ; \; placement B)$.

When no object is grasped, *grasp* will be noted "$_-$", for example $(\; _- \; ; \; p_A^1 \; ; \; p_B^2)$. We call such a state a *transit state*.

When object X is held by the robot, *placement X* will be noted "$_-$", for example $(G_A^2 \; ; \; _- \; ; \; p_B^4)$. We call such a state an *transfer state*.

Let $C(grasp \; ; \; placement A \; ; \; placement B)$ denote the set of configurations in CS associated to a given task state. $C(\; _- \; ; \; p_A^i \; ; \; p_B^j)$ and $C(G_A^i \; ; \; _- \; ; \; p_B^j)$ correspond to n-dimensional "slices" of CS where n is the dimension of CSR.

For a transit state:

$$C(\; _- \; ; \; p_A^i \; ; \; p_B^j) = \mathsf{Free}(CSR \times \{p_A^i\} \times \{p_B^j\})$$

i.e. all configurations such that the robot does not overlap object A in placement p_A^i nor object B in placement p_B^j nor the obstacles.

For a transfer state:

$$C(G_A^i \; ; \; _- \; ; \; p_B^j) = \mathsf{Free}(CSR \times G_A^i(CSR) \times \{p_B^j\})$$

i.e. all configurations such that the robot holding object A in grasp G_A^i does not overlap object B in placement p_B^j nor the obstacles.

Remark: When $C(G_A^i \; ; \; _- \; ; \; p_B^j) = \emptyset$ the corresponding state is invalid.

A node models a transition between a transit state and a transfer state. We say that the node "belongs" to these states. For example, a node in $I(G_A^i \; ; \; p_A^j \; ; \; p_B^k)$ "belongs" to the transit state $(\; _- \; ; \; p_A^j \; ; \; p_B^k)$ and to the transfer state $(G_A^i \; ; \; _- \; ; \; p_B^k)$.

Transit edges: A transit edge can be built between a node N and any node N' which belongs to the same transit state and which is "directly" reachable. This simply means that N and N' represent configurations that are in a same connected component of $C(\; _- \; ; \; p_A^i \; ; \; p_B^j)$.

Transfer edges: A transfer edge can be built between a node N and any node N' that belongs to the same transfer state and that is "directly" reachable. This simply means that N and N' represent configurations that are in a same connected component of $C(G_A^i \; ; \; _- \; ; \; p_B^j)$ or $C(G_B^i \; ; \; p_A^j \; ; \; _-)$

We have then to construct two CS slices for any given node. However, a given CS will be used for a great number of nodes.

Figure 3 represents several nodes in the manipulation graph that corresponds to the example. The drawing at the center of the figure represents the node $I(G_B^2 \; ; \; p_A^3 \; ; \; p_B^2)$. Transit and transfer edges are built using $C(_- \; ; \; p_A^3 \; ; \; p_B^2)$ and $C(G_B^1 \; ; \; _- \; ; \; p_B^2)$.

Figure 4 illustrates the "links" between several configuration space slices that are traversed by the system when it executes the sequence represented in Figure 1. Several states are represented; for each state, the regions in white represent the projection of the connected components of its CS slice onto the robot configuration space. In the initial state of Figure 1, the robot is in the "left" connected component of $C(\; _- \; ; \; p_A^2 \; ; \; p_B^4)$. The only possible transition (arc 11) is to move the robot until it is able to grasp object A in G_A^2. The transitions sequence is 11-10-7-9-6-5...Note that the solution involves state $(\; _- \; ; \; p_A^2 \; ; \; p_B^4)$ twice, but it traverses only once a given connected component of $C(\; _- \; ; \; p_A^2 \; ; \; p_B^4)$.

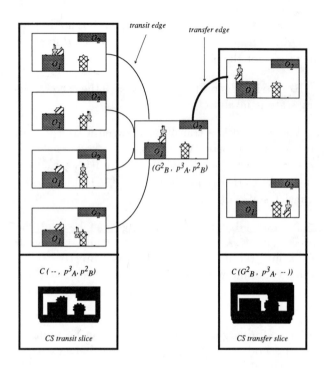

Figure 3: *A partial representation of a manipulation graph*

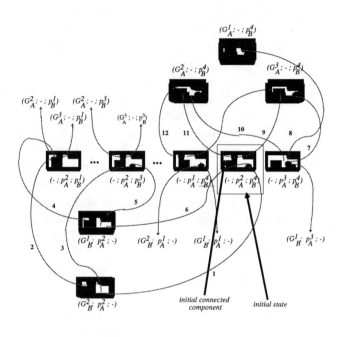

Figure 4: *Links between Configuration Space slices*

2.3 Search Strategies in the Manipulation Graph

The size of the graph grows rapidly depending on the number of grasps and placements. The number of nodes corresponds to *number of grasps* × *number of placements* (in our simple example, there are $(3+2) \times (4 \times 4) = 80$ nodes). The number of transit slices is equal to the number of legal placements $((4 \times 4) - 4 = 12$ in the example). The number of transfer slices is equal to the number of combinations of grasps for an object and placements for the other objects $(3 \times 4 + 2 \times 4 = 20$ in the example).

The cost of building an edge is expensive and depends mainly on the cost of computing a n-dimensional CS slice (where n is the number of degrees of freedom of the robot). However, a CS slice is used several times; for example $C(\_ ; p_A^j ; p_B^k)$ will be used for all nodes in $I(G_A^i ; p_A^j ; p_B^k)$ and in $I(G_B^i ; p_A^j ; p_B^k)$. The first time, it has to be computed; and then, it will only be used in order to find a path.

MG has not to be built completely before execution. It can be explored and built incrementally. Powerful heuristics remain to be explored in order to "drive" the

system towards the goal. However, even simple heuristics based only on the distance between the positions of objects allow to limit substantially the construction of the graph.

Note that, if several objects have the same shape and the same grasps and placements, the number of different CS slices to build can be considerably reduced. In the case, similar to the example, of two identical objects with 3 different grasps and 4 placements, we have only 12 transfer slices and 6 transit slices. Then, another way to limit the complexity, when exploring the graph edges, is to consider only a gross approximation of objects shape (by classifying them into a limited number of classes: small, elongated, big...) in order to use a same CS slice for a great number of nodes.

2.4 Implementation

We have implemented a system based on the method described above. It is composed of two modules: a *Manipulation Task Planner* and a *Motion Planner*.

The *Manipulation Task Planner* builds incrementally the manipulation graph and searches solution paths in it. It makes use of the Motion Planner in order to

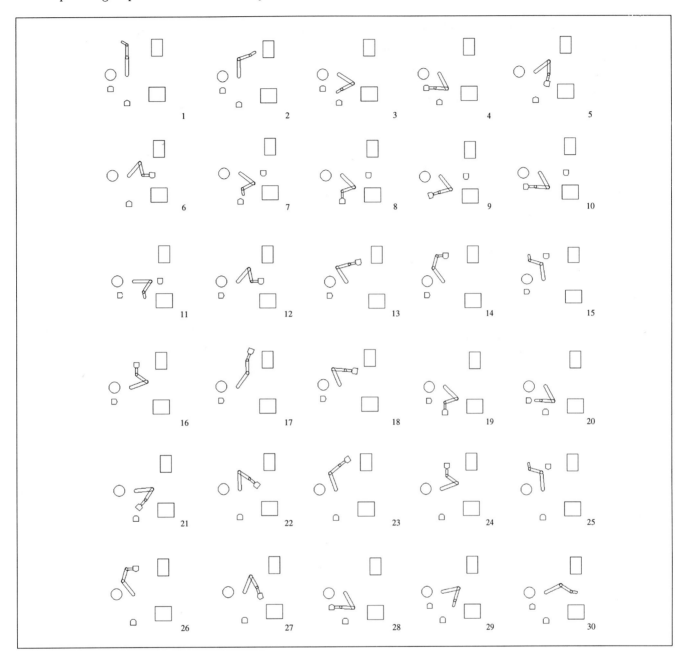

Figure 5: *A manipulation task.*

build the different *CS* slices corresponding to transfer and transit states, to structure them into connected components and to find paths between two robot configurations.

The search in the manipulation graph is performed using a A* algorithm. The cost functions currently used are based on the length of the movable bodies trajectory. The incremental graph construction allows a nice feature: for the first plans, the system is quite slow but it becomes more efficient progressively as it re-uses parts of the graph already developed.

The *Manipulation Task Planner* is implemented in order to be used for an arbitrary number of objects and does not depend on a specific motion planner module.

For the *Motion Planner*, we have first implemented a method working for a polygonal body in translation amidst polygonal obstacles (the obstacles are grown using Minkowsky sum and the trajectory is built using a visibility graph). This has been done mainly in order to demonstrate the feasibility of the approach. Figure 1 shows the plan produced for the example.

A second implementation is based on a general motion planner [17] which works for manipulators is a 3-dimensional workspace. It allows to solve manipulation planning problems as complicated as the example of Figure 5. Note that the sub-sequences 14-15-16 and 24-25-26 correspond to re-grasping operations of an object.

3   The polygonal case for one movable object with an infinite set of grasps

This section describes a method for solving the manipulation problem in the case of a polygonal robot and a polygonal object moving in translation amidst polygonal obstacles. It has been implemented in the case where both polygons are convex. In order to illustrate the different steps, we will rely on the simple example of Figure 6.

Let us consider $CS = CSR \times CSO$ the configuration space of the robot and the object together. In order to simplify the notations, we denote by :

- ACS the admissible (i.e. without any collision between the bodies) configurations space ($ACS = \mathsf{Free}(CS)$),

- $ACSR(co)$ the admissible configuration space of the robot when the movable object is placed at $co \in CSO$, and

- $ACSO(cr)$ the admissible configuration space of the movable object when the robot lies at $cr \in CSR$.

We assume that the robot has to avoid any contact with the obstacles (hypothesis H, see Appendix). We define $GRASP$ as the subset of all the configurations verifying hypothesis H and such that the

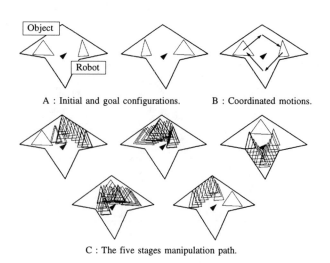

A : Initial and goal configurations.   B : Coordinated motions.

C : The five stages manipulation path.

Figure 6: *An illustration of manipulation constraints.*

robot touches the object. Finally, the movable object may be placed anywhere in the environment, as long as the bodies do not collide. As a consequence, $GRASP \cap PLACEMENT = GRASP$.

Let $\mathcal{C}$ be the set of connected components of $GRASP$. Let us consider the graph whose nodes are the elements of $\mathcal{C}$ and whose edges correspond to the existence of a transit-path between two configurations of the associated nodes. Thanks to the reduction property (see Appendix), two configurations in $GRASP$ are connected by a manipulation path if and only if they belong to two elements of $\mathcal{C}$ which are in the same connected component of the graph.

Therefore, according to the resolution scheme stated in Section 1, a manipulation graph can be built by :

1. computing the connected components of $GRASP$,

2. and linking them by transit-paths.

In a first step, we compute a cell decomposition of ACS (which solves the coordinated motion problem); then a retraction on the boundary of ACS gives a cell decomposition of $GRASP$. Finally, the connectivity by transit-paths between the various connected components of $GRASP$ is given by a study of the connectivity of $ACSR$, whose structure can be extracted from ACS cells.

3.1 *ACS* cell decomposition and coordinated motions

To compute the cell decomposition of *ACS*, we have chosen to use an adaptation of the projection method developed by Schwartz and Sharir in [15] for the case of two discs. Any other, and perhaps better ([7, 14, 16]) method could have been used. However, the purpose of this paper is not to give some optimal algorithm, but to demonstrate the feasibility of our approach.

Let us consider an object position *co*. $ACSR(co)$ is obtained by removing, from *ACSR* (robot admissible configurations without the object), the set $COL(co)$ of all configurations where the robot and the object collide [3].

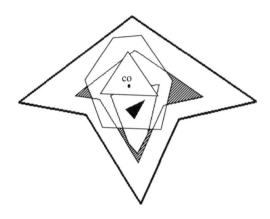

Figure 7: *The hatched areas represent the connected components of* $ACSR(co)$.

Let us observe now the evolution of $ACSR(co)$ with respect to *co*. In a neighborhood of most object positions, $ACSR(co)$ varies only quantitatively, keeping the same structure. However, at some object positions, some $ACSR(co)$ connected components may appear, disappear, be split or merged. More precisely, the geometrical structure of the connected components of $ACSR(co)$ are modified when some vertices appear or disappear. These changes correspond to specific values of *co* which constitute a set of *critical curves* (see [15] for a proof). Figure 8 gives an example of a critical curve along which a connected component of $ACSR(co)$ is divided into two separate components.

---

[3] $COL(co)$ is the polygon obtained by Minkowski difference between the robot and the object, and placed in *co*.

The critical curves provide a decomposition of *ACSO* into *non-critical regions*.

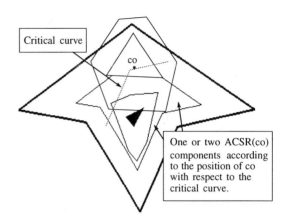

Figure 8: *A critical curve*

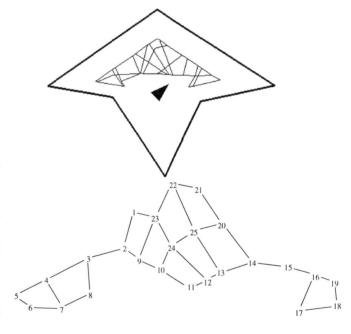

Figure 9: *Cell decomposition of ACSO and the associated graph*

Let us consider a non-critical region *R*. $\{\{co\} \times ACSR(co) \mid co \in R\}$ constitutes cells of *ACS* (there are as many cells as the number of connected components of $ACSR(co)$). The set of all such cells is the expected cell decomposition.

In order to compute the critical curves, we introduce a symbolic description of $ACSR(co)$. Let us recall that the boundary of $ACSR(co)$ is constituted by $ACSR$ edges and $COL(co)$ edges. We assign a numerical label to all the vertices of $ACSR$ and a literal symbol to the edges of $COL(co)$. We denote by $b[8,9]$ the intersection between the edge b of $COL(co)$ and the segment $[8,9]$ of $ACSR$. Therefore, the bottom connected component in the example of Figure 10 is labeled by the sequence $(3,4,[4,5]b,b[8,9],9,[9,10]b,b[1,2],2)$.

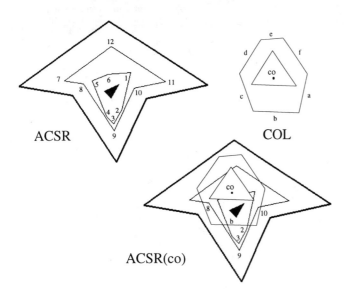

Figure 10: *An illustration of the labels used to characterize the connected components of $ACSR(co)$.*

To compute the critical curves, we do not consider all the possible changes in this sequence. We just need to consider the changes *on the letters* (i.e. the changes induced by $COL(co)$ and not by $ACSR$ vertices). In Figure 10 for instance, when co moves to the bottom, the disappearance of vertex 2 in the sequence above does not induce a critical curve, while the fusion of the two b labels (when edge b meets vertex 3) does.

The critical curves of our example are shown in Figure 9, together with the graph of non-critical regions of $ACSO$.

Let us recall that each non-critical region induces as many cells in ACS as the number of connected components in $ACSR(co)$ when co belongs to the region.

Now we structure all these cells into a *coordinated motion graph*. Two cells are adjacent in this graph if and only if :

- the associated non-critical regions are adjacent in $ACSO$,

- and the symbolic descriptions of the associated $ACSR(co)$ components just differ by a letter.

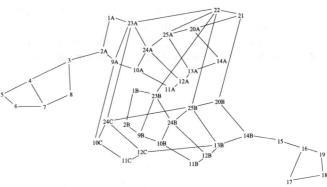

Figure 11: *The coordinated motion graph.*

Figure 11 shows the coordinated motion graph of our example. In Figure 10, co belongs to the non-critical region 10 (Figure 9). This region gives rise to three ACS cells numbered $10A$, $10B$ and $10C$ in Figure 11. $10C$ is the node corresponding to the label mentioned above.

The fundamental property is : there exists a coordinated motion between two configurations in ACS if and only if they belong to nodes of a same connected component of the coordinated motion graph. The proof is exactly the same as in [15].

3.2 GRASP cell decomposition and contact motion

Let us consider the above $ACSR(co)$ component labeled by $(3,4,[4,5]b,b[8,9],9,[9,10]b,b[1,2],2)$. There are two edges labeled by b in it. This means that there are two connected sets of configurations where the robot is in contact with the object. It is not possible to go from one set to the other one without leaving the contact. These connected sets are easily extractable from the symbolic description of the $ACSR(co)$ components. In our example,

$([4, 5]b, b[8, 9])$ and $([9, 10]b, b[1, 2])$ are the symbolical descriptions of the two classes of contact. By definition, they always contain two terms.

Now, let us consider a non-critical region R. The set $\{\{co\} \times COL(co) \cap ACSR(co) \mid co \in R\}$ constitutes cells of $GRASP$ (there are as many cells as the number of connected components of $COL(co)$ along the $ACSR(co)$ boundary).

We follow the same method as for the coordinated motion problem. We structure the $GRASP$ cells into a *contact graph*. Two cells are adjacent in this graph if and only if :

- the associated non-critical regions are adjacent in $ACSO$,

- and their symbolic descriptions just differ by one term.

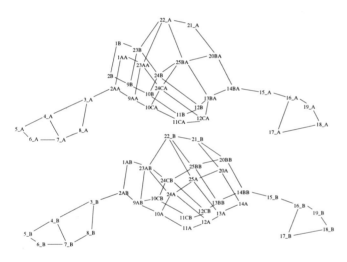

Figure 12: *The contact graph.*

Figure 12 shows the contact graph of our example. Region 10 gives rise to three ACS cells. The frontier of two of them ($10A$ and $10B$) contain one connected component in $GRASP$. They then give rise to two $GRASP$ cells (which keep the same name in the contact graph). The frontier of $10C$ contains two connected components in $GRASP$; they give rise to $GRASP$ cells $10CA$ and $10CB$.

This graph verifies the following property : there exists a motion keeping the contact between the robot

and the object, between two configurations in $GRASP$, if and only if both configurations belong to nodes of a same connected component of the contact graph. The proof is exactly of the same kind as for the coordinated motion graph.

3.3 The manipulation graph

Let us come back to our manipulation planning problem. At this time we have captured the connectivity of $GRASP$. We know that two configurations in the same connected component of $GRASP$ may be linked by a manipulation path (Reduction Property).

Now we have to study the existence of transit-paths between $GRASP$ components. This study is very easy from the above labeling. Indeed let us consider $ACSR(co)$ in Figure 10. There are two grasp classes in the bottom component. Nevertheless, it is possible for the robot to move *alone* in this component; that means that the robot can go from a position in the first grasp class to any other one in the second grasp class. Then, these two classes are linked by transit-paths.

The existence of such transit-paths is very easy to compute from the labeling of $ACSR(co)$. Indeed, two $GRASP$ cells are connected by a transit-path if and only if they belong to the frontier of the same ACS cell. In our current example, only the cells $10CA$ and $10CB$ (which come from the same ACS cell $10C$) are linkable by a transit path.

Computing the connectivity of $GRASP$ components by transit-path is equivalent to adding to the contact graph edges between nodes defined from a same ACS cell. These additional edges are referred as "transit edges" in Figure 13. With our notations, two nodes in the contact graph whose "names" contain the same number and the same first letter are linked by a transit edge. The resulting graph is the *manipulation graph*.

3.4 Manipulation path finding

Let us consider an initial configuration c_i and a final one c_f, defining the initial and final positions of the robot and the object. According to the property of the manipulation graph, we use a three-steps procedure :

1. First, we compute the ACS cells C_i and C_f containing c_i and c_f. Then, we compute the set $\mathcal{G}_i$ (resp. $\mathcal{G}_f$) of $GRASP$ cells reachable from c_i (resp. c_f). This computation is a 2-dimensional problem

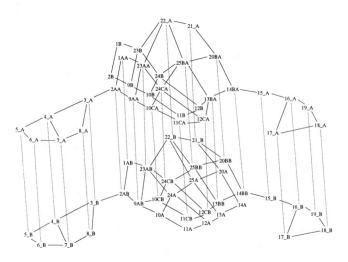

Figure 13: *The manipulation graph, where dotted lines describe transit edges.*

since the transit-paths have to lie in $ACSR(co_i)$ (resp. $ACSR(co_f)$).

2. The second step consists in searching a path in the manipulation graph between a cell in $\mathcal{G}_i$ and a cell in $\mathcal{G}_f$. If no such path exists, the procedure stops. There is no solution. Otherwise, we obtain a sequence $\mathcal{P}ath$ of $GRASP$ cells.

3. Finally the complete path is built from elementary manipulation paths lying in the $GRASP$ cells of $\mathcal{P}ath$ and from transit-paths associated to the transit edges contained in $\mathcal{P}ath$.

Comments on Step 1 : The computation of C_i (resp. C_f) is a 2-dimensional location problem performed in $ACSR(co_i)$ (resp. $ACSR(co_f)$) : we have to determine the connected component of $ACSR(co_i)$ (resp. $ACSR(co_f)$ containing cr_i (resp. cr_f). This fully characterizes C_i (resp. C_f). Then the computation of $\mathcal{G}_i$ (resp. $\mathcal{G}_f$) is very easy, since these $GRASP$ cells belong to the frontier of C_i (resp. C_f) : with our notations, a $GRASP$ cell belonging to the frontier of some ACS cell appears in the contact graph with the same numerical label and the same first letter as the ACS cell appears in the coordinated motion graph.

Comments on Step 2 : The second step is performed using a A^* algorithm. Several cost criteria can be introduced : the length of the complete path, the number of

grasping changes... The experimental results will give an illustration of the influence of the cost definition.

Comments on Step 3 : Step 3 consists in computing the complete manipulation path. $\mathcal{P}ath$ is a sequence of $GRASP$ cells. Two consecutive cells in this sequence are linked either by an edge already appearing in the contact graph, or by a transit edge. Moving inside a $GRASP$ cell needs a specific procedure : we have implemented the method presented in the proof of the reduction property . Finally, there remains to compute the transit-path associated to a transit edge : this is a 2-dimensional problem solved in some $ACSR(co)$ slice by a visibility graph method for instance (which is the method we have implemented).

Remark : The algorithm can be adapted in order to take into account partial goals : in this case the goal configuration describes only a goal position for the object (resp. the robot) without specifying a goal position for the robot (resp. the object). Such an extension is easy when the robot goal is unspecified and the object goal is known (the identification of the goal node is given by the non-critical region containing the object). The case of an unspecified object goal is also tractable, but would require some tedious algorithmic details.

3.5   Complexity

In order to evaluate the complexity of the algorithm, we have to distinguish the *decision* part of the manipulation problem (i.e. proving the existence of a solution) from the *complete* problem (i.e. the computation of a solution if any).

In our algorithm, the decision problem is solved by building and searching the graph. The complexity of this part is clearly dominated by the construction of the non-critical regions. Let us denote by n_e, n_r and n_o the number of vertices of the environment, the robot and the object respectively. We assume that both object and robot are convex.

All the critical curves lie in $ACSO$ (Figure 9), whose computation can be done in $O(n_e n_o \log(n_e n_o))$; moreover, the complexity of the admissible configuration space of the object is in $\Omega(n_e n_o)$ (see for instance [6]). Similarly the complexity of the admissible robot configurations subset $ACSR$ is in $\Omega(n_e n_r)$ and can be computed in $O(n_e n_r \log(n_e n_r))$.

The critical curves are defined by the coincidence between an edge (resp. a vertex) of $ACSR$ and a vertex (resp. an edge) of $COL(co)$ boundary (i.e. the set of contacts between the object and the robot). The edge number of $COL(co)$ is exactly $(n_o + n_r)$. Then, the number of critical curves is in $\Omega(n_e n_r(n_o + n_r))$.

The cell decomposition of $ACSO$ is given by the computation of the intersections between the $\Omega(n_e n_r(n_o + n_r))$ critical curves, and the $\Omega(n_e n_o)$ edges of $ACSO$ boundary. This can be done in $O((n_e n_r(n_o + n_r) + n_e n_o)^2 \log(n_e n_r(n_o + n_r) + n_e n_o))$ and gives $\Omega((n_e n_r(n_o + n_r) + n_e n_o)^2)$ non-critical regions.

Finally, in each non-critical region, the number of connected components of $ACSR(co)$ is in $O(n_e n_r)$. They are computed by intersection of the $(n_r + n_o)$ edges of $COL(co)$, and the $\Omega(n_e n_r)$ edges of $ACSR$, for each non-critical region, each time placing the object on some point of the region. Such an intersection is computed in $O((n_r + n_o)n_e n_r \log(n_r + n_o + n_e n_r))$. The manipulation graph is then obtained in $O((n_e n_r(n_o + n_r) + n_e n_o)^2(n_r + n_o)n_e n_r \log(n_r + n_o + n_e n_r))$.

The number of robot and object edges may be considered to be small comparing with n_e. Then, defining n as the environment complexity, previously called n_e, our algorithm runs in $O(n^3 \log(n))$ time. We did not try to optimize it at this time. Perhaps more sophisticated cell-decomposition (like [16]) could be used in the same framework.

Finally, the complexity of the complete problem (i.e. the computation of a solution path), is dominated in the worst case by the number of elementary manipulation paths (see Appendix). Therefore the complexity of the complete problem not only depends on the complexity of the environment but also on the "clearance" of the robot in the environment.

3.6   Experimental results

Figures 14 to 17 show results obtained with the above described implementation.

The solution given by the planner to our illustrative example is shown in Figure 6C. The initial and final object positions are respectively in the non-critical regions 6 and 19 (Figure 9). The initial and final configurations of the manipulation problem are respectively in the ACS cells 5 and 18 (Figure 11).

Let us consider now the solution path (5_B, 4_B, 3_B, 2AB, 9AB, 23AB, 22_B, 22_A, 25BA, 13BA, 14BA, 15_A, 16_A, 19_A, 18_A). Figure 6Ca shows the transit-path along which the robot first reaches the object. In Figure 6Cb, the object is moved with a sequence of transit and transfer paths along the same connected component of $GRASP$. If we refer to the manipulation graph (Figure 13), the manipulation path is built, at this time, from the sequence (5_B, 4_B, 3_B, 2AB, 9AB, 23AB, 22_B) of $GRASP$ cells. A transit edge appears between vertices 22_B and 22_A; the associated transit path is shown in Figure 6Cc. Figure 6Cd describes the subsequence (22_A, 25BA, 13BA, 14BA, 15_A, 16_A, 19_A, 18_A). A last transit-path allows to reach the robot goal configuration.

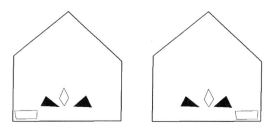

Figure 14: *Initial and goal configurations.*

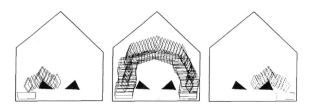

Figure 15: *Solution using a cost function that minimizes the number of grasps.*

Figures 16 to 15 illustrate the influence of heuristics used to find a path through the manipulation graph.

In Figure 16, the weights of transit and transfer links of manipulation graph are nearly equivalents, each referring to the straight line distance between reference points of the $GRASP$ cells. Then the path found by the search algorithm involves numerous transit-paths and re-grasps.

On the contrary, Figure 15 describes the obtained

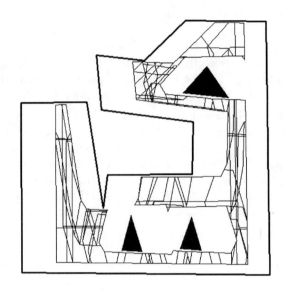

Figure 18: *Non-critical regions.*

Figure 16: *Solution using a cost function that minimizes the length of the object path.*

solution when transit-paths are heavier, due to a simple multiplying coefficient. A path is then found that allows to keep the same grasp along the whole path.

Finally, Figures 17 and 19 show the solution given to a more intricate situation, involving numerous re-grasps. The second and third images detail a transit and a transfer path at the beginning of a contact motion. The heuristic used there takes into account the average length of available grasp frontier for each grasp cell. Then it avoids narrow grasps which could involve numerous re-graspings.

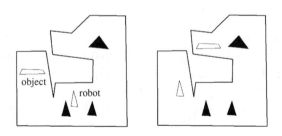

Figure 17: *A more intricate situation.*

The algorithm has been implemented in C on a Sun SPARC Station 2. The computation of the manipulation graph of Figure 13 (including the computation of the non-critical regions) takes 3.4 seconds. The search for a path takes 0.25 seconds for the example in Figure 6. In this simple case, the cell decomposition of the admissible configurations of the object gives 25 non-critical regions, and the manipulation graph consists of 69 nodes and 126 edges. In the example of Figure 14, we obtain 146 non-critical regions, 257 nodes and 475 edges. The case of Figure 17 gives 288 non-critical regions (shown in Figure 18) and its manipulation graph consists of 619 nodes and 1115 edges.

4   Related work and open problems

The first paper that attacks motion planning in presence of movable obstacles is [18]; in this paper, Wilfong gave the first results on the complexity of the problem : he proved that the problem is $PSPACE$-hard (resp. NP-hard) in two dimensional environments where only translations are allowed and when the final configuration specifies (resp. does not specify) the final positions of all the movable objects. In the same reference, Wilfong gives a solution in $O(n^3 \log^2 n)$ (where n is the number of vertices of a polygonal environment) for the case of a convex polygonal robot moving in translation

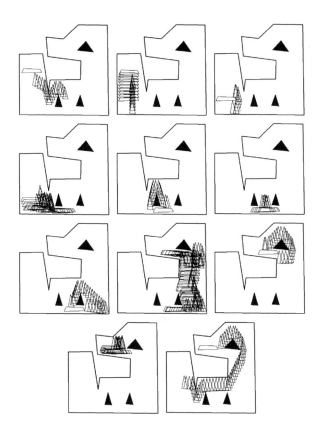

Figure 19: *A manipulation path.*

amidst one polygonal object (with a finite grasp set) and obstacles.

In [12] we presented a general algorithm for the case of one robot and one movable object. We showed [4] how to decompose the space of grasping configurations into a finite number of cells in order to make the problem tractable. However the method makes use of general tools from algebraic geometry, leading to an inefficient algorithm in practice.

More recently, several authors attacked the manipulation planning problem in various contexts.

Koga and Latombe [8, 9] addressed the case of multi-arm manipulation planning where several robots have to cooperate to carry a single movable object amidst obstacles. In this context, re-grasps of the object are often required along the path to avoid collisions and

---

[4]The principles are similar to those presented in Section 3, but they are established for general robot systems.

may involve changing the arms grasping the object. In [8] they propose several implemented planners to deal with problems of increasing difficulty for two identical arms in a 2D workspace. For problems involving many degrees of freedom, which is usually the case in multi-arm manipulation, they use an adapted version of the randomized potential field planner [2]. An extension to a robot system made of three 6 DOF arms in a 3D workspace is also presented in [9]. In this work, the number of legal grasps of the object is finite and the algorithms require the object to be held at least by one robot at any time during a re-grasp operation. The approach relies on several simplifications, but it yields to impressive results for complex and realistic problems.

The approach developed by Barraquand and Ferbach [3] consists in translating the manipulation planning problem into convergent series of less constrained sub-problems increasingly penalizing the motions that do not satisfy the constraints. Each subproblem is solved using variational dynamic programming.

[4] describes a heuristic algorithm for a circular robot and where all the obstacles can be moved by the robot in order to find its way to its goal.

Finally, let us pinpoint the interesting extension of the manipulation planning problem attacked by Lynch and Mason [13]. In their context, grasping is replaced by pushing. The space of stable pushing directions imposes a set of nonholonomic constraints on the robot motions, which opens issues of controllability.

All the above mentioned studies contribute to establish the manipulation planning problem as a specific and challenging instance of the motion planning problem with constraints. However, several open questions remain, ranging from a theoretical analysis of the problem to the investigation of new practical instances. One key theoretical aspect concerns the conditions under which the reduction property can be extended to the case of several objects and robots. On a practical point of view, the problem represents also a challenge to motion planning techniques because of its additional complexity.

Acknowledgement : This work has been partially supported by the Esprit Programme through Basic Research Action 6546 PROMotion. We are grateful to B. Dacre-Wright who implemented the algorithm described in Section 3.

Appendix : Reduction property

In this appendix, we consider only one movable object. In this case, two configurations belonging to a same connected component of $GRASP \cap PLACEMENT$ are m-connected. In fact, this property holds up to a precise definition of what is a grasping configuration.

Let us consider $CS = CSR \times CSO$ the configuration space of the robot and the object together. Let us denote by $CONTACT$ the domain of the configurations where the robot and the object are in contact. $CONTACT$ is a subset of $\mathsf{Free}(CS)$ boundary.

Hypothesis H: We assume that the robot has to avoid any contact with the obstacles.

We then define $GRASP$ as the subset of $CONTACT$ of all the configurations verifying hypothesis H.

Property 1 *Two configurations of a same connected component of* GRASP $\cap$ PLACEMENT *are connected by a manipulation path.*

Proof: Let a and b be two configurations in a connected component of $GRASP \cap PLACEMENT$. There exists a path $p : [0,1] \mapsto GRASP \cap PLACEMENT$ linking these two configurations[5] ($p(0) = a$, $p(1) = b$). We define p_r and p_o as the projections of p onto CSR and CSO respectively.

Let $c = p(t)$ be any configuration on the path. Thanks to the hypothesis H, $p_r(t)$ lies in an open set of $\mathsf{Free}(CSR)$. We then can find an open disc $D_\epsilon \subset \mathsf{Free}(CSR)$ centered on $p_r(t)$ and with a radius $\epsilon > 0$.

Since p is continuous, there exists $\eta_1 > 0$ such that :

$$\forall \tau \in]t - \eta_1, t + \eta_1[, p_r(\tau) \in D_{\epsilon/2}.$$

Similarly, $p_r - p_o$ is a continuous function. Then there exists $\eta_2 > 0$ such that, for any $\tau \in]t - \eta_2, t + \eta_2[$,

$$\| (p_r(\tau) - p_o(\tau)) - (p_r(t) - p_o(t)) \|_{\mathbf{R}^2} < \epsilon/2.$$

This last assertion means that the relative grasp configuration does not vary more than $\epsilon/2$ along the path p between $p(t - \eta_2)$ and $p(t + \eta_2)$.

Let us consider $\eta = min\{\eta_1, \eta_2\}$:

$$\forall \tau, \sigma \in]t - \eta, t + \eta[, \quad p_o(\sigma) + (p_r(\tau) - p_o(\tau)) \in D_\epsilon. \quad (1)$$

[5] p designates a path as well as the associated function.

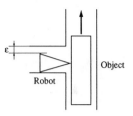

Figure 20: *A case with finite (but great) number of transit and transfer paths.*

Let $c_1 = p(\tau_1)$ and $c_2 = p(\tau_2)$ be any two configurations on p, with τ_1 and τ_2 in $]t - \eta, t + \eta[$ (we assume that $\tau_1 < \tau_2$). We prove now that c_1 and c_2 can be linked by one transfer path followed by one transit path.

Let us consider the path $(p_o(\tau), p_o(\tau) + (p_r(\tau_1) - p_o(\tau_1)))$, with $\tau \in [\tau_1, \tau_2]$. This path is clearly a transfer path with constant grasp $(p_r(\tau_1) - p_o(\tau_1))$, between $p(\tau_1)$ and $(p_o(\tau_2), p_o(\tau_2) + (p_r(\tau_1) - p_o(\tau_1)))$. According to relation 1, this path is admissible. Let us consider the path $(p_o(\tau_2), p_o(\tau_2) + (p_r(\tau) - p_o(\tau)))$, with $\tau \in [\tau_1, \tau_2]$. This path is clearly a transit path between $(p_o(\tau_2), p_o(\tau_2) + (p_r(\tau_1) - p_o(\tau_1)))$ and $p(\tau_2)$. Again, according to relation (1), this path is admissible. The concatenation of both paths constitute a manipulation between $p(\tau_1)$ and $p(\tau_2)$.

As path p_r is a compact set included in an open set of $\mathsf{Free}(CSR)$, we can apply this local transformation on a *finite* covering of $[0, 1]$. We have then a finite number of elementary manipulation paths which constitutes a manipulation path linking a and b. $\square$

Remark : the number of elementary manipulation paths used in the proof of the reduction property depends on the clearance of p_r in $\mathsf{Free}(CSR)$. The worst case is reached in the example Figure 20 where the number of elementary paths is clearly in $O(\frac{1}{\epsilon})$.

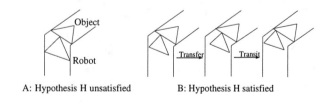

A: Hypothesis H unsatisfied   B: Hypothesis H satisfied

Figure 21: *Illustration of hypothesis H.*

Note that the reduction property does not hold when hypothesis *H* is not satisfied. Figure 21 illustrates this fact. Figure 21A shows an example where both the robot and the object touch the environment. There exists a coordinated path, but no feasible manipulation path (the robot cannot move the object with a constant grasp). Figure 21B shows how the robot can "manipulate" the object even when this one is in contact.

References

[1] R. Alami, T. Siméon, J.P. Laumond, "A geometrical Approach to planning Manipulation Tasks. The case of discrete placements and grasps." International Symposium on Robotics Research, Tokyo, August 1989.

[2] J. Barraquand and J.C. Latombe, "Robot motion planning : a distributed representation approach," in *Int. Journal of Robotics Research*, 10 (6), 1991.

[3] J. Barraquand and P. Ferbach, "A penalty function method for constrained motion planning," in *IEEE* Conf. on Robotics and Automation, 1994.

[4] P.C. Chen, Y.K. Hwang, "Practical path planning among movable obstacles." in *IEEE* Conf. on Robotics and Automation, 1991.

[5] B. Dacre-Wright, J.P. Laumond, R. Alami, "Motion planning for a robot and a movable object amidst polygonal obstacles," in *IEEE* Conf. on Robotics and Automation, 1992.

[6] S.J. Fortune, "A fast algorithm for polygon containment by translation", ICALP, 1985.

[7] S. Fortune, G. Wilfong, C. Yap, "Coordinated motion of two robot arms." in *IEEE* Conf. on Robotics and Automation, 1986.

[8] Y. Koga and J.C. Latombe, "Experiments in dual-arm manipulation planning," in *IEEE* Conf. on Robotics and Automation, 1992.

[9] Y. Koga and J.C. Latombe, "On multi-arm manipulation planning," in *IEEE* Conf. on Robotics and Automation, 1994.

[10] J.-C. Latombe, *Robot Motion Planning*, Kluwer Press, 1991.

[11] J.P. Laumond, R. Alami, "A new geometrical approach to manipulation task planning: the case of a circular robot amidst polygonal obstacles and a movable circular object." Technical Report LAAS/CNRS 88314, Toulouse, Octobre 1988.

[12] J.P. Laumond, R. Alami, "A geometrical approach to planning manipulation tasks in Robotics.", Technical Report LAAS/CNRS 89261, Toulouse, August 1989.

[13] K. M. Lynch, "Stable pushing : mechanics, controllability and planning," in this volume.

[14] D. Parsons, J. Canny, "A motion planner for multiple mobile robots." in *IEEE* Conf. on Robotics and Automation, 1990.

[15] J.T. Schwartz, M. Sharir, "On the piano mover problem III : Coordinating the motion of several independant bodies : the special case of circular bodies amidst polygonal barriers." Int. J. of Robotics Research, 2 (3), 1983.

[16] M. Sharir, S. Sifrony, "Coordinated motion planning for two independent robots." ACM Symposium on Computational Geometry, 1988.

[17] T. Siméon, "Planning collision-free trajectories by a configuration space approach," in *Geometry and Robotics*, J.D. Boissonnat and J.P. Laumond Eds, Lecture Notes in Computer Science, 391, Springer Verlag, 1989.

[18] G. Wilfong, "Motion planning in the presence of movable obstacles." in *ACM* Symp. on Computational Geometry, 1988.

Natural Sets in Manipulation Tasks

Randy C. Brost, *Sandia National Laboratories, Albuquerque, NM, USA*

A key feature distinguishing robotics from traditional computer science is its connection to the physical world. Robot planning software may use elegant algorithms supported by ironclad analytic proofs, but ultimately nature will decide whether the software output is correct in the sense of accomplishing the task goal. Thus a chief goal of robotics research is to understand and capture this nature in a way that allows algorithmic analysis to produce robust physical results. This is made particularly difficult by the presence of uncertainty, which arises from the inevitable discrepancy between the real task and its idealized computer model. This paper reviews fundamental sets of states, forces, and actions that exist for a broad class of robot manipulation tasks, and ties these sets to past and future approaches to developing robust manipulation planning and execution systems.

1 Robotics as a Study of Nature

Robots differ from ordinary computers in that they must execute actions that affect the external world. This difference imposes a harsh criterion on the capability of a given robot system: Nature is the ultimate judge of whether the robot has chosen an appropriate action to accomplish a given goal. Consequently, algorithms developed to automatically plan and execute robot actions must adequately capture the inherent nature of the task, or face the consequence of having nature point out the error of their ways.

This paper examines the thesis that there are natural sets that characterize robot tasks, and that these sets determine the likelihood that a given robot action will succeed. These sets may be expressed in the state space, force space, or action space of the task, or in some other coordinate system. One way or another, an effective robot planning and execution system must encode these sets in order to reliably perform successful

actions. Similarly, these sets may be exploited by robot systems to solve difficult problems that may otherwise elude solution. In this paper we will focus on manipulation tasks, but we expect that these ideas will apply to other robot tasks as well.

Let's begin by examining a typical robot task, with an eye toward making observations of nature. We will study data from an experiment that Alan Christiansen and I performed for another paper [6], but which are also useful in the context of this paper. Our example task is a grasping operation that appears as one step in a larger assembly procedure; this task is shown in Figure 1. The goal of the task is to grasp the gear in the configuration shown at the bottom of Figure 2, using a SCARA robot with a parallel-jaw gripper. To perform a grasping operation, the robot opens the gripper fingers, moves the gripper to an initial position near the gear, and then closes the fingers. Figure 2 shows the motion of the gear that occurs during a typical successful operation. Figure 3 shows examples of the types of failure that can occur. Clearly, the success of this operation depends on the chosen initial gripper position relative to the gear, as well as other factors. In this experiment, we ask the simple question:

> "Which commanded initial positions will succeed, and which will not?"

We answer this question by repetitively executing grasping operations at different initial positions, checking success after each operation. This is performed by an automatic procedure that allows the execution of a large number of trials. For each desired grasping operation, this procedure initially locates the gear using computer vision, calculates the gripper position in world coordinates, moves to the appropriate initial position, executes the grasping operation, and then checks the success of the operation using computer vision. The

127

Figure 1: *An example task chosen for empirical study.*

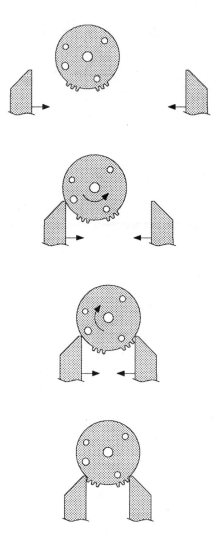

Figure 2: *A typical grasping action, showing four phases of contact: No contact, single-finger pushing, two-finger squeezing, final grasp.*

outcome of the operation is then recorded, and the next desired grasping operation is executed.

A number of precautions were taken to improve the accuracy of the collected data. Lighting sources were designed to provide even illumination with minimal robot shadows, and the support surface was chosen to provide a crisp high-contrast image of the gear. Edge-based feature localization algorithms were used to measure the gear position, and these were found to be much more robust and precise than typical blob analysis techniques. Measured image locations were adjusted to compensate for lens distortion using a radially-symmetric distortion model; the coefficients of this model were determined from calibration data obtained from a 21×21 grid of dots. We estimate the accuracy of the gear position measurements to be within ± 1mm and $\pm 3°$. The gear diameter is 41.48mm.

This apparatus was used to study the reliability of grasping operations throughout a region of initial positions. The region of interest was a 20mm square with the upper left corner at the center of the gear; the operations corresponded to a 1mm grid covering this region. Each grid point corresponds to the initial placement of the center point between the fingertips relative to the gear center. The initial orientation of the gear relative to the gripper was the same in all of the trials; only the initial (x, y) position was varied. Thus each experimental run consisted of $21 \cdot 21 = 441$ grasping operations, each with a different initial position.

We performed thirty experimental runs, for a total of 13,230 grasping operations. The results were then tabulated to form a histogram of successful grasping operations associated with each initial position. These data are shown in Figure 4, in two forms. At the top of the figure, the gear outline is drawn actual size, with the region of sampled points shown dashed. The shaded region outlines the set of points that succeeded thirty times out of thiry; this is the *observed strong backprojection* of this task. Surrounding the shaded region is another outline, which delineates the set of points that

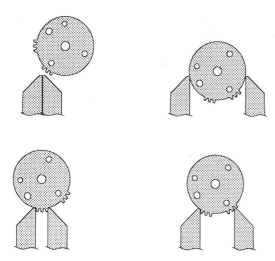

Figure 3: *Example grasp failures.*

were observed to succeed at least once; this is the *observed weak backprojection* of this task. Points outside this outline failed thirty times out of thirty. At the bottom of the figure, the histogram of observed successes is drawn as a three-dimensional plot, showing the variation from complete failure to reliable success. The shaded boxes of this plot identify squares where all four corners succeeded thirty times out of thirty. The plot axes correspond to an (x, y) coordinate system whose origin is at the center of the gear.

This entire experiment was repeated three times, under varying conditions. The resulting data are shown in Figure 5. In part (a), a small amount of sand was added to the table surface, simulating the effect of grit or a less structured environment. In part (b), the finger speed was doubled, with no sand present. In part (c) the finger speed was doubled and sand was present. All four of the conditions studied represent reasonable simulations of practical situations that may be encountered in industry. As can be seen from the figure, these varying conditions had a dramatic effect on the observed strong and weak backprojections for the task. For a more detailed discussion of this experiment and its results, see [6].

These experiments demonstrate that there are sets of actions that succeed, and sets of actions that fail, and that these sets are a natural artifact of the task conditions. Note that our experimental data do not exactly characterize these sets — we would expect the true strong backprojection to be smaller than the ob-

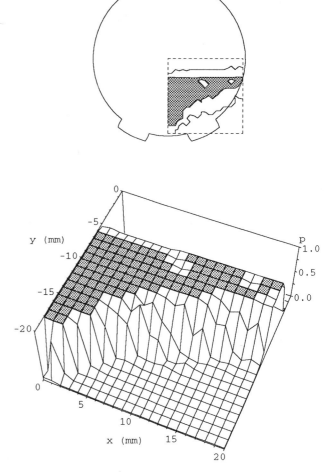

Figure 4: *The result of exploring the space of grasping actions. In this experiment, the support surface was clean, and the fingers closed slowly.*

served strong backprojection, and the true weak backprojection to be larger than the observed weak backprojection. Nonetheless, the data strongly support the existence of these sets, except in Figure 5(c), where it seems likely that the true strong backprojection is null.

These sets characterize the combined effect of the geometry, mechanics, and uncertainty of the task, and reflect the constraints on successful actions that are imposed by nature. In a sense, these sets are what nature gives us to work with as designers of robot algorithms. If we wish to build reliable grasp planners, then we must somehow encode these sets in our planning and execution systems.

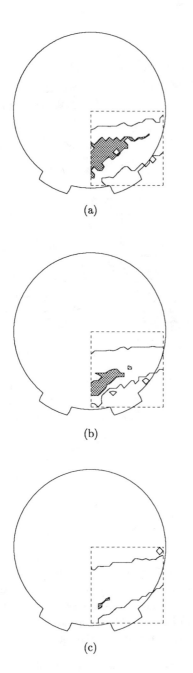

(a)

(b)

(c)

Figure 5: *The results of the grasping experiment repeated under varying conditions. (a) With sand contaminating the support surface, and slow finger closing speed. (b) With a clean support surface and fast finger closing speed. (c) With sand contaminating the support surface, and fast finger closing speed.*

How can we accomplish this? One approach would be to take the classical scientific method of making observations of nature, forming a hypothetical model of nature, conducting an experiment to validate the model, and then using the model to develop practical systems. Can this method be applied to robotics?

2 Previous Examples

In fact, we can see several past examples where this approach has already been undertaken:

Compliant Motion

In 1974 Hirochika Inoue implemented a demonstration of an assembly task using the Purbrick arm [14]. This arm was equipped with a force-sensing wrist, which Inoue used to implement both guarded moves and compliant motions. Inoue immediately noticed the influence of task uncertainty, which had a dramatic impact on the reliability of some of the tight-tolerance insertion operations included in the task. Inoue developed several clever strategies to cope with this difficulty; these included tilting the peg relative to the hole to expand the effective target region of the insertion operation, and applying an intentional offset to allow sliding in a known direction to find the hole.

Later researchers studied this operation in an attempt to understand in general terms why these strategies were successful. This led to a number of important insights. First, Tomás Lozano-Pérez observed that the relationship between task geometry and uncertainty may be cleanly expressed in the configuration space of the task [17]. Using techniques initially developed for path planning, Lozano-Pérez converted the peg-in-hole problem to the problem of moving a point into a narrowed hole in configuration space. When the peg is tilted, the mouth of the c-space hole widens substantially, making it easier to accomplish the insertion operation. This work was later extended to include motion dynamics by Lozano-Pérez, Mason, and Taylor [19]. This work, often referred to as the *LMT formalism*, proposed the notion of strong and weak preimages, and showed how these sets could be used to synthesize Inoue's offset-slide insertion strategy.

In a related vein, a group at Draper Laboratories studied the mechanics of the peg-insertion operation, and identified constraints on the peg's compliance that would assure reliable insertion. These constraints were expressed in a *jamming diagram,* which was used to identify the stiffness parameters of a remote center of compliance device for peg insertion [29]. Using the observations of Inoue and Lozano-Pérez, this group later extended this device to tilt the peg at the beginning of the insertion operation, increasing the range of allowable initial insertion positions [24]. These devices are in wide use in industry today.

Grasping and Pushing Operations

At the same time as Inoue's assembly demonstration, Pingle, Paul, and Bolles developed a demonstration of automatic hinge assembly using multiple robot arms [26]. In viewing the film of their demonstration, Matt Mason observed a subtle operation that was used to grasp a hinge plate. The hinge plate was grasped by simultaneously closing the fingers and translating the gripper, which had the ultimate effect of locating the plate very precisely between the fingers. Pingle, *et al.* undoubtedly chose this operation because it provided a robust method of obtaining a precise grasp of the plate for subsequent assembly operations. Mason's observations ultimately led to his seminal work addressing the relationship between pushing mechanics and robot tasks [21]. This work led directly to implemented planners for constructing robust pushing and grasping operations for robot fingers that are infinitely long [20, 4]. These planners operated by using Mason's mechanics model to identify sets of reliable operations, which were then shrunk to account for uncertainty. Balorda later applied similar techniques to identify robust two-finger pushing actions [1].

Analyzing Friction

In 1781 Coulomb studied the intrinsic properties of dry friction, and proposed the familiar Coulomb friction model [9]. Somewhat later, Mosely developed the graphical abstraction of Coulomb's law, which is now known as the friction cone [23]. In his Master's thesis, Erdmann explored the implications of Coulomb's law applied to multiple-contact manipulation tasks, and developed the notion of the *configuration-space friction cone,* which extends the ordinary friction cone to include induced moments and multiple contacts [11]. Later, Brost and Mason simplified Erdmann's constructions to develop graphical techniques for analyz-

ing multiple-contact situations with friction [7]. Erdmann's method and Brost and Mason's method both focus on the construction of sets of forces that are possibly consistent with a given contact situation. Mason later applied these models to characterize the set of forces that could be applied to reliably push a block along a wall, and then used this characterization to synthesize a collection of actions that push the block while retaining contact with the wall [22]. See Figure 6. Mason demonstrated the efficacy of these synthesized operations through a series of simple physical experiments.

This list is by no means exhaustive, but does provide insight into how the classical scientific method can be applied to solve robotics problems. Each of these examples began with fundamental observations of nature, followed by insights that were gained from those observations, models developed from the insights, and manipulation strategies developed using the models. In some cases the final utility of these models has not been fully established, but work continues toward resolving these questions.

3   State Spaces and Action Spaces

Past robot planning systems have often constructed sets in the task state space, the robot's action space, or both. The distinction between these spaces is sometimes confusing. The *state space* of a task is defined by the range of the variables used to describe the system's state. These may include configuration, velocity, or force variables, or other state parameters. Points in the state space correspond to the true states of the task system.

The *action space* is defined by the range of the parameters used to define a robot action. Let's suppose that our robot is controlled by a computer that attempts to accomplish a series of goals. Further, let's suppose that this computer has a repertoire of actions that may be commanded; each action corresponds to a subroutine that drives the robot's actuators to perform some action. These actions may be simple, such as "Move in a straight line to position $\mathbf{x}$," or complex, such as "Follow the potential field $f(\mathbf{x})$, responding to contacts with a generalized-spring compliant motion defined by the time-varying stiffness matrix $\mathbf{K}(t)$, and avoid all objects in motion." Whether simple or complex, we will think of actions as procedures that may

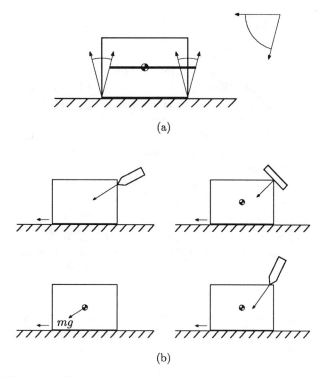

Figure 6: *The result of Mason's block-along-wall analysis. A rectangular block is resting on a horizontal surface, in contact with a vertical wall. (a) In order to push the block along the wall, one must apply a line of force that passes through the bold segment, in a direction inside the interval shown. This corresponds to a convex set of forces in the $(F_x, F_y, \tau/\rho)$ space of planar forces. (b) Four strategies that provide such a force. The fourth strategy only works if there is more friction between the pusher and the block than between the block and the wall.*

be called, with certain unbound parameters. The position $\mathbf{x}$ is the parameter for the first example, while the second example has the parameters $f(\mathbf{x})$ and $\mathbf{K}(t)$. These parameters define the action space for each action. Given a database of such actions, the planner's job is to select the appropriate action to execute, and to choose appropriate values of the unbound action parameters.

It is the action space that is ultimately of interest to the robot, since the robot can only directly control the commands that it sends to its actuators. The robot cannot command task states, so these are of secondary interest. However, the state space can provide a useful mechanism for analyzing the task to identify the

appropriate action to command. This is because the state space allows the integration of differential task mechanics, as well as reasoning about the set of possible outcomes of a given action.

A useful technique may be applied when the robot can execute both compliant and position-controlled motions, and the accuracy of this position control obeys a history-independent bound. The technique is appropriate when the task goal is a contact state that must be achieved by a compliant motion. In this technique, we perform a trajectory analysis of compliant motions in the task state space, forming the strong backprojection of the goal. We then shrink the resulting backprojection by the position control uncertainty; if any points remain, we formulate a two-step plan comprised of first executing a position-controlled motion to a point in the shrunk backprojection, and then executing the compliant motion. Figure 7 shows how this method may be used to synthesize the offset-slide strategy used by Inoue.

This method exploits useful properties of both the state space and the action space. For a given commanded motion direction, the state space allows identification of initial states that will reliably reach a desired final state, even if the robot cannot directly command a position-controlled motion to establish the state. After the state-space analysis is complete, the shrinking operation transforms the set of reliable initial states into a set of commanded initial positions in the action space. Assuming that an appropriate path planner is available, the robot may directly utilize any of these positions as action commands. If we had augmented the state space in Figure 7 by adding a dimension describing the compliant motion direction, then a similar shrinking procedure could be applied to directly identify robust commanded motion directions. Variants of this basic method were applied in [4, 5, 1] to plan pushing and grasping operations; see [15] for similar constructions applied to compliant motion synthesis.

The distinction between state space and action space is perhaps most keenly highlighted by Sanjiv Singh and Reid Simmons' recent work in robot excavation [28]. Singh and Simmons began by defining a family of canonical digging actions, parameterized by a few control variables such as digging depth, shovel angle, etc. Using these action definitions and a description of the aggregate soil properties, they developed constraints in

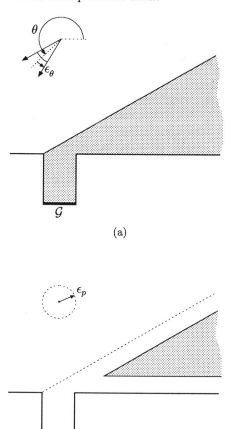

Figure 7: *A principled approach to landfall navigation, using the LMT formalism. This is the standard peg-in-hole task, where the peg has been shrunk to a point, and the walls of the hole have been grown accordingly. The goal of the task is to move the point to the goal G at the bottom of the hole, using a generalized damper motion. The robot may execute a position-controlled motion to move the peg to an initial position in free space before initiating the compliant motion. The uncertainty in motion direction and position control are ϵ_θ and ϵ_p, respectively. (a) First the strong backprojection is formed in the state space; this is the set of initial states that will reliably achieve the goal. (b) The strong backprojection is shrunk by the position control uncertainty, yielding a set of commanded initial positions that will reliably succeed.*

the action space delineating actions that obeyed force and kinematic limits, and then used numerical search to find optimal points in this space. This planner an-

alyzes sets of successful actions in the action space, even though the state space of the task is impractical to represent due to the large number of soil particles.

4 Advantages of Set-Based Methods

In our model of a robot with a repertoire of actions, it becomes apparent that set-based approaches to robot task planning may be overkill — after all, the planner only needs to choose a single point in the action space. Why incur the work of constructing an entire set of successful actions when only a single point is needed? This is a legitimate question, and in some cases the set-based approach will not be justified. The primary advantage of set-based models is that they support powerful, general-purpose analysis techniques. This may be seen through three particular advantages:

- *Reasoning about uncertainty.* Analysis in the state space allows direct construction of the set of states that may occur for a given action, or the set of initial states that may accomplish the task goal. This provides a principled approach to landfall navigation. For centuries navigators have intentionally diverted their course slightly to assure that they would encounter landfall or some other constraint on a known side of their ultimate target. We have seen two examples of this strategy for manipulation tasks: The offset-slide peg insertion strategy employed by Inoue, and the gear-grasping task studied in Section 1. How can such strategies be synthesized in general? Synthesizing these strategies is more difficult than planning ordinary landfall navigation strategies, because of the nuances of the task geometry and mechanics. However, these strategies may be automatically synthesized by constructing the strong backprojection of the task goal, and then shrinking this set by the uncertainty in the robot's control of the initial state. This general construction may be applied for tasks with complex task mechanics and geometry, which are difficult to solve using other techniques.

- *Satisfaction of multiple constraints.* Manipulating sets of successful actions allows the identification of actions that simultaneously satisfy multiple constraints. This can be important in tasks that are part of a larger procedure such as an assembly task, or if there are certain critical states that must be avoided at all times (such as contact with a sensitive surface). Examples include the geometric constraint propogation methods used to identify grasp configurations in [16, 18], and the constraints imposed in the contact-cement example in [5].

- *Identification of optimal strategies.* When the entire set of acceptable solutions is known, it is generally possible to find the optimal or near-optimal solution given a desired optimality metric. In scenarios where a given action will be executed a large number of times (such as in large-volume manufacturing), this benefit may far outweigh the extra analysis cost.

In addition to these pragmatic considerations, the set-based approach has an appeal from a scientific perspective. As a model of nature, this approach seems to capture salient aspects of the task, as evidenced by the grasping experimental results given above.

5   Implementation Strategies

Implementing systems that exploit natural sets in manipulation tasks may involve a variety of techniques. Some may attempt to produce an exact representation of these sets, while others may construct crude approximations. Still other techniques may completely avoid set constructions, but implicitly represent the relevant sets through the structure of the algorithm. Here are some past examples:

Direct Geometric Construction. Mike Erdmann and Bruce Donald both implemented planners that constructed strong and weak backprojections directly from a representation of the task geometry and generalized damper mechanics [12, 10].

Critical Value Analysis. A number of planning algorithms have been produced that identify critical conditions that separate regions of qualitatively similar points in a continuous space. Examples include the push-stability and squeeze-grasp diagrams in [4], the motion-direction analysis in [2], the static equilibrium analysis in [5], and the non-directional blocking graph in [30].

Generate-and-Test. Mani and Wilson developed a planner that identified sequences of pushing operations to orient planar parts [20]. Their planner utilized pushing analysis diagrams with exactly the same semantics

as Brost's push-stability diagrams, except Mani and Wilson's diagrams were generated using a discretized generate-and-test computation.

Numerical Integration. For some tasks we do not have closed-form equations that describe the task trajectory, but we do have methods for identifying the set of possible instantaneous motions. One example is the task of pushing with an uncertain support pressure distribution [25]. For these tasks, it is possible to generate the strong backprojection and other LMT sets using numerical integration to characterize the set of possible trajectories in the task state space [5].

Constraint-Posting and Numerical Search. Singh's excavation planner constructs an implicit representation of the set of successful actions by posting constraints in the action space and then applying numerical search techniques to find a near-optimal action in the constraint-satisfying subspace [28].

Invariant Identification. In Ken Goldberg's planner for orienting planar polygonal parts through grasping operations, the implicit set representation is buried deep in the algorithm design [13]. A given polygon shape has a diameter function, which is a transformation of the configuration-space obstacle of the part with respect to the gripper. This diameter function has certain local minima and maxima, which define capture regions for the frictionless squeeze-grasping operation. These capture regions may be expressed as a monotonic transfer function from initial states to final states, which forms the basis of the algorithm.

Inferences Gained from Empirical Observation. Alan Christiansen developed a planning and execution system that began with almost no information, and developed the ability to solve difficult manipulation problems based on empirical observation [8]. Given a simple bounding-box inductive bias, Christiansen's planner built up internal models of strong and weak backprojections by cataloging observations of initial and final action outcomes. These approximations to the true backprojections were then used to construct multi-step plans.

This list provides us with a number of possible approaches to implementing planners based on natural sets; there are undoubtedly other approaches not listed above. This menagerie of implementation methods underscores the importance of separating the issues of developing models of manipulation tasks and developing systems that exploit those models to automatically plan robust actions.

6   Discussion

This paper has made the case that manipulation planning is fundamentally and unavoidably tied to natural sets. Observations such as the gear-grasping experiment in Section 1 and Mason's block-along-wall experiment suggest that we have no choice but to somehow encode these sets in our planning algorithms for robot manipulation tasks. Thus, I'll claim that these sets by their very nature must form a key cornerstone in the foundation of successful algorithms in robotics.

A number of open problems remain. What are these sets really like for practical manipulation tasks? What are good models of these sets? What are the most effective techniques for computing them? Is there some way to obtain the benefits of set-based analysis while incurring only the cost of identifying a single point?

One of the most interesting open problems is how to merge these ideas with more reactive techniques of sensing and control. In order to emphasize the role of natural sets in manipulation tasks, this paper has intentionally taken a somewhat narrow view. For example, the idea of centering robot planning algorithms on models of natural sets may seem incompatible with the reactive architecture ideas of Rod Brooks [3], or the cyclic process control schemes developed in Marc Raibert's laboratory [27]. This need not be the case. Ultimately, I expect that all of these ideas will play a prominent role in the development of robust, flexible autonomous systems. Perhaps we will discover some way to integrate these seemingly different approaches into a single unifying framework. In a way, it seems that we must: How else will we create a robot that can determine how to assemble parts with complicated shapes, respond appropriately to unexpected situations, and maintain stable dynamic locomotion?

Acknowledgements

Many thanks to Randy Wilson and Pang Chen for helpful comments on drafts of this paper, and to Alan Christiansen, Bruce Donald, Mike Erdmann, Ken Goldberg, Tomás Lozano-Pérez, Matt Mason, Marc Raibert, and Sanjiv Singh for numerous discussions that led to the ideas presented here. The author is supported by US DOE contract DE-AC04-94AL85000.

References

[1] Z. Balorda. Automatic planning of robot pushing operations. In *Proceedings, IEEE International Conference on Robotics and Automation*, pages (1)732–737, May 1993.

[2] A. J. Briggs. An efficient algorithm for one-step planar compliant motion planning with uncertainty. Technical Report TR 89-980, Cornell University Department of Computer Science, March 1989.

[3] R. A. Brooks. A robust layered control system for a mobile robot. *IEEE Journal of Robotics and Automation*, 2(1):14–23, March 1986.

[4] R. C. Brost. Automatic grasp planning in the presence of uncertainty. *International Journal of Robotics Research*, 7(1):3–17, February 1988.

[5] R. C. Brost. *Analysis and Planning of Planar Manipulation Tasks*. PhD thesis, Carnegie Mellon University School of Computer Science, January 1991.

[6] R. C. Brost and A. D. Christiansen. Probabilistic analysis of manipulation tasks: A computational framework. Technical Report SAND Report 92-2033, Sandia National Laboratories, January 1994.

[7] R. C. Brost and M. T. Mason. Graphical analysis of planar rigid-body dynamics with multiple frictional contacts. In H. Miura and S. Arimoto, editors, *Robotics Research: The Fifth International Symposium*, pages 293–300, Cambridge, Massachusetts, 1990. MIT Press.

[8] A. D. Christiansen. Manipulation planning from empirical backprojections. In *Proceedings, IEEE International Conference on Robotics and Automation*, pages 762–768, April 1991.

[9] C. A. Coulomb. Théorie des machines simples en ayant égard au frottement de leurs parties et à la roider des cordages. In *Mémoires de Mathématique et de Physique présentés à l'Academie Royale des Sciences*, Paris, 1781.

[10] B. R. Donald. A geometric approach to error detection and recovery for robot motion planning with uncertainty. *Artificial Intelligence*, 37:223–271, 1988.

[11] M. A. Erdmann. On motion planning with uncertainty. Master's thesis, MIT Department of Electrical Engineering and Computer Science, August 1984.

[12] M. A. Erdmann. Using backprojections for fine motion planning with uncertainty. *International Journal of Robotics Research*, 5(1):19–45, Spring 1986.

[13] K. Y. Goldberg. *Stochastic Plans for Robotic Manipulation*. PhD thesis, Carnegie Mellon University School of Computer Science, November 1990.

[14] H. Inoue. Force feedback in precise assembly tasks. Memo 308, MIT Artificial Intelligence Laboratory, August 1974.

[15] J. C. Latombe. *Robot Motion Planning*. Kluwer Academic Publishers, Boston, 1991.

[16] T. Lozano-Pérez. Automatic planning of manipulator transfer movements. *IEEE Transactions on Systems, Man, and Cybernetics*, SMC-11(10):681–698, October 1981. Reprinted in M. Brady, *et al.*, editors, *Robot Motion: Planning and Control*, MIT Press, Cambridge, Massachusetts, 1982.

[17] T. Lozano-Pérez. Task planning. In M. Brady, J. M. Hollerbach, T. L. Johnson, T. Lozano-Pérez, and M. Mason, editors, *Robot Motion: Planning and Control*, pages 473–498. MIT Press, Cambridge, Massachusetts, 1982.

[18] T. Lozano-Pérez, J. L. Jones, E. Mazer, P. A. O'Donnell, W. E. L. Grimson, P. Tournassoud, and A. Lanusse. Handey: A task-level robot system. In R. Bolles and B. Roth, editors, *Robotics Research: The Fourth International Symposium*, pages 29–36, Cambridge, Massachusetts, 1988. MIT Press.

[19] T. Lozano-Pérez, M. T. Mason, and R. H. Taylor. Automatic synthesis of fine-motion strategies for robots. *International Journal of Robotics Research*, 3(1):3–24, Spring 1984.

[20] M. Mani and R. D. W. Wilson. A programmable orienting system for flat parts. In *North American Manufacturing Research Institute Conference XIII*, 1985.

[21] M. T. Mason. Mechanics and planning of manipulator pushing operations. *International Journal of Robotics Research*, 5(3):53–71, Fall 1986.

[22] M. T. Mason. How to push a block along a wall. In *NASA Conference on Space Telerobotics*, February 1989.

[23] H. Moseley. On the equilibrium of the arch. *Transactions of the Cambridge Philosophical Society*, V:293–314, 1835.

[24] J. L. Nevins and D. E. Whitney. *Concurrent Design of Products and Processes*. McGraw-Hill, New York, 1989.

[25] M. A. Peshkin and A. C. Sanderson. The motion of a pushed, sliding workpiece. *IEEE Journal of Robotics and Automation*, 4(6):569–598, December 1988.

[26] K. Pingle, R. Paul, and R. Bolles. *Programmable Assembly: Three Short Examples*. Stanford Artificial Intelligence Laboratory, 1974. Film.

[27] M. H. Raibert. *Legged Robots that Balance*. MIT Press, Cambridge, Massachusetts, 1986.

[28] S. Singh and R. Simmons. Task planning for robotic excavation. In *Proceedings, IEEE/RSJ International Conference on Intelligent Robots and Systems*, pages 1284–1291, July 1992.

[29] D. E. Whitney. Quasi-static assembly of compliantly supported rigid parts. *Journal of Dynamic Systems, Measurement, and Control*, 104:65–77, March 1982. Reprinted in M. Brady, *et al.*, editors, *Robot Motion: Planning and Control*, MIT Press, Cambridge, Massachusetts, 1982.

[30] R. H. Wilson. *On Geometric Assembly Planning*. PhD thesis, Stanford University Department of Computer Science, March 1992.

Grasp Metrics: Optimality and Complexity

B. Mishra, *Courant Institute, New York University, NY, USA*

In this paper, we discuss and compare various metrics for goodness of a grasp. We study the relations and trade-offs among the goodness of a grasp, geometry of the grasped object, number of fingers and the computational complexity of the grasp-synthesis algorithms. The results here employ the techniques from convexity theory first introduced by the author and his colleagues [14,21].

1 Introduction

In robotics, the hand models usually consist of a small number (four or five) of articulated fingers, possibly with a palmar surface. A variety of finger arrangements have been suggested: for example, an anthropomorphic arrangement consisting of an opposable 'thumb' and several more fingers arranged to work cooperatively.

Many problems in *dextrous manipulation* involving such a hand model has been studied by mapping an object to be manipulated into a low dimensional hypersurface in a higher-dimensional wrench space by the so-called "*wrench map* [21]." By studying the convexity geometric properties of the image of this wrench map, one can answer several interesting questions in this field.

- Lower and upper bounds on the number of fingers to grasp an object under a variety of models [18,21]. The arguments employ various Helly-type theorems: namely, Carathéodory's theorem [5] and Steinitz's theorem [30].

- Properties of *force* and *form closures* and a related geometric notion of *immobility*. Relation between force and form closures [18,22,23].

- Characterization of ungraspable ("exceptional") objects [21].

- Linear time (thus, optimal) grasp synthesis algorithms [21].

- Grasp control: updating the finger forces in $O(1)$ time in response to a time-varying external wrench [18].

- Study of the strength (also called, efficiency) of a grasp [9,14]. The algorithms and the bounds are based on a quantitative version of the Steinitz's theorem. Also, see [2]

- Analysis and planning algorithms for fixturing and workholding [3,19,20,35].

Several other approaches to these problems have emerged, contemporaneously or subsequently. Notable among these approaches are purely geometric techniques [4,10,16,25,31], topological techniques [12] and algebraic techniques [28,29]. However, while in certain instances these other techniques have proven to be more powerful, they seem to lack the depth and elegance of the convexity theoretic approach.

In this paper, we take a fresh look at the problem of formulating "*grasp metrics*" and their relation with the quantitative versions of Helly-type theorems. Parts of the results described here are based on some old results (jointly with Kirkpatrick and Yap) and some work in progress (jointly with Teichmann).

2 Terminology and Overview

Formally, we consider an idealized robot hand, consisting of several independently movable force-sensing fingers; this hand is used to grasp a rigid object B. Furthermore, we make the following simplifying assumptions:

- (*Smooth Body*) B is a full-bodied (i.e. no internal holes) compact subset of the Euclidean 3-space. Furthermore, B has a piece-wise smooth boundary ∂B.

- (*Point Contact*) For each finger-contact on the body, we may associate a nominal point of contact, $\mathbf{p} \in \partial B$. By convention, we pick $\mathbf{n}(\mathbf{p})$ to be the unit normal pointing into the interior of B.

  For each such point $\mathbf{p}$, we can define a wrench system $\{\Gamma^{(1)}(\mathbf{p}), \Gamma^{(2)}(\mathbf{p}), \ldots, \Gamma^{(\ell)}(\mathbf{p})\}$, $(0 \leq \ell \leq 6)$, where the number and screw-axes of the wrench system depend on the contact type. Some of these wrenches can be *bisense* (i.e. can act in either sense) and the remaining wrenches, *unisense*. (For a discussion of screw theory, and in particular, wrenches and twists, see [13] and [26]. Also, see the appendix.)

- (*Compliance*) We will consider the case when the fingers are stiff—the force/torques applied at the fingers are generated by some actuators whose mechanics need not concern us.

Many interesting special cases occur, depending on how we model the *static friction* and the *stiction* between the fingers and the body B. In the case, where the contacts are frictionless, a finger can only apply force $\mathbf{f}$ on the body in the direction $\mathbf{n}(\mathbf{p})$ at the point $\mathbf{p}$. Also if the fingers are non-sticky, then the force $\mathbf{f}$ has a non-negative magnitude, $f = \mathbf{f} \cdot \mathbf{n}(\mathbf{p}) \geq 0$. Such grips are also known as 'positive grips'. In this case, the wrench system associated with each point is:

$$\Gamma(\mathbf{p}) = \{[\mathbf{n}(\mathbf{p}), \mathbf{p} \times \mathbf{n}(\mathbf{p})]\}$$

Thus, corresponding to a set of finger-contacts, we have a system of n wrenches,

$$\{\mathbf{w}_1, \ldots, \mathbf{w}_k, \mathbf{w}_{k+1}, \ldots, \mathbf{w}_n\},$$

the first k of which are bisense and the remaining last $n - k$ of the wrenches are unisense. Let us assume that the magnitudes of these wrenches are given by the scalars f_i's

$$\{f_1, \ldots, f_k, f_{k+1}, \ldots, f_n\},$$

where $f_1, \ldots, f_k \in \mathbb{R}$ and $f_{k+1}, \ldots, f_n \in \mathbb{R}_{\geq 0}$, and not all the magnitudes are zero. We call such a system of wrenches and the wrench-magnitudes, a *grip*, $\mathcal{G}$, and say that this grip $\mathcal{G}$ generates an external wrench $\mathbf{w} = [F_x, F_y, F_z, \tau_x, \tau_y, \tau_z] \in \mathbb{R}^6$, if

$$\mathbf{w} = \sum_{i=1}^{n} f_i \, \mathbf{w}_i.$$

In matrix notation, the above equation is expressed as

$$\mathbf{w} = \mathcal{W} \begin{bmatrix} f_1 \\ f_2 \\ \vdots \\ f_n \end{bmatrix},$$

where $\mathcal{W}$ is a $6 \times n$ matrix whose columns are the corresponding n wrenches of the system $\{\mathbf{w}_1, \ldots, \mathbf{w}_k, \mathbf{w}_{k+1}, \ldots, \mathbf{w}_n\}$, associated with the contact points of the grip. The matrix $\mathcal{W}$ is called a *grip matrix* of the grip defining the system of wrenches

$$\{\mathbf{w}_1, \ldots, \mathbf{w}_k, \mathbf{w}_{k+1}, \ldots, \mathbf{w}_n\}.$$

2.1 Closure Grasp

Next, we consider the concept of a *closure grasp*: A system of wrenches $\mathbf{w}_1, \ldots, \mathbf{w}_n$ (as before) is said to constitute a *force/torque closure grasp* if and only if any arbitrary external wrench can be generated by varying the magnitudes of the wrenches (subject to the constraints imposed by the senses of the wrenches). A necessary and sufficient condition for a closure grasp is that the (module) sum of the linear space spanned by the vectors $\mathbf{w}_1, \ldots, \mathbf{w}_k$ and the positive space spanned by the vectors $\mathbf{w}_{k+1}, \ldots, \mathbf{w}_n$ is the entire $\mathbb{R}^6$:

$$\text{lin}\,(\mathbf{w}_1, \ldots, \mathbf{w}_k) + \text{pos}\,(\mathbf{w}_{k+1}, \ldots, \mathbf{w}_n) = \mathbb{R}^6.$$

Let us denote, by L, the linear space $\text{lin}\,(\mathbf{w}_1, \ldots, \mathbf{w}_k)$, and, by $L^\perp$, the orthogonal complement of L in $\mathbb{R}^6$. Let π be the linear projection function of $\mathbb{R}^6$ onto $L^\perp$ whose kernel is L. Then it can be shown that a necessary and sufficient condition for a closure grasp is

$$\text{lin}\,(\mathbf{w}_1, \ldots, \mathbf{w}_k) + \text{pos}\,(\pi\mathbf{w}_{k+1}, \ldots, \pi\mathbf{w}_n) = \mathbb{R}^6.$$

The above equation in turn is equivalent to the following conditions:

$$\mathbf{0} \in \text{int conv}\,(\pi\mathbf{w}_{k+1}, \ldots, \pi\mathbf{w}_n)$$

in $L^\perp$. Here, if $k = 0$ (i.e. positive grip) then the above condition reduces to the following:

$$\mathbf{0} \in \text{int conv}\,(\mathbf{w}_1, \ldots, \mathbf{w}_n).$$

Let us assume that $\dim(L) = d$. Then there is a linear basis W of L

$$W = \{\mathbf{w}_{j_1}, \ldots, \mathbf{w}_{j_d}\} \subseteq \{\mathbf{w}_1, \ldots, \mathbf{w}_k\}$$

which when adjoined with a set of vectors W',

$$W' = \{\mathbf{w}_{j_{d+1}}, \ldots, \mathbf{w}_{j_6}\} \subseteq \{\mathbf{w}_{k+1}, \ldots, \mathbf{w}_n\}$$

yields $\widehat{W} = W \cup W'$, a linear basis of $\mathbb{R}^6$. Thus under the condition that we have a closure grasp, we can find $\mathbf{g} \in \mathbb{R}^k \times \mathbb{R}_{\geq 0}^{n-k}$ such that

$$-(\mathbf{w}_{k+1} + \cdots + \mathbf{w}_n) = \sum_{i=1}^{n} g_i \, \mathbf{w}_i,$$

and thus

$$\sum_{i=1}^{n} f_{h,i} \, \mathbf{w}_i, = \mathbf{0},$$

$$\text{where } \mathbf{f}_h \in \mathbb{R}^k \times \mathbb{R}_{\geq 0}^{n-k},$$
$$\text{and } f_{h,k+1} > 0, \ldots, f_{h,n} > 0.$$

In other words,

$$\mathbf{f}_h \in \text{null}(\mathcal{W}) \cap \mathbb{R}^k \times \mathbb{R}_{>0}^{n-k};$$

i.e., $\mathbf{f}_h$ is a null vector of the grip matrix and all its unisense components are strictly positive.

Now any external wrench $\mathbf{w}$ can be expressed as a linear combination of the vectors in the basis $\widehat{W}$. Thus there is a vector $\mathbf{f}_p \in \mathbb{R}^n$, whose non-zero entries are in the positions $j_1, \ldots, j_6$, and

$$\mathbf{w} = \sum_{k=1}^{6} f_{p,j_k} \, \mathbf{w}_{j_k} = \sum_{i=1}^{n} f_{p,i} \, \mathbf{w}_i.$$

Now consider a vector $\mathbf{f} = \mathbf{f}_p + \lambda \, \mathbf{f}_h \in \mathbb{R}^k \times \mathbb{R}_{\geq 0}^{n-k}$, where λ is chosen to be of a sufficiently large positive value that ensures that the negative components in $\mathbf{f}_p$ are dominated by the positive components of $\mathbf{f}_h$. Thus,

$$\mathbf{w} = \sum_{i=1}^{n} f_i \, \mathbf{w}_i = \sum_{i=1}^{n} (f_{p,i} + \lambda \cdot f_{h,i}) \, \mathbf{w}_i.$$

These arguments yield a simple algorithm to find *at least* one set of force targets that can generate a given external wrench. Also, as the external wrench is varied in the course of a manipulation task, a slight variation of this algorithm updates the force targets in $O(1)$ time.

Observe that the above formulation has turned a problem in mechanics into a combinatorial geometric problem, now amenable to many interesting techniques in convexity theory and computational and combinatorial geometry.

2.2 Grasp Metric: Desiderata

Given a grasp $\mathcal{G}$ described by a wrench system

$$\{\mathbf{w}_1, \ldots, \mathbf{w}_n\},$$

we would frequently like to be able to say how good this grasp is as compared to another grasp producing a different wrench system. Clearly, such a "measure of goodness" must possess some physical intuitions that correspond to how we normally view a grasp—e.g., a closure grasp should be preferable to a non-closure grasp (i.e., immobile grasp) or a force/torque closure grasp should be better than one achieving only force closure but not torque closure, etc. However, the situation becomes more complicated as we begin to compare two grasps that are both force-torque closure—here, we must rely on the underlying physics. Independent of the physical considerations, however, it seems appropriate that the grasp metrics possess some simple mathematical properties enumerated below.

In what follows, we shall refer to the value of a grasp under a grasp metric by the terms *quality*, *strength* and/or *efficiency*, in view of the nonstandard and often confusing arrays of terminologies that have appeared in the literature, and denote such a metric by r with a distinguishing subscript. By abuse of notation, sometimes we will call a "measure" a grasp metric, even if it does not satisfy the following properties. In this paper, a grasp metric will yield a positive scalar real value and the grasps will be totally ordered, although, we foresee situations where a grasp metric may be given by a higher dimensional vector (e.g., a pair) with some partial ordering on the grasps. The generalizations are straightforward. The grasp metric can also be generalized to include certain other parameters (e.g, friction, stiction, tasks under considerations, scaling in various force/torque dimensions, hand geometry, some universality criteria, etc.). We simply avoid these added complications at this point.

Let the grasp, $\mathcal{G}$, be given by the wrench system

$$\{\mathbf{w}_1, \ldots, \mathbf{w}_n\},$$

and let the associated grasp strength be given by

$$r = r(\mathcal{G}) = r(\mathbf{w}_1, \ldots, \mathbf{w}_n).$$

Then the following properties hold.

1. **Positivity:** For any grasp $\mathcal{G}$ with the wrench system
$$\{\mathbf{w}_1, \ldots, \mathbf{w}_n\},$$
$r = r(\mathcal{G}) \geq 0$, the inequality being strict if and only if $\mathcal{G}$ is a force/torque closure grasp.

2. **Boundedness:** If the cardinality of the wrench system, $n < \infty$ is bounded (i.e., the number of contact points for $\mathcal{G}$ is bounded), then so is $r = r(\mathcal{G}) < \infty$. This condition is weaker than the corresponding condition that requires r to be bounded, independent of n.

3. **Subadditivity:** For any two grasps $\mathcal{G}_1$ and $\mathcal{G}_2$, with the wrench systems:
$$\{\mathbf{w}_1, \ldots, \mathbf{w}_m\},$$
and
$$\{\mathbf{w}_{m+1}, \ldots, \mathbf{w}_n\},$$
respectively, we have
$$r(\mathbf{w}_1, \ldots, \mathbf{w}_m, \mathbf{w}_{m+1}, \ldots, \mathbf{w}_n)$$
$$\geq r(\lambda \mathbf{w}_1, \ldots, \lambda \mathbf{w}_m)$$
$$+ r((1-\lambda)\mathbf{w}_{m+1}, \ldots, (1-\lambda)\mathbf{w}_n),$$
where $0 \leq \lambda \leq 1$.

An immediate consequence of the above axioms is that, for a given object and a given hand with fixed number of fingers (with associated contact types), if the object allows a force/torque closure grasp then there is an optimal grasp of the object with that hand with a grasp quality r^*, when the grasp metric satisfies the first two properties.

Additional consequence of the above axioms are as follows:

1. **Scaling:** For any $\lambda > 1$,
$$r(\lambda \mathbf{w}_1, \ldots, \lambda \mathbf{w}_n) \geq r(\mathbf{w}_1, \ldots, \mathbf{w}_n).$$
However, many of our grasp metrics can be shown to actually satisfy the equality condition under scaling.

2. **Monotonicity:**
$$r(\mathbf{w}_1, \ldots, \mathbf{w}_n, \mathbf{w}_{n+1}) \geq r(\mathbf{w}_1, \ldots, \mathbf{w}_n).$$
Also, note that
$$r(\mathbf{w}_1, \ldots, \mathbf{w}_m, \mathbf{w}_{m+1}, \ldots, \mathbf{w}_n)$$
$$\geq \max(r(\mathbf{w}_1, \ldots, \mathbf{w}_m), r(\mathbf{w}_{m+1}, \ldots, \mathbf{w}_n)).$$
Thus the last condition tells us that a hand with large number of fingers is better than another with smaller number of fingers, assuming that they allow same contact types.

2.3 Grasp Metrics Based on Resistable Wrench

Note that in the above descriptions of closure grasps, we have made an implicit unrealistic assumption that *the magnitudes of finger forces are no way constrained.* In particular, it is quite likely that a force/torque closure grasp may resist any arbitrary external wrench; but it may only do so by applying an unrealistically large force at a finger in response to a fairly small external wrench in some direction.

In order to alleviate this problem, we may assume that certain additional constraint is imposed on the magnitudes of the finger forces—the *"finger force constraint"* being expressible as

$$\chi \quad : \quad \mathbb{R}^n \to \{0, 1\}$$
$$: \quad (f_1, f_2, \ldots, f_n) \mapsto \begin{cases} 1, & \text{if the} \\ & \text{``constraint''} \\ & \text{holds;} \\ 0, & \text{otherwise.} \end{cases}$$

The characteristic function naturally defines the set

$$S_\chi \quad = \quad \Big\{ (f_1, f_2, \ldots, f_n) \in \mathbb{R}^n : $$
$$\chi(f_1, f_2, \ldots, f_n) = 1 \Big\}$$
$$\subseteq \quad \mathbb{R}^n.$$

We say that χ (or equivalently, S_χ) is *faithful* if
$$\mathbb{R}^k \times \mathbb{R}^{n-k}_{\geq 0} \subseteq \text{pos}(S_\chi),$$
and that χ is *strongly faithful* if
$$\mathbb{R}^k \times \mathbb{R}^{n-k}_{\geq 0} = \text{pos}(S_\chi),$$

Thus the set of external wrenches that can be generated by the grasp, subject to the finger force constraint, χ, is given by G_χ, called the *"feasible wrench set:"*

$$G_\chi(\mathbf{w}_1, \ldots, \mathbf{w}_n)$$
$$= \quad \Big\{ \mathbf{w} = \sum_{i=1}^n f_i \mathbf{w}_i : \chi(f_1, f_2, \ldots, f_n) = 1 \Big\}$$
$$\subseteq \quad \mathbb{R}^6.$$

We also use the notation $R_\chi = -G_\chi$ to denote the *"resistable wrench set,"* the set of external wrenches that can be resisted by the grasp.

Note that

- If S_χ is convex, closed and compact, then so is G_χ.

- If $\{\mathbf{w}_1, \ldots, \mathbf{w}_n\}$ forms a force/torque closure grasp and if χ is faithful then

$$\mathbf{0} \in \text{int } G_\chi(\mathbf{w}_1, \ldots, \mathbf{w}_n).$$

- If $S_{\chi_1} \subset S_{\chi_2}$ then $G_{\chi_1} \subset G_{\chi_2}$.

Some natural finger force constraints that one may impose are of the following kinds:

- **Convex Constraint**:

$$\chi_{con} \quad : \quad f_{k+1} \geq 0, \ldots, f_n \geq 0$$
$$\text{and } \sum_{i=1}^{n} |f_i| \leq 1.$$

Note that χ_{con} is convex, closed, compact and strongly faithful.

$G_{\chi_{con}}$ is given by the convex hull of the vectors $\mathbf{w}_1, -\mathbf{w}_1, \ldots, \mathbf{w}_k, -\mathbf{w}_k, \mathbf{w}_{k+1}, \ldots, \mathbf{w}_n$:

$$G_{\chi_{con}} = \text{conv}\Big(\{\mathbf{w}_i, -\mathbf{w}_i : 1 \leq i \leq k\}$$
$$\cup \{\mathbf{w}_i : k+1 \leq i \leq n\}\Big).$$

- **Max Constraint**:

$$\chi_{max} \quad : \quad f_{k+1} \geq 0, \ldots, f_n \geq 0$$
$$\text{and } \max_{i \in \{1, \ldots, n\}} |f_i| \leq 1.$$

Clearly, χ_{max} is convex, closed, compact and strongly faithful.

$G_{\chi_{max}}$ is given by the Minkowski sum of the vectors $\mathbf{w}_1, -\mathbf{w}_1, \ldots, \mathbf{w}_k, -\mathbf{w}_k, \mathbf{w}_{k+1}, \ldots, \mathbf{w}_n$:

$$G_{\chi_{max}} = \bigoplus\Big(\{\mathbf{w}_i, -\mathbf{w}_i : 1 \leq i \leq k\}$$
$$\cup \{\mathbf{w}_i : k+1 \leq i \leq n\}\Big).$$

- **Hybrid Constraint**:
Let $P_1, P_2, \ldots, P_l$ be a partition of the indices $\{1, \ldots, n\}$. Then

$$\chi_{hyb} \quad : \quad f_{k+1} \geq 0, \ldots, f_n \geq 0$$
$$\text{and } \sum_{i \in P_j} |f_i| \leq 1, \quad 1 \leq j \leq l.$$

Again, χ_{hyb} is convex, closed, compact and strongly faithful.

$G_{\chi_{hyb}}$ is given by the Minkowski sum of the convex hulls of the vectors corresponding to each partition P_j:

$$G_{\chi_{hyb}} = \bigoplus_{j=1}^{l} \text{conv}\Big(\{\mathbf{w}_i, -\mathbf{w}_i : i \in P_j'\}$$
$$\cup \{\mathbf{w}_i : i \in P_j''\}\Big),$$

where

$$P_j' = P_j \cap \{1, \ldots, k\},$$

and

$$P_j'' = P_j \cap \{k+1, \ldots, n\}.$$

Now consider the largest ball of radius $r = r_\chi(\mathbf{w}_1, \ldots, \mathbf{w}_n)$ in $\mathbb{R}^6$ centered at $\mathbf{0}$ and contained in the corresponding feasible wrench set, $G_\chi(\mathbf{w}_1, \ldots, \mathbf{w}_n)$. We shall refer to this r as the "*residual radius*" of G_χ. Then it is trivial to see that there exists an external wrench of magnitude only infinitesimally larger than r that cannot be generated or resisted by the grasp under consideration, if it must respect the finger force constraint χ. This value r may thus be used to define a grasp metric.

Note that since

$$S_{\chi_{con}} \subseteq S_{\chi_{hyb}} \subseteq S_{\chi_{max}} \subseteq n \, S_{\chi_{con}},$$

we have

$$G_{\chi_{con}} \subseteq G_{\chi_{hyb}} \subseteq G_{\chi_{max}} \subseteq n \, G_{\chi_{con}}$$

and

$$r_{\chi_{con}} \leq r_{\chi_{hyb}} \leq r_{\chi_{max}} \leq n \, r_{\chi_{con}}.$$

Note that, since the underlying geometric problem remains largely unchanged irrespective of the finger force constraint chosen, we shall often focus only on the simplest situation represented by the constraint χ_{con}.

2.4 Grasp Metrics: Variations

(1) One may consider a finger force constraint of the following kind:

$$\tilde{\chi}_2 : \sum_{i=1}^{n} (f_i)^2 \leq 1.$$

Note that $\tilde{\chi}_2$ is convex, closed, compact and faithful, but not strongly faithful, as $S_{\tilde{\chi}_2} = \mathcal{B}^n$, the n-dimensional ball, and

$$\text{pos}(S_{\tilde{\chi}_2}) = \mathbb{R}^n.$$

Then $G_{\tilde{\chi}_2}$ is the image of the n-dimensional ball, $\mathcal{B}^n$, under the linear map defined by the grip matrix $\mathcal{W}$, and thus a 6-dimensional ellipsoid. Since the lengths of the principal axes of the resulting ellipsoid are given by the singular values of the grip matrix $\mathcal{W}$ (i.e., the nonnegative square roots of the eigenvalues of the real positive definite square matrix $\mathcal{W}^T \mathcal{W}$), its residual radius is given by the smallest singular value of the grip matrix $\mathcal{W}$, $r_{\tilde{\chi}_2} = \sigma_{min}(\mathcal{W}) > 0$. Note that, since $\tilde{\chi}_2$ is faithful but not strongly faithful, this grasp metric may be highly misleading. However, note that

$$r_{\chi_2} < r_{\tilde{\chi}_2},$$

where χ_2 is a corresponding strongly faithful constraint of the following form:

$$\chi_2 \quad : \quad f_{k+1} \geq 0, \ldots, f_n \geq 0 \text{ and}$$
$$\text{and} \sum_{i=1}^{n} (f_i)^2 \leq 1.$$

Note, however that r_{χ_2} can be arbitrarily small in relation to $r_{\tilde{\chi}_2}$.

(2) Another grasp metric was suggested by Jeff Trinkle [33]. This metric will be denoted here by r_{null} and the motivations for it are described below. Note that earlier we had observed that, given a grasp with the corresponding system of n wrenches,

$$\{\mathbf{w}_1, \ldots, \mathbf{w}_k, \mathbf{w}_{k+1}, \ldots, \mathbf{w}_n\},$$

that satisfy the closure condition, we always have a nontrivial null vector $\mathbf{f}_h$ of the grip matrix $\mathcal{W}$, $\mathbf{f}_h = \{f_{h,1}, \ldots, f_{h,k}, f_{h,k+1}, \ldots, f_{h,n}\} \in \mathbb{R}^k \times \mathbb{R}^{n-k}_{>0}$ (all unisense components are strictly positive) such that

$$0 = \sum_{i=1}^{n} f_{h,i} \mathbf{w}_i = \mathcal{W} \begin{bmatrix} f_{h,1} \\ f_{h,2} \\ \vdots \\ f_{h,n} \end{bmatrix} = \begin{bmatrix} 0 \\ 0 \\ \vdots \\ 0 \end{bmatrix}.$$

Following Trinkle's suggestion, we now consider the grasp metric defined as follows:

$$r_{null} = \max_{\mathbf{f} \in \text{null}(\mathcal{W})} \left\{ \min_{k+1 \leq i \leq n} f_i : \right.$$
$$\sum_{i=1}^{n} f_i \mathbf{w}_i = 0$$
$$\left. \& \quad \mathbf{f} \in \mathbb{R}^k \times \mathbb{R}^{n-k}_{>0} \right\}.$$

The quality r_{null} can be computed efficiently by the following linear programming formulation:

- GIVEN: A $6 \times n$ grip matrix $\mathcal{W}$ and a linear, compact, closed, convex and strongly faithful finger force constraint condition χ.
- SOLVE:

$$\begin{aligned} \text{maximize} \quad & \lambda \\ \text{subject to} \quad \mathcal{W}\mathbf{f} &= \mathbf{0} \\ -\lambda &\leq 0 \\ \lambda - f_i &\leq 0, \\ & (k+1 \leq i \leq n) \\ \chi(\mathbf{f}) &= 1, \end{aligned}$$

where $\mathbf{f} = (f_1, \ldots, f_n)^T$ and λ is a real number.

For instance, if the finger force constraint χ is χ_{con}, then the last condition is simply

$$f_{k+1} \geq 0, \ldots, f_n \geq 0,$$

and

$$|f_1| + \cdots + |f_k| + |f_{k+1}| + \cdots + |f_n| \leq 1.$$

For a closure grasp there is a positive λ satisfying the feasibility conditions, and thus, since the feasible set is bounded, there is an optimal solution: $r_{null} = \lambda^*$, $\mathbf{f}_h^* = (f_{h,1}^*, \ldots, f_{h,k}^*, f_{h,k+1}^*, \ldots, f_{h,n}^*)$.

Note that r_{null} satisfies the positivity and boundedness conditions but not the monotonicity (thus, subadditivity) property. That is adding a new finger can actually make an existing grasp worse! For instance, in the Figure 1, only considering the force-closure, it is easily seen that r_{null} for both the grasps are same (i.e., the largest possible value $1/4$).

Here, we shall compute certain relations that explicitly exhibit certain problems with the grasp metric r_{null}. Recall that, we have a subset $\widehat{W} \subset \{\mathbf{w}_1, \ldots, \mathbf{w}_n\}$, linearly spanning the wrench space $\mathbb{R}^6$. Among all such $\widehat{W}$'s, let $\widehat{W}^*$ be a basis of $\mathbb{R}^6$ that is *maximally orthonormal* in the sense that it maximizes the following positive real-valued function:

$$d(\widehat{W}) = \left(\frac{|\det \widehat{W}| \, (\min_{\mathbf{w} \in \widehat{W}} |\mathbf{w}|)}{\prod_{\mathbf{w} \in \widehat{W}} |\mathbf{w}|} \right),$$

where $\det \widehat{W}$ is the determinant of a 6×6 square matrix whose columns are the vectors of $\widehat{W}$. Note that by Hadamard inequality, we have

$$0 \leq d(\widehat{W}^*) \leq \min_{\mathbf{w} \in \widehat{W}} |\mathbf{w}|.$$

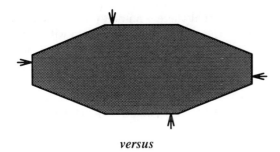

versus

Both grasps are equally good under r_{null}

Figure 1: *The grasp metric r_{null} does not distinguish between the above two grasps.*

We show that

$$r_{\chi_{con}} \geq d(\widehat{W}^*) \left(\frac{r_{null}}{1 + 6r_{null}} \right),$$

where r_{null} is computed with the linear condition χ_{con}.

We proceed as follows: Consider an external wrench $\mathbf{w}$ oriented arbitrarily but of magnitude

$$|\mathbf{w}| = d(\widehat{W}^*) \left(\frac{r_{null}}{1 + 6r_{null}} \right).$$

It suffices to show that $\mathbf{w}$ can be generated by the wrenches in the system of wrenches associated with the grasp, subject to the finger force constraint. Let $\widetilde{\mathbf{w}} = (1 + 6r_{null})\mathbf{w}$. Clearly, $\widetilde{\mathbf{w}}$ can be expressed as a linear combination of the vectors in $\widehat{W}^*$ as follows:

$$\widetilde{\mathbf{w}} = \sum_{k=1}^{6} f_{p,j_k} \mathbf{w}_{j_k},$$

such that by Cramer's rule,

$$f_{p,j_k} = \frac{\det \left[\mathbf{w}_{j_1}, \ldots, \mathbf{w}_{j_{k-1}}, \widetilde{\mathbf{w}}, \mathbf{w}_{j_{k+1}}, \ldots, \mathbf{w}_{j_6} \right]}{\det \widehat{W}^*},$$

and

$$|f_{p,j_k}| \leq \frac{|\mathbf{w}_{j_1}| \cdots |\mathbf{w}_{j_{k-1}}| \, |\widetilde{\mathbf{w}}| \, |\mathbf{w}_{j_{k+1}}| \, \cdots |\mathbf{w}_{j_6}|}{|\det \widehat{W}^*|}.$$

$$= r_{null} \frac{\min_{\mathbf{w} \in \widehat{W}^*} |\mathbf{w}|}{|\mathbf{w}_{j_k}|}$$

$$\leq r_{null}.$$

Thus, we can express $\widetilde{\mathbf{w}}$ as

$$\widetilde{\mathbf{w}} = \sum_{k=1}^{6} f_{p,j_k} \mathbf{w}_{j_k} + \sum_{i=1}^{n} f_{h,i}^* \mathbf{w_i}$$

$$= \sum_{i=1}^{n} f_i \mathbf{w_i},$$

where $f_{k+1} \geq 0, \ldots, f_n \geq 0$, since $|f_{h,i}^*| \geq r_{null}$ and $|f_{p,j_k}| \leq r_{null}$. However,

$$\sum |f_i| \leq 1 + 6r_{null},$$

since $\sum |f_{h,i}^*| \leq 1$ and $\sum |f_{p,j_k}| \leq 6r_{null}$. Thus by scaling, we have

$$\mathbf{w} = \frac{\widetilde{\mathbf{w}}}{1 + 6r_{null}} = \sum_{i=1}^{n} \frac{f_i}{1 + 6r_{null}} \mathbf{w_i},$$

satisfying all the necessary conditions for χ_{con}. Thus

$$r_{\chi_{con}} \geq |\mathbf{w}| = d(\widehat{W}^*) \left(\frac{r_{null}}{1 + 6r_{null}} \right).$$

However, in general, a large value for r_{null} does not imply a good value for $r_{\chi_{con}}$, if we do not have a good value for $d(\widehat{W}^*)$. For instance, for any dimension d (e.g., $d = 6$), there is a positive $\epsilon_d > 0$, such that for any $0 < \epsilon < \epsilon_d$, we can always find $d + 1$ unit vectors, $\{\mathbf{w}_0, \mathbf{w}_1, \ldots, \mathbf{w}_d\}$, such that the corresponding $r_{\chi_{con}} = \epsilon$, but $r_{null} > 1/(2d)$. Choose $\mathbf{w}_0$ arbitrarily, and place the remaining d vectors closely in a cluster such that their centroid is on the ray $\lambda(-\mathbf{w}_0)$ (where $\lambda > 0$) and the d simplex contains a residual d-ball of radius ϵ—this can always be accomplished. Then there is a small $0 < \delta < 1 - 1/d$ such that

$$\left(\frac{1 - \delta}{2} \right) \mathbf{w}_0 + \left(\frac{1 + \delta}{2d} \right) (\mathbf{w}_1 + \cdots + \mathbf{w}_d) = \mathbf{0}.$$

Conversely, for any dimension d, and any sufficiently small $\epsilon > 0$, we can find $n = \lceil 1/\epsilon \rceil$ d-dimensional unit vectors $\{\mathbf{w}_1, \ldots, \mathbf{w}_n\}$, such that $r_{null} \leq \epsilon$ but $r_{\chi_{con}} \geq 1/d$. Assume that $n = m(d + 1)$ is a multiple of $d + 1$. Choose $d + 1$ clusters of m unit vectors each and place each cluster closely about the vertices of a regular d-simplex Δ inscribed within the unit sphere. Since the convex hull of these vectors contains Δ, it has a residual radius of value no smaller than $1/d$, but since $r_{null} \leq 1/n$, r_{null} is arbitrarily small.

(3) Yet another variation on this theme may be obtained by considering some different "geometric object" inscribed in the feasible wrench set, G_χ. Easy variations can be obtained by either considering full-dimensional spheres with respect to different norms (i.e. L_1 or L_∞ norms, instead of L_2), ellipsoids (the parameters being dependent on the task that the hand is supposed to perform) or by considering lower dimensional spheres or ellipsoids (i.e., 3-spheres spanning only the force vectors, and/or 3-spheres spanning only the torque vectors.)

(4) Another special case arises as follows: Suppose we know that the grasp is required to resist a set of external wrenches, each of which can be expressed as a sum of a fixed external wrench $\overline{w}$ and an additional arbitrarily varying external wrench $\widetilde{w}$, whose magnitude and orientation are unknown. We wish to maximize the magnitude of this unknown component to the extent possible; the associated grasp metric is then given by this maximal value. It is easily seen that the grasp metric is given by the radius of the largest sphere centered at $-\overline{w}$ and contained in the feasible wrench set G_χ. An ubiquitous example of such a constant external wrench is given by the wrench generated by the weight of the object being grasped. Sometimes, we may further generalize this concept by requiring that the constant external wrenches are known only to the extent that they belong to a set $\overline{W}$. Then we wish to maximize the parameter r such that $\overline{W} \oplus r\mathcal{B} \subseteq G_\chi$, where $\overline{W} \oplus r\mathcal{B}$ is the Minkowski sum of the set $\overline{W}$ and a 6-dimensional ball $\mathcal{B}$ of radius r.

(5) One important special case is as follows: Recall that the map Γ applied to a point $p \in \partial B$ produces a system of wrenches in $\mathbb{R}^6$ that would result if a unit force is applied at the contact point p. Also recall that the map Γ is uniquely defined by the contact type and the point on the boundary. Let us now consider the set $\Gamma(\partial B) \subseteq S^3 \oplus \mathbb{R}^3$ (the unit cylinder). Let $\lambda \in \mathbb{R}_{\geq 0}$ be a maximal positive real number such that

$$\lambda \Gamma(\partial B) \subseteq G_\chi.$$

Then it is clear that there is a point $p \in \partial B$ such that if one "pushes" the object B at the point p with a "*nasty finger*" with a force of magnitude only infinitesimally larger than λ, such a finger will be able to break the grasp. Thus $r_{\chi_{nasty}} = \lambda$ defines a grasp metric.

2.5 Grasp Metrics Based on Virtual Coefficients

Yet another formulation of a closure grasp is via *form closure*. Recall that with each nominal point of contact we can also associate a twist system; they describe the degrees of freedom of the body local to that contact point. Thus a system with a set of contacts is free to move by a twist if and only if the virtual coefficient of any wrench and the twist is nonnegative, since otherwise the virtual work done by some wrench would be negative. This situation occurs when the twist d is *reciprocal* to the bisense wrenches

$$w_i \odot d = 0, \quad (\forall i = 1, \ldots, k)$$

and *reciprocal* or *repelling* to the unisense wrenches

$$w_i \odot d \geq 0, \quad (\forall i = k+1, \ldots, n).$$

A set of twists (associated with the contacts) is said to constitute a *form closure* if and only if any arbitrary twist is resisted by the set of contacts. That is, the object is *totally constrained* with no degree of freedom left. Thus if d is an arbitrary twist then it must be *non-reciprocal* to some bisense wrench w_i ($i = 1, \ldots, k$) or must be *contrary* to some unisense wrench w_i ($i = k+1, \ldots, n$):

$$w_i \odot d \;\neq\; 0, \quad (\exists i = 1, \ldots, k)$$
$$\text{or}$$
$$w_i \odot d \;<\; 0, \quad (\exists i = k+1, \ldots, n).$$

Put another way, this is equivalent to saying that, for any arbitrary vector $d' \in \mathbb{R}^6$, we have

$$w_i \cdot d' \;=\; 0, \quad (\forall i = 1, \ldots, k)$$
implies that
$$w_i \cdot d' = \pi w_i \cdot d' \;<\; 0, \quad (\exists i = k+1, \ldots, n),$$

which, in turn is equivalent to the condition that

$$0 \in \text{int conv } (\pi w_{k+1}, \ldots, \pi w_n)$$

Thus, *force/torque closure* and *form closure* are equivalent.

Also note that, one can use the definition of a form closure to define a grasp metric in terms of the virtual coefficients determined by the system of wrenches of the grasp and a unit twist. For instance, one may propose the following:

1. Let $\mathbf{d}$ be an arbitrary twist, $|\mathbf{d}|_2 = 1$. Then, define the *minimal virtual coefficient* of the grasp $\{\mathbf{w}_1, \ldots, \mathbf{w}_n\}$ with respect to $\mathbf{d}$ to be

$$\mu_{con}(\mathbf{d}) = \max_{i \in I} |\mathbf{w}_i \odot \mathbf{d}|$$

where $i \in I$ if and only if $\mathbf{w}_i$ is contrary to $\mathbf{d}$ $(1 \le i \le n)$, (i.e., $\mathbf{w}_i \odot \mathbf{d} < 0$). Note that $I \neq \emptyset$ if the grasp $\{\mathbf{w}_1, \ldots, \mathbf{w}_n\}$ is a closure grasp. Now define the grasp metric to be

$$\mu_{con} = \min\Big(\mu_{con}(\mathbf{d}) : \mathbf{d} \in \mathbb{R}^6 \text{ is a unit twist}\Big)$$

This represents the smallest amount of virtual work an adversary may have to perform to "break the grasp." A little thought will show that this is exactly the grasp metric defined by the residual radius $r_{\chi_{con}}$ of the set $G_{\chi_{con}}$.

2. As earlier, let $\mathbf{d}$ be an arbitrary twist, $|\mathbf{d}|_2 = 1$. Then, define the *minimal virtual coefficient* of the grasp $\{\mathbf{w}_1, \ldots, \mathbf{w}_n\}$ with respect to $\mathbf{d}$ to be

$$\mu_{max}(\mathbf{d}) = \sum_{i \in I} |\mathbf{w}_i \odot \mathbf{d}|$$

where $i \in I$ if and only if $\mathbf{w}_i$ is non-reciprocal to $\mathbf{d}$ $(1 \le i \le k)$ or contrary to $\mathbf{d}$ $(k+1 \le i \le n)$, (i.e., $\mathbf{w}_i \odot \mathbf{d} \neq 0$ or < 0, depending respectively on whether $1 \le i \le k$ or $k+1 \le i \le n$). Now define the grasp metric to be

$$\mu_{max} = \min\Big(\mu_{max}(\mathbf{d}) : \mathbf{d} \in \mathbb{R}^6 \text{ is a unit twist}\Big)$$

One sees that this is exactly the grasp metric defined by the residual radius $r_{\chi_{max}}$ of the set $G_{\chi_{max}}$.

One can also easily devise a grasp metric by means of virtual coefficients that corresponds exactly to one's favorite finger force constraint.

2.6 Some Remarks

We note that most of the grasp metrics considered are dependent on the coordinate system chosen—namely, on the choice of the torque origin. This can be addressed by either asking that the torque origin is always chosen at the *center of mass* of the object, or by considering different measure of the feasible wrench set, e.g., volume. But such solutions seem ad hoc and without an immediate physical interpretation. Another problem is that the torque and force dimensions are not comparable. The scalings chosen in either dimension is clearly artificial, but do affect the grasp metric. A simple solution is to leave the two dimensions separate and define the grasp metrics by a pair of numbers. While we avoid these issues for the time being, we hope to come back to these problems in the future.

2.7 Synthesis of a Grasp

Let us next consider the problem of grasp synthesis with total disregard for the condition of optimality and in the simplest possible situation, where there is no friction at the contact points—the so-called "positive grip."

In order to obtain a particular grasp on an object, it must be determined if that grasp is achievable. It is for this reason, researchers have studied the question of how many fingers (wrenches) are required to obtain certain grasps on the object.

Reuleaux and Somoff determined that the closure grasp of a two dimensional object requires at least *four* wrenches and of a three dimensional object requires at least *seven*, where the wrenches are normal to the surface of the object.

Mishra, Schwartz and Sharir [21] gave general bounds on the number of fingers in the case of a *positive grip*; they also provided an algorithm that finds *at least one* such grip on a polyhedral object and their algorithm runs in time linear in the number of faces of the object. Here, we briefly describe the techniques of Mishra, Schwartz, and Sharir.

Recall that a non-frictional grip is called a *positive grip*. Note that, in this case, the fingers are assumed to be point fingers, a finger can only apply a force on the object along the surface-normal at the point of contact, directed inward. In this situation, we have a *wrench*

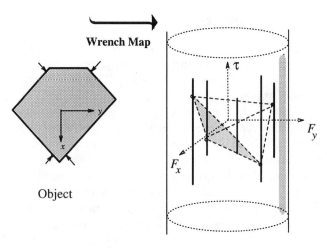

Wrench Map

Object

Its Image under the Wrench Map

Figure 2: *A pictorial explanation of the techniques of Mishra, Schwartz and Sharir.*

map, Γ, mapping ∂B into the six-dimensional *wrench space* $\mathbb{R}^6$ as follows:

$$\Gamma: \partial B \to \mathbb{R}^6 \; : \mathbf{p} \mapsto [\mathbf{n}(\mathbf{p}), \mathbf{p} \times \mathbf{n}(\mathbf{p})].$$

Essentially, Γ maps $\mathbf{p}$ to the point $\Gamma(\mathbf{p})$ ($\in S^3 \oplus \mathbb{R}^3$, a unit radius cylinder) in the wrench space that represents the effects of applying a unit force at $\mathbf{p}$ in the direction $\mathbf{n}(\mathbf{p})$.

Now, it can be shown that

Theorem 1 *Given an arbitrary compact rigid object B whose piece-wise smooth surface ∂B is not a "surface of revolution," B can be held with a closure grasp by a positive grip of at most twelve fingers.* □

The assumption that ∂B is not a surface of revolution is essential, since, in the absence of friction, no matter how many fingers are employed the object can always be rotated about its axis of symmetry. A general description of such exceptional objects is given in [21]. Also, see [29].

By virtue of the discussion of the previous section, we only need to show that

$$\Big(\exists \, \mathbf{p}_1, \dots, \mathbf{p}_m \in \partial B, \; m \leq 12 \Big)$$

$$\Big[\mathbf{0} \in \text{int conv} \Big(\Gamma(\mathbf{p}_1), \dots, \Gamma(\mathbf{p}_m) \Big) \Big].$$

The proof is in two steps:

Step 1. $\Rightarrow$ We claim that

$$\mathbf{0} \in \text{int conv} \, \Gamma(\partial B).$$

Using Gauss' divergence theorem, we see that

$$\left[\int_{\partial B} \mathbf{n}(\mathbf{p}) \, dS, \int_{\partial B} (\mathbf{p} \times \mathbf{n}(\mathbf{p})) \, dS \right] = \mathbf{0},$$

i.e.

$$\mathbf{0} \in \text{conv} \, \Gamma(\partial B).$$

Thus, it remains to be shown that the origin is indeed an interior point of the convex hull; assume to the contrary. Then it can be shown that there is a nonzero vector $\mathbf{g} = [F, \tau] \in \mathbb{R}^6$ orthogonal to the linear subspace spanned by $\Gamma(\partial B)$. Thus, for each function $c(\mathbf{p})$ we have

$$\int_{\partial B} c(\mathbf{p}) \, [(F + \tau \times \mathbf{p}) \cdot \mathbf{n}(\mathbf{p})] \, dS = 0.$$

In particular, substitute the function $c(\mathbf{p}) = (F + \tau \times \mathbf{p}) \cdot \mathbf{n}(\mathbf{p})$ into the last equation, to deduce that $(F + \tau \times \mathbf{p}) \cdot \mathbf{n}(\mathbf{p})$ must be identically 0 over ∂B. But, this is possible only in the case when the surface of B is a surface of revolution.

Step 2. $\Rightarrow$ The rest follows by an application of the following theorem from combinatorial geometry:

Theorem 2 (Steinitz's Theorem[30]) *If $X \subseteq \mathbb{R}^d$ and $\mathbf{p} \in \text{int conv} \, X$, then $\mathbf{p} \in \text{int conv} \, Y$ for some $Y \subseteq X$ with $|Y| \leq 2d$.* □

Our main result then follows from Steinitz's Theorem, once we identify X with $\Gamma(\partial B)$, and Y with the set $\{\Gamma(\mathbf{p}_1), \dots, \Gamma(\mathbf{p}_m)\}$, with $m \leq 2 \times 6 = 12$.

Note that here we see how to pick one grasp. There is a simple linear time algorithm to find such a grasp with 12 fingers. An interesting related algorithmic question is to understand the complexity of the problem of choosing an optimal grasp (say, with a fixed number $m \geq 12$ of fingers) such that

$$G_\chi(\Gamma(\mathbf{p}_1), \dots, \Gamma(\mathbf{p}_m))$$

contains as a large a residual ball as possible, under one's favorite faithful, convex and compact finger force constraint. This leads us to the study of the Quantitative Steinitz's Theorem (Q.S.T.).

An argument as above for one- or two-dimensional objects yields a theory and results with appropriate changes in the dimension of the wrench space (one and three, respectively, instead of six) and the number of fingers sufficient for positive closure grasps (two and six, respectively, instead of twelve). Also note that the same line of reasoning leads to a much more tight calculation of number of fingers necessary and sufficient for equilibrium grasps (two for 1-dimensional objects, four for 2-dimensional objects and seven for 3-dimensional objects). Other results for other contact types may be obtained in a similar manner.

3 Q.S.T.

Following the discussion of the earlier sections, we see that the selection of an optimal grasp, with respect to any of the grasp metrics of choice, leads to the study of a stronger *quantitative version* of the Steinitz's theorem. We start with few notations, as follows.

For any convex set $X \subseteq \mathbb{R}^d$, let the *residual ball* of X refer to the maximal ball $\mathcal{B}(X)$ centered at the origin $\mathbf{0}$ such that $\mathcal{B}(X)$ is fully contained inside the convex set X. The *residual radius* of X, denoted $r(X)$ is the radius of this residual ball $B(X)$. By an abuse of notation, we write $r(X)$ instead of $r(\text{conv } X)$, if X is not convex. Let

$$
\begin{aligned}
r_d(m, X) &= \max\{r(Y): \\
&\quad Y \subseteq X \text{ and } |Y| \le m\},
\end{aligned}
$$

$$
\begin{aligned}
r_d(m) &= \min\{r_d(m, X): \\
&\quad X \subseteq \mathbb{R}^d \text{ and } r(X) \ge 1\}.
\end{aligned}
$$

In the notation, we shall omit the subscript d, if the dimension is clear from the context. (In this paper, the interesting case is $d = 6$.) Thus the original Steinitz's theorem can now be interpreted to say that

$$
r_d(2d) > 0.
$$

A quantitative version of Steinitz's theorem provides more precise bounds for the number $r_d(m)$, when $m \ge 2d$.

Now the optimal closure grasp with m fingers can be expressed in terms of the residual radius values given by a quantitative Steinitz's theorem. For the sake of simplicity consider the finger force constraint given by χ_{con}. Then the grasp metric for an optimal closure grasp with an m-fingered positive grip for a body B

can be seen to be the value, $r_6(m, \text{conv } \Gamma(\partial B))$. To see this, note that if we choose m points in $G_{\chi_{con}} = \text{conv } \Gamma(\partial B)$ with residual radius r then any external wrench vector $\mathbf{v}$ of magnitude at most r can be written as a convex combination of the m chosen points. So if $\mathbf{v}$ is any external wrench that is applied to the body B, and $\mathbf{v}$ lies in the residual ball of radius r, we can resist this external wrench by applying suitable forces (of magnitude at most 1) at the grasp points such that these forces sum to $-\mathbf{v}$; hence, we maintain the body in equilibrium. Thus, we see that the quantity $r_6(m)$ gives a universal measure for the quality of a closure grasp with $m \ge 12$ fingers.

3.1 B.K.P. Bounds

The special case, where $m = 2d$ had been studied by Bárány, Katchalski and Pach [2]; they showed that

$$
r_d(2d) > \frac{c}{(2ed)^{\lfloor d/2 \rfloor} d^2}.
$$

These results seem to indicate that a twelve-finger positive grip, while sufficient to provide a closure grasp, may not be adequate to achieve a desirable grasp quality.

Theorem 3 (Q.S.T.)
Bárány, Katchalski and Pach [2]
For any positive d there is a constant $r = r_d(2d) > d^{-2d}$ such that given any set $X \subseteq \mathbb{R}^d$ of points in d-space whose convex hull contains the unit ball $\mathcal{B}^d$ centered at the origin $\mathbf{0}$, there is a subset $Y \subseteq X$ with at most $2d$ points whose convex hull contains a ball centered at $\mathbf{0}$ with radius r.

Proof

The proof is constructive. We first choose $(d + 1)$ rays, placed regularly as follows: Let Δ be a regular d-simplex inscribed in the unit ball $\mathcal{B}$. By assumption, Δ is contained in the conv X. The desired rays R_i ($1 \le i \le d + 1$) are the ones joining the origin to the vertices of Δ. Let $p'_i = R_i \cap \partial \text{conv } X$. Each such p'_i lies on a face of the conv X and thus can be expressed as a convex combination of at most d points of X. The totality of these $d(d + 1)$ points $Y' \subseteq X$ contains Δ and thus a ball of radius $1/d$ in its convex hull. Let $P = \text{conv } Y'$ and $P_1, P_2, \ldots, P_l$ be the facets of P, each of which may be assumed to a simplex (otherwise triangulate nonsimplicial facets). Clearly, $l \le \binom{d(d+1)}{d}$. Also note that pos P_i's cover the sphere S^{d-1}. Choose a facet, say P_1, such that the surface area of $S \cap \mathrm{pos}\, P_1$ is as large as possible and thus greater than

$$\frac{\int_{S^{d-1}} dA}{l}.$$

Now choose two rays $R_1' \in \mathrm{pos}\, P_1$ and $R_2' = -R_1'$ such that the minimal angle between R_1' and the facets of P_1 is maximal and equal to α. Again, choose d points of X such that their convex hull contains the point $R_2' \cap \partial \mathrm{conv}\, X$. These points and the vertices of P_1 is the desired Y. Clearly $|Y| \leq 2d$ and has residual radius r, where

$$r = \frac{\sin \alpha}{\alpha} \left(1 + \frac{1}{d^2} + \frac{2 \cos \alpha}{d}\right)^{1/2}$$

$$\alpha > \arctan \frac{\int_{S^{d-1}} dA}{l d \int_{S^{d-1}} dA}.$$

Simple calculation then shows that $r_d(2d) > d^{-2d}$. Additional efforts lead to the improvement mentioned earlier and uses Upper Bound Theorem for a tighter estimation of the number of facets of P. □

3.2   K.M.Y. Bounds

Kirkpatrick, Mishra and Yap [14] have provided more general bounds. Here, we consider the d-dimensional case for $d > 2$. The techniques are slightly weaker than the 2-dimensional case considered in greater entails in [14].

3.2.1   Lower Bound

We first give a lower bound for $r_d(m)$ for sufficiently large m (in particular, for all $m \geq 13^d d^{\frac{d+3}{2}}$). Thus, m is chosen to be large enough to guarantee that

$$k = \left\lfloor \left(\frac{m}{2d^2}\right)^{1/(d-1)} \right\rfloor$$

takes integral values, greater than $\lceil 11\sqrt{d} \rceil$.

Lemma 1 *For any set $X \subseteq \mathbb{R}^d$ whose convex hull contains the unit ball $\mathcal{B}^d$ centered at the origin $\mathbf{0}$, we can find a set $Y \subseteq X$ of at most m points with residual radius*

$$r(Y) \geq 1 - 3d \left(\frac{2d^2}{m}\right)^{\frac{2}{d-1}}, \quad \text{for all } m \geq 13^d d^{\frac{d+3}{2}}.$$

Proof

Let k be defined as a function of d and m, as before. It suffices to show that

$$r(Y) \geq 1 - \frac{97}{48} \frac{d}{(k-1)^2} > 1 - \frac{3d}{(k+1)^2},$$

in the given range for k.

Henceforth, P will stand for the convex hull of X. Let C be the d-dimensional cube whose faces are normal to the appropriate coordinate axes, of side-length 2 and containing the unit ball $\mathcal{B}^d$. On each face of C we place a $k \times k \times \cdots \times k$ ($d-1$ times) grid (so the grid points have coordinates that are integer multiples of $\frac{1}{k-1}$ and two adjacent grid points are $\frac{2}{k-1}$ apart). Note that there are fewer than $2dk^{d-1} \leq m/d$ 'grid cubes' on the union of the $2d$ faces of C. Through each grid point p, we pass a ray R from the origin. Let R intersect the unit sphere S^{d-1} at $y(R)$. For each such ray R, we choose at most d vertices of P (the convex hull of X) as follows. If the ray passes through an i-face of P, we choose $i+1$ vertices of P whose convex span intersects that ray and is contained in that i-face. Thus the set Y of chosen vertices has at most m points. The convex hull of Y contains the set Y' of all points of the form $y(R)$ where R is a ray passing through the grid point.

Let R be any ray originating from $o = \mathbf{0}$ and suppose it intersects some face of C at a point a where a lies inside a grid cube S. Consider the triangle oab where b is any other point on the boundary of S.

$$\sin \angle(aob) = |ab| \cdot \frac{\sin \angle(oab)}{|ob|}$$

$$\leq \frac{2\sqrt{d}}{k-1} \cdot \frac{1}{1} \leq \frac{1}{5}$$

Choose α to be

$$\alpha = \arcsin \frac{2\sqrt{d}}{k-1}.$$

Let q_0 be any point at distance $\cos \alpha$ from the origin. We show that q_0 lies in the convex hull of Y'. Let R_0 be the ray from $o = \mathbf{0}$ through q_0 and suppose R_0 intersects the grid cube S_0. Let K_0 be the cone bounded by the set of rays originating from o that makes an angle of α with R_0. Hence each ray that passes through a vertex of S_0 is contained in K_0. There is a unique hyperplane H_0 containing $\partial(K_0) \cap S^{d-1}$. Note that $q_0 = R_0 \cap H_0$. Let

$$T_0 = \{y(R) : R \text{ passes through a vertex of } S_0\}$$

and

$$T_1 = \{R \cap H_0 : R \text{ passes through a vertex of } S_0\}.$$

By definition, $T_0 \subseteq Y'$. Note that each point in T_0 lies on the side of H_0 not containing the origin. This means that the convex hull of Y' contains the set T_1. But the convex span of the set T_1 contains the point $q_0 = R_0 \cap H_0$. This proves $r(Y) \geq r(Y') \geq \cos \alpha$.

$$
\begin{aligned}
\cos \alpha &= (1 - \sin^2 \alpha)^{1/2} \\
&> 1 - \frac{\sin^2 \alpha}{2} - \frac{\sin^4 \alpha}{8} \sum_{i=0}^{\infty} \sin^{2i} \alpha \\
&= 1 - \frac{\sin^2 \alpha}{2} - \frac{\sin^2 \alpha}{8} \left[\frac{\sin^2 \alpha}{1 - \sin^2 \alpha} \right] \\
&\geq 1 - \frac{97 \sin^2 \alpha}{192} \\
&\qquad (\text{since } \sin^2 \alpha \leq 1/25) \\
&\geq 1 - \frac{97d}{48(k-1)^2}.
\end{aligned}
$$

This proves the lower bound lemma. $\qquad\square$

3.2.2 Upper Bound

Next, we derive an upper bound for $r_d(m)$. For this purpose, we let X be all the points on the unit sphere and then bound the largest radius of a ball contained in the convex hull of m points on the unit sphere. The convex hull of any such m points forms a polytope. The proof relies on the facts that (1) any "long" edges of this polytope bound the radius of the contained ball and (2) since the polytope has only m vertices it must have some "long" edges. The detailed calculations provide an appropriate numerical bound.

Lemma 2 *Let $X \subseteq \mathbb{R}^d$ be the set of all points on the surface of the d-dimensional unit ball centered at the origin $\mathbf{0}$. Thus, the convex hull of X contains the unit ball $\mathcal{B}^d$ centered at the origin $\mathbf{0}$. Then any set $Y \subseteq X$ of at most m points has a residual radius*

$$r(Y) \leq 1 - \frac{1}{17}\left(\frac{2d^2}{m}\right)^{\frac{2}{d-1}}, \quad \text{for all } m \geq 3^d d^2.$$

Proof

The proof proceeds in two steps: We first show that for all $m > 0$ and for all $0 < \Theta < \pi/4$,

$$r(Y) \leq \max\left(\cos\frac{\Theta}{2}, 1 - \frac{1 - \tan^2 \Theta}{16}\left(\frac{2d^2}{m}\right)^{\frac{2}{d-1}}\right).$$

Then by an appropriate choice of the parameter Θ ($\Theta = 4\pi/53$), we obtain the claimed bound.

(1) Let Y be a set of m points in $\mathbb{R}^d$ all lying on the surface of a unit ball and $P = \text{ConvexHull}(Y)$. Let P' be the polyhedron obtained from P by triangulating the nonsimplicial facets of P. Let pq be an edge of the polyhedron P'. Then

$$r(Y) \leq \cos\frac{\angle(poq)}{2}.$$

Thus, if

$$\alpha = \max_{pq = \text{edge of } P'} \angle(poq),$$

is the maximum of all such angles, then

$$r(Y) \leq \cos\frac{\alpha}{2}.$$

If $\alpha \geq \Theta$ then

$$r(Y) \leq \cos\frac{\Theta}{2}.$$

Henceforth, we assume that $\alpha < \Theta$. Let t stand for $\tan \Theta$; thus $0 < t < 1$.

(2) Let $p \in Y$ be any point, and define its *truncated cone* K_p as follows:

$$K_p = \{x : \angle(xop) \leq \alpha \text{ and } x \cdot p \leq 1\}.$$

Now, if q is an arbitrary point on the surface of the unit ball, then the line segment oq belongs to K_p, for each vertex p of some (simplicial) facet of P'. As each such simplex facet has d vertices, the collection of truncated cones cover each point in the unit ball at least d times. Thus, we see that

$$m \cdot \text{Volume}(K_p) \geq d \cdot \text{Volume}(\text{unit ball}).$$

Let $V_d(r)$ stand for the volume of a d-dimensional ball of radius r.

$$V_d(r) = V_d(1)r^d.$$

Thus, the volume of the d-dimensional unit ball is given by

$$
\begin{aligned}
V_d(1) &= 2\int_0^{\pi/2} V_{d-1}(\sin\theta)\,\sin\theta\,d\theta \\
&= 2V_{d-1}(1)\int_0^{\pi/2} \sin^d\theta\,d\theta \\
&= K(d)V_{d-1}(1),
\end{aligned}
$$

where $K(d)$ is defined by the last equation. The volume of each K_p is given by

$$
\begin{aligned}
\text{Volume}(K_p) &= \int_0^1 V_{d-1}(r \tan \alpha) \, dr \\
&= V_{d-1}(\tan \alpha) \int_0^1 r^{d-1} \, dr \\
&= \frac{V_{d-1}(\tan \alpha)}{d}.
\end{aligned}
$$

Substituting the volumes into the preceding inequality, we get

$$
m \frac{\tan^{d-1} \alpha V_{d-1}(1)}{d} \geq d K(d) V_{d-1}(1).
$$

Hence,

$$
\begin{aligned}
1 \quad > \quad & t = \tan \Theta > \tan \alpha \\
\geq \quad & \left(\frac{d^2 K(d)}{m} \right)^{\frac{1}{d-1}} = c(d, m),
\end{aligned}
$$

where $c(d, m)$ is defined in the last equation. Using the inequality $c(d, m)^2 < t^2$, we get

$$
\begin{aligned}
\cos^2 \alpha \quad &\leq \quad \frac{1}{1 + c(d, m)^2} \\
&\leq \quad 1 - c(d, m)^2 + c(d, m)^4 \\
&\leq \quad 1 - (1 - t^2) c(d, m)^2.
\end{aligned}
$$

Hence,

$$
\begin{aligned}
\cos \alpha \quad &= \quad 2 \cos^2 \frac{\alpha}{2} - 1 \\
&\leq \quad \left(1 - (1 - t^2) c(d, m)^2 \right)^{\frac{1}{2}} \\
&\leq \quad 1 - \frac{1 - t^2}{2} c(d, m)^2
\end{aligned}
$$

and

$$
\cos^2 \frac{\alpha}{2} \leq 1 - \frac{1 - t^2}{4} c(d, m)^2.
$$

Finally, we get

$$
\begin{aligned}
\cos \frac{\alpha}{2} \quad &\leq \quad \left(1 - \frac{1 - t^2}{4} c(d, m)^2 \right)^{\frac{1}{2}} \\
&\leq \quad 1 - \frac{1 - t^2}{8} c(d, m)^2.
\end{aligned}
$$

Hence,

$$
r(Y) \leq 1 - \frac{1 - t^2}{8} \left(\frac{d^2 K(d)}{m} \right)^{\frac{2}{d-1}}.
$$

(3) Note that (e.g., [11], page 369)

$$
\begin{aligned}
K(d) \quad &= \quad 2 \int_0^{\pi/2} \sin^d \theta \, d\theta \\
&= \quad \begin{cases} \frac{(2k-1)!!}{(2k)!!} \pi, & \text{if } d = 2k = \text{even}; \\ \frac{(2k)!!}{(2k+1)!!} 2, & \text{if } d = 2k + 1 = \text{odd}. \end{cases} \\
&\geq \quad 2 \left(\frac{1}{2} \right)^{\frac{d-1}{2}}.
\end{aligned}
$$

Here $k!!$ stands for $k(k-2)(k-4) \cdots (\ell+4)(\ell+2)\ell$ (terminating in $\ell = 1$ or 2, depending on whether k is odd or even). Thus

$$
r(Y) \leq 1 - \frac{1 - \tan^2 \Theta}{16} \left(\frac{2d^2}{m} \right)^{\frac{2}{d-1}}.
$$

(4) The stated bound follows with appropriate choice of the parameter Θ, as shown below: Let $m \geq 3^d d^2$; then

$$
\left(\frac{2d^2}{m} \right)^{\frac{2}{d-1}} < \frac{1}{9}.
$$

Choose the parameter $\Theta = 4\pi/53$, and observe that

$$
\cos \frac{2\pi}{53} < 1 - \frac{1}{17 \times 9} \leq 1 - \frac{1}{17} \left(\frac{2d^2}{m} \right)^{\frac{2}{d-1}}.
$$

Since $1 - \tan^2 \frac{4\pi}{53} > 16/17$,

$$
\begin{aligned}
1 - \frac{1 - \tan^2(4\pi/53)}{16} &\left(\frac{2d^2}{m} \right)^{\frac{2}{d-1}} \\
\leq \quad 1 - \frac{1}{17} &\left(\frac{2d^2}{m} \right)^{\frac{2}{d-1}}.
\end{aligned}
$$

$\square$

If we choose $\Theta = \pi/5$ in the preceding proof, we can show that: for all $m > 0$,

$$
r(Y) \leq 1 - \frac{15}{512} \left(\frac{d}{m} \right)^{\frac{2}{d-1}}.
$$

If $m \leq d$, then

$$
0 = r(Y) < 1 - \frac{15}{512} \left(\frac{d}{m} \right)^{\frac{2}{d-1}}.
$$

On the other hand, if $m > d$, we get the result since $\cos \frac{\pi}{10} < 1 - \frac{15}{512}$ and $1 - \tan^2 \frac{\pi}{5} > 15/32$.

Summarizing the preceding lemmas, we have

Theorem 4 *For all* $m \geq 13^d d^{\frac{d+3}{2}}$,

$$\frac{1}{17}\left(\frac{2d^2}{m}\right)^{\frac{2}{d-1}} \leq 1 - r_d(m) \leq 3d\left(\frac{2d^2}{m}\right)^{\frac{2}{d-1}}.$$

$\square$

The results in this section seem highly pessimistic for moderately small number of fingers. However, if one allows large number of fingers, or allows frictional and/or soft contact models, there is a possibility for synthesizing moderately efficient closure grasps. Kirkpatrick et. al. [14] have provided certain approximation algorithms for these problems and some related computational geometric problems. However, still much research is needed to provide practical algorithms.

4 Nasty Finger Model

Here, we present a result due to Walter Meyer ("A Seven Finger Robot Hand is Weak," unpublished manuscript, Adelphi University), which shows that there is a family of boxes (rectangular parallelo pipeds), "skinny boxes," all of unit length, where the optimal grasp (under r_{nasty} model) has a grasp metric value bounded by a linear function of the skinny dimension. These results have been motivated by the model proposed by Kirkpatrick, Mishra and Yap [14] who left the problem for the small number (≥ 7) of fingers unanswered. Note that the problem of quantifying the trade-offs between the number of fingers and the best achievable grasp metric values remain largely open in the most general context.

Observe that as the dimension of the object to be grasped become smaller, the ability of a finger to generate large torques become weaker. However, it is also equally harder for the external ("hostile" or "nasty") environment to impose a large counteracting torque to break a grasp on a small body. That is, the robot hand and its opponent (adversarial environment) are equally balanced. This balance is made explicit, in our discussion, by always assuming that the object, B, to be grasped is scaled in such a manner that $R_\chi(\partial B)$ contains a unit residual sphere in its interior. That is, if we are allowed to use arbitrarily large number of fingers (in the limit going to infinity) then we can resist any external wrench of unit magnitude. (Note, however, for this argument to work out, one has to rule out such finger force constraints as χ_{max} and replace

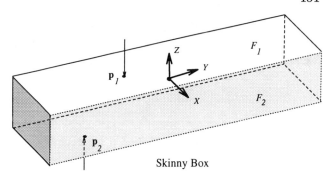

Figure 3: *The "skinny box" example.*

it by an appropriate version of χ_{hyb} with some bound on the number of the partitions. With χ_{max}, as the number of fingers increase so does the bound on the volume of the resistable wrench set.)

Now consider a family of rectangular parallelopipeds—"skinny boxes"—each of which has one side of unit length and the remaining two sides equal and of length $\varepsilon < 1$. We denote such a box by Box_ε. Then the following result holds:

Theorem 5 (Meyer) *For any seven fingered positive grip of a skinny box,* Box_ε

$$r_{nasty} \leq 5\sqrt{2}\varepsilon.$$

Proof

Note that any closure grasp of the object must place at least one finger on each of the six faces of the box. Thus there can be exactly one face with two fingers. Thus there are at least two opposing parallel long faces F_1 and F_2 with one finger on each. The contact points are $\mathbf{p}_1 \in F_1$ and $\mathbf{p}_2 \in F_2$. Let us now assign a coordinated system with the origin an the center of mass of the skinny box, the Z axis normal to the faces F_1 and F_2 and the X and Y axes normal to the other four faces. That is, X-Y plane is parallel to the faces F_1 and F_2. Note that the torques generated by the contacts $\mathbf{p}_1$ and $\mathbf{p}_2$ are in the X-Y plane. Because of the choice of the dimensions of the skinny box, the torque component due to the other five points of contact in either the X or Y direction is no more than $\varepsilon/2$; thus, making the total torque component in the X-Y plane due to these five fingers no more than $5\sqrt{2}\varepsilon/2$.

Now assume that the magnitude of the forces at the points of contact $\mathbf{p}_1$ and $\mathbf{p}_2$ are respectively f_1 and

f_2. Let us now place the "nasty finger" close to one of the corners of the faces F_1 and F_2 at a point $\mathbf{p}_{nasty}$, exerting a force of magnitude λ. Without loss of generality, assume that $\mathbf{p}_{nasty} \in F_1$. Then, by the force and torque equilibria conditions, we have

$$|(f_1\mathbf{p}_1 - f_2\mathbf{p}_2 + \lambda\mathbf{p}_{nasty}) \times Z| \;\leq\; 5\sqrt{2}\varepsilon/2$$
$$\lambda \;=\; |f_1 - f_2|$$

Since

$$\max(|\mathbf{p}_1 - \mathbf{p}_{nasty}|, |\mathbf{p}_2 - \mathbf{p}_{nasty}|) \geq 1/2,$$

by design, we see that $\lambda \leq 5\sqrt{2}\varepsilon$. $\square$

Note however, that the argument above fails for nine fingers. That is with nine fingers, one can devise a grasp (for every pair of opposing faces, place two fingers on a face and one on the opposing face) that has a good value for this grasp metric, independent of the skinny dimension of the box. Also, note that this argument fails to generalize to the other grasp metrics directly. These issues need to be explored further.

5 Optimal Three-finger Planar Grasps

For the sake of concreteness, we now consider some special cases, where we wish to obtain the best three-finger grasp of a planar polygonal object assuming non-frictional contacts. Note that in this case, since it is not possible to guarantee that the resulting grasp will have the force/closure properties, we are willing to sacrifice the condition requiring torque-closure. In other words, we wish only to achieve a three-finger grasp such that the smallest external force such a grasp and resist is as large as possible.

More formally given a simple n-gon P, we wish to choose three distinct points $\mathbf{p}_1$, $\mathbf{p}_2$ and $\mathbf{p}_3$ on the interior of the edge segments of P such that the following properties hold:

1. The unit inner normals $\mathbf{n}(\mathbf{p}_1)$, $\mathbf{n}(\mathbf{p}_2)$ and $\mathbf{n}(\mathbf{p}_3)$ are concurrent.

2. The unit inner normals $\mathbf{n}(\mathbf{p}_1)$, $\mathbf{n}(\mathbf{p}_2)$ and $\mathbf{n}(\mathbf{p}_3)$ positively spans the two-dimensional force space, i.e.,

$$(\forall \mathbf{w} \in \mathbb{R}^2)\,(\exists f_i \geq 0, 1 \leq i \leq 3)\; \mathbf{w} = \sum_{i=1}^{3} f_i \mathbf{n}(\mathbf{p}_i).$$

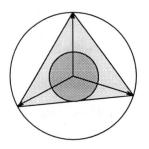 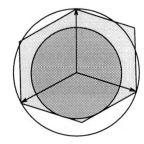

Residual circle with respect to χ_{con} Residual circle with respect to χ_{max}

Figure 4: *Grasp metrics associated with χ_{con} and χ_{max}.*

3. The unit normals are "well-balanced" in the sense that

$$\min\Big\{|\mathbf{w}| : \mathbf{w} \in \mathbb{R}^2,$$
$$(\exists f_i \geq 0, 1 \leq i \leq 3)\, \chi(f_1, f_2, f_3) = 1$$
$$\mathbf{w} = \sum_{i=1}^{3} f_i \mathbf{n}(\mathbf{p}_i)\Big\},$$

is as large as possible (among all choices of $\mathbf{p}_1$, $\mathbf{p}_2$ and $\mathbf{p}_3$). Here, $\chi(f_1, f_2, f_3)$ denotes a finger force constraint condition on the magnitude of the forces applied at the points of contact. For instance,

$$\chi_{con} : f_i \geq 0, \sum f_i \leq 1,$$

or

$$\chi_{max} : f_i \geq 0, \max f_i \leq 1.$$

Thus the first property denotes the trivial torque equilibrium condition; the second property denotes the force closure condition and the third property measures the goodness of the grasp. In English, the third property says: under the condition χ_{con}, we wish to maximize the radius of a disk, centered at origin and contained in the triangle formed by (convex hull of) the points (on the unit circle) corresponding to the vectors $\mathbf{n}(\mathbf{p}_1)$, $\mathbf{n}(\mathbf{p}_2)$ and $\mathbf{n}(\mathbf{p}_3)$. Similarly, under the condition χ_{max}, we wish to maximize the radius of a disk, centered at origin and contained in the Minkowski sum of the points (on the unit circle) corresponding to the vectors $\mathbf{n}(\mathbf{p}_1)$, $\mathbf{n}(\mathbf{p}_2)$ and $\mathbf{n}(\mathbf{p}_3)$—a convex hexagon.

Let the corresponding radii be denoted as $\rho_{con}(\mathbf{p}_1$, $\mathbf{p}_2$, $\mathbf{p}_3)$ and $\rho_{max}(\mathbf{p}_1, \mathbf{p}_2, \mathbf{p}_3)$, respectively. Note that, if the angle α_i's $(1 \leq i \leq 3)$ denote the angles between the inner normals then $\alpha_{max} = \max(\alpha_1,\ \alpha_2,\ \alpha_3) \geq 2\pi/3$ completely determines the radii

$$\rho_{con} = \cos(\alpha_{max}/2), \quad \text{and}$$
$$\rho_{max} = \sin\alpha_{max}.$$

Thus both these metrics are monotonically decreasing functions of $2\pi/3 \leq \alpha_{max} \leq \pi$, and it suffices to minimize α_{max}. However, for the sake of the ease of exposition, we will frequently use $\rho = \rho_{con}$, and refer to it as the "*residual radius*" of $\mathbf{n}(\mathbf{p}_1)$, $\mathbf{n}(\mathbf{p}_2)$ and $\mathbf{n}(\mathbf{p}_3)$. The optimal value of residual radius is denoted by ρ^*.

Note that given an edge $e = ab$ of the polygon P, for every point $p \in ab$, $\mathbf{n}(\mathbf{p})$ defines a unique point $q(e)$ on the unit circle in $\mathbb{R}^2$. Thus we may simply refer to this point on the unit circle by $q(e)$. Henceforth, let the edges of the n-gon be given as $E = \{e_1,\ e_2,\ \ldots,\ e_n\}$ and the corresponding points on the unit circle be $Q = \{q_1,\ q_2,\ \ldots,\ q_n\}$, where $q_i = q(e_i)$ $(1 \leq i \leq n)$.

We may note at this point that there is a trivial $O(n^3)$ time algorithm to find an optimal grasp of a simple n-gon, P, by exhaustively enumerating all edge triples of P ad by examining each triple successively. In order for an edge triple $(e_i,\ e_j,\ e_k)$ to produce three necessary optimal contact points, it must be the case that $(q_i,\ q_j,\ q_k)$ form a triangle with a positive residual radius of ρ^*—a condition that can be checked easily in $O(1)$ time. However, this is not sufficient—since we must check that there are three points $\mathbf{p}_i \in e_i$, $\mathbf{p}_j \in e_j$ and $\mathbf{p}_k \in e_k$ satisfying the torque equilibrium condition; namely, that $\mathbf{n}(\mathbf{p}_1)$, $\mathbf{n}(\mathbf{p}_2)$ and $\mathbf{n}(\mathbf{p}_3)$ are concurrent meeting at some point c.

This is not hard but requires some thought. We proceed as follows: Consider an edge ab of P. Let $HP(a, ab)$ be the open half plane containing ab and delimited by a line containing a and normal to ab and similarly, let $HP(b, ab)$ be the open half plane containing ab and delimited by a line containing b and normal to ab. Let

$$\text{slab}(e) = HP(a, ab) \cap HP(b, ab),$$

where $e = ab$.

Then it is easy to see that for a triple of edges $(e_i,\ e_j,\ e_k)$ to satisfy the torque equilibrium condition, it is necessary and sufficient that

$$\text{slab}(e_i) \cap \text{slab}(e_j) \cap \text{slab}(e_k) = C \neq \emptyset.$$

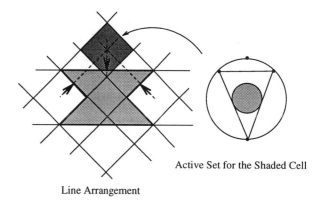

Line Arrangement   Active Set for the Shaded Cell

Figure 5: *The line arrangement associated with an object.*

The point of concurrency $c \in S$, and the contact points $\mathbf{p}_i$, $\mathbf{p}_j$ and $\mathbf{p}_k$ are determined by the normals from c onto the edges e_i, e_j and e_k.

Thus our previous arguments can be summarized to be saying that an edge triple $(e_i,\ e_j,\ e_k)$ defines an optimal grasp if $\text{slab}(e_i) \cap \text{slab}(e_j) \cap \text{slab}(e_k)$ is nonempty and that the triangle formed by the corresponding points on the unit circle has a positive residual radius of ρ^*, maximal among all choices of edge triples. These considerations yield an $O(n^3)$-time algorithm.

5.1 An Improved Algorithm

Next, we ask if it is possible to improve upon the trivial $O(n^3)$-time algorithm. Here, we present an $O(n^2 \log n)$-time algorithm for finding the optimal three fingered planar grasp for an arbitrary simple polygon and some simple improvements for convex polygon. We first describe the algorithm assuming that the polygon P is nondegenerate (in the sense that will be made precise later) and then remark on how the nondegeneracy can be eliminated by a simple modification to the algorithm.

The algorithm can be described as follows: First we create the two-dimensional line arrangement formed by a collection of lines consisting of three lines per edge, where the triplet of lines associated with an edge ab are: (1) the line containing the edge ab, (2) the line normal to ab, containing a and (3) the line normal to ab, containing b. Now consider a nonempty cell C of this arrangement: we say a point $q = q(e)$ on the unit circle is *active* for this cell, if $\text{slab}(e) \supseteq C$. The subset of points on the unit circle (among the points q_1, q_2,

..., q_n of Q) that are active for this cell C, is called its active set and denoted by active(C) $\subseteq Q$. Now, if we find three points q_i, q_j and $q_k \in$ active(C), whose residual radius $\rho(C)$ is as large as possible (and positive), then it is seen that ρ^* is simply the maximum of all $\rho(C)$'s taken over all cells of the arrangement.

Note that there are at most $O(n^2)$ cells altogether and as we go from one cell C to its adjacent cell C' then the active(C') can be computed from the active(C) by adding or deleting a point on the unit circle, depending on the line containing the $C \cap C'$. Of course, here we have tacitly assumed that the polygon is nondegenerate, in the sense that the all the lines on the arrangement are distinct, since otherwise $C \cap C'$ may belong to more than one line of the arrangement and thus require addition and deletion of more than one point of the set Q. Clearly, the active sets for all the cells can be computed in $O(n^2)$ time by visiting the cells of the arrangement, starting from a cell with an empty active set (such a cell exists sufficiently far away from the polygon P). However, computing the $\rho(C)$ for each cell may still take $O(n)$ time, thus forcing the entire procedure to take $O(n^3)$ time.

We circumvent this problem by the following simple trick: First of all we maintain the active(C)'s in a clockwise order in a dynamic balanced binary search tree. Since each update operation on this data structure takes $O(\log n)$ time, this increases the complexity of computing the active sets of all the cells to $O(n^2 \log n)$-time.

At any instant, we only remember $\tilde{\rho}$—the maximal residual radius seen so far. That is, $\tilde{\rho}$ is simply the maximum of those $\rho(C)$'s corresponding to only those cells C that have been visited so far. We also remember the edge triple associated with the radius value $\tilde{\rho}$. When we go from a visited cell C to an adjacent unvisited cell C', we do one of two things: If going to the next cell entails deletion of a point, q_i, on the unit circle, then we only have to update the active(C'); the maximal residual radius of C' cannot be larger than that of C and thus $\tilde{\rho}$ remains unchanged. If going to the next cell, on the other hand, entails addition of a point, q_i on the unit circle, then we have to both update the active(C') and check if $\tilde{\rho}$ can be improved. If the maximal residual radius of C', $\rho(C') > \tilde{\rho}$, then the associated triplet from active(C') must involve the new point q_i and two of the old points. How can we do this

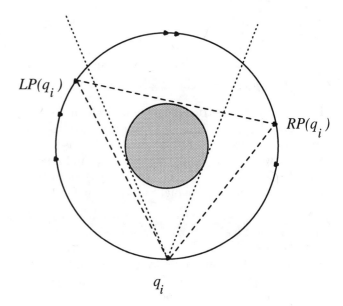

Figure 6: *Test involving q_i and a possible residual radius value of ρ_k.*

operation quickly?

First note that residual radii cannot take all possible values but only one of $\binom{n}{2}$ values, each value being determined by a pair of distinct points q_l and q_m and is equal to the radius of the circle that is centered at the origin and has the line containing q_l and q_m as tangent. All these radii can be sorted in $O(n^2 \log n)$ time and are denoted by

$$0 \leq \rho_1 \leq \rho_2 \leq \cdots \leq \rho_i \leq \cdots < 1$$

Suppose before visiting the cell C' the maximal residual radius seen so far is $\tilde{\rho} = \rho_i$. When we go to the cell C' (which requires adding the point q_i), we will successively test if it has a residual radius no smaller than ρ_{i+1}, ρ_{i+2}, etc. until we fail for some value ρ_j $(j > i)$. Each such test can be performed in $O(\log n)$ time as explained below.

Let $i < k \leq j$, and we wish to test if active(C') has three points involving q_i and of residual radius $\geq \rho_k$. Consider a circle $C(\rho_k)$ of radius ρ_k and centered at the origin. Two distinct points of active(C') are said to be *mutually visible* if the line segment connecting these two points do not intersect the interior of $C(\rho_k)$. Thus our test succeeds if we can find a pair of mutually visible distinct points among the active(C'), each of which is also mutually visible with q_i. Let the *left-most partner* of q_i be the *last* mutually visible point of q_i encountered, visiting the points of active(C') in clockwise order starting from q_i. We call this point $LP(q_i)$. Similarly, we define the the *rightmost partner* of q_i by visiting the points of active(C') in anticlockwise order, and call it $RP(q_i)$. Since the active points of C' are kept in their sorted order in a balanced search structure, both $LP(q_i)$ and $RP(q_i)$ can be computed in $O(\log n)$ time. Then it only remains to check that $LP(q_i)$ and $RP(q_i)$ are mutually visible, a step that can be accomplished in $O(1)$ time.

Thus, we can keep track of $\tilde{\rho}$ by performing a sequence of tests per each new cell, each of which takes $O(\log n)$ time. Note that while there is no a priori bound on the number of tests we may need to perform for a new cell, it should be obvious that all but the last test succeeds and the last test fails. Thus there are at most one test per cell that fails, and the totality of all such failed tests incur a cost of $O(n^2 \log n)$. On the other hand, if we have a successful test involving a radius value ρ_k, then *we shall never perform another successful test involving ρ_k, subsequently*. Thus, the total number of successful tests are bounded by the number possible radii values ($\binom{n}{2}$ of those) and altogether they incur a cost of $O(n^2 \log n)$. Clearly, when we are done visiting all the cells, we have the global maximal residual radius ρ^* together with the edge triple, which readily give the three contact points, and we have spent $O(n^2 \log n)$ time.

If the polygon P is degenerate then the resulting arrangement may force us to add and delete many points of Q while going from a cell to its adjacent cell. If we enforce the discipline that all the deletions are performed before all the additions and each update is performed sequentially then the correctness of the algorithm still holds and the performance analysis goes through *mutatis utandis*. In summary, we have

Theorem 6 *Given an arbitrary simple n-gon P, in $O(n^2 \log n)$ time, we can compute a three finger optimal grasp of P.*

Proof

The complexity analysis follows from the discussion preceding the theorem: The possible radii values can be computed and sorted in $O(n^2 \log n)$ time; the cells can all be visited with the active sets computation taking

$O(n^2 \log n)$ time; the tests involved in going from cell to cell are no more than the sum of the possible radii values and the number of cells in the arrangement with each test taking $O(\log n)$ time and thus contributing only $O(n^2 \log n)$ cost to the total cost.

To see the correctness of the algorithm, note that if the computed value at the end is $\tilde{\rho}$ then clearly

$$\tilde{\rho} \leq \rho^*.$$

Conversely, consider the set of edge triples that lead to the maximal residual radius ρ^* and all the cells that are contained in the intersection of three slabs associated with each such edge triple. Among all such cells consider the one that was visited the earliest, say C'. Let the preceding visited cell be denoted C. Let the maximal residual value seen up to the time C was visited be $\tilde{\rho}'$. Thus $\tilde{\rho}' \leq \rho(C) < \rho^* = \rho(C')$. Thus active($C'$) must have been obtained by addition of a point $q_i \in Q$. Thus, active(C') has two other points q_j and q_k such that a residual circle of radius ρ^* touches an edge of the triangle formed by q_i, q_j and q_k. Thus ρ^* is a possible residual radius and the tests at the cell C' involving possible residual radii values larger than $\tilde{\rho}'$ will all succeed up to ρ^*. Thus if the computed value at the end is $\tilde{\rho}$ then

$$\tilde{\rho} \geq \rho^*.$$

Hence they must be both equal and our algorithm correctly determines the optimal three finger grasp for P. □

5.2 Some Related Open Questions

There are several open questions related to the problem of finding optimal planar grasps. We briefly discuss these problems.

(1) Consider a variation on the above problem: Suppose we are given a simple polygon P with certain subset of ∂P designated as "forbidden" and its complement, "feasible." Assume that the feasible parts of the polygon consists of at most K segments (the edge segment ab being allowed to be a point a ($a = b$), in the degenerate case). We are asked to find an optimal three-finger grasp of the polygon with none of the fingers on a forbidden region. Using a small variation of the above algorithm, we can solve this problem in $O(k^2 \log k)$ time—only modify the line arrangement to consist of the following triple of lines per feasible edge segment $ab \subset e$, where e is an edge of P: (1) the line

containing e, (2) the line normal to e and containing a, and (3) the line normal to e and containing b. If the edge segment is a point $a \in e$ then the above situation degenerates to two lines, one containing e and the other normal e at a.

(2) We do not know whether there is a better solution for the above problem with improved complexity. For instance, it is not even clear whether there are $O(n)$ time algorithms for objects with simpler geometry, e.g., convex objects. We have an $O(n)$ solution only for what we shall refer to as *circular convex* polygonal objects. A convex polygon P will be called *circular* if there is a point c in its interior (its *center*) so that the line segment from c to a line containing an edge e and normal to e is entirely within P. For instance, the convex hull of a set of points on a circle defines a circular polygon (thus, the name). Note that in this case,

$$\bigcap_{e \in E} \text{slab}(e) \neq \emptyset,$$

where E is the set of edges of P. Clearly, we can find a small neighborhood U of the center c such that active(U) is all of Q. In this case, the problem reduces to simply finding three points q_i, q_j and $q_k \in Q$ such that the residual radius of the resulting triangle is as large as possible.

We need to extend the notion of residual radius as follows: The residual radius of a triangle Δ is the *signed* radius of the largest disk centered at the origin that is either fully outside or fully inside Δ, the sign being positive or negative depending on whether the disk is inside or outside Δ respectively.

Assume that the points of Q are ordered in the anti-clockwise order as

$$q_1 > q_2 > \cdots > q_n.$$

for any point $q \in Q$ its successor, succ(q), is the point immediately following it in the clockwise order.

We start with three arbitrarily chosen distinct points, say $u_0 = q_1$, $u_1 = q_2$ and $u_2 = q_3$, for instance. At any instance, assume that, we have three points u_0, u_1 and u_2, at least two of which are distinct, and

$$u_0 \geq u_1 \geq u_2 \geq u_0$$

There are two cases to consider: (1) they are not all distinct, i.e., $u_i = u_{i+1}$ and (2) they are all distinct and the residual disk touches the edge $u_{i-1}u_i$.

In the first case, we advance the "forward point" u_i (i.e, replace u_i by succ(u_i)) with the hope of making the points distinct (this may not succeed and lead to further advancements of this kind). In the second case, we advance the "backward point u_i (i.e, replace u_i by succ(u_i)) with the hope of releasing the "limiting edge" $u_{i-1}u_i$ and thus possibly (but not always) increasing the residual radius.

The algorithm keeps advancing the forward or the backward point (as the case may be) while recording the maximal residual radius seen so far until u_0 returns to its initial position, at which point it halts and outputs the edge triple corresponding to the maximal residual radius. Since the polygon is a circular convex polygon, one can easily determine the contact points by taking the normals from the center of the polygon to each edge of the edge triple. The correctness and the complexity analysis of the algorithm can be shown in a manner similar to the discussions in section 5 of the paper by Kirkpatrick et. al. [14] and is omitted here. Note, however, that the above technique fails for arbitrary convex polygons if we relax the condition of circularity.

Note that the above technique can be easily adapted to the following problem: Given a simple polygon P and a center $c \in \mathbb{R}^2$, find a 3-finger optimal grasp of P such that the inner normals at the contact points go through c. This problem is solved by simply running the above algorithm starting with an active set, active(U) of a small open neighborhood of c. The resulting algorithm takes $O(n)$ time.

(3) Sometimes, we wish to determine not just one optimal three finger grasp but all of them. Then we may use any one of this class of optimal grasps, depending on the task at hand. Clearly, the brute force $O(n^3)$ time algorithm will succeed to do so. Note that the algorithm of the previous section cannot be easily modified into a two pass algorithm, since addition of a new point (in the process of going from one cell to an adjacent cell) may create an $O(n)$ edge triplets of residual radius ρ^*. Here, we describe an $O(n^2 \log n)$ algorithm for the special case when the object is convex.

Let P be a convex n-gon and let the possible residual radii (as in the preceding subsection) be given as

$$0 \leq \rho_1 \leq \rho_2 \leq \cdots \leq \rho_i \leq \cdots < 1.$$

We shall find the optimal residual radius ρ^* by performing a binary search on the sequence of possible residual radii. For a given value of ρ_i, we can enumerate all the edge triples that lead to a residual radius of ρ_i in $O(n^2)$ as follows: Corresponding to the possible radius value ρ_i, there are at most $O(n)$ edge pairs (e_i, e_j)'s such that the corresponding points q_i and $q_j \in Q$ on the unit circle satisfy the property that the line determined by $q_i q_j$ is tangent to a circle $C(\rho_i)$ centered at the origin and of radius ρ_i. Now for each such edge pair, we need to check in $O(n)$ time if there is another edge e_k such that $q_k \in Q \setminus \{q_i, q_j\}$ is mutually visible (with respect to $C(\rho_i)$) to both q_i and q_j and that

$$\text{slab}(e_i) \cap \text{slab}(e_j) \cap \text{slab}(e_k) \neq \emptyset.$$

We can thus enumerate all the e_k's that succeed this test. The binary search only considers $O(\log n)$ different values of ρ_i's and terminates with success with the largest possible value ρ^* and enumerating all edge triples corresponding to ρ^*. It is then trivial to describe all possible three finger optimal planar grasps. Thus the algorithm has a time complexity of $O(n^2 \log n)$.

However, the algorithm applied to a nonconvex polygon leads to an $O(n^3)$-time algorithm, as in a pathological case, there may be $O(n^2)$ edge pairs to be considered for a given value of ρ_i. It is noteworthy that this algorithm is rather simple to implement and may perform well in practice. For instance, if one performs binary search on the real interval $[0,1]$ (instead of the possible radii values), then for a random polygon this algorithm can compute in $O(n \log n \log(1/\epsilon))$ all three finger grasps whose corresponding residual radii lie in the range $[\rho, \rho^*]$ of size $< \epsilon$, for sufficiently small positive ϵ.

(4) We still do not know how to find optimal m-finger planar grasp $(m \geq 4)$ in time better than what can be obtained by the brute force algorithm taking time $O(n^{O(m)})$. For instance, it is not even clear if there is an algorithm to compute such an optimal grasp in time $O(n^{m-1}\text{polylog } n)$. The complication arises by virtue of the torque components that one has to consider in the case when $m \geq 4$.

Some progress has been made, by modifying the problem to that of choosing an optimal set of m-finger contact points out of a preselected $O(n)$ points on the boundary of the polygon, ∂P. For instance, we have an $O(n^3 \log n)$ time algorithm to find such an optimal four finger grasp in this case. The technique employed for this case is a generalization of the preceding algorithm involving binary search. We suspect that the

algorithm generalizes to m fingers $(m > 4)$ and has a time complexity of $O(n^{m-1} \log n)$.

(5) In case of parallel jaw grippers and three-jaw grippers grasping an n-gon, one can compute the optimal grasps in time $O(n)$. The algorithms in these cases involve simply going around the object and trying all possible grasps [9]. It is not clear, if these grippers are comparable to multi-fingered hands in terms of how well they optimize various grasp metrics.

(6) Another problem of interest is to study the similar optimality problem in the case of "fixturing," where a polygonal object has to be fixtured by a set of toe-clamps that can be placed only at places designated by a set of toe-slots (which are usually arranged on a regular square grid) [3,35,19,20]. The added difficulty arises because of the geometric constraints imposed by the toe-slots. It is easily seen that for a rectilinear object the optimal fixel (fixture element) placement can be determined in $O(n)$ time. However, the problem seems quite difficult even when we consider a convex polygonal object.

(7) Recently, we have been able to design "reactive hands" for grasping. These algorithms operate by determining a sensor-dependent binary vector and then actuating a small set of actuators by a simple table-lookup procedure[31,32]. It remains an intriguing open question whether it is possible to design general multi-fingered reactive hands that always find an optimal grasp.

6 Computational Issues

The study of grasp metrics suggest two kinds of algorithmic (and the related complexity) problems:

1. Computing the quality of a given grasp under the chosen grasp metric.

2. Computing the optimal grasp of an object by an m-fingered hand under the chosen grasp metric.

Henceforth, we shall refer to these two problems as 1) *Computation Problem* and 2) *Optimization Problem*, respectively. In general, the computation problem seems to have received far more attention than the optimization problem.

6.1   Computation Problem

Given: A grasp $\mathcal{G}$ by means of its associated system of wrenches: $\{\mathbf{w}_1, \ldots, \mathbf{w}_k, \mathbf{w}_{k+1}, \ldots, \mathbf{w}_n\}$, where $\mathbf{w}_1, \ldots, \mathbf{w}_k$ are bisense and the remaining $\mathbf{w}_{k+1}, \ldots, \mathbf{w}_n$ are unisense. Also given is a grasp metric, described by the finger force constraint χ, where χ is assumed to be convex, closed, compact and faithful.

Compute: The quality of the given grasp $r_\chi(\mathcal{G})$. Recall that r_χ is the radius of the largest sphere centered at the origin and contained in the feasible wrench set G_χ.

$$G_\chi = \left\{ \mathbf{w} = \sum f_i \mathbf{w}_i : \chi(f_1, \ldots, f_n) = 1 \right\}.$$

Thus the given grasp has a quality of $r_\chi(\mathcal{G})$ if and only if

$$\left(\forall r \leq r_\chi(\mathcal{G}) \right) \left(\forall \mathbf{w} = (w_1, \ldots, w_6) \in \mathbb{R}^6 \right)$$
$$\left[|\mathbf{w}| = r \ \Rightarrow \ \mathbf{w} \in G_\chi \right],$$

or equivalently,

$$\left(\forall r \leq r_\chi(\mathcal{G}) \right) \left(\forall \mathbf{w} \in \mathbb{R}^6 \right) \left(\exists f_1, \ldots, f_n \right)$$
$$\left[(|\mathbf{w}| = r) \ \Rightarrow \ \mathbf{w} = \sum f_i \mathbf{w}_i \right.$$
$$\left. \wedge \, (\chi(f_1, \ldots, f_n) = 1 \right].$$

Thus we need to maximize $r_\chi(\mathcal{G})$ subject to the conditions described above to get the quality of the grasp.

Thus, if we assume that $\chi(f_1, \ldots, f_n)$ is given by a set of algebraic equations and inequalities of some bounded degree, then the above problem can be solved by using efficient algorithms in Tarski's elementary geometry in exponential time:

$$O\left(2^{O(n \log n)} \right),$$

in the cardinality of the wrench system n. (See Renegar [27].)

(1) Note that some simple improvements can be obtained in the special cases as follows: For instance, consider the finger force constraint defined by χ_{con}, in this case $G_{\chi_{con}}$ can be expressed as the intersection of upto $\binom{n}{6} = O(n^6)$ half spaces which define the convex hull of

$$\{\mathbf{w}_i, -\mathbf{w}_i : i = 1, \ldots, k\} \cup \{\mathbf{w}_i : i = k+1, \ldots, n\}.$$

Now considering the set of all normals from the origin to the boundary hyperplanes of these half spaces, one can compute $r_{\chi_{con}}$ by simple brute force. Note that computing these half spaces is quite trivial. For every six vectors, consider the hyperplane corresponding to their affine hull. If all the remaining points lie on one side (say positive) of the hyperplane, then we include the associated half space in our collection. The resulting algorithm takes $O(n^7)$ time and is thus polynomial. The exponent can be slightly improved by introducing additional data structure or randomization, but since we suspect that the improvement is not substantial, we shall not discuss this any further.

A similar consideration for χ_{max}, however, gives a $O(2^{O(n)})$-time algorithm, but we do not know if for the special cases where χ is given by set of linear equations and inequalities, the problem has a better complexity than the one with trivial $O(2^{O(n \log n)})$ bound. Note that since n is usually small (i.e., between 4 and 12), this complexity may be deemed acceptable.

(2) Another approach to this problem would be to probe G_χ in m distinct directions, each direction being given by a ray R going through the origin. Also assume that there is an oracle, which given R, returns $R \cap \partial G_\chi$. Thus after m probes, we can choose as an approximation for r_χ, a function of the probe values returned by the oracle (for instance, the minimum of the magnitudes). Here, we have tacitly assumed that we have no knowledge of G_χ other than what is provided by the oracle. Standard argument (similar to the one in [1]) shows that the number of probes will be exponentially large in d in order to obtain a good approximation, in general. Here, of course, $d = 6$, the dimension of G_χ.

The argument is as follows: We operate in dimension d. When the oracle is given a d-dimensional ray R, the oracle returns a unit vector $\mathbf{v}_R$, $|\mathbf{v}_R| = 1$, in the direction of R. We perform m such probes, $m \geq 13^d d^{\frac{d+3}{2}}$. Choose an ϵ

$$\epsilon \ll \frac{1}{68} \left(\frac{2d^2}{m} \right)^{\frac{2}{d-1}}.$$

If we compute an approximate residual radius of some value less than $1 - 2\epsilon$ then the oracle reveals G_χ to be a unit d-ball. If we compute an approximate residual radius of some value larger than

$$1 + 2\epsilon - \frac{1}{17} \left(\frac{2d^2}{m} \right)^{\frac{2}{d-1}}$$

then the oracle reveals G_χ to be the convex hull of the points it had produced, which has a residual radius

$$\leq 1 - \frac{1}{17}\left(\frac{2d^2}{m}\right)^{\frac{2}{d-1}},$$

thus contradicting again. By the choice of our ϵ value, our approximation must fall into either category. Thus the ϵ-approximation, for a given ϵ requires m probes such that

$$m \geq \frac{2d^2}{(c\epsilon)^{d-1/2}}.$$

Thus improving the approximation by a single bit will require increasing the number of probes by a factor of $2^{d-1/2}$.

(3) Before we leave this discussion, we note that as $r_{\tilde{\chi}_2}$ and r_{null} are given by simple linear algebraic formulations, they can be computed rather easily by matrix computation and linear programming techniques, respectively. However, $r_{\tilde{\chi}_2}$ and r_{null} are not of much help in computation (or, even in providing an approximation) involving some general χ.

6.2 Optimization Problem

Given: An object B and a robot hand with exactly m fingers. The contact type of each finger is given so that given a point $\mathbf{p} \in \partial B$, the corresponding system of wrenches

$$\Gamma(\mathbf{p}) = \{\mathbf{w}_1, \ldots, \mathbf{w}_\ell\}$$

($\ell \leq 6$) can be computed easily. Also given is a grasp metric, described by the finger force constraint χ, where χ is assumed to be convex, closed, compact and faithful.

Compute: A set of m contact points $\mathbf{p}_1, \ldots, \mathbf{p}_m \in \partial B$ such that the resulting grasp $\mathcal{G}$ involving the wrench system

$$\bigcup_{i=1}^{m} \Gamma(\mathbf{p}_i)$$

has the optimal grasp strength $r^* = r^*_\chi(B) = r_\chi(\mathcal{G})$. Thus we wish to

- maximize r^*
- subject to the following condition

$$\left(\exists \mathbf{p}_1, \ldots, \mathbf{p}_m\right)\left(\forall r \leq r^*\right)\left(\forall \mathbf{w} \in \mathbb{R}^6\right)$$

$$\left[(\mathbf{p}_1, \ldots, \mathbf{p}_m \in \partial B \, \wedge \, |\mathbf{w}| = r) \Rightarrow\right.$$

$$\left.\mathbf{w} \in G_\chi(\bigcup_{i=1}^{m} \Gamma(\mathbf{p}_i))\right].$$

Thus, if we assume that $\chi(f_1, \ldots, f_n)$ is given by a set of algebraic equations and inequalities of some bounded degree and ∂B is given by piecewise algebraic surfaces, each of bounded degree and of cardinality n then the above problem can be solved by using efficient algorithms in Tarski's elementary geometry in time $O(n^{O(m)})$.

Note that if we simply try to place the fingers on each of $\binom{n}{k}$ ($1 \leq k \leq m$) portions of the surface ∂B and exhaustively try each possibility then the complexity of the algorithm can be improved to

$$O\left(n^m 2^{O(m \log m)}\right).$$

However, it is unclear whether any further improvement for this problem can be obtained in a very general setting.

(1) When m is substantially large, however one can obtain a good approximate algorithm for the case of χ_{con}. In this case, we first choose a large number of candidate points on the surface of the object ∂B such that the points are placed fairly closely and choose as our set X the image of these points under the wrench map Γ. Thus conv $X = \tilde{G}_{\chi_{con}}$ is a good approximation of the feasible wrench set $G_{\chi_{con}}$ in the sense that the residual radius of $\tilde{G}_{\chi_{con}}$ is a close approximation of that of $G_{\chi_{con}}$:

$$r(\tilde{G}_{\chi_{con}}) \geq r(G_{\chi_{con}})(1 - \epsilon_n),$$

where ϵ_n depends on the number of candidate points n and ∂B. The rest follows from the discussion below. We start with the following algorithmic problem:

> Given a set X of n points in d-dimensional Euclidean space, whose residual radius $r(X)$ is positive, find a subset $Y \subseteq X$ of at most m points such that the following inequality holds:
>
> $$\frac{r(Y)}{r(X)} \geq \tilde{r}_d(m) = 1 - 3d\left(\frac{2d^2}{m}\right)^{\frac{2}{d-1}}.$$
>
> Here m and n are assumed to be sufficiently large, i.e. $n \geq m \geq 13^d d^{\frac{d+3}{2}}$.

We see that this problem can be solved by essentially following the ideas outlined in the proof of KMY bounds: We first choose a set Y' of at most m/d points on the surface of the unit ball such that the residual radius of Y' is no smaller than $\tilde{r}_d(m)$. We can then determine a set $Y \subseteq X$ of at most m points such that for some $\lambda_{\min} \geq r(X)$, the convex hull of Y contains the set of points

$$\lambda_{\min} Y' = \{\lambda_{\min} \, q \colon q \in Y'\}.$$

Thus

$$r(Y) \geq r(\lambda_{\min} Y') \geq \lambda_{\min} \tilde{r}_d(m) \geq r(X)\tilde{r}_d(m).$$

The points of Y' are chosen as follows: Let C be the d-dimensional cube comprising the points $(y_1, \ldots, y_d)$ with $|y_i| \leq 1$ for $i = 1, \ldots, d$. On each face of C, we place a $k \times k \times \cdots \times k$ ($(d-1)$ times) grid, with k taking the value

$$\left\lfloor \left(\frac{m}{2d^2}\right)^{\frac{1}{(d-1)}} \right\rfloor.$$

Let

$$Y' = \left\{ \overrightarrow{op} \cap \mathcal{S}^{d-1} \colon p \text{ is a grid point} \right\}.$$

Thus $|Y'| \leq 2dk^{d-1} \leq m/d$. For each $q \in Y'$, we determine an appropriate set $X_q \subseteq X$ of at most d points such that

$$\overrightarrow{oq} \cap \partial \left(\mathrm{ConvexHull}(X)\right) \in \mathrm{ConvexHull}(X_q);$$

thus for some λ_q,

$$\lambda_q q \in \mathrm{ConvexHull}(X_q).$$

Let Y be

$$Y = \bigcup_{q \in Y'} X_q,$$

with $\lambda_{\min}$ taking the value $\min_{q \in Y'} \lambda_q$. Evidently, $\lambda_{\min} \geq r(X)$.

Note that $|Y| \leq m$, and

$$\lambda_{\min} Y' \subseteq \mathrm{ConvexHull}(Y).$$

This demonstrates the correctness of the algorithm, since we know that the residual radius of Y' is bounded from below by $\tilde{r}_d(m)$. (See the proof for the KMY bounds).

In order to complete the algorithm, we show how to efficiently compute the set X_q (for any point q) using the following linear programming formulation. Let $X = \{p_1, p_2, \ldots, p_n\}$. Without loss of generality, we assume that the points of X are in *general position*, i.e., *at most d points of X may lie on any $(d-1)$ dimensional hyperplane*. If not, the original points of X may be perturbed using generic perturbation methods (see, for example, [34]); the following discussions still apply

mutatis mutandis. Define the $d \times n$ matrix $\mathbf{A}$ whose j^{th} column consists of the coordinates of the point p_j. Corresponding to the point q, define a column d-vector $\mathbf{b}$. The linear programming problem (LP) is given as follows:

- GIVEN: A $d \times n$ matrix $\mathbf{A}$ and a column d-vector $\mathbf{b}$.

- SOLVE:

$$
\begin{aligned}
\text{minimize} \quad & -\lambda \\
\text{subject to} \quad \mathbf{A}\mathbf{x} &= \lambda \mathbf{b} \\
\mathbf{e}^{\mathrm{T}}\mathbf{x} &= 1 \\
\mathbf{x} &\geq \mathbf{0} \\
\lambda &\geq 0,
\end{aligned}
$$

where $\mathbf{x} = (x_1, \ldots, x_n)^{\mathrm{T}}$, $\mathbf{e} = (1, \ldots, 1)^{\mathrm{T}}$ and $\mathbf{0} = (0, \ldots, 0)^{\mathrm{T}}$ are column n-vectors.

Let $\mathbf{x}^*, \lambda^*$ be an optimal solution of (LP). Then $\lambda^* > 0$ is the maximum value of λ such that

$$\lambda^* q = \sum_{i=1}^{n} x_i^* p_i,$$

with $\sum_{i=1}^{n} x_i^* = 1$, and $x_i^* \geq 0$.

Now consider the following *dual* of the (LP), which will be referred to as (DLP):

$$
\begin{aligned}
\text{maximize} \quad & y_{d+1} \\
\text{subject to} \quad a_{1,1}y_1 + \cdots + a_{d,1}y_d + y_{d+1} &\leq 0 \\
a_{1,2}y_1 + \cdots + a_{d,2}y_d + y_{d+1} &\leq 0 \\
&\vdots \quad \vdots \\
a_{1,n}y_1 + \cdots + a_{d,n}y_d + y_{d+1} &\leq 0 \\
-b_1 y_1 - \cdots - b_d y_d &\leq -1
\end{aligned}
$$

This problem can be solved in $O(3^{d^2} n)$ time by using Clarkson-Dyer's improvement on Megiddo's multidimensional search technique [7,8,17]. Let us now see how to recover the solution to the original problem.

Clearly both (LP) and (DLP) have optimal solutions. Let an optimal solution for (DLP) be

$$\mathbf{y}^* = (y_1^*, \ldots, y_d^*, y_{d+1}^*).$$

Let $I_q \subseteq \{1..n\}$ be the set of all the indices j such that

$$\mathbf{a}_j \cdot \mathbf{y}^* = a_{1,j}y_1^* + \cdots + a_{d,j}y_d^* + y_{d+1}^* = 0.$$

where $\mathbf{a}_j = (a_{1,j}, \cdots, a_{d,j}, 1)^{\mathrm{T}}$. By the Complementary Slackness Theorem (see [6]), this implies that for all $i = 1, \ldots, n$, if $x_i^* > 0$ then $i \in I_q$. By virtue of our non-degeneracy hypothesis about the points of X, we see that $|I_q| \leq d$. We now claim that $X_q = \{p_j : j \in I_q\}$ can serve as a desired solution. Clearly, $X_q \subseteq X$, has at most d points and

$$\vec{oq} \cap \partial(\text{ConvexHull}(X)) \in \text{ConvexHull}(X_q)).$$

Note that even if the original set had been perturbed (by a sufficiently small amount) the set X_q chosen from the unperturbed set X still provides the desired solution.

To summarize:

Theorem 7 *For* $n \geq m \geq 13^d d^{(d+3)/2}$*, we can find an* m*-fingered grasp* $\mathcal{G}$ *of an object* B *such that*

$$\frac{r_{\chi_{con}}(\mathcal{G})}{r^*} \geq (1 - \epsilon_n)\left(1 - 3d\left(\frac{2d^2}{m}\right)^{\frac{2}{d-1}}\right),$$

in time $O(3^{d^2}mn)$. *Here* n *is the number of candidate points and* ϵ_n *is a small number that depends on* n *and* ∂B. $\square$

7 Bibliographic Notes

The general framework describing the connection between grasping and convexity theory is due to Mishra, Schwartz and Sharir [21]. The results relating form and force/torque closure appear in the paper by Mishra and Silver [22] and the results relating immobility and force/torque closure appear in the paper by Mishra and Teichmann [23]. Some of the other related results are taken from our earlier work [18,23].

The grasp metric of section 2 is partly based on our earlier work [14]. For the sake of concreteness, only the finger force constraint χ_{con} was discussed. Ferrari and Canny [9] subsequently suggested that in some situations χ_{max} condition may be of more practical interest. Independently, Li and Sastry [15] proposed a metric that corresponds to the conditions referred to here as $\tilde{\chi}_2$ and χ_2. Trinkle [33] studied the problem of computing with the grasp metric, r_{null}, for small number of fingers. The relations between r_{null} and $r_{\chi_{con}}$ described here has not appeared elsewhere. The grasp metric based on the nasty finger model and the result proving the weakness of a seven finger grasp is due to Walter Meyer and was motivated by our original (KMY) grasp metric.

The results involving Q.S.T. are due to Bárány, Katchalski and Pach [2] and the generalizations motivated by the study of grasp metrics are due to Kirkpatrick, Mishra and Yap [14]. The approximate algorithm in the general case

and appearing in the last section is based on the results of Kirkpatrick, Mishra and Yap [14]. The results related to optimal three finger planar grasp are jointly with Marek Teichmann of N.Y.U.

An algorithmic study of the modular fixturing problem in manufacturing was initiated by the present author [19,20]. Recently, some important further progress in this area, specially with an emphasis on the quality of the fixturing, has been made by Brost, Goldberg, Wong and Zhuang [3,35]. The study of reactive robotics algorithm in general as well as in the context of grasping, has been initiated by Teichmann and Mishra [31,32].

Appendix: Geometric Terminology

A.1 Linear Spaces and Convexity

A d-dimensional space, $\mathbb{R}^d$, equipped with the standard linear operations, is said to be a *linear space*.

1. A *linear combination* of vectors $\mathbf{x}_1, \ldots, \mathbf{x}_n$ from $\mathbb{R}^d$ is a vector of the form
   $$\lambda_1 \mathbf{x}_1 + \cdots + \lambda_n \mathbf{x}_n,$$
   where $\lambda_1, \ldots, \lambda_n$ are in $\mathbb{R}$.

2. An *affine combination* of vectors $\mathbf{x}_1, \ldots, \mathbf{x}_n$ from $\mathbb{R}^d$ is a vector of the form
   $$\lambda_1 \mathbf{x}_1 + \cdots + \lambda_n \mathbf{x}_n,$$
   where $\lambda_1, \ldots, \lambda_n$ are in $\mathbb{R}$, with $\lambda_1 + \cdots + \lambda_n = 1$.

3. A *positive (linear) combination* of vectors $\mathbf{x}_1, \ldots, \mathbf{x}_n$ from $\mathbb{R}^d$ is a vector of the form
   $$\lambda_1 \mathbf{x}_1 + \cdots + \lambda_n \mathbf{x}_n,$$
   where $\lambda_1, \ldots, \lambda_n$ are in $\mathbb{R}_{\geq 0}$.

4. A *convex combination* of vectors $\mathbf{x}_1, \ldots, \mathbf{x}_n$ from $\mathbb{R}^d$ is a vector of the form
   $$\lambda_1 \mathbf{x}_1 + \cdots + \lambda_n \mathbf{x}_n,$$
   where $\lambda_1, \ldots, \lambda_n$ are in $\mathbb{R}_{\geq 0}$ with $\lambda_1 + \cdots + \lambda_n = 1$.

By convention, we allow the empty linear combination (with $n = 0$) to take the value $\mathbf{0}$. We also assume that the empty linear combination is neither an affine combination nor a convex combination.

Note that affine, positive and convex combinations are all linear combinations, and a convex combination is both affine and positive combinations.

A nonempty subset $L \subseteq \mathbb{R}^d$ is said to be a

1. *linear subspace*: if it is closed under linear combinations;

2. *affine subspace (or, flat)*: if it is closed under affine combinations;

3. *positive set (or, cone)*: if it is closed under positive combinations; and

4. *convex set*: if it is closed under convex combinations.

The intersection of any family of linear subspaces of $\mathbb{R}^d$ is again a linear subspace of $\mathbb{R}^d$. For any subset M of $\mathbb{R}^d$, the intersection of all linear subspaces containing M (i.e. the smallest linear subspace containing M) is called the *linear hull* of M (or, the linear subspace *spanned* by M), and is denoted by lin M.

Similarly, the intersection of any family of affine subspaces, or positive sets or convex sets of $\mathbb{R}^d$ is again, respectively, an affine subspace or positive set or convex set. Thus for any subset M of $\mathbb{R}^d$, we can define

1. the *affine hull* (denoted by aff M) to be the smallest affine subspace containing M,

2. the *positive hull* (denoted by pos M) to be the smallest positive set containing M, and

3. the *convex hull* (denoted by conv M) to be the smallest convex set containing M.

They are also called, respectively, the affine subspace, positive set and convex set *spanned* by M.

Equivalently, the linear hull lin M can be defined to be the set of all linear combinations of vectors from M. Similarly, the affine hull aff M (respectively, the positive hull pos M, the convex hull conv M) can be defined to be the set of all affine (respectively, positive, convex) combinations of vectors from M.

A set $\mathbf{x}_1, \ldots, \mathbf{x}_n$ of n vectors from $\mathbb{R}^d$ is said to be *linearly independent* if a linear combination

$$\lambda_1 \mathbf{x}_1 + \cdots + \lambda_n \mathbf{x}_n$$

can only have the value $\mathbf{0}$, when $\lambda_1 = \cdots = \lambda_n = 0$; otherwise, the set is said to be *linearly dependent*.

A set $\mathbf{x}_1, \ldots, \mathbf{x}_n$ of n vectors from $\mathbb{R}^d$ is said to be *affinely independent* if a linear combination

$$\lambda_1 \mathbf{x}_1 + \cdots + \lambda_n \mathbf{x}_n \quad \text{with } \lambda_1 + \cdots + \lambda_n = 0$$

can only have the value $\mathbf{0}$, when $\lambda_1 = \cdots = \lambda_n = 0$; otherwise, the set is said to be *affinely dependent*.

A *linear basis* of a linear subspace L of $\mathbb{R}^d$ is a set M of linearly independent vectors from L such that $L = \text{lin } M$. The dimension dim L of a linear subspace L is the cardinality of any of its linear basis.

An *affine basis* of an affine subspace A of $\mathbb{R}^d$ is a set M of affinely independent vectors from L such that $A = \text{aff } M$. The dimension dim A of an affine subspace A is one less than the cardinality of any of its affine basis.

Let C be any convex set. Then by *d-interior* of C, denoted $\text{int}_d\, C$, we mean the set of points p such that, for some d-dimensional affine subspace, A, p is interior to $C \cap A$ relative to A. If c is the dim aff C, then by an abuse of notation, we write int C to mean $\text{int}_c\, C$.

For subsets A and B of $\mathbb{R}^d$ and λ real define the *(Minkowski) sum* of A and B by

$$A + B = \{a + b : a \in A, b \in B\},$$

and let λA be

$$\lambda A = \{\lambda a : a \in A\}.$$

We shall write $A \oplus B$ instead of $A + B$ if A and B are contained in subspaces of $\mathbb{R}^d$ for which the usual direct sum exists: $A \oplus B$ is then called the *direct sum* of A and B. Call C *directly irreducible* if there is no representation of C of the form $A \oplus B$ where both A and B are different from the origin. By a decomposition theorem of Gruber, we have the result that each convex body C can be represented in the form $C_1 \oplus \cdots \oplus C_m$ where $C_1, \ldots, C_m$ are directly irreducible. Such a representation is unique modulo the order of the summands.

A.2 Screw Theory

A *screw* is defined by a straight line in three-dimensional Euclidean space, called, its *screw-axis* and an associated *pitch, p*. A screw is represented by a six-dimensional vector, $\mathbf{s} = (S_1, S_2, S_3, S_4, S_5, S_6)$, known as the *screw coordinates*. The screw coordinates are interpreted in terms of the *Plücker line coordinates*, (L, M, N, P, Q, R), of the screw axis, as follows:

$$
\begin{aligned}
L &= S_1, \\
M &= S_2, \\
N &= S_3, \\
P &= S_4 - pS_1, \\
Q &= S_5 - pS_2, \\
R &= S_6 - pS_3,
\end{aligned}
$$

where L, M and N are proportional to the direction cosines of the screw axis, and P, Q and R are proportional to the moment of the screw axis about the origin of the reference frame (i.e. the cross product of a vector from the origin to a point on the axis and a unit vector, directed along the screw axis). The pitch of the screw is then given by

$$p = \frac{S_1 S_4 + S_2 S_5 + S_3 S_6}{S_1^2 + S_2^2 + S_3^2},$$

and the magnitude of the screw is given by

$$|\mathbf{s}| = \begin{cases} \sqrt{S_1^2 + S_2^2 + S_3^2}, & \text{if } p < \infty; \\ \sqrt{S_4^2 + S_5^2 + S_6^2}, & \text{if } p = \infty. \end{cases}$$

A *unit screw* is a screw with unit magnitude. Scalar multiplication and vector addition are valid for infinitesimal screws, and the screws are closed under these operations. Thus the six-dimensional space of infinitesimal screws forms a vector space.

Sometimes, we simply consider the 2-norms of a screw (as a six-dimensional vector), disregarding its pitch:

$$|\mathbf{s}|_2 = \sqrt{\sum_{i=1}^{6} S_i^2}.$$

Given two screws $\mathbf{s}' = (S_1', S_2', S_3', S_4', S_5', s_6')$ and $\mathbf{s}'' = (S_1'', S_2'', S_3'', S_4'', S_5'', S_6'')$, we define their *virtual coefficient* as

$$\mathbf{s}' \odot \mathbf{s}'' = S_1'S_4'' + S_2'S_5'' + S_3'S_6'' + S_4'S_1'' + S_5'S_2'' + S_6'S_3''.$$

Note that the operation '$\odot$' is a commutative operation from $\mathbb{R}^6 \times \mathbb{R}^6$ into $\mathbb{R}$.

Two screws $\mathbf{s}'$ and $\mathbf{s}''$ are said to be

1. *reciprocal*: if their virtual coefficient is zero, i.e. $\mathbf{s}' \odot \mathbf{s}'' = 0$,

2. *repelling*: if their virtual coefficient is strictly positive, i.e. $\mathbf{s}' \odot \mathbf{s}'' > 0$, and

3. *contrary*: if their virtual coefficient is strictly negative, i.e. $\mathbf{s}' \odot \mathbf{s}'' < 0$.

An ensemble of screws is known as a *screw system*, and is defined by a set of $n \leq 6$ independent *basis screws*. The *order* of a screw system is equal to the number of basis screws required to define it; such a system is also called an *n-system*. The order of a screw system reciprocal to an *n*-system is $(6 - n)$.

With an infinitesimal rigid motion of an object in three-dimensional Euclidean space there is an associated screw called *twist* such that the body rotates about and translates along its screw axis. The screw coordinates of a twist are given by $\mathbf{t} = (T_1, T_2, T_3, T_4, T_5, T_6)$, where the first three components T_1, T_2 and T_3 correspond to the angular displacement (or angular velocity), $\overline{\omega}$, of the body and the last three components T_4, T_5 and T_6 correspond to the translational displacement (or translational velocity), $\overline{v}$, of a point fixed in the body and lying at the origin of the coordinate system. The pitch of the twist is given by

$$p = \frac{\overline{\omega} \cdot \overline{v}}{\overline{\omega} \cdot \overline{\omega}}.$$

The pitch of the twist is the ratio of the magnitude of the velocity of a point on the twist axis to the magnitude of the angular velocity about the twist axis. If the pitch of a twist is zero then the twist corresponds to a pure rotation, and if the pitch of a twist is infinite then the twist corresponds to a pure translation. The magnitude of the twist is given by

$$|\mathbf{t}| = \begin{cases} \|\overline{\omega}\|_2, & \text{if } p < \infty; \\ \|\overline{v}\|_2, & \text{if } p = \infty. \end{cases}$$

Similarly, with any system of forces and torques acting on a rigid object in three-dimensional Euclidean space there is an associated screw called *wrench* such that the system of forces and torques can be replaced by an equivalent system of single force along the wrench axis and a torque about the same wrench axis. The screw coordinates of a wrench are given by $\mathbf{w} = (W_1, W_2, W_3, W_4, W_5, W_6)$, where the first three components W_1, W_2 and W_3 correspond to the resultant force, $\overline{f}$, acting on the body along the wrench axis and the last three components W_4, W_5 and W_6 correspond to the resultant torque, $\overline{\tau}$, acting on the body about the wrench axis. The pitch of the wrench is given by

$$p = \frac{\overline{f} \cdot \overline{\tau}}{\overline{f} \cdot \overline{f}}.$$

The pitch of the wrench is the ratio of magnitude of the torque acting about a point on the axis to the magnitude of the force acting along the axis. If the pitch of a wrench is zero then the wrench corresponds to a pure force, and if the pitch of a wrench is infinite then the wrench corresponds to a pure moment. The magnitude of the wrench is given by

$$|\mathbf{w}| = \begin{cases} \|\overline{f}\|_2, & \text{if } p < \infty; \\ \|\overline{\tau}\|_2, & \text{if } p = \infty. \end{cases}$$

Note that the virtual coefficient of a twist $\mathbf{t} = (\overline{\omega}, \overline{v})$ and a wrench $\mathbf{w} = (\overline{f}, \overline{\tau})$ is

$$\mathbf{w} \odot \mathbf{t} = \overline{f} \cdot \overline{v} + \overline{\tau} \cdot \overline{\omega},$$

the rate of change of work done by the wrench $\mathbf{w}$ on a body moving with the twist $\mathbf{t}$.

If a twist $\mathbf{t}$ is reciprocal to a wrench $\mathbf{w}$, then the wrench does no work when the body is displaced infinitesimally by the twist. Thus for two reciprocal screws, a twist about one of the screws is possible while the body is being constrained about the other screw. Similarly, if $\mathbf{t}$ is repelling to $\mathbf{w}$, then positive work is done by the constraining wrench when the body is displaced infinitesimally by the twist. This implies that the twist can be accomplished, but then the contact of the wrench will be definitely broken. Lastly, if $\mathbf{t}$ is contrary to $\mathbf{w}$, then negative (virtual) work must be done by the constraining wrench when the body is displaced infinitesimally by the twist. This implies that such a displacement is impossible, if we assume that the objects being considered are all rigid.

For a given wrench system acting on a body, we say that the body has *total freedom*, if the body can undergo all possible twists, without breaking the contacts associated with the wrenches; we also say that the body has *total constraint*, if the body cannot undergo any twist, without breaking the contacts; otherwise, we say that the body has *partial constraint*.

References

[1] I. Bárány and Z. Füredi. "Computing the Volume is Difficult", in *Proc. 18th Annual ACM Symposium on Theory of Computing*, pp. 442–447, 1986.

[2] I. Bárány, M. Katchalski and J. Pach. "Quantitative Helly-type Theorems", *Proc. AMS*, pp. 109–114, **86**, 1982.

[3] R.C. Brost and K.Y. Goldberg. "A Complete Algorithm for Synthesizing Modular Fixtures for Polygonal Parts," Submitted to the *1994 IEEE International Conference on Robotics and Automation*, 1994.

[4] R.C. Brost, *Analysis and Planning of Planar Manipulation Tasks*, Ph.D. Thesis, School Computer Science, Carnegie Mellon University, Pittsburgh, PA. 1991.

[5] C. Carathéodory, "Über den Variabilitätsbereich der Koeffizienten von Potenzreihen, die gegebene Werte nicht annehmen", *Math. Ann.*, **64**, 95–115, 1907.

[6] V. Chvátal. *Linear Programming*, W.H. Freeman and Company, New York, 1983.

[7] K.L. Clarkson. "Linear Programming in $O(n3^{d^2})$ Time," *Information Processing Letters*, **22**, pp. 22–24, 1988.

[8] M.E. Dyer. "On a Multidimensional Search Technique and Its Application to the Euclidean One Center Problem," Tech Report, Teeside Polytechnic, United Kingdom, 1984.

[9] C. Ferrari and J. Canny. "Planning Optimal Grasps," *Proceedings of the 1992 IEEE International Conference on Robotics and Automation*, Nice (France), pp. 2290–2295, 1992.

[10] K.Y. Goldberg, *Stochastic Plans for Robotic Manipulation*, Ph.D. Thesis, School Computer Science, Carnegie Mellon University, Pittsburgh, PA. 1991.

[11] I.S. Gradshteyn and I.M. Ryzhik, *Table of Integrals, Series and Products*, Academic Press, New York, Corrected and Enlarged Edition, 1980.

[12] J.-W. Hong, G. Lafferriere, B. Mishra and X. Tan. *Proceedings of the 1990 IEEE International Conference on Robotics and Automation*: ICRA'90, pp. 1568–1573, Cincinnati, Ohio, May 13–18, 1990.

[13] K.H. Hunt. *Kinematic Geometry of Mechanisms*, Clarendon Press, Oxford, 1978.

[14] D. Kirkpatrick, B. Mishra and C.-K. Yap. "Quantitative Steinitz's Theorem with Applications to Multifingered Grasping," *Discrete & Computational Geometry*, Springer-Verlag, New York, pp. 295–318, Volume **7**, Number 3, 1992.

[15] Z-X. Li and S. Sastry. "Task Oriented Optimal Grasping by Multifingered Robot Hands," *Proceedings of the 1987 IEEE International Conference on Robotics and Automation*, pp. 389–394, 1987.

[16] X. Markenscoff, L. Ni and C.H. Papadimitriou. "The Geometry of Grasping," *International Journal of Robotics Research*, pp. 61–74, Volume **9**, Number 1, 1990.

[17] N. Megiddo. "Linear Programming in Linear Time when the Dimension is Fixed," *Journal of the ACM*, **31**(1), pp. 114–127, 1984.

[18] B. Mishra. "Dexterous Manipulation: A Geometric Approach," *Advances in Robot Kinematics*, (With Emphasis on Symbolic Computation), Vol. **XIV**, (Edited by S. Stifter and J. Lenarčič), pp. 17–27, Springer-Verlag, Wien, New York, 1991.

[19] B. Mishra. "Workholding—Analysis and Planning," *Proceedings: IEEE/RSJ International Workshop on Intelligent Robots and Systems*: IROS'91, pp. 53–57, Vol. **1**, Osaka, Japan, November 3–5, 1991.

[20] B. Mishra. "An Algorithmic Approach to Fixturing," *1994 NSF Design and Manufacturing Grantees Conference*, M.I.T., Massachusetts, Cambridge, January, 1994 (To appear).

[21] B. Mishra, J.T. Schwartz and M. Sharir, "On the Existence and Synthesis of Multifinger Positive Grips," Special Issue: Robotics, *Algorithmica*, Springer-International, pp. 541–558, Volume **2**, Number 4, 1987.

[22] B. Mishra and N. Silver. "Some Discussion of Static Gripping and Its Stability," *IEEE Transactions on Systems, Man and Cybernetics*, pp. 783–796, Volume **19**, Number 4, July/August, 1989.

[23] B. Mishra and M. Teichmann. "On Immobility," Special Issue: Robot Kinematics, *Laboratory Robotics and Automation*: International Journal, VCH Publisher, pp. 145–153, Volume **4**, 1992.

[24] B. Mishra and M. Teichmann. "Three Finger Optimal Planar Grasp," Submitted to the *1994 IEEE/RSJ International Workshop on Intelligent Robots and Systems*: IROS'94, Grenoble, (France), 1994.

[25] V. Nguyen. "Constructing Force-Closure Grasps," *Proceedings of the 1986 IEEE International Conference on Robotics and Automation*, San Francisco, California (USA), pp. 1368–1373, 1986.

[26] M.S. Ohwovoriole. *An Extension of Screw theory and its Applications to the Automation of Industrial Assemblies*, Stanford Artificial Intelligence Laboratory Memo AIM-338, April, 1980.

[27] J. Renegar "Recent Progress on the Complexity of the Decision Problem for the Reals," **6**, *Discrete and Computational Geometry: Papers from the DIMACS Special Year*, (Edited by J.E. Goodman, R. Pollack and W. Steiger), pp. 287–308, AMS & ACM, 1991.

[28] F. Reuleaux. *Theoretische Kinematic: Gunzüge einer Theorie des Maschinwesens*, Braunschweig, Vieweg; Also, Dover, New York, 1963.

[29] J.M. Selig and J. Rooney. "Reuleaux Pairs and Surfaces That Cannot Be Gripped" *The International Journal of Robotics Research*, pp. 79—86, **8**(5), 1989.

[30] E. Steinitz "Bedingt Konvergente Reihen und Konvexe Systeme", *J. reine angew. Math.*, (I) 143:128–175, 1913; (II) 144:1–48, 1914 and (III) 146:1–52, 1916.

[31] M. Teichmann and B. Mishra. "Reactive Algorithms for Grasping Using a Modified Parallel Jaw Gripper," Submitted to the *1994 IEEE International Conference on Robotics and Automation*: ICRA'94, San Diego, California, 1994.

[32] M. Teichmann and B. Mishra. "Reactive Algorithms for 2 and 3 Finger Grasping," Submitted to the *1994 IEEE/RSJ International Workshop on Intelligent Robots and Systems*: IROS'94, Grenoble, (France), 1994.

[33] J.C. Trinkle. "A Quantitative Test for Form Closure Grasps," Department of Computer Science, Texas A&M University, College Station, TX, 1992.

[34] C.-K. Yap. "A Geometric Consistency Theorem for a Symbolic Perturbation Scheme," *Journal of Computers and Systems Sciences* **40**(1), pp. 2–18, 1990.

[35] Y. Zhuang, K.Y. Goldberg and Y-C. Wong. "On the Existence of Modular Fixtures," Submitted to the *1994 IEEE International Conference on Robotics and Automation*, 1994.

Algorithms for Computing Force-Closure Grasps of Polyhedral Objects

Jean Ponce, Attawith Sudsang, and Steve Sullivan, *University of Illinois, Urbana, IL, USA*
Bernard Faverjon, *ALEPH Technologies, Saint Martin d'Hères, France*
Jean-Daniel Boissonnat and Jean-Pierre Merlet, *INRIA Sophia-Antipolis, Valbonne, France*

We present an overview of our work on the synthesis of force-closure grasps of two- and three-dimensional polyhedral objects. We consider the case of a gripper equipped with three or four hard fingers and assume point contact with friction. We give several necessary and several sufficient conditions for equilibrium and force closure, and present various algorithms for computing force-closure grasps. We show some results and outline a number of extensions.

1 Introduction

We address the problem of computing stable three- and four-finger grasps of polygonal and polyhedral objects. More precisely, we consider force-closure grasps, such that arbitrary forces and torques exerted on the grasped object can be balanced by the contact forces exerted by the fingers. We assume hard-finger point contact with Coulomb friction.

Reulaux first used the related notions of force and form closure to study the properties of mechanisms [59]. In the robotics context, conditions for force closure have been given by Salisbury and Roth [61, 62], Ji and Roth [26, 27], Nguyen [45] and Trinkle [66] for example. Bounds on the minimum number of contacts necessary to achieve force closure have been established by Lakshminarayana [31] and Mishra, Schwartz, and Sharir [41] in the frictionless case. When Coulomb friction is taken into account, Markenscoff, Ni, and Papadimitriou [35] have shown that, under very general conditions, three (resp. four) fingers are sometimes necessary and always sufficient to achieve a force-closure grasp for a non-exceptional two-dimensional (resp. three-dimensional) object. This result and the recent availability of dextrous grippers such as the Salisbury hand [61] or the Utah/MIT Dextrous Hand [24] are practical motivations for studying three- and four-finger grasping.

Several algorithms for synthesizing force-closure grasps of polyhedral and curved objects have also been proposed: for example, Kerr and Roth [28, 29] give a numerical search algorithm for optimizing the equilibrium forces applied by the fingers, while Ji and Roth give an analytical method for computing these forces in the three-finger case [26, 27]. Mishra, Schwartz, and Sharir [41] and Markenscoff, Ni, and Papadimitriou [35] present linear-time algorithms for computing at least one force-closure grasp of a polyhedral object. Nguyen gives a geometric method for finding maximal independent two-finger grasps of polygons [45]. Pollard and Lozano-Pérez plan three-finger grasps of polyhedral objects by first intersecting the object with some plane and then optimizing the choice of a focus point within this plane [51]. Omata uses fractional programming methods for manipulating polygonal objects with two fingers in the plane [48]. Ponce and Faverjon use linear programming to compute maximal independent three-finger grasps of polygonal objects [52, 53]. Faverjon, Ponce, and Stam [17, 55] and Chen and Burdick [11] use global non-linear optimization methods to compute two-finger grasps of curved objects. Ferrari and Canny [19] and Mirtich and Canny [40] compute optimal force-closure grasps of polygonal objects by maximizing the set of external wrenches that can be balanced by the contact wrenches. Other approaches to grasp analysis and synthesis include [1, 3, 4, 7, 8, 12, 13, 18, 20, 21, 25, 29, 32, 33, 34, 44, 49, 63, 65]. See [42, 50] for recent surveys.

In this paper, we present several methods for synthesizing three-finger force-closure grasps of 2D polygonal objects and four-finger force-closure grasps of 3D polyhedral objects. We give necessary and sufficient conditions for equilibrium and force closure that are linear in the unknown position of the fingers on the polygonal faces and in the position of an additional point where the friction cones at the fingers intersect. This reduces the problem of computing the force-closure grasp regions in configuration space to the problem of constructing the projection of a polytope defined in a high-dimension space onto a lower-dimension subspace. We present efficient projection algorithms based on linear programming and variable elimination among linear constraints. Once the grasp regions have been computed, maximal object segments where fingers can be positioned independently while ensuring force closure are found by linear optimization. We have implemented the proposed approaches and present several examples.

Mathematical details have been omitted for the sake of conciseness and can be found in [54, 56, 57].

2 Force-Closure Grasps

We characterize the stable grasps of a two- or three-dimensional object by a hand equipped with d hard fingers, assuming point contact with friction. A grasp is defined geometrically by the position $\mathbf{x}_i$ of the fingers on the boundary of the object, with $i = 1, .., d$. We assume that the points $\mathbf{x}_i$ are all different, that no three of them are aligned, and (in the three-dimensional case) that no four of them are coplanar.

Each finger exerts in $\mathbf{x}_i$ a force $\mathbf{f}_i$ and a moment $\mathbf{x}_i \times \mathbf{f}_i$ with respect to the origin. Force and moment are combined into a zero-pitch wrench $\mathbf{w}_i = (\mathbf{f}_i, \mathbf{x}_i \times \mathbf{f}_i)$ [6, 38]. This wrench is a three-dimensional vector in the case of planar grasps or a six-dimensional vector in the case of three-dimensional grasps.

2.1 Equilibrium and Force Closure

Definition 1 *A grasp is said to achieve force closure when any external wrench can be balanced by the wrenches at the fingertips.*

In other words, a grasp is force-closure iff, for any external wrench $\mathbf{w}$, the fingers can exert a set of wrenches $\mathbf{w}_i$ such that

$$\mathbf{w} + \sum_{i=1}^{d} \mathbf{w}_i = 0.$$

Again, this definition is independent of the choice of coordinate system.

Force closure is sometimes called *force/torque closure* [41, 42]. A related notion is *form closure*: a grasp

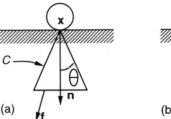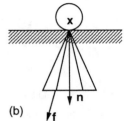

Figure 1: *Coulomb friction: (a) continuous friction cone; (b) discrete approximation by an m-sided pyramid.*

is said to achieve form closure when it prevents all motions (including infinitesimal ones) of the object held by the hand (this is also related to the notions of reciprocal, contrary, and repelling screws [2, 6, 23, 46, 47]). Force and form closure are dual from each other, in the same sense as forces and moments are dual from angular and linear velocities in mechanics [60] and, as noted in [66] for example, force-closure grasps are form-closure and vice versa. We should also note that there is unfortunately no general agreement on terminology in the grasping literature (see [40, 66] for discussions of this problem). Our definitions match the ones given in [40, 45, 53, 55].

From now on, we assume point contact with Coulomb friction. The force exerted by each finger lies in the internal friction cone of half-angle θ at the contact point (Figure 1(a)). Note that writing that a force $\mathbf{f}$ lies in the friction cone is quadratic in $\mathbf{f}$. This condition can be expressed instead as a set of linear constraints in $\mathbf{f}$ by approximating the friction cone by an m-sided pyramid (Figure 1(b)). This will prove important in the sequel.

Definition 2 *A grasp is said to achieve equilibrium when there exists a set of forces (not all of them being zero) in the friction cones at the fingertips such that the sum of the corresponding wrenches is zero.*

Intuitively, equilibrium is a somewhat weaker condition than force closure. Formally, it is shown in [41, 42, 45] for example that force closure implies equilibrium. More interestingly, the converse is also true for *non-marginal equilibrium* grasps, i.e., grasps such that the forces achieving equilibrium lie strictly inside the friction cones at the fingertips.

Proposition 1 *Non-marginal equilibrium is sufficient*

for force closure in the two-dimensional, three-finger case and the three-dimensional, four-finger case.

See [45] for a proof in the two-dimensional, three-finger case, and [57] for a proof in the three-dimensional, four-finger case.

Since for a given friction coefficient, a grasp achieving equilibrium with non-zero forces trivially achieves non-marginal equilibrium for any strictly greater friction coefficient, it also achieves force closure for this friction coefficient. In practice, it should be noted that Proposition 1 does not tell us anything about the actual wrenches that the hand should exert to balance a given external load. In particular, a real hand can only exert bounded forces. See [19, 26, 27, 33, 36, 40] and Section 8 for a discussion of related issues.

In the next two sections, we will give necessary and sufficient condition for equilibrium, hence for force closure. We will first attack the problem of characterizing three-finger grasps of polygonal objects, which can be treated in an elementary fashion and illustrates key problems in the characterization of equilibrium and force-closure grasps. We will then switch to the problem of characterizing four-finger grasps of polyhedral objects. This will require using some elementary results of Grassmann geometry [14, 39].

3 Three-Finger Grasps of Polygonal Objects

Proposition 2 *A necessary condition for three contact points to form an equilibrium grasp is that the intersection of the corresponding double-sided friction cones is not empty.*

See [54] for an elementary proof. As shown by Figure 2, this condition is not sufficient.

It is easy to prove the following condition, which is sufficient, but not necessary, in the case where the intersection point lies in the *internal* friction cones

Proposition 3 *A sufficient condition for three contact points to form an equilibrium grasp is that the intersection of the corresponding open internal friction cones with the triangle formed by these points is not empty.*

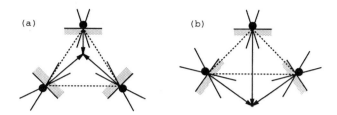

Figure 2: *Three-finger grasps: (a) achieves equilibrium, (b) does not.*

Note that, for a polygonal object, writing that some point $\mathbf{x}_0$ belongs to the friction cone at $\mathbf{x}_i$ is a linear condition in both $\mathbf{x}_0$ and $\mathbf{x}_i$: the vector $\mathbf{x}_0 - \mathbf{x}_i$ must lie on the left side of the right edge of the friction cone, and on the right side of its left edge (this would not be true in the three-dimensional case). On the other hand, writing that $\mathbf{x}_0$ belongs to the triangle formed by the contact points is a bilinear constraint on the coordinates of $\mathbf{x}_0$, $\mathbf{x}_1$, $\mathbf{x}_2$, and $\mathbf{x}_3$.

Next, we will give two sufficient conditions which are linear in all the variables, but first we must give another definition.

Definition 3 *A set of vectors positively span $\mathbb{R}^n$ if any vector in $\mathbb{R}^n$ can be written as a strictly positive combination of the given vectors (Figure 3).*

Figure 3: *(a) Three edges whose internal normals positively span the plane. (b) Three edges whose internal normals do not positively span the plane.*

Proposition 4 *When the surface normals at three contact points positively span the plane and the intersection of the three internal friction cones at these points is not empty, either two of the points form a two-finger equilibrium grasp, or the three points form a three-finger equilibrium-closure grasp.*

The proof of this proposition is essentially a case analysis of the possible contact configurations [54].

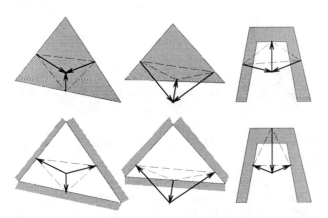

Figure 4: *Different types of three-finger equilibrium grasps.*

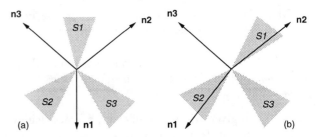

Figure 5: *Normals positively spanning the plane: (a) the pairwise angles are all smaller than $\pi - 2\theta$; (b) the angle between $\mathbf{n}_1$ and $\mathbf{n}_2$ is greater than $\pi - 2\theta$. For each normal, a forbidden sector of vectors making angles greater than $\pi - 2\theta$ is shown as a grey triangle.*

Considering only the case of intersection points inside the internal friction cones may be overly restrictive: as demonstrated by Figure 4, there are many types of three-finger grasps; sometimes the intersection points lie in the internal friction cones, while other times they lie in the external ones.

Under a slightly restrictive condition on the normals, the following proposition allows us to give another sufficient condition for force closure which is more general from the point of view of the intersection points.

Proposition 5 *A sufficient condition for three contact points to form an equilibrium grasp is that the three internal normals at the contact points positively span the plane with pairwise angles less than $\pi - 2\theta$ (Figure 5), and the intersection of the three open double-sided friction cones is not empty.*

See again [54] for a proof. Note that this condition is a disjunction of linear constraints (the common point may lie in the internal or in the external friction cone at each of the contact points).

In Section 5, we will give an algorithm for computing all configurations of three contact points satisfying the conditions of Propositions 4 or 5. We will assume that all contacts occur in the interior of the polygonal edges. Of course, force-closure grasps can occur with one finger at a concave vertex: the force exerted by this finger may lie anywhere in the friction cones of the incident edges or in the region separating them [9] (see [8] for a general discussion of friction models at vertex contacts). The corresponding force-closure grasps can

be characterized by modifying the force-closure constraints to take into account this larger friction cone. This case can be easily treated as a sub-case of the one we are looking at by considering concave vertices as zero-length edges with normal aligned with the bisector of the edges' normals.

4 Four-Finger Grasps of Polyhedral Objects

4.1 Line Geometry

A zero-pitch wrench $\mathbf{w} = (\mathbf{f}, \mathbf{x} \times \mathbf{f})$ can also be seen as a representation of the line of action of the force $\mathbf{f}$ applied at the point $\mathbf{x}$; the six coordinates $(w_1, .., w_6)$ of $\mathbf{w}$ are called the Plücker coordinates of the line.

More precisely, the Plücker coordinates are homogeneous coordinates for a projective space of dimension 5: the wrenches $\mathbf{w} = (\mathbf{f}, \mathbf{x} \times \mathbf{f})$ and $\lambda\mathbf{w}$, with $\lambda \neq 0$, both represent the line with direction $\mathbf{f}$ passing through the point $\mathbf{x}$ (note that $\mathbf{w}$ is independent of the choice of $\mathbf{x}$ along the line). The set of lines form a quadric surface, called the Grassmannian, defined by $w_1 w_4 + w_2 w_5 + w_3 w_6 = 0$ in this projective space. Hence they form a variety of dimension 4.

Equilibrium implies that the lines of action of the forces are linearly dependent. Grassmann geometry characterizes the varieties of various dimensions formed by sets of dependent lines [14, 39], and it can be used to characterize equilibrium geometrically. For example, three linearly dependent lines from a flat pencil. In other words, a necessary condition for three-finger

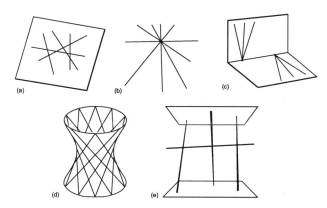

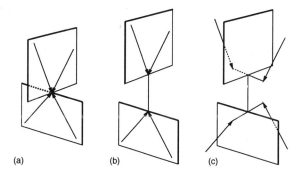

Figure 6: *Configurations of linearly-dependent lines: (a) four coplanar lines, (b) four intersecting lines, (c) two flat pencils of lines having a line in common, (d) a regulus. As shown in (e), each line in the regulus intersects three skew lines. (After [22].)*

Figure 7: *Examples of grasps achieving equilibrium: (a) the contact forces intersect in one point; (b) they form two non-coplanar pencils; (c) they form a regulus. The contact faces are not shown.*

equilibrium in three dimensions is that the lines of action of the three forces be coplanar and intersect in a point. This condition is similar to the one of Proposition 2, and can be used to synthesize three-finger force-closure grasps of polyhedral objects [26, 56]. Here, we concentrate on the four-finger case, as illustrated by the following proposition, which is another classical result from Grassmann geometry [14, 39].

Proposition 6 *A set of four linearly-dependent lines either lie in a single plane, intersect in a single point, form two flat pencils having a line in common but lying in different planes, or form a regulus (Figure 6).*

The lines in a regulus lie on a doubly-ruled hyperboloid of one sheet (Figure 6(d)). A regulus can also be defined as the set of lines intersecting a fixed set of three skew lines (Figure 6(e)).

Note that this result is attributed by Ball [2, pp. 186-187, p. 511] to Möbius [43, p. 177] in the following form: if four forces are in equilibrium they must be generators of the same hyperboloid. (The hyperboloids associated with four intersecting lines and two pencils lying in different planes and having one line in common are degenerate.)

From now on we restrict our attention to non-planar grasps, i.e., to sets of four contact forces whose lines of action do not all lie in the same plane.

4.2 A Necessary and Sufficient Condition for Equilibrium

Can the three types of linearly-dependent non-coplanar lines yield equilibrium grasps? The answer is yes, as demonstrated graphically by Figure 7. If we group the contact forces into pairs whose directions lie in two planes, we see that the three types of grasps form a hierarchy characterized by how the lines of action of the forces do or do not intersect in these planes: in Figure 7(a), the four lines intersect in a point that lies in both planes, while in Figure 7(b) the two pairs of lines have been pulled apart, each pair of lines intersecting on the line formed by the intersection of the two planes. Finally, in the regulus case, the lines in each pair do not intersect anymore; they have been pulled away from their original plane and now lie parallel to it (Figure 7(c)). Clearly the forces shown in Figure 7(b) balance each other and exert no torque about the line common to the two planes; they achieve equilibrium. Likewise, the forces shown in Figure 7(c) balance each other, but this time the two corresponding pairs of forces exert opposite torques about the line; this grasp also achieves equilibrium.

The main result of this section is the following necessary and sufficient condition for forces whose lines of action are linearly dependent to achieve equilibrium in the case where the contact points are not all coplanar.

Proposition 7 *A necessary and sufficient condition for four non-coplanar points to form an equilibrium grasp with four non-zero contact forces is that*

(P1) there exist four lines in the corresponding double-sided friction cones that either intersect in a single point, form two flat pencils having a line in common but lying in different planes, or form a regulus, and

(P2) the vectors parallel to these lines and lying in the internal friction cones at the contact points positively span $\mathbb{R}^3$.

The proof of this proposition is given in [57].

4.3 A Sufficient Condition for Equilibrium

As before, we seek conditions for equilibrium that are linear in the unknown grasp parameters (the finger positions and the contact forces). Unfortunately, (P2) is definitely non-linear: it can be written as

$$\sum_{i=1}^{4} \alpha_i \mathbf{f}_i = 0, \quad \text{with } \alpha_i \geq 0 \quad \text{for } i = 1, \dots, 4,$$

which is a bilinear constraint on the unknown coefficients α_i and forces $\mathbf{f}_i$.

In this section, we give a sufficient condition for equilibrium, using a condition on the surface normals which ensures that (P2) is satisfied. Since, for a given choice of four contact faces, the surface normals are fixed, this replaces the non-linear condition (P2) by a simple test on these normals.

We first need a definition.

Definition 4 *We say that four vectors θ-positively span $\mathbb{R}^3$ when, for any triple $\mathbf{u}_1, \mathbf{u}_2, \mathbf{u}_3$ of these vectors, the cones C_1, C_2, C_3 of half-angle θ centered on $\mathbf{u}_1, \mathbf{u}_2,$ and $\mathbf{u}_3$ lie in the interior of the same half-space and the cone $-C_4$ of half-angle θ centered on the direction opposite to the fourth vector $\mathbf{u}_4$ lies in the interior of the intersection of the trihedra formed by all triples of vectors belonging to $C_1, C_2,$ and C_3 (Figure 8).*

The following proposition is a simple corollary of Proposition 7.

Proposition 8 *A sufficient condition for four non-coplanar points to form an equilibrium grasp with four non-zero contact forces is that*

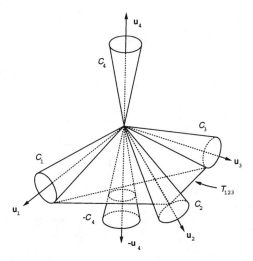

Figure 8: *Four vectors θ-positively spanning $\mathbb{R}^3$. T_{123} is the intersection of the trihedra formed by all triples of vectors belonging to $C_1, C_2,$ and C_3.*

(P1) there exist four lines in the corresponding double-sided friction cones that either intersect in a single point, form two flat pencils having a line in common but lying in different planes, or form a regulus, and

(P3) the surface normals at the four contact points θ-positively span $\mathbb{R}^3$.

The condition for equilibrium given by Proposition 8 is similar to the condition given by Proposition 5 in the two-dimensional, three-finger case [52, 53]. Note that Proposition 8 also provides a sufficient condition for three-finger equilibrium in the case where gravity acts as a fourth finger [56].

The main advantage of Proposition 8 over Proposition 7 is that it replaces the condition (P2) —which depends on the actual contact forces' directions— by condition (P3) —which depends on the normals to the grasped faces only.

In the case of grasps whose contact forces intersect in a single point, Proposition 8 allows us to compute equilibrium grasps in two steps: we first select faces whose normals satisfy (P3), then compute the grasp configurations satisfying (P1). (P1) can itself be decomposed into sixteen elementary conditions (the common point may lie in the internal or the external friction cone at each contact point), each of them being a conjunction of linear constraints (see Section 5).

Proposition 8 does not yield obvious linear conditions for equilibrium in the case of forces lying in two flat pencils or in a regulus. In the rest of this discussion, we will focus on *concurrent* grasps, whose contact forces intersect in a point; this will allow us to use linear programming as a basis for grasp planning. We will briefly come back to the general case in Section 8.

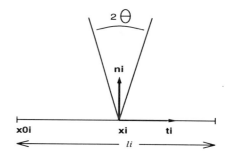

Figure 9: *The parameters of a three-finger grasp.*

5 The Basic Algorithm

We now present an algorithm, based on linear programming, for computing the maximal regions of the grasp configuration space that yield force-closure grasps. We use the case of three-finger grasps of polygonal objects to illustrate the basic algorithm. Its generalization to four-finger grasps of polyhedral objects is straightforward, and it is omitted for the sake of conciseness. A vital part of the algorithm is the projection of a polytope from a high-dimension space onto a lower-dimension subspace, which is detailed separately in Section 6.

5.1 Linear Constraints

Let us rewrite the sufficient condition of Proposition 8 in algebraic form. Consider three polygonal edges defined by an end point $\mathbf{x}_{0i}$, a unit direction $\mathbf{t}_i$, and a length l_i (Figure 9), and assume that the normals $\mathbf{n}_i$ to the edges positively span the plane. We suppose that the edges are oriented counterclockwise and that the $\mathbf{n}_i$ are the internal normals. A point on edge i is defined by $\mathbf{x}_i = \mathbf{x}_{0i} + u_i \mathbf{t}_i$, with $u_i \in [0, l_i]$. Writing that a point $\mathbf{x}_0 = (x_0, y_0)$ belongs to the intersection of the friction cones at the points $\mathbf{x}_i$, we obtain the following twelve constraints in the five variables u_1, u_2, u_3, x_0, y_0:

$$\begin{cases} \left[\mathbf{x}_0 - \mathbf{x}_{0i} - u_i \mathbf{t_i}\right] \times \left[\mathbf{n_i} - \tan\theta \mathbf{t_i}\right] \geq 0, \\ \left[\mathbf{x}_0 - \mathbf{x}_{0i} - u_i \mathbf{t_i}\right] \times \left[\mathbf{n_i} + \tan\theta \mathbf{t_i}\right] \leq 0, \\ u_i \geq 0, \\ u_i \leq l_i, \end{cases} \quad (1)$$

for $i = 1, 2, 3$.

A sufficient condition for the existence of a three-finger force-closure grasp for the triple of edges considered is that the polytope defined in $\mathbb{R}^5$ by the above twelve constraints is not empty.

The force-closure grasps satisfying the hypotheses of Proposition 5 can be found by considering in turn all possible combinations of internal and external friction cones at each contact point. The equations are the same as before except that the inequalities defining the friction cone in (1) may have to be reversed. The set of solutions is the union of the polytopes corresponding to the different combinations.

It is important to realize that, in either case, linear programming can readily be used to assess whether a set of linear inequalities such as (1) admits a solution, or to find representative grasp configurations optimizing some linear merit function. This is true even though (1) includes the x_0, y_0 unknowns besides the variables of interest, u_1, u_2, u_3.

Nonetheless, it is convenient to characterize the force-closure regions in the three-dimensional grasp configuration space, independently of the extra variables x_0, y_0. For example, this yields a trivial test for force closure. More importantly, we will see in the next section that, while it is theoretically possible to find independent grasp regions without eliminating x_0 and y_0, this is much less efficient than working directly with the three-dimensional force-closure regions. Several algorithms for performing the elimination of these extra variables will be presented in Section 6.

5.2 Finding Independent Contact Regions

After eliminating the variables x_0, y_0, we obtain the polytope representing all force-closure grasps for each triple of edges. Because of the uncertainty in robotics systems, we would like to minimize the sensitivity of a grasp to positioning errors. A way of achieving this is to seek triples of *independent contact regions* [45]. These regions are such that for any triple of contact points chosen in them, the corresponding grasp exhibits force closure. In the grasp configuration space, these regions are obviously represented by parallelepipeds with sides aligned with coordinate axes and contained in the polytope of force-closure grasps.

Among all the solutions to this problem, we define the *maximal independent contact regions* as those maximizing a criterion depending on their size and location. A reasonable criterion is to maximize the minimum of the lengths of the three contact regions. Empirically, we have observed that there is not, in general, a unique solution to this problem, and that, for sufficiently long edges, the size of the contact regions depends only on the size of the friction cones. In this case, there are two parallel faces in the polytope representing the set of force-closure grasps and a one- or two-dimensional set of maximal parallelepipeds can be found. For this reason, we add a secondary criterion in order to select a unique solution: we try to center as well as possible the center of mass of the object in the triangle formed by the contact points. This enables us to decrease the effect of gravity and inertial forces during the motion of the robot. It should be noted that this is only a heuristic choice for taking into account the three-dimensional nature of the problem even though we are dealing with planar objects.

We now show how to translate the problem of finding maximal independent contact regions into a linear programming problem. A parallelepiped with sides aligned with the coordinate axes in the grasp configuration space is defined by three intervals $[u_i^-, u_i^+], i = 1, 2, 3$ corresponding to its projection on each axis. Because of convexity, we only need to write that the eight vertices of the parallelepiped are contained in the set of force-closure grasps in order to guarantee that the entire parallelepiped is also contained in it. Let l be the minimum of the lengths of the three intervals. We add to the existing set of linear constraints new constraints expressing that, for each i, $0 \leq l \leq u_i^+ - u_i^-$. Maximizing the minimum length criterion thus reduces to maximizing l under all the constraints.

The second criterion can also be expressed linearly by introducing an additional variable d measuring the L^∞ distance between the center of mass $\mathbf{g}_p = (x_p, y_p)^T$ of the polygon and the center of mass $\mathbf{g}_c = (x_c, y_c)^T$ of the contacts corresponding to the mid-points of the intervals $[u_i^-, u_i^+]$. We write that d is greater than $x_p - x_c$, $x_c - x_p$, $y_p - y_c$, and $y_c - y_p$. Since $\mathbf{g}_p$ is fixed and x_c, y_c depend linearly on the unknowns u_i^-, u_i^+, these constraints on d are themselves linear in all the unknowns. Minimizing the distance criterion amounts to maximizing $-d$.

Finding the maximal independent contact regions amounts to solving a linear program with eight variables, the function to maximize being a weighted combination of the two above criteria. We also rank the maximal independent contact regions we have found for different triples of edges by using the output value of the simplex (i.e., the weighted sum of the criteria after optimization). For each triple of edges, we choose the mid-points of the maximal independent contact regions as representative grasps.

It should be noted that the independent contact regions can also be found without projecting the original polytope. In this case, we use the original inequalities (1) instead of those defining the projection. In general, the number of constraints will therefore be about the same, sometimes smaller. However, we must also add sixteen variables corresponding to the values of x_0, y_0 at each vertex of the parallelepiped. The corresponding optimization program has more than twice as many variables as the original one, and is correspondingly much slower. As shown in Section 7, this is empirically confirmed by our experiments.

5.3   Four-Finger Case

The generalization of the algorithm presented above to the four-finger case is theoretically straightforward. Writing that a contact point $\mathbf{x}_i$ lies inside a convex polygonal face and that the line joining it to the intersection point $\mathbf{x}_0$ lies in the corresponding polyhedral friction cone yields a set of linear constraints similar to (1). Likewise, the computation of a maximal box inscribed in the grasp configuration space can be done through linear programming. The main difference is in the number of variables and constraints, which is much larger for four-finger grasps of polyhedral objects (the configuration space is now eleven-dimensional). This makes the elimination of the coordinates of $\mathbf{x}_0$ an even more crucial step of the algorithm: as discussed in Section 7, the algorithm simply does not run to completion (in a reasonable time) without it.

6   Projecting Polytopes

We now attack the problem of eliminating the coordinates of $\mathbf{x}_0$ among the force-closure constraints. This is equivalent to projecting a five- or eleven-dimensional polytope onto a three- or eight-dimensional grasp configuration space. We solve this problem with a general algorithm for projecting a d-dimensional polytope onto some $(d-k)$-dimensional subspace.

Consider a polytope P, defined in a Euclidean space E of dimension d as the intersection of n half-spaces

$$H_i = \{\mathbf{x} : \mathbf{A}_i\mathbf{x} + b_i \leq 0\}, \quad i = 1, \ldots, n. \quad (2)$$

Here $\mathbf{A}_i = (A_{i1}, .., A_{id})$ is a $1 \times d$ real matrix, $\mathbf{x} = (x_1, \ldots, x_d)^T$ is the vector of coordinates of a point in E, and b_i is a real number. The $(d-1)$-faces f_i of P lie in the hyperplanes bounding the half-spaces H_i.

We address the problem of constructing the orthogonal projection Q of P onto a $(d-k)$-dimensional subspace F of E. Several approaches are possible: for example, observing that Q itself is a polytope, one can first compute the whole facial structure of P, and thus its vertices, and then construct Q as the convex hull of their projections onto F. In the worst case, P and Q have size $O(n^{\lfloor\frac{d}{2}\rfloor})$, and they can be constructed in time $O(n^{\lfloor\frac{d}{2}\rfloor})$ [5, 10].

Instead, we propose to compute directly the equations of the hyperplanes bounding Q in F. From now on, we assume without loss of generality that F is defined by $x_i = 0$ for $1 \leq i \leq k$.

6.1 A Gaussian-Elimination Approach

We present a two-step algorithm based on Gaussian elimination [53]: we first eliminate the variables $x_1, \ldots, x_k$ among the n inequalities defining P, which yields a set of half-spaces defining a cylindrical polytope C in E; we then discard the redundant half-spaces through linear programming and construct Q by (trivially) projecting the remaining ones onto F.

Before detailing the algorithm, let us give its geometric interpretation (Figure 10). We will say that a hyperplane is *vertical* when it contains the directions $x_1, \ldots, x_k$. For any j in $1, .., k+1$, a $(d-j)$-face of P lies in the intersection of j hyperplanes bounding P. The *apparent contour* is formed by the $(d-j)$-faces that are contained in vertical hyperplanes. The boundary of Q, or *silhouette*, is the projection of the apparent contour. Here, we are interested in constructing the hyperplanes of F that are part of the silhouette, or equivalently, the vertical hyperplanes supporting the apparent contour.

The adjacency relationships between the $(d-1)$-faces of P are a priori unknown. The first step of the algorithm constructs a set of candidate half-spaces bounded

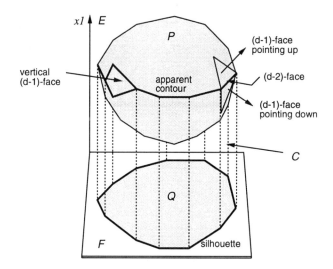

Figure 10: *The geometry of the projection algorithm in the case $d = 3$, $k = 1$.*

by vertical hyperplanes that may contain a $(d-k-1)$-face of the apparent contour. The second step rejects the redundant candidates and constructs the projection.

Step 1. In principle, any $(k+1)$-tuple of $(d-1)$-faces may yield a $(d-k-1)$-face of the apparent contour. The inequalities defining the corresponding half-spaces can be written as

$$\begin{cases} A_{i_1,1}x_1 + \ldots + A_{i_1,k}x_k \\ \quad + A_{i_1,k+1}x_{k+1} + \ldots + A_{i_1,d}x_d + b_{i_1} \leq 0, \\ \ldots \\ A_{i_k,1}x_1 + \ldots + A_{i_k,k}x_k \\ \quad + A_{i_k,k+1}x_{k+1} + \ldots + A_{i_k,d}x_d + b_{i_{j-1}} \leq 0, \\ A_{i_{k+1},1}x_1 + \ldots + A_{i_{k+1},k}x_k \\ \quad + A_{i_{k+1},k+1}x_{k+1} + \ldots + A_{i_{k+1},d}x_d + b_{i_{k+1}} \leq 0. \end{cases}$$

We eliminate the variables $x_1, \ldots, x_k$ in k Gaussian-elimination steps. It is very important to remark that, because we deal with inequalities instead of equalities, the multiplicative coefficients used during Gaussian elimination must all be positive. In other words, we can only eliminate the variable x_1 if there exists some l in $\{1, \ldots, k+1\}$ such that, for all $m \neq l$ in $\{1, \ldots, k+1\}$, we have $A_{i_l,1}A_{i_m,1} \leq 0$ (intuitively, the normal to one of the hyperplanes under consideration must face "up" while the others face "down", see Figure 11). Eliminating x_1 yields a new system of k inequalities in $x_2, \ldots, x_k$. After k such steps, checking each

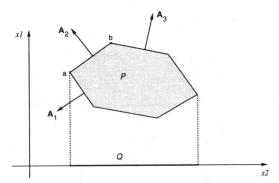

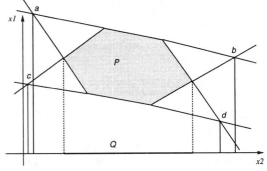

Figure 11: *Sign constraints in the case $d = 2, k = 1$ (the $(d-2)$-faces are vertices): the signs of the x_1 coordinates of the normals $\mathbf{A}_1$ and $\mathbf{A}_2$ are different: the $(d-2)$-face a obtained by eliminating x_1 between the equations of the corresponding hyperplanes belongs to the apparent contour. The signs of the x_1 coordinates of the normals $\mathbf{A}_2$ and $\mathbf{A}_3$ are the same: the $(d-2)$-face b obtained by eliminating x_1 between the equations of the corresponding hyperplanes projects inside the silhouette.*

Figure 12: *Potential $(d-2)$-faces a, b, c, d actually project outside the silhouette. In this example, $d = 2$, $k = 1$, and the $(d-2)$-faces are vertices.*

time that one of the coefficients of the leading variable has a sign different from all the other ones, we obtain a single linear inequality

$$A_{k+1}^* x_{k+1} + \ldots + A_d^* x_d + b^* \leq 0 \qquad (3)$$

in $x_{k+1}, \ldots, x_d$ only. This inequality defines the desired half-space.

Step 2. Some of the vertical hyperplanes bounding the half-spaces found in the previous step actually contain a $(d-k-1)$-face of the apparent contour, but since we do not know the adjacency relationships between the hyperplanes bounding P, others may actually lie outside the polytope (Figure 12). We must identify the former and discard the latter. We compute the signed distance between the original polytope and each candidate hyperplane by maximizing (3) under the original constraints (2). This is done once through linear programming, and hyperplanes such that the distance found is strictly negative are rejected.

What is the cost of this algorithm? We must consider $\binom{n}{k+1}$ $(k+1)$-tuples of $(d-1)$-faces, so for a fixed k we have to consider a total of $O(n^{k+1})$ tuples. For a fixed dimension, the cost of the best linear programming algorithms proposed so far is linear in the number of constraints [15], so rejecting redundant faces can be done in $O(n^{k+2})$ time.

An alternative to the previous method is to eliminate the k variables $x_1, \ldots, x_k$ one at a time, rejecting at each iteration the hyperplanes lying outside the apparent contour. In this case, we must consider at each projection step $n(n-1)/2$ pairs of $(d-1)$-faces. In the worst case, the size (number of facets) of the projection will be $O(n^2)$. If k projection steps must be performed, the worst-case size of the output will be $O(n^{2^k})$, and the worst-case complexity of the algorithm will be $O(n^{2^k+1})$. Note however that the average cost will be much smaller since the input to each step is the output of the previous step. While such a traditional asymptotic analysis of the algorithm will predict its worst-case behavior for large values of n, we must keep in mind that the original number of inequalities can be very small depending on the application: in the case of three-finger grasps of polygonal objects, the original 5-dimensional polytope defined by (1) has only 6 non-vertical 4-faces, and only 15 pairs of faces must be considered during the first projection. Typically, about 7 or 8 of those will correspond to faces whose normals point in opposite directions, and an additional couple may be rejected as redundant. This suggests that, in general, we may expect that the number of constraints defining the projection will in fact be quite small. This is what we have observed in practice, as shown in Section 7.

6.2 A Contour-Tracking Approach

In this section, we propose an algorithm that computes the projection Q of P onto F in an output-sensitive way. We restrict the discussion to the case

$k > 1$. The case $k = 1$ can be easily solved using linear programming. We will assume that the hyperplanes $A_i X + b_i = 0$, $i = 1, \ldots, n$, are in general position, i.e. no $(d + 1)$-tuples of hyperplanes have a common intersection and no two vertices of P project onto the same vertex of Q. General techniques such as the simulation of simplicity of [16] can be used to make this hypothesis valid.

The general outline of the algorithm is the following :

1. Find a first vertex of P which projects onto a vertex of Q , e.g., an extremum in some direction orthogonal to the projection direction (this is a linear program). Put it in a stack S.

2. Repeat the following steps until S is empty

   (a) Take a vertex V out of S and put it in a dictionary D of already considered vertices. Construct the facial structure of the cone incident to V and find all the edges (rays) of the cone that contain an edge of the silhouette of P, i.e. that project onto edges of Q.

   (b) Each such ray r is considered in turn. We compute the vertex W distinct from V of the edge e contained in the ray. If W has not been already considered (it does not belong to D), add W to S, and report the corresponding edge of P, VW.

3. Compute the vertical hyperplanes of E bounding P (hence the hyperplanes of F bounding Q) from the edge set.

A simple implementation of this algorithm takes $O(tn)$ time where t is the size of Q.

It should be noted that the time complexity can be improved by preprocessing the hyperplanes H_i so as to answer the ray shooting queries in sub-linear time. In [57], we present a solution to that problem, based on a recent result by Matoušek and Schwarzkopf [37]. The idea of our algorithm is to start with little memory and do ray shootings until the cumulated time of all the queries performed so far exceeds the preprocessing time and the memory storage. At that point, we can allocate, restart preprocessing, and repeat that process until all the faces have been constructed. We obtain the following result.

Proposition 9 *The projection of a d-polytope of E onto a $(d-k)$-dimensional subspace F can be computed*

in time and space

$$T = O\left((d - k)n^{\gamma + \varepsilon} t^{\gamma + \varepsilon}\right)$$

where ε is any positive constant, t is the size of Q, and

$$\gamma = \frac{1}{1 + \frac{1}{\lfloor \frac{d}{2} \rfloor}}$$

In our case, $d = 5$ or 11, so $\gamma = 2/3$ or $5/6$.

7   Implementation and Results

7.1   Three-Finger Case

The implementation has been written in Common Lisp with calls to C functions from *Numerical Recipes in C* [58] for the simplex implementation of linear programming. In this case we have used the iterative elimination method to perform the projection of the grasp convex.

We first show results obtained using the conditions of Proposition 4. Figure 13 shows a polygonal object and its four best grasps. The half-angle of the friction cones in this case is 10 degrees. The circle near the center of the figure is positioned at the center of mass of the object, while the small black disc is the centroid of the three contact points. For each region, the representative positions of the fingers are shown as solid discs. We have implemented a simple feasibility test by testing finger-finger and finger-edge intersection. The contact regions are shown as thick segments along the polygonal boundary, and the internal friction cones corresponding to the representative points for each region are drawn as dotted lines. In this example, our test object has 21 edges, there are 1140 different triples of edges. 326 triples of edges have normals that positively span the plane, and, among them, 105 grasps are found in a total of 27.6s of CPU time on a SUN SPARCstation 1. This includes the projection stage as well as the grasp optimization. Among the grasps found, the average number of constraints after projection is 8.4. As noted in the previous section, we have also run the optimization using directly the original inequalities (1), without the projection stage. The performance is, as expected, much worse in this case, with a total computing time of 108.7s. It is also interesting to note that the relatively expensive pruning stage of the projection algorithm where redundant inequalities are discarded

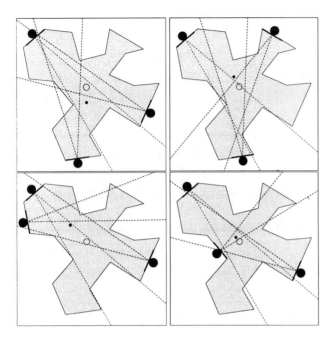

Figure 13: *The four best grasps of a polygon.*

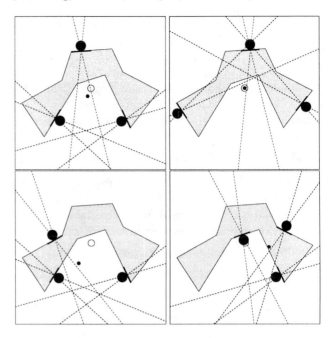

Figure 14: *The four best grasps of another polygon.*

is also worth the extra effort: without it, the total computing time is 100.6s.

Figure 14 shows some grasps found using the condition of Proposition 5 for another object. In this case, the double-sided friction cones are also shown. The friction half-angle has been set to 10 degrees, and the polygon has 10 edges. Among the 120 possible triples of edges, 16 satisfy the test of Proposition 5, i.e., their normals positively span the plan with pairwise angles smaller than $\pi - 2\theta$, and a total of 41 grasps are found. The four best grasps are shown in the figure, along with the double-sided friction cones. Note that, in the first grasp for example, two of the fingers are moving away from each other instead of squeezing. The total computing time is 12.2s. Among the grasps found, the average number of constraints after projection is 10.0. When the projection stage is omitted, the total computing time is 52.5s.

7.2   Four-Finger Case

The implementation has been entirely written in C, using once again the simplex implementation from *Numerical Recipes in C* [58] for linear programming. In this case we have used both the Gaussian elimination method and the apparent contour tracking methods to perform the projection of the grasp convex. We briefly compare the performances of these methods in practice, then show some results. (It should be noted that we also tried the algorithm without the projection step, but the program simply never ran to completion –we interrupted it after a few hours– probably due to the very large number of constraints and variables involved.)

A four-finger grasp may be defined by eleven parameters: two local coordinates (u_i, v_i) defining the contact point for each face, plus a point $\mathbf{x}_0$ which lies inside the friction cones centered at each finger. Each (convex) face contributes n_i hyperplanes expressing the shape of the face (linear constraints on u_i and v_i), and m hyperplanes approximating its friction cone –involving u_i, v_i, and $\mathbf{x}_0 = (x_0, y_0, z_0)^T$– to the grasp convex, for a total of $\sum_{i=1}^{4} n_i + 4m$ constraints.

In general, we wish to eliminate three variables among all sets of four hyperplanes, checking that the resulting vertical hyperplane bounds the convex. For even a simple problem (three-sided faces and four-sided friction cones, giving 28 constraints), a direct application of this approach would be expensive due to the $\binom{28}{4} = 20475$ combinations to be checked. However, note that a large fraction of the input hyperplanes are already vertical; these can be set aside and removed from further consideration, since they contain their intersection with non-vertical hyperplanes. Furthermore, there are some sets which may be ruled out by simply examining the signs of the hyperplane coefficients: for an edge to project onto the silhouette, at least one constraint must be violated by moving a point off the edge parallel to a projection direction. This can only happen if there is some disagreement in sign between the hyperplane coefficients in projection directions. We can thus consider only those sets of four non-vertical planes with sign differences in each of the x_0, y_0 and z_0 coordinates. This restricts the number of four-tuples considerably, requiring only about 1400 comparisons for a problem of the size mentioned above.

For each set of four non-vertical hyperplanes passing the above test, eliminating the three variables x_0, y_0, z_0 yields a vertical hyperplane. We decide whether to keep or reject this hyperplane by maximizing its equation under the original constraints. If the distance is zero, the hyperplane bounds the convex and we keep it. For small problems like the one mentioned above, there are typically 100 to 150 hyperplanes at the end of this process, which takes approximately 5 seconds on a SPARCstation 10.

We have also implemented the contour-tracking method described earlier. At each successive vertex of the apparent contour, we must test all subsets of ten incident hyperplanes. Finding the contour edges and the corresponding vertical hyperplanes typically takes 5 seconds on a SPARCstation 10, yielding approximately 50,000 to 75,000 edges and, of course, the same number of hyperplanes as before. This means that for our problem the Gaussian-elimination and contour-tracking methods have comparable performances. This is probably due to the relatively small number of hyperplanes bounding the initial polytope: preliminary experiments involving random polytopes confirm that contour tracking is better than Gaussian elimination for polytopes with many faces. For example, when projecting a random eight-dimensional polytope onto a five-dimensional sub-space, contour tracking and Gaussian elimination have comparable performances when the polytope only has 40 faces, but contour tracking is already four times faster for 50 faces, and eight times faster for 60 faces.

Once the projection has been computed, we proceed to find the corresponding maximal independent

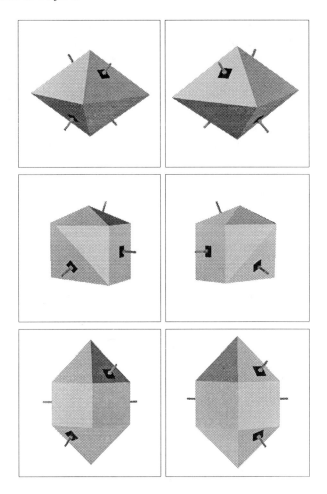

Figure 15: *Some four-finger grasps.*

grasp regions. We wish to inscribe a box into an eight-dimensional convex. The shortest edge of the box should be as long as possible. The box itself may be defined as a set of intervals for the face coordinates $[u_i^-, u_i^+], i = 1, .., 8$. Because of convexity, the entire box will be contained in the convex if each of the 256 vertices are valid grasps; we need versions of the grasp constraints for each of the 256 combinations of interval endpoints. However, since most of the constraints do not involve all of the face variables, we do not end up with 256 times the number of original constraints (for a typical case of 30 grasp constraints, the new problem has approximately 2000 constraints). As in the two-dimensional case, we maximize the length of the shortest side, and since there may be a family of solutions with the same minimum length, we also position the centroid of the contact points as close as possible to the center of mass of the polyhedron. Finding the optimal box thus amounts to solving a linear program with 25 variables and approximately 2000 constraints, which takes approximately 70 seconds on a SPARCstation 10.

We illustrate our results for three polyhedra. Figure 15 (top) shows a diamond shape, and the grasp regions found for the two possible four-finger grasps. Figure 15 (middle) shows two views of one of the five possible grasps for an 11-sided object. The other four grasps come from placing the top finger on each of the remaining four top triangular faces. Finally, Figure 15 (bottom) shows two of the 40 grasps of a 15-sided object.

8   Discussion and Future Work

We have presented algorithms for computing three- and four-finger force-closure grasps of polygonal and polyhedral objects and demonstrated efficient implementations of these algorithms. Let us conclude by discussing some of the issues raised by our implementation and by sketching some future research directions.

The algorithms presented in this paper rely on the existence of a point $\mathbf{x}_0$ in the intersection of the friction cones at the contact points. We have recently implemented an alternative method using the equivalent condition that the *inverted* friction cones at the point $\mathbf{x}_0$ must intersect the corresponding faces. Writing that the inverted friction cone of half-angle θ in $\mathbf{x}_0$ intersects a face f is equivalent to writing that this point projects inside the face grown by $d \tan \theta$, where d is the distance between $\mathbf{x}_0$ and f. The grown face is bounded by straight lines and circular arcs. By approximating the latter by polygonal arcs (which is equivalent to approximating the friction cones by polyhedral pyramids), we can thus express the force-closure conditions as linear constraints on the position of $\mathbf{x}_0$. An advantage of this method is that the configuration space corresponds to the position of the point $\mathbf{x}_0$ and is only three-dimensional. This approach bypasses the projection step but does not yield directly maximal grasp regions. Good grasps can be computed through linear programming by maximizing the size of a circle inscribed in the overlap regions between the inverted friction cones and the object faces. Any grasp within

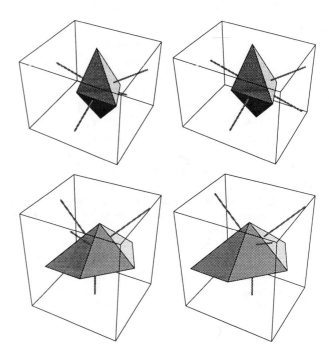

Figure 16: *Four-finger grasps computed using inverted friction cones.*

these overlap regions yields a force-closure grasp. Figure 16 shows some four-finger grasps computed with this method.

Even though Propositions 7 and 8 hold for all types of non-planar four-finger grasps, the algorithm described in Section 5 is restricted to to equilibrium forces whose lines of action all pass through some point. It is clearly important to generalize our approach to the other types of four-finger grasps. Grasps involving forces that lie in two flat pencils having a line in common may prove particularly important in practice, for example for grasping an elongated object with two cooperating robots equipped with simple two-finger grippers. We have developed a linear method using inverted friction cones to compute this type of grasps when the direction of the line common to the two pencils is constrained to lie in a prescribed cylinder [64], and Figure 17 shows an example. This is only a first step: as mentioned in Section 4, the equations that characterize equilibrium for these more general grasps are normally non-linear, and new methods will have to be developed.

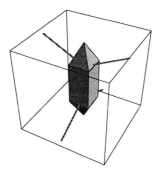

Figure 17: *A grasp formed by forces lying in two flat pencils having a line in common but lying in different planes.*

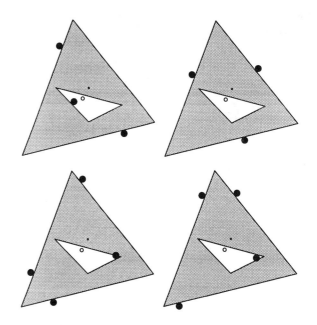

Figure 18: *Optimal three- and four-finger grasps of a polygonal object (in this case feasibility tests have not been implemented yet).*

geometry does not.

So far, we have ranked grasps according to the size of the independent grasp regions and how well the center of mass of the object is centered among the contact points. It would also be interesting to compute grasps that are optimal according to some functional consideration [19, 30, 33]: for example, given a fixed set of contact points and some bound on the magnitude of the contact forces, Ferarri and Canny [19] propose to compute a maximal ball centered at the origin and contained in the convex formed by the contact wrenches; clearly, the contact forces can generate any wrench contained in this ball, and its radius provides a measure of the grasp's efficiency. Computing a grasp configuration which is optimal according to this criterion is more challenging; it is a non-linear problem since the moment of the forces depends bilinearly on the finger positions and the force directions. We have recently developed an iterative method for computing multi-finger grasps of polygonal objects [64]. It uses a linear approximation of the force-closure constraints and linear programming to construct a good initial guess for the optimal grasp configuration. This guess is then refined through non-linear optimization. Figure 18 shows three- and four-finger optimal grasps found by a preliminary implementation of this approach.

Although we have implemented (at least in the two-dimensional case) a simple a posteriori test to check whether the computed grasps are feasible, we do not enforce feasibility during the actual grasp computation. We plan to address this problem and the associated issue of reachability [51] in the future. We will also explore the case of soft fingers that can exert pure torques in addition to pure forces, a more difficult setting in which general screw theory does apply, but simple line

References

[1] B.S. Baker, S.J. Fortune, and E.H. Grosse. Stable prehension with a multi-fingered hand. In *IEEE Int. Conf. on Robotics and Automation*, pages 570–575, Saint-Louis, Missouri, 1985.

[2] R.S. Ball. *A treatise on the theory of screws.* Cambridge University Press, 1900.

[3] J. Barber, R.A. Volz, R. Dessai, R. Rubinfeld, B. Schipper, and J. Wolter. Automatic two-fingered grip selection. In *IEEE Int. Conf. on Robotics and Automation*, pages 890–896, San Francisco, CA, 1986.

[4] A. Blake. Computational modelling of hand-eye coordination. *Phil. Trans. R. Soc. Lond. B*, 337:351–360, 1992.

[5] J.-D. Boissonnat, O. Devillers, R. Schott, M. Teillaud, and M. Yvinec. Applications of random sampling to on-line algorithms in computational geometry. *Discrete Comput. Geom.*, 8:51–71, 1992.

[6] O. Bottema and B. Roth. *Theoretical Kinematics.* North Holland Publishing Co., 1979.

[7] R.C. Brost. Automatic grasp planning in the presence of uncertainty. *International Journal of Robotics Research*, 7(1):3–17, 1988.

[8] R.C. Brost. *Analysis and planning of planar manipulation tasks.* PhD thesis, Carnegie-Mellon University School of Computer Science, January 1991.

[9] S.J. Buckley. *Planning and teaching compliant motion strategies.* PhD thesis, MIT Dept. of Electrical Engineering and Computer Science, January 1987.

[10] B. Chazelle. An optimal convex hull algorithm for point sets in any fixed dimension. Technical Report CS-TR-336-91, Dept. Comput. Sci., Princeton Univ., Princeton, NJ, 1991.

[11] I.M. Chen and J.W. Burdick. Finding antipodal point grasps on irregularly shaped objects. In *IEEE Int. Conf. on Robotics and Automation*, pages 2278–2283, Nice, France, June 1992.

[12] M.R. Cutkosky. Mechanical properties for the grasp of a robotic hand. Technical Report CMU-RI-TR-84-24, Carnegie-Mellon University Robotics Institute, 1984.

[13] M.R. Cutkosky. *Grasping and fine manipulation for automated manufacturing.* PhD thesis, Carnegie-Mellon University Department of Mechanical Engineering, January 1985.

[14] A. Dandurand. The rigidity of compound spatial grid. *Structural Topology*, 10, 1984.

[15] H. Edelsbrunner. *Algorithms in Combinatorial Geometry.* EATCS Monographs in Theoretical Computer Science. Springer-Verlag, 1987.

[16] H. Edelsbrunner and E. P. Mücke. Simulation of simplicity: a technique to cope with degenerate cases in geometric algorithms. *ACM Trans. Graph.*, 9:66–104, 1990.

[17] B. Faverjon and J. Ponce. On computing two-finger force-closure grasps of curved 2D objects. In *IEEE Int. Conf. on Robotics and Automation*, pages 424–429, Sacramento, CA, April 1991.

[18] R.S. Fearing. Simplified grasping and manipulation with dextrous robot hands. MIT AI Memo 809, MIT Artificial Intelligence Lab, 1984.

[19] C. Ferrari and J.F. Canny. Planning optimal grasps. In *IEEE Int. Conf. on Robotics and Automation*, pages 2290–2295, Nice, France, June 1992.

[20] K. Goldberg and M.T. Mason. Bayesian grasping. In *IEEE Int. Conf. on Robotics and Automation*, pages 1264–1269, May 1990.

[21] H. Hanafusa and H. Asada. Stable prehension by a robot hand with elastic fingers. In *Seventh International Symposium on Industrial Robots*, Tokyo, Japan, 1977.

[22] D. Hilbert and S. Cohn-Vossen. *Geometry and the Imagination.* Chelsea, New York, 1952.

[23] K.H. Hunt. *Kinematic Geometry of Mechanisms.* Clarendon, Oxford, 1978.

[24] S.C. Jacobsen, J.E. Wood, D.F. Knutti, and K.B. Biggers. The Utah-MIT Dextrous Hand: Work in progress. *International Journal of Robotics Research*, 3(4):21–50, 1984.

[25] J.W. Jameson. *Analytic techniques for automated grasps.* PhD thesis, Stanford University, Stanford, CA, 1985.

[26] Z. Ji. *Dexterous hands: optimizing grasp by design and planning.* PhD thesis, Stanford University, Dept. of Mechanical Engineering, 1987.

[27] Z. Ji and B. Roth. Direct computation of grasping force for three-finger tip-prehension grasps. *Journal of Mechanics, Transmissions, and Automation in Design*, 110:405–413, December 1988.

[28] J.R Kerr. *An Analysis of multi-fingered hands.* PhD thesis, Stanford University, Stanford, CA, 1984.

[29] J.R. Kerr and B. Roth. Analysis of multi-fingered hands. *International Journal of Robotics Research*, 4(4), Winter 1986.

[30] D.G. Kirkpatrick, B. Mishra, and C.K. Yap. Quantitative Steinitz's theorems with applications to multifingered grasping. In 20th *ACM Symp. on Theory of Computing*, pages 341–351, Baltimore, MD, May 1990.

[31] K. Lakshminarayana. Mechanics of form closure. Technical Report 78-DET-32, ASME, 1978.

[32] C. Laugier. A program for automatic grasping of objects with a robot arm. In *Eleventh International Symposium on Industrial Robots*, 1981.

[33] Z. Li and S. Sastry. Task-oriented optimal grasping by multifingered robot hands. In *IEEE Int. Conf. on Robotics and Automation*, pages 389–394, 1987.

[34] H. Liu, T. Iberall, and G. A. Bekey. The multidimensional quality of task requirements for dextrous robot hand control. In *IEEE Int. Conf. on Robotics and Automation*, pages 452–457, 1989.

[35] X. Markenscoff, L. Ni, and C.H. Papadimitriou. The geometry of grasping. *International Journal of Robotics Research*, 9(1):61–74, February 1990.

[36] X. Markenscoff and C.H. Papadimitriou. Optimum grip of a polygon. *International Journal of Robotics Research*, 8(2):17–29, April 1989.

[37] J. Matoušek and O. Schwarzkopf. Linear optimization queries. In *Proc. 8th Annu. ACM Sympos. Comput. Geom.*, pages 16–25, 1992.

[38] J.M. McCarthy. *An Introduction to Theoretical Kinematics*. MIT Press, 1990.

[39] J.P. Merlet. Singular configurations of parallel manipulators and grassmann geometry. In J.D. Boissonnat and J.P. Laumont, editors, *Geometry and Robotics*, volume 391 of *Lecture Notes in Computer Science*, pages 194–212. Springer-Verlag, 1988.

[40] B. Mirtich and J.F. Canny. Optimum force-closure grasps. Technical Report ESRC 93-11/RAMP 93-5, Robotics, Automation, and Manufacturing Program, University of California at Berkeley, July 1993.

[41] B. Mishra, J.T. Schwartz, and M. Sharir. On the existence and synthesis of multifinger positive grips. *Algorithmica, Special Issue: Robotics*, 2(4):541–558, November 1987.

[42] B. Mishra and N. Silver. Some discussion of static gripping and its stability. *IEEE Systems, Man, and Cybernetics*, 19(4):783–796, 1989.

[43] A.F. Möbius. *Lehrbuch der Statik*. Leipzig, 1837.

[44] D.J. Montana. Contact stability for two-fingered grasps. *IEEE Transactions on Robotics and Automation*, 8(4), August 1992.

[45] V-D. Nguyen. Constructing force-closure grasps. *International Journal of Robotics Research*, 7(3):3–16, June 1988.

[46] M.S. Ohwovoriole. *An extension of screw theory and its application to the automation of industrial assemblies*. PhD thesis, Stanford University, Stanford, CA, 1980.

[47] M.S. Ohwovoriole. An extension of screw theory. *Journal of Mechanical Design*, 103:725–735, 1981.

[48] T. Omata. Fingertip positions of a multifingered hand. In *IEEE Int. Conf. on Robotics and Automation*, pages 1562–1567, Cincinatti, OH, 1990.

[49] J. Pertin-Troccaz. On-line automatic programming: A case study in grasping. In *IEEE Int. Conf. on Robotics and Automation*, pages 1292–1297, Raleigh, NC, 1987.

[50] J. Pertin-Troccaz. Grasping: a state of the art. In O. Khatib, J. Craig, and T. Lozano-Pérez, editors, *The Robotics Review 1*. MIT Press, 1989.

[51] N.S. Pollard and T. Lozano-Pérez. Grasp stability and feasibility for an arm with an articulated hand. In *IEEE Int. Conf. on Robotics and Automation*, pages 1581–1585, Cincinatti, OH, 1990.

[52] J. Ponce and B. Faverjon. On computing three-finger force-closure grasps of polygonal objects. In *International Conference on Advanced Robotics*, pages 1018–1023, Pisa, Italy, June 1991.

[53] J. Ponce and B. Faverjon. On computing three-finger force-closure grasps of polygonal objects. *IEEE Transactions on Robotics and Automation*, 1994. Submitted.

[54] J. Ponce and B. Faverjon. On computing three-finger force-closure grasps of polygonal objects. *IEEE Transactions on Robotics and Automation*, 1994. Submitted.

[55] J. Ponce, D. Stam, and B. Faverjon. On computing force-closure grasps of curved two-dimensional objects. *International Journal of Robotics Research*, 12(3):263–273, June 1993.

[56] J. Ponce, S. Sullivan, J-D. Boissonnat, and J-P. Merlet. On characterizing and computing three- and four-finger force-closure grasps of polyhedral objects. In *IEEE Int. Conf. on Robotics and Automation*, pages 821–827, Atlanta, Georgia, May 1993.

[57] J. Ponce, S. Sullivan, J-D. Boissonnat, and J-P. Merlet. On characterizing and computing three- and four-finger force-closure grasps of polyhedral

objects. Technical report, Beckman Institute, University of Illinois, 1994. In preparation.

[58] W. Press, B. Flannery, S. Teukolsky, and W. Vetterling. *Numerical Recipes in C*. Cambridge University Press, 1988.

[59] F. Reulaux. *The kinematics of machinery*. MacMillan, NY, 1876. Reprint, Dover, NY, 1963.

[60] B. Roth. Screws, motors, and wrenches that cannot be bought in a hardware store. In *Proc. International Symposium on Robotics Research*, pages 679–693. MIT Press, 1984.

[61] J.K. Salisbury. *Kinematic and force analysis of articulated hands*. PhD thesis, Stanford University, Stanford, CA, 1982.

[62] J.K. Salisbury and B. Roth. Kinematic and force analysis of articulated hands. *ASME Journal of Mechanisms, Transmissions, and Automation in Design*, 105:33–41, 1982.

[63] S.A. Stansfield. Reasoning about grasping. In *Proc. AAAI Nat. Conf. Artif. Intell.*, pages 768–773, Saint Paul, MN, 1988.

[64] A. Sudsang. On computing multi-finger force-closure and optimal grasps of two- and three-dimensional objects. Master's thesis, Department of Computer Science, University of Illinois, 1994. Beckman Institute Tech. Report. In preparation.

[65] R. Tomovic, G. A. Bekey, and W. J. Karplus. A strategy for grasp synthesis with multifingered robot hands. In *IEEE Int. Conf. on Robotics and Automation*, pages 83–89, 1987.

[66] J.C. Trinkle. On the stability and instantaneous velocity of grasped frictionless objects. *IEEE Transactions on Robotics and Automation*, 8(5):560–572, October 1992.

Solving Complex Motion Planning Problems by Combining Geometric and Physical Models: The Cases of a Rover and of a Dextrous Hand

Christian Laugier, Christian Bard, Moëz Cherif, Ammar Joukhadar
LIFIA-INRIA Rhône-Alpes, 46 avenue Félix Viallet, 38031 Grenoble Cedex 1, France

This paper addresses motion planning for robots subjected to strong physical interaction constraints. More precisely, we will show how to solve two particular instances of this general problem: the case of a rover moving on a hilly three dimensional terrain, and the case of a dextrous hand executing a grasping operation. In addition to the high dimension of the associated search spaces, such motion planning problems exhibit new difficulties coming from the fact that the dynamics of the robot/ environment interactions may strongly modify the characteristics of the generated motion. Indeed, parameters like friction, inertia, or accelerations generate physical phenomena like sliding or skidding which may obviously affect the final behaviour of the robot. Therefore, such physical characteristics have to be carefully modeled and integrated into the motion planning paradigm. The purpose of this paper is to show how this goal can be achieved by appropriately combining geometric and physical models. The paper introduces first the concept of "physical model" before describing the algorithmic constructions which have been developed for the purpose of modeling the robot, the environment, and the robot/environment physical interactions. Then, it is shown how such models have been combined with more classical geometric techniques for solving the two above mentioned instances of the problem (the case of the rover and the case of the dextrous hand). In both cases, we successively describe the modeling of the task, the connection which has been performed between geometric and physical constructions, the characteristics of the motion planning process, and the experimental results which have been obtained.

1 Introduction

1.1 Overview of the problem

This paper addresses motion planning for robots subjected to *strong physical interaction constraints*. More precisely, we will show how to solve two particular instances of this general problem: the case of a rover moving on a hilly three dimensional terrain, and the case of a dextrous hand executing a grasping operation. In addition to the high dimension of the associated search spaces, such motion planning problems exhibit new difficulties coming from the fact that the dynamics of the robot/ environment interactions may strongly modify the characteristics of the generated motion. Indeed, parameters like friction, inertia, or accelerations generate physical phenomena like sliding or skidding which may obviously affect the final behaviour of the robot. Therefore, such motion planning problems cannot be solved using purely geometric approaches. This means that new constructions allowing the characterization of the main physical properties of the robot, of the task, and of the robot/environment interactions have to be developed and integrated into the motion planning paradigm. As we will see further down, such constructions have been implemented using an appropriate instantiation of the concept of "physical model" which has initially been developed in the field of Computer Graphics for the purpose of solving animation problems [14] [26] [27].

Let A be the robot, and q_{start} and q_{goal} be respectively the initial configuration and the goal configuration of A. The problem to solve can be roughly stated as follows: find a *"safe"* and *"executable"* motion of A starting at q_{start} and ending at q_{goal}. The motion is said to be *"safe"* if it verifies the geometric constraints of the task (no-collision, contact relations to achieve or to maintain); it is considered as *"executable"* if the kinematic and the dynamic constraints of A are verified all along the motion (this includes the non-holonomic constraints of A and the control constraints imposed by the robot/environment interactions). A more complete formulation of this problem is given in section 2.1.

1.2   Related work

Until now, very few methods have been devised for solving motion planning problems requiring to explicitly take into account the dynamic characteristics of the task or the robot/environment interactions. Some researchers have introduced simple dynamic models within the motion planning process in order to generate safe and executable motions for a vehicle; some other researchers have used simple geometric formulations of basic stability criteria for planning robust grasping strategies. In an other domain of Computer Sciences, the production of realistic movements in animated graphical scenes has forced some researchers to develop adequate models based upon the application of the main principles of physics (the "physical models"). Even if the goal is different, we will show further down that such models are of some interest for us.

Vehicle motion planning.   During the last decade, most of the work in the field of motion planning for mobile robots has focused on the problem of the automatic generation of collision-free trajectories in two- dimensional workspaces. This problem can be formulated in terms of geometric and kinematic constraints, and it can be solved using computational geometry techniques (see [17]). Very few results have been obtained yet, when additional dynamic constraints have to be processed. Generally, all the implemented motion planners apply some restrictive hypotheses to reduce the algorithmic complexity and to make the problem tractable (see [9]).

More recently, some interesting contributions addressing motion planning for a vehicle moving on a rough terrain have been reported [23][24][10][18][25]. In these approaches, the robot dynamics is generally reduced to some basic constraints which are associated to a mobile point, and the robot/terrain interactions considered are mainly restricted to the evaluation of some simple stability criteria based on the Coulomb law. A first approach in this framework is to reason on a continuous three dimensional B-patch model of the terrain, using a two steps method for searching an optimal trajectory on this terrain (determination of an optimal path and optimization of the trajectory on each involved patch) [23][24]. In this approach, only some basic dynamic constraints associated to the center of mass of the vehicle are modeled, and only robot/terrain interactions which can be represented using elementary

stability criteria are explicitly considered (such criteria characterize simple no-sliding and no tip-over conditions). In order to simplify the computation of the reaction and friction forces, the contact points are assumed to be planar and all the interaction forces are transferred to the center of mass of the vehicle. As mentioned in [24], such assumptions restrict the terrain to be smooth and the obstacles to be large compared to the robot size.

An other approach consists in applying graph search techniques operating on a discretized model of the surface of the terrain [10][18][25]. In this case, the contact between a wheel and the ground is reduced to a single point, and the stability criteria are evaluated with respect to the planar faces which approximate the surface of the terrain. For instance, [25] assumes that the robot does not tip-over if the projection of its center of mass belongs to the substantiation polygons; it also considers that the stability conditions can be evaluated using simple criteria based on static. Such an approach works fearly well when the effects of dynamics do not significantly modify the vehicle trajectory. It generally fails when more realistic dynamic phenomena have to be taken into account (e.g. sliding or skidding of the wheels, local ground deformations ...).

Automatic grasping.   The major part of the work done in the field of automatic grasping has focused on the problem of the automatic generation of accessible and potentially stable grasping configurations for classical grippers (see [28] for an overview of this work). Additional work has also been done for extending the capabilities of the previous approaches, in order to be able to deal with dextrous hands. The related solutions are generally based upon the concept of "hand preshaping" (see [28] and [2]). But, all these geometric based approaches do not completely solve the problem because the stability conditions are partially evaluated using some heuristics and very simple geometric criteria.

Some other contributions have focused on the stability problem. In this case, the solutions which have been developed rely on an evaluation of the forces and torques which are associated to the contact points. Nguyen [22] constructs force-closure grasps on polygonal objects using an explicit representation of the positions which generate stable situations onto the edges of the object. Hanafusa and Asada [12] make use of an appropriate potential function to find stable grasping configurations for a robot hand having three elastic fingers. A similar approach including the processing of some slipping phenomena has been reported in [1]. A more recent work aimed at integrating stability and slipping criteria within the planning process using a representation of some elementary physical properties of the hand/object system (mass and stiffness) has also been reported in [20]. As for the study of friction phenomena, one can find in the literature some interesting contributions which use friction and contacts as a physical guide for the execution of some manipulation tasks: Mason [21] studied the friction phenomena in order to generate appropriate sequences of motions and contacts in the plane for achieving a given positioning task; Fearing [8] applied a similar approach to re-orient an object using the fingers of a dextrous hand.

In spite of the ability of the above-mentioned approaches to solve some identified subproblems of automatic grasping, the generation of complete solutions taking into account dynamic effects and hand/object interactions is far to be accomplished. Because of the intrinsic complexity of the general problem, all the previous contributions rely on the application of simplifying hypotheses (polygonal or polyhedral world, punctual contacts, quasi-static conditions ...). Dealing with more realistic situations involving complex dynamic phenomena (produced by the hand/object interactions), requires to make use of appropriate models and algorithms. It will be shown in the sequel how "physical models" have been used to solve this problem.

Object animation in Computer Graphics. Researchers in the field of Computer Graphics have developed the concept of "physical model" for providing objects with realistic behaviours in animated scenes. Several approaches have been developed for the purpose of animating articulated rigid bodies [14] or deformable objects [26] [27]. Among these approaches, the model of Cordis-Anima [19] which basically represents the physical world using elementary particles and spring connectors is the first system of this type which has been used for solving robotics problems [15][16]. As we will see further down, the concept of physical model which has been developed for solving these animation problems has strongly influenced our approach.

1.3 Contribution of the paper

Despite the ability of the above-mentioned approaches to solve some instances of the path planning problem (i.e. instances mainly involving geometric/kinematic constraints), the automatic generation of "*executable motions*" for a robot subjected to strong physical interaction constraints is far to be fully accomplished. This comes from the fact that the behaviour of the robot results in this case from the combination of various geometric and physical parameters: the non-collision and kinematic criteria, the mechanical architecture of the robot, the characteristics of the control law to apply, and the robot/environment interactions.

The main contribution of this paper is to propose a motion planning method which takes into account such non-trivial features. It is based upon a two-stage approach combining a discrete search strategy and a continuous motion generation technique. The discrete search strategy (called the "*Explore*" step) is applied at each planning step to determine a pertinent set of potential intermediate safe configurations for the robot; it is based upon the application of classical search strategies and of geometric/kinematic models. The continuous motion generation technique (called the "*Move*" step) is applied at each planning step to compute an executable motion for the robot; this motion generation step is constrained by both the safe configurations which have been selected by the "Explore" step and by the dynamic parameters of the motion (i.e. the dynamic characteristics of the robot, the physical robot/environment interactions and the control strategy to apply for moving the robot towards the next subgoal). It is executed using appropriate models which have been derived from the concept of "physical model" initially developed in the field of Computer Graphics. The basic underlying idea is to consider that motions and/or deformations of physical objects result from the application of some basic physical laws which involve forces whose application points depends on the intrinsic structure of the interacting objects. In our approach, which applies most of the basic Cordis-Anima principles [19], a set of interacting objects is modeled by a set of representative punctual masses combined within a network of interaction including linear and non-linear terms. An important property of such a model relies in the fact that any component obeys the Newtonian dynamics.

The addressed motion planning problem and the principles of our approach are respectively described in sections 2.1 and 2.2. Section 3 presents the concept of physical model and the basic constructions which have been developed for modeling the robot, the workspace, and the physical robot/environment interactions. The way geometric models and physical models have been used for modeling the tasks to achieve in the case of the rover application and of the dextrous hand application is respectively described in section 4.1 and 4.2. Sections 5.1 and 5.2 respectively show how the addressed motion planning problem has been solved in both cases using the proposed "Explore & Move" strategy. Finally, section 6 addresses implementation and experimentation issues.

2   Outline of the approach

2.1   Statement of the problem

Let $\mathcal{A}$ be the robot, W be the workspace, C_A be the configuration space of $\mathcal{A}$, ST_A be the state space of $\mathcal{A}$, q_{start} be the initial configuration of $\mathcal{A}$, q_{goal} be the goal configuration to reach, and CO_g and CO_d be respectively the set of geometric constraints and the set of kinematic and dynamic constraints to satisfy during the execution of the task. CO_g can be formulated in C_A; it includes classical non-collision constraints and contact relation constraints (for instance a contact relation to achieve or to maintain during the motion). CO_d is generally expressed in ST_A; it includes non-holonomic kinematic constraints and dynamic constraints coming from both the robot mechanical characteristics and the robot/environment interactions[1].

The problem to solve is to find a *"safe"* and *"executable"* motion $\Gamma(q_{start}, q_{goal})$ allowing to move $\mathcal{A}$ from q_{start} to q_{goal} while respecting the constraints CO_g and CO_d. $\Gamma(q_{start}, q_{goal})$ is said to be "safe" if it verifies the constraints CO_g, and it is considered as "executable" if it verifies the constraints CO_d. The constraints CO_g to consider when dealing with the rover motion planning problem, express the fact that the vehicle has to avoid collisions, to maintain a contact

---

[1]Some non-holonomic kinematic constraints can also been represented in CA using appropriate neighbourhood constraints to consider at search time (see for instance the Dubins [7] curves or the discrete search strategies described in [17] in the case of a car-like robot).

between several wheels and the ground, and to avoid to tip-over. In the case of grasping tasks, the robot has to avoid collisions and to achieve potentially stable contact relations between the fingers and the object to be grasped (in this case, q_{goal} is expressed in terms of contact relations involving features of the object and of the hand). Processing such constraints can be done using geometric approaches (see for instance [25], [4], and [2]). The constraints CO_d to consider when generating executable motions are the non-holonomic kinematic constraints, the dynamic constraints, and the constraints imposed by the forces and the torques which are created by both the vehicle/terrain interactions and the control strategy to apply. In the case of grasping tasks, the forces and torques generated by the hand/object interactions must also respect some given equilibrium conditions depending on the task to perform. Processing such constraints obviously requires to make use of appropriate models having the ability to evaluate dynamic equations. As we will see further down, this is done in our approach using the concept of "physical model".

2.2   The motion planning approach

The motion planning approach which is proposed in this paper to solve the previous problem is based upon a two-stage approach combining a discrete search strategy and a continuous motion generation technique. The discrete search strategy (called the *"Explore"* step) is applied at each planning step to determine a pertinent set of potential intermediate safe configurations for the robot; it is based upon the application of classical search strategies and of geometric/kinematic models. In the case of the rover, the intermediate safe configurations represent the next subgoals to try to reach using the continuous motion generation technique; in the case of the dextrous hand, the intermediate configurations represent some selected preshapes of the hand. The continuous motion generation technique (called the *"Move"* step) is applied at each planning step to compute an executable motion for the robot; this motion generation step is constrained by both the safe configurations which have been selected by the "Explore" step and by the dynamic parameters of the motion (i.e. the dynamic characteristics of the robot, the physical robot/environment interactions and the control strategy to apply). In the case of the rover, the vehicle tries to reach the selected next subgoal by applying an appropriate motion strategy which integrates the wheel/ground interactions; in the case of the dextrous hand, the motions of the fingers are generated using an appropriate representation of the selected hand preshape and a model of the physical hand/object interactions. In both cases, the "*Move*" function makes use of representations which have been derived from the concept of "physical model" initially developed in the field of Computer Graphics for the purpose of producing realistic animation scenes [14] [26] [27] [19]. The main characteristics of the physically based model that we have developed for implementing the previous "Explore & Move" algorithm are described in the next section. The instantiations of this approach in the case of the rover application and of the dextrous hand application are respectively described in sections 4 and 5.

3 Physical modeling

3.1 The basic idea

Geometric models cannot been used to process complex physical interactions. This comes from the fact that the effects of such interactions result from the integration of differential equations involving distributed force and position parameters. Modeling such features requires to make use of appropriate constructions which obey the Newtonian physics. Some approaches exhibiting this property have already been developed in the field of Computer Graphics for the purpose of producing realistic animation scenes (see above). The basic idea in such approaches, is to consider that motions and/or deformations of physical objects result from the application of physical laws involving a set of forces whose application points depend on the intrinsic structure of the interacting objects. A practical way to obtain such a behaviour is to discretize the objects of the scene appropriately, and to associate each model component with a set of differential equations combining two dual variables: the force F and the position P (or the velocity V). Both uniform discretization techniques [26] [27] and task- oriented discretization approaches [19] have been proposed in the literature. But we have shown in [15] and in [16] that a task-oriented discretization approach is better adapted to the processing of robot/environment interactions.

In a task-oriented discretization approach, any object is represented by a set of interconnected particles having the following properties (see for instance the Cordis-Anima model [19]): (1) each particle is seen as a punctual mass m which obeys the Newtonian dynamics and which is surrounded by a spherical non-penetration "elastic" area; (2) the set of particles respects the mass, the inertia, and the spatial occupancy characteristics of the modeled object; (3) the particles are interconnected using interaction components (referred as the "connectors") which associate pairs of interacting particles with appropriate physical laws —as it is shown further down, interaction laws like viscous/elastic behaviours, elastic collisions, or dry friction are modeled by combining "spring" and "damper" components appropriately—.

Consequently, the physical model $\Phi(W)$ of a set W of interacting objects is represented by a network in which each node n_i defines an object component (i.e. a punctual mass and its associated non-penetration area), and each arc a_{ij} represents a physical interaction law. In this model, a node n_i is characterized by a pair (p_i, m_i), where p_i and m_i are respectively the position and the weight of the associated punctual mass; an arc a_{ij} defines a particular interaction equation of type $F_i = -F_j = \phi_{ij}(p_i, p_j)$, where ϕ_{ij} represents an elementary interaction law or a composite non-linear interaction. ¿From a computational point of view, a node n_i can be seen as an algorithm which computes a position p_i according to the laws of the Newtonian Physics and to the current forces provided by the associated connectors $a_{ij}, a_{ik}...$, and an arc a_{ij} represents a function which produces a force according to the characteristics of ϕ_{ij} and to the current value of the positions p_i and p_j, figure 3 illustrates.

3.2 Basic constructions for modeling robotic applications

Robotics applications mainly differ from Computer Graphics applications by the following characteristics: complex quasi-rigid and articulated objects have to be processed while object deformations are generally limited, collision and friction phenomena are of prime importance for the motion generation process, internal control forces and torques have to be considered, and motion planning problems require to optimize the computation time. These characteristics have necessitated the development of more appropriate constructions for modeling quasi-rigid components (e.g. the robot links), articulated objects (e.g. the robot joints), locally deformable objects (e.g. the ground in rover applications), and complex contact interactions like dry friction. The basic idea is to make use of fine physical models for representing contact and joint interactions, and to apply more global representations for modeling the behaviour of the other components of the scene (see section 4.1). In order to implement this idea, we have developed some basic constructions derived from the concepts proposed in [19] and in [16]: the linear and torsion spring/damper connectors, the contact interaction models, and the motion generation mechanisms.

The linear and torsion spring/damper connectors. A *linear spring/damper connector* (*LS*) defines a viscous/elastic interaction law between two particles by generating a force which tries to bring back these particles to a given equilibrium state, figure 1a illustrates. The force involved in this relation can be expressed by the following linear equation: $\vec{F} = \lambda \Delta \vec{P} - \mu \vec{P}$, where λ is the stiffness of the "spring", $\Delta \vec{P}$ is the variation of its length (p_0 the initial length), μ is the viscosity factor, and $\vec{P}$ is the speed of the motion. The term $-\mu \vec{P}$ is a "damper" used to prevent the oscillation of the system near the equilibrium position of the pair of particles (μ has a large value for a quasi-rigid object and a small one for an elastic body). Such a connector is useful for representing traction behaviours and prismatic joints, but it is not adapted to the representation of flexion phenomena and of revolute joints[2], because only one linear degree of freedom is constrained by *LS*. This is why we have developed the *torsion spring/damper connector* (*TS*) for associating an angular constraint to a set of three particles. In *TS*, the forces which are generated onto the non central particles can be defined by the expression: $\vec{F_i} = (\lambda \Delta \theta - \mu \dot{\theta}) \vec{k_i}$, where $\Delta \theta$ is the variation of the angle formed by the three particles, $\dot{\theta}$ is the angular speed, and $\vec{k_i}$ is the unitary vector normal to the direction defined by the involved pair of particles, figure 1b illustrates. Because of the action/reaction principle, the force $\vec{F}$ applied to the central particle is equal to

the negative sum of $\vec{F_1}$ and $\vec{F_2}$. An interesting property of *TS* is to constrain a set of three particles to only move on a straight line when the equilibrium value θ_0 is equal to 180°. This property may be used to model a prismatic joint as shown on figure 1c: the *LS* relations define a viscous/elastic prismatic joint (the joint value is l), and the *TS* relations constrain the particles to only move along the joint axis. Figure 1d shows how to model a revolute joint using *TS* and *LS* relations (in this case, the joint value θ is associated to the *TS* relations).

The contact interactions. Contacts generate surface-surface interactions whose characteristics depend on several parameters: the distribution of the contact points, the relative velocity of the interacting objects, the involved forces, and the physical characteristics of the surfaces in contact. Some approaches for studying vehicle/ground interactions have already been developed in the field of Mechanics. These approaches generally consist in combining partial theoretical models with empirical models derived from experimentations, in order to elaborate a set of elementary behavioural laws (i.e. laws which associate a particular type of dynamic behaviour to a given type of ground [3]). But these approaches cannot been applied in physical modeling, because they cannot been expressed in terms the numerical integration of a set of discretized differential equations.

The basic idea which has been applied in our model, is to represent such interactions using a combination of simple mechanical models of collision and of friction phenomena. *Collision forces* are defined by the following expression: $\vec{F_c} = -\lambda(\vec{d} - \vec{d_0})$ if $d < d_0$ and $\vec{F_c} = 0$ else, where λ is the rigidity factor of the collision, d is the distance between the colliding particles, and d_0 is the nominal distance defining the contact relation. *Friction forces* are defined using the Coulomb law: $\vec{F_f} = -k|\vec{F_n}|\vec{u}$, where k is the static or the kinetic friction parameter, $\vec{F_n}$ is the normal component of the force $\vec{F}$ applied at the contact point, and $\vec{u}$ is the direction defined by the tangential force component $\vec{F_t}$ or by the relative velocity $\vec{V}$ of the interacting particles. Complete surface-surface interactions can be modeled using these basic constructions: the viscous/elastic law associated to a non-penetration threshold characterizes the contact formation (such a contact may involve several pairs of particles), and surface adherence law

---

[2]Such properties may be represented by appropriately combining several *LS* connectors (see for instance [16]), but this necessitates to introduce additional fictitious punctual masses and connectors in the model. A consequence of this approach is to artificially modify the mass distribution and to increase the complexity of the model.

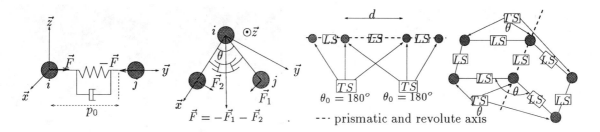

Figure 1: *Basic constructions of the model: (a) the LS connector, (b) the TS connector, (c) modeling of a prismatic joint, (d) modeling of a revolute joint.*

that produces sliding and gripping effects are modeled using the previous equations. Dealing with more complex surface-surface interactions requires to make use of a *finite state automaton*. Indeed, phenomena like dry friction basically involves three different states: no contact, gripping, and sliding under friction constraints. The commutation from one particular state to another is determined by conditions involving gripping forces, sliding speed, and relative distances (see figure 2). In this representation, each state is characterized by a specific interaction law. For instance, a viscous/elastic law is associated to the gripping state, and a Coulomb equation is associated to the sliding state [16].

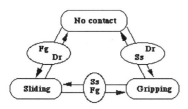

Figure 2: *A finite state automaton to model dry friction phenomena*

The motion generation mechanisms. Let W be a set of interacting objects and F be an external force which is applied to some components of $\Phi(W)$. The dynamic behaviour of W after having applied F can be simulated at a given frequency (this frequency depends on the type of phenomenon to simulate) by applying the following algorithm iteratively:

1. Compute the forces to associate with each arc a_{ij} in $\Phi(W)$ by evaluating the interaction function ϕ_{ij}.

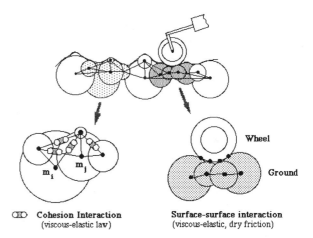

Figure 3: *A physical model of the terrain and of the vehicle/terrain interactions.*

2. Compute the new position p_i to associate with each node n_i in $\Phi(W)$ using the Newtonian law: $\sum_{\forall a_{ik}} F_{ik} = m_i \cdot \gamma_i = m_i \cdot (d^2 p_i / dt^2)$[3]

Let $\Phi(\mathcal{A}, W)$ be the physical model which is associated to the system composed of the robot $\mathcal{A}$ and of the environment W. $\Phi(\mathcal{A}, W)$ is obtained by combining $\Phi(\mathcal{A})$ and $\Phi(W)$ using appropriate contact interaction laws, figure 3 illustrates. Generating a motion of $\mathcal{A}$ while reacting with W can be performed by associating a "propelling force" F to $\mathcal{A}$ and by applying the previous algorithm onto $\Phi(\mathcal{A}, W)$. In practice, F re-

---

[3]In practice, this expression is evaluated using a finite-difference formulation: $\vec{P}_t = \frac{\Delta T^2}{m} * \vec{F}_{ext} + 2\vec{P}_{t-\Delta T} - \vec{P}_{t-2\Delta T}$, where $\vec{P}_t$ is the position of the particle at time t, ΔT is the time step, and $\vec{F}_{ext}$ is the sum of the external forces applied on P.

sults from the application of a given control onto $\mathcal{A}$. A straightforward solution to do that is to generate a constant propelling force F to apply to some given particles of $\Phi(\mathcal{A})$ (e.g. the punctual mass located at the center of the front axle of a vehicle). This solution has already been used to perform some simple simulations (see [16]), but it does not enable us to apply more realistic control strategies. An other solution which has already been proposed in [16], is to associate an internal source of kinetic energy to some appropriate components of the physical model of the mechanical structure of $\mathcal{A}$. Such a construction is called a "physical effector". For instance, the physical effector of a propelling wheel of a vehicle will be represented by the coupling of a controlled torque with a pair of rigidly connected punctual masses belonging to the model of the wheel.

4   The rover motion planning problem

4.1   Task modeling

As it has already been mentioned in section 2, our motion planning approach requires to combine geometric and physical models. It is the purpose of this section to describe the models which have been used to solve the rover motion planning problem.

Geometric/kinematic models. Let $G(\mathcal{A})$ and $G(\mathcal{T})$ be respectively the geometric/kinematic models of the rover $\mathcal{A}$ and of the terrain $\mathcal{T}$. $G(\mathcal{A})$ is represented as an articulated object involving a set of rigid bodies Ω_i (e.g. the main body, a link of the chassis, an axle, or a wheel) and a set of active or passive joints δ_k (e.g. a joint connecting an axle to a wheel, or a peristaltic joint for modifying the length of the chassis), figure 4 illustrates. $G(\mathcal{A})$ is represented using a traditional tree structure, in which the leaves represent the Ω_i and the arcs represent spatial relationships or joints.

A configuration q of $\mathcal{A}$ is given by $6 + n_\delta$ parameters: *six* parameters $(x, y, z, \theta, \phi, \psi)$ specifying the position/orientation of the main body in the reference frame of W, and n_δ parameters specifying the values of the joints δ_k.

The terrain $\mathcal{T}$ is initially described using an elevation map defined from a regular grid (x, y) expressed in the reference frame of W (such an elevation map is generally provided by a vision system). For making

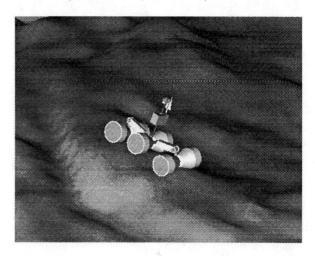

Figure 4: A six-wheeled vehicle on a terrain

the collision and the contact computations possible, we have represented $G(\mathcal{T})$ as a connected set of patches.

Physical model of the vehicle. As explained in section 3, the physical model $\Phi(\mathcal{A})$ of the rover $\mathcal{A}$ is basically represented by a network of interacting particles. A first way to obtain this network is to appropriately discretized the components Ω_i of $\mathcal{A}$ and to construct the appropriate connectors. Such an approach has already been used in [15] and in [16] for simulating the dynamic behaviour of a rover moving on a terrain. In spite of the efficiency of this approach to simulate the vehicle/terrain interactions, it is not directly applicable to the motion planning problem because it is time consuming and task dependent[4]. In order to reduce the computation time, we have developed an other instance of the model [5] [6] in which each sub-network associated to a rigid component of $\mathcal{A}$ is represented by a single node. Such a node is connected to its neighbours in the network using viscous/elastic connectors as described in section 3, but it differs from the basic particles by the fact that it obeys the laws of the physics of the solids (it is characterized in practice by position/orientation and by force/torque parameters). In this model, the joints δ_k of $\mathcal{A}$ are represented using the approach described in section 3, figure 6 illustrates.

---

[4]On one hand, the required processing time increases with the number of particles and of connectors. On the other hand, the processing of contact interactions requires to make use of a fine discretization of the interacting objects (and consequently to increase the number of particles).

This new type of node in $\Phi(\mathcal{A})$ requires to also consider the classical laws of the dynamic when propagating force and torques parameters in the network. Let $r_i(t)$ and $\alpha_i(t)$ be respectively the position and the orientation parameters at time t of a rigid body Ω_i of $\mathcal{A}$. The motions of Ω_i are specified by the Euler/Newton equations:

$$F_i = m_i \, \ddot{r}_i(t) \,, \qquad T_i = \dot{L}_i(t) = I_{\Omega_i} \, \ddot{\alpha}_i(t) \qquad (1)$$

where F_i and T_i are respectively the sums of forces and torques applied on Ω_i, $L_i(t)$ is the angular moment about the center of mass G_{Ω_i}, and I_{Ω_i} is the inertia tensor of Ω_i about the frame axes. $\dot{L}_i(t)$ is also related to Coriolis and centrifugal terms (see [11]), but we will make the assumption that such terms are negligible (because the robot is supposed to move with a low velocity). Then, F_i and T_i can be calculated using the *Euler*'s principle of superposition:

$$F_i = F_d + \sum_{force\, j} F_{i,j} \qquad (2)$$

$$T_i = \sum_{force\, j} (G_{\Omega_i} P_{i,j} \times F_{i,j}) + \sum_{torque\, k} T_{i,k} \qquad (3)$$

where $T_{i,k}$ are the torques acting on Ω_i, $P_{i,j}$ are the points where the forces $F_{i,j}$ are applied, $G_{\Omega_i} P_{i,j}$ is the vector from G_{Ω_i} to $P_{i,j}$, and $\times$ is the cross product. The term F_d includes the forces generated by the gravity and additional forces such as the viscosity of the environment. The set of forces $F_{i,j}$ results from the evaluation of the physical interactions of Ω_i with the other components of $\Phi(\mathcal{A})$ —through the joints of $\mathcal{A}$— and with the the involved components of $\Phi(\mathcal{T})$ —through the wheel/ground contact interactions—. The torques $T_{i,k}$ are generated by the control mechanisms (i.e. the "physical effector") associated to the wheels of $\mathcal{A}$.

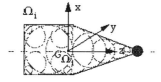

Figure 5: Representing a wheel of $\mathcal{A}$ by a set of spheres.

Physical model of the Terrain. According to our approach (see section 3), the physical model $\Phi(\mathcal{T})$ of the terrain $\mathcal{T}$ is represented by a network of interacting particles. Constructing this network requires to first determine an appropriate set of particles for representing $\mathcal{T}$, and then to define the type of the interaction

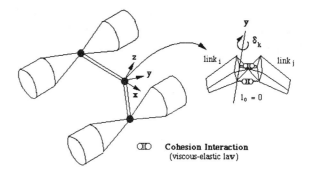

Figure 6: Modeling a 3D revolute mechanism.

laws to associate to these particles. The first step of the modeling process consists in converting the geometric model $G(\mathcal{T})$ of the terrain $\mathcal{T}$ into a set $\mathcal{ST}$ of spheres whose profile approximates the surface of $\mathcal{T}$ (as shown in Figure 3). The sizes of the spheres which are used for constructing $\mathcal{ST}$ depend on the geometry of $\mathcal{T}$, on the global physical characteristics of $\mathcal{T}$ (modeling deformable area requires more particles than rigid area), and on the characteristics of the wheel/ground interactions to process (the sizes of the spheres used to represent $\mathcal{T}$ must be consistent with the number of particles and with the distribution factor used for modeling the wheels of $\mathcal{A}$). In $\mathcal{ST}$, each sphere (p, r) represents a particle (m, r, p), where m is a punctual mass, r is the radius of the associated non-penetration area, and p is the position of the particle. It should be noticed that this approach allows the representation of an hilly terrain using a reasonable number of elements, since the selected particles are not uniformly distributed onto the surface of $\mathcal{T}$.

The next step of the modeling process consists in connecting the previous particles using appropriate interaction connectors. Cohesion properties are obtained using spring/damper connectors, and surface-surface interaction characteristics are modeled using the contact interaction constructions described in section 3.2, figure 3 illustrates.

Physical model of the vehicle/terrain interactions. The most difficult problem to solve when dealing with a vehicle moving on a terrain, is to define realistic vehicle/terrain interaction laws. In our approach, such interaction laws are modeled using the basic contact interaction constructions described in section 3.2. But, the ability of the model to represent as accurately as possible the wheel/ground interactions depends on two criteria: the distribution of the contact points and the type of the surface-surface interactions to be considered. Thanks to our representation of $\Phi(\mathcal{A})$ and of $\Phi(\mathcal{T})$, the distribution of the wheel/ground contact points can easily been determined using a fast distance computation algorithm involving a structured set of spheres [13]: these spheres represents the pairs (particle, non-penetration area) which model the wheels of $\mathcal{A}$ and the terrain $\mathcal{T}$. In order to make the contact interaction computation possible, a set of elementary particles is distributed onto the contact area defined by the contact points (see figure 3). The number of particles involved in this representation depends on both the size of the contact area and the accurary the model.

As explained in section 3.2, the surface-surface interactions are processed using two types of constructions: a viscoos-elastic law associated to a non-penetration threshold, and a surface adherence law that produces sliding and gripping effects. In order to process dry friction phenomena, we also make use of the finite state automaton described in figure 2.

4.2 Motion Planning

Instanciation of the "Explore & Move" approach. As mentioned in 2.1, the problem to solve is to find a *"safe" and "executable"* motion $\Gamma(q_{start}, q_{goal})$ allowing to move the rover $\mathcal{A}$ from a configuration q_{start} to a given configuration q_{goal}, while respecting a set of geometric/kinematic and of dynamic constraints CO_g and CO_d. Our approach to solve this problem is to apply an "Explore & Move" strategy. The "explore" step is executed using a discrete graph search strategy operating on the geometric model $G(W)$ of the task; the "move" step is performed using a continuous motion generation technique operating on the physical model $\Phi(W)$ of the task. The "explore" function is applied at each planning step to determine a ranked set of potentially reacheable safe configurations of $\mathcal{A}$. These configurations are then considered as the next subgoals to try to reach using the "move" function. In case of failure of the motion generation step, the system backtracks in order to process an unexplored solution.

Let q be a configuration[5] of $\mathcal{A}$, and $\tilde{q} = (x, y, \theta)$ be the 2D position/orientation parameters of q. A safe

---

configuration of $\mathcal{A}$ is a configuration which satisfies the following constraints COg: the wheels of $\mathcal{A}$ are in contact with the terrain, there is no collision between $\mathcal{A}$ and $\mathcal{T}$, and some basic static equilibrium conditions are satisfied (no tip-over and non sliding geometric criteria). A safe configuration q of $\mathcal{A}$ is called a *"stable placement"* $P(\tilde{q})$ of $\mathcal{A}$, where $\tilde{q}$ represents the 2D position/orientation parameters of q. Because of the presence of the constraints CO_g, there exists obvious inter- dependencies between the parameters of a stable placement $P(\tilde{q})$. This property allows us to reduce the dimension of the search space when searching for stable placements of $\mathcal{A}$ [4]: given a sub-configuration $\tilde{q} = (x, y, \theta)$ of $\mathcal{A}$, it is possible to compute the other configuration parameters of q which verifies the constraints CO_g. The related computation is performed by an iterative algorithm called the "placement algorithm". The basic idea in this algorithm is to simulate the falling of $\mathcal{A}$ onto the terrain $\mathcal{T}$, at the 2D position/orientation specified by (x, y, θ). This can be done by minimizing a "potential energy function" including a gravity term (applied to each component of $\mathcal{A}$) and a spring energy term (applied to each wheel/axle joint), under a set of constraints expressing the fact that the contact points must stay on the surface of the ground (see [4] for more details). It should be noticed that this formulation of stability is based on simple geometric criteria, and that it does not guarantee the real stability of $\mathcal{A}$. It is the purpose of the next planning step to evaluate more complete stability criteria (see below).

The "Explore" function. A practical consequence of the previous property, is to make it possible to apply the discrete search strategy of the "Explore" function onto the subspace $\mathcal{C}_{search}$ defined by $\tilde{q}$ —i.e. the configuration space $E(x, y, \theta)$—. This approach has already been used in [25]. It basically consists in applying an A^* algorithm to find a near-optimal path in an incrementally generated graph $\mathcal{G}$ representing the explored configurations of $\mathcal{C}_{search}$. Let $\tilde{q}^i$ be the current sub-configurations of $\mathcal{A}$ in $\mathcal{C}_{search}$. The associated complete configuration of $\mathcal{A}$ in CA is given by the "placement" relation $q^i = P(\tilde{q}^i)$, where $P(\tilde{q}^i)$ is computed using the previous "placement algorithm". Then, the expansion of the node $N_i \in \mathcal{G}$ associated to $\tilde{q}^i$ is obtained by

---

evaluating a cost function and by applying a simple control u during a fixed period of time ΔT. In the current implementation, the cost function estimates the length of the generated trajectory and the control u defines 8 types of motion: 3 forward movements satisfying the non-holonomic constraints of $\mathcal{A}$ (left turn, straight, and right turn), 3 similar backward movements, and 2 "pure turns" (left and right) executed without moving $\mathcal{A}$ in the forward or in the backward direction (such turns are executed by applying appropriate controls onto the right and left wheels of $\mathcal{A}$). The successors of $\tilde{q}^i$ obtained using this technique are considered as candidates for potential stable placements of $\mathcal{A}$, and only the successors which generates stable placements are stored into the ranked list which is considered for the next planning step. Let $\tilde{q}^i_{next}$ be the best successor (according to the cost function) of $\tilde{q}^i$. The next planning step (i.e. the "Move" step) consists in trying to generate a continuous motion allowing $\mathcal{A}$ to move from $q^i = P(\tilde{q}^i)$ to $q^i_{next} = P(\tilde{q}^i_{next})$ (see below).

The "Move" function. The purpose of this motion generation step is to check for the existence of an *executable* motion allowing $\mathcal{A}$ to move from q^i to q^i_{next}, i.e. a motion which verifies the kinematic/dynamic constraints CO_d of the task (i.e. the kinematic/dynamic constraints of $\mathcal{A}$, the constraints imposed by the vehicle/terrain interactions, and the constraints coming from the applied control strategy). Solving this problem requires to cope with differential equations involving position, velocity and acceleration terms, and consequently to formulate the problem into the state-space $ST_{\mathcal{A}}$ of $\mathcal{A}$. As it has already been shown in section 4.1, this can be done using our physical model $\Phi(\mathcal{A}, \mathcal{T})$ of the task. But the main difficulty which remain to overcome is to find an appropriate way to generate a motion which allows $\mathcal{A}$ to move towards the next sub-goal represented by q^i_{next}. Let δt be the time increment of the motion generation process ($\delta t = \Delta T/p$), s^i be the state of $\mathcal{A}$ corresponding to the configuration q^i, and let $s^i(n\delta t)$ be the state of $\mathcal{A}$ obtained after having applied n elementary motion step from s^i (i.e. after having applied a sequence of n controls on the "physical effectors" of $\Phi(\mathcal{A})$). Determining the required sequence of controls to apply to $\mathcal{A}$ can be done by executing an iterative algorithm involving two complementary functions: a function which hypothesizes a nominal sub-path Γ^i_k between the current sub-configuration $\tilde{q}(n\delta t)$ and the next sub-goal represented by $\tilde{q}^i_{next}$, and a function which tracks Γ^i_k while processing the physical vehicle/terrain interactions, figure 7 illustrates.

In the current implementation of the algorithm, Γ^i_k is constructed using a technique derived from the Dubins approach [7]. Then, the obtained sub-path is a smooth curve made of straight line segments and of circular arcs (the choice of the gyration radius to apply at a given step of the algorithm depends on the length of the involved path[6] and on the mechanical characteristics of $\mathcal{A}$, see [6]).

The tracking function operate on the model $\Phi(\mathcal{A}, \mathcal{T})$. It takes as input the velocity controls applied on each controlled wheel during a time increment δt. These controls are computed from the linear and steering speeds which are associated to the reference point of $\mathcal{A}$ when moving on Γ^i_k. They are converted into a set of torques $u(t)$ to apply to the "physical effectors" of $\Phi(\mathcal{A})$, in order to be processed by the motion generator (see section 3.2).

Since the motion generation step integrates physical phenomena like sliding or skidding, the state $s^\star_i(n\delta t)$ of $\mathcal{A}$ obtained after having applied n successive controls may be different from the nominal state $s_i(n\delta t)$. The processed motion generation step will be considered as a failure when the sub-configuration $\tilde{q}^\star_i(n\delta t)$ is too far from its nominal value $q_i(n\delta t)$. The previous algorithm is iterated until the neighbourhood of s^i_{next} is reached or until a failure is detected (see figure 7).

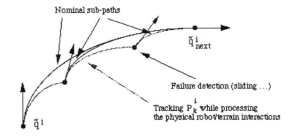

Figure 7: The motion generation scheme.

---

[6]If this length is large enough, the velocities of the controlled wheels are positive; otherwise, the velocities of the opposite controlled wheels have opposite signs and the vehicle will skid and use more energy to execute the motion.

5 The dextrous grasping problem

5.1 Task modeling

As it has already been mentioned in section 2, our motion planning approach requires to combine geometric and physical models. It is the purpose of this section to describe the models which have been used to solve the dextrous grasping problem.

Geometric/kinematic models. Let $G(\mathcal{A})$ and $G(\mathcal{O})$ be respectively the geometric/kinematic models of the hand $\mathcal{A}$ and of the object $\mathcal{O}$ to be grasped. $G(\mathcal{A})$ is represented as an articulated object having 9 revolute joints for moving the fingers and 6 degrees of freedom for positioning the hand in the $3D$ space. This model is constructed using classical CAD models, and a configuration q of $\mathcal{A}$ is characterized by 15 parameters: *six* parameters $(x, y, z, \theta, \phi, \psi)$ specifying the position/orientation of the hand in the reference frame of W, and 9 parameters specifying the values of the finger joints.

The object $\mathcal{O}$ is initially represented by a $3D$ occupancy grid which is either obtained by a vision reconstruction process or derived from a classical CAD model. For the purpose of the shape analysis process on which the "hand preshaping" step relies, $G(\mathcal{O})$ has been converted into an octree representation (see [2]).

Physical model of the hand/object system. According to our approach (see section 3), the physical model $\Phi(\mathcal{A}, \mathcal{O})$ of the hand/object system is represented by a network of interacting particles. Constructing this network requires to first determine an appropriate set of particles for representing $\mathcal{A}$ and $\mathcal{O}$, and then to define the type of the interaction laws to associate to these particles.

The first step of the modeling process consists in converting the geometric models of $G(\mathcal{O})$ and of the rigid components of $G(\mathcal{A})$ into sets $\mathcal{ST}_i$ of spheres whose profiles approximate the shapes of the associated components. As in the case of the rover application (see section 4.1), the sizes of the spheres which are used for constructing the sets $\mathcal{ST}_i$ depend on both geometric and physical criteria (shape, size, deformability factor, hand/object interaction characteristics). Each sphere $(p, r) \in \mathcal{ST}_i$ represents a particle (m, r, p), where m is a punctual mass, r is the radius of the associated non-penetration area, and p is the position of the particle.

The next step of the modeling process consists in connecting the previous particles using appropriate interaction connectors. Cohesion properties are obtained using spring/damper connectors, revolute joints are modeled by appropriately combining linear and torsion spring/damper connectors, and surface-surface interactions are characterized using the basic contact interaction constructions (see section 3.2). In this model, external forces are generated by the gravity, by the hand/object interactions, and by the forces/torques produced by the control mechanisms (i.e. the "physical effectors") which are associated to the finger joints. As we will see in the next section, these control mechanisms are modeled using appropriate sets of linear spring/damper connectors (the global characteristics of the associated network depend on the type of task to perform).

5.2 Motion planning

Instantiation of the "Explore & Move" approach. As mentioned in 2.1, the problem to solve is to find a "safe" and "executable" motion for the robot hand $\mathcal{A}$, starting from a given configuration q_{start} and ending at a partially specified goal configuration q_{goal} representing a "*stable grasp*" (i.e. a hand/object configuration which is accessible to the hand and which involves grasping forces allowing the robot to robustly maintain the object into the hand). q_{goal} is generally expressed in terms of contact relations involving features of the robot hand and of the object to be grasped. The first step of the planning process (the "*Explore*" step) is to determine q_{goal} using an appropriate discrete search strategy operating upon the geometric model $G(W)$ of the task. In our approach, a partially specified goal configuration q_{goal} is represented by a "*hand preshape*" $\mathcal{P}$ which characterizes an infinite set of functionally equivalent grasps (see below). The second step of the planning process consists in searching for a particular stable grasp, by generating appropriate motion strategies constrained by both the selected "hand preshape" $\mathcal{P}$ and the hand/object physical interactions. This planning step is performed using the "*move*" function. In case of failure of the motion generation step (i.e. no stable grasp has been found), the system backtracks in order to process an unexplored solution represented by a new "hand preshape".

Hand preshaping using an appropriate "Explore" function. Because of the high dimension of the hand configuration space C_A, a classical exhaustive analysis of the set of possible grasp configurations is not applicable when dealing with a dextrous hand. This is why we have chosen to apply another approach motivated by studies of human grasping: the human hand "preshapes" to fit an object's form while reaching for it, before selecting a particular grasp using tactile and force feedback during grasp closure. Consequently, a *"hand preshape"* represents a class of hand configurations from which the final grasp is chosen. In this approach, the "hand preshaping" step relies on a task-oriented exploration of a subset of C_A, and the grasp selection step makes use of a reactive motion generator. The task-oriented exploration of C_A is performed in our system using two complementary steps: (1) construct an appropriate task-oriented model of $\mathcal{O}$ which cleanly represents the object proportions and the accessibility relations, and (2) select a preshape $\mathcal{P}$ among the obtained set of candidates. An interesting property of this approach is to drastically reduce the complexity of the search step, by constructing a symbolic representation of the task which combines qualitative information (global shape, protuberances, relative size of components ...) with a limited amount of quantitative data such as the center of mass or the principal dimensions. In our approach, such a representation is constructed using the concept of "elliptical cylinder" (see [2]). Then, the task-oriented model of the task is obtained using an efficient heuristic reconstruction technique that is well- adapted to the needs of grasp planning [2]: a topographic analysis of the object shape is performed by slicing the octree model $G(\mathcal{O})$, and the result of this analysis is used to reconstruct $\mathcal{O}$ as a hierarchy of *elliptical cylinders and associated approach directions*, figure 8 illustrates.

Once the task-oriented model has been constructed, a particular hand preshape $\mathcal{P}$ and its associated goal region G on $G(\mathcal{O})$ is heuristically selected. This choice mainly depends on the characteristics of the task to perform. It is perform using an heuristic function which combines hand/object proportion criteria, stability criteria, and functional knowledge about grasping. For instance, a preshape $\mathcal{P}$ representing a class of *three-digits* grasps will be chosen for robustly grasping an almost spherical object whose dimensions fits with the dimensions of the hand fingers. A preshape $\mathcal{P}$

produced by this planning step is characterized by the following parameters [2]: the relative hand/object location which is considered as the starting point for the closure step, the approach direction for the hand, and the intended configuration and contact parameters for the fingers (i.e. the number of finger involved, the type of contact to achieve −tip, pad, or palm−, and the relative placement of the fingers when executing the grasp operation −opposition between the thumb and the other fingers, opposition between the three fingers ...−). It has been shown in [2] that a small number of different types of preshapes are needed (for instance, only 12 types of preshapes have to considered for our three-fingered hand).

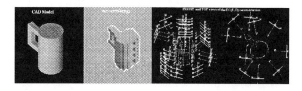

Figure 8: 3D reconstruction of a mug as a hierarchy of elliptical cylinders EC_i and their associated approach directions D_i^k. Each pair (EC_i, D_i^k) represents a candidate for a "hand preshape".

Grasp determination using a "Move" function. The purpose of this motion generation step is to check for the existence of an *executable motion of $\mathcal{A}$ ending at a configuration $q - goal$ corresponding to a stable grasp of $\mathcal{O}$*. The constraints CO_d to satisfy during this motion are the kinematic/dynamic constraints of $\mathcal{A}$, the constraints imposed by the hand/object interactions that must respect some given equilibrium conditions, and the constraints coming from the applied control strategy. As in the case of the rover, this problem has to be formulated in the state space ST_A of $\mathcal{A}$ for being solved. This is done using our physical model $\Phi(\mathcal{A}, \mathcal{O})$ of the task (see section 5.1). Then, the equilibrium conditions $SG(\mathcal{O})$ can easily be expressed as follows: $\vec{V} = \vec{0}$ and $\sum \vec{F_{ext}} = \vec{0}$, where $\vec{V}$ is the speed of $\mathcal{O}$ and each $\vec{F_{ext}}$ is an external force applied on $\mathcal{O}$ (i.e. a gravity term or a force applied by a finger of $\mathcal{A}$)/footnoteGenerating a force-closure grasp allowing the execution of a task requiring to apply a force F_{task} onto the object $\mathcal{O}$ can easily been done by introducing this new term within the expression $\sum \vec{F_{ext}} = \vec{0}$..

Let q_i and s_i be respectively the hand configuration

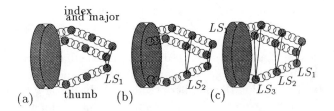

Figure 9: *Defining grasping strategies using appropriate sets of LS relations for connecting the joints of $\mathcal{A}$: (a) executing a tip contact based grasp; (b) executing a pad contact based grasp; (c) executing a palm contact based grasp.*

and the hand state associated to the selected preshape $\mathcal{P}$. Then, the problem to solve is to generate an executable motion $\Gamma(q_i, q_{goal})$ of $\mathcal{A}$ (i.e. a motion which verifies the constraints CO_d) starting at s_i and ending at a state s_{goal} characterized by the equilibrium condition $SG(\mathcal{O})$. Since CO_d includes contact conditions imposed by $\mathcal{P}$, $\Gamma(q_i, q_{goal})$ obviously depends on the type of grasp to achieve: a grasp consisting in surrounding the objet with the articulated fingers will not involve the same control strategy that a tip contact based grasp. In our approach, a particular grasping strategy is achieved using an appropriate set of linear spring/damper connectors (LS connectors) for connecting the joints and the finger tips of $\mathcal{A}$, figure 9 illustrates. Let (P_i, P_j), (λ, μ), and (lo, l) be respectively the pair of particles (joints or finger tips) to connect, the rigidity/viscosity factors, and the final/initial length of the considered LS connector. The values to associate to these parameters depend on the type of the involved hand/object contacts (tip, pad, or palm contacts) and on the characteristics of the grasp to achieve: pairs (P_i, P_j) are defined by the hand/contact characteristics, (λ, μ) parameterizes the motion generation process, and pairs (l, l_0) are defined by the preshape $\mathcal{P}$ and by the local morphological characteristics of $\mathcal{O}$. For instance, a "spherical grasp" involves an l_0 value depending on the global size of $\mathcal{O}$, while a "cylindrical grasp" will generates a small l_0 value for the pair (index, major), and a l_0 value depending on the size of $\mathcal{O}$ for the pair (index, thumb). It should be noticed that this approach consisting in connecting sets (P_1, P_2, P_3) of "coordinated" points of the fingers (finger tips or $i - th$ joints of the fingers) permits the generation of a set of forces having a null resulting value (such a property is obviously very interesting for achieving a stable

grasp).

Complete motion strategies can also been generated using a similar approach where the goal to achieve is represented using a "ghost" of the moving object associated to a set of particles having an "infinite mass", figure 10 illustrates.

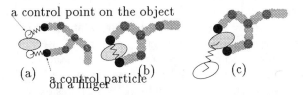

Figure 10: *Defining motion strategies using a set of LS relations for connecting the control points of the moving object $\mathcal{A}$ with the associated "infinite mass particles" belonging to a "ghost" of $\mathcal{A}$ which is located at the goal configuration: (a) $\mathcal{A}$ is attracted towards the selected preshape; (b) the fingers of $\mathcal{A}$ attract each others in order to achieve the required grasp; (c) the goal configuration of the object $\mathcal{O}$ attracts the hand/object system.*

6 Implementation and Experiments

The approach presented in this paper has been implemented on a Silicon Graphics workstation within the robot programming system ACT[7]. Several experiments have been successfully processed in simulation for both the rover application and the dextrous grasping application.

The Rover Application. The rover $\mathcal{A}$ considered for these experiments is a non-holonomic vehicle having an articulated mechanical structure and a locomotion system composed of six motorized wheels, figure 4 illustrates. The central pair of wheels is connected to the main body of $\mathcal{A}$ and the two others pairs of wheels are respectively connected to the front and the rear axles of $\mathcal{A}$. These axles are connected to the chassis by joints allowing roll and yaw movements, and the front and the rear links of the chassis are articulated in order to enable a "peristaltic" transformation of $\mathcal{A}$.

Figure 11 shows the trajectory generated by the system when only forward and backward motions respecting the non-holonomic constraints of $\mathcal{A}$ are allowed.

[7] The ACT system is developed and currently marketed by the Aleph Technologies Company.

Figure 12 shows the trajectory obtained when "maneuvers" are allowed (i.e. when "pure turns" that locally violate the non-holonomic constraints can be executed by $\mathcal{A}$, see section 4.2). In this case, the produced maneuver is executed by applying controls having opposite signs onto the wheels of each axle; the smooth part of the trajectory is executed by applying positive controls of various magnitudes onto the wheels of each axle. Figure 13 shows the profiles of the forces applied on the six wheels of the rover when executing the last left turn of the trajectory shown in figure 11. In these profiles, the location of the forces having a large magnitude corresponds to the place where the wheels are crossing an irregular area.

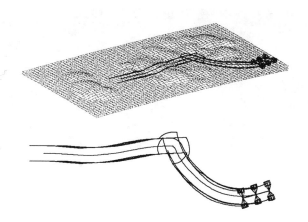

Figure 12: Trajectories executed by the main body and the six wheels of $\mathcal{A}$ when maneuvers are allowed.

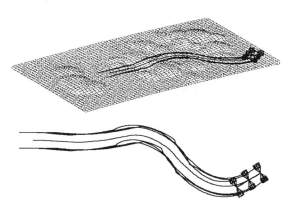

Figure 11: Trajectories executed by the main body and the six wheels of $\mathcal{A}$ when no maneuver is allowed.

The dextrous grasping application. The articulated hand $\mathcal{A}$ used for these experiments is composed of three fingers, where each finger has three links connected by two revolute joints, see figure 14. In the physical model $\Phi(\mathcal{A})$, the links of the fingers are modeled by n particles connected by $(n-1)$ LS relations and by $(n-2)$ TS relations. As explained in section 5.1, the number n of particles depends on the characteristics of the task to perform (and in particular on the deformability properties to consider). In the shown experiments, only 3 particles have been used to model each rigid link of the fingers of $\mathcal{A}$. Several grasping experiments have been processed with this simple hand, by considering tip or pad contacts involving various friction parameters. Figure 14 shows how a grasping operation may fail when friction forces are not suffi-

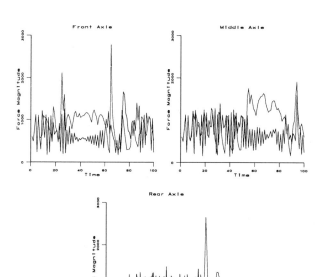

Figure 13: The forces applied on the wheels of the three axles of $\mathcal{A}$ during the execution of the last left turn of the motion shown in the previous figure.

cient to generate a stable grasp. Some other experiments have also been processed using a physical model of the Salisbury hand (this model is composed of 300 particles). Despite the increasing number of particles, these experiments have shown that the motion genera-

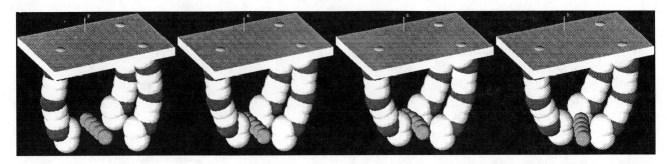

Figure 14: (A simple articulated hand trying to grasp a bar: (1) hand preshaping step; (2) first closure step ending when the finger/object contacts have been achieved; (3) next closure step consisting in applying the computed forces at the contact points (the object is subjected to a deformation); (4) the bar slips out of the hand because the friction forces are not sufficient to generate a stable grasp.

tion process could be still performed in "real time" (i.e. a motion step is generated in less than one second).

Acknowledgment

The work presented in this paper has been partly supported by the CNES (Centre National des Etudes Spatiales) through the RISP national project, the MRE (Ministère de la Recherche et de l'Espace) and by the Esprit-BRA project SECOND. It has been also partly supported by the Rhône-Alpes Region through the IMAG/INRIA Robotics project SHARP.

References

[1] B.S. Baker and S. Fortune and E. Grosse: *"Stable prehension with a multi-fingered hand"*, IEEE Int. Conf. on Robotics and Automation, 1985, March, St Louis.

[2] C. Bard , C. Laugier , C. Milesi-Bellier, J.Troccaz, B. Triggs , G. Vercelli: *"Achieving dextrous grasping by integrating planning and vision based sensing"*, International Journal of Robotics Research. To appear in 1994.

[3] M.G.Bekker: *"Introduction to Terrain-Vehicle Systems"*, The University of Michigan Press, 1969.

[4] Ch. Bellier, Ch. Laugier, B. Faverjon, *"A kinematic Simulator for Motion Planning of a Mobile Robot on a Terrain"*, IEEE/RSJ Int. Workshop on Intelligent Robot and Systems, Yokohama, Japan, July 1993.

[5] M. Cherif, Ch. Laugier, *"Using Physical Models to Plan Safe and Executable Motions for a Rover Moving on a Terrain"*, First Workshop on Intelligent Robot Systems, J.L.Crowley and A. Dubrowsky (Eds), Zakopane, Poland, July 1993.

[6] M. Cherif, Ch. Laugier, Ch. Milési-Bellier, B. Faverjon, *"Combining Physical and Geometric Models to Plan Safe and Executable Motions for a Rover Moving on a Terrain"*, Third Int. Symposium on Experimental Robotics, Kyoto, Japan, October 1993.

[7] L.E. Dubins, *"On Curves of Minimal Length with a Constraint on Average Curvature, and with Prescribed Initial and Terminal Positions and Tangents"*, in American Journal of Mathematics, 79:497-516, 1957.

[8] R.S. Fearing: *"Implementing a force strategy for objects re-orientation"*, IEEE Int. Conf. on Robotics and Automation, San Francisco, April, 1986.

[9] T. Fraichard and C. Laugier: *"Dynamic Trajectory Planning, Path-Velocity Decomposition and Adjacent Paths"*, Int. Joint Conf. on Artificial Intelligence, 1993, August, Chambery, France.

[10] D. Gaw, A. Meystel, *"Minimum-time Navigation of an Unmanned Mobile Robot in a 2-1/2D World with Obstacles"*, IEEE Int. Conf. on Robotics and Automation, April 1986.

[11] H. Goldstein, *"Classical Mechanics"*, 2nd edition, Addison-Wesley, Massachusetts, 1983.

[12] H. Hanafusa and H. Asada: *"Stable prehension by a robot hand with elastic fingers"* 7th Int. Symp. on Industrial Robots, 1977, October, Tokyo.

[13] J.E. Hopcroft, J.T. Schwartz, M. Sharir, *"Efficient Detection of Intersections among Spheres"*, Int. Journal on Robotics Research, Vol. 2, No. 4, 1983.

[14] P.M. Isaacs, M.F. Cohen: *"Controlling dynamic simulation with kinematic constraints, behaviour functions and inverse dynamics"*, Computer Graphics(SIGGRAPH'87), 21(4), July 1987.

[15] S. Jimenez, A. Luciani, C. Laugier, *"Physical Modeling as an Help for Planning the Motions of a Land Vehicle"*, IEEE/RSJ Int. Workshop on Intelligent Robot and Systems, Osaka, Japan, November 1991.

[16] S. Jimenez, A. Luciani, C. Laugier, *"Predicting the Dynamic Behaviour of a Planetary Vehicle Using Physical Models"*, IEEE/RSJ Int. Workshop on Intelligent Robot and Systems, Yokohama, Japan, July 1993.

[17] J.C. Latombe, *"Robot Motion Planning"*, Kluwer Academic Publishers, Boston, USA, 1990.

[18] A. Liegeois, C. Moignard, *"Minimum-time Motion Planning for Mobile Robot on Uneven Terrains"*, in Robotic Systems, Kluwer 1992, pp. 271-278.

[19] A. Luciani and al: *"An unified view of multiple behaviour, flexibility, plasticity and fractures: balls, bubbles and agglomerates"*, IFIP WG 5.10 on Modeling in Computer Graphics, Springer Verlag, 1991, pp.55-74.

[20] H. Maekawa: *"Stable grasp and Manipulation of 3D Object by Mutifingered Hands"*, International Symposium on Measurement and Control in Robotics (ISMCR'92), 1992, November, Tsukuba, Japan.

[21] M.T. Mason: *"Manipulator grasping and pushing operations"*, Massachusetts Institute of Technology, 1982, June.

[22] V. Nguyen: *Constructing force-closure grasps in 3D*, IEEE Int. Conf. on Robotics and Automation, 1987, March/April, Raleigh.

[23] Z. Shiller, J.C. Chen, *"Optimal Motion Planning of Autonomous Vehicles in Three Dimensional Terrains"*, IEEE Int. Conf. on Robotics and Automation, May 1990.

[24] Z. Shiller, Y.R. Gwo, *"Dynamic Motion Planning of Autonomous Vehicles"*, IEEE Trans. on Robotics and Automation, vol 7, No 2, April 1991.

[25] Th. Simeon, B. Dacre Wright, *"A Practical Motion Planner for All-terrain Mobile Robots"*, IEEE/RSJ Int. Workshop on Intelligent Robot and Systems, Yokohama, Japan, July 1993.

[26] D. Terzopoulos, J. Platt, A. Barr, K. Fleischer: *"Elastically deformable models"*, Computer Graphics, 21(4), July 1987.

[27] D. Terzopoulos, K. Fleischer: *"Modeling inelastic deformations: Visco-elasticity, plasticity, fracture"*, Computer Graphics, 22(4), August 1988.

[28] J. Pertin-Troccaz: *"Grasping: a state of the art"*, The Robotics Review, MIT Press, Spring, 1989.

Geometric Reasoning About Mechanical Assembly[*]

Randall H. Wilson, *Sandia National Laboratories, Albuquerque, NM, USA*
Jean-Claude Latombe, *Stanford University, Stanford, CA, USA*

In which order can a product be assembled or disassembled? How many hands are required? How many degrees of freedom? What parts should be withdrawn to allow the removal of a specified subassembly? To answer such questions automatically, important theoretical issues in geometric reasoning must be addressed. This paper investigates the planning of assembly algorithms specifying (dis)assembly operations on the components of a product and the ordering of these operations. It also presents measures to evaluate the complexity of these algorithms and techniques to estimate the inherent complexity of a product. The central concept underlying these planning and complexity evaluation techniques is that of a "non-directional blocking graph," a qualitative representation of the internal structure of an assembly product. This representation describes the combinatorial set of parts interactions in polynomial space. It is obtained by identifying physical criticalities where geometric interferences among parts change. It is generated from an input geometric description of the product. The main application considered in the paper is the creation of smart environments to help designers create products that are easier to manufacture and service. Other possible applications include planning for rapid prototyping and autonomous robots.

1 Introduction

Reasoning about mechanical assembly (and disassembly) is an important research topic which has attracted interest from researchers in both artificial intelligence [38] and robotics [27]. In which order can a product be assembled or disassembled? How many hands are required? How many degrees of freedom?

What parts should be removed from the assembled product to allow replacement of a specified subassembly? To automatically answer questions like these, interesting theoretical issues in geometric reasoning must be addressed and computational techniques of broad interest must be developed. This investigation will also benefit many applications. The one considered in this paper is the creation of smart interactive environments to help designers create products that are easier to manufacture and service. Other important applications include planning for rapid prototyping and autonomous robots.

Given a mechanical assembly product (such as a toaster, an automobile, or a plane), an *assembly algorithm* specifies assembly and/or disassembly operations on its components and the ordering of these operations. The manufacturing, maintenance, and repair procedures for the product are all instances of assembly algorithms. These algorithms may be executed by humans (where the algorithm takes the form of an instruction sheet), robots (here expressed as a robot program), or specific machines (here an input to the design of these machines). Assembly algorithms are the key link between design and manufacturing/servicing. If we can measure the cost of executing an assembly algorithm, then the product's "inherent complexity", with respect to manufacturing and servicing, is given by the lower bound cost of all the assembly algorithms that are possible for this product.

Here we investigate the following two related problems: (1) the automatic generation of assembly algorithms from CAD data, and (2) the characterization of the complexity of assembly designs. Our main goal is to provide efficient computational support for the "concurrent engineering" approach to design, in which constraints arising from manufacturing and servicing are taken into account at design time [9]. As products are designed with more parts of various sorts (e.g.,

---

[*]This article appeared previously in the journal *Artificial Intelligence*, and is being republished by permission of the publisher, Elsevier.

machined, composite, electrical, electronic) densely packed to provide more functions per cubic inch, the need for powerful assembly support tools will increase dramatically.

A typical CAD model of an assembly describes the geometry of the parts composing the assembly and the spatial relations among these parts. It is appropriate for graphic rendering and some man-machine interaction, but it does not directly provide the information that is needed to easily plan assembly algorithms. Indeed, to synthesize such algorithms one must first analyze how the various components in an assembly constrain their respective motions. However, there is a combinatorial set of potential interactions among parts. This suggests that the CAD model be converted into another representation making these constraints explicit. We propose one such representation, called the *non-directional blocking graph*, or NDBG, which describes the potential interactions among parts in polynomial space. Its construction derives from the observation that infinite families of motions can be partitioned into finite collections of subsets such that the interferences among the parts are constant over every subset. Once computed, the NDBG can be exploited for a variety of purposes, including the efficient (polynomial) generation of assembly algorithms.

There are often many algorithms to assemble or disassemble a product, and we may not want to produce all. To deal with this issue, we propose to measure the complexity of an algorithm along various axes, e.g., the number of hands it requires, the longest sequence of operations that cannot be parallelized, the total number of elementary motions. Along each axis, we define the algorithmic complexity of a product as the lower bound on the complexity of all assembly algorithms for this product. This definition immediately yields various classes of products, for instance the class of products that can be (dis)assembled with single translations only. The complexity of a product and its class provide concise pertinent information to feed back to the designers during design; it can also be used as a reference to select the most interesting assembly sequences. We will see that the NDBG is directly exploitable to estimate various complexity measures of a design.

The NDBG is a qualitative representation of the internal structure of an assembly product, which is obtained by identifying physical criticalities where interferences

among parts change. Constructing the NDBG of an assembly design is a precomputation step allowing subsequent computations (synthesis of assembly algorithms, evaluation of complexity measures) to be performed in efficient query time. We will describe several types of NDBGs (for different families of motions) whose computation takes polynomial time in the size of the CAD input (number and complexity of parts). This computation can be done from a full assembly model, or incrementally, while the assembly is being designed. The exploitation of the NDBGs can be aimed at answering specific requests from the designers (e.g., how many parts must be removed from the assembled product before a specified subassembly can itself be extracted?) and/or posting warnings to the designers (e.g., when the product changes from one complexity class to another).

Notation: The following notation will be used throughout the paper:
- n: number of parts (polygons or polyhedra) in an assembly A,
- v: total number of vertices of the parts in A,
- r: number of pairs of parts in contact in A,
- c: number of contacts (edge-edge or plane-plane) in A,
- k: total number of vertices in the convex hulls of the c contacts in A.
All other notations are defined and used locally.

2 Related Work

The automatic planning of assembly and disassembly operations has attracted the interest of AI researchers for a long time. The classical blocks world can be seen as a primitive assembly planning domain [15, 38]. Moreover, some AI planners have considered more complex domains. For instance, NOAH [42] was originally aimed at supplying instructions to a human apprentice to repair an air compressor, including disassembly and assembly plans. However, because they are usually interested in more general planning issues than just assembly planning, virtually all AI planners make use of a very abstract geometric description of the objects and their relations expressed in logical notation, e.g. ON(A,B). Additional geometric knowledge is implicitly coded in the operators representing the actions that can be executed (e.g., stacking a block on top of another). A noticeable exception is BUILD [13], which includes a simple treatment of such notions as stability and friction. The interest of AI in general planning is still very high [35].

The geometric approach to assembly planning originated in robotics with the work reported in [29] (AUTOPASS), [31] (LAMA), and [45]. It is more limited in scope than traditional AI planning and focuses specifically on issues raised by the manipulation of physical objects. It has motivated various research in basic path planning, motion planning with uncertainty, manipulation planning with movable objects, and grasp planning [27]. However, the high complexity of assembly planning when viewed as a general motion planning problem has led researchers to turn their attention toward a simpler subproblem known as *assembly sequence planning*, or simply *assembly sequencing* [1, 22]. In this problem, only the constraints (mainly geometric ones) arising from the assembly itself are considered; the manipulation system (e.g., the robots) executing this plan is simply ignored by assuming that the parts composing the assembly are flee-flying objects.

The early assembly sequencers were mainly sequence editors. Geometric reasoning was supplied by a human who answered questions asked by the computer systems; the assembly sequences were inferred from the answers to these questions [7, 10]. Automated geometric reasoning was later added to answer these questions automatically [5, 19, 20, 28, 30, 47, 51]. This development resulted in generate-and-test assembly sequencers, with a module guessing candidate sequences and generating questions to check their feasibility, and geometric reasoning modules answering these questions. This approach tends to repeat the same geometric computations many times. Mechanisms for saving and reusing previous computations, such as the "precedence expressions" [47], have been proposed to overcome this drawback, but with limited success. In practice, the generate-and-test paradigm remains relatively inefficient (it takes time exponential in the number of parts) and is applicable only to assemblies with few parts. The NDBG avoids this combinatorial trap. Though there is a combinatorial set of potential parts interactions, the NDBG represents them in polynomial space, allowing valid operations and assembly sequences to be directly generated in polynomial time.

Originally, the research in assembly planning was aimed at assisting process planning in order to reduce delays between design and manufacturing, and eventually produce better plans [10]. This goal is still valuable, especially for rapid prototyping and even mass production. Recently, however, the interest has shifted toward generating assembly sequences to evaluate assembly designs and help designers create products that are easier to manufacture [44, 48]. In this new context, automated geometric reasoning and computational efficiency of assembly planning are critical issues that must be thoroughly explored. The synthesis of pertinent information to feed back to designers is another important issue. The concept of the algorithmic complexity of an assembly design presented in this paper directly addresses this issue. It derives in part from informal complexity measures currently in use in several companies (e.g., see [6]).

The field of computational geometry has also explored issues relevant to assembly planning, for example, set separation problems [46]. Given a 2D polygonal assembly A, the problem of deciding whether there is a direction d and a subassembly $S \subset A$ such that a translation along d separates S from the rest of A is addressed in [4]. An algorithm to construct a sequence of translations separating two polygonal parts is given in [39]. Several techniques presented in this paper have been influenced by the work in computational geometry.

The construction of an NDBG is based on the identification of physical criticalities to decompose a continuous set into a finite number of regions that are treated as single entities. This approach relates to the general interests of qualitative physics [11] and, more specifically, qualitative kinematics [14, 24], which studies the internal motions of parts in an operational device. It yields more meaningful decompositions than blind discretizations not based on any sort of criticality (discontinuity, singularity, or event). The notion of an "aspect graph" used in computer vision also has the same qualitative flavor as the NDBG. The aspect graph of an object describes the appearance of the object from all possible points of view. Aspect graphs were first computed by discretizing the set of viewing directions. Recent algorithms take advantage of the fact that, except at critical viewing directions, the occluding contours of an object remain qualitatively (i.e., topologically) the same for small changes in the viewpoint (e.g., see [25]).

In [17] a criticality-driven approach makes it possible to plan a sensorless sequence of squeezing operations to achieve some specified orientation of a polygonal part independent of its initial orientation.

3   Virtual Manipulation Systems

Assume that we are given the description of a robot system and the design of an assembly product. We wish to plan an algorithm to make the robot system assemble the product from its individual parts. This problem can be formulated as a manipulation problem whose solution is a path in a large-dimensional space, the composite configuration space of the robot and all the movable parts [27]. A point (configuration) in this space fully represents a spatial placement of the robot and the other parts. A solution path describes all the motions that are necessary to construct the assembly. It is a continuous curve connecting an initial configuration where the parts are separated, to a goal configuration where they are assembled together according to the product design. The curve must satisfy certain physical constraints (e.g., the parts cannot move on their own). This view, however, makes assembly planning a highly complex motion planning problem. It also raises several conceptual issues that are still poorly understood, for example, the interaction between stability, fixturing, and grasping. Moreover, the manipulation system may not be known in advance. In fact, assembly planning is often used to help specify this system.

This leads to approaching assembly planning hierarchically, by solving a succession of simplified, but increasingly more realistic planning problems. The assembly algorithms for one problem are used to prune large subsets of the solution spaces of subsequent problems and to guide the search of these spaces. Each problem can then be seen as a planning problem for an abstraction of the (possibly not yet known) real manipulation system. We call this abstraction a *virtual manipulation system*. For example, at a high level of abstraction the parts composing the assembly are considered free-flying geometric objects; the corresponding constraints on assembly algorithms arise only from the product itself. At lower levels of abstraction, grippers, fixtures, and machines are introduced, along with uncertainties. As the virtual manufacturing system becomes more realistic, more detailed as-

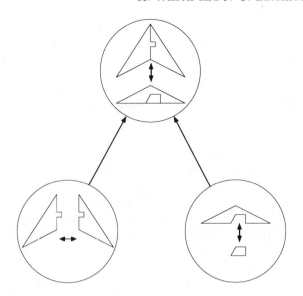

Figure 1: *Tree of an assembly algorithm for a simple product*

sembly algorithms can be generated. Just as one can write computer algorithms without knowing the details of their implementation, one can plan assembly algorithms without knowing which machines (or humans) will perform the manipulation.

Throughout the rest of this paper we will assume a very abstract virtual manufacturing system, in which individual parts are free-flying rigid geometric objects. This assumption corresponds to the most frequent situation in the early phase of the geometric design of a product and also underlies previous work in assembly sequencing (see Section 2). At this level of abstraction, we define an *assembly instruction* as the specification of the relative motions of m subsets of assembly parts, with $m \in \{2, 3, \ldots\}$, between an initial and a goal geometric arrangement of the parts. The m subsets are called the *moved sets* of the instruction. All the parts in the same moved set remain in constant relative position during the motion specified by the instruction; usually, each moved set forms a connected composite object, but this is not required.

An algorithm to assemble a product from its parts is a partial ordering of assembly instructions in time. The graph of this relation is a tree whose maximal element (the root) is the instruction generating the product itself (see Fig. 1). Furthermore, no two instructions that are not comparable in the ordering move the same

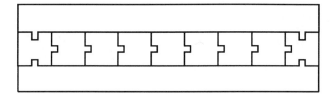

Figure 2: *An assembly with two feasible decompositions*

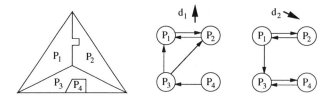

Figure 3: *A simple assembly and two* DBGs

assembly component. In the virtual assembly system considered here, an algorithm to disassemble a product is obtained by simply inverting the ordering of an algorithm assembling the product. An algorithm to service a product usually combines instructions disassembling and re-assembling subsets of a product. For simplification, but without loss of generality, in the rest of this paper we will only consider algorithms to assemble products, with initial arrangements where all the parts are separated. Such an algorithm is *correct* if all its instructions specify collision-free motions of the parts (contacts between parts are allowed, however) and the final relative positions of the parts are those specified in the geometric description of the product. An assembly algorithm of this form is often called an *assembly sequence*.

4   Blocking Structure of an Assembly

In assembly planning the goal state (the assembled product) is considerably more constrained than the initial state (the separated parts). This suggests that planning should proceed backward from the goal state, i.e., by disassembling the product. Indeed, the contacts among the parts in the assembled product can be used to quickly filter out many impossible motions.

Contact analysis is the basis of much previous work in assembly sequencing (e.g., [20]), where it was part of an overall generate-and-test planning approach. This approach recursively partitions an assembly into subassemblies and uses contact analysis to check the feasibility of each decomposition. However, the number of candidate decompositions is exponential in the number of parts, even though there may exist very few feasible ones. For example, in the assembly of Fig. 2, which consists of $n-2$ interlocking parts sandwiched between two plates, only two decompositions are feasible. The inherent inefficiency of generate-and-test led many authors to restrict their planners to assembly sequences

in which a product is built by adding a single part at a time. With this restriction, generate-and-test has a lower-order exponential dependence on the number of parts. But it is often preferable to build subassemblies that are later merged into larger ones.

The inefficiency of generate-and-test derives from the fact that essentially the same contact analysis is done many times. This repetition can be avoided by performing a complete contact analysis first. The results, recorded in a compact data structure, can then be used to directly determine which assembly decompositions are feasible. This approach yields the concept of an NDBG. For clarity, we introduce a first simple type of NDBG below (2D with infinitesimal translations). More sophisticated NDBGs will be presented in Section 5. In particular, we will show that NDBGs are not limited to contact analysis.

4.1   Directional Blocking Graph

Consider a planar assembly A made of n polygonal parts $P_1, \ldots, P_n$. The interiors of any two parts in A are disjoint. If the boundaries of two parts intersect, the two parts are said to be in contact.

Suppose that we wish to remove one part, say P_i, by translating it along a direction defined by the unit vector d. We say that a part P_j of A ($j \neq i$) *blocks* the translation of P_i along d if an arbitrarily small translation of P_i along d leads the interiors of P_i and P_j to intersect. Hence, if P_j blocks the translation of P_i, the two parts are necessarily in contact. A subassembly S of A is *locally free* to translate in direction d iff no part in $A \setminus S$ blocks the translation of any part of S along d.

The *directional blocking graph*, or DBG, $G(d, A)$ of A for an *infinitesimal translation* along d is a directed graph with nodes representing the parts of A. An arc connects P_i to P_j iff P_j blocks the translation of P_i along d. Fig. 3 shows a simple assembly made of 4 polygonal parts and its DBGs for infinitesimal translations along d_1 and d_2.

A subassembly S of A is locally free to translate in direction d iff no arcs in $G(d, A)$ connect parts in S to parts in $A \setminus S$. If $G(d, A)$ is strongly connected,[1] no such subassembly exists. Otherwise, at least one strong component of $G(d, A)$ must have no outgoing arcs. For example, in Fig. 3 the subassemblies $\{P_1, P_2\}$ and $\{P_1, P_2, P_3\}$ are locally free to translate in direction d_1; only the subassembly $\{P_3, P_4\}$ is locally free in direction d_2.

For a subassembly S to be assemblable with $A \setminus S$ by a translation along $-d$, it must be locally free to translate in direction d. This condition is necessary but not sufficient, since global accessibility is not considered. Also, merging S and $A \setminus S$ may require rotation. For those reasons, in Section 5 we will extend the above notions to non-infinitesimal translations and motions with rotation.

4.2   Non-Directional Blocking Graph

We represent the set of all translation directions by the unit circle $\mathcal{S}^1$. This circle is the locus of the extremity of the vector d when its origin is fixed.

Let P_i and P_j be two parts in contact. The set of directions in which P_i is locally free to translate relative to P_j is a closed cone (possibly a half-space or a single ray) [20, 48]. For every pair of parts P_i and P_j in contact in A, we draw the diameters of $\mathcal{S}^1$ parallel to the two sides of the cone characterizing the local freedom of P_i relative to P_j. The drawn diameters partition $\mathcal{S}^1$ into an arrangement of regions, the endpoints of the diameters and the open circular arcs between them. Every such region is *regular* in the sense that the DBG $G(d, A)$ remains constant when d varies over it. We denote the DBG of A for any direction in a regular region R by $G(R, A)$.

The arrangement of points and intervals on $\mathcal{S}^1$ and the associated DBGs form the *non-directional blocking graph* $\Gamma(A)$ of A for infinitesimal translations. It represents the blocking structure of A for infinitesimal translations in all directions.

---

[1] A *strongly connected component* (or *strong component*) of a directed graph is a maximal subset of nodes such that for any pair of nodes (X_1, X_2) in this subset, a path connects X_1 to X_2. A graph is strongly connected if it has only one strong component.

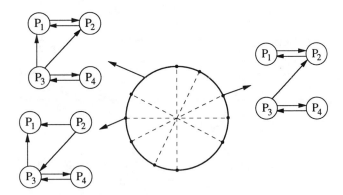

Figure 4: *Part of the NDBG of the assembly of Fig. 3*

Part of the NDBG of the assembly of Fig. 3 is shown in Fig. 4. The partition of $\mathcal{S}^1$ in this example consists of 20 regular regions.

4.3   Computation

Let us assume for simplification that the contacts between any two parts in A are edge-edge contacts (hence, there are no isolated contact points). Assume also that if a concave vertex of a part coincides with a convex vertex of another part, then the edges abutting these vertices are in pairwise contact. These assumptions eliminate cases requiring specific attention; however, these cases can be easily treated within the complexity bounds given below.

Let the input to the computation consist of the geometric model of the parts in the assembly and the specification of the contacting edges between every two parts. If these contacts are not given explicitly they can be easily inferred from the spatial relations among the parts in quadratic time in the total number of vertices of the parts [48]. Let n be the number of parts in A and c the total number of edge contacts. We represent each DBG as an $n \times n$ adjacency matrix.

Under the above assumptions, the partition of $\mathcal{S}^1$ is obtained by drawing the diameters parallel to all edge contacts. It contains $O(c)$ regular regions and is computed in $O(c \log c)$ time.

For each regular region R, we select an arbitrary direction d in R and compute the DBG $G(d, A)$. After clearing the adjacency matrix for the DBG $G(d, A)$, each edge contact is considered separately. For every edge contact between parts P_i and P_j, the inner product of d with the outer normal to P_i in the contacting edge with P_j is computed. If the inner product is strictly positive, an arc from P_i to P_j is added to $G(d, A)$ (if it does not already exist). The computation of $G(d, A)$ takes $O(n^2 + c)$ time. We can avoid the initial clearing of the adjacency matrix by keeping track of the updated cells in a stack of pointers (see [2], page 71). This modification reduces the computing complexity of $G(d, A)$ to $O(c)$. By repeating the computation for all regular regions, the NDBG $\Gamma(A)$ is constructed in $O(c^2)$ time. The size of $\Gamma(A)$ is $O(cn^2)$.

The above computation can be refined as follows. For any pair of parts in contact, if there are more than two edge contacts, we only retain the two diameters of $\mathcal{S}^1$ which bound the cone of directions in which one part is free to translate relative to the other. This yields a total of $O(r)$ diameters, where r is the number of pairs of parts in contact in A. The arrangement of $\mathcal{S}^1$ is computed in $O(c + r \log r)$ time. The resulting NDBG is computed in $O(c + r^2)$ time and has $O(rn^2)$ size. We always have $r \leq c$ and for many assemblies $r \ll c$. While c depends on both the number and the complexity of the parts of A, r only depends on their number. Although in the worst case $r \in O(n^2)$, it is often smaller.

The NDBG can also be incrementally updated as the product is being designed. Each modification can be treated as a combination of part deletions and part additions. The deletion of a part P is obtained by removing the node corresponding to P and all its adjacent arcs from every DBG, and merging regular regions that were previously separated by contacts involving P. It can be done in $O(r)$ time. Adding a part contacting r' parts already in the assembly can be done in $O(r'r)$ time.

4.4 Generating Candidate Algorithms

The above NDBG $\Gamma(A)$ is an implicit representation of a set of assembly algorithms in which every instruction merges exactly two subassemblies without rotation. This set contains all correct assembly algorithms; it may also contain incorrect algorithms, since global accessibility constraints were not considered to compute the NDBG. Hence the assembly algorithms given by $\Gamma(A)$ must be validated by further tests (not described here; see [48]). These tests operate on assembly operations; therefore each instruction in an assembly algorithm should be validated before constructing

the rest of the algorithm. Since $\Gamma(A)$ allows the direct generation of instructions satisfying local freedom constraints, the number of additional tests required is usually relatively small.

Let a candidate algorithm be any algorithm contained in $\Gamma(A)$, and a candidate partitioning of A be two subassemblies, S and $A \setminus S$, such that one is locally free to translate in some direction d. Each DBG contains n nodes and at most $2r$ arcs. Hence, finding the strong components of any DBG takes $O(r)$ time [3] and generating a candidate partitioning of A, given $\Gamma(A)$, takes $O(r^2)$ time. Furthermore, if a candidate algorithm exists, one candidate algorithm also exists for each of the two subassemblies in any candidate partitioning. Thus, generating a candidate algorithm takes $O(r^2 n)$ time. Let u be the total number of candidate partitionings of A; $u \in O(2^n)$, but for most assemblies it is much smaller. The set of all candidate partitionings of A for some direction d is computed in $O(ru)$ output-sensitive time by reducing $G(d, A)$ to the acyclic graph of its strong components (see [48]). The set of all candidate partitionings of A is computed in $O(r^2 u)$ time.

4.5 Computing Variant

One can notice that there is little or no change between the DBGs of two adjacent regular regions. Thus, once we have computed the DBG for one region, call it R_1, we can incrementally modify this graph to get the DBG for the next region in the NDBG list, instead of computing this DBG from scratch. We can proceed in this same way until all the regions have been considered.

To do this, we slightly modify the DBG of a region by attaching a *weight* to each arc of the graph. In $G(R_1, A)$, this weight is the number of inner products that were strictly positive in the above computation. The absence of an arc from P_i to P_j is treated as an arc of weight 0, and conversely. Let R_1 be a circular arc. The next region R_2 in the NDBG is necessarily a singleton. Let D be the diameter of $\mathcal{S}^1$ that ends at R_2, and $\{E_1, \ldots, E_s\}$, $s \geq 1$, be the set of all contact edges in A parallel to D. $G(R_2, A)$ can be derived from $G(R_1, A)$ by applying the following *crossing rule*:

> Initialize G to $G(R_1, A)$. For every contact edge E_k ($k = 1$ to s), let P_i and P_j be the two parts sharing this edge. If the inner product of any direction in R_1 and the outgoing

normal to P_i in E_k is strictly positive, then retract 1 from the weight of the arc connecting P_i to P_j in G.

The graph G obtained at the end of the loop is $G(R_2, A)$. (Again, every arc weighted 0 is interpreted as no arc.)

If R_1 is a singleton and R_2 a circular arc, the crossing rule is similar:

Initialize G to $G(R_1, A)$. For every contact edge E_k ($k = 1$ to s), let P_i and P_j be the two parts sharing this edge. If the inner product of any direction in R_2 and the outgoing normal to P_i in E_k is strictly positive, then add 1 to the weight of the arc connecting P_i to P_j in G.

Using these crossing rules and representing only one DBG at any one time allow us to successively compute all the DBGs in $O(r)$ amortized time after the $O(c + r \log r)$ computation of the partition of $\mathcal{S}^1$. Indeed, the cost of computing a DBG by applying the crossing rule is proportional to the number s of contact edges involved in the computation. This number is in $O(r)$, but throughout the computation of the entire NDBG, each edge is considered only twice. Hence, the time complexity of the construction of *all* the remaining DBGs is only $O(r)$.

A candidate partitioning can be computed in $O(r^2)$ time, and a candidate algorithm in $O(r^2 n)$ time by constructing the DBG of every generated subassembly. Compared to the algorithm of the previous two subsections, this computing variant allows substantial space saving, since it does not require representing all $O(r)$ DBGs at any given time. Moreover, if a computation (for example, evaluating a complexity measure) uses the DBGs, one at a time, in the same sequence as they are generated, this incremental technique also has greater time efficiency.

5 Other Blocking Graphs

The notion of an NDBG introduced in the previous section admits several extensions and variants. We present some of them below. See [48] for more detail.

5.1 3D Assemblies

The NDBG for infinitesimal translations can be easily extended to 3D assemblies made of polyhedral parts. As in the 2D case, we simplify our analysis and eliminate cases requiring specific, but straightforward attention, by making the following assumptions: (1) the contacts between any two parts in A are face-face contacts; (2) if a concave edge of a part coincides with a convex edge of another part, then the faces bounded by these edges are in pairwise contact; (3) if a non-convex vertex of a part coincides with a vertex of another part, then the edges abutting these vertices are in pairwise contact. Let c be the total number of face-face contacts between parts of A.

The set of translation directions in 3D is represented by the unit sphere $\mathcal{S}^2$. The c plane-plane contacts induce arcs of great circles partitioning $\mathcal{S}^2$ into an arrangement of $O(c^2)$ regular regions of dimensions 2, 1 and 0 (faces, edges, and vertices, respectively). This arrangement is computed in $O(c^2)$ time using a topological sweep [12]. The NDBG is computed in $O(c^3)$ time and has size $O(c^2 n^2)$. If only one DBG is represented at any one time, we can use crossing rules between regions and successively compute all the DBGs in $O(c^2)$ time. Generating a candidate partitioning of A into two subassemblies, given the NDBG, takes $O(rc^2)$ time, where r is the number of pairs of parts in contact. Generating a candidate assembly algorithm takes $O(rc^2 n)$ time.

As in the 2D case, slightly lower complexity bounds can be obtained by considering each of the r pairs of parts in contact in sequence. For each pair, the set of directions in which one part is locally free to translate relative to the other is a convex cone. The intersection of this cone with the sphere $\mathcal{S}^2$ is a "polygon" bounded by arcs of great circles. Only the arrangement of regions created by these arcs need be used to compute the NDBG.

5.2 Infinitesimal Generalized Motions

One important extension is to allow motions in rotation. Fig. 5 shows a simple case where a part blocks another part for any infinitesimal translation, while a rotation is feasible.

Let us consider the 3D case only (the 2D case is just simpler), with polyhedral parts. The direction of an infinitesimal generalized motion (combining translation and rotation) is given by a unit vector in 6D.

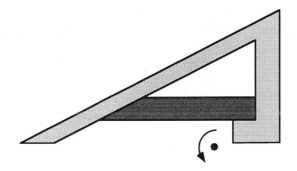

Figure 5: *Need for rotation*

Hence, the set of all possible directions of motion make up the unit 5D sphere $\mathcal{S}^5$.

A Cartesian frame is attached to each part in A. For any two parts P_i and P_j, the configuration (position and orientation) of P_i relative to P_j is defined as the position and orientation of the frame attached to P_i relative to the frame attached to P_j. An infinitesimal generalized motion of P_i with respect to P_j is described by a 6D vector $dX = (dx, dy, dz, d\alpha, d\beta, d\gamma)$. The components dx, dy, and dz are those of the translation of the origin of the frame of P_i along the axes of the frame of P_j. The components $d\alpha$, $d\beta$, and $d\gamma$ are the infinitesimal angles by which the frame of P_i rotates about the axes of the frame of P_j.

Let V be a vertex of P_i. The motion described by dX causes V to undergo a translation $J_V\, dX$, where J_V is the 3×6 Jacobian matrix of the transform that gives the coordinates of V in the frame of P_j as a function of the configuration of P_i relative to P_j. Assume that P_i and P_j are in contact such that the vertex V of P_i is contained in the face F of P_j. Let n_F be the outgoing normal vector to F. The motion dX causes V to penetrate F when $n_F J_V dX < 0$, to break the contact with F when $n_F J_V dX > 0$, and to slide in F when $n_F J_V dX = 0$. The set of motions dX *allowed* by the contact between V and F are those satisfying $n_F J_V dX \geq 0$.

Now let P_i and P_j be two parts in contact such that a face F_i of P_i and a face F_j of P_j intersect at a piece of planar surface (plane-plane contact). The set of motions dX allowed by this contact is the intersection of all the closed half-spaces $n_{F_j} J_{V_k} dX \geq 0$ computed for the vertices V_k of the convex hull of the intersection of F_i and F_j [18, 49]. For example, in Fig. 6, the vertices

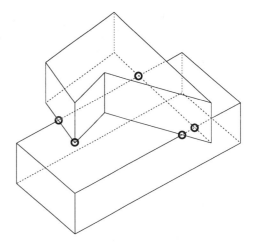

Figure 6: *Plane-plane contact expressed as point-plane contacts*

V_k are circled.

We make the same simplifying assumptions about contacts as in Subsection 5.1. For each vertex V_k of the convex hull of the intersection of two parts, the equation $n_F J_{V_k} dX = 0$ defines a 5D hyperplane in the 6D space of infinitesimal generalized motions, which partitions $\mathcal{S}^5$ into two open half-spheres and a great circle. The set of all such hyperplanes, determined by the vertices of the convex hulls of the planar contacts, induces an arrangement of regions of dimensions $5, \ldots, 1, 0$ on $\mathcal{S}^5$. It is plain to see that the DBG is constant over each such region; hence, the regions of the arrangement are regular.

The intersection of two non-convex faces F_i and F_j having v_i and v_j vertices, respectively, has $O(v_i v_j)$ vertices and can be computed in $O(v_i v_j)$ time. Its convex hull is constructed in $\Theta(v_i v_j \log v_i v_j)$ time [40] and has $O(v_i + v_j)$ vertices. Indeed, all the intersection vertices of $F_i \cap F_j$ lie on edges of F_i and F_j, but each particular edge can contribute at most two vertices of the convex hull. Let k be the total number of vertices in the convex hulls of the c contacts. We have $k \in O(cv)$, where v is the total number of vertices in the parts of A; in practice, however, k is much smaller.

Since a singleton in the above arrangement arises whenever five hyperplanes intersect, the arrangement contains $O(k^5)$ regions. It is constructed in $O(k^5)$ time by a multi-dimensional topological sweep [12]. Constructing a DBG in any region is done in time $O(k)$. A crossing rule similar to the pure translational case can be established, yielding an $O(rk^5)$ time algorithm to build the NDBG for a 3D assembly, where $r \in O(n^2)$ is the number of pairs of parts in contact.

A necessary condition for a subassembly S to be assemblable with $A \setminus S$ is that there exists a DBG G such that no arcs in G connect parts in S to parts in $A \setminus S$. However, this condition is not sufficient. The above NDBG is thus an implicit representation of a set of assembly algorithms in which every instruction merges exactly two subassemblies. This set contains all correct assembly algorithms; it may also contain incorrect algorithms due to the fact that global accessibility constraints have not been considered.

Given the above NDBG, computing a candidate partitioning of A into two subassemblies takes $O(rk^5)$ time. Generating a candidate assembly algorithm takes $O(rk^5 n)$ time.

5.3 Infinite Translations

To address global accessibility, we now present a variant of the NDBG that derives from the analysis of the interferences among parts for a family of non-infinitesimal motions: infinite translations. Given any two parts P_i and P_j in A (these two parts might not be in contact), we say that P_j blocks the infinite translation of P_i along d if the volume that P_i sweeps out when it translates along d from its initial position in A to infinity intersects P_j. The notions of a DBG and an NDBG for infinite translations follow from this blocking relation.

Let P_i and P_j be any two parts in the assembly A. The set $\mathcal{B}_{ij}$ of directions d along which P_j blocks P_i is identical to the set of directions along which the "grown" object

$$P_j \ominus P_i = \{a_j - b_i \mid a_j \in P_j, b_i \in P_i\},$$

i.e., the Minkowski difference of the two sets of points P_j and P_i at their positions in A, blocks the infinite translation of the coordinate origin O. If P_i and P_j are polygons in 2D (resp. polyhedra in 3D), then $P_j \ominus P_i$ is also a polygon (resp. a polyhedron) [27, 32].

Again, let us consider the 3D case with polyhedral parts. $\mathcal{B}_{ij}$ is the intersection of $\mathcal{S}^2$ and the polygonal cone of all rays erected from O and intersecting $P_j \ominus P_i$. This intersection is a region of $\mathcal{S}^2$ bounded by segments of great circles. The great circles supporting these segments create an arrangement of regular regions overwhich the DBG for infinite translations remains constant. This arrangement and the associated DBGs form the NDBG of the assembly for infinite translations.

A sufficient condition for a subassembly S to be directly assemblable with $A \setminus S$ is that there exists a DBG G in the above NDBG such that no arcs in G connect parts in S to parts in $A \setminus S$. This condition is not necessary since, for instance, a path with multiple extended translations may be required. The above NDBG is an implicit representation of all correct assembly algorithms in which every instruction merges two subassemblies by a single translation.

Let v be the total number of vertices in the parts of A. The arrangement on $\mathcal{S}^2$ has size $O(v^4)$ and can be computed in $O(v^4)$ time. The NDBG is computed in $O(n^2 v^4)$ time. Each DBG has $O(n^2)$ arcs, so that finding its strong components takes $O(n^2)$ time. Hence, deciding whether A can be broken down into two subassemblies that can be merged by a single translation (and constructing one such partitioning, if one exists) takes $O(n^2 v^4)$ time. Generating an assembly algorithm takes $O(n^3 v^4)$ time.

5.4 Discussion

One can easily imagine other NDBGs. For example, one may consider the case where an assembly is constructed by merging more than two, say $m + 1$, subassemblies at a time. The corresponding NDBG for infinitesimal translations in 3D would be obtained by partitioning the $(3m - 1)$-dimensional sphere $\mathcal{S}^{3m-1}$, representing m simultaneous translation directions, into an arrangement of regular regions.

NDBGs for infinite generalized motions and, more generally, sequences of extended motions along different directions could be investigated as well. The investigation for sequences of extended translations is under way in [33]. It leads to partitioning the composite configuration space $\mathcal{C}_A$ of the parts in A into regions over which all the relative placements of the parts induce the same NDBGs for infinitesimal translations. This construct yields a collection of NDBGs distributed over regions of $\mathcal{C}_A$. A careful representation of $\mathcal{C}_A$ yields a polynomial collection of NDBGs. From there, the generation of a partitioning of A for a bounded-length sequence of extended translations takes polynomial time. The construct can be generalized to sequences of extended generalized motions.

Each NDBG is defined for a limited family of motions and describes only the class of assembly algorithms for these motions. NDBGs for infinitesimal motions are underconstrained and may contain incorrect algorithms. NDBGs for infinite motions or bounded-length sequences of extended motions are overconstrained and may not include all possible algorithms. Trying to construct and exploit a unique NDBG or collection of NDBGs covering all possible motions would bring us back to the general planning problem of finding a coordinated path for a set of parts. This problem is known to be PSPACE-hard [23, 37] and is strongly believed to require exponential time in the number of parts.

Polynomial NDBGs such as those described above should be regarded as efficient filters to quickly identify feasible assembly operations and algorithms (for instance, using the NDBG for infinite translations) and eliminate infeasible ones (with the NDBG for infinitesimal rigid motions). When required, the candidate algorithms generated by the latter can be analyzed further using more general, but also more expensive, motion planning techniques.

Allowing more than two subassemblies to be merged simultaneoulsy and/or accepting sequences of extended motions yield NDBGs that are considerably more expensive to compute. An interesting development would therefore be to first construct the simplest NDBGs and then extend them locally (that is, for the subsets of parts that require the extension) when they do not allow the generation of assembly algorithms.

6   Implementation

We have implemented the algorithms constructing the NDBGs for infinitesimal and infinite translations both in 2D and 3D [48]. In 3D our implementation allows parts with planar, cylindrical, and some helicoidal faces. For infinitesimal generalized motions, we have implemented a hybrid algorithm that has the same time complexity as the translational version. This algorithm considers all pure translations, plus some "suggested generalized motions" inferred from nonplanar contacts in the assembly. For example, a cylindrical contact suggests pure rotation about its axis; a threaded contact

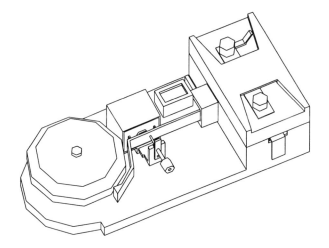

Figure 7: *Electric bell*

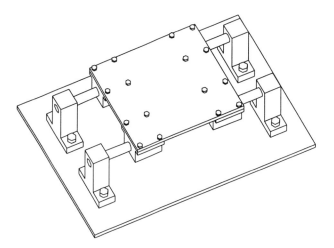

Figure 8: *Friction-testing machine*

between two helicoidal surfaces suggests a screwing motion; etc. The set of suggested generalized motions is incomplete but accounts for most motions required by actual assembly products.

Our programs are written in CommonLisp as part of a larger assembly sequencing system [48] and run on a DEC 5000 workstation. We have used the NDBGs to compute candidate partitionings of assemblies and construct candidate assembly algorithms. We ran the programs on a variety of assemblies including a 22-part electric bell (Fig. 7), a friction-testing machine with 36 parts (Fig. 8), and the 42-part model-aircraft combustion engine (Fig. 9). Table 1 gives the running times

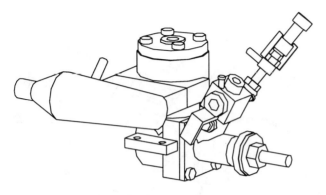

Figure 9: *Model-aircraft engine*

Assembly	Bell	Friction	Engine
Parts	22	36	42
Gen-and-Test	4.7	—	74
NDBG	1.4	3.4	8.5

Table 1: *Times to identify candidate assembly instructions, in seconds*

of both the generate-and-test approach and the NDBG-based methods to identify all partitionings (hence, final assembly instructions) that satisfy local freedom constraints. The generate-and-test algorithm was stopped after failing to find any candidate instructions for the friction-testing machine in two days.

Assembly sequences generated for the electric bell were converted into robot programs executed by a RobotWorld system [43].

7   Complexity of an Assembly

The outcome of assembly planning, i.e., assembly algorithms, can be used to specify, select, design, and/or program the manipulation systems that will execute these algorithms. Another goal which currently attracts increasing interest is to use this outcome to help designers create products that are easier to manufacture and service. The problem, however, is that there may be many possible assembly algorithms, especially when the manipulation system is loosely defined. Showing the most promising algorithms (for some criterion), e.g., by displaying a graphic simulation of their execution, would certainly be useful. But it would have to be limited to a few milestones of the design process in order to avoid taking too much of the designers'

time.

Instead, we propose to measure the inherent complexity of a product (at its current stage of design) as the lower-bound complexity of all the assembly algorithms that are feasible for this product. This idea is analogous to the computational complexity of a problem, which is defined as the lower-bound complexity (in terms of time and space) of all the computer algorithms that solve the problem [16]. The estimated complexity measures for a product can be used to provide concise, pertinent information to the designers.

To be useful, however, the complexity of an assembly product must be measured along many more axes than that of a computational problem. We propose several such axes below, but by no means is this list exhaustive. Moreover, assembly algorithms apply to a number of parts that is small relative to the amount of data processed by typical computer algorithms. As a result, a much tighter evaluation of assembly complexity is required than the classical asymptotic analysis applied to computer algorithms. We will show how the NDBGs make it possible to perform this analysis in many cases.

7.1   Complexity Measures

Let an assembly product be *admissible* if it admits at least one correct assembly algorithm. In the following discussion we consider only admissible assemblies.

An interesting measure of the complexity of an assembly algorithm is the number of hands it requires. Let us say that an assembly instruction is ℓ-handed if it has $\ell + 1$ moved sets ($\ell \geq 1$). An assembly algorithm is m-handed if all its assembly instructions are ℓ-handed, with $\ell \leq m$, and at least one is m-handed. A product is *p-handed* if all correct assembly algorithms for this product are m-handed, with $m \geq p$, and at least one is p-handed. A p-handed product requires p moving "hands" to be (dis)assembled, in addition to a fixed one (e.g., a vise). It is shown in [37] that an assembly made of n parts may require up to $n - 1$ moving hands to be assembled (i.e., the assembly can only be built by moving every part relative to all the other parts simultaneously). However, it is in general much more cost-effective to manufacture and service a 1-handed product than a multi-handed one. The class of 1-handed products is thus an important one.

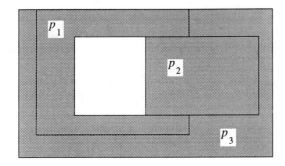

Figure 10: *Three-part latch assembly*

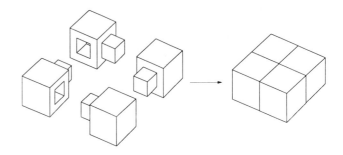

Figure 11: *Nonlinearizable 1-handed assembly*

Let a *subassembly* be any subset of parts in their final relative positions. An assembly algorithm is *monotonic* if each of its instructions merges the corresponding moved sets into a subassembly [48, 51]. A product is *p-handed monotonic* if it admits a p-handed monotonic assembly algorithm. For example, the 3-part latch assembly shown in Fig. 10 is not 1-handed monotonic, but is 2-handed monotonic. It is clear that any admissible n-part assembly is $(n-1)$-handed monotonic; however, there exist p-handed n-part products, with $p < n - 1$, that are not $(p+i)$-handed monotonic, for any $i \in [0, n - p - 2]$. A non-monotonic assembly (for some given number of hands) requires bringing parts into intermediate relative positions (where they are ungrasped) that have to be changed later (which requires regrasping parts). It may require many more motions than a monotonic assembly with the same number of parts. So, the class of 1-handed monotonic products is an important subclass of the 1-handed products.

The nature and the number of degrees of freedom required to perform the motions specified by the assembly instructions are another key measure of product complexity. Let an assembly instruction be *m-prismatic* if the motion of every moved set is a sequence of at most m extended translations. Let a product be *p-prismatic* if it admits an assembly algorithm in which all instructions are m-prismatic, with $m \leq p$, for some initial arrangement of its parts in which every two parts are separated, and one instruction is p-prismatic. Since single translations are much more cost-effective to generate than multiple translations and generalized motions, the class of 1-prismatic products is another important one. One can further characterize the complexity of a product within this class by the minimal number of translation directions

that are needed. A *stack* product is a 1-prismatic 1-handed monotonic product that admits an algorithm in which all instructions specify translations along the same direction. Large subassemblies of many small consumer electronic products are stack products. The concepts of prismatic and stack products can be easily extended to allow for screws in the products.

The length of the longest sequence of instructions in an assembly algorithm, the *length* of the algorithm, is another important attribute of this algorithm. Indeed, assuming that the time taken by any assembly instruction is lower and upper bounded regardless of the moved sets, it directly impacts the total execution time. The length of the shortest correct algorithm for a product is therefore a pertinent measure of the complexity of this product. It characterizes the extent to which the product can be broken down into subassemblies that can be manufactured concurrently.

However, some mass-produced assemblies are more cost-effective to manufacture by bringing one part at a time. This concept can be captured as follows. Let us say that an instruction is *linear* if it is 1-handed and one of its moved sets is a single part. A linear algorithm consists of linear instructions only, hence is a total ordering. A product that admits a linear algorithm is said to be *linearizable*. Many 1-handed products are not linearizable (see one example in Fig. 11) and are in general less amenable to mass-production assembly lines. A refinement of this concept is to measure the minimal number p of nonlinear instructions over all possible 1-handed assembly algorithms; the minimal number of separate assembly lines needed for the product is $p + 1$.

Moved sets need to be grasped or fixtured in a way that parts in the same set cannot move relative to each other under a variety of forces (gravity, acceleration,

contact). This is achieved by positioning fingers or fixture elements on their boundary. A grasp of a moved
set achieves *form closure* if, when the fingers are locked
relative to each other, no part in the moved set can
move relative to any other and the fingers [26, 27].
(Another concept, involving friction and forces, is that
of force closure; we will not discuss it here, though it
is perhaps more practical than form closure.) Given
some abstract dimensionless model of a finger, e.g., a
point-plane contact, a measure of the complexity of an
assembly instruction is the number of fingers required
to achieve a form-closure grasp of all its moved sets.
Weaker notions can also be considered. For instance, a
grasp achieves *prismatic form-closure* if it prevents every part to move in translation relative to any other. A
grasp achieves *form closure of order q* if it prevents up
to $q + 1$ subassemblies to move relative to each other.
Thus, a form-closure grasp of order 0 assumes that no
internal motion in the moved set is possible; a form-
closure grasp of order 1 assumes that a single internal motion (along any direction) is possible; etc. The
higher the degree of form-closure, the safer the grasp.
The number of fingers to achieve form-closure grasps
for a *single* rigid part (equivalently, form-closure of order 0 for a set of parts) is investigated in several papers,
including [34, 36].

There exist other pertinent measures of complexity.
For example, uncertainties may also play an important
role in assembly instructions, requiring sensors to be
used. The number of elementary sensors (e.g., plane
probe, diameter sensors) may also be used as a measure
of product complexity. However, since we have ignored
physical and uncertainty issues in this paper, we will
not consider such a measure further.

7.2   Evaluation of Complexity Measures

We now show how NDBGs can be used to estimate complexity measures of a product. We mainly focus on the
NDBG for infinitesimal translations in 3D, but a similar development can be done with other NDBGs. We
evaluate the time complexity for evaluating some complexity measures assuming the NDBG as the input to
this computation. However, several measures can be
computed while the NDBG is constructed, within the
time complexity required by this construction; some
can even be more efficiently obtained without computing the NDBGs.

Let us say that a 1-handed assembly algorithm is
correct for infinitesimal translations if, for every instruction, one moved set is locally free to translate in
some direction d in the final arrangement specified by
the instruction. The set of assembly algorithms that
can be extracted from the NDBG $\Gamma(A)$ of an assembly
A for infinitesimal translations includes all correct 1-
handed monotonic prismatic assembly algorithms, the
importance of which was discussed above. If no assembly algorithm can be extracted from $\Gamma(A)$, it is guaranteed that the assembly is not 1-handed monotonic
prismatic.

If a candidate algorithm can be extracted from $\Gamma(A)$,
one may wish to know if the assembly is linearizable
for translations. Checking the existence of a candidate
partitioning of A into S and $A \setminus S$, where S is a singleton, and finding one such partitioning, if one exists,
take the same time, $O(rc^2)$, as checking the existence
of any candidate partitioning (see Subsection 5.1). In
addition, if an assembly is linearizable, all its subassemblies are also linearizable. Hence, checking that a candidate linear algorithm exists takes $O(rc^2n)$ time. If
none exists, the product is not linearizable for translations. (This check is a case where using the NDBG
is not very pertinent. Indeed, a candidate linear algorithm can be directly computed from the description
of the c contacts in only $O(nc \log c)$ time.)

The minimal length of all the candidate algorithms
for a product is a lower bound on the minimal length
of all correct one-handed monotonic prismatic assembly algorithms. The set of all candidate algorithms can
be represented as an AND/OR graph [21] and searched
for the shortest one using alpha-beta pruning [38]. But
in general, the size of this graph is exponential in the
number n of parts. Efficiently computing the minimal
length of candidate algorithms, or a bounded approximation of this length, is still an open problem.

The NDBG $\Gamma(S)$ of any subassembly $S \subseteq A$ for infinitesimal translations can also be used to generate
a prismatic form-closure grasp of order 1 for S. Let
each finger be modeled by a point-plane contact. We
scan all the DBGs in $\Gamma(S)$. For every DBG, we consider
the reduced graph of its strong components; for every
component C with no outgoing arc (i.e., not blocked by
any other component), we place a dimensionless finger
f in a face F of C whose external normal vector has
positive inner product with d and we update $\Gamma(S)$ by

adding f to the parts of S, that is, by adding the plane parallel to F to the partition of $\mathcal{S}^2$. At the end, the arrangement in $\mathcal{S}^2$ must be such that, in every DBG, each strongly connected component with no outgoing arc contains one finger. In general, the number of fingers in the generated grasp is not minimal to achieve prismatic form-closure of order 1; it nevertheless gives a good indication of the cost of grasping or fixturing S.

The set of assembly algorithms that one can extract from the NDBG of A for infinitesimal generalized motions includes all correct 1-handed monotonic assembly algorithms. Various complexity measures can be extracted as above.

The set of assembly algorithms that can be extracted from the NDBG of A for infinite translations is the set of all the 1-handed monotonic 1-prismatic assembly algorithms for A. This NDBG allows the computation of various complexity measures for this family of algorithms. In particular, A is a stack product exactly when an algorithm can be extracted from a single DBG. Such a DBG must be an acyclic graph (this is a necessary and sufficient condition), which can be checked in $O(n^2)$ time. Hence, determining whether A is a stack product takes $O(n^2v^4)$ time.

Notice that no combination of the previous NDBGs allows one to decide in all cases whether a product is admissible.

7.3 Example

We have implemented several algorithms to evaluate complexity measures of a product from its NDBG for infinitesimal translations and suggested rotations (see Section 6). We applied these algorithms to a 23-part simplified version of the engine shown in Fig. 9. The results are as follows:

Number of hands and monotonicity. The NDBG allows the generation of candidate algorithms. Hence, the engine admits 1-handed monotonic algorithms that are correct for infinitesimal motions.

Is it prismatic? The engine contains a number of threaded contacts, so it is not prismatic. If these threaded contacts are replaced by cylindrical ones, the existence of candidate algorithms in the modified NDBG suggests that there may exist 1-handed monotonic prismatic algorithms.

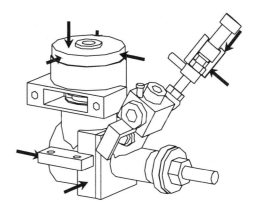

Figure 12: *Form-closure grasp of order 1 of the engine*

Is it linearizable? As far as infinitesimal motions are concerned, the product is linearizable. This computation was performed with the faster method mentioned above, not directly using the NDBG.

Minimal length. The minimum-length algorithm contained in the NDBG has depth 6. This was computed by searching the AND/OR graph representing all 1-handed monotonic assembly algorithms for the engine. The graph has 1027 nodes (each distinct subassembly is represented by a single node) and required 14 minutes to build and search. Better algorithms are needed for products having many more parts.

Is it a stack product? The engine requires infinitesimal translation in several directions to assemble, so it cannot be a stack product.

Number of fingers. Fig. 12 shows a prismatic form-closure grasp of order 1 computed for the simplified engine. The grasp consists of 10 fingers. The contact points of 8 fingers are shown as arrows; two additional fingers are required to hold a loose part inside the engine. It may seem that an additional finger is needed to support the assembly from below. In fact, the finger pointing up and left at the top-right achieves that function.

The implemented algorithms illustrate the kind of information which one may extract from NDBGs. Some of them, however, are based on inefficient brute-force methods that are not practical for complex products. Furthermore, no algorithms have been implemented yet to exploit NDBGs for infinite translation or any other family of extended motions. It is not obvious that polynomial NDBGs will allow all interesting complexity measures to be computed in polynomial time. Very likely, for some measures, we will have to content ourselves with approximate algorithms. Incremental algorithms operating in marginal time while the product is being designed will also be useful.

8 Conclusion

This paper presented techniques to reason about mechanical assembly. We investigated the following two problems: (1) the automatic generation of assembly algorithms from CAD data, and (2) the characterization of the complexity of assembly designs. Our main goal was to provide efficient computational support to help designers create products that are easier to manufacture and service. The proposed techniques nevertheless have broader application. They could also be used, for example, to plan the actions of autonomous maintenance robots.

One contribution of this paper consists of the notion of an NDBG and the algorithms to construct it. The construction of an NDBG is based on an analysis of the interferences (blocking relations) among the parts in an assembly. Several variants were presented for different families of motion inducing different blocking relations. Unlike previous methods, the NDBG allows assembly sequences to be computed in time polynomial in the complexity of the parts of the assembly.

Another contribution of the paper is the notion of the algorithmic complexity of an assembly product, the set of complexity measures proposed, and the techniques to estimate them from precomputed NDBGs. Our goal here was to concisely assess the difficulty, hence the cost, of (dis)assembling a product.

Computational complexity has been instrumental to identify the strengths and weaknesses of particular computer algorithms, and classify problems. It has contributed enormously to the progress of computer technology. We think that algorithmic complexity measures for assembly products could have a similar impact on assembly products and processes, bringing improvements that would outpace any advances in the underlying manufacturing technologies themselves. We also expect that it will contribute to the emergence of a completely new type of manufacturing technology

where a large number of very simple, mass-produced actuators and sensors are combined in highly modular manufacturing systems. The RISC approach to robotics recently proposed in [8] takes a similar view.

One of our current research goals is to construct polynomial NDBGs for more complex families of motion and investigate the effect of additional constraints (e.g., uncertainty, grasp accessibility) on the construction of NDBGs. Another goal is to investigate further the exploitation of NDBGs and design a more comprehensive set of efficient complexity evaluation methods. On the experimental side, we are currently integrating our algorithms into a highly interactive design-assistant system.

Acknowledgments: This research was funded by ARPA contract N00014-88-K-0620, NSF grant IRI-9306544-001, and a grant of the Stanford Integrated Manufacturing Association (SIMA). The authors also thank Leo Guibas, Rajeev Motwani, and Achim Schweikard for their encouragement and comments.

References

[1] —— 1990. Special Issue on Robotic Assembly and Task Planning. *AI Magazine*, AAAI Press, 11(1).

[2] Aho, A.V., Hopcroft, J.E., and Ullman J.D. 1974. **The Design and Analysis of Computer Algorithms.** Reading, MA: Addison-Wesley.

[3] Aho, A.V., Hopcroft, J.E., and Ullman J.D. 1983. **Data Structures and Algorithms.** Reading, MA: Addison-Wesley.

[4] Arkin, E.M., Connelly, R., and Mitchell, J.S. 1989. On Monotone Paths among Obstacles with Applications to Planning Assemblies. *ACM Symp. on Computational Geometry*, 334–343.

[5] Baldwin, D.F. 1990. Algorithmic Methods and Software Tools for the Generation of Mechanical Assembly Sequences, Master's thesis, MIT.

[6] Boothroyd, G. 1991. **Assembly Automation and Product Design.** Marcel Dekker, Inc., New York, NY.

[7] Bourjault, A. 1984. Contribution à une Approche Méthodologique de l'Assemblage Automatisé: Elaboration Automatique des Séquences Opératoires. PhD thesis, Faculté des Sciences et des Techniques de l'Université de Franche-Comté, France.

[8] Canny, J.F. and Goldberg, K.Y. 1992. RISC Robotics. in preparation.

[9] Cutkosky, M.R. and Tenenbaum, J.M. 1990. A Methodology and Computational Framework for Concurrent Product and Process Design. *ASME J. of Mechanism and Machine Theory*, 25(3):365–381.

[10] De Fazio, T.L. and Whitney, D.E. 1987. Simplified Generation of All Mechanical Assembly Sequences. *IEEE J. of Robotics and Automation*, RA-3(6):640–658. Errata in RA-4(6):705-708.

[11] de Kleer, J. and Williams, B.C. (guest editors). 1991. Special Volume, Qualitative Reasoning about Physical Systems II. *Artificial Intelligence*, 51(1–3).

[12] Edelsbrunner, H. and Guibas, L. 1989. Topologically Sweeping an Arrangement. *J. of Computer and System Sciences*, 38:165–194.

[13] Fahlman, S.E. 1974. A Planning System for Robot Construction Tasks. *Artificial Intelligence*, 5(1):1–49.

[14] Faltings, B. 1992. A Symbolic Approach to Qualitative Kinematics. *Artificial Intelligence*, 56(2–3), 139–170.

[15] Fikes, R.E. and Nilsson, N.J. 1971. STRIPS: A New Approach to the Application of Theorem Proving to Problem Solving. *Artificial Intelligence*, 2(3-4):189–208.

[16] Garey, M.R. and Johnson, D.S. 1979. Computers and Intractability: A Guide to the Theory of NP-Completeness. New York, NY:W.H. Freeman and Co.

[17] Goldberg, K.Y. 1990. Stochastic Plans for Robotic Manipulation. PhD Thesis, Dept. of Computer Science, Carnegie Mellon University.

[18] Hirukawa, H., Matsui, T., and Takase, K. 1991. A General Algorithm for Derivation and Analysis of Constraint for Motion of Polyhedra in Contact. *IEEE/RSJ Int. Workshop on Intelligent Robots and Systems*, Osaka, 38–43.

[19] Hoffman, R.L. 1991. A Common Sense Approach to Assembly Sequence Planning. In Homem de Mello, L.S. and Lee, S., editors, Computer-Aided Mechanical Assembly Planning, Boston, MA: Kluwer Academic Publishers, 289–314.

[20] Homem de Mello, L.S. 1989. Task Sequence Planning for Robotic Assembly. PhD Thesis, Carnegie Mellon University.

[21] Homem de Mello, L.S. and Sanderson, A.C. 1991. Representations of Mechanical Assembly Sequences. *IEEE Tr. on Robotics and Automation*, 7(2):211–227.

[22] Homem de Mello, L.S. and Lee, S. (editors) 1991. Computer-Aided Mechanical Assembly Planning. Boston, MA: Kluwer Academic Publishers.

[23] Hopcroft, J.E., Schwartz, J.T., and Sharir, M. 1984. On the Complexity of Motion Planning for Multiple Independent Objects: PSPACE-Hardness of the 'Warehouseman's Problem', *The Int. J. of Robotics Research*, 3(4):76–88.

[24] Joskowicz, L. and Sacks, E. 1991. Computational Kinematics. *Artificial Intelligence*, 51(1–3):381–416.

[25] Kriegman, D.J. and Ponce, J. 1990. Computing Exact Aspect Graphs of Curved Objects: Solids of Revolution. *Int. J. of Computer Vision*, 5(2):119–135.

[26] Lakshminarayana, K. 1978. The Mechanics of Form Closure. *ASME Paper*, No. 78-DET-32.

[27] Latombe, J.C. 1991. Robot Motion Planning. Boston, MA: Kluwer Academic Publishers.

[28] Lee, S. and Shin, Y.G. 1990. Assembly Planning Based on Geometric Reasoning, *Computation and Graphics*, 14(2):237–250.

[29] Liebermann, L.I. and Wesley, M.A. 1977. AUTOPASS: An Automatic Programming System for Computer Controlled Mechanical Assembly. *IBM J. of Research and Development*, 21(4):321–333.

[30] Liu, Y. 1990. Symmetry Groups in Robotic Assembly Planning. PhD Thesis, COINS Tech. Rep. 90-83, Computer and Information Science Dept., University of Massachusetts, Amherst.

[31] Lozano-Pérez, T. 1976. The Design of a Mechanical Assembly System. Tech. Rep. AI-TR 397, AI Lab., MIT.

[32] Lozano-Pérez, T. 1983. Spatial Planning: A Configuration Space Approach, *IEEE Tr. on Computers*, 32(2):108–120.

[33] Lozano-Pérez, T. and Wilson, R.H. 1992. Assembly Sequencing. In preparation.

[34] Markenscoff, X., Ni, L., and Papadimitriou, C.H. 1990. The Geometry of Grasping. *The Int. J. of Robotics Research*, 9(1):61–74.

[35] McDermott, D.V. 1992. Robot Planning. *AI Magazine*, AAAI Press, 13(2):55–79.

[36] Mishra, B., Schwartz, J.T., and Sharir, M. 1987. On the Existence and Synthesis of Multifinger Positive Grips. *Algorithmica*, 2:541–558.

[37] Natarajan, B.K. 1988. On Planning Assemblies. *ACM Symp. on Computational Geometry*, 299–308.

[38] Nilsson, N.J. 1980. Principles of Artificial Intelligence, Los Altos, CA: Morgan Kaufmann.

[39] Pollack, R., Sharir, M., and Sifrony, S. 1988. Separating Two Simple Polygons by a Sequence of Translations, *Discrete and Computational Geometry*, 3:123–136.

[40] Preparata, F.P. and Shamos, M.I. 1985. Computational Geometry: An Introduction. New York, NY:Springer-Verlag.

[41] Reif, J.H. 1979. Complexity of the Mover's Problem and Generalizations. *20th IEEE Symp. on Foundations of Computer Science*, 421–427.

[42] Sacerdoti, E.D. 1977. A Structure for Plans and Behavior. Amsterdam: Elsevier.

[43] Scheinman, R. 1988. RobotWorld: A Multiple Robot Vision Guided Assembly System. In Bolles, R.C. and Roth, B., editors, Robotics Research, Cambridge, MA: MIT Press, 23–27.

[44] Subramani, A. 1992. Development of a Design for Service Methodology. PhD Thesis, Dept. of Industrial and Manufacturing Engineering, The University of Rhode Island, Kingston, RI.

[45] Taylor, R.H. 1976. Synthesis of Manipulator Control Programs from Task-Level Specifications. PhD Thesis, Dept. of Computer Science, Stanford University.

[46] Toussaint, G.T. 1985. Movable Separability of Sets. In Toussaint, G.T., editor, Computational Geometry. Amsterdam: Elsevier.

[47] Wilson, R.H. and Rit, J.F. 1990. Maintaining Geometric Dependencies in an Assembly Planner. *IEEE Int. Conf. on Robotics and Automation*, Scottsdale, AZ, 890–895.

[48] Wilson, R.H. 1992. On Geometric Assembly Planning. PhD Thesis, Dept. of Computer Science, Stanford University.

[49] Wilson, R.H. and Matsui, T. 1992. Partitioning an Assembly for Infinitesimal Motions in Translation and Rotation. *IEEE Int. Conf. on Intelligent Robots and Systems*, Raleigh, NC, 1311–1318.

[50] Wilson, R.H. and Latombe, J.C. 1992. On the Qualitative Structure of a Mechanical Assembly. *AAAI-92 Nat. Conf. on Artificial Intelligence*, San Jose, CA, 697–702.

[51] Wolter, J.D. 1988. On the Automatic Generation of Plans for Mechanical Assembly. PhD thesis, The University of Michigan.

[52] Wolter, J., Chakrabarty, S., and Tsao, J. 1992. Mating Constraint Languages for Assembly Sequence Planning. *IEEE Int. Conf. on Robotics and Automation*, Nice, France, 2367–2374.

The Hand Complexity of Assembly

B.K. Natarajan, *Hewlett Packard Labs, Palo Alto, CA 94304, USA*

We present an overview of results on the hand-complexity of assembly. We find that it is difficult to categorize the number of hands required to assemble a composite in terms of geometric restrictions on its components—even for composites of convex components, more than two hands are necessary.

1 Introduction

An important engineering problem is that of design for manufacturability—designing a composite part so that it can be quickly and cheaply assembled during mass-production. Prior to the advent of the robot in the assembly line, solutions to this problem centered on designs that enabled increased use of jigs and fixtures, in order to minimize the physical demands on the human operator. After the introduction of the robot to the assembly line, solutions to the problem have expanded to include designs that permit improved automated handling. For instance, since it is difficult to obtain compliant motions from robots, we see the increased use of flexible plastics and composite materials. Yet, broadly speaking, the focus of design for manufacture has remained on improving conventional designs rather than on unconventional designs approached from radical starting points. One such starting point would be to relinquish the tenet that all composites must be such that they can be assembled by two-handed creatures. When robots are used in assembly, we should consider composites that cannot be assembled with two hands, but can be assembled relatively quickly with several hands. In this paper, we review some results obtained in this context.

2 Preliminaries

We avoid formal definitions where intuition suffices. We are concerned with two and three dimensional space.

A *component* is any closed and bounded subset of space.

A *composite* is an arrangement of a set of components in a specific configuration.

Rather than study the assembly of composites, we will study their disassembly, with the understanding that the two are simply inverses of each other. This makes for technical simplicity.

A pair of components are said to be *separated* if they are arbitrarily far apart.

An *assembly sequence* for a composite is a sequence of motions of the components of the composite that leaves them pairwise separated.

The *number of hands* required by an assembly sequence is the maximum number of simultaneous and distinct motion vectors occurring in the sequence.

In the above definition, we assume that one hand is sufficient to move several unconnected components with the same velocity. Furthermore, we ignore practical considerations such as gravity. Hence, the definition is but a lower bound on the number of physical hands or robots that the assembly sequence might require in the practical situation.

The *number of hands* required by a composite is the minimum number of hands required over all possible assembly sequences for the composite.

3 Results

We first consider the computational complexity of deciding the existence of an assembly sequence for a given composite.

Proposition 1 *Deciding the existence of an assembly sequence for a given composite is PSPACE-hard, even if each component in the composite is limited to be a polyhedron of at most 44 vertices.*

Proof: Natarajan (1988). A modification of the proof of Reif (1979) that the Piano Mover's Problem is PSPACE-hard. □

In light of the above result, we must consider composites limited to simpler classes of objects and inquire into the existence and nature of assembly sequences for them.

Proposition 2 *Every composite of convex polygons in the plane has a two-handed assembly sequence of translations in the plane.*

Proof: Guibas and Yao (1980) showed that in every collection of convex polygons, and for every direction in the plane, there is some polygon that can be moved to infinity along that direction without colliding with the others. In particular, their proof is as follows. Without loss of generality assume that the direction of choice is the positive X-direction. Consider all the polygons whose uppermost vertex is visible from infinity. Of these, consider the one with the lowest uppermost vertex. Clearly this polygon can be moved to infinity without collision. It follows that every composite of convex polygons has a two-handed assembly sequence consisting only of translations in the plane. □

A set S is said to be *star-shaped* if there is a point $x \in S$ such that for all points $y \in S$, the line xy is completely contained in S. The point x is called the *star-center* of S.

Proposition 3 *There exists a composite of N star-shaped polygons in the plane for which N hands are necessary for assembly.*

Proof: Natarajan (1988). Figure 1 shows a composite of star-shaped objects consisting of a triangular block and $N - 1$ daggers arranged radially about the apex of the triangle. In order to pull the daggers out of the block, each dagger must be moved radially outward in a distinct direction, requiring N hands. □

Proposition 4 *N hands suffice for a composite of N star-shaped polygons in the plane.*

Proof: Dawson (1984). Let $P_1, P_2, P_3, \cdots P_N$ be the N polygons and let x_i be the star-center of P_i. Pick any point o as the origin. Translate P_i with velocity vector $\vec{x_i}$. That is, translate P_i along the line joining o and x_i with speed equal to the distance between o and

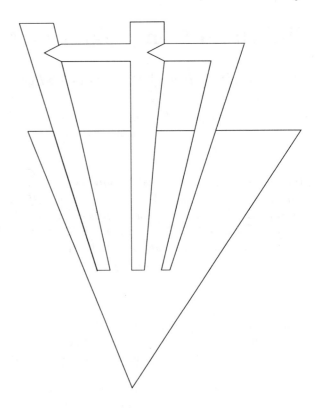

Figure 1: *A composite of 4 star-shaped polygons that requires 4 hands for assembly.*

x_i. We now argue that doing so will mutually separate the polygons without collisions. To do so, it suffices to point out that such a sequence of translations is equivalent to a global dilation, i.e., a simple scaling of distances in the plane, followed by a shrinking of each polygon about its star-center. Clearly a global dilation cannot cause collisions. Also, since each polygon is star-shaped, shrinking about the star-center cannot cause collisions. The claim follows. □

We now seek to extend the above propositions to three dimensional space.

Proposition 5 *N hands are necessary and sufficient for composites of N star-shaped polyhedra.*

Proof: Sufficiency follows from the proof of Proposition 4. To prove necessity, it is easy to see that we can extend the components of Figure 1 to three dimensions by making the block a cone, and fashioning each of the daggers out of two cylindrical pieces. □

As it happens, Proposition 2 on convex polygons does not extend smoothly to convex polyhedra. In particular, in Proposition 2 we showed that in every composite of convex polyhedra, we can always move one to infinity without colliding with others. In three dimensions, such is not the case.

Proposition 6 *There exist composites of convex polyhedra where no single polyhedron can be translated without collision.*

Proof: There are three examples of such composites. Snoeyink and Stolfi (1993) note that Fejes-Toth and Heppes gave one comprising 13 polyhedra—twelve tetrahedra arranged around a dodecahedron. In the same paper, Snoeyink and Stolfi point out that the dodecahedron is not essential to the construction. Dawson (1984) gave a composite comprised of twelve convex tiles arranged to form a rhombic dodecahedron. In both of the above examples, the components interlock so that no one component can be translated or rotated. To prohibit just translations, a simpler construction is possible. Natarajan (1988) gave an example comprised of 16 thin triangular tiles arranged to form a twisted octahedron, as shown in Figure 2. To obtain such a composite, start with eight triangular tiles arranged to form an octahedron. Shrink each tile by a small amount, and then twist it clockwise about its normal. This will prevent the tiles from moving in their plane, or inward into the octahedron. Now place eight more tiles in the untwisted position atop the twisted tiles. These cannot be moved outward as they are obstructed by the twisted tiles. □

In light of the above proposition, it is clear that we cannot expect as strong a result for convex polyhedra as we had for convex polygons. Yet, each of the examples of Proposition 6, involved tiling the sphere with interlocking pieces, so that no single piece can be moved. However, the composite can still be taken apart by a two-handed sequence of translations in it can be partitioned into two sets of components each tiling a hemisphere and the two sets can be separated. This led to the following conjecture.

Conjecture 1 *Natarajan (1988). Every composite of convex polyhedra has a two-handed assembly sequence of translations.*

As we see below, Snoeyink and Stolfi (1993) disproved the conjecture.

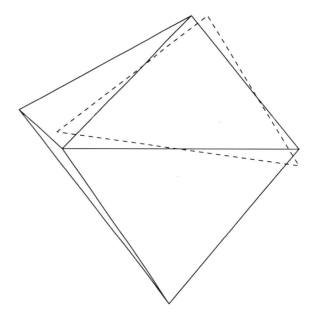

Figure 2: *An octahedron composed of sixteen triangular tiles where none can be translated without collision. Only one of the eight tiles in the twisted position is shown.*

Proposition 7 *There exists a composite of six tetrahedra without a two-handed assembly sequence of translations, but with a three-handed sequence of translations. Furthermore, the counterexample is minimal in that every composite of five or fewer convex polyhedra can be taken apart with a two-handed sequence of translations.*

Proof: Figure 3 shows six tetrahedra interlaced to form the edges of a tetrahedron. Snoeyink and Stolfi (1993) argue that no subset of the six tetrahedra can be translated without collision with the remainder. Hence, no two-handed sequence of translations exists. To show that a three-handed sequence exists, simply pair off the tetrahedra into pairs that share a contact point. Each of the three pairs is star-shaped about its contact point and by the arguments of Proposition 5, we see that there exists a three-handed sequence of translations.

To see that the counterexample is minimal, we first consider composites of four or fewer convex polyhedra. Take any one of the four, and construct a set of three separating planes between it and the remainder. These three planes cannot form a closed and bounded region, since at least four planes are necessary for such. Thus, there exists a direction along which the chosen polyhedron can be moved to infinity without collision. For collections of five polyhedra a similar but more involved argument is given in Snoeyink and Stolfi (1993). □

per contact normal across the partition. The variables stands for the normal contact forces. The constraints in the linear program are that the forces and torques should sum to zero and that all contact forces must be in the direction of the outward normals at the contact surfaces. If a basic feasible solution exists and has 7 non-zero values, then the tricontahedron cannot be partitioned with two-hands. By exhaustive enumeration on a computer, Snoeyink and Stolfi (1993) verify such for each of the 2^{30} partitions. □

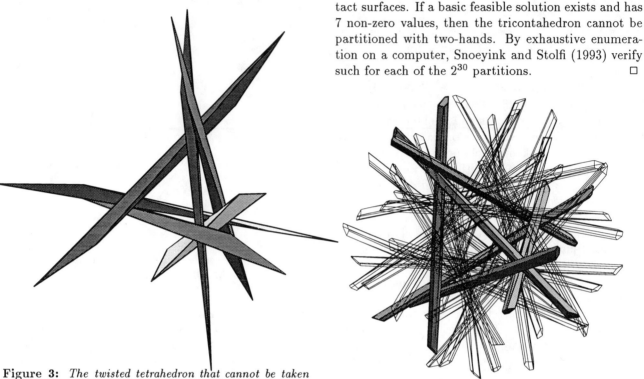

Figure 3: *The twisted tetrahedron that cannot be taken apart with a two-handed sequence of translations, Snoeyink and Stolfi (1993).*

Figure 4: *The tricontahedron that cannot be taken apart with two hands. Snoeyink and Stolfi (1993).*

Furthermore, Snoeyink and Stolfi (1993) show that there exists composite of convex polyhedra that cannot be assembled with two hands, even if rotations are allowed.

Proposition 8 *There exists a composite of thirty tetrahedra that cannot be assembled with two hands.*

Proof: Figure 4 shows thirty tetrahedra interlaced to form a tricontahedron. The ends of the tetrahedron are cut off in the figure for the sake of clarity. Essentially, the tricontahedron is comprised of five copies of the twisted tetrahedron of Figure 3, laced together. One copy is shown shaded in Figure 4 and the other four copies are shown in outlines. The thirty tetrahedra can be partitioned into 2^{30} subsets. For each partition we can set up a linear program with one variable

4   Conclusion

We examined a number of results on the hand complexity of assembly. Several important open problems remain. Among them are

(1) Find a minimal composite of convex polyhedra that cannot be assembled with two hands.

(2) How many hands are required for composites of convex polyhedra?

(3) What is the computational complexity of computing assembly sequences for composites of convex polyhedra?

(4) Is is possible to prove useful lower bounds on the tradeoff between the number hands and the length of an assembly sequence for various classes of composites?

5 Acknowledgements

Thanks to J. Snoeyink for the discussions and for providing Figures 3 and 4.

6 References

Dawson, R., *On removing a ball without disturbing the others*. Mathematics Magazine, 57(1):27-30, Jan. 1984.

Guibas, L.J., and Yao, F.F., *On translating a set of rectangles*, in Proc. 12th ACM Symposium on Theory of Computing, pp.154-160, 1980.

Natarajan, B.K., *On planning assemblies*, in Proc. 4th ACM Symposium on Computational Geometry, pp.299-308, 1988.

Reif, J.H., *Complexity of the mover's problem and generalizations*, in Proc. 20th IEEE Foundations of Computer Science, pp. 421-427, 1979.

Snoeyink, J. and Stolfi, J., *Objects that cannot be taken apart with two hands*, in Proc. 9th ACM Symposium on Computational Geometry, pp.247-256, 1993.

Nonholonomic Motion Planning via Optimal Control

J-P. Laumond, *LAAS/CNRS, 7 Avenue du Colonel Roche*
31077 Toulouse, France

This paper presents a survey of some recent results in nonholonomic motion planning. A general planner is presented : it consists of three steps. The first one computes a collision-free geometric path without taking into account the kinematic constraints; this path is approximated by a sequence of optimal paths and then smoothed. From a theoretical view point, using optimal control provides original results in complexity analysis. Moreover we present results from three different planners working respectively for a polygonal car-like robot, a circular car-like robot and a mobile robot with trailers. As a conclusion we show how nonholonomic motion planning via optimal control opens very challenging issues.

1 Introduction

Nonholonomic systems are characterized by constraint equations involving the time derivatives of the system's configuration variables. These equations are non integrable; they typically arise when the system has less controls than configuration variables. For instance a car-like robot has two controls (linear and angular velocities) while its moves in a 3-dimensional configuration space. As a consequence, any path in the configuration space does not necessarily correspond to a feasible path for the system. This is basically why the purely geometric techniques developed in motion planning for holonomic systems (see [19]) do not apply directly to nonholonomic ones.

Nevertheless, for controllable nonholonomic systems without drift, the existence of a feasible and collision-free path is characterized by the existence of any collision-free path. This is a consequence of the Lie algebra rank condition (LARC) [40].

Several authors developed techniques for steering nonholonomic systems [18, 34, 37, 14, 32, 44] in the absence of obstacle.

In the presence of obstacles, the LARC implies that any collision-free path can be approximated by a feasible and collision-free one.

Therefore, the existing collision-free nonholonomic motion planning algorithms consists of two steps. The first one computes a collision-free path without taking into account the nonholonomic constraints. If such paths do not exist, then there is no solution to the problem. Else, the second step consists in approximating any path computed at the first step by a sequence of feasible ones.

Such a scheme has been completely instantiated in [21] for the car-like system. The method consists in approximating a collision-free path (computed by using computational geometry techniques) by a sequence of shortest paths. The main advantage of the method is to compute quite easily the combinatorial complexity of the problem. Nevertheless, the extension of the algorithm depends on the knowledge of the optimal paths.

Recently Sussmann and Liu [41] propose an algorithm that provides a sequence of feasible paths that converge *uniformly* to any given path. This guarantees that one can choose a feasible path which is *arbitrarily close* to a given collision-free path. The method uses high frequency sinusoidal inputs. It is general and quite hard to implement in practice. In [43], Tilbury *et al* exploit the idea in the framework of a mobile robot with two trailers. The authors show how the general method by Sussmann and Liu can be simplified in this case. Nevertheless, experimental results show that the approach cannot be applied in practice, mainly because the convergence is very slow.

Another approach to nonholonomic motion planning in the presence of obstacles has been developed by Barraquand and Latombe [3] and applied to the tractor-trailer system : it consists of a search in a grid representing the configuration space; at each iteration, the procedure computes six successors corresponding to a discretized set of controls; then the choice of the successor is based on a criterion minimizing the number of reversals.

This paper presents an overview of the nonholonomic motion planning approach using optimal control introduced in [21] and enlightens its theoretical issues and practical advantages. As a conclusion we pinpoint issues in optimal control related to nonholonomic motion planning : the method developed by Mirtich and Canny [28, 29] for a car-like robot uses a computation of the metric induced by the optimal paths; such computations in general cases are very challenging.

2 Algorithm, scope, convergence, completeness, complexity, limitations

2.1 Algorithm

The nonholonomic motion planning approach using optimal control consists of three steps :

Step 1 : Plan a collision-free path $Path_1$ for the robot without taking into account the kinematic constraints.

Step 2 : Approximate $Path_1$ by a sequence $Path_2$ of optimal feasible and collision-free paths.

Step 3 : Smooth $Path_2$ by reducing the length of the sequence.

The first step corresponds to the classical "Piano Mover" problem. It is solved by using any geometric planner. The only property required by the approach is to guaranty that $Path_1$ lies in an *open* component of the collision-free configuration space. This means that the geometric planner has to avoid configurations in contact. Therefore the cell decompositions of the free-space or the Voronoï diagrams are very well adapted. The methods using visibility graphs or a retraction on the boundary of the free-space generate paths which are piece-wise in contact; applying these geometric planners in the context of our approach requires some adaptations, such as the growing of the robot by some fixed value.

The second step of the algorithm consists of a recursive procedure which splits $Path_1$ into pieces until the endpoints of each piece can be linked by a collision-free optimal feasible path. The computational efficiency of

this step depends mainly on the efficiency of the optimal path computation and of the collision-checker.

In the final step, $Path_2$ is smoothed by an iterative procedure which selects randomly two configurations lying on the current path and computes an optimal path between them; if this optimal path is collision-free then it replaces the corresponding part of the current path. The cost of the path is then decreasing. The procedure stops when one does not obtain any significant improvement after a fixed number of iterations.

2.2 Scope, convergence and completeness

Property : The algorithm stops in finite time for small-time controllable systems.

Proof : If there is no solution for the sub-problem solved at Step 1, then there is no solution for the problem. Deciding the existence of $Path_1$ is a decidable problem.

Now, let us consider a small-time controllable system (e.g., a symmetric system verifying the LARC condition at every point). For every point c, there exists a neighborhood whose all the points are reachable from c by a path included in the neighborhood; the corresponding optimal paths are also obviously included in the neighborhood. Then $Path_1$ can be covered by a *finite* number of such neighborhoods lying in the collision-free configuration space (indeed, $Path_1$ is a *compact* set lying in an *open* connected component of the free space).

By the recursive procedure of Step 2, one tries to link by an optimal path pairs of points such as the distance between them is strictly decreasing. Therefore, in the worst case, all the pairs of points would lie in the previous covering; all of them can be linked by an optimal collision-free path. The number of such pairs is then finite, and the algorithm stops.

Finally, Step 3 stops after a finite number of iterations. □

Notice that this property does not hold for any controllable systems. For instance, a car-like robot moving only forwards is controllable (i.e., each point can be reached from any other one in the absence of obstacles). This system is not symmetric and no longer small-time controllable. For such a system the existence of a feasible collision-free path is not characterized by the existence of $Path_1$, and then our algorithm does not work; special approaches have to be defined in this case (see [11, 15]).

For small-time controllable systems, if $Path_1$ exists, the recursive procedure to find a feasible path always converges in finite time. Then, the completeness of the algorithm clearly depends on the completeness of the geometric planner computing $Path_1$, or deciding that such a path does not exist.

2.3 Complexity

Each of the three steps of the algorithm requires a special combinatorial complexity analysis.

The first step corresponds to the classical piano mover problem. In analytical approaches (see [19] for a bibliography), the complexity of the geometric planners is a function of the geometric complexity of the robot and the obstacles (e.g., number of vertices, number of algebraic patches describing the environment, degrees of the patches...). In distributed representation approaches [2], the environment and the configuration space are modelled as bitmaps; in this case the complexity of the geometric planner depends on the resolution of the bitmaps.

The second step computes a sequence of optimal paths. Each elementary optimal path is checked to be collision free. The computation of an optimal path does not depends on the environment; its complexity is $O(1)$. The complexity of the collision-checker depends on the same parameters as in Step 1; for instance, the collision-checker for a polygonal car-like robot moving amidst polygonal obstacles along shortest paths may be implemented in $O(nm)$, where m and n are the number of vertices of the robot and the environment respectively.

Therefore the complexity of the second step is dominated by the length of the output sequence $Path_2$. The number of iterations in Step 2 depends on the "size" of the free-space around $Path_1$. The worst case is reached when $Path_1$ lies in a narrow "corridor" of the free-space, whose direction corresponds to the most "difficult" direction to follow for the system. Sub-Riemannian geometry allows to make these intuitive notions precise. We just present here a short account of the theory (see [4, 23, 5] for details).

Let us consider a control system defined by a set of vector fields; let us assume that the tangent space at

every point can be spanned by a finite family of these vector fields together with their Lie brackets (i.e., the system verifies the LARC at every point). The minimal length of the Lie bracket required to span the tangent space at a point is said to be the degree of nonholonomy of the system at this point. The degree of nonholonomy of the system is the upper bound d of all the degrees of nonholonomy defined locally. Let us assume that this bound is finite (it may be infinite [23]).

The cost of the optimal paths induces a metric in the configuration space of the system. A ball of radius r corresponding to this metric is the set of all the points in the configuration space reachable by a path of cost lesser than r. The balls grow faster in the directions given by the vector fields directly controled than in the directions defined by the Lie brackets of these vector fields. A powerfull result from sub-Riemannian geometry shows that the growing law depends on the degree of nonholonomy [6, 12, 45, 31] : when r is small enough, the ball grows as r in the directions directly controled; it grows as r^d in the directions spanned by Lie brackets of length d.

The complexity of the second step of our algorithm (i.e., the length of the sequence $Path_2$) is upper bounded by the minimum number of collision-free balls centered on $Path_1$. Therefore :

Property : For small-time controllable systems with a finite degree of nonholonomy d, the complexity of Step 2 is in $O(\epsilon^{-d})$ where ϵ is the smallest distance of $Path_1$ to an obstacle.

A direct and self-contained proof of this property appears in [21] for the case of a car-like robot.

Finally, there is no way to model the complexity of Step 3 of the algorithm. Since all the elementary optimal paths have to lie in the free-configuration space (i.e., an open domain), this procedure can not converge. Indeed, the optimal paths in the presence of obstacles belong to the closure of the free-space. This is why Step 3 stops when the smoothing procedure does not make a current path "significantly" shorter.

2.4 Limitations and extensions

The algorithm is based on the computation of the optimal paths. Such analytical computations are very challenging in optimal control theory. Only few results exist. At this moment, only the case of the car-like robot has been completely solved [36, 42, 7, 39]. The case of a two-driving wheels mobile robots is addressed in [16]. Its complete solution is still an open problem.

This means that the scope of the algorithm seems quite limited. Nevertheless, one may use numerical methods to compute optimal paths (see [10]).

Moreover, optimal paths may be replaced in the algorithm by some canonical paths. The only property required by the approach is the following : at each point c, for any neighborhood $N(c)$ and any point $c' \in N(c)$, there is a canonical path linking c and c' and lying in $N(c)$.

The following section presents three implementations of the algorithm done at LAAS. The first two planners come from [21] (Section 3.1) and the third one from [24] (Sections 3.2 and 3.3).

3 Implementations

3.1 Car-like robots

Let us condider a car. We denote by (x, y) the coordinates of the midpoint of the rear wheels. θ is the direction of the car. We define the car-like robot as the following control system :

$$\begin{pmatrix} \dot{x} \\ \dot{y} \\ \dot{\theta} \end{pmatrix} = \begin{pmatrix} \cos \theta \\ \sin \theta \\ 0 \end{pmatrix} u_1 + \begin{pmatrix} 0 \\ 0 \\ 1 \end{pmatrix} u_2 = X u_1 + Y u_2$$

with $|u_2| \leq |u_1| \leq 1$.

The direction of the robot is always tangent to its trajectory. Moreover the curvature of the path (when it is defined) is always lesser than 1. It appears that $\{X, Y, [X, Y]\}$ spans the tangent space at every point in $\mathbf{R}^2 \times S^1$. The system has no drift; it is small-time controllable at any point. Its degree of nonholonomy is 2.

The optimal paths used at Step 2 are the minimal length paths. There are computed from a *finite* family of canonical paths that has been proved to be sufficient to always contain a shortest path by Reeds and Shepp [36].

Figure 1 from [21] shows an example of a solution provided by a planner working for a polygonal robot moving amidst polygonal obstacles.

Step 1 of the algorithm is solved by the geometric planner developed in [1]. This geometric planner

is based on the computation of the boundary of the collision-free configuration space. A retraction on the boundary and a path search on the boundary provides collision-free paths whose some pieces are in contact. To avoid this inconvenience, the robot is first grown with some small value. This adaptation guarantees that $Path_1$ lies in an open domain of the configuration space. Nevertheless the path lies very close to the obstacles and Step 2 is then time consuming (see the length of the sequence $Path_2$ in Figure 1 (top, right)).

Figure 2 (also from [21]) shows another example provided by a planner working for a circular robot. In this case the geometric planner compute $Path_1$ from a Voronoï diagram. In this case $Path_1$ is as far from the obstacles as possible and the length of the sequence $Path_2$ is much shorter than in the previous example (see Figure 2 (middle, left)).

In [20], Latombe presents another implementation with a geometric planner using a distributed representation approach.

According to the value of the degree of nonholonomy of the system, for all these planners, Step 2 runs in $O(\epsilon^{-2})$. This bound is reached in the case of a parallel parking task.

3.2 Mobile robot with a trailer

The system is defined by 4 parameters (x, y, θ, φ), where x, y and θ have the same meaning as in the previous example, and φ is the angle of the trailer with respect to the direction of the mobile robot. We assume that the system is submitted to the geometric constraint $|\varphi| < \frac{\pi}{4}$.

The control system is :

$$\begin{pmatrix} \dot{x} \\ \dot{y} \\ \dot{\theta} \\ \dot{\varphi} \end{pmatrix} = \begin{pmatrix} \cos \theta \\ \sin \theta \\ 0 \\ -\sin \varphi \end{pmatrix} u_1 + \begin{pmatrix} 0 \\ 0 \\ 1 \\ 1 \end{pmatrix} u_2$$
$$= X \quad u_1 + Y \quad u_2$$

One may check that $\{X, Y, [X, Y], [X, [X, Y]]\}$ spans the tangent space at every point in $\mathbf{R}^2 \times (S^1)^2$. The system is small-time controllable at any point. Its degree of nonholonomy is 3 (see [23] for details).

Figure 3 shows an example of path provided by the planner presented in [24]. The planner has been implemented on a Sun Sparc 10. The geometric planner is

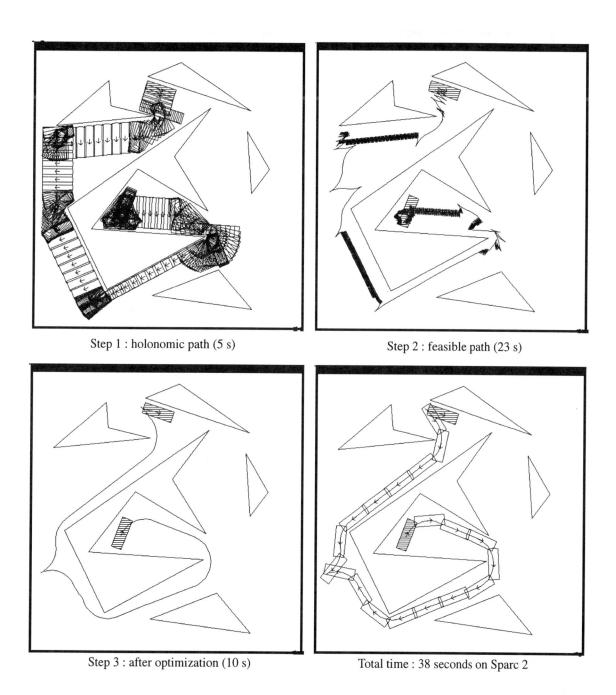

Step 1 : holonomic path (5 s)

Step 2 : feasible path (23 s)

Step 3 : after optimization (10 s)

Total time : 38 seconds on Sparc 2

Figure 1: *A solution for a polygonal car-like robot*

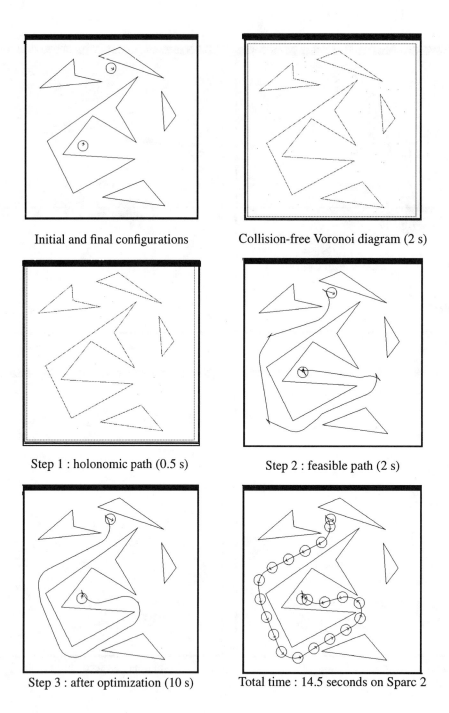

Initial and final configurations

Collision-free Voronoi diagram (2 s)

Step 1 : holonomic path (0.5 s)

Step 2 : feasible path (2 s)

Step 3 : after optimization (10 s)

Total time : 14.5 seconds on Sparc 2

Figure 2: *The same problem as in Figure 1 for a circular robot*

derived from the distributed representation approach defined in [2]. The geometric constraint $|\varphi| < \frac{\pi}{4}$ is taken into account at this level. In the example, the size of the bitmap modeling the workspace is 300×250. $Path_1$ (Figure 3 (top, left)) has been computed in 10 seconds[1].

For Step 2 we use the numerical method developed in [10]; we compute near-optimal paths by expending the controls from a Fourier basis,

$$\begin{pmatrix} u_1 \\ u_2 \end{pmatrix} = \sum_{i=1}^{\infty} \alpha_i e_i$$

and minimizing the cost

$$\tilde{J} = \sum_{i=1}^{r} \alpha_i^2 + \beta \|x(T) - x_{goal}\|.$$

Computing an elementary optimal path takes 2 seconds in average.

In the example, the length of the sequence $Path_2$ (top, right) is 10; it has been computed in 90 seconds (notice that this time includes the computation of the optimal paths and the collision-checker). The solution shown in the figure (bottom) required 4 iterations of the smoothing procedure and took 90 seconds.

Again, according to the value of the degree of nonholonomy of the system, the algorithm runs in $O(\epsilon^{-3})$. As for the car-like system, this bound is reached in the case of a parallel parking task.

3.3 Mobile robot with two trailers

The corresponding 5-dimensionated system is :

$$\begin{pmatrix} \dot{x} \\ \dot{y} \\ \dot{\theta} \\ \dot{\varphi}_1 \\ \dot{\varphi}_2 \end{pmatrix} = X u_1 + Y u_2$$

with

$$X = \begin{pmatrix} \cos\theta \\ \sin\theta \\ 0 \\ -\sin\varphi_1 \\ \sin\varphi_1 - \cos\varphi_1 \sin\varphi_2 \end{pmatrix}, \quad Y = \begin{pmatrix} 0 \\ 0 \\ 1 \\ 1 \\ 0 \end{pmatrix}$$

and where φ_1 and φ_2 are the angles between the robot and the first trailer and between the first trailer and the second one respectively. We assume the geometric constraints $|\varphi_1| < \frac{\pi}{4}$ and $|\varphi_2| < \frac{\pi}{4}$.

Figure 4 from [24] shows an example of a path provided by an extension of the planner above. $Path_1$ (top, left) has been computed in 2 seconds (including the preprocessing). The length of the sequence $Path_2$ (top, right) is 4; it has been computed in 151 seconds. The solution shown in the figure (bottom) required 4 iterations of the smoothing procedure and took 73 seconds.

In [23] one checks that

$$\{X, Y, [X, Y], [X, [X, Y]], [X, [X, [X, Y]]]\}$$

spans the tangent space at every point in $\mathbf{R}^2 \times (S^1)^3$ verifying $\varphi_1 \not\equiv \frac{\pi}{2}$ (regular points), while

$$\{X, Y, [X, Y], [X, [X, Y]], [X, [X, [X, [X, Y]]]]\}$$

spans the tangent space elsewhere (i.e., at singular points). The system is small-time controllable at any point. Its degree of nonholonomy is 4 at regular points and 5 at singular points. Therefore the algorithm runs in $O(\epsilon^{-5})$ in the worst case. This bound is reached in the case of a special parking task where the goal to reach would verify $\varphi_1 \equiv \frac{\pi}{2}$ in a very constrained space, while the complexity of the classical parallel parking task (i.e., $\theta = \varphi_1 = \varphi_2 = 0$) is in $O(\epsilon^{-4})$.

Very recently, it has been proven [26] that the degree of nonholonomy for a mobile robot with n trailers is $Fib(n+3) - 1$ where $Fib(i)$ is the i-th number of Fibonacci (defined by $Fib(i) = Fib(i-1) + Fib(i-2)$, with $Fib(0) = 0$ and $Fib(1) = 1$). This means that the complexity of a collision-free path for that system may be in $O(\epsilon^{-Fib(n+3)-1})$ in the worst case [2].

4 Conclusion and open problems

The nonholonomic motion planning presented in this paper is general. It can be applied to any small-time controllable systems so long as one gets a geometric planner and a method to compute optimal paths.

---

[1]This time includes the preprocessing; the path search itself takes 1 second.

[2]This complexity is directly related to the number of maneuvers; one better understands why the luggage carrier drivers in the airports do not play at parking their vehicles between two other ones...

Figure 3: *A first geometric path (top, left), its approximation before smoothing (top, right) and after (bottom).*

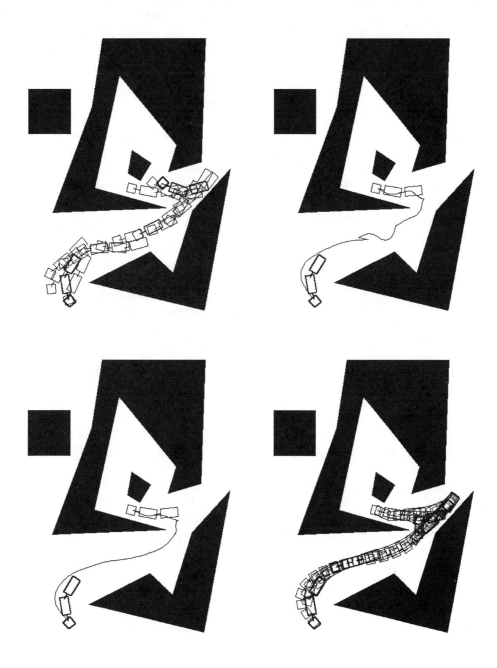

Figure 4: *A first geometric path (top, left), its approximation before smoothing (top, right) and after (bottom).*

From a practical point of view, the geometric planners based on an analytical representation of the free-space (e.g., those presented in Section 3.1) are limited to low dimensional spaces. Dealing with n-dimensional systems, for $n \geq 4$, often requires to apply alternative methods such as random methods [2, 35, 17] (e.g., the method presented in Section 3.2).

On the other hand, there is no general way to compute analytically optimal paths for nonholonomic robots. Reeds and Shepp's result is an exception. Nevertheless the alternative proofs of their results [42, 7] and the extensions [39, 8] use a combination of optimal control techniques and geometric tools; they open a way which seems to be very promising. In the absence of analytical results, one gets numerical methods to compute approximated optimal paths (see Section 3.2); making these methods more efficient constitutes another research issue.

Finally, the performance of our planner mainly depends on the geometric path computed at the first step. It could be interesting to introduced at this first step a heuristic search taking into account the kinematic constraints of the system.

The most advanced result that gathers in a first step the collision avoidance constraints and the kinematic ones has been developed by Mirtich and Canny for the car-like system [28]. Their clever idea consists in computing some Voronoi diagram; rather than using the Euclidean metric, they use the metric induced by the cost of the optimal paths. The associated skeleton may lead in infeasible directions. A second stage then consists in computing a feasible path by moving roughly "along" the skeleton through a sequence of optimal paths. Due to the property of the skeleton, this method tends to generate paths of low complexity (i.e., the length of the sequence is short). The method has been implemented for a point moving among polygonal obstacles [29]. Extending the approach to a polygonal robot requires the computation of the optimal paths which "crash" the robot against an obstacle; the problem is solved in [30].

To generalize the approach to any nonholonomic system, the main obstacle is the knowledge of the global shape of the singular metrics. At this moment, the only existing results either hold locally (see the literature in sub-Riemannian geometry, e.g., [6, 12, 31, 45]) or are known for special systems [25].

Acknowledgments

This work has been supported by the European Esprit 3 Program within the Project 6546 PROMotion. The idea and the results presented in this paper come from joint works with A. Bellaïche, P. Jacobs, S. Sekhavat, P. Souères and M. Taïx.

References

[1] F. Avnaim, J. Boissonnat, and B. Faverjon, "A practical exact motion planning algorithm for polygonal objects amidst polygonal obstacles," in *IEEE International Conference on Robotics and Automation*, Philadelphia, pp. 1656–1661, 1988.

[2] J. Barraquand, and J.C. Latombe, "Robot motion planning : a distributed representation approach," *International Journal of Robotics Research*, 10 (6), 1991.

[3] J. Barraquand, and J.C. Latombe, "Nonholonomic multibody mobile robots : controllability and motion planning in the presence of obstacles," *Algorithmica*, 10, pp. 121-155, 1993.

[4] A. Bellaïche, J.P. Laumond, and P. Jacobs, "Controllability of car-like robots and complexity of the motion planning problem," in *International Symposium on Intelligent Robotics*, pp. 322–337, Bangalore, India, January 1991.

[5] A. Bellaïche, J.P. Laumond, and M. Chyba, "Canonical nilpotent approximation of control systems : application to nonholonomic motion planning," in *32nd IEEE Conf. on Decision and Control*, December 15-17, 1993.

[6] A. Bellaïche, "Basic sub-Riemannian geometry," in *Spring School on Robot Motion Planning*, Rodez, 1993.

[7] J.D. Boissonnat, A. Cerezo and J. Leblond, "Shortest paths of bounded curvature in the plane," in *IEEE Conf. on Robotics and Automation*, pp. 2315–2320, Nice, France, 1992.

[8] X-N. Bui, P. Souères, J-D. Boissonnat, and J-P. Laumond, "Shortest path synthesis for Dubins nonholonomic robot," in *IEEE International Conference on Robotics and Automation*, pp. 2–7, San Diego, May 1994.

[9] G. Desaulniers, and F. Soumis, "An efficient algorithm to find a shortest path for a car-like robot," Les Cahiers du GERAD, G-93-18, Montréal, June 1993.

[10] C. Fernandes, L. Gurvitz, and Z.X. Li, "Optimal nonholonomic motion planning for a falling cat," in *Nonholonomic Motion Planning*, Z. Li and J.F. Canny Eds, Kluwer Academic Pub., 1993.

[11] S.J. Fortune and G.T. Wilfong, "Planning constrained motions," in *ACM STOCS*, pp. 445–459, Chicago IL, May 1988.

[12] V. Y. Gershkovitch, "Estimates for ϵ-balls of nonholonomic metrics," in *Global Analysis, Studies and Applications IV*, Y.G. Borisovitch, Y.E. Gliklikh Eds, Lecture Notes in Mathematics, Vol. 1453.

[13] I. Girard, "Etude d'un système de planification et de suivi de chemin pour un robot mobile articulé," M.Sc.A. Report, Ecole Polytechnique de Montréal, June 1993.

[14] G. Jacob, "Lyndon discretization and exact motion planning," in *European Control Conference*, pp. 1507–1512, Grenoble, France, 1991.

[15] P. Jacobs and J. Canny, " Planning smooth paths for mobile robots," in *IEEE International Conference on Robotics and Automation*, pp. 2–7, 1989.

[16] P. Jacobs, A. Rege, and J.P. Laumond, "Non-holonomic motion planning for Hilare-like robots," in *International Symposium on Intelligent Robotics*, pp. 338–347, Bangalore, India, January 1991.

[17] L. Kavraki, and J-C. Latombe, "Randomized preprocessing of configuration space for fast path planning," in *IEEE International Conference on Robotics and Automation*, San Diego, May 1994.

[18] G. Lafferriere and H. J. Sussmann, "Motion planning for controllable systems without drift: A preliminary report," Tech. Rep. SYSCON-90-04, Rutgers Center for Systems and Control, June 1990.

[19] J.C. Latombe, *Robot Motion Planning*, Kluwer Academic Pub., 1990.

[20] J.C. Latombe, "A Fast Path Planner for a Car-Like Indoor Mobile Robot," in *Ninth National Conference on Artificial Intelligence, AAAI*, pp. 659–665, Anaheim, CA, July 1991.

[21] J.P. Laumond, P. Jacobs, M. Taïx, and R. Murray, "A motion planner for nonholonomic mobile robot", LAAS/CNRS Report 92413, Toulouse, October 1992 (to appear in *IEEE Trans. on Robotics and Automation*).

[22] J.P. Laumond, "Controllability of a Multibody Mobile Robot," in *IEEE Transactions on Robotics and Automation*, Vol. 9, N° 6, pp. 755–763, December 1993.

[23] J.P. Laumond, "Singularities and topological aspects in nonholonomic motion planning," in *Nonholonomic Motion Planning*, Z. Li and J.F. Canny Eds, Kluwer Academic Pub., 1993.

[24] J.P. Laumond, S. Sekhavat, and M. Vaisset, "Collision-free motion planning for a nonholonomic mobile robot with a trailer", in *IFAC Symposium on Robot Control*, Capri, September 1994.

[25] J.P. Laumond, and P. Souères, "Metric induced by the shortest paths for a car-like mobile robot," in *IEEE/RSJ International Conference on Intelligent Robots and Systems*, pp. 1299–1303, July 1993.

[26] F. Luca, and J.J. Risler, "The maximum degree of nonholonomy for the car with n trailers," in *IFAC Symposium on Robot Control*, Capri, September 1994.

[27] *Nonholonomic Motion Planning*, Z. Li and J.F. Canny Eds, Kluwer Academic Pub., 1993.

[28] B. Mirtich, and J. Canny, "Using skeletons for nonholonomic motion planning among obstacles," in *IEEE Conf. on Robotics and Automation*, pp. 2533–2540, Nice, France, 1992.

[29] B. Mirtich, "Using skeletons for nonholonomic motion planning," Report ESRC 92-16/RAMP 92/6, Berkeley, June 1992.

[30] P. Moutarlier, B. Mirtich, and J. Canny, "The best way to crash a mobile robot," Working Paper, Berkeley, 1993.

[31] J. Mitchell, "On Carnot-Caratheodory metrics," *J. of Differential Geometry*, Vol. 21, pp. 35–45, 1985.

[32] S. Monaco and D. Normant-Cyrot, "An introduction to motion planning under multirate control,"in *Proc. of the CDC*, pp. 1780–1785, 1992.

[33] R.M. Murray and S. Sastry, "Steering nonholonomic systems using sinusoids," in *Proc. of the CDC*, pp. 2097–2101, 1990.

[34] R.M. Murray and S. Sastry, "Steering nonholonomic systems in chained forms," in *Proc. of the CDC*, pp. 1121–1126, 1991.

[35] M. H. Overmars, "A random approach to motion planning," Tech. Rep. RUU-CS-92-32, Utrecht Univ., October 1992.

[36] J. A. Reeds and R. A. Shepp, "Optimal paths for a car that goes both forward and backwards," *Pacific Journal of Mathematics*, 145 (2), pp. 367–393, 1990.

[37] P. Rouchon, M. Fliess, J. Lévine and P. Martin, "Flatness and motion planning : the car with n trailers," in *European Control Conference*, pp. 1518–1522, 1993.

[38] O.J. Sordalen, "Conversion of the kinematics of a car with n trailers into a chained form," in *IEEE Conf. on Robotics and Automation*, 1993.

[39] P. Souères, J.P. Laumond, "Shortest path synthesis for a car-like robot," in *European Control Conference*, Groningen, June 1993.

[40] H. Sussmann, "Lie brackets, real analyticity and geometric control," in *Differential Geometric Control Theory* (R. Brockett, R. Millman, and H. Sussmann, eds.), vol. 27 of *Progress in Mathematics*, pp. 1–116, Michigan Technological University, Birkhauser, June 28 – July 2 1982.

[41] H. J. Sussmann and W. Liu, "Limits of highly oscillatory controls and the approximation of general paths by admissible trajectories," Tech. Rep. SYSCON-91-02, Rutgers Center for Systems and Control, February 1991.

[42] H.J. Sussmann and W. Tang, "Shortest paths for the Reeds-Shepp car : a worked out example of the use of geometric techniques in nonlinear optimal control," Report SYCON-91-10, Rutgers University, 1991.

[43] D. Tilbury, J.P. Laumond, R. Murray, S. Sastry and G. Walsh, "Steering car-like systems with trailers using sinusoids," in *IEEE Conf. on Robotics and Automation*, pp. 1993–1998, Nice, France, 1992.

[44] D.Tilbury, R. Murray and S. Sastry, "Trajectory generation for the n-trailer problem using Goursat normal form," Memo No UCB/ERL M93/12, Berkeley, Feb. 1993.

[45] A.M. Vershik and V.Y. Gershkovich, "Nonholonomic problems and the theory of distributions," *Acta Applicandae Mathematicae*, Vol. 12, pp. 181–209, 1988.

Stable Pushing: Mechanics, Controllability, and Planning

Kevin M. Lynch, *Carnegie Mellon University, Pittsburgh, PA, USA*
Matthew T. Mason, *Carnegie Mellon University, Pittsburgh, PA, USA*

We would like to give robots the ability to position and orient parts in the plane by pushing, particularly when the parts are too large or heavy to be grasped and lifted. Unfortunately, the motion of a pushed object is generally unpredictable due to unknown support friction forces. With multiple pushing contact points, however, it is possible to find pushing directions which cause the object to remain fixed to the manipulator. These are called stable pushing directions. In this paper, we consider the problem of planning pushing paths using stable pushes. The space of stable pushing directions imposes a set of nonholonomic constraints on the motion of the manipulator, and we study the resulting issues of local and global controllability of the configuration of the object. We describe a planner for finding stable pushing paths among obstacles, and experimental results are given.

1 Introduction

Pushing provides a simple and practical solution to the problem of positioning and orienting objects in the plane, particularly when the manipulator lacks the size, strength, or dexterity to grasp and lift them. For instance, we often use pushing to rearrange furniture. We would like to give robots the ability to position and orient parts in the plane by pushing.

Unfortunately, the precise motion of a pushed object is generally unpredictable due to the indeterminacy of the distribution of support forces and ambiguities arising from the Coulomb model of friction. If there are two or more pushing contacts, however, there may exist a space of pushing directions which admit only a single solution to the motion of the object: the motion which causes the object to maintain its configuration relative to the pusher. The object is effectively rigidly attached to the pusher, and the push is called a stable push (Lynch [31]). A pushing path is formed by stringing together stable pushes, as in Figure 1.

Our goal is to develop algorithms to automatically find pushing plans to position and orient parts in the plane. Toward this end, in this paper we study the following three issues in pushing:

Mechanics. How does an object move when it is pushed? We are particularly interested in identifying the set of stable pushing directions when the pusher makes line contact with the object.

Controllability. The directions an object can move during pushing are limited due to the limited set of forces that can be applied by the pusher. Given these limitations, our study of controllability is motivated by questions of whether or not it is possible to push the object to the goal configuration, with and without obstacles. We examine the local and global controllability of objects pushed with either point contact or stable line contact.

Planning. Pushing paths consist of sequences of stable pushes, and the space of stable pushing directions imposes nonholonomic constraints on the motion of the manipulator. We draw on recent work on path planning for nonholonomic mobile robots by Barraquand and Latombe [6] to construct a planner to find stable pushing paths among obstacles.

1.1 Related work on pushing

Mason [37] identified pushing as an important manipulation process for manipulating several objects at once, for reducing uncertainty in part orientation, and as a precursor to grasping. Building on some early work by Prescott [49] and MacMillan [35], Mason implemented a numerical routine to find the motion of an object with a known support distribution being pushed at a single

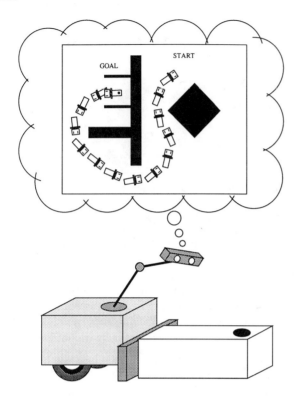

Figure 1: *A mobile robot pushing a box using stable pushes with line contact.*

point of contact. Recognizing that the support distribution is usually unknown, Mason derived a simple rule for determining the rotation sense of the pushed object which depends only on the center of mass of the object. Mason and Brost [39] and Peshkin and Sanderson [46] followed this work by finding bounds on the rotation rate of the pushed object. Goyal *et al.* [17, 18] studied the relationship between the motion of the sliding object and the associated support friction when the support distribution is completely specified. Alexander and Maddocks [3] considered the other extreme, when only the geometric extent of the support area is known, and describe techniques to bound the possible motions of the pushed object.

These results have been used by to plan manipulator pushing operations. Mason [37] used pushing and grasping to reduce uncertainty. Mani and Wilson [36] built a system for orienting a part in an initially unknown orientation by executing a series of linear pushes with a fence. Peshkin and Sanderson [47], and later Brokowski *et al.* [7], considered a similar problem,

where the part is carried on a conveyor belt and reoriented by interactions with fences suspended above the belt. Brost [8] developed algorithms to identify stable parallel-jaw grasping motions for polygonal objects in the presence of uncertainty, and Goldberg [16] developed a planner to find a sequence of parallel-jaw grasps to orient a polygonal part. Brost [9] finds the linear pushing motions resulting in a goal pusher/object equilibrium configuration by integrating the object's motion with respect to the pusher. This is like "catching" the object by pushing it. Balorda [4, 5] has investigated catching by pushing with two points of contact. Mason [38] has shown how to synthesize robot pushing motions to slide a block along a wall, a problem later studied by Mayeda and Wakatsuki [40] considering pushing forces out of the plane.

Feedback control of the motion of an object pushed with a single point of contact has been studied by many researchers, including Gandolfo *et al.* [15], Inaba and Inoue [22], Latombe and colleagues [57], Lynch *et al.* [34], Okawa and Yokoyama [45], and Salganicoff *et al.* [51]. A control strategy for pushing by two cooperating mobile robots is described by Donald *et al.* [11]. Learning (Miura [42], Salganicoff *et al.* [52], Zrimec [59]) and friction parameter estimation (Yoshikawa and Kurisu [58], Lynch [32]) have also been proposed to improve control.

Particularly relevant to the topic of this paper is work by Akella and Mason [1] and Narasimhan [43]. Akella and Mason have considered the problem of planning pushing sequences to reconfigure polygonal parts in the obstacle-free plane. Each pusher motion is a linear motion, and the object can rotate without slipping on the straight-edge of the pusher. The precise motion of the object during the push is unknown, but at the end of each push, a known edge of the object will be aligned with the pusher. Between pushes, the pusher breaks contact and recontacts the object. The approach is guaranteed to find an open-loop plan for any polygonal object and any initial and goal configuration.

Narasimhan [43] has studied the problem of moving an object among obstacles by pushing with point contact. The first step is to find a holonomic path connecting the start and goal configurations. Then the robot guides the object along the path using visual feedback. At each control step, the robot chooses the best push from its model. The model, built up from simulated or experimental data, attempts to describe the motion of the object as a function of the contact location and pushing direction.

The results of this paper build on earlier work reported in (Lynch [33]).

1.2 Assumptions

1. Friction is assumed to conform to Coulomb's law. The frictional force at a sliding contact opposes the motion with magnitude $\mu_k f_n$, where f_n is the magnitude of the normal contact force and μ_k is the kinetic coefficient of friction. At a sticking contact, the frictional force can act in any tangential direction with any magnitude less than or equal to μf_n, where μ is the static coefficient of friction. For simplicity, we will assume that the static and kinetic coefficients of friction are equal.

2. All pushing forces lie in the horizontal support plane, and gravity acts along the vertical.

3. The pusher and slider move in the horizontal plane.

4. Friction properties are uniform over the support plane.

5. Pushing motions are slow enough that inertial forces are negligible. This is the quasi-static assumption. Pushing forces are always balanced by the support frictional forces acting on the object.

1.3 Definitions

The slider $\mathcal{S}$ is a rigid object in the plane $\mathcal{W} = \mathbf{R}^2$, and its configuration space $\mathcal{C}$ is $\mathbf{R}^2 \times S^1$. The slider is pushed by a rigid pusher $\mathcal{P}$ at a point or set of points on a closed, piecewise smooth curve Γ, which typically forms the perimeter of the slider $\mathcal{S}$. A world frame $\mathcal{F}_\mathcal{W}$ with origin $O_\mathcal{W}$ is fixed in the plane, and a slider frame $\mathcal{F}_\mathcal{S}$ with origin $O_\mathcal{S}$ is attached to the center of friction of the slider $\mathcal{S}$. For a uniform coefficient of support friction, the center of friction of the slider is the point in the support plane beneath the center of mass (MacMillan [35]). Configurations measured in the slider frame $\mathcal{F}_\mathcal{S}$ have coordinates $(x, y, \theta)^T$. The configuration $\mathbf{q} = (x_w, y_w, \theta_w)^T$ describes the position and orientation of the slider frame $\mathcal{F}_\mathcal{S}$ relative to the world frame $\mathcal{F}_\mathcal{W}$. See Figure 2.

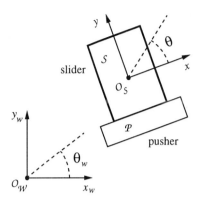

Figure 2: *The world frame $\mathcal{F}_\mathcal{W}$ and the slider frame $\mathcal{F}_\mathcal{S}$.*

Forces $\mathbf{f}$ and velocities $\mathbf{v}$ are always defined with respect to the slider frame $\mathcal{F}_\mathcal{S}$. A force $\mathbf{f} \in \mathbf{R}^3$ is given by its force and moment components $(f_x, f_y, m)^T$. A nonzero force $\mathbf{f}$ is the product of its magnitude f and its direction $\hat{\mathbf{f}}$. A force direction is a three-dimensional unit vector and may be represented as a point on the unit sphere ($\hat{\mathbf{f}} \in S^2$). The sphere of force directions is called the force sphere. Similarly, a nonzero velocity $\mathbf{v} = (v_x, v_y, \omega)^T$ is given by the product of its magnitude v and direction $\hat{\mathbf{v}}$, and the sphere of velocity directions is called the velocity sphere. We will sometimes represent a velocity direction by its corresponding center of rotation. The function $\mathrm{COR}(\cdot)$ maps velocity directions to rotation centers in the slider frame $\mathcal{F}_\mathcal{S}$, such that $\mathrm{COR}(\hat{\mathbf{v}})$ returns the point about which the velocity direction $\hat{\mathbf{v}}$ is a pure rotation, along with the sense of rotation. The domain of the function $\mathrm{COR}(\cdot)$ is the velocity sphere, and the range consists of two copies of the plane, one for each rotation sense, and a line at infinity for translations. Figure 3 illustrates the mapping from velocity directions to rotation centers.

For the quasi-static pushing problem, we are only concerned with force and velocity directions, not their magnitudes. We assume only that the manipulator is strong enough to move the slider, and that it moves slowly enough to satisfy the quasi-static assumption. A pushing plan generated by the planner in Section 4 may be properly thought of as a pushing path, not a trajectory. To generate a manipulator trajectory from this path, times must be assigned to each point along the path such that the quasi-static assumption is satisfied.

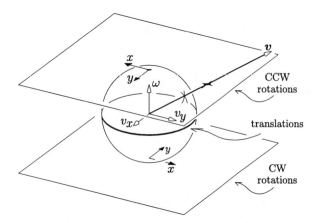

Figure 3: *The mapping COR(·) from velocity directions on the unit sphere to rotation centers in the slider frame $\mathcal{F}_S$.*

1.4 Overview

In the next section we define the pushing control system, some basic definitions of controllability, and their application to the pushing control system. Armed with these tools, in Section 3 we study the mechanics of pushing and the controllability of objects pushed with either point contact or line contact. Finally, Section 4 presents an approach to synthesizing manipulator plans for positioning and orienting objects among obstacles using stable pushes with line contact.

2 Controllability of planar objects with velocity constraints

The set of velocity directions which the slider can follow during pushing is limited due to the limited set of force directions which can be applied by the pusher. These limitations constitute a set of nonholonomic constraints on the motion of the slider. In addition, it is clear that a slider that can be pushed in one direction cannot be pulled in the opposite direction by simply reversing the motion of the pusher. Despite these constraints on the motion, we know by experience that it is often possible to move objects to desired configurations by pushing. In this section we formalize these ideas using tools from nonlinear control theory.

2.1 The pushing control system

The pushing control system can be described abstractly by the autonomous nonlinear control system

$$\dot{\mathbf{q}} = F(\mathbf{q}, \mathbf{c}),$$

where $\mathbf{c}$ is the control input describing the pushing contact configuration and the velocity of the pusher in the slider frame $\mathcal{F}_S$. The motion of the slider in the world frame $\mathcal{F}_W$ is a function F of the control input and the configuration of the slider. For the rest of this paper, we will use the following more concrete description of the control system:

$$\Sigma : \quad \dot{\mathbf{q}} = F(\mathbf{q}, u) = X_u(\mathbf{q}),$$

$$\mathbf{q} \in \mathcal{C} = \mathbf{R}^2 \times S^1, \quad u \in \{0, \ldots, n\},$$

$$X_u(\mathbf{q}) = \begin{cases} (0,\ 0,\ 0)^T & \text{if } u = 0 \\ \begin{pmatrix} \cos\theta_w & -\sin\theta_w & 0 \\ \sin\theta_w & \cos\theta_w & 0 \\ 0 & 0 & 1 \end{pmatrix} \begin{pmatrix} \hat{v}_{ux} \\ \hat{v}_{uy} \\ \hat{\omega}_u \end{pmatrix} & \text{otherwise.} \end{cases}$$

A nonzero control u chooses one of n distinct combinations of contact configurations and pushing velocities in the slider frame $\mathcal{F}_S$. Associated with each control u is a vector field X_u describing the motion of the slider in the world frame $\mathcal{F}_W$. The tangent vector $X_u(\mathbf{q})$ is the (unit) velocity of the slider in the world frame $\mathcal{F}_W$, and $\hat{\mathbf{v}}_u$ is the (unit) velocity of the slider in the slider frame $\mathcal{F}_S$. The set of nonzero vector fields X_u is denoted $\mathcal{X}$, and the set of nonzero velocity directions $\hat{\mathbf{v}}_u$ in the slider frame $\mathcal{F}_S$ is $\hat{\mathcal{V}}$. Each of the n nonzero controls results in a distinct velocity direction.

Three aspects of the control system Σ bear mentioning. (1) The absence of a drift vector field (a vector field which is only a function of the state of the slider) implies that the slider will not move when it is not pushed ($u = 0$). This is a consequence of the quasi-static assumption: the slider's kinetic energy is instantly dissipated by support friction, and the slider's state is merely its configuration. (2) For any constant control, the slider's velocity direction is constant in the slider frame $\mathcal{F}_S$. (3) The control system is not necessarily symmetric. In general, it is not possible to follow a vector field X_u backwards. This is due to the unilateral nature of frictional contact: pushing forces must have a nonnegative component in the direction of the contact normal, and the ability to push the slider forward does not guarantee the ability to pull the slider backward. (Of course, with a proper choice of pushing contact configurations and velocities, it may be the case that both X_u and $-X_u$ are in the set of n vector fields for all u, making the system effectively symmetric, but this is not the case in general.)

2.2 Definitions of controllability

We will adopt Sussmann's terminology [54] to describe some basic concepts of reachability and controllability as they apply to the pushing control system. For a pushing control system Σ, a slider configuration $\mathbf{q}$, and a time $T \geq 0$, the set of configurations which the slider can reach from $\mathbf{q}$ at time T is denoted $A_\Sigma(T, \mathbf{q})$. The set $A_\Sigma(\leq T, \mathbf{q})$ is the union of the sets of reachable configurations $A_\Sigma(t, \mathbf{q})$ for $0 \leq t \leq T$. The set of all reachable configurations from $\mathbf{q}$ is denoted simply $A_\Sigma(\mathbf{q})$, the union of $A_\Sigma(t, \mathbf{q})$ for $0 \leq t < \infty$.

The pushing control system Σ, or equivalently the configuration of the slider $\mathcal{S}$, is *controllable from* $\mathbf{q}$ if $A_\Sigma(\mathbf{q}) = \mathcal{C}$. The configuration is *locally controllable from* $\mathbf{q}$ if $A_\Sigma(\mathbf{q})$ contains a neighborhood of $\mathbf{q}$, and the configuration is *small-time locally controllable from* $\mathbf{q}$ if $A_\Sigma(\leq T, \mathbf{q})$ contains a neighborhood of $\mathbf{q}$ for all $T > 0$. (The local controllability property has been called *global-local controllability* by Haynes and Hermes [19], and small-time local controllability has been referred to as *local-local controllability* by Haynes and Hermes [19] and *local controllability* by Hermann and Krener [20] and Sussmann [53].) The configuration is *accessible from* $\mathbf{q}$ if $A_\Sigma(\mathbf{q})$ has nonempty interior in the configuration space $\mathcal{C}$, and it is *small-time accessible from* $\mathbf{q}$ if $A_\Sigma(\leq T, \mathbf{q})$ has nonempty interior for all $T > 0$. (The accessibility property has been called *weak controllability* by Hermann and Krener [20], and small-time accessibility has been referred to as *local weak controllability* by Hermann and Krener [20], *accessibility* by Sussmann and Jurdjevic [56], and *local accessibility* by Nijmeijer and van der Schaft [44].)

The definitions of small-time local controllability and small-time accessibility are applied to the pushing control system under the assumption that the control u can be changed instantaneously from any control to any other control. We can drop this assumption if we use the following essentially equivalent definitions. The slider is small-time locally controllable from $\mathbf{q}$ if, for any neighborhood U of $\mathbf{q}$, the set of accessible configurations without leaving U contains a neighborhood of $\mathbf{q}$. The slider is small-time accessible from $\mathbf{q}$ if, for any neighborhood U of $\mathbf{q}$, the set of accessible configurations without leaving U has nonempty interior.

If a controllability property holds for all $\mathbf{q} \in \mathcal{C}$, the phrase "from $\mathbf{q}$" can be omitted. Because the configuration space $\mathcal{C}$ is everywhere the same and the set of feasible slider velocities is fixed in the slider frame $\mathcal{F}_\mathcal{S}$, any property which holds for any $\mathbf{q} \in \mathcal{C}$ also holds for all $\mathbf{q} \in \mathcal{C}$. Similarly, any property which does not hold for some $\mathbf{q} \in \mathcal{C}$ does not hold for any $\mathbf{q} \in \mathcal{C}$.

By definition, controllability implies local controllability and accessibility, and small-time local controllability implies small-time accessibility. By the connectivity of the configuration space $\mathcal{C}$, small-time local controllability implies controllability and small-time accessibility implies accessibility. For autonomous linear systems, all of these concepts are equivalent (Hermann and Krener [20]).

Of these properties, small-time accessibility can be established by an algebraic test on the set of vector fields $\mathcal{X}$. The Lie algebra $L(\mathcal{X})$ of this set of vector fields is the space of linear combinations of these vector fields and the vector fields created by repeated Lie bracket operations. The Lie bracket of the vector fields X and Y is denoted $[X, Y]$. Defining $B_0(\mathcal{X}) = \mathcal{X}$ and $B_{k+1}(\mathcal{X}) = B_k(\mathcal{X}) \cup \{[X, Y]$ for all $X, Y \in B_k(\mathcal{X})\}$, then the Lie algebra $L(\mathcal{X})$ is spanned by vector fields in $B_\infty(\mathcal{X})$. A control system is small-time accessible from $\mathbf{q}$ if it satisfies the *Lie Algebra Rank Condition*, which states that the dimension of the space spanned by the tangent vectors at $\mathbf{q}$ of vector fields in $L(\mathcal{X})$ must equal the dimension of the state manifold (see, for example, Hermann and Krener [20]).

The Lie bracket $[X, Y]$ of the vector fields X and Y in local coordinates is

$$[X, Y](\mathbf{q}) = \frac{\partial Y(\mathbf{q})}{\partial \mathbf{q}} X(\mathbf{q}) - \frac{\partial X(\mathbf{q})}{\partial \mathbf{q}} Y(\mathbf{q}).$$

(See Nijmeijer and van der Schaft [44] for a derivation.) For the pushing control system Σ, $\partial X(\mathbf{q})/\partial \mathbf{q}$ is given by

$$\begin{pmatrix} \partial X_{x_w}(\mathbf{q})/\partial x_w & \partial X_{x_w}(\mathbf{q})/\partial y_w & \partial X_{x_w}(\mathbf{q})/\partial \theta_w \\ \partial X_{y_w}(\mathbf{q})/\partial x_w & \partial X_{y_w}(\mathbf{q})/\partial y_w & \partial X_{y_w}(\mathbf{q})/\partial \theta_w \\ \partial X_{\theta_w}(\mathbf{q})/\partial x_w & \partial X_{\theta_w}(\mathbf{q})/\partial y_w & \partial X_{\theta_w}(\mathbf{q})/\partial \theta_w \end{pmatrix},$$

where

$$X = (X_{x_w}, X_{y_w}, X_{\theta_w})^T.$$

Using the definition of X_u from Section 2.1, $\partial X_u(\mathbf{q})/\partial \mathbf{q}$ evaluates simply to

$$\begin{pmatrix} 0 & 0 & -\hat{v}_{ux}\sin\theta_w - \hat{v}_{uy}\cos\theta_w \\ 0 & 0 & \hat{v}_{ux}\cos\theta_w - \hat{v}_{uy}\sin\theta_w \\ 0 & 0 & 0 \end{pmatrix}.$$

For the control system Σ, the Lie algebra $L(\mathcal{X})$ is spanned by vector fields in $B_1(\mathcal{X})$: it is only necessary to look at the vector fields $\mathcal{X}$ and their Lie brackets in order to decide small-time accessibility.

It is easily shown that if the control system is small-time accessible and symmetric (all vector fields can be followed forward and backward), then it is also small-time locally controllable. Sussmann [55] proves this and other sufficient conditions for small-time local controllability.

2.3 Controllability of the pushing control system

Here we study the controllability properties of the control system Σ. We would like to find necessary and sufficient conditions for the various forms of controllability. We begin by studying accessibility.

If $n = 1$ for the control system Σ, then the slider is confined to a one-dimensional integral curve of X_1, and the control system Σ is not accessible. If $n = 2$, a basis of the Lie algebra $L(\mathcal{X})$ is given by X_1, X_2, and $X_3 = [X_1, X_2]$, where

$$X_1 = (\hat{v}_{1x}\cos\theta_w - \hat{v}_{1y}\sin\theta_w, \hat{v}_{1x}\sin\theta_w + \hat{v}_{1y}\cos\theta_w, \hat{\omega}_1)^T,$$

$$X_2 = (\hat{v}_{2x}\cos\theta_w - \hat{v}_{2y}\sin\theta_w, \hat{v}_{2x}\sin\theta_w + \hat{v}_{2y}\cos\theta_w, \hat{\omega}_2)^T,$$

$$X_3 = [X_1, X_2] =$$

$$(\hat{\omega}_1(-\hat{v}_{2x}\sin\theta_w - \hat{v}_{2y}\cos\theta_w) + \hat{\omega}_2(\hat{v}_{1x}\sin\theta_w + \hat{v}_{1y}\cos\theta_w),$$

$$\hat{\omega}_1(\hat{v}_{2x}\cos\theta_w - \hat{v}_{2y}\sin\theta_w) + \hat{\omega}_2(-\hat{v}_{1x}\cos\theta_w + \hat{v}_{1y}\sin\theta_w),$$

$$0)^T.$$

The dimension of $L(\mathcal{X})$ is given by $\text{rank}(X_1 \ X_2 \ X_3)$. If the determinant of this matrix is nonzero, then its rank is three. A simple calculation yields

$$\det(X_1 \ X_2 \ X_3) = (\hat{\omega}_2\hat{v}_{1x} - \hat{\omega}_1\hat{v}_{2x})^2 + (\hat{\omega}_2\hat{v}_{1y} - \hat{\omega}_1\hat{v}_{2y})^2.$$

The determinant is only zero if (1) $\hat{\omega}_1 = \hat{\omega}_2 = 0$ or (2) $\hat{\mathbf{v}}_1$ and $\hat{\mathbf{v}}_2$ are multiples of each other. (Since $\hat{\mathbf{v}}_1$ and $\hat{\mathbf{v}}_2$ are distinct unit vectors, this condition is equivalent

to $\hat{\mathbf{v}}_1 = -\hat{\mathbf{v}}_2$.) If condition (1) holds, the slider cannot rotate. If condition (2) holds, the slider is confined to a one-dimensional curve of the configuration space $\mathcal{C}$. If neither of these conditions hold, the Lie bracket operation has essentially created a new linearly independent control vector field, the dimension of $L(\mathcal{X})$ is three, and the Lie Algebra Rank Condition is satisfied.

Proposition 1 *The control system Σ is small-time accessible if and only if the set of velocity directions $\hat{\mathcal{V}}$ of the slider $\mathcal{S}$ contains two velocity directions, $\hat{\mathbf{v}}_1$ and $\hat{\mathbf{v}}_2$, such that they are not both translations ($\hat{\omega}_1 \neq 0$ or $\hat{\omega}_2 \neq 0$) and $\hat{\mathbf{v}}_1 \neq -\hat{\mathbf{v}}_2$.*

Remark: As noted earlier, small-time accessibility implies accessibility on the configuration space $\mathcal{C}$. Here we note that any control system Σ which is accessible must also be small-time accessible. Thus the conditions of Proposition 1 are also necessary and sufficient for accessibility.

Proposition 1 is a straightforward generalization of a result due to Barraquand and Latombe [6] which states that the Lie Algebra Rank Condition is satisfied for any car-like mobile robot which can take at least two steering angles. A car-like mobile robot can drive both forward and backward, and this symmetry, coupled with small-time accessibility, implies small-time local controllability. As we have already noted, however, the pushing control system Σ may not be symmetric, so small-time local controllability does not follow from small-time accessibility.

For some systems, however, accessibility may imply controllability even when the system is not small-time locally controllable. Several authors, including Jurdjevic [24] and Jurdjevic and Sussmann [25], have identified control systems of this type. (Sussmann [54] gives other references.) Consider, for example, the task of setting the minute-hand of a watch if it can only be rotated in a clockwise direction. The configuration of the minute-hand is not small-time locally controllable, but the topology of its configuration manifold S^1 renders the minute-hand's configuration controllable.

For control systems Σ, it is easily shown that accessibility implies controllability on the configuration space $\mathcal{C}$, and the conditions of Proposition 1 are also necessary and sufficient for controllability.

Proposition 2 *For a control system Σ, accessibility implies controllability. The slider $\mathcal{S}$ may be moved*

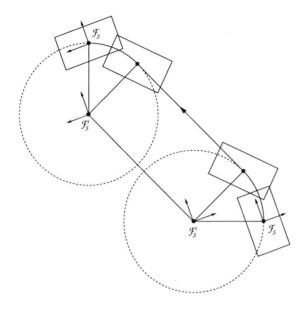

Figure 4: *An example path using one rotational and one translational velocity direction. The frames $\mathcal{F}_{\mathcal{S}}$ and $\mathcal{F}'_{\mathcal{S}}$ are drawn at the initial and final locations.*

from any configuration to any other configuration in the obstacle-free plane if and only if the set of velocity directions $\hat{\mathcal{V}}$ contains two velocity directions, $\hat{\mathbf{v}}_1$ and $\hat{\mathbf{v}}_2$, such that they are not both translations ($\hat{\omega}_1 \neq 0$ or $\hat{\omega}_2 \neq 0$) and $\hat{\mathbf{v}}_1 \neq -\hat{\mathbf{v}}_2$.

Proof: We first note that accessibility is a necessary condition for controllability. The following argument shows that it is also sufficient. First consider the case $\hat{\omega}_1 \neq 0, \hat{\omega}_2 = 0$. The slider may reach any configuration $\mathbf{q}_{goal}$ from any other configuration $\mathbf{q}_{init}$ by following X_1, then X_2, then X_1. The rotation center $\mathrm{COR}(\hat{\mathbf{v}}_1)$ is located at a point $R_1 = (-\hat{v}_{1y}/\hat{\omega}_1, \hat{v}_{1x}/\hat{\omega}_1)$ in the slider frame $\mathcal{F}_{\mathcal{S}}$. We define a new frame $\mathcal{F}'_{\mathcal{S}}$ with its origin at R_1. The frame $\mathcal{F}'_{\mathcal{S}}$ is aligned with and fixed in the slider frame $\mathcal{F}_{\mathcal{S}}$. In the frame $\mathcal{F}'_{\mathcal{S}}$, the two velocity directions are a pure rotation $\hat{\mathbf{v}}'_1 = (0, 0, sgn(\hat{\omega}_1))^T$ and a pure translation $\hat{\mathbf{v}}'_2 = \hat{\mathbf{v}}_2$, and the problem is to transfer the frame $\mathcal{F}'_{\mathcal{S}}$ from $\mathbf{q}'_{init}$ to $\mathbf{q}'_{goal}$. This is achieved by simply rotating the frame $\mathcal{F}'_{\mathcal{S}}$ in place, translating it to the final position, and rotating it to the final orientation. Any of these steps could have zero length. An example is shown in Figure 4.

If both $\hat{\mathbf{v}}_1$ and $\hat{\mathbf{v}}_2$ have nonzero angular components,

with rotation centers at R_1 and R_2 in the slider frame $\mathcal{F}_{\mathcal{S}}$, respectively, controllability can be demonstrated by showing that the slider can always translate to a configuration from which it can rotate to the goal. A translation is obtained by following X_1 and X_2 such that the total rotation is an integral multiple of 2π. The set of paths that follow X_1 and then X_2 and satisfy this condition defines a circle of final positions of the origin $O_{\mathcal{S}}$ of the slider frame $\mathcal{F}_{\mathcal{S}}$. The radius r of the circle is the distance between R_1 and R_2, and the center of the circle is a distance r from $O_{\mathcal{S}}$ on a ray from $O_{\mathcal{S}}$ parallel to the ray from R_2 through R_1. Translations consisting of paths following X_2 and then X_1 define a similar circle (see Figure 5). The locus of points to which the slider frame $\mathcal{F}_{\mathcal{S}}$ can translate with only a single control change is given by these two circles. Each point on the locus is the origin of a similar, translated locus, and the slider frame $\mathcal{F}_{\mathcal{S}}$ can translate to any point in the plane by concatenating translations with one control change. □

For a controllable system Σ with two velocity directions ($n = 2$), the slider may have to travel a long distance to reach nearby configurations. Thus the conditions of Proposition 2 are not sufficient for small-time local controllability. Before addressing the conditions for small-time local controllability, we establish the following fact.

Proposition 3 *Consider a set of velocity directions $\hat{\mathcal{V}}$ and its convex hull $\hat{\mathcal{V}}_{CH}$. Any path from $\mathbf{q}_1$ to $\mathbf{q}_2$ using velocity directions in $\hat{\mathcal{V}}_{CH}$ can be followed arbitrarily closely by another path, also from $\mathbf{q}_1$ to $\mathbf{q}_2$, using only velocity directions in $\hat{\mathcal{V}}$.*

Proof: This proposition states that bang-bang control using extremal velocity directions suffices to follow any feasible path between two configurations arbitrarily closely. The proof proceeds by showing that bang-bang control is sufficient when the velocity direction set is the set of convex combinations of two velocity directions. By induction we can then show that bang-bang control is sufficient for the set of convex combinations of any number of velocity directions. We will not give the full proof here, but we point the reader to the proof of Proposition 5 in Appendix B of (Barraquand and Latombe [6]), which demonstrates the sufficiency of bang-bang control for the set of convex combinations of velocity directions $\hat{\mathbf{v}}_1$ and $\hat{\mathbf{v}}_2$ such that at least one of these is not a translation ($\hat{\omega}_1 \neq 0$). The proof for

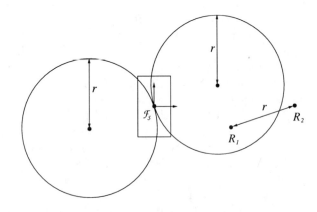

Figure 5: *The locus of points to which the slider frame $\mathcal{F}_S$ can translate, using one control change and rotation centers R_1 and R_2 in the slider frame $\mathcal{F}_S$. The left circle results from rotating about R_1 and then R_2 in the slider frame $\mathcal{F}_S$, and the right circle is obtained by reversing the order.*

the case of two translation directions ($\hat{\omega}_1 = \hat{\omega}_2 = 0$) is trivial.   □

On any open set of the configuration space $\mathcal{C}$, Proposition 3 says that we can consider the available velocity directions to be the convex hull of the velocity direction set $\hat{\mathcal{V}}$. Therefore, if the set of velocity directions $\hat{\mathcal{V}}$ contains four velocity directions that positively span the velocity sphere, the configuration of the slider $\mathcal{S}$ is small-time locally controllable. This also follows from the following theorem:

Theorem 1 (Sussmann [53]) *Let $\mathcal{X}$ be a finite set of vector fields on an open set of the state manifold containing $\mathbf{q}$. The set of nonzero tangent vectors at $\mathbf{q}$ is denoted $\mathcal{X}(\mathbf{q})$. Then*
(1) If $\mathbf{0}$ is in the interior of the convex hull of $\mathcal{X}(\mathbf{q})$, the system is small-time locally controllable from $\mathbf{q}$.
(2) If $\mathbf{0}$ does not belong to the convex hull of $\mathcal{X}(\mathbf{q})$, the system is not small-time locally controllable from $\mathbf{q}$.

To apply this theorem, we define the convex hull in $\mathbf{R}^3$ of a set of unit velocities $\mathcal{X}(\mathbf{q})$ to be $CH_{\mathbf{R}^3}(\mathcal{X}(\mathbf{q}))$ with boundary $\partial CH_{\mathbf{R}^3}(\mathcal{X}(\mathbf{q}))$.

The first half of Theorem 1 indicates that if the velocity directions $\hat{\mathcal{V}}$ of the control system Σ positively span the velocity sphere, the control system Σ is small-time locally controllable. The second half of the theorem says that if the velocity directions $\hat{\mathcal{V}}$ are confined

to any open hemisphere of the velocity sphere, then Σ is not small-time locally controllable. Theorem 1 does not completely answer the question of small-time local controllability at $\mathbf{q}$, however, as it does not address the case when $\mathbf{0}$ lies in $\partial CH_{\mathbf{R}^3}(\mathcal{X}(\mathbf{q}))$. To resolve this case, we must consider the derivatives of $\mathcal{X}(\mathbf{q})$ (Sussmann [53]).

If $\mathbf{0}$ lies in $\partial CH_{\mathbf{R}^3}(\mathcal{X}(\mathbf{q}))$, then there is a unique linear subspace Z of maximum dimension such that $\mathbf{0}$ lies in the interior of $\partial CH_{\mathbf{R}^3}(\mathcal{X}(\mathbf{q}))$ in this space Z. The set $\mathcal{X}_Z$ is the subset of vector fields $X \in \mathcal{X}$ such that the tangent vector $X(\mathbf{q})$ lies in the subspace Z. The dimension of the subspace Z is one if the portion of $\partial CH_{\mathbf{R}^3}(\mathcal{X}(\mathbf{q}))$ through $\mathbf{0}$ is a line segment, and $\mathcal{X}_Z$ consists of two opposite vector fields. The dimension of the subspace Z is two if the portion of $\partial CH_{\mathbf{R}^3}(\mathcal{X}(\mathbf{q}))$ through $\mathbf{0}$ is a planar region, and $\mathcal{X}_Z$ consists of at least three vector fields such that the associated velocity directions span a great circle of the velocity sphere.

The set of all vector fields $[X, Y]$ such that $X, Y \in \mathcal{X}_Z$ is denoted $\mathcal{X}_Z^1$. Applying Sussmann's sufficient condition for small-time local controllability [53], the control system Σ is small-time locally controllable if the convex hull of $\mathcal{X}(\mathbf{q})$ and $\mathcal{X}_Z^1(\mathbf{q})$ contains $\mathbf{0}$ in its interior. For the control system Σ, Sussmann's sufficient condition is also necessary.

When the dimension of Z is one, the set of vector fields $\mathcal{X}_Z$ consists of two opposite vector fields. The Lie bracket of opposite vector fields is zero, so $\mathcal{X}_Z^1$ consists of the zero vector field. The convex hull of $\mathcal{X}(\mathbf{q})$ and $\mathcal{X}_Z^1(\mathbf{q})$ does not contain $\mathbf{0}$ in its interior, and the control system Σ is not small-time locally controllable.

Now consider the case where the dimension of the subspace Z is two. In this case, the velocity directions corresponding to $\mathcal{X}_Z$ positively span a great circle of the velocity sphere. Provided that this great circle does not lie in the $\omega = 0$ plane, the control system Σ is small-time locally controllable. To see this, recall that two nonopposite velocity directions which are not both translations are sufficient for small-time accessibility. If both velocity directions can be reversed (a total of four velocity directions), then the system is small-time locally controllable. These velocity directions positively span a great circle of the velocity sphere such that $\hat{\omega}$ is not identically zero. By Proposition 3, on any open set of the configuration space $\mathcal{C}$, any set of velocity directions which span the same great circle is equivalent.

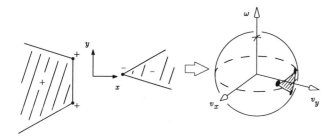

Figure 6: *The convex hull of three velocity directions, represented in the rotation center space and on the velocity sphere.*

Proposition 4 *The control system Σ is small-time locally controllable if and only if the set of velocity directions $\hat{\mathcal{V}}$ positively spans a great circle of the velocity sphere which does not lie in the $\omega = 0$ plane.*

This result is stated in terms of velocity directions on the velocity sphere, but it can just as easily be stated in terms of rotation centers. The positive linear combination of two rotation centers of the same sense is given by the line segment of rotation centers of the same sense between the points. The positive linear combination of rotation centers of opposite senses is given by all points on the line through the two rotation centers but not between them. The sense of the rotation centers changes at infinity, which corresponds to a translation (see Figure 6). A great circle on the velocity sphere is equivalent to a line of rotation centers with both rotation senses. Figure 7 gives examples of rotation center sets which yield small-time local controllability.

From Propositions 2 and 4, the following corollary follows immediately.

Corollary 1 *The number of distinct combinations n of pushing contact configurations and pushing directions must be (1) at least two for the configuration of the slider $\mathcal{S}$ to be controllable by pushing, and (2) at least three for the configuration of the slider $\mathcal{S}$ to be small-time locally controllable by pushing. These bounds are tight.*

3 Mechanics and controllability of pushed objects

In this section we study the mechanics problem of determining the motion of a pushed object. Using these

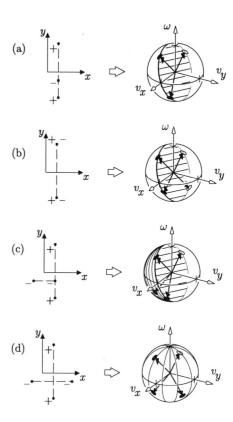

Figure 7: *Examples of rotation center sets which yield small-time local controllability. (a) These three rotation centers positively span a great circle of the velocity sphere. (b) These four rotation centers positively span a great circle of the velocity sphere. (c) These four rotation centers positively span a closed hemisphere of the velocity sphere. (d) These four rotation centers positively span the velocity sphere.*

results and those of the previous section, we elucidate the controllability properties of objects pushed with either point contact or stable line contact.

3.1 Mechanics of pushing

During quasi-static pushing, the force applied by the pusher is balanced by the frictional forces applied by the support surface to the slider. Equivalently, the force applied by the pusher is equal to the frictional force which the slider applies to the support plane. The frictional force $\mathbf{f}$ which the slider applies to the support plane when moving with velocity $\mathbf{v}$ will be expressed in terms of the following:

x	point of contact $(x, y, 0)^T$ between the slider $\mathcal{S}$ and the support surface in the slider reference frame $\mathcal{F}_\mathcal{S}$
$d\mathbf{x}$	differential element of support area
$p(\mathbf{x})$	support pressure at $\mathbf{x}$
$\mu_s(\mathbf{x})$	support friction at $\mathbf{x}$
$s(\mathbf{x})$	the product of $p(\mathbf{x})$ and $\mu_s(\mathbf{x})$, referred to as the support friction distribution
$\mathbf{v}(\mathbf{x})$	the linear velocity of $\mathbf{x}$, given by $(v_x - \omega y, v_y + \omega x, 0)^T$, where the slider velocity $\mathbf{v}$ is $(v_x, v_y, \omega)^T$
$\mathbf{f}_{xy}$	the linear components $(f_x, f_y, 0)^T$ of $\mathbf{f}$

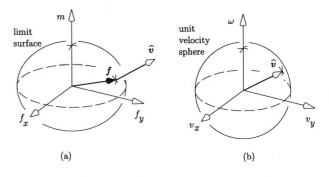

Figure 8: *Mapping a force $\mathbf{f}$ through a limit surface to a velocity direction $\hat{\mathbf{v}}$.*

The origin $O_\mathcal{S}$ of the slider frame $\mathcal{F}_\mathcal{S}$ is located at the center of friction of the slider, which is the unique point of the slider in the support plane such that $\int_\mathcal{S} \mathbf{x} s(\mathbf{x}) d\mathbf{x} = \mathbf{0}$. When $\mu_s(\mathbf{x})$ is constant, $O_\mathcal{S}$ is located directly beneath the center of mass (MacMillan [35]). All forces and velocities are expressed with respect to the slider frame $\mathcal{F}_\mathcal{S}$.

The force $\mathbf{f}_{xy}$ and moment m components of the force $\mathbf{f}$ applied to the support plane by a slider $\mathcal{S}$ are given by Mason [37]:

$$
\begin{aligned}
\mathbf{f}_{xy} &= \int_\mathcal{S} \frac{\mathbf{v}(\mathbf{x})}{|\mathbf{v}(\mathbf{x})|} s(\mathbf{x}) d\mathbf{x} \\
m\hat{\mathbf{k}} &= \int_\mathcal{S} \mathbf{x} \otimes \frac{\mathbf{v}(\mathbf{x})}{|\mathbf{v}(\mathbf{x})|} s(\mathbf{x}) d\mathbf{x},
\end{aligned}
$$

where $\hat{\mathbf{k}}$ is the unit vector $(0, 0, 1)^T$. These expressions simply state that the differential frictional force applied by the slider at each support point $\mathbf{x}$ acts in the direction of the velocity of $\mathbf{x}$ with magnitude $s(\mathbf{x})$.

3.1.1 The limit surface

As the slider's velocity direction $\hat{\mathbf{v}}$ moves over the velocity sphere, the force $\mathbf{f}$ moves on a two-dimensional surface in the three-dimensional force space. This surface is called the *limit surface*, and Goyal *et al.* [17] describe several of its important properties. This surface is closed, convex, and it encloses the origin of force space. The limit surface encloses the set of all forces which can be statically applied to the slider, and during quasi-static motion the applied force lies on the limit surface. The slider's velocity direction vector $\hat{\mathbf{v}}$ is normal to the limit surface at the force $\mathbf{f}$. If the force $\mathbf{f}$ lies

on the limit surface with an associated velocity direction $\hat{\mathbf{v}}$, then the force $-\mathbf{f}$ also lies on the limit surface with an associated velocity direction $-\hat{\mathbf{v}}$.

If the support friction distribution $s(\mathbf{x})$ is finite everywhere, the limit surface is smooth and strictly convex, defining a continuous one-to-one mapping from the set of force directions $\hat{\mathbf{f}} \in S^2$ to the set of velocity directions $\hat{\mathbf{v}} \in S^2$. Mason [37] also showed that if the applied force has zero moment about the center of friction, the resulting slider velocity is translational and parallel to the applied force.

If the support friction distribution $s(\mathbf{x})$ becomes infinite at any point, however, the situation is somewhat different. For example, if a support point $\mathbf{x}_0$ supports a finite force, the pressure $p(\mathbf{x}_0)$ is infinite, and if the coefficient of friction $\mu_s(\mathbf{x}_0)$ is nonzero, then $s(\mathbf{x}_0)$ is infinite. (Similarly if $\mu_s(\mathbf{x}_0)$ is infinite.) In these cases, the limit surface contains flat facets. At these facets, a set of force directions maps to the same velocity direction.

If the support friction distribution $s(\mathbf{x})$ is infinite only at points on a line, and $s(\mathbf{x})$ integrates to zero over the rest of the support surface, then the limit surface contains vertices. The normals to the limit surface at these vertices are not uniquely defined: the same force maps to a set of velocity directions.

The simplest limit surface with vertices and facets is that of a barbell: a rod with two points of equal support force at either end (Figures 9 and 10). This example has been used by Mason [37] and Goyal *et al.* [17]. If the support points lie on the y-axis at $(0, -1)$ and $(0, 1)$, then an applied force on the x-axis maps nondeterministically through a vertex of the limit surface

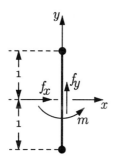

Figure 9: *A barbell with two support points.*

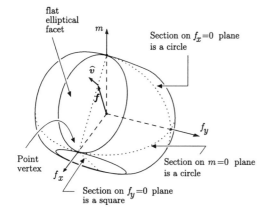

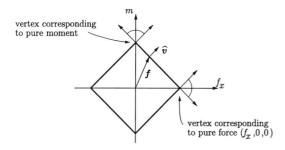

Figure 10: *The limit surface for the barbell, taken from (Goyal et al. [17]).*

to a rotation center at $(0, y)$, for any $y \leq -1$ or $y \geq 1$. If the velocity is a pure rotation about one of the supports, the force applied to the support plane lies on a facet of the limit surface. The force is indeterminate due to the static indeterminacy of the friction force at the stationary support point.

3.1.2 Solving for the motion of a pushed object

Each contact point between the pusher and the slider may be sticking, breaking free, or sliding to the left or right. The *contact mode* describes the qualitative behavior of each contact point between the pusher and the slider. For each possible contact mode i, there is a space of possible slider velocities $\mathcal{V}_{k,i}$ which are kinematically consistent with that contact mode and the known pusher velocity. By Coulomb's law, each contact mode also specifies a polyhedral cone of possible pushing forces in the three-dimensional force space. This cone is the convex hull of the individual friction cones at the sticking contacts and the friction cone edges at the sliding contacts (Erdmann [13, 14]). This composite friction cone is intersected with the limit surface to find a cone of possible velocities $\mathcal{V}_{f,i}$ (Figure 11). If $\mathcal{V}_{k,i} \cap \mathcal{V}_{f,i} = \emptyset$, contact mode i cannot occur; otherwise, contact mode i is feasible and any of the velocities in the intersection set is a possible solution to the motion of the slider.

3.2 Pushing with point contact

3.2.1 Mechanics of point contact pushing

When the slider is pushed at a single point of contact, there are only three possible contact modes: sticking contact, left sliding, and right sliding. Mason [37] developed search procedures to find the slider motion resulting in force-moment balance (for a known support friction distribution $s(\mathbf{x})$) for sticking and sliding contact. The actual motion is given by the contact mode which is consistent with all kinematic and force constraints. Peshkin and Sanderson [46] and Alexander and Maddocks [3] derived similar search procedures which find the slider motion by minimizing the power loss due to frictional sliding. Peshkin and Sanderson [48] proved that the power minimization approach is equivalent to Mason's force-moment balance approach.

In general, the support friction distribution $s(\mathbf{x})$ is indeterminate. For this reason, several authors have investigated the use of weaker models of the support friction distribution. With these weaker models, it is generally impossible to find the exact motion of the slider, but it is possible to find constraints on the motion. Mason [37] derived a simple rule for determining the rotation sense of an object pushed with point contact when the support friction distribution is unknown

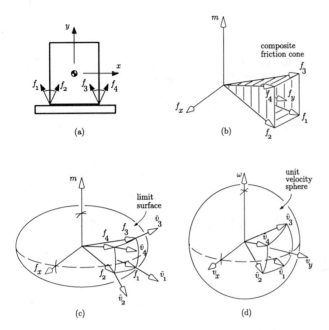

Figure 11: *(a) The forces that the pusher can apply to the slider during sticking contact are represented by the two friction cones. (b) The convex hull of these friction cones in the three-dimensional force space. The result is a composite friction cone of possible pushing forces. (c) Mapping these forces through the limit surface for the slider. (d) The slider velocity directions corresponding to forces inside the composite friction cone.*

but the center of friction is known. This work was extended by Mason and Brost [39] to find bounds on the rotation rate using information about the extent of the slider support area. Peshkin and Sanderson [46] also assumed a known center of friction and modeled the support region as the smallest disk centered at the center of friction and enclosing the support area. For a given push at a point contact, their method returns the set of all possible rotation centers for all support friction distributions $s(\mathbf{x})$ consistent with the disk and the center of friction. Alexander and Maddocks [3] derived constraints on the possible rotation centers when only the geometry of the slider support area is known.

3.2.2 Controllability with point contact pushing

Open-loop pushing with point contact is inherently unstable, and many researchers have constructed point contact pushing control systems using visual (Inaba and Inoue [22], Gandolfo *et al.* [15], Salgani-

coff *et al.* [51], Narasimhan [43]), tactile (Okawa and Yokoyama [45], Lynch *et al.* [34]), and infrared ranging feedback (unpublished work by Latombe and colleagues [57]). Here we examine the controllability of such a control system.

Proposition 5 *The configuration of a slider $\mathcal{S}$ with a bounded support friction distribution $s(\mathbf{x})$ is controllable by pushing if and only if the pusher can apply two pushing force directions, $\hat{\mathbf{f}}_1$ and $\hat{\mathbf{f}}_2$, such that they do not both pass through the center of friction ($\hat{m}_1 \neq 0$ or $\hat{m}_2 \neq 0$) and $\hat{\mathbf{f}}_1 \neq -\hat{\mathbf{f}}_2$.*

Proof: Because the support friction distribution $s(\mathbf{x})$ is bounded, the limit surface has no facets, and therefore the two force directions map through the limit surface to two distinct velocity directions. At least one of the force directions has nonzero moment, so at least one of the velocity directions has a nonzero angular component. Because the two force directions are not opposite, the two velocity directions are also not opposite. Therefore, by Proposition 2, the slider is controllable. □

A slider which is uncontrollable by pushing is a disk centered at its center of friction with a pushing friction coefficient of zero. All pushing forces pass through the center of friction, creating zero moment about the center of friction. The slider cannot be rotated. On the other hand, the barbell of Section 3.1.1 is, in some sense, nondeterministically controllable using a single pushing force on a vertex of the limit surface. Such a force maps nondeterministically to a set of velocity directions which satisfy Proposition 2.

If the slider is polygonal, a natural question is whether the slider is controllable by pushing with point contact on a single edge. Theorem 2 is a direct application of Proposition 5.

Theorem 2 *The configuration of a slider $\mathcal{S}$ with a bounded support friction distribution $s(\mathbf{x})$ is controllable by pushing on a straight edge if and only if the edge (1) has nonzero length or (2) has nonzero friction and is not a point at the center of friction.*

A slider which is controllable by pushing may have to be pushed a long distance to reach nearby configurations. If the object is small-time locally controllable, however, it can be maneuvered in tight spaces. In order to find conditions for small-time local controllability of a slider, first recall that the set of available pushing contacts is given by Γ, a closed, piecewise smooth curve. At each point of Γ which is not a vertex, the curve Γ has a unique inwardly-pointing contact normal. At a vertex, we assume that the contact normal can take any direction in the range specified by the contact normals adjacent to the vertex. Each contact point and contact normal specifies a pushing force direction which can be applied to the frictionless slider $\mathcal{S}$, and the curve of all such force directions is denoted $\hat{\mathbf{f}}(\Gamma)$. Because Γ is a closed curve, $\hat{\mathbf{f}}(\Gamma)$ is also a closed curve of force directions on the force sphere.

Theorem 3 *The configuration of any slider $\mathcal{S}$ with a closed, piecewise smooth curve Γ of available pushing contact points is small-time locally controllable by pushing with point contact, unless the pushing contact is frictionless and Γ is a circle centered at the slider's center of friction (i.e., a frictionless disk).*

Proof: By reasoning similar to that of Hong *et al.* [21], Γ always admits at least two pairs of noncollinear antipodal points. A pair of antipodal points of Γ is two points whose contact normals are opposite and lie on the same line, giving rise to opposite pushing force directions. By the limit surface mapping, these force directions yield opposite velocity directions. Because there are at least two pairs of noncollinear antipodal points, there are also at least two distinct pairs of opposing velocity directions $\{\hat{\mathbf{v}}_1, -\hat{\mathbf{v}}_1\}$ and $\{\hat{\mathbf{v}}_2, -\hat{\mathbf{v}}_2\}$, and these four velocity directions positively span a great circle of the velocity sphere. Therefore, by Proposition 4, the slider is small-time locally controllable, unless this great circle lies in the $\omega = 0$ plane.

Now consider the case where the two pairs of opposing velocity directions $\{\hat{\mathbf{v}}_1, -\hat{\mathbf{v}}_1\}$ and $\{\hat{\mathbf{v}}_2, -\hat{\mathbf{v}}_2\}$ span a great circle in the $\omega = 0$ plane. If Γ is not a circle, Mishra *et al.* [41] show that $\hat{\mathbf{f}}(\Gamma)$ positively spans the force sphere. This implies that $\hat{\mathbf{f}}(\Gamma)$ must contain force directions with positive and negative moment, and therefore the slider must be capable of rotating in either direction. The $\omega = 0$ great circle, any clockwise velocity direction, and any counterclockwise velocity direction positively span the velocity sphere, and the slider is small-time locally controllable.

The only remaining case is when Γ is a circle. If the center of friction is offset from the center of the circle, then only one pair of antipodal points can give zero

moment about the center of friction. Therefore the great circle positively spanned by $\{\hat{\mathbf{v}}_1, -\hat{\mathbf{v}}_1, \hat{\mathbf{v}}_2, -\hat{\mathbf{v}}_2\}$ does not lie in the $\omega = 0$ plane, and the slider is small-time locally controllable by Proposition 4. If the center of friction is at the center of the circle and there is nonzero friction at the pushing contact, the slider can be translated in any direction and rotated in either direction (using tangential frictional forces to create moment about the center of friction). The set of velocity directions positively spans the velocity sphere, and the slider is small-time locally controllable. If the center of friction is at the center of the circle and the pushing contact is frictionless, however, the object cannot be rotated, as all pushing forces create zero moment about the center of friction. A frictionless disk centered at its center of friction is the only type of slider which is not small-time locally controllable by point contact pushing. (Such a disk *could* rotate if its limit surface contains vertices, but there is no guarantee that it *will* rotate.)   □

3.3 Stable pushing with line contact

3.3.1 Mechanics of line contact pushing

In the previous section we examined the controllability of sliders pushed with point contact, but we would also like to synthesize pushing controllers. Unfortunately, the motion of a slider pushed with a single point of contact is unpredictable, because it depends on the unknown support friction distribution $s(\mathbf{x})$. If there are two or more pushing contacts, however, there may exist a space of pushing directions which, despite some uncertainty in the support friction distribution $s(\mathbf{x})$, result in a predictable motion of the slider: sticking at all contact points. If the slider is sticking at all contact points, it maintains its configuration relative to the pusher, and the push may be continued indefinitely. The slider is effectively rigidly attached to the pusher. We call such a push a stable push, and we will use these stable pushes to execute open-loop pushing plans.

In this paper, we focus on stable pushing with line contact: all contacts between the pusher and the slider are collinear with contact normals perpendicular to the line. A straight-edge pusher always makes line contact. We also assume that the coefficient of friction at all pushing contacts is the same. Figure 12 shows examples of line contact. A line contact is equivalent to two

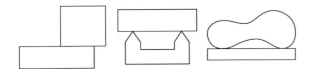

Figure 12: *Examples of line contact.*

contact points at the ends of the line.

For a given line contact, we use the following definitions:

$\hat{V}_{stable}$: The set of pushing directions such that the slider is guaranteed to remain fixed to the pusher during the motion.

$\hat{V}_{\mathcal{F}}$: The set of velocity directions such that the frictional force **f** which the slider applies to the support during the motion is contained in the composite friction cone $\mathcal{F}$ from the line pushing contact. This set of velocity directions is found by intersecting $\mathcal{F}$ with the limit surface (Figure 11). Membership in $\hat{V}_{\mathcal{F}}$ is a necessary but not sufficient condition for a velocity direction to belong to $\hat{V}_{stable}$.

If a velocity direction $\hat{\mathbf{v}}$ is in $\hat{V}_{stable}$, then it is also in $\hat{V}_{\mathcal{F}}$, but the converse is not necessarily true. Although stable contact is always a possible solution if the pushing direction $\hat{\mathbf{v}}$ is in $\hat{V}_{\mathcal{F}}$, there may be other solutions. In this section, we describe a procedure for finding a conservative approximation to $\hat{V}_{\mathcal{F}}$ and provide a theorem for identifying a subset belonging to $\hat{V}_{stable}$.

Usually the support friction distribution $s(\mathbf{x})$ is unknown, and therefore we cannot determine $\hat{V}_{\mathcal{F}}$ exactly. Instead, we assume that the center of friction and the shape of the slider are known. In a previous paper (Lynch [31]), we presented an algorithm for finding an approximation to $\hat{V}_{\mathcal{F}}$ for a slider with a known center of friction. This algorithm utilizes results due to Peshkin and Sanderson [46] on the possible support friction distributions of a disk slider.

Here we describe a simple procedure for finding a conservative approximation to $\hat{V}_{\mathcal{F}}$. We will illustrate the procedure in the rotation center space. Without loss of generality, assume that the line contact is horizontal with an upward pointing contact normal, as in Figure 13.

(a) The coefficient of friction defines two friction cone edges, at angles $\tan^{-1}\mu$ to the contact normal,

where μ is the coefficient of friction. For each edge of the friction cone, draw two lines perpendicular to the friction cone edge such that the entire slider is contained between the two lines. For an applied force at an edge of the friction cone, the resulting rotation center must lie in its respective band (Mason and Brost [39], Alexander and Maddocks [3]). Counterclockwise (clockwise) rotation centers between the two bands and to the left (right) of the slider correspond to force angles inside the angular limits of the friction cone. See Figure 13(a).

(b) For each endpoint of the line contact, draw two lines perpendicular to the line through the center of friction and the endpoint. One of these lines is the perpendicular bisector between the contact point and the center of friction (Mason and Brost [39]). The other is a distance r^2/p from the center of friction and on the opposite side from the endpoint, where p is the distance from the endpoint to the center of friction and r is the distance from the center of friction to the most distant support point of the slider (Peshkin and Sanderson [46]). (Actually, this second line should be slightly more distant from the center of friction; see [46] for details.) The rotation center from pushing on this endpoint must lie in the band between these two lines (Mason and Brost [39]). All rotation centers between the two bands correspond to forces that pass between the two endpoints. See Figure 13(b).

(c) The intersection of the closed regions found in (a) and (b) yield a set of rotation centers corresponding to forces which are guaranteed to lie on or inside the composite friction cone $\mathcal{F}$ from the line pushing contact. This is a conservative approximation to $\hat{V}_{\mathcal{F}}$. See Figure 13(c).

This simple conservative approximation procedure misses some tight turning rotation centers (velocity directions with a large angular component) but finds all translations belonging to $\hat{V}_{\mathcal{F}}$. If the composite friction cone $\mathcal{F}$ contains any pure force (zero moment; a force through the center of friction) in its interior, the procedure will find a convex set of velocity directions with nonempty interior and a range of translation directions. If the composite friction cone $\mathcal{F}$ contains only a single pure force direction (necessarily on the boundary of $\mathcal{F}$), the procedure finds only a single translation in the di-

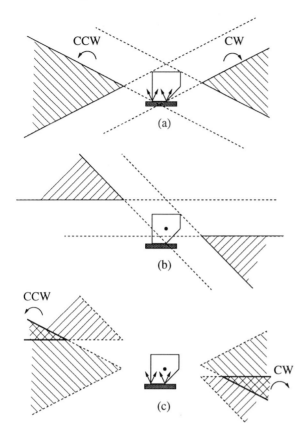

Figure 13: *(a) Rotation centers which can be achieved by forces inside the friction cone angular limits. (b) Rotation centers which can be achieved by forces passing between the line contact endpoints. (c) The intersection of the closed regions found in (a) and (b) correspond to rotation centers which can be achieved by forces inside the composite friction cone $\mathcal{F}$ defined by the line contact.*

rection of this pure force. If the composite friction cone $\mathcal{F}$ does not contain a pure force, this procedure finds no velocity directions belonging to $\hat{\mathcal{V}}_{\mathcal{F}}$. In fact, with no information about the support friction distribution $s(\mathbf{x})$ other than the center of friction, no velocity directions are guaranteed to be in $\hat{\mathcal{V}}_{\mathcal{F}}$ if the line contact cannot apply a force through the center of friction.

Proposition 6 *If the pusher $\mathcal{P}$ can apply a set of forces $\mathcal{F}$ to the slider $\mathcal{S}$ such that $\mathcal{F}$ does not include a force through the center of friction, then, in the absence of any other information about the support friction distribution $s(\mathbf{x})$, no single velocity direction is guaranteed to be achievable by a force in $\mathcal{F}$.*

Proof: With no information about the support friction distribution $s(\mathbf{x})$, we can always choose the support friction distribution to be concentrated arbitrarily close to the center of friction. For all rotation centers not located at the center of friction, this corresponds to pushing forces passing through or arbitrarily close to the center of friction, which are not included in the composite friction cone $\mathcal{F}$. If the rotation center is located at the center of friction, any support friction distribution $s(\mathbf{x})$ which is symmetric about the center of friction corresponds to a pushing force which is a pure moment. A pure moment cannot be applied by pushing on a finite object. □

Unfortunately, the necessary force condition for stability is not also a sufficient condition for stability. While stable contact is always a solution when the pusher follows a velocity direction in $\hat{\mathcal{V}}_{\mathcal{F}}$, there may be other solutions. In order to show that stable contact is the only solution, it is necessary to show that every other contact mode is inconsistent (Lynch [31]). The following theorem gives a sufficient condition for a pushing velocity direction to result in stable contact. The proof is somewhat lengthy and therefore is omitted.

Theorem 4 *Let $\hat{\mathcal{V}}_{\mathcal{F}}$ be the set of rotation centers which satisfy the necessary force condition for stability for a pusher $\mathcal{P}$ and slider $\mathcal{S}$ in line contact. Draw two lines perpendicular to the line contact such that the entire slider $\mathcal{S}$ is contained between the two lines. All rotation centers in the interior of the set $\hat{\mathcal{V}}_{\mathcal{F}}$ and outside the two lines are guaranteed to belong to the set of stable pushing directions $\hat{\mathcal{V}}_{stable}$. Therefore, all velocity directions interior to the set found by the procedure above belong to $\hat{\mathcal{V}}_{stable}$.*

Remark: The conditions of this theorem are sufficient but not necessary. In particular, velocity directions in $\hat{\mathcal{V}}_{\mathcal{F}}$ which do not satisfy the conditions of Theorem 4 may indeed be provably stable.

All velocity directions $\hat{\mathbf{v}}$ belong to one of the following sets:

1. $\hat{\mathbf{v}} \in \hat{\mathcal{V}}_{stable}$. The slider maintains its position and orientation relative to the pusher during the push.

2. $\hat{\mathbf{v}} \in \hat{\mathcal{V}}_{unstable}$. The pushing direction $\hat{\mathbf{v}}$ does not belong to $\hat{\mathcal{V}}_{\mathcal{F}}$. The slider will slip and/or roll relative to the pusher.

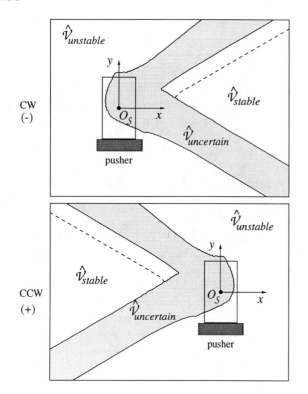

CW
(-)

CCW
(+)

Figure 14: *Stable and unstable pushing directions, represented as clockwise and counterclockwise rotation centers in the slider frame $\mathcal{F}_S$, for a box with a center of friction at its center, pushing contact on the bottom edge, and a coefficient of friction of 1.0. The dotted line indicates the boundary of $\hat{\mathcal{V}}_{stable}$ found using the conservative approximation described in the text.*

3. $\hat{\mathbf{v}} \in \hat{\mathcal{V}}_{uncertain}$. The pushing direction $\hat{\mathbf{v}}$ may or may not be stable. The information in the model is insufficient to predict the resulting contact mode.

The algorithm described in (Lynch [31]) was used to generate Figure 14, which shows the sets $\hat{\mathcal{V}}_{stable}$, $\hat{\mathcal{V}}_{unstable}$, and $\hat{\mathcal{V}}_{uncertain}$ for a rectangular box pushed at one edge with a coefficient of friction of 1.0. If the rotation center of the pusher motion is in $\hat{\mathcal{V}}_{stable}$, the box will remain fixed to the pusher. If the rotation center of the pusher motion is in $\hat{\mathcal{V}}_{unstable}$, the box will slip and/or roll with respect to the pusher. If the rotation center is in $\hat{\mathcal{V}}_{uncertain}$, the contact may or may not be stable. In the sequel, we will only use pushing directions that are guaranteed to be stable.

3.3.2 Controllability using stable pushes with line contact pushing

If the composite friction cone $\mathcal{F}$ from the line pushing contact contains a pure force in its interior, then the procedure described above finds a convex set of velocity directions $\hat{\mathcal{V}}_{\mathcal{F}}$ with nonempty interior (including a range of translations). Theorem 4 tells us that all velocity directions in the interior must be stable. By Proposition 2, we get the following.

Theorem 5 *If a slider $\mathcal{S}$ is pushed with line contact with a composite friction cone $\mathcal{F}$ such that $\mathcal{F}$ contains a pure force (zero moment about the center of friction of the slider) in its interior, then the configuration of the slider is controllable by provably stable pushes. The procedure outlined in Section 3.3.1, along with Theorem 4, finds a set of stable pushing directions that yield controllability.*

If Theorem 5 is satisfied, the slider can be pushed to any configuration in the obstacle-free plane using provably stable pushes. The proof of Proposition 2 suggests possible approaches to planning pushing paths using only two velocity directions in $\hat{\mathcal{V}}_{stable}$.

We can apply Theorem 5 to prove the following result regarding polygonal sliders.

Theorem 6 *For a polygonal slider $\mathcal{S}$ and a sufficiently long straight-edge pusher $\mathcal{P}$, any edge of the convex hull of $\mathcal{S}$ can be used as the line pushing contact. If the center of friction of the slider $\mathcal{S}$ is in the interior of its convex hull, and there is nonzero friction at the line contacts, then there is at least one line contact from which the slider is controllable by provably stable pushes.*

Proof: For any center of friction in the interior of the convex hull of $\mathcal{S}$, there is at least one normal to the interior of an edge of the convex hull that passes through the center of friction. A force $\mathbf{f}$ along that normal is in the interior of the composite friction cone for that edge, and therefore, by Theorem 5, the slider is controllable from that edge by provably stable pushes.
□

It is intuitively clear that a slider is not small-time locally controllable by pushing a single edge. It is impossible to "pull" the slider backward, and therefore it cannot be maneuvered in tight spaces. If we allow the pusher to change contact configurations, however, in some cases the configuration of the slider may be made small-time locally controllable by stable pushes with line contact. The following theorem gives a sufficient condition for a slider to be small-time locally controllable by line contact pushing.

Theorem 7 *Given a set of line pushing contacts with composite friction cones $\mathcal{F}_i$, find the set of all pure forces (zero moment about the center of friction of the slider) interior to at least one of the composite friction cones $\mathcal{F}_i$. If these pure forces positively span the plane of pure forces, then the configuration of the slider $\mathcal{S}$ is small-time locally controllable by provably stable pushes. The procedure outlined in Section 3.3.1, along with Theorem 4, finds a set of stable pushing directions that yield small-time local controllability.*

Proof: If the conditions of Theorem 7 are satisfied, then the procedure of Section 3.3.1, along with Theorem 4, will find a set of translation directions interior to $\hat{\mathcal{V}}_{stable}$ that positively span the plane of translations ($\omega = 0$). Because each of these translation directions has a neighborhood of velocity directions also in $\hat{\mathcal{V}}_{stable}$, the stable velocity directions $\hat{\mathcal{V}}_{stable}$ positively span the velocity sphere.   □

Figure 15 shows the stable pushing directions if the box of Figure 14 can be pushed on two opposite edges. The stable pushing directions positively span the space of velocity directions, and therefore the box is small-time locally controllable by stable pushing with these two contact configurations.

4   Planning stable pushing paths among obstacles

The stable pushing directions impose nonholonomic constraints on the motion of the pusher, and the problem is to plan free pushing paths among obstacles subject to these nonholonomic constraints. This problem is similar to the problem of planning paths for car-like mobile robots, the subject of much recent robotics research. The configuration space of a car-like mobile robot is also $\mathbf{R}^2 \times S^1$, but its feasible velocity direction set is only one-dimensional, corresponding to the angle of the steering wheel. Despite this, Laumond [28] showed that a car-like robot which can reverse can reach any configuration in any open connected subset of its free configuration space. Barraquand and

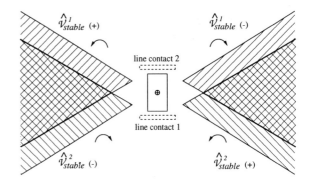

Figure 15: *Stable pushing directions, represented as rotation centers, for a box that can be pushed on either of two opposite edges with a coefficient of friction of 1.0. These pushing directions positively span the space of velocity directions, and therefore the box is small-time locally controllable by stable pushing with line contact.*

Latombe [6] showed that only two steering directions are required.

Many path planners for car-like mobile robots have been proposed; see Chapter 9 of Latombe's textbook [27] for a survey. In our work on planning pushing paths, we chose to adapt an algorithm by Barraquand and Latombe [6] due to its simplicity and its adaptability to two-dimensional velocity direction sets. The resulting algorithm closely parallels that described by Barraquand and Latombe. We provide only a brief description below. We refer the interested reader to (Barraquand and Latombe [6]) for details.

4.1   Planner description

In order to plan stable pushing paths, we first choose a discrete set of m possible line contact pushing configurations. We then calculate $\hat{\mathcal{V}}^i_{stable}$, the set of provably stable pushing directions, for contact configurations $i = 1, \ldots, m$. These are the feasible pushing directions. In light of Proposition 3, the planner uses a discrete set of velocity directions $\hat{\mathcal{V}}^i_{extremal}$, $i = 1, \ldots, m$, where $\hat{\mathcal{V}}^i_{extremal}$ is a subset of $\hat{\mathcal{V}}^i_{stable}$, and the convex hull of $\hat{\mathcal{V}}^i_{extremal}$ on the velocity sphere is a polygonal approximation of $\hat{\mathcal{V}}^i_{stable}$. We will denote the union of $\hat{\mathcal{V}}^i_{stable}$ for all i to be $\hat{\mathcal{V}}_{stable}$ and the union of $\hat{\mathcal{V}}^i_{extremal}$ for all i to be $\hat{\mathcal{V}}_{extremal}$.

The workspace is an open rectangular subset of $\mathbf{R}^2$, and obstacles are represented as closed polygons. For a particular pushing contact configuration, $\mathcal{PS}$ is the closed region of the workspace occupied by the pusher $\mathcal{P}$ and the slider $\mathcal{S}$. $\mathcal{PS}(\mathbf{q})$ is the closed region of the workspace occupied by the pusher and the slider at the configuration $\mathbf{q}$. The obstacle space $\mathcal{C}_{obs}$ is the closed set of configurations $\mathbf{q}$ such that $\mathcal{PS}(\mathbf{q})$ intersects an obstacle or the walls bounding the workspace. The free space $\mathcal{C}_{free}$ is the open set of the configuration space $\mathcal{C}$ complementary to $\mathcal{C}_{obs}$.

The free space $\mathcal{C}_{free}$ changes with the contact configuration, but for simplicity the planner uses a single representation of free space for all contact configurations. This simplification is appropriate when the pusher $\mathcal{P}$ is much smaller or much larger than the slider $\mathcal{S}$. The region occupied by the pusher and the slider is treated as the smallest single disk that encloses them for all pushing contact configurations. The obstacles are then grown by the radius of this disk into generalized polygons (Laumond [29]) and $\mathcal{PS}$ is treated as a point at the center of the disk. A collision occurs when this point intersects the grown obstacles. This representation of the free space $\mathcal{C}_{free}$ is not exact, but the resulting code is very simple and the collision-detection routine is very fast.

The planner keeps a tree T of configurations reached in the search and a list OPEN of pointers to configurations in T whose successors have not yet been generated. The pointers in the list OPEN are sorted by the costs of the paths to the associated configurations. The planner begins by making the slider's initial configuration $\mathbf{q}_{init}$ the root of the tree T and initializing the list OPEN with a pointer to this configuration. The main loop of the planner is a simple best-first search. Roughly, the planner sets the current configuration to that indicated by the minimum-cost pointer in OPEN, and it removes this pointer from OPEN. Subsequent configurations are generated by integrating each velocity direction in $\mathcal{V}_{extremal}$ forward a short distance, and each new collision-free configuration is added to the tree T with a record of the velocity direction and a pointer to the previous configuration. Pointers to these new configurations are inserted into the sorted list OPEN. This continues until one of the three termination conditions is satisfied: (1) the list OPEN becomes empty, (2) the number of configurations generated exceeds some user-specified value, or (3) the planner reaches a configuration in a user-specified neighborhood $\mathcal{G}(\mathbf{q}_{goal})$ of the goal configuration $\mathbf{q}_{goal}$. Note

that the planner is not exact, as it only finds a path to a goal neighborhood.

The cost of a path is a function of the number of pushing steps, the number of changes of the control, and the number of changes of the contact configuration. In the current implementation, the cost is the sum of an integer a times the number of pushing steps, an integer b times the number of control changes, and an integer c times the number of changes of the contact configuration. Typically, $a = 1$ and $b = c = 0$, or $a = b = 0$ and $c = 1$. In the former case, the planner will tend to find short paths with many contact changes, and in the latter case, the planner will tend to find longer paths with fewer contact changes. The values of each of a, b, and c can be nonzero to weight the relative importance of these factors.

The planner checks for collisions at each new configuration in the search, not along the path. For this reason, the disk approximation to $\mathcal{PS}$ is grown by the maximum distance any point in $\mathcal{PS}$ can move in a single step. This ensures that if a new configuration is free, then the path which transferred it there is also free. This approach is simple and fast, but somewhat conservative; the planner may not find paths through tight spots where paths exist.

The planner also discards configurations which are sufficiently close to configurations which have been reached with a lower cost and the same pushing contact configuration. Two configurations are considered sufficiently close if they occupy the same cell of a predefined grid on the configuration space.

The user must specify the parameters defining the size of the goal neighborhood $\mathcal{G}(\mathbf{q}_{goal})$, the length of the control steps, and the resolution of the configuration space grid used to check for prior occupancy. These parameters are interdependent. The resolution of the grid should be sufficiently fine that the application of any control step moves the configuration to a new grid cell, and the goal neighborhood should be large enough that $\mathcal{PS}$ cannot jump over it. The user must also specify the maximum number of configurations that the planner will explore before returning failure.

We assume that the pusher $\mathcal{P}$ can change pushing contact configurations at any time. If the slider is pushed by a manipulator capable of moving above the obstacles, such as a robot arm pushing objects on a table, this is a reasonable assumption. If the pusher is constrained to move in the plane, such as a mobile robot, however, this planner does not address the problem of finding paths between pushing contact configurations.

4.2 Properties of the planner

This planner inherits properties of Barraquand and Latombe's planner [6]. Changing contact configurations in the pushing case is essentially equivalent to shifting between forward and reverse in the car-like robot case. (Note that we assume there is always a path for the pusher from one feasible pushing contact configuration to another.) If the convex hull of the extremal velocities $\hat{\mathcal{V}}^i_{extremal}$ completely covers $\hat{\mathcal{V}}^i_{stable}$ for all i, then, with an exact collision-detection routine, it is possible to choose search parameters such that the following properties hold:

- Completeness. *If there is a feasible pushing path from the initial configuration* $\mathbf{q}_{init}$ *to the goal neighborhood* $\mathcal{G}(\mathbf{q}_{goal})$ *using the stable pushing direction set* $\hat{\mathcal{V}}_{stable}$, *then the planner will find a pushing path.*

- Optimality. *If the cost of the pushing path is given by the number of changes of the pushing contact configuration, then the planner will find a pushing path with* λ *changes of the pushing contact configuration, where* λ *is the minimum number of contact changes for any feasible pushing path connecting the initial configuration* $\mathbf{q}_{init}$ *to the goal neighborhood* $\mathcal{G}(\mathbf{q}_{goal})$ *using velocity directions in* $\hat{\mathcal{V}}_{stable}$.

Now assume that the extent of the pusher $\mathcal{P}$ is negligible so that $\mathcal{PS}$ is equivalent to the slider $\mathcal{S}$. If $\hat{\mathcal{V}}_{extremal}$ (and therefore $\hat{\mathcal{V}}_{stable}$) satisfies the conditions for small-time local controllability, then, with an exact collision-detection routine, it is possible to choose search parameters such that the following property holds:

- *If the initial configuration* $\mathbf{q}_{init}$ *and the goal configuration* $\mathbf{q}_{goal}$ *lie in the same connected component of the slider's free configuration space* $\mathcal{C}_{free}$, *then the planner will find a pushing path connecting* $\mathbf{q}_{init}$ *to the goal neighborhood* $\mathcal{G}(\mathbf{q}_{goal})$.

The proofs of these properties do not suggest how to choose the search parameters. See (Barraquand and Latombe [6]) for a more detailed discussion of the properties of the planner.

4.3 Experimental results

The planner is implemented in C on a Sun 4/30. This section presents some pushing paths generated for the box of Figure 14. For the line contact of Figure 14, the set of stable pushing directions $\hat{\mathcal{V}}_{stable}$ is closely approximated by the convex combination of four pushing directions—two rotations and two translations. The planner was given these four extremal velocities, plus a translational velocity along the length of the box. This pushing direction is also in $\hat{\mathcal{V}}_{stable}$ and was included to simplify straight paths. Figure 16 shows a pushing path for an omnidirectional mobile robot pusher which cannot change its contact configuration. This path minimizes the number of pushing steps and took 50 seconds to generate.

A car-like mobile robot pusher imposes additional constraints on the possible pushing directions. Figure 17 shows the intersection of $\hat{\mathcal{V}}_{stable}$ with the space of feasible robot velocity directions, limited by the rolling constraint of parallel wheels and a minimum turning radius constraint. Figure 18 shows the path generated for the same problem as above, using only translational motion and the fastest feasible clockwise and counterclockwise turning motions in $\hat{\mathcal{V}}_{stable}$. The robot is forced to take a longer path due to its kinematic constraints. The planner took 30 seconds to find this path.

In the next example, a robot arm pushes the box on a table, and the arm is capable of lifting up and changing the contact configuration. The pusher can contact the box at two opposite edges, as shown in Figure 15. The resulting sets of stable pushing directions yield small-time local controllability. For each contact configuration, the planner uses the four extremal velocity directions and a translational velocity direction along the length of the box. The area of the pusher $\mathcal{P}$ in the plane is assumed to be negligible.

Figures 19, 20, and 21 show pushing paths which solve the same problem using different cost functions. In Figure 19, the planner minimizes the number of pushing steps, and it finds a path which takes 39 pushing steps and two changes in the contact configuration. In Figure 20, the planner minimizes the number of changes in the contact configuration, and it finds a path which takes 111 pushing steps and involves no changes in the contact configuration. Figure 21 shows the path found for a cost function with $a = 0$, $b = 1$, and $c = 10$. This cost function indicates that we would

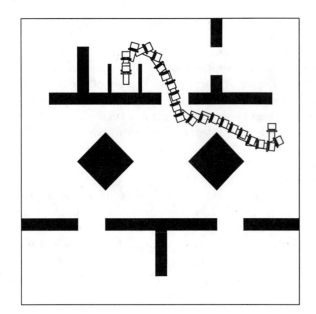

Figure 16: *The box being pushed to the goal by an omni-directional mobile robot. Four control steps separate each snapshot.*

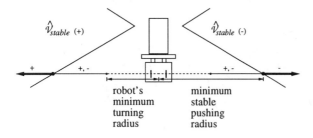

Figure 17: *Intersecting the feasible velocity directions of the car-like mobile robot pusher (represented as rotation centers) with the stable pushing directions for the box. The rays of intersection are indicated in bold.*

like to find a path involving as few control changes as possible without changing the contact configuration. The path takes 120 pushing steps with five control changes. The goal neighborhood for these examples is approximately ±10% the length of the box and ±3 degrees. Each of these paths took about two minutes to generate.

4.4 Variations on the planner

In order to shrink the size of the goal neighborhood without sacrificing much speed in planning, the con-

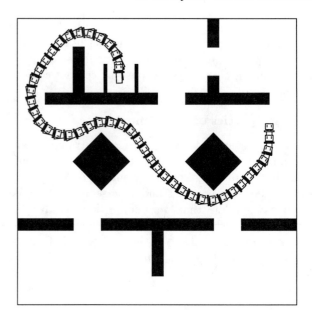

Figure 18: *The box being pushed to the goal by a car-like mobile robot. Four control steps separate each snapshot.*

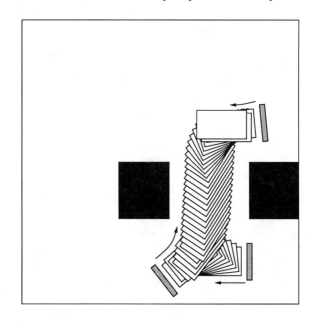

Figure 19: *A pushing path that minimizes the number of pushing steps.*

trol step and grid cell sizes could be decreased in the neighborhood of the goal. This would allow fine positioning near the goal.

The planner could be modified to find a path to a sin-

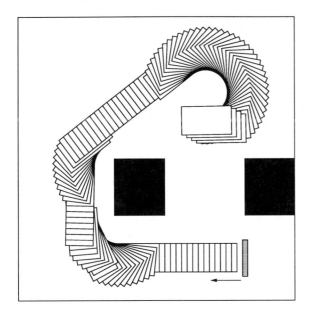

Figure 20: *A pushing path that minimizes the number of changes in the contact configuration.*

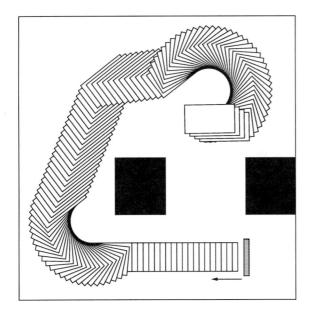

Figure 21: *A pushing path that minimizes the number of control changes without changing the contact configuration.*

gle goal configuration instead of a goal neighborhood. A very simple way to implement this is to plan a path to a goal neighborhood as before, then back up along the path until a configuration is found from which $\mathcal{PS}$

can move exactly to the goal configuration using a single pushing direction in $\hat{\mathcal{V}}_{stable}$. This is possible due to the fact that the space of stable pushing directions often has nonempty interior on the velocity sphere. The preimage of the goal configuration under all constant stable pushing directions therefore encloses a volume of the configuration space $\mathcal{C}$ with nonempty interior. (This is not true for the case of a car-like mobile robot pusher.)

The examples presented in Section 4.3 use stable translational velocity directions which border $\hat{\mathcal{V}}_{unstable}$. These pushing paths may be made robust to bounded uncertainty in the coefficient of contact friction and the location of the center of friction by shrinking the set of stable pushing directions $\hat{\mathcal{V}}_{stable}$. Brost [8, 9] has applied this idea to several manipulation tasks.

4.5 Open problems

This planner ignores the issue of finding the path for the pusher from one feasible contact configuration to another, but assumes that a free path always exists. If the pusher is constrained to move in the plane of the obstacles, this problem must be addressed. In the terminology of Alami *et al.* [2], a path between contact configurations is called a *transit path*, and a pushing path is called a *transfer path*. Alami *et al.* describe an approach to planning both the transit and transfer paths using *manipulation graphs*. In their problem, the robot and the manipulated object translate in a planar workspace with obstacles, and the robot can grasp the object at a discrete set of locations and place the object at a discrete set of configurations. The planner of Dacre-Wright *et al.* [10] addresses the case of a convex planar robot and object where the robot can grasp the object at any contact. Koga and Latombe [26] use manipulation graphs to find manipulation plans for a dual-arm robot system manipulating an object among obstacles in the plane.

The push planner is asymptotically optimal in the sense that if a feasible pushing path exists, then with an appropriate choice of search parameters, it will find a path which minimizes the number of changes in the pushing contact configuration. We can make no claim, however, that the resulting path will be "short." An open problem is how to find shortest paths for a polygonal set of feasible velocity directions. Dubins [12] found a set of canonical paths in the obstacle-free plane which is guaranteed to include the minimum arclength path for a car-like mobile robot which can only go forward. This result has been utilized by Jacobs and Canny [23] in planning paths among obstacles. Reeds and Shepp [50] enumerated a set of canonical paths, guaranteed to include the shortest path in the obstacle-free plane, for a car-like mobile robot which can reverse. This result has also been used in planning paths among obstacles (Laumond *et al.* [30]). As far as we are aware, there have been no similar results for more general sets of velocity directions.

We have assumed that the set of possible contact configurations is specified by the user. An interesting problem is how to choose the set of contact configurations based on knowledge of the pusher, the slider, the environment, and the initial and goal configurations. We have focused on line contact configurations in this paper, but a similar approach can be used with other types of pushing contacts.

In order to manipulate an unknown object, a robot could estimate the friction parameters of the object (Yoshikawa and Kurisu [58], Lynch [32]), perform the mechanics analysis, and use the results in the planner described here. Alternatively, the robot could empirically determine a set of stable pushing directions and plug these directly into the planner.

The pushing paths described in this paper are stabilized by using more than one contact point between the pusher and the slider. No sensing is required: the inherent mechanics of the task essentially close a tight feedback loop. Nevertheless, unmodeled effects could cause the pusher to lose control of the motion of the object. The planner described here could be considered the feedforward component of a pushing control system.

5   Conclusion

Although there has been a wealth of previous research on pushing, there has been relatively little work toward answering the fundamental question "Can the object be pushed there from here?" This paper begins to answer that question by elucidating some of the controllability properties of objects pushed with point or line contact. For the case of line contact, we have presented an algorithm for finding open-loop pushing plans among obstacles.

Acknowledgments

We thank Mike Erdmann for a discussion on Theorem 3. This work was supported by NSF Grant IRI-9114208.

References

[1] S. Akella and M. T. Mason. Posing polygonal objects in the plane by pushing. In *IEEE International Conference on Robotics and Automation*, pages 2255–2262, Nice, France, 1992.

[2] R. Alami, T. Simeon, and J.-P. Laumond. A geometrical approach to planning manipulation tasks. The case of discrete placements and grasps. In *International Symposium on Robotics Research*, pages 453–459, Tokyo, Japan, Aug. 1989.

[3] J. C. Alexander and J. H. Maddocks. Bounds on the friction-dominated motion of a pushed object. *International Journal of Robotics Research*, 12(3):231–248, 1993.

[4] Z. Balorda. Reducing uncertainty of objects by robot pushing. In *IEEE International Conference on Robotics and Automation*, pages 1051–1056, Cincinnati, OH, 1990.

[5] Z. Balorda. Automatic planning of robot pushing operations. In *IEEE International Conference on Robotics and Automation*, pages 1:732–737, Atlanta, GA, 1993.

[6] J. Barraquand and J.-C. Latombe. Nonholonomic multibody mobile robots: Controllability and motion planning in the presence of obstacles. *Algorithmica*, 10:121–155, 1993.

[7] M. Brokowski, M. Peshkin, and K. Goldberg. Curved fences for part alignment. In *IEEE International Conference on Robotics and Automation*, pages 3:467–473, Atlanta, GA, 1993.

[8] R. C. Brost. Automatic grasp planning in the presence of uncertainty. *International Journal of Robotics Research*, 7(1):3–17, Feb. 1988.

[9] R. C. Brost. Dynamic analysis of planar manipulation tasks. In *IEEE International Conference on Robotics and Automation*, pages 2247–2254, 1992.

[10] B. Dacre-Wright, J.-P. Laumond, and R. Alami. Motion planning for a robot and a movable object amidst polygonal obstacles. In *IEEE Inter-*

national Conference on Robotics and Automation, pages 2474–2480, Nice, France, 1992.

[11] B. R. Donald, J. Jennings, and D. Rus. Information invariants for cooperating autonomous mobile robots. In *International Symposium on Robotics Research*, Hidden Valley, PA, 1993.

[12] L. E. Dubins. On curves of minimal length with a constraint on average curvature and with prescribed initial and terminal positions and tangents. *American Journal of Mathematics*, 79:497–516, 1957.

[13] M. A. Erdmann. On motion planning with uncertainty. Master's thesis, Massachusetts Institute of Technology, Aug. 1984.

[14] M. A. Erdmann. On a representation of friction in configuration space. *International Journal of Robotics Research*, 13(3):240–271, 1994.

[15] F. Gandolfo, M. Tistarelli, and G. Sandini. Visual monitoring of robot actions. In *IEEE/RSJ International Conference on Intelligent Robots and Systems*, pages 269–275, Osaka, Japan, 1991.

[16] K. Y. Goldberg. Orienting polygonal parts without sensors. *Algorithmica*, 10:201–225, 1993.

[17] S. Goyal, A. Ruina, and J. Papadopoulos. Planar sliding with dry friction. Part 1. Limit surface and moment function. *Wear*, 143:307–330, 1991.

[18] S. Goyal, A. Ruina, and J. Papadopoulos. Planar sliding with dry friction. Part 2. Dynamics of motion. *Wear*, 143:331–352, 1991.

[19] G. W. Haynes and H. Hermes. Nonlinear controllability via Lie theory. *SIAM Journal on Control*, 8(4):450–460, Nov. 1970.

[20] R. Hermann and A. J. Krener. Nonlinear controllability and observability. *IEEE Transactions on Automatic Control*, AC-22(5):728–740, Oct. 1977.

[21] J. Hong, G. Lafferiere, B. Mishra, and X. Tan. Fine manipulation with multifinger hands. In *IEEE International Conference on Robotics and Automation*, pages 1568–1573, Cincinnati, OH, 1990.

[22] M. Inaba and H. Inoue. Vision-based robot programming. In *International Symposium on Robotics Research*, Tokyo, Japan, 1989.

[23] P. Jacobs and J. Canny. Planning smooth paths for mobile robots. In *IEEE International Conference on Robotics and Automation*, pages 2–7, Scottsdale, AZ, 1989.

[24] V. Jurdjevic. Certain controllability properties of analytic control systems. *SIAM Journal on Control*, 10(2):354–360, May 1972.

[25] V. Jurdjevic and H. J. Sussmann. Control systems on Lie groups. *Journal of Differential Equations*, 12:313–329, 1972.

[26] Y. Koga and J.-C. Latombe. Experiments in dual-arm manipulation planning. In *IEEE International Conference on Robotics and Automation*, pages 2238–2245, Nice, France, 1992.

[27] J.-C. Latombe. *Robot Motion Planning*. Kluwer Academic Publishers, 1991.

[28] J.-P. Laumond. Feasible trajectories for mobile robots with kinematic and environment constraints. In *International Conference on Intelligent Autonomous Systems*, pages 346–354, 1986.

[29] J.-P. Laumond. Finding collision-free smooth trajectories for a non-holonomic mobile robot. In *International Joint Conference on Artificial Intelligence*, pages 1120–1123, 1987.

[30] J.-P. Laumond, P. E. Jacobs, M. Taix, and R. M. Murray. A motion planner for nonholonomic mobile robots. Technical Report 92413, LAAS, Oct. 1992. To appear in *IEEE Transactions on Robotics and Automation*, 1994.

[31] K. M. Lynch. The mechanics of fine manipulation by pushing. In *IEEE International Conference on Robotics and Automation*, pages 2269–2276, Nice, France, 1992.

[32] K. M. Lynch. Estimating the friction parameters of pushed objects. In *IEEE/RSJ International Conference on Intelligent Robots and Systems*, pages 186–193, Yokohama, Japan, 1993.

[33] K. M. Lynch. Planning pushing paths. In *International Conference on Advanced Mechatronics*, pages 451–456, Tokyo, Japan, 1993.

[34] K. M. Lynch, H. Maekawa, and K. Tanie. Manipulation and active sensing by pushing using tactile feedback. In *IEEE/RSJ International Conference on Intelligent Robots and Systems*, pages 416–421, Raleigh, NC, 1992.

[35] W. D. MacMillan. *Dynamics of Rigid Bodies.* Dover, New York, 1936.

[36] M. Mani and W. Wilson. A programmable orienting system for flat parts. In *North American Manufacturing Research Institute Conference XIII*, 1985.

[37] M. T. Mason. Mechanics and planning of manipulator pushing operations. *International Journal of Robotics Research*, 5(3):53–71, Fall 1986.

[38] M. T. Mason. Compliant sliding of a block along a wall. In *International Symposium on Experimental Robotics*, pages 568–578. Springer-Verlag, June 1989.

[39] M. T. Mason and R. C. Brost. Automatic grasp planning: An operation space approach. In *Sixth Symposium on Theory and Practice of Robots and Manipulators*, pages 321–328, Cracow, Poland, Sept. 1986. Alma-Press.

[40] H. Mayeda and Y. Wakatsuki. Strategies for pushing a 3D block along a wall. In *IEEE/RSJ International Conference on Intelligent Robots and Systems*, pages 461–466, Osaka, Japan, 1991.

[41] B. Mishra, J. T. Schwartz, and M. Sharir. On the existence and synthesis of multifinger positive grips. *Algorithmica*, 2(4):541–558, 1987.

[42] J. Miura. Integration of problem solving and learning in intelligent robots. In *International Symposium on Robotics Research*, pages 13–20, Tokyo, Japan, 1989.

[43] S. Narasimhan. Task-level strategies for robot tasks. Forthcoming PhD thesis, Department of Computer Science and Electrical Engineering, Massachusetts Institute of Technology, 1994.

[44] H. Nijmeijer and A. J. van der Schaft. *Nonlinear Dynamical Control Systems.* Springer-Verlag, 1990.

[45] Y. Okawa and K. Yokoyama. Control of a mobile robot for the push-a-box operation. In *IEEE International Conference on Robotics and Automation*, pages 761–766, Nice, France, 1992.

[46] M. A. Peshkin and A. C. Sanderson. The motion of a pushed, sliding workpiece. *IEEE Journal of Robotics and Automation*, 4(6):569–598, Dec. 1988.

[47] M. A. Peshkin and A. C. Sanderson. Planning robotic manipulation strategies for workpieces that slide. *IEEE Journal of Robotics and Automation*, 4(5):524–531, Oct. 1988.

[48] M. A. Peshkin and A. C. Sanderson. Minimization of energy in quasi-static manipulation. *IEEE Transactions on Robotics and Automation*, 5(1):53–60, Feb. 1989.

[49] J. Prescott. *Mechanics of Particles and Rigid Bodies.* Longmans, Green, and Co., London, 1923.

[50] J. A. Reeds and L. A. Shepp. Optimal paths for a car that goes both forwards and backwards. *Pacific Journal of Mathematics*, 145(2):367–393, 1990.

[51] M. Salganicoff, G. Metta, A. Oddera, and G. Sandini. A direct approach to vision guided manipulation. In *International Conference on Advanced Robotics*, Tokyo, Japan, 1993.

[52] M. Salganicoff, G. Metta, A. Oddera, and G. Sandini. A vision-based learning method for pushing manipulation. In *AAAI Fall Symposium on Machine Learning in Computer Vision*, Raleigh, NC, 1993.

[53] H. J. Sussmann. A sufficient condition for local controllability. *SIAM Journal on Control and Optimization*, 16(5):790–802, Sept. 1978.

[54] H. J. Sussmann. Lie brackets, real analyticity and geometric control. In R. W. Brockett, R. S. Millman, and H. J. Sussmann, editors, *Differential Geometric Control Theory*. Birkhauser, 1983.

[55] H. J. Sussmann. A general theorem on local controllability. *SIAM Journal on Control and Optimization*, 25(1):158–194, Jan. 1987.

[56] H. J. Sussmann and V. Jurdjevic. Controllability of nonlinear systems. *Journal of Differential Equations*, 12:95–116, 1972.

[57] M. H. Yim, personal communication.

[58] T. Yoshikawa and M. Kurisu. Identification of the center of friction from pushing an object by a mobile robot. In *IEEE/RSJ International Conference on Intelligent Robots and Systems*, pages 449–454, Osaka, Japan, 1991.

[59] T. Zrimec. *Towards Autonomous Learning of Behavior by a Robot.* PhD thesis, University of Ljubljana, Computer and Information Science, 1990.

The Geometry of a Robot Programming Language

Daniel E. Koditschek *University of Michigan, Ann Arbor, MI, USA*

This paper explores the problem of building robot navigation plans via scalar valued functions in the face of incomplete information about the configuration space such as might be available from onboard sensors. It seems as though syntactical aspects of navigation function construction may play an important role. This problem provides an important concrete instance of the need for intelligent control.

1 Introduction

Let a robot be conceived rather broadly as an electro-mechanical system whose input signals take the form of commands to actuators — motors that deliver torque or force — and for which all goals of interest involve *work* in the physical sense — the application of specified forces over specified motions. If we seek to program such systems then we are bound eventually to compile our more abstractly expressed goals into their "machine language:" forced Lagrangian dynamics.

Automated reasoning about the static properties of the physical environment has been a central concern of computational geometry for at least a decade. The dynamics of physical settings also admits a geometric representation. This has been understood for several decades [1], but has not generally been exploited in engineering practice. One presumes that geometry should offer a natural language for expressing abstract goals relating to the physics of manipulation [6].

The possibility of addressing both the syntactic and dynamical aspects of robotics in the same formalism has nowhere seemed to me as great as in path planning. For here, the goal set is a point located in a set with a boundary, hence, in dynamical terms the task may be encoded as a point attractor with maximal domain of attraction. Such a

dynamical setup can be re-expressed algebraically via a special Lyapunov function — a geometric construct that we have termed a *navigation function* in earlier work [9]. In turn, scalar valued functions may be interpreted as expressions of rules. [1] Thus, there is at least a hint of a formal connection between robot rules and robot controls in this setting.

This paper offers a very sketchy and speculative rumination on the possible use and shape of such a connection. There are many syntactical or rule based planners and controllers in the AI world, and various efforts have been made to project these onto the domain of real-time control. It seems certain that this would be a profitable enterprise. It is my prejudice that such projections cannot have wider applicability than case by case tinkering absent a more formal characterization. I manage in this paper merely to offer a glimpse of some questions whose answers might set the stage for such a formalism in the specific context of robot navigation. Thus, I mean by the word "syntax," some formal automaton that outputs symbol strings — "sentences" — with the properties

Tuning: any string represents (and can be "compiled" into) a navigation function for a known (suspected) configuration space

Switching: when a new configuration space is presented for which the present symbol string is not valid, the automaton recognizes this fact, switches state and generates a new valid string

Identifying the nature of the "syntax" that might emerge from this point of view of path planning has two motivations. The first is abstract. A recent study [5] has revealed that there are mathematical obstructions to the kind of on-

---

[1] This notion seems to be the starting point of fuzzy control.

line smooth tuning one might wish for even when the topology of the configuration space is fixed — apparently, switching is built in to this problem. Yet, on the other hand, we have recently begun to realize that certain "formal" structures seem to yield valid navigation functions for topologically diverse model spaces. It seems important to understand the relation of the algebraic structure to the analytic properties of these constructions. Second, apart from the intrinsic intellectual interest, understanding how to devise programs that yield flexibly re-configurable navigation plans becomes critically important when faced with incomplete information about the environment. The navigation functions offer a concrete means of parametrizing plans according to the information available. As that information about the environment changes, their adjustment will define a hybrid dynamical system.

2 Navigation Functions

In this section I explore the relationship between the syntactic and the analytical aspects of navigation problems. Here, attention is focussed on the tuning functions and symbol strings themselves rather than possible tuning dynamics and switching automata that might generate them.

2.1 Analytical Aspects

Let the workspace, $\mathcal{W}$, be some compact connected subset of $\mathbb{R}^3$ with non-empty interior. Let $\mathcal{R}$ be the set of all placements of a specified kinematic chain (formed from a finite number of compact connected subsets of $\mathbb{R}^3$ with non-empty interior) in $\mathbb{R}^3$ and let $\mathcal{Q}(\mathcal{W}, \mathcal{R})$ denote the resulting configuration space — the set of all free placements of the chain in the workspace. Suppose there is a distinguished "destination," $q^* \in \mathcal{Q}$. For the entire paper we shall work within the generalized damper model of dynamics

$$\dot{q} = u$$

where input force, u, is proportional to a velocity rather than acceleration. The extension of these ideas to the true Lagrangian case is quite straightforward [8].

The navigation functions were defined in [9] to be scalar valued maps, Morse functions, taking their maximum value (unity) on the entire bound-

ary of $\mathcal{Q}$, their minimum value (zero) at the destination q^*, and possessing no other local minima. If $\varphi(q)$ is a navigation function then call the vector field $f(q)$ a *plan* if it admits φ as a Lyapunov function. For example, note that the gradient vector field defined by φ is a plan. Given a plan, f, the feedback,

$$u = f(q)$$

results in a closed loop system whose trajectories never pass through the boundary and converge to q^* from all but a set of initial conditions of zero measure.

We have shown that every compact connected smooth manifold with boundary admits a *navigation function* and that a navigation function for one manifold, $\varphi : \mathcal{M} \to \mathbb{R}$, composed with a diffeomorphism from another, $h : \mathcal{Q} \approx \mathcal{M}$, can serve as a navigation function for the second manifold, $\varphi \circ h : \mathcal{Q} \to \mathbb{R}$ [9]. We have furnished computationally effective recipes for such diffeomorphisms in the case that $\mathcal{M}$ is a punctured sphere and all the information about the boundaries of $\mathcal{Q}$ is available [12].

2.2 Algebraic Aspects

I have just now outlined a "theory" of navigation functions considered as being topologically rigid but geometrically extensible. That is as much about them as I know how to say formally. But in fact, we have employed the same general algebraic structure for several topologically very different models, $\mathcal{M}$. It seems compelling to try to understand what can be formalized about this algebraic structure, how it might be improved, and over what class of models it and its descendants may yield a navigation function.

2.2.1 An Algebraic Construction

Consider the function

$$\varphi = \delta_l \circ \gamma / \beta$$

where, δ_l is a member of a parametrized family of diffeomorphism from $[0, \infty]$ to $[0, 1]$, γ is the Euclidean distance to the goal in $\mathcal{M}$ and β goes to 0 on the boundary, $\partial \mathcal{M}$. Setting the boundary value is achieved by taking β to be the algebraic product of functions that vanish on each connected component of the boundary

$$\beta = \beta_0 \cdot \beta_1 \cdot \ldots \cdot \beta_k.$$

Consider now a family of loci $\{b_i\}_{i=0,k} \subset \mathbf{E}^n$, each point, b_j forming the center of a disk, $\mathcal{D}_j$ of radius ρ_j. Suppose all the "smaller" disks are in one larger disk, $\mathcal{D}_j \subset \mathcal{D}_0, j > 0$ but that otherwise they are pairwise disjoint. Take for the model space, the "geometric sphere world"

$$\mathcal{M} = \mathcal{D}_0 - (\mathcal{D}_1 \cup \ldots \cup \mathcal{D}_k).$$

This is the configuration space for a point robot in a punctured sphere — that is, the workspace is $\mathcal{W} = \mathcal{M}$, and $\mathcal{R} \approx \mathbf{E}^2$ is the set of all translations of a point, so that $\mathcal{Q}(\mathcal{M}, \mathbf{E}^2) \approx \mathcal{M}$. It can be shown that if β_j vanishes on $\partial \mathcal{D}_j$ then a function of the form φ is a navigation function on $\mathcal{M}$ when δ_l is chosen properly [9]. Implicit in the proof of this result is that more disk punctures can be accommodated in an essentially "automatic" manner. One just adds a new factor, β_{k+1}, to the recipe for β and adjusts δ_l. Thus, relatively simple adjustments to the structure of φ handle topological changes in the the configuration space. Somehow, this should not seem surprising — why?

What does seem surprising is that the same construction works in the case of a "spider robot" in a punctured sphere. That is, keeping $\mathcal{W} = \mathcal{M}$, take $\mathcal{R}$ to be the rigid placements of a locked overlapping pair of disks, $\mathcal{S}$, into $\mathcal{M}$,

$$\mathcal{R} \approx SE(\mathcal{S}, \mathcal{M})$$

so that the configuration space is a solid torus punctured by toral tubes (that may split and join as they are traced through the interior). Rimon [13] has shown that (after a bit more adjustment) the structure remains valid here too — φ is a navigation function on $\mathcal{Q}(\mathcal{M}, SE(\mathcal{S}, \mathcal{M}))$.

Moreover, recent work on potential fields for multiple robots [14] and for robotic assembly [7] suggests that the same construction remains valid when $\mathcal{W} = \mathbf{E}^n$ and $\mathcal{R}$ is the simultaneous placement of m disks,

$$\mathcal{R} \approx \prod_{i=1}^{m} SE(\mathcal{D}_i, \mathbf{E}^n).$$

The condition for adjusting δ_l to get a rigorous proof in the case $n = 1$ amounts to a convexity condition for φ [7], and this may offer some hint of what is going on. But note that in general, $\mathcal{Q}$ will not itself be a convex space, so there is no chance of φ satisfying a convexity property.

In each of these cases, the metric features of the objects being placed figures prominently. Each is

formed from disks in some essential manner. Thus, there seems to be some geometric (rather than topological) property of $\mathcal{R}$ that is favored by φ. What property is this?

2.2.2 A Fuzzy Interpretation

Imagine building a formal syntax of scalar valued expressions in the following manner. Let the product of two functions denote their "conjunction." Represent "negation" of a function by its reciprocal. This yields a sentential logic, $\mathcal{L}$. Is there some model for $\mathcal{L}$?

One might loosely interpret the "sentence"

$$\frac{\gamma}{\beta_0 \cdot \beta_1 \cdot \ldots \cdot \beta_k}$$

to mean "move toward the zero of γ and do not move toward the zero of β_0 and do not move toward the zero of β_1 and..." and so on. This would be an anthropo-centric semantics for $\mathcal{L}$. On the other hand, for some non-trivial range of model spaces, $\mathcal{M} = \mathcal{Q}(\mathcal{W}, \mathcal{R})$, the construction is a mere "shaping", δ_l, away from a navigation function φ. Since possessing the navigation property is equivalent to a closed loop plan, one might intepret this as a "robo-centric" semantics for $\mathcal{L}$.

Can these two semantics — the anthropo- and robo- centered — be formalized as models (in the sense of mathematical model theory) of $\mathcal{L}$? Are there other sentences that "mean" the same thing in both semantic systems?

3 Sensor Based Navigation Functions

This section introduces a setting wherein the tuning of diffeomorphisms and switching between symbol strings must be done online and in a coordinated manner. A (perhaps *the*) central problem in robot navigation concerns the incorporation of sensor derived information about the configuration space. Heretofore, we have assumed full a priori knowledge of $\mathcal{Q}$ and precise online information concerning $q \in \mathcal{Q}$. But real sensors provide merely filtered views of the workspace and it is a major problem to devise a means of recovering from the evolving history of sensor readings the configuration space and the robot's state therein.

There are really two hard problems. First, how can one infer configuration space properties from the highly filtered views of the workspace given by sensors? Second, how can one successfully adjust the plan in an online manner to be consistent with the sensor derived adjustments in the presumed configuration space. I speculate briefly on the manner in which navigation functions relate to the first problem and then move on to consider their role in addressing the second.

3.1   Navigation from Sensory Data

Loosely following Donald [3], we will suppose that a "sensor," s, provides reports to a "sensorium," S, concerning the classes of "perceptually equivalent" configuration space locations

$$s : \mathcal{Q} \to S,$$

so that given a report, $\sigma \in S$ we "know"

$$q \in \mathcal{Q}_\sigma := s^{-1}(\sigma)$$

only up to an equivalence class.

3.1.1   Projections of a Known a Configuration Space

Assume complete information about $\mathcal{Q}$. We will loosely follow Erdmann [4] in proposing an adaptation of navigation function methods to this situation. Since $\mathcal{R}$ is always a Lie Group, it is parallelizable [2], that is,

$$T\mathcal{R} \approx \mathcal{R} \times \mathbf{E}^k$$

where it is understood that the Euclidean vector space $\mathbf{E}^k$ represents the Lie Algebra. Directions of infinitesimal motion in $\mathcal{R}$ can thus be represented (uniformly over $\mathcal{R}$) by $\mathbf{S}^{k-1}$ — unit ball of of $\mathbf{E}^k$. Consider the "cone of progress" generated over a perceptual equivalence class, $\mathcal{Q}_\sigma$, by a navigation function

$$\mathcal{V}_\sigma := \left\{ \frac{grad \, \varphi \, (q)}{\|grad \, \varphi \, (q)\|} : \exists q \in \mathcal{Q}_\sigma \right\} \subset \mathbf{S}^{k-1}$$

If $\mathcal{V}_\sigma$ is contained in a hemisphere, $\mathcal{V}_\sigma \subset \mathcal{H}^+$, then any direction in the opposing, hemisphere, $w_\sigma \in \mathcal{H}^-$ will result in progress. That is to say, the output feedback rule

$$u(\sigma) = w_\sigma$$

generates a plan: it results in a closed loop system for which φ is a Lyapunov function.

Hemispheric containment is a very strong condition. It seems to imply, for example that there

be a "beacon" at the destination $q^* \in \mathcal{Q}$ — that is to say, we would require a privileged sensor value, $\sigma^* \in S$ whose perceptual equivalence class is the singleton at the destination, $\mathcal{Q}_{\sigma^*} = \{q^*\}$. Such restrictive assumptions need to be replaced with a filtering theory.

In general, even in linear control theory problems, output feedback is rarely sufficient to achieve stabilization, and what is really needed here is some "structural observer theory" that would use the time history of the sensor reports, $\sigma(t)$ to refine the "raw" partition induced by single measurements. That, of course, is what many researchers are presently trying to work out, and I will say no more about such efforts here.

3.1.2   Estimation From Projections of an Unknown Configuration Space

As the previous discussion shows, when a sensor provides information about a known environment, there is, in control theoretic parlance, an observer problem. When the sensor provides information about an unknown environment there arises an adaptive observer problem.

Suppose there is a parametrized family of possible workspaces, $\{\mathcal{W}_p\}_{p \in \mathcal{P}}$, into any one of which the robot, $\mathcal{R}$ may have been placed. Denote by $\mathcal{Q}^p$ the configuration space $\mathcal{Q}(\mathcal{W}_p, \mathcal{R})$. There is now a different sensor map for each workspace, $s_p : \mathcal{Q}^p \to S$. We require an estimation procedure for adjusting p by comparing the sensor reports, σ, with the estimates, $s_p(q)$. This is an "adaptive" problem in the sense that both q and p are unknown and must be each adjusted online. I will now simply presume the choice of some parameter tuning method that results in a trajectory of estimates over time, $p(t)$.

3.2   Tuning and Switching

We have just seen that sensor based navigation problems give rise to evolving views of the true configuration space, $\mathcal{Q}^{p(t)}$ resulting from a motion in parameter space, $p : \mathbb{R} \to \mathcal{P}$, consequent upon some sensor interpretation scheme applied over the course of exploration. For some portions of the motion, we expect geometric changes in the configuration space, and for others we may expect topological changes.

3.2.1 Tuning: Adaptation to Changing Geometry

In those situations where we have succeeded in constructing them, our diffeomorphisms are defined in terms of the specific geometric features of the particular configuration space [12]. Thus for a selected p^* we may think of an associated diffeomorphism, $h_{p^*} : Q^{p^*} \to \mathcal{M}$ and, for some neighborhood of $\mathcal{N}(p^*) \subset \mathcal{P}$ one might imagine an associated family diffeomorphisms, $\{h_p\}_{p \in \mathcal{N}(p*)}$. Then a planner might "track" the changing sensory views with the navigation function sequence

$$\varphi(t, q) =: \varphi \circ h_{p(t)}(q).$$

3.2.2 Switching: Adaptation to Changing Topology

But it must be expected that local changes in the shape of the workplace (or the robot's shape) will cause topological changes in the configuration space.

Say that $(\mathcal{W}_1, \mathcal{R}_1)$ is *equivalent* to $(\mathcal{W}_2, \mathcal{R}_2)$ if $\mathcal{Q}(\mathcal{W}_1, \mathcal{R}_1)$ is diffeomorphic to $\mathcal{Q}(\mathcal{W}_2, \mathcal{R}_2)$. Using the Hausdorf "shape" metric, the set of all robot-workspace pairs has a natural topology. Say that a pair $(\mathcal{W}, \mathcal{R})$ is *structurally stable* if it lies in some open neighborhood of equivalent pairs; otherwise say that it is a *bifurcation pair*.

I now conjecture that the structurally stable pairs form an open dense subset of the space of pairings. This would impose a cellular partition on $\mathcal{P}$, the parameter space that summarizes the a priori knowledge of the workspace structure. Each open cell would correspond to a structurally stable class separated from its neighboring class by an empty interior bifurcation set. If so, then it is to be expected that a one-parameter family of pairings — a sequence of estimates of the workspace returned over time as the robot directs its sensors through the environment — will generally consist of intervals of equivalence punctuated by bifurcations.

Thus, there is a recognition problem and an action problem for the syntax generating automaton. Will it be possible using local clues from the sensor to deduce a bifurcation? There is a discrete sequence of models, $\{\mathcal{M}_k\}_{k \in K}$, to be determined. How might it be possible to choose the model, a model function, φ_k, a nominal geometry, p_k^* and a tuned family $\{h_p\}_{p \in \mathcal{N}(p_k^*)}$?

3.2.3 Tuning and Switching Must be Mixed Together

Topologically equivalent (smooth) configuration spaces admit the same navigation function up to diffeomorphism. Thus, we might have imagined that $\{h_p\}_{p \in \mathcal{N}(p_k^*)}$ is a smooth parametrization. Unfortunately, this may not always be possible.

In recently reported work, Hirsch and Hirsch [5] have examined the fiber bundle, $\pi : \bar{\mathcal{E}} \to \mathcal{B}$, of navigation functions to the family of two-spheres with a finite number of punctures. Recall that a two-sphere, $\mathbf{S}^2$, with k punctures is a topological model for the configuration space of a (rotationally symmetric) mobile robot in a closed room with some number of obstacles. The location of the punctures defines a point in

$$\mathcal{B} = \underbrace{\mathbf{S}^2 \times ... \times \mathbf{S}^2}_{k \text{ times}},$$

and for each $b = (b_1, ..., b_k) \in \mathcal{B}$, the fiber $\bar{\mathcal{E}}_b$ denotes the set of Morse functions that take their only minimum on b_1, their maxima at $b_2, ...b_k$ and have some number of other saddle points as well.

It shouldn't matter where the punctures are on the sphere since all the sets parametrized by $\mathcal{B}$ are diffeomorphic. Thus, one imagines the possibility of assigning to each b a distinguished member of its fiber, $\varphi_b : \mathbf{S}^2 \to I\!\!R^2$ such that a smooth trajectory $b(t) \in \mathcal{B}$ lifts to a smoothly parametrized family of navigation functions, $\varphi_{b(t)}$. Apparently, this is not possible — the Hirsch $\bar{\mathcal{E}}$ admits no smooth section.

4 Conclusion: Is there an "Intelligent" Dynamics

Recently, Narendra and I have asserted that an "intelligent" controller will require the ability to switch and tune [11]. Surely, sensor based robot navigation represents an example of the need for such intelligence. There seem to be at least two significant gaps in the ideas explored by this paper.

First, as discussed in Section 3.1.2, one requires an "estimation dynamics" to move around in $\mathcal{P}$ on the basis of the reports from s_p. This may be addressed by standard algorithms or it may give rise to presently unsolved problems, depending upon the parametrization — that is, how p enters Q^p and, thereby, enters s_p and h_p. At any rate, there is a large existing body of theory and practice to inform such investigations [10].

Second, as discussed immediately above in Section 3.2, one requires a means of switching between models and model functions. Here, the problem seems less well delimited. There is a presumption that the manner in which one equivalence class of $(\mathcal{W}, \mathcal{R})$ pairs bifurcates into another class may be captured by some discrete dynamics. This is indeed presumptuous, but it is perhaps not impossible to imagine for restricted classes of problems. One would then need to understand how the "dynamics" of configuration space bifurcations could be captured in the "syntax" generating automaton that produces the sequence φ_k.

Acknowledgements

This work was supported in part by the National Science Foundation under grants IRI-9123266 and IRI-9216823.

References

[1] Ralph Abraham and Jerrold E. Marsden. *Foundations of Mechanics.* Benjamin/Cummings, Reading, MA, 1978.

[2] William M. Boothby. *An Introduction to Differentiable Manifolds and Riemannian Geometry.* Academic Press, New York, 1975.

[3] Bruce Donald and Jim Jennings. Constructive recognizability for task-directed robot programming. In *IEEE Int. Conf. Rob. Aut.,* 1992.

[4] Michael Erdmann. Towards task-level planning: Action based sensor design. Technical Report CMU-CS-92-116, Carnegie Mellon University, Pittsburgh, PS, Feb. 1992.

[5] Michael Hirsch and Morris Hirsch. On the non-existence of certain morse functions. Talk presented at the 1993 Mid-West Dynamical Systems Conference, Oct. 1993.

[6] D. E. Koditschek. Task encoding: Toward a scientific paradigm for robot planning and control. *Journal of Robotics and Autonomous Systems,* 9:5–39, 1992.

[7] D. E. Koditschek. An approach to autonomous robot assembly. *Robotica,* (to appear), 1994.

[8] Daniel E. Koditschek. The Application of Total Energy as a Lyapunov Function for Mechanical Control Systems. In J. Marsden, Krishnaprasad, and J. Simo, editors, *Control Theory and Multibody Systems,* volume 97, pages 131–158. AMS Series in Contemporary Mathematics, 1989.

[9] Daniel E. Koditschek and Elon Rimon. Robot navigation functions on manifolds with boundary. *Advances in Applied Mathematics,* 11:412–442, 1990.

[10] K. S. Narendra and A. Annaswamy. *Stable Adaptive Systems.* Prentice-Hall, NY, 1988.

[11] K. S. Narendra and D. E. Koditschek. Intelligent control systems design. (Funded Proposal: National Science Foundation), 1991.

[12] E. Rimon and D. E. Koditschek. The construction of analytic diffeomorphisms for exact robot navigation on star worlds. *Transactions of the American Mathematical Society,* 327(1):71–115, Sep 1991.

[13] Elon Rimon. A navigation function for a simple rigid body. In *Proc. IEEE Int. Conf. Rob. and Aut.,* pages 2–7, Sacramento, CA, April 1991. IEEE Computer Society.

[14] Louis L. Whitcomb, Alfred Rizzi, and Daniel E. Koditschek. Comparative experiments with a new adaptive controller for robot arms. In *Proc. IEEE Int. Conf. Rob. and Aut.,* pages 2–7, Sacramento, CA, April 1991. IEEE Computer Society.

The Robot Localization Problem

Leonidas J. Guibas, *Stanford University, Palo Alto, CA, USA*

Rajeev Motwani, *Stanford University, Palo Alto, CA, USA*

Prabhakar Raghavan, *IBM TJ Watson Research Center, Yorktown Heights, NY, USA*

We consider the following problem: given a simple polygon $\mathcal{P}$ and a star-shaped polygon $\mathcal{V}$, find a point (or the set of points) in $\mathcal{P}$ from which the portion of $\mathcal{P}$ that is visible is congruent to $\mathcal{V}$. The problem arises in the localization of robots using a range-finder — $\mathcal{P}$ is a map of a known environment, $\mathcal{V}$ is the portion visible from the robot's position, and the robot must use this information to determine its position in the map. We give a scheme that preprocesses $\mathcal{P}$ so that any subsequent query $\mathcal{V}$ is answered in optimal time $O(m + \log n + A)$, where m and n are the number of vertices in $\mathcal{V}$ and $\mathcal{P}$, and A is the number of points in $\mathcal{P}$ that are valid answers (the output size). Our technique allows us to trade off smoothly between the query time and the preprocessing time or space. We also devise a data structure for output-sensitive determination of the visibility polygon of a query point inside a polygon $\mathcal{P}$. We then consider a variant of the localization problem in which there is a maximum distance to which the robot can "see" — this is motivated by practical considerations, and we outline a similar solution for this case. We also show that a single localization query $\mathcal{V}$ can be answered in time $O(mn)$ with no preprocessing.

1 Introduction

We consider the following problem: a robot is at an unknown position in an environment for which it has a map. It "looks" about its position, and based on these observations must infer the place (or set of places) in the map where it could be located. This is known as the *localization problem* in robotics [7, 22].

Aside from being an interesting and fundamental geometric problem, the problem has several practical applications. As described in [7], localization eliminates the need for complex position-guidance equipment to be built into factories and buildings. Unmanned spacecraft require localization for the following reason [18, 20]: a rover lands on Mars, a map of whose terrain is available to it. It looks about its position, and then infers its exact position on the Martian surface. Another application comes from robots that follow a planned path through a scene: the control systems that guide a robot along the planned path gradually accumulate errors due to mechanical drift. Thus it is desirable to use localization from time to time to verify the actual position of the robot in the map, and apply corrections as necessary to return it to the planned path [22].

We assume that the robot is in an environment such as an office or factory, with vertical walls and a flat floor. The subject of this paper is localization using a *range finder* [19], a device commonly used in real robots [7, 8, 9, 19]. A range finder is a device that emanates a beam (laser or sonic), and determines the distance to the first point of contact with any object in that direction. This is similar to the *finger probe* model [6, 21] studied in computational geometry. In practice a robot sends out a series of beams spaced at small angular intervals about its position, measuring the distance to points at each of these angles. The discrete "points of contact" are joined together to obtain a *visibility polygon* $\mathcal{V}$ with m vertices.

The robot has a map of its environment: a polygon $\mathcal{P}$ (possibly containing holes) having n vertices. We assume that the robot has a compass: its representations of $\mathcal{P}$ and $\mathcal{V}$ have a common reference direction (say North). We wish to solve the following problem: given $\mathcal{P}$ and $\mathcal{V}$, determine all the points $p \in \mathcal{P}$ such that the visibility polygon of p is exactly $\mathcal{V}$.

Because the map is likely to be fixed for a given environment, our main interest is in preprocessing the map so that subsequent queries can be answered quickly. Our main contribution is a scheme for preprocessing a simple polygon $\mathcal{P}$ so that any query $\mathcal{V}$ can be answered in time $O(m + \log n + A)$, where A is the size of the output (the number of places in $\mathcal{P}$ at which the visibility polygon would match $\mathcal{V}$); this query time is the best possible. Our preprocessing takes $O(n^5 \log n)$ time and $O(n^5)$ space (Section 5). We also exhibit a smooth trade-off between the query time and the preprocessing cost. The preprocessing time and space complexity achieve their worst-case bound only for instances of map polygons that contain degeneracies. Let the number of mutually visible pairs of vertices in $\mathcal{P}$ be v and let r be the number of reflex vertices in $\mathcal{P}$, then the bound can be expressed as $O(n^3 v)$ and this is always $O(n^4 r)$.

The development of our scheme involves the study of a fundamental property of simple polygons — the *visibility cell decomposition* — that has other potential applications. Sections 2, 3 and 4 study some properties of this decomposition. An interesting application of these ideas is to the construction of a data structure for the output-sensitive determination of the visibility polygon of a query point (see Section 6.4). This requires a query time of $O(\log n + m)$ using $O(n^2 r)$ preprocessing and space, where m is the size of the visibility polygon. Once again we can trade-off between the query time and the space requirement.

In Section 6 we consider variants of the basic problem. We first describe the effect of holes in the map polygon. Next, we consider a variant motivated by a property of some range-finders — that a distance measurement is obtained only if there is a wall within a certain maximum distance D, and otherwise no reading is obtained (indicating only that the distance to the nearest wall is greater than D in that direction). We show how our approach can be modified to deal with this feature without increasing the query time, but with additional preprocessing. We then address the following question: given no preprocessing, what is the complexity of answering a single query? We provide an algorithm running in time $O(mn)$ by applying results on ray-shooting to our problem.

Independently of our work, Bose, Lubiw and Munro [1] have obtained some of the results presented here. In particular, they provide a scheme for preprocessing a simple polygon so as to compute the vertices of the polygon that are visible from a query point in time $O(\log n + m)$ time. The machinery developed for this purpose includes some of the structure theorems described below.

1.1 Overview of our scheme.

We now give a brief overview of our scheme to motivate the study of the visibility cell decomposition in Sections 2, 3 and 4. We discretize the problem by partitioning the map polygon into regions such that within a region the visibility polygon of any point is roughly the same — in Section 2 we call this rough view a *skeleton*. An intuitive definition of the skeleton of $\mathcal{V}$ is that it is a contraction of $\mathcal{V}$ whose boundary contains exactly those vertices from $\mathcal{V}$ that can be certified to be vertices of $\mathcal{P}$. We provide a data structure which quickly identifies all the regions that have the same skeleton as the query $\mathcal{V}$. We then check the candidate regions to see if they contain any points that have exactly the same view as $\mathcal{V}$. Some difficulties that arise are:

a) Due to occlusions by reflex vertices, an edge of the map polygon may have neither or only one of its endpoints visible from a point inside the polygon. Our characterization of a skeleton must cater to these incomplete edges.

b) If the line segments forming several edges of the polygon are collinear, it is possible that a "window" in the map allows the robot to see only an interior portion of one of these edges. Further, it cannot easily identify which of these collinear edges it sees. The problem is compounded when there are several such windows and collections of collinear edges. In fact, this is a main source of complexity in our preprocessing. In the case where the map $\mathcal{P}$ has no collinear edges the preprocessing time can be improved to $O(n^3)$.

c) There can be regions that match the skeleton but contain no point whose visibility polygon is congruent to V. Thus we must still pinpoint those visibility regions (from all the ones that share this skeleton) that contain a point whose visibility polygon exactly matches $\mathcal{V}$. We must do so in time proportional to A, so we cannot check each candidate region individually. We reduce this problem to a form of point-location in a planar subdivision.

2 Visibility Polygons and Skeletons

Let $\mathcal{P}$ denote a polygon with n sides. We will refer to $\mathcal{P}$ as the `map polygon`. Let P denote the boundary of $\mathcal{P}$. We first assume that $\mathcal{P}$ has no holes, deferring this general case to Section 6.1. From now on all polygons will be assumed to be oriented with respect to a common reference direction.

Two points in P are <u>visible</u> to each other if the straight line joining them meets P only at these endpoints. The <u>visibility polygon</u> $\mathcal{V}(p)$, for any point $p \in \mathcal{P}$, is the polygon consisting of all points in $\mathcal{P}$ that are visible from p. Assume that the number of vertices in $\mathcal{V}(p)$ is m.

Let $V(p)$ denote the boundary of $\mathcal{V}(p)$. Assume that p does not lie on P, and hence it does not lie on $V(p)$. In general, there may be edges and vertices in $V(p)$ which are not in P but lie in the interior of $\mathcal{P}$. To deal with such cases we define the notion of spurious edges and vertices. Informally, an edge or a vertex is non-spurious if the view from p provides a guarantee that this edge or vertex is on the boundary of $\mathcal{P}$.

Definition 1 *An edge of $V(p)$ is* <u>spurious</u> *if it is collinear with p.*

Definition 2 *A vertex $v \in V(p)$ is* <u>spurious</u> *if it lies on a spurious edge $(u, v) \in V(p)$ and the other end-point u is closer to p.*

This definition may label as spurious an edge (or a vertex) which actually lies on P (see Fig. 1). This only happens if that edge is collinear with p. In that case, the closer of the two end-points of the edge may be visible from p but it will then block the view of any other point on the edge. Thus, although the edge (u, v) is in $V(p)$, the robot sitting at point p cannot infer this from its localized view. Similarly, the definition of a spurious vertex assumes that if a ray from p goes through vertices u and v of P, in that order, then u is an obstacle to the visibility of v from p. As the next lemma shows, the non-spurious components of $V(p)$ are invariant under any modifications to $\mathcal{P} \setminus \mathcal{V}(p)$. The proof is an easy consequence of the above definitions.

Lemma 1 *An edge e, or a vertex v, in V is non-spurious if and only for each choice of $\mathcal{P}$ and $p \in \mathcal{P}$ such that $\mathcal{V}(p) = V$, e and v lie on P.*

A <u>reflex vertex</u> in P is a vertex which subtends an angle greater than 180^0 inside $\mathcal{P}$. It is the existence of reflex vertices which creates obstacles to viewing the points inside $\mathcal{P}$. Note that a spurious vertex can never be a reflex vertex in $V(p)$.

Definition 3 *A reflex vertex v in $V(p)$ is a* <u>blocking vertex</u> *if at least one edge incident on v in P does not intersect $\mathcal{V}(p)$.*

It is now easy to establish the following lemma.

Lemma 2 *If a vertex in $V(p)$ is a non-blocking reflex vertex, then both its incident edges in $V(p)$ must be non-spurious.*

The next lemma follows from the observation that each spurious edge can be extended to pass through p. Thus, no two spurious edges of $V(p)$ can meet each other except at p, and by assumption p does not lie on the boundary of $\mathcal{V}(p)$.

Lemma 3 *No two spurious edges can be adjacent in $V(p)$.*

For each spurious edge, the end-point closer to p is a blocking reflex vertex and the other end-point is a spurious vertex.

Lemma 4 *Let $e \in V(p)$ be a non-spurious edge and e' the edge of P on which it lies. Then e is the only portion of e' visible from p and the edge e is of one of the following three types.*

- <u>full edge</u>: *the end-points of e are the same as those of e';*

- <u>half edge</u>: *one end-point of e is spurious, the other is an end-point of e';*

- <u>partial edge</u>: *both end-points of e are spurious vertices.*

We now conclude that the spurious vertices and edges in $V(p)$ can only occur in certain specific patterns. Consider a clockwise traversal of $V(p)$ starting with an arbitrary blocking vertex. (If no such vertex exists then $\mathcal{V}(p) = \mathcal{P}$, trivializing the whole problem.) The sequence of vertices seen in this traversal can be decomposed into chains of consecutive non-spurious vertices alternating with chains of consecutive spurious vertices – call this the <u>chain decomposition</u> of $V(p)$.

Lemma 5 *The* <u>chain decomposition</u> *of $V(p)$ has the following properties.*

1. *A non-spurious chain can contain blocking vertices only as its end-points.*

2. *A spurious chain is of length at most 2.*

3. *Consider a spurious chain with only one vertex v. Let x be the last vertex of the preceding chain and y the first vertex of the succeeding chain. Then one of x and y is a blocking vertex joined to v by a spurious edge, while the other is non-blocking and is joined to v by a half-edge.*

4. *Consider a spurious chain with two vertices u and v, in that order. Let x be the last vertex of the preceding chain and y the first vertex of the succeeding chain. Then both x and y are blocking vertices with spurious edges going to u and v, respectively, and the edge (u,v) is a partial edge.*

Fix a unique non-spurious vertex v_o of $V(p)$ as the origin with reference to which we specify all other points. Let $V^*(p)$ be the polygon induced by the non-spurious vertices of $V(p)$ ordered by a clockwise traversal of starting at v_o. In $V^*(p)$, if two adjacent vertices are from the same chain in $V(p)$, then their edge is the same as in $V(p)$, and this must be a non-spurious edge. Otherwise, the two vertices are end-points of neighboring chains in $V(p)$ and their edge in $V^*(p)$ is a newly introduced artificial edge.

Every artificial edge e' of $V^*(p)$ corresponds to some half or partial edge e of $V(p)$. The edge e is one of the edges of $V(p)$ which connect the two chains whose endpoints are the vertices of e'. We will label each artificial edge e' with a characterization of the line on which the corresponding edge e lies. This line-characterization will be the coefficients of the linear equation that defines the line containing e, with the origin at v_o.

Definition 4 *The skeleton $V^*(p)$ of a visibility polygon $V(p)$ is the polygon induced by the non-spurious vertices of $V(p)$. Each artificial edge of $V^*(p)$ is labeled with the line equation and the type of the corresponding half or partial edge.*

The skeleton of a visibility polygon can also be looked upon as a polygon induced by all the full edges in $V(p)$, such that the chains of edges are tied together by artificial edges. It is important to keep in mind that a skeleton is a *labeled* polygon as described above.

Definition 5 *The embedding of a skeleton $V^*(p)$ is a 1-1 mapping h from the vertices in the skeleton into the vertices of P such that:*

1. *For each vertex v in $V^*(p)$, the location of $h(v)$ relative to $h(v_o)$ is identical to the location of v relative to v_o in $V(p)$.*

2. *There is a full edge between vertices u and v in $V^*(p)$ if and only if there is an edge of P with $h(u)$ and $h(v)$ as end-points.*

3. *Let l' be the line labeling an artificial edge between vertices u and v in $V^*(p)$. Then there is an edge e of P lying on a line l whose equation (with $h(v_o)$ as origin) is that of l', and a point of e visible from both u and v.*

Does $V^*(p)$ have enough information to uniquely determine the point p? Unfortunately not: the information about the end-points of a half or partial edge e in $V(p)$ is absent from the labels of the corresponding edge e' in $V^*(p)$. (The reason for this imprecise labeling will become clear later when we describe our search mechanism.) Thus a single embedding may have several candidate edges in P for the edge e. These candidate edges must all be collinear. For instance, see Fig. 1. The partial edge on the left visible from p, and that visible from q, are collinear. Note also that the skeletons of p and q come out to be the same.

Let r denote the number of reflex vertices in $\mathcal{P}$.

Theorem 1 *Given a visibility polygon $V(p)$, its skeleton $V^*(p)$ has at most r embeddings in $\mathcal{P}$ and this bound is the best possible.*

Proof: Let us label any embedding h of $V^*(p)$ in $\mathcal{P}$ by the vertex $h(v_o)$, where v_o is the origin vertex in $V^*(p)$. We claim that there can be at most one embedding with the label v, for any vertex v in P. This follows from the observation that the location of every vertex in $V^*(p)$ is fixed with reference to the origin vertex. Having specified the location of the origin as being at v immediately fixes the location of every other vertex and thus uniquely determines the embedding h itself.

Since v_o is a reflex vertex, there are at most r distinct labels for the embeddings of $V^*(p)$. It follows that number of distinct embeddings cannot exceed r.

To see this bound is tight to within constant factors, consider the polygon $\mathcal{P}$ and the visibility polygon $V(p)$ in Fig. 2. It is clear that the origin of $V^*(p)$ can be mapped to any of the vertices marked with o in P. Clearly, there are $\Omega(n)$ such embeddings.   □

The location of the point p is fixed with reference to the origin vertex v_o of $V^*(p)$. Thus, the only possible locations of the point p are the (at most r) locations corresponding to the different embeddings of the skeleton. Notice that an embedding of $V^*(p)$ may not correspond to an embedding of $V(p)$, although the converse is always true.

Corollary 1 *The number of solutions in $\mathcal{P}$ to a localization query $\mathcal{V}$ is at most r.*

3 Visibility Cell Decomposition

We now describe the subdivision of the map polygon into visibility cells such that the points in each cell have essentially the same visibility polygon. The subdivision is created by introducing lines into the interior of the map polygon. Each line partitions $\mathcal{P}$ into two regions, one where a vertex v is not visible due to the obstruction created by vertex u and another region where u cannot block the view of v. Each such line starts at a reflex vertex u and ends at the boundary of $\mathcal{P}$. It is collinear with a vertex v which is either visible from u or is adjacent to it in P. This line is said to be *emanating* from v and *anchored* at u.

It is convenient to give each of the interior lines a direction and consider the interior of the map polygon to be dissected by a collection of such rays. A ray determined by vertex v as emanating vertex and vertex u as anchor vertex proceeds *from u into P away* from v. It forms the boundary between regions that can see v and others that cannot. We will classify this ray as a *left* or *right* ray for v according to whether the obstruction defining the anchor u is to the left or the right of the ray, from the point of view of an observer sitting on v and looking along the ray. Note again that a ray starts at the anchor vertex and proceeds away from the emanating vertex.

Theorem 2 *In the visibility cell decomposition of the map polygon $\mathcal{P}$:*

1. *The number of lines introduced in the interior of $\mathcal{P}$ is $O(nr)$, and this bound is the best possible.*

2. *Each cell in the decomposition is a convex polygonal region inside $\mathcal{P}$.*

Proof: It easy to see that there are at most $O(nr)$ interior lines in the decomposition since each interior line is generated by a pair of anchor and emanating vertices, and each pair of vertices generates at most two interior lines. It is also fairly easy to construct examples where the upper bound is achieved, see Fig. 2.

For the second part, let $\mathcal{C}$ be a visibility cell which is non-convex. Let w be a reflex vertex on the boundary of $\mathcal{C}$ (if the cell is non-convex, its boundary must have at least one reflex vertex). We first claim that w cannot be a vertex from the boundary of $\mathcal{P}$. Otherwise, the edges incident on w would be extended into interior lines, and these interior lines would sub-divide $\mathcal{C}$. On the other hand, no interior vertex can be a reflex vertex for its bordering visibility cells since it is formed by the intersection of two interior lines which start and end at the polygon boundary. This gives a contradiction. $\square$

Since we can have $\Theta(nr)$ lines forming the cell decomposition, an obvious bound on the total complexity of the cells in the decomposition is $O(n^2r^2)$. However, the structure of our problem can be exploited to obtain the following tight bound.

Theorem 3 *The number of visibility cells in a given map polygon, as well as their total complexity, is $O(n^2r)$, and this bound is the best possible.*

Proof: We will show that the number of subdivision vertices is $O(n^2r)$. There is a total of $O(nr)$ rays, and each ray gives rise to one boundary vertex only. Therefore it suffices to count non-boundary vertices for the asymptotic bound in the theorem.

Consider a vertex of the subdivision which does not lie on the boundary P. There must be two rays whose intersection gives rise to this vertex v. Let the first ray emanate from the vertex a with the anchor vertex A, and the second ray emanate from b with the anchor B. Consider the two lines containing the given rays, they divide the plane into four "quadrants". Notice that if both rays are of the same orientation (i.e. both left or both right) then the corresponding anchors must lie in adjacent quadrants. In the case where the two rays have different orientations, the corresponding anchors lie in either the same quadrant or in opposite quadrants.

Label each vertex of the subdivision that does not lie on the boundary of P by the pair of vertices that the two rays determining the vertex emanate from, as well as by the two bits specifying the relative placement of the two anchors in the quadrants defined above. The two bits classify the subdivision vertices into four types. Type 1 vertices are determined by two left rays and have the anchors in adjacent quadrants; Type 2 vertices are similar except that the rays are both of the right orientation. Type 3 vertices have the anchors in the same quadrant, while Type 4 have the anchors in opposite quadrants.

The hardest case to deal with is that of the Type 4 vertices – when the two anchors lie in opposite quadrants. Consider a Type 4 vertex v determined by two rays, say a ray from a with right anchor A, and a ray from b with left anchor B. Suppose we take the portion of the boundary of P from A to B not containing a or b and replace it with the interior segments Av and vB. We have not eliminated any vertices of our subdivision with the same label as v because in the eliminated part of P at least one of a of b is not visible. In the remaining part of $\mathcal{P}$, there cannot be any other Type 4 vertex labeled by a and b which uses either A or B as an anchor for the two rays. Thus, there can be at most r vertices with this label. Since the total number of labels is $O(n^2)$, we have the desired bound.

There are three other cases to consider. In each case, a simple topological argument shows that only a *unique* vertex can possess that label. Thus there can be a total of $O(n^2r)$ interior vertices of our subdivision and so the number of cells is similarly bounded.

For a lower bound, consider a polygon P that has n small bays (each say of three sides) lying roughly along a straight line and facing a convex chain of n sides that is visible from all of them (see Fig. 2). Then within each bay we can get $\Theta(n^2)$ regions corresponding to the visible subchain of the convex chain. This gives a total of $\Theta(n^3)$ subdivision vertices in total.   □

4   Visibility from a Cell

We now examine the visibility polygons for all the points in a particular cell and extract some common features from these views. We must be careful about the assignment of the points on the interior lines to the cells in the decomposition. Consider an interior ray emanating from vertex u and anchored at vertex v. The cells bordering this ray are divided into classes, those which can see u and those which cannot. The boundary edges determined by this ray are assumed to be a part of the latter kind of cells only. Using this

rule, each interior vertex gets assigned to a unique cell also.

The following theorem ties together the notions of a skeleton and a visibility cell.

Theorem 4 *For any visibility cell $\mathcal{C}$, and points p, $q \in \mathcal{C}$, $V^*(p) = V^*(q)$.*

Proof: Consider the straight line joining p and q, call this line l. Clearly, l is totally contained in $\mathcal{C}$, even when p and q are boundary points. Further, no interior line of the decomposition intersects l. Suppose s is the point on l which is the closest to q such that it has a different visibility skeleton than $V^*(q)$.

Consider first the case where $V^*(s)$ and $V^*(q)$ do not have the same underlying polygon, i.e. they have a different set of vertices. Assume, without loss of generality, that the difference between the two skeletons is that a vertex of P, say x, is visible from p but is not visible from q. Then there must exist a reflex vertex $y \in V^*(q)$ such that it is an obstruction for q viewing x. Then the ray emanating from x and anchored at y must intersect l between s and q, giving a contradiction. Therefore, the underlying polygon for the skeleton is invariant over the entire cell.

Any difference in the skeletons $V^*(q)$ and $V^*(s)$ must then be in the labeling of the edges. It is fairly easy to verify that the location of full edges and artificial edges must be the same in both cases, as also the labeling of the artificial edges as corresponding to either half or partial edges. The difference, if any, must be in the line-characterizations that label some artificial edge. In that case, there must exist two edges of P, say e_s and e_q, such that s can see some portion of e_s but cannot see e_q at all, and q can see some portion of e_q but cannot see e_s at all. We then conclude that there is some point between s and q on l which sees a vertex not visible from q. This contradicts the assumption that s was the closest point of l to q which has a different skeleton.

Thus, there can be no point s on l which has different visibility skeleton from q.   □

This theorem allows us to make the following definition.

Definition 6 *For any visibility cell $\mathcal{C}$, we define the* visibility skeleton of the cell *$V^*(\mathcal{C})$ as the common visibility skeleton for all points contained in $\mathcal{C}$.*

The exact choice of an edge e of P for any half or partial edge is also invariant over the cell, although the portion of e that can be seen varies from point to point in the cell. However, Theorem 4 does not guarantee that if two points have the same visibility skeleton then they are in the same visibility cell. In fact, a visibility skeleton can have $\Theta(r)$ embeddings and could have several distinct visibility cells in its kernel, all with the same skeleton (see Fig. 2). This is because there could be several collinear edges of P that are all candidates for being the half or partial edge corresponding to a particular artificial edge of the skeleton.

Definition 7 *The binary relation "$\equiv$" over the visibility cells in the decomposition is such that for any two cells $\mathcal{C}_1 \equiv \mathcal{C}_2$ if and only if $V^*(\mathcal{C}_1) = V^*(\mathcal{C}_2)$.*

It is easy to verify that this is an equivalence relation over the cells such that each equivalence class of cells is associated with a unique visibility skeleton. Let EC_i, for $1 \leq i \leq T$, be the equivalence classes for $\mathcal{P}$, and denote by V_i^* the visibility skeleton of the cells in the ith equivalence class.

Finally, we would like to characterize the complexity of a visibility cell in terms of the complexity of its visibility skeleton.

Theorem 5 *Let $\mathcal{C}$ be a visibility cell whose visibility skeleton has s artificial edges. Then the complexity of $\mathcal{C}$ is $O(s)$.*

Proof: Each side of the polygon $\mathcal{C}$ is determined by a ray anchored at one of the blocking reflex vertices. Each such blocking vertex can have at most 2 rays which bound $\mathcal{C}$. Since the number of blocking reflex vertices is $O(s)$, the result follows. □

Basically, this is so because each side of the cell arises from a ray anchored at an end-point of an artificial edge.

5 Data Structures and Search Algorithms

We now describe the construction of the data structures and algorithms for query processing. In the first step of preprocessing we compute the cell decomposition of the map polygon, and the visibility skeleton for each cell. The skeleton of each cell is represented as an $O(m)$-dimensional real vector. These vectors are stored in a multi-dimensional search tree each of whose leaves indexes an equivalence class. Given $\mathcal{V}(p)$, we extract the visibility skeleton $V^*(p)$ and query this data structure to identify the equivalence class of cells where p must be located. The last, and most non-trivial, stage of the search is concerned with identifying the possible locations of p within the equivalence class. This reduces to a search problem in a planar subdivision.

5.1 Computing Cells and their Skeletons.

The visibility cells and their skeletons are computed by the following steps: (i) for each reflex vertex, identify the vertices of $\mathcal{P}$ that are visible from it — each such vertex can give rise to a line in the arrangement with one end-point at that reflex vertex; (ii) compute the arrangement of all these lines; (iii) compute, for each vertex in the arrangement, the visibility polygon (and hence the skeleton). This last computation can be done in an "incremental" fashion as we walk along a line of the arrangement — the visibility polygon incurs only one change from one vertex on a line to the next vertex. Let N denote the number of visibility cells; note that N is always $O(n^2 r)$.

Theorem 6 *The preprocessing time and space are $O(nr \log n + nN)$.*

Proof: By standard results in ray-shooting [15] and the construction of arrangements [10], steps (i) and (ii) above can together be completed in $O(nr \log n + nN)$ time and space. To bound the complexity of step (iii), we note that the visibility skeleton of adjacent regions differ in at most a constant number of contiguous edges. Thus we may generate the N visibility skeletons by the following procedure that walks along the lines forming the arrangement. We assume for simplicity of description that there are no degeneracies in the arrangement, i.e., at most two lines intersect at any point of the arrangement except at vertices of $\mathcal{P}$. Assume that every vertex of the arrangement that is not a vertex of $\mathcal{P}$ is labeled by the emanating and anchor vertices of the two lines that intersect to form that vertex.

We first compute the visibility polygons of the r reflex vertices in time $O(nr \log n)$ [13]. Call this set ρ.

We begin at the (reflex) anchor vertex of the first line (call it ℓ) in the arrangement; given ρ, we can in time $O(n)$ obtain the visibility skeleton of the cell of the arrangement that contains ℓ and its anchor and lies to the right of ℓ. We then walk on the arrangement along ℓ; at each new vertex w of the arrangement a vertex v of $\mathcal{P}$ either becomes visible or invisible. Indeed, v is the emanating vertex of the other line that intersects ℓ at w. From this we can in $O(n)$ time and space read off and write down the visibility skeleton of the region bounded by ℓ, w and lying to the right of ℓ. We repeat this for all the lines forming the arrangement, in a total of $O(Nn)$ time and space. $\square$

Let $\mathcal{C}_i$, $1 \le i \le N$, be the cells in the subdivision. Define c_i to be the complexity of $\mathcal{C}_i$, m_i to be the complexity of the $V^*(\mathcal{C}_i)$ and s_i to be the number of blocking vertices in these skeletons. Note that c_i is $O(s_i)$, and that $s_i \le m_i$. By Theorem 3 we obtain the following.

Corollary 2 *In the visibility cell decomposition,* $\sum_{i=1}^{N} c_i = O(n^2 r)$.

5.2   Locating the Equivalence Class.

Assume that we have obtained a subdivision of $\mathcal{P}$ into N visibility cells, and have computed a visibility polygon for one point in each of the cells. We start by showing how to compute the equivalence classes.

Given a visibility polygon of complexity m, the corresponding skeleton can be computed in $O(m)$. We will encode each of the N skeletons as an M-dimensional real vector, where $M = O(m)$. The encoding fixes an origin of the skeleton (as described earlier) and specifies the position of every other vertex relative to the origin using only $2m$ real numbers. Further, we store the edge labels using another $2m$ components. The ordering of the vertices and edges in the skeleton is specified implicitly in the ordering of the components of the encoding. This entire process takes $O(m)$ time.

A crucial property of the representation is that two cells have the same skeleton if and only if their representations are identical. This motivated the definition of a skeleton and is vital to constructing and searching the equivalence classes.

We first partition the cells according to the number of vertices in their visibility skeletons. Consider now the N_m cells whose skeletons have complexity m. We construct a multi-dimensional search tree [17] for the vectors corresponding to the skeleton representations of these cells. These search trees can be constructed

in $O(mN_m + N_m \log N_m)$ time and space, and support exact match queries in time $O(m + \log N_m)$. The various skeletons whose vector representations are identical will reach the same leaf of this search tree. Thus the leaves are in 1-1 correspondence with the equivalence classes of cells in this collection. (In practice, it would be more efficient to recompute the equivalence search trees using one representative from each equivalence class, after having computed the equivalence classes as above.)

We have at most n different search trees corresponding to the different values of m. Given a query $\mathcal{V}$, we can easily determine which tree to search and thence the correct equivalence class.

Theorem 7 *The T equivalence classes, EC_i, can be computed in $O(nN)$ time using the equivalence search trees. The n search trees can all be constructed in $O(nN)$ time and space, and answer queries in time $O(m + \log n)$.*

5.3   Searching within an Equivalence Class.

It remains to specify how we search within an equivalence class of visibility cells for all the possible locations of the query point p. We will have one data structure for each equivalence class, associated with the corresponding leaf of the equivalence tree.

We now fix our attention on any one equivalence class EC. Let V^* be the skeleton corresponding to EC. Each cell in EC can be identified with a distinct embedding of V^*. The cell must lie in the kernel of that embedding. For each embedding there could be several cells in the kernel, but these must be disjoint convex polygonal regions. Let the class EC consist of cells from k different embeddings. The complexity of any one cell in EC is at most m. The following theorem bounds the overall complexity $c(EC)$ of all the cells in EC.

Theorem 8 *The total number of cells in any equivalence class EC, as well as their total complexity $c(EC)$, is $O(n^2)$. This bound is the best possible.*

It turns out to be much easier to prove a stronger version of the above theorem.

Definition 8 *The class $\mathcal{C}_m$ of cells in the decomposition consists of all cells whose skeletons contain m non-spurious vertices.*

We will show that the total complexity of the cells in the class $\mathcal{C}_m$ is $O(n^2)$. This implies that the total complexity of the cells in an equivalence class is also $O(n^2)$ since these cells must all belong to $\mathcal{C}_m$ for some m.

Theorem 9 *The overall complexity of the cells in $\mathcal{C}_m$ is $O(n^2)$.*

Proof: We start by giving an alternate proof of Theorem 3 which will provide some intuition about the current proof. Consider any ray R in the subdivision anchored at a and emanating from b. Suppose we walk along this ray, starting at a and going away from b. During this walk there can be at most $2n$ changes in the set of visible vertices. This is because, in a simple polygon, once a vertex disappears from view it will never be visible again. This implies that the ray R has $O(n)$ vertices on it. Since the number of rays is $O(nr)$, we get the desired bound.

The original proof of Theorem 3 will also prove useful here. Recall that each vertex of the subdivision has a label determined by the two rays on which it lies. The label consists of the names of the two emanating map vertices, as well as two bits specifying the layout of the anchor vertices with respect to these rays. There are four types of vertices corresponding to the four possible layouts. There could be as many as n vertices of Type 4 which carry the label of a particular pair of map vertices. It is the latter kind of vertices which can be large in number. Overall there are only $O(n^2)$ vertices which are of Types 1-3 or lie on the boundary of the map polygon.

We now have to do a careful assignment of the edges and vertices of the subdivision to the visibility cell. The rays determining a vertex create four quadrants, one of which is the region where the two emanating vertices are blocked from view by the anchors. The vertex is assigned to the incident cell which lies in this quadrant. Similarly, each edge lies on a ray on side of which the emanating vertex is hidden from view by the anchor vertex. The edge is assigned to the adjacent cell which lies on that side of the ray. Thus each edge/vertex is assigned to only one cell, but it still bounds all its neighboring cells.

This assignment of edges and vertices is motivated by our definition of visibility. Consider a ray emanating at v and anchored at w. From the view of any point

p on this ray, the vertex v is not visible. Thus in the skeleton $V^*(p)$ there will be a spurious edge from w to v, and v will be considered a spurious vertex.

Consider now the edges which bound the cells in $\mathcal{C}_m$. Some of these edges could be portions of the boundary of the map polygon. There are at most nr rays in the subdivision and each ray will create one additional subdivision vertex on the boundary P. The number of subdivision edges lying on P cannot exceed the number of subdivision vertices on P, and these are at most $nr + n$ in number. We will ignore all such edges from now on since our goal is to prove a bound of $O(n^2)$.

Consider any particular ray R emanating at the vertex v and anchored at the vertex w. This ray is collinear with v and, starting from w it goes away from v until it first hits the boundary P. We will assume that the ray is directed away from v. There may be several subdivision vertices on the ray, besides its endpoints, and this will divide R into a sequence of edges which bound some cells. We will think of these edges as open intervals separated by the vertices on R.

Assume that from the point of view of an observer sitting at v and looking in the direction of the ray, the anchor w lies to the left. We will argue only for the case of such "leftist" rays, the case of the rightist rays being similar. Each edge on the ray will be assigned to the cell lying to its left. However, we will have to count each such edge as contributing to the complexity of both the cell to its right as well as the one to its left. We will provide the argument only for the cells to the right – the other side can be handled similarly.

As we sweep down the ray from w to the boundary of P, we will traverse the edges on the ray in order. If the cell bounded by an edge (i.e. the cell to its right) is in $\mathcal{C}_m$, then the edge will contribute to the complexity of $\mathcal{C}_m$. Consider the vertices on this ray. Think of the ray as a vertical line which is directed towards the north. Each vertex x on this ray is caused by another ray, call it R', whose anchor may lie to the left or right of R and this doesn't really matter. The anchor for R' may lie below it – in this case the vertex is called an *incrementing* vertex. This means that as we cross v, a new vertex comes into view, and this vertex is the emanating vertex for R'. Similarly, a vertex is called *decrementing* if its anchor lies above R'. Upon crossing such a vertex, the emanating vertex falls out of view. The key observation is that a Type 4 vertex must be a decrementing vertex, and moreover its anchor must lie to the right of R.

Consider the sequence of edges we see that bound cells (to the right) which are in $\mathcal{C}_m$. Note that since all these edges are not assigned to the cells to the right, the cells will have v in their skeleton. Thus any two consecutive $\mathcal{C}_m$ edges will be separated by a collection of vertices (including their end-points) which contains an equal number of incrementing and decrementing vertices. We will label each $\mathcal{C}_m$ edge with the nearest incrementing vertex which lies below it. Since no two $\mathcal{C}_m$ edges are adjacent on R, each such edge gets assigned a unique label except possibly the first such edge on R. The crucial point here is that none of the labels can be a vertex of Type 4.

Each vertex can label at most four $\mathcal{C}_m$ edges since it lies on two rays and can be used as a label for at most two $\mathcal{C}_m$ edges on each of these two rays – one each when we are considering the cells to left or to the right of R. Since none of the labels is of Type 4, the number of distinct labels is $O(n^2)$. Moreover, the number of unlabeled $\mathcal{C}_m$ edges is at most nr, i.e. one per ray. We conclude that the number of edges contributing to the complexity of cells in $\mathcal{C}_m$ is $O(n^2)$.

It is easy to see that this bound can be reached in the example which proves the tightness of the $O(n^3)$ bound on the complexity of the subdivision. □

Consider any one embedding h of V^* and all the cells in the kernel of the embedded skeleton which belong to EC. Let $h(p)$ denote the point in P which has the same location relative to $h(v_o)$ as does p to v_o. The following theorem states that if $h(p)$ lies in one of the cells with skeleton V^* then $h(p)$ is a valid answer to the query $\mathcal{V}(p)$.

Theorem 10 *Let h be any embedding of $V^*(p)$ and $h(p)$ the corresponding location of p. Then, $\mathcal{V}(h(p)) = \mathcal{V}(p)$ if and only if $V^*(h(p)) = V^*(p)$.*

Proof: Clearly, if $h(p)$ and p have the same visibility polygon then they have the same skeleton. The non-trivial part is to show that if they have the same skeleton then they have the same visibility polygon.

Since the two points have the same skeleton, all the non-spurious vertices are identically laid out in the two visibility polygons, as are all the full edges. Consider now any fixed artificial edge e' of their skeleton. Suppose that e' is labeled as being in correspondence with

a half edge. Then in the visibility polygon there is a spurious edge se which is adjacent to this half edge. The spurious edge starts at the same reflex vertex in both visibility polygons and is collinear with p. Thus the only difference between the two visibility polygons could be in the location of the spurious vertex where se meets the half edge. However, one end-point of the half edge is non-spurious and has the same location in both visibility polygons. Moreover, the line-characterization of the half edge is the same in both cases. Therefore, the location of the spurious vertex must also be identical in the two cases.

The other case to be considered is where the artificial edge e' is in correspondence with a partial edge. In this case there are two spurious edges in the visibility polygon which meet the partial edge. By an argument similar to the one above, it is easy to see that the location of the two spurious vertices must also be identical in the two visibility polygons.

We conclude that the relative location of all the spurious vertices must be identical in $\mathcal{V}(h(p))$ and $\mathcal{V}(p)$, and hence the two polygons must be identical.   □

Thus, to verify if p could have been in any fixed embedding, it suffices to check if $h(p)$ lies in a cell with the same visibility skeleton. This reduces to the following problem: given a collection of disjoint convex polygons of total complexity $c(EC)$, we wish to identify the polygon (if any) where a point q is located. When k is large, searching independently in each embedding's cells would require time at least k, which may be much larger than A (the output size). However, observe that all the embeddings are congruent and have the same location of p with respect to the origin. The only difference between the embeddings is in the visibility cells therein which have V^* as their skeleton. The cells in different embeddings are totally unrelated to each other.

Our solution is to consider all embeddings at once by overlaying all their cells into one embedding of the skeleton. Thus, with reference to the origin of V^*, we have k collections of convex polygons of total complexity $c(EC)$ that are overlaid to create a planar subdivision. The problem is now that of performing a point location in this subdivision: each region is labeled by the set of visibility cells that intersect to create it. The overall complexity of the subdivision is at most quadratic in $c(EC)$. Using data structures for point location in planar subdivisions due to Kirkpatrick [16] or Edelsbrunner, Guibas and Stolfi [11], we obtain:

Theorem 11 *The localization problem can be solved with $O(n^2N)$ space, $O(n^2N \log n)$ preprocessing and a query time of $O(m + \log n + A)$.*

A problem that we face in implementing this approach is that we need to associate with each region in the subdivision a list of the visibility cells whose intersection create that region. This could blow up the space requirement by as much as a factor of n. It is possible to avoid the blow-up in space by using the technique outlined in Section 6.4. An important point is that these data structures will locate the point p as being in the intersection of some number α of the original polygons. To actually enumerate all these polygons (or the corresponding locations of p) would take time proportional to α, and this quantity may be as large as n. Thus, if we want any one solution, the query time drops to $O(m + \log n)$.

We also show that the space can be reduced at the expense of increased query time. Moreover, we can smoothly trade-off the query time with the preprocessing time and space.

Theorem 12 *Let $1 \le f(n) \le n$. The localization problem can solved in $O\left(n^2N \log n/f(n)\right)$ space and preprocessing, and query time $O(m + f(n) \log n + A)$.*

Proof: Consider any particular equivalence class EC. Pick $\gamma = n^2/f(n)$ and consider the embeddings whose complexity is at most γ each. These can be partitioned into groups of embeddings such that each group has complexity roughly γ. The idea is to overlay the cells from the embeddings in a particular group, using a total space of $O(\gamma^2)$ for each such group. The number of such groups cannot exceed $O(n^2/\gamma)$ since the overall complexity of all the embeddings is bounded by $O(n^2)$. The total space required by these groups is $O(n^2\gamma)$. Searching independently in each group's planar subdivision requires time $O(n^2 \log n/\gamma)$.

The embeddings of complexity at least γ each are searched independently also. Their total space requirement is $O(n^2)$. Moreover, since they cannot be more than $O(n^2/\gamma)$ in number, it requires $O(n^2 \log n/\gamma)$ time to search these too. Thus, our total space requirement for this equivalence class is $O(n^2\gamma)$ and the query time is $O(n^2 \log n/\gamma)$. This implies the desired result. $\square$

The next result is obtained when we perform an independent point location in each distinct embedding.

Theorem 13 *The localization problem can be solved with $O(nN)$ space, $O(nN \log n)$ preprocessing and a query time of $O(m + r \log n + A)$.*

6 Extensions and Variants

6.1 Map polygons with holes.

When the map polygon has holes, the size and complexity of the visibility cell decomposition can be higher. In this case, we have a tight bound of $O(n^2r^2)$ on the number of visibility cells. When all the holes are convex, the increase in the number of visibility cells can be bounded in terms of the *number of holes*. This also applies to the increase in the preprocessing and space bounds. We omit the details here, and the concomitant increase in preprocessing time and space.

6.2 The limited range version.

We now consider a feature of range-finders that arises in practice — they can reliably obtain range readings only up to some distance D [19]. Beyond this distance, the noise levels are too high to measure the distance, and we only learn that the distance is greater than D. Our approach to preprocessing $\mathcal{P}$ can be modified to work even in this case. The set of points within distance D of an edge of $\mathcal{P}$ is an oval region. Consider now the arrangement of the oval regions defined by all the edges of $\mathcal{P}$; since any two ovals intersect at most at four places, this arrangement partitions the plane into $O(n^2)$ subdivisions. Within each subdivision, the set of edges of $\mathcal{P}$ that are within distance D is invariant. Intersecting this arrangement with our visibility cell decomposition, we obtain a modified decomposition with the property that in each subdivision, the (redefined) skeletons are the same. Our search process is now applied to this modified decomposition to obtain the same query time.

6.3 The single-shot query problem.

Consider now the problem of answering a single query $\mathcal{V}$. Here the cost of any preprocessing must be included in the cost of answering the query. We present an algorithm running in time $O(nm)$ based on some results in *ray-shooting*. Suppose we wish to determine, at each vertex of $\mathcal{P}$, where the ray going in a fixed direction first hits $\mathcal{P}$ again. This problem is equivalent to trapezoidalizing $\mathcal{P}$ using lines parallel to the shooting direction. This can be done in linear-time [5, 12] given a triangulation of the polygon $\mathcal{P}$. The recent result of Chazelle [4] shows that the triangulation itself can be computed in linear time.

Theorem 14 *Given a map polygon $\mathcal{P}$ and a visibility polygon $\mathcal{V}$, the set of valid locations of p can be determined in time $O(nm)$.*

Proof: First, for each of the $O(m)$ spurious and artificial edges in $\mathcal{V}$ we determine the answer to the ray-shooting query from each vertex of $\mathcal{P}$. This requires a total of $O(nm)$ time. Now, we try each possible embedding of $\mathcal{V}$ in $\mathcal{P}$, and there are at most n of these. In each potential embedding, we can determine its validity by using the information from the ray-shooting queries in time $O(m)$. Some additional processing and details are involved, and omitted here. $\square$

When the range-finder has a limited range D, we can answer a single query in $O(n^2)$ time.

6.4 Visibility query processing.

Consider the problem of preprocessing a polygon $\mathcal{P}$ so that a visibility query can be efficiently answered. A visibility query is a point $p \in \mathcal{P}$, and we are required to compute the vertices of P visible from p. An interesting side effect of our results is the construction of an efficient data structure for this problem. The obvious approach is to compute the visibility cell decomposition and the skeleton for each cell. A query can now be answered by performing a point location in this cell decomposition. This is inefficient in its use of space since we must store the skeletons for each cell.

We now sketch an idea for making the space requirement linear in the complexity of the cell decomposition. The idea is to avoid storing the full list of visible vertices at each cell. Instead, at each interior line we store the change in the visibility as we cross it. Borrowing on an idea of Chazelle [2], we actually store the visibility skeleton only at those cells where the visibility is a "local" minima. These are the cells where crossing any boundary edge leads to an increase in the number of visible vertices. To compute the visiblity in any cell, we perform a walk to a minimal visibility cell, keeping track of the changes in visibility as we cross interior

lines. If we are not at a minimal cell, there is always a cell boundary such that crossing it will cause the number of visible vertices to decrease. The length of the walk cannot exceed the size of the output.

Theorem 15 *Using $O(N)$ space and preprocessing, a visibility query can be answered in time $O(\log n + m)$ where m is the size of the output visibility polygon.*

Using the results from [3], we can achieve linear space at the cost of increasing the query time to $O(n \log n)$.

7 Further Work

- The single-shot problem resembles a classic string-matching problem, and it is likely that an algorithm running in time $o(mn)$ can be devised using techniques from that field.

- We assumed that the robot has a compass and thus $\mathcal{V}$ is oriented with respect to $\mathcal{P}$. What if this assumption were removed?

- A hard but natural extension of our problem is to the case of 3-dimensional polyhedral terrains. Here the robot's viewing mechanism would be a camera which would deliver two-dimensional images.

Acknowledgements

We gratefully acknowledge useful discussions with Bernard Chazelle, John Hershberger and Emo Welzl.

Rajeev Motwani's research was supported by Mitsubishi Corporation, NSF Grant CCR–9010517 and NSF Young Investigator Award CCR-9357849, with matching funds from IBM, Schlumberger Foundation, Shell Foundation, and Xerox Corporation.

This paper is a revised version of a preliminary report which appeared in the proceedings of the Symposium on Discrete Algorithms [14].

References

[1] P. Bose, A. Lubiw, and J. I. Munro. Efficient visibility queries in simple polygons. In *Proceedings of the 4th Canadian Conference on Computational Geometry*, pages 23–28, 1992.

[2] B. Chazelle. An improved algorithm for the fixed-radius neighbor problem. *Information Processing Letters*, 16:193–198, 1983.

[3] B. Chazelle, H. Edelsbrunner, M. Grigni, L. Guibas, J. Hershberger, M. Sharir, and J. Snoeyink. Ray Shooting in Polygons Using Geodesic Triangulations. In *Proceedings of the ICALP Conference*, pages 661–673, 1991.

[4] B. Chazelle. Triangulating a simple polygon in linear time. *Discrete and Computational Geometry*, 6:485–524, 1991.

[5] B. Chazelle and J. Incerpi. Triangulation and shape-complexity. *ACM Transactions on Graphics*, 3:135–152, 1984.

[6] R. Cole and C-K. Yap. Shape from probing. *Journal of Algorithms*, 8:19–38, 1987.

[7] I.J. Cox. Blanche — an experiment in guidance and navigation of an autonomous robot vehicle. *IEEE Transactions on Robotics and Automation*, 7:193–204, 1991.

[8] I.J. Cox and J.B. Kruskal. On the congruence of noisy images to line segment models. In *Proceedings of the 2nd International Conference on Computer Vision*, pages 252–258, 1988.

[9] I.J. Cox and J.B. Kruskal. Determining the 2- or 3-dimensional similarity transformation between a point set and a model made of lines and arcs. In *Proceedings of the 28th IEEE Conference on Decision and Control*, pages 1167–1172, 1989.

[10] H. Edelsbrunner. *Algorithms in Combinatorial Geometry*, volume 10 of *EATCS Monographs on Theoretical Computer Science*. Springer-Verlag, Heidelberg, West Germany, 1987.

[11] H. Edelsbrunner, L. J. Guibas, and J. Stolfi. Optimal point location in a monotone subdivision. *SIAM Journal on Computing*, 15:317–340, 1986.

[12] A. Fournier and D. Y. Montuno. Triangulating simple polygons and equivalent problems. *ACM Transactions on Graphics*, 3(2):153–174, 1984.

[13] L. Guibas, J. Hershberger, D. Leven, M. Sharir, and R. E. Tarjan. Linear time algorithms for visibility and shortest path problems inside simple polygons. In *Proceedings of the 2nd Annual ACM Symposium on Computational Geometry*, pages 1–13, 1986.

[14] L. Guibas, R. Motwani, and P. Raghavan. The Robot Localization Problem in Two Dimensions. In *Proceedings of the Third Annual ACM-SIAM Symposium on Discrete Mathematics*, pages 259–268, 1992.

[15] L. Guibas, M. Overmars, and M. Sharir. Intersecting line segments, ray shooting, and other applications of geometric partitioning techniques. In *Proceedings of the 1st Scandanavian Workshop on Algorithm Theory*, volume 318 of *Lecture Notes in Computer Science*, pages 64–73. Springer-Verlag, 1988.

[16] D. G. Kirkpatrick. Optimal search in planar subdivisions. *SIAM Journal on Computing*, 12:28–35, 1983.

[17] K. Mehlhorn. *Data Structures and Algorithms 3: Multi-dimensional Searching and Computational Geometry*. Springer-Verlag, 1984.

[18] D.P. Miller, D.J. Atkinson, B.H. Wilcox, and A.H. Mishkin. Autonomous navigation and control of a Mars rover. In *Automatic Control in Aerospace: Selected papers from the IFAC Symposium, Tsukuba, Japan, 1989*, pages 111–114. Pergamon, Oxford, UK, 1989.

[19] G.L. Miller and E.R. Wagner. An optical rangefinder for autonomous robot cart navigation. In I.J. Cox and G.T. Wilfong, editors, *Autonomous Robot Vehicles*. Springer-Verlag, Berlin, 1990.

[20] C.N. Shen and G. Nagy. Autonomous navigation to provide long distance surface traverses for Mars rover sample return mission. In *Proveedings of the IEEE International Symposium on Intelligent Control*, pages 362–367, 1989.

[21] S.S. Skiena. *Geometric Probing*. PhD thesis, University of Illinois, Urbana-Champaign, April 1988. Also available as Computer Science Dept. Report UIUCDCS-R-88-1425.

[22] C. Ming Wang. Location estimation and uncertainty analysis for mobile robots. In I.J. Cox and G.T. Wilfong, editors, *Autonomous Robot Vehicles*. Springer-Verlag, Berlin, 1990.

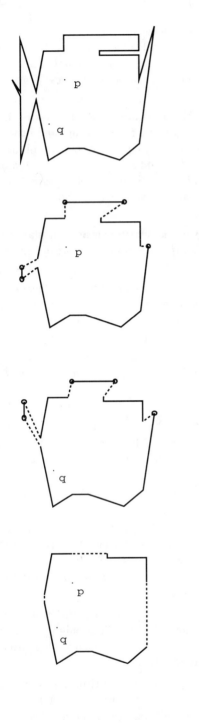

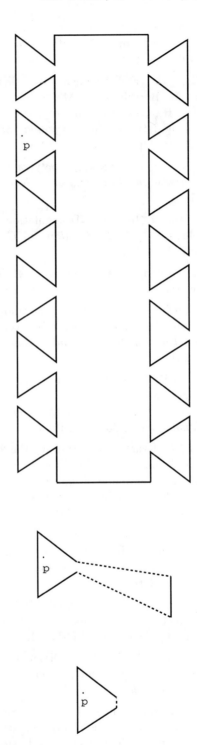

Fig. 1. A map polygon, the visibility polygons from p and q, and their common skeleton.

Fig. 2. A map polygon, the visibility polygon from p and its skeleton.

The Complexity of Sensing by Point Sampling

Yan-Bin Jia, *Carnegie Mellon University, Pittsburgh, PA*

Michael Erdmann, *Carnegie Mellon University, Pittsburgh, PA*

In assembly tasks it is often necessary to recognize parts arriving via a conveyor belt or a parts feeder at some robot work cell. Generally the parts feeder will have reduced the number of possible poses of the parts to a small finite set. In order to distinguish between the remaining poses of the parts some simple sensing or probing operation may be used. In this paper we consider the problem of finding the minimum number of sensing points required to distinguish between a finite set of polygonal shapes. For instance, we might imagine embedding a series of point light detectors in a feeder tray. Then we would be interested in the question "What is the minimum number of light detectors that can fully distinguish between all the possible shapes?" Or we might imagine a set of mechanical probes that touches the feeder at a finite number of predetermined points. Then we would ask "What are the minimum number of probing points and where should the probes be located in order to distinguish all the possible shapes?" We address these questions in this paper.

Intuitively, each sensing point can be regarded as a binary bit that has two values 'contained' and 'not contained'. So the robot senses a shape by reading out the binary representation of the shape, that is, by checking which points are contained in the shape and which are not. The formalized sensing problem: Given n polygons with a total of m edges in the plane, locate the fewest points such that each polygon contains a distinct subset of points in its interior. We show that this problem is equivalent to an NP-complete set-theoretic problem introduced as Discriminating Set. By a reduction to Hitting Set (and hence to Set Covering), an $O(n^2 m^2)$ approximation algorithm is presented to solve the sensing problem with a ratio of $2 \ln n$. Based on a reverse reduction, we prove that one can use an algorithm for Discriminating Set with ratio $c \log n$ to construct an algorithm for Set Covering with ratio

$c \log n + O(\log \log n)$. Thus approximating Discriminating Set exhibits the same hardness as that of approximating Set Covering recently shown in [24] and [4]; this result implies that the ratio $2 \ln n$ is asymptotically optimal unless $\text{NP} \subset \text{DTIME}(n^{\text{poly} \log n})$. Finally we analyze the complexity of subproblems of Discriminating Set, based on their relationship to a generalization of Independent Set called 3-Independent Set.

1 Introduction

One of the fundamental tasks in automatic assembly is for robots to efficiently determine the positions and orientations, termed the *poses*, of the individual parts to be assembled. The geometric shapes of these parts are designed early in the manufacturing process, so they are known in advance. Often the possible poses in which a part settles on the assembly table are of a small number, either reduced by a parts feeder or limited by the mechanical constraints imposed by a sequence of planned manipulations. (See Erdmann and Mason's tray-tilting method [14] and Brost's squeeze-grasp method [6] for examples of the latter case.) These facts together allow the implementation of effective sensing mechanisms, which usually take the form of simple and fast hardware systems coupled with efficient geometric algorithms [7]. The efficiency of such sensing mechanisms depends on both the time cost of the physical operations and the time complexity of the algorithms involved. Consequently, minimizing one or both of these two factors has become an important aspect of sensor design.

In order to illustrate the goals of this paper, consider a polygonal part resting on a horizontal assembly table. The table is bounded by vertical fences at its bottom-left corner, as shown in Figure 1. Pushing the part towards that corner will eventually cause the part to settle in one of the 12 stable poses listed in the fig-

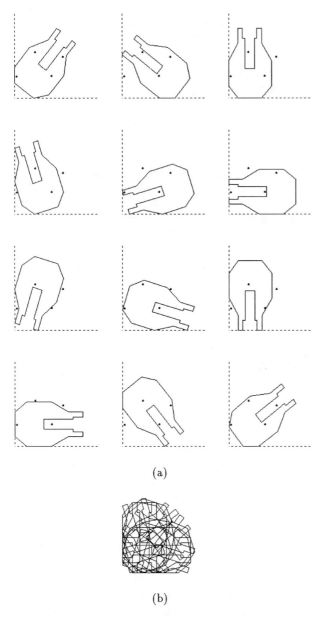

(a)

(b)

Figure 1: *Sensing by point sampling. (a) The 12 possible stable poses of an assembly part after pushing, along with 4 sampling points (optimal by Lemma 1) found by our implementation of the approximation algorithm to recognize these poses, where dashed line segments are fences perpendicular to the plane. (b) The planar subdivision formed by these poses which consists of 610 regions.*

ure. (Note to reach a stable pose both fences must be in contact with some vertices of the part while at least

one fence must be in contact with no less than two vertices.) In order to distinguish between these 12 poses, the robot has marked 4 points on the table beforehand, so it can infer the pose from which marks are covered by the part and which are not[1].

The above "shape recovery" method is named *sensing by point sampling*, as a loose analogy to the reconstruction of band limited functions by sampling on a regular grid in signal processing. To save the expense of sampling, the robot wants to mark as *few* points as possible. The problem: *How to compute a minimum set of points to be marked so that parts of different types and poses can be distinguished from each other by this method?*

1.1 Related Work

Natarajan [25] examined a similar strategy of detecting the orientations of polygonal and polyhedral objects with an analysis of the numbers of sensors sufficient and necessary for the task. More recent related work includes [5] and [2]. [5] shows that the problem of deciding whether k line probes are sufficient to distinguish a convex polygon from a collection of n convex polygons is NP-complete. This result is very similar to our Theorem 2. A variation of the line-probing result in [5] would give us the point sampling result of Theorem 2. [2] proves a similar result as well, namely that the problem of constructing a decision tree of minimum height to distinguish among n polygons using point probes is NP-complete. This result holds even if all the polygons are convex. [2] also exhibits a greedy approximation algorithm for constructing such a decision tree. This result is similar to our approximation algorithm of Section 4, with a similar ratio bound. The difference is that our greedy algorithm seeks to minimize the total number of probe points rather than the tree height.[2]

---

[1]This can be implemented in multiple ways, such as placing light detectors in the table, probing at the points, or if the robot has a vision system, taking a scene image and checking the corresponding pixel values.

[2]It is easy to give an example for which a minimum height decision tree uses more than minimum number of total probes, while a decision tree with minimum number of total probes does not attain the minimum height. Consider the problem of discriminating sets $\{a, b, a', e'\}$, $\{a, a'\}$, $\{b, b', d'\}$, $\{c, b'\}$, $\{d, c'\}$ and $\emptyset$ which can be viewed as probing a collection of polygons by the later transformation

Closely related work includes the research by Romanik and others on geometric testability (see, for example, [27], [28], and [1]). Their research develops strategies for verifying a given polygon using a series of point probes. Moreover, the research examines the testability of more general geometric objects, such as polyhedra, and develops conditions that determine whether a class of objects is (approximately) testable.

A number of researchers have looked at the problem of determining or distinguishing objects using finger probes. Finger probing is closely related to sensing by point sampling, as indicated by our discussion of [5]. For a more extensive survey of probing problems and solutions see the paper by Skiena [29].

There would seem to be connections between our work and the concept of VC-dimension often used in learning theory. For instance, in this paper we develop the notion of a "discriminating set" to distinguish different polygons. The concept of a discriminating set bears some resemblance to the idea of shattered sets associated with VC-dimension. However, discriminating sets and shattered sets are different. A minimum discriminating set is the smallest set of points that uniquely identifies every object in a set of objects, whereas VC-dimension is the size of the largest set of points shattered by the set of objects. Thus, the VC-dimension of a finite class gives a lower bound on the size of a minimum discriminating set. For dense polygon distributions, the two cardinalities will be the same, namely $\log n$, where n is the number of polygons. For sparsely distributed polygons, the two cardinalities are different. For instance, the VC-dimension can be 1, while the minimum discriminating set has size $n-1$. See Figure 2.

Finally, the work described in this paper is part of our larger research goal to understand the information requirements of robot tasks. Related work includes the sensor design methodology of Erdmann [13] and the information invariants of Donald et al. [11]. [13] proposes a method for designing sensors based on the particular manipulation task at hand. The resulting sensors satisfy a minimality property with respect to the given task goal and the available robot actions. [11] investigates the relationship between sensing, action, distributed resources, communication paths, and computation, in the solution of robot tasks. That work provides a method for comparing disparate sensing strategies, and thus for developing minimal or redundant strategies, as desired.

1.2 The Formal Problem

Consider n simple polygons $P_1, \ldots, P_n$ in the plane, not necessarily disjoint from each other. We wish to locate the minimum number of points in the plane such that no two polygons P_i and P_j, $i \neq j$, contain exactly the same points. In order to avoid ambiguities in sensing, we require that none of the located points lie on any edge of $P_1, \ldots, P_n$. The planar subdivision formed by $P_1, \ldots, P_n$ divides the plane into one unbounded region, some bounded regions outside $P_1, \ldots, P_n$, called the "holes", and some bounded regions inside. (For example, the 12 polygons in Figure 1(a) form the subdivision in Figure 1(b) which consists of 610 regions, none of which is a hole.) Immediately we make two observations: (1) Points on the edges of the subdivision or in the interior of the unbounded region or in a "hole" do not need to be considered as locations; (2) for each bounded (open) region inside some polygon only one point needs to be considered.

Let Ω denote the set of bounded regions in the subdivision which are contained in at least one of $P_1, \ldots, P_n$. Each polygon P_i, $1 \leq i \leq n$, is partitioned into one or more such regions; we write $\omega \sqsubseteq P_i$ when a region ω is contained in polygon P_i. A *region basis* for polygons $P_1, \ldots, P_n$ is a subset $\Delta \subseteq \Omega$ such that

$$\{\, \omega \mid \omega \in \Delta \text{ and } \omega \sqsubseteq P_i \,\} \neq \{\, \omega \mid \omega \in \Delta \text{ and } \omega \sqsubseteq P_j \,\},$$

for $1 \leq i \neq j \leq n$; that is, each P_i contains a distinct collection of regions from Δ. A region basis Δ^* of minimum cardinality is called a *minimum region basis*. Thus the problem of sensing by point sampling becomes the problem of finding a minimum region basis Δ^*. We will call this problem *Region Basis* and focus on it throughout the paper. The following lemma gives the upper and lower bounds for the size of such Δ^*.

Lemma 1 *A minimum region basis Δ^* for n polygons $P_1, \ldots, P_n$ satisfies $\lceil \log n \rceil \leq |\Delta^*| \leq n-1$.*

Proof. To verify the lower bound $\lceil \log n \rceil$, note that each of the n polygons must contain a distinct subset

---

technique in the proof of Theorem 2. The decision tree using minimum probes a, b, c, d always has height 4, and the decision tree using probes a', b', c', d', e' can achieve minimum height 3.

of Δ^*; so $n \leq 2^{|\Delta^*|}$, the cardinality of the power set 2^{Δ^*}.

To verify the upper bound $n - 1$, we incrementally construct a region basis Δ of size at most $n - 1$. This construction is similar to Natarajan's Algorithm 2 [25]. Initially, $\Delta = \emptyset$. If $n > 1$, without loss of generality, assume P_1 has the *smallest area*. Then there exists some region $\omega_1 \in \Omega$ outside P_1. Split $\{P_1, \ldots, P_n\}$ into two *nonempty* subsets, one including those P_i containing ω_1 and the other including those not; and add ω_1 into Δ. Recursively split the resulting subsets in the same way, and at each split, add into Δ its defining region (as we did with ω_1) if this region is not already in Δ, until every subset eventually becomes a singleton. The Δ thus formed is a region basis. Since there are $n - 1$ splits in total and each split adds at most one region into Δ, we have $|\Delta| \leq n - 1$. $\qquad\square$

Figure 2 gives two examples for which $|\Delta^*| = \lceil \log n \rceil$ and $|\Delta^*| = n - 1$ respectively. Therefore these two bounds are tight.

We can view all the bounded non-hole regions as elements of Ω, and all the polygons $P_1, \ldots, P_n$ as subsets of Ω. Then a region basis Δ is a subset of Ω that can discriminate subsets $P_1, \ldots, P_n$ by intersection. Hence the Region Basis problem can be rephrased as: Find a subset of Ω of minimum size whose intersections with any two subsets P_i and P_j, $1 \leq i \neq j \leq n$, are not equal. The general version of this set-theoretic problem, in which Ω stands for an arbitrary finite set and $P_1, \ldots, P_n$ stand for arbitrary subsets of Ω, we call *Discriminating Set*. We have thus reduced Region Basis to Discriminating Set, and the former problem will be solved once we solve the latter one.

Let us analyze the amount of computation required for the geometric preprocessing to reduce Region Basis to Discriminating Set. Let m be the total *size* of $P_1, \ldots, P_n$, i.e., the sum of the number of vertices each polygon has; trivially $m \geq 3$. Then the planar subdivision these polygons define has at most s vertices, where $3 \leq s \leq \binom{m}{2}$. By Euler's relation on planar graphs, the number of regions and the number of edges are upper bounded by $2s - 4$ and $3s - 6$ respectively. So we can construct the planar subdivision either in time $O(m \log m + s)$ using an optimal algorithm for intersecting line segments by Chazelle and Edelsbrunner [8], or in time $O(s \log m)$ using a simpler plane sweep version

by Nievergelt and Preparata [26]. To obtain the set of regions each polygon contains, we only need to traverse the portion of the subdivision bounded by that polygon, which takes time $O(s)$. It follows that the reduction to Discriminating Set can be done in time $O(m \log m + ns)$, or $O(nm^2)$ in the worst case.

Here is a short summary of the structure of the paper: Section 2 proves the NP-completeness of Discriminating Set; based on this result, Section 3 establishes an equivalence between Discriminating Set and Region Basis, hence proving the latter problem NP-complete; Section 4 presents an $O(n^2 m^2)$ approximation algorithm for Region Basis with ratio $2 \ln n$ and shows that further improvements on this ratio are hard; and Section 5 closes up with a complexity analysis of various subproblems of Discriminating Set, along with the definition of a family of related NP-complete problems called k-Independent Sets. We have implemented our approximation algorithm and have tested it on both real data taken from mechanical parts and random data extracted from the arrangements of random lines. The algorithm works very well in practice.

2   Discriminating Set

Given a collection C of subsets of a finite set X, suppose we want to identify these subsets just from their *intersections* with some subset $D \subseteq X$. Thus D must have distinct intersection with every member of C, that is,

$$D \cap S \neq D \cap T, \qquad \text{for all } S, T \in C \text{ and } S \neq T.$$

We call such a subset D a *discriminating set* for C with respect to X. From a different point of view, each element $x \in D$ can be regarded as a binary "bit" that, to represent any subset $S \subseteq X$, gives value '1' if $x \in S$ and value '0' otherwise. In such a way D represents an encoding scheme for subsets in C.

Below we show that the problem of finding a minimum discriminating set is NP-complete. As usual, we consider the decision version of this minimization problem:

DISCRIMINATING SET (D-SET)
Let C be a collection of subsets of a finite set X and $l \leq |X|$ a non-negative integer. Is there a discriminating set $D \subseteq X$ for C such that $|D| \leq l$?

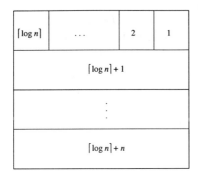

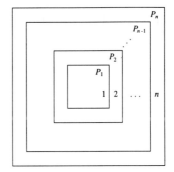

(a)

(b)

Figure 2: *Two examples whose minimum region basis sizes achieve the lower bound $\lceil \log n \rceil$ and the upper bound $n - 1$ respectively. Bounded regions in the examples are labelled with numbers. (a) For $1 \leq i \leq n$ polygon P_i is defined to be the boundary of the union of regions $\lceil \log n \rceil + 1, \ldots, \lceil \log n \rceil + i$, and all regions k with $1 \leq k \leq \lceil \log n \rceil$ such that the kth bit of the binary representation (radix 2) for $i - 1$ is 1. Thus $\Delta^* = \{1, 2, \ldots, \lceil \log n \rceil\}$. (b) The polygons $P_1, \ldots, P_n$ contain each other in increasing order: $\Delta^* = \{2, 3, \ldots, n\}$.*

Our proof of the NP-completeness for D-Set uses a reduction from Vertex Cover (VC) which determines if a graph $G = (V, E)$ has a *cover* of size not exceeding some integer $l \geq 0$, i.e., a subset $V' \subseteq V$ that, for each edge $(u, v) \in E$, contains either u or v. The reduction is based on a key observation, that for any three finite sets S_1, S_2 and S_3,

$$S_1 \cap S_2 \neq S_1 \cap S_3 \quad \text{if and only if} \quad S_1 \cap (S_2 \triangle S_3) \neq \emptyset,$$

where '$\triangle$' denotes the operation of symmetric difference, i.e., $S_2 \triangle S_3 = (S_2 \setminus S_3) \cup (S_3 \setminus S_2)$.

Theorem 1 *Discriminating Set is NP-complete.*

Proof. That D-Set $\in$ NP is trivial.

Next we establish VC $\propto_P$ D-Set, that is, there exists a polynomial-time reduction from VC to D-Set. Let $G = (V, E)$ and integer $0 < l \leq |V|$ be an instance of VC. We need to construct a D-Set instance (X, C) such that the collection C of subsets of X has a discriminating set of size l' or less if and only if G has a vertex cover of size l or less.

The construction uses the component design technique described by Garey and Johnson [16]. It's rather natural for us to begin by including every vertex of G in set X, and assigning each edge $e = (u, v)$ a subset $S(e)$ in C which contains at least u and v; in other

words, we have $V \subset X$ and

$$\{u, v\} \subset S(e) \in C, \qquad \text{for all } e = (u, v) \in E.$$

In order to ensure that any discriminating set D for C contains at least one of u and v from subset $S(e)$, we add an *auxiliary* subset A_e into C that consists of some new elements *not* in V, and in the meantime define

$$S(e) = \{u, v\} \cup A_e.$$

Hence $S(e) \triangle A_e = \{u, v\}$; and $D \cap \{u, v\} \neq \emptyset$ follows directly from $D \cap S(e) \neq D \cap A_e$. Since any discriminating set D' for $\{A_e \mid e \in E\}$ can also distinguish between $S(e_1)$ and $S(e_2)$, and between $S(e_1)$ and A_{e_2}, for any $e_1, e_2 \in E$ and $e_1 \neq e_2$, D' unioned with a vertex cover for G becomes a discriminating set for C. Conversely, every discriminating set D for C can be split into a discriminating set for $\{A_e \mid e \in E\}$ and a vertex cover for G.

The $m = |E|$ auxiliary subsets should be constructed in a way such that we can easily determine the size of their minimum discriminating sets in order to set up the entire D-Set instance. There is a simple way: We introduce m elements $a_1, a_2, \ldots, a_m \notin V$ into X, and define subsets A_e, for $e \in E$, to be

$$\{a_1\}, \{a_2\}, \ldots, \{a_m\},$$

where the order of mapping does not matter. It's clear that there are m minimum discriminating sets for the above subsets: $\{a_1, \ldots, a_m\} \setminus \{a_i\}$, $1 \leq i \leq m$.

Setting $l' = l + m - 1$, we have completed our construction of the D-Set instance as

$$X = V \cup \{a_1, \ldots, a_m\}; \qquad a_1, \ldots, a_m \notin V;$$
$$C = \{S(e) \mid e \in E\} \cup \{A_e \mid e \in E\}.$$

The construction can be carried out in time $O(|V| + |E|)$. We omit the remaining task of verifying that G has a vertex cover of size at most l if and only if C has a discriminating set of size at most $l + m - 1$. $\square$

One thing about this proof is worthy of note. All subsets in C constructed above have at most three elements. This reveals that D-Set is still NP-complete even if $|S| \leq 3$ for all $S \in C$, a stronger assertion than Theorem 1. The subproblem where all $S \in C$ have $|S| \leq 1$ is obviously in P, for an algorithm can simply count $|C|$ in linear time and then answer "yes" if $l \geq |C| - 1$ and "no" if $0 \leq l < |C| - 1$. For the remaining case in which all $S \in C$ have $|S| \leq 2$, we will prove in Section 5 that the NP-completeness still holds. However, the proof will be a bit more involved than the one we just gave under no restriction on $|S|$.

At the end of this section, we give a problem that is equivalent to D-Set:

Row-Differing Submatrix
Given an $m \times n$ matrix A of 0's and 1's and integer $0 \leq l \leq n$, is there an $m \times l$ matrix B formed by l columns of matrix A such that no two rows of B are identical?

3   Region Basis

Now that we have shown the NP-completeness of D-Set, the minimum region basis cannot be computed in polynomial time through the use of an efficient algorithm for D-Set, because no such algorithm would exist unless P = NP. This conclusion, nevertheless, leads us to conjecture that the minimization problem Region Basis is also NP-complete. Again we consider the decision version:

Region Basis (RB)
Given n polygons $P_1, \ldots, P_n$ and integer $0 \leq l \leq n - 1$, does there exist a region basis Δ for the planar subdivision Ω formed by $P_1, \ldots, P_n$ such that $|\Delta| \leq l$?

The condition $0 \leq l \leq n - 1$ above is necessary because we already know from Lemma 1 that a minimum region basis has size at most $n - 1$.

Consider a mapping $\mathcal{F}$ from the set of RB instances to the set of D-Set instances that maps regions to elements and polygons to subsets in a one-to-one manner. Every RB instance is thus mapped into an equivalent D-Set instance, as pointed out in Section 1. We claim that $\mathcal{F}$ is not onto. Suppose $\mathcal{F}$ were onto. Then the elements of each subset in a D-Set instance must correspond to regions in some RB instance. The union of these regions must be a polygon, and this polygon must map to the subset given in the D-Set instance. However, this is not always possible. Consider a D-Set instance generated from a nonplanar graph such that each edge is a subset containing its two vertices as only elements. No RB instance can be mapped to such a D-Set instance. For if there were such an RB instance, the geometric dual of the planar subdivision it defines would contain a planar embedding for the original nonplanar graph. This is an impossibility, hence we have a contradiction.

Thus the set of RB instances constitutes a *proper* subset of the set of D-Set instances; in other words, RB is isomorphic to a subproblem of D-Set. Therefore, the NP-completeness of RB does not follow directly from that of D-Set established earlier. Fortunately, however, D-Set has an equivalent subproblem which is isomorphic to a subproblem of RB under $\mathcal{F}$. That isomorphism provides us with the NP-completeness of Region Basis.

Theorem 2 *Region Basis is NP-complete.*

Proof. That RB $\in$ NP is easy to verify, based on the fact mentioned in Section 1 that the number of regions in the planar subdivision is at most quadratic in the total size of the polygons.

Let (X, C) be a D-Set instance, where

$$X = \{x_1, x_2, \ldots, x_m\};$$
$$C = \{S_1, S_2, \ldots, S_n\} \subseteq 2^X.$$

Without loss of generality, we make two assumptions

$$\bigcup_{i=1}^{n} S_i = X \qquad \text{and} \qquad \bigcap_{i=1}^{n} S_i = \emptyset,$$

because elements contained in none of the subsets or contained in all subsets can always be removed from any discriminating set of (X, C). Now add in a new element $a \notin X$ and consider the D-Set instance $(X \cup \{a\}, C')$, where $C' = \{ S_i \cup \{a\} \mid 1 \le i \le n \}$. Clearly $(X \cup \{a\}, C')$ and (X, C) have the same set of irreducible discriminating sets[3] and hence they are considered equivalent.

The planar subdivision defined by the constructed RB instance for $(X \cup \{a\}, C')$ takes the configuration shown in Figure 3(a): A rectangular region is divided by a horizontal line segment into two identical regions of which the bottom region is named $\omega(a)$; the top region is further divided, this time by vertical line segments, into $2m - 1$ identical regions of which the odd numbered ones, from left to right, are named $\omega(x_1), \ldots, \omega(x_m)$ respectively. Remove those $m - 1$ unnamed regions on the top. For $1 \le i \le n$ define polygon P_i to be the boundary of the union of all regions $\omega(x)$, $x \in S_i \cup \{a\}$. It should be clear that P_i is indeed a polygon; and the two assumptions guarantee that $P_1, \ldots, P_n$ form the desired subdivision. Note the subdivision consists of $m + 1$ rectangular regions and $4m + 2$ vertices. All can be computed in time $\Theta(m)$, given the coordinates of the four vertices of the bounding rectangle. Thus the reduction takes time $\Theta(\sum_{i=1}^{n} |S_i|)$.

It is clear that C has a discriminating set of size l or less if and only if there is a region basis of the same size for $P_1, \ldots, P_n$. Hence we have proved the NP-completeness of RB. $\qquad \square$

The above proof implies that we may regard Discriminating Set and Region Basis as equivalent problems. Note that the polygons $P_1, \ldots, P_n$ in Figure 3 are not convex; will Region Basis become P when all the polygons are convex? This question is answered by the following corollary.

Corollary 1 *Region Basis remains NP-complete even if all the polygons are convex.*

Proof. Same as the proof of Theorem 2 except that we use the planar subdivision shown in Figure 3(b). (The vertices of the subdivision partition an imaginary circle (dotted) into $2n$ equal arcs.) $\qquad \square$

---

[3] A discriminating set D is said to be *irreducible* if no subset $D' \subset D$ can be a discriminating set.

4   Approximation

Sometimes we can derive a polynomial-time approximation algorithm for the NP-complete problem at hand from some existing approximation algorithm for another NP-complete problem, by reducing one problem to the other. In fewer cases, where the reduction *preserves* the solutions, namely, every instance of the original problem and its reduced instance have the same set of solutions, any approximation algorithm for the reduced problem together with the reduction will solve the original problem. The problem to which we will reduce Discriminating Set is Hitting Set:

HITTING SET
Given a collection C of subsets of a finite set X, find a minimum *hitting set* for C, i.e., a subset $H \subseteq X$ of minimum cardinality such that $H \cap S \ne \emptyset$ for all $S \in C$?

Karp [20] shows Hitting Set to be NP-complete by a reduction from Vertex Cover. The reducibility from D-Set to Hitting Set follows a key fact we observed when proving Theorem 1: The intersections of a finite set D with two finite sets S and T are not equal if and only if D intersects their symmetric difference $S \triangle T$. Given a D-Set instance, the corresponding Hitting Set instance is constructed simply by replacing all the subsets with their *pairwise* symmetric differences. Thus every discriminating set of the original D-Set instance is also a hitting set of the constructed instance, and vice versa.

The approximability of Hitting Set can be studied through another problem, Set Covering:

SET COVERING
Given a collection C of subsets of a finite set X, find a minimum *cover* for X, i.e., a subcollection $C' \subseteq C$ of minimum size such that $\bigcup_{S \in C'} S = X$?

This problem is also shown to be NP-complete by Karp by a reduction from Exact Cover by 3-Sets [20]. A greedy approximation algorithm for this problem due to Johnson [18] and Lovász [23] guarantees to find a cover $\hat{C}$ for X with ratio

$$\frac{|\hat{C}|}{|C^*|} \le H(\max_{S \in C} |S|) \quad \text{or simply} \quad \frac{|\hat{C}|}{|C^*|} \le \ln |X| + 1,$$

where C^* is a minimum cover and $H(k) = H_k = \sum_{i=1}^{k} \frac{1}{i}$, known as the kth harmonic number. The algorithm works by selecting, at each stage, a subset from

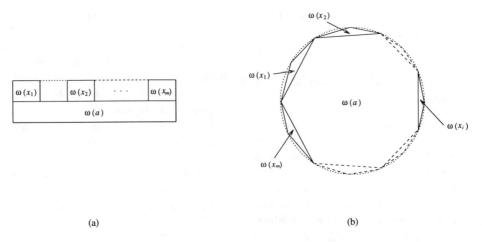

(a)   (b)

Figure 3: *Two reductions from Discriminating Set to Region Basis.*

C that covers the most remaining uncovered elements of X. We refer the reader to [9] for a general analysis of the greedy heuristic for Set Covering.

Hitting Set and Set Covering are *duals* to each other—the roles of set and element in one problem just get switched in the other. More specifically, let a Hitting Set instance consist of some finite set X and a collection C of its subsets; its dual Set Covering instance then consists of a set $\bar{C}$ and a collection of its subsets $\bar{X}$ where

$$\bar{C} = \{\, \bar{S} \mid S \in C \,\} \quad \text{and} \quad \bar{X} = \{\, \bar{x} \mid x \in X \,\},$$

and where each subset $\bar{x}$ is defined as[4,5]

$$\bar{x} = \{\, \bar{S} \mid S \in C \text{ and } S \ni x \,\}.$$

Intuitively speaking, the element $x \in X$ "hits" the subset $S \in C$ in the original instance if and only if the subset $\bar{x}$ "covers" the element $\bar{S} \in \bar{C}$ in the dual instance. Thus it follows that $H \subseteq X$ is a hitting set for C if and only if $\bar{H} = \{\, \bar{x} \mid x \in H \,\}$ is a cover for $\bar{C}$. Hence the corresponding greedy algorithm for Hitting Set selects at each stage an element

that "hits" the most remaining subsets. It is clear that the approximation ratio for Hitting Set becomes $H(\max_{x \in X} |\{\, S \mid S \in C \text{ and } S \ni x \,\}|)$ or $\ln |C| + 1$.[6]

As a short summary, the greedy heuristic on a Discriminating Set instance (X, C) works by finding a hitting set for the instance $(X, \{\, S \triangle T \mid S, T \in C \,\})$. Since an element can appear in at most $\lfloor \frac{n^2}{4} \rfloor$ such pairwise symmetric differences, where $n = |C|$, the approximation ratio attained by this heuristic is $\ln \lfloor \frac{n^2}{4} \rfloor + 1 < 2 \ln n$. The same ratio is attained for Region Basis by the heuristic that selects at each step a region discriminating the most remaining pairs of polygons, where n is now the number of polygons.

The greedy algorithm for Set Covering (dually for Hitting Set) can be carefully implemented to run in time $O(\sum_{S \in C} |S|)$ [10]. The reduction from a D-Set instance (X, C) to a Hitting Set instance takes time $O(|C|^2 \max_{S \in C} |S|)$. Combining the time complexity of the geometric preprocessing in Section 1, we can easily verify that Region Basis can be solved in time $O(nm^2 + n^2 m^2) = O(n^2 m^2)$, where n and m are the number and size of polygons respectively.

In the remainder of this section we establish the hardness of approximating D-Set and hence Region Basis. Both problems allow the same approximation ratio since the reductions from one to another do not

---

[4]According to this definition, $\bar{x} = \bar{y}$ may hold for two different elements $x \neq y$. In this case only one subset is included in $\bar{X}$.

[5]This definition also establishes the duality between D-Set and a known NP-complete problem called Minimum Test Set (see [16]). Given a collection of subsets of a finite set, Minimum Test Set asks for a minimum subcollection such that exactly one from each pair of distinct elements is contained in some subset from this subcollection.

[6]Kolaitis and Thakur [22] syntactically define a class of NP-complete problems with logarithmic approximation algorithms that contains Set Covering and Hitting Set, and show that Set Covering is complete for the class.

change the number of subsets (or polygons) in an instance. First we should note that the ratio bound $H(\max_{S \in C} |S|)$ of the greedy algorithm for Set Covering is actually tight; an example that makes the algorithm achieve this ratio for arbitrarily large $\max_{S \in C} |S|$ is given in [18].

Next we present a reverse reduction from Hitting Set to D-Set to show that an algorithm for D-Set with approximation ratio $c \log n$ can be used to obtain an algorithm for Hitting Set with ratio $c \log n + O(\log \log n)$, where $c > 0$ is any constant and n is the number of subsets in an instance. Afterwards, we will apply some recent results on the hardness of approximating Set Covering (and thus Hitting Set).

Lemma 2 *For any $c > 0$, if $c \log n$ is the approximation ratio of Discriminating Set, then Hitting Set can be approximated with ratio $c \log n + O(\log \log n)$.*

Proof. Suppose there exists an algorithm $\mathcal{A}$ for D-Set with approximation ratio $c \log n$. Let (X, C) be an arbitrary instance of Hitting Set, where $C = \{S_1, \ldots, S_n\} \subseteq 2^X$, and let $n = |C|$. To construct a D-Set instance, we first make $f(n)$ isomorphic copies $(X_1, C_1), \ldots, (X_{f(n)}, C_{f(n)})$ of (X, C) such that $X_i \cap X_j = \emptyset$ for $1 \leq i \neq j \leq f(n)$. Here f is an as yet undetermined function of n upper bounded by some polynomial in n. Now consider the enlarged Hitting Set instance $(X', C') = (\bigcup_{i=1}^{f(n)} X_i, \bigcup_{i=1}^{f(n)} C_i)$. Every hitting set H' of (X', C') has $H' = \bigcup_{i=1}^{f(n)} H_i$, where H_i is a hitting set of (X_i, C_i), $1 \leq i \leq n$; so from H' we can obtain a hitting set H of (X, C) with $|H| \leq |H'|/f(n)$ merely by taking the smallest one of $H_1, \ldots, H_{f(n)}$.

Next we introduce a set A consisting of new elements $a_1, a_2, \ldots, a_{\log(nf(n))} \notin X'$; and for $1 \leq i \leq nf(n)$ define auxiliary sets A_i:

$$A_i = \{ a_j \mid 1 \leq j \leq \log(nf(n)) \text{ and the } j\text{th bit of}$$
$$\text{the binary representation of } i - 1 \text{ is } 1 \}.$$

It is not hard to see $\{a_1, \ldots, a_{\log(nf(n))}\}$ must be a subset of any discriminating set for $A_1, \ldots, A_{nf(n)}$; therefore it is the minimum one for these auxiliary sets. The constructed D-Set instance is then defined to be (X'', C''), where

$$X'' = X_1 \cup \cdots \cup X_{f(n)} \cup \{a_1, a_2, \ldots, a_{\log(nf(n))}\};$$
$$C'' = \{ T \cup A_{(i-1)n+j} \mid T \in C_i \text{ and } T \cong S_j \} \cup$$
$$\{ A_1, \ldots, A_{nf(n)} \}.$$

It is easy to verify that every discriminating set of (X'', C'') is the union of A and a hitting set of (X', C').

Now run algorithm $\mathcal{A}$ on the instance (X'', C'') and let D be the discriminating set found. Then

$$\frac{|D|}{|D^*|} \leq c \log(|C''|) = c \log(2nf(n)),$$

where D^* is a minimum discriminating set. From the construction of (X'', C'') we know that $D = H_1 \cup \cdots \cup H_{f(n)} \cup A$ and $D^* = H_1^* \cup \cdots \cup H_{f(n)}^* \cup A$, where for $1 \leq i \leq n$, H_i and H_i^* are some hitting set and some minimum hitting set of (X_i, C_i), respectively. Let H_k satisfy $|H_k| = \min_{i=1}^{f(n)} |H_i|$ and thus let $H \cong H_k$ be a hitting set of (X, C). Also, let H^* with $|H^*| = |H_1^*| = \cdots = |H_{f(n)}^*|$ be a minimum hitting set of (X, C). Then

$$\frac{|D|}{|D^*|} = \frac{\sum_{i=1}^{f(n)} |H_i| + |A|}{\sum_{i=1}^{f(n)} |H_i^*| + |A|}$$
$$\geq \frac{f(n) \cdot |H| + \log(nf(n))}{f(n) \cdot |H^*| + \log(nf(n))}.$$

Combining the two inequalities above generates:

$$\frac{|H|}{|H^*|} \leq c \log(2nf(n)) + \frac{(c \log(2nf(n)) - 1) \cdot \log(nf(n))}{f(n) \cdot |H^*|}$$
$$\leq c \log(2nf(n)) + \frac{(c \log(2nf(n)) - 1) \cdot \log(nf(n))}{f(n)}$$
$$= c \log n + \Big[c + c \log f(n) +$$
$$\frac{(c \cdot (1 + \log n + \log f(n)) - 1) \cdot (\log n + \log f(n))}{f(n)} \Big].$$

Setting $f(n) = \log^2 n$, all terms in the brackets can be absorbed into $O(\log \log n)$ after simple manipulations on asymptotics [17]; thus we have

$$\frac{|H|}{|H^*|} \leq c \log n + O(\log \log n).$$

$\square$

Though Set Covering has been extensively studied since the mid 70's, essentially nothing on the hardness of approximation was known until very recently. The results of [3] imply that no polynomial approximation scheme exists unless P = NP. Based on recent results from interactive proof systems and probabilistically checkable proofs and their connection to approximation, several asymptotic improvements on the hardness of approximating Set Covering have been made. In particular, Lund and Yannakakis [24] showed that Set Covering cannot be approximated with ratio $c \log n$ for any $c < \frac{1}{4}$ unless $\text{NP} \subset \text{DTIME}(n^{\text{poly} \log n})$; Bellare et al. [4] showed that approximating Set Covering within any constant is NP-complete, and approximating it within $c \log n$ for any $c < \frac{1}{8}$ implies $\text{NP} \subset \text{DTIME}(n^{\log \log n})$. Based on their results and by Lemma 2, we conclude on the same hardness of approximating D-Set and Region Basis:

Theorem 3 *Discriminating Set and Region Basis cannot be approximated by a polynomial-time algorithm with ratio bound $c \log n$ for any $c < \frac{1}{4}$ unless $\text{NP} \subset \text{DTIME}(n^{\text{poly} \log n})$, or for any $c < \frac{1}{8}$ unless $\text{NP} \subset \text{DTIME}(n^{\log \log n})$.*

Following the above theorem, the ratio $2 \ln n \approx 1.39 \log n$ of the greedy algorithm for D-Set remains asymptotically optimal if NP is not contained in $\text{DTIME}(n^{\text{poly} \log n})$.

5   More on Discriminating Set

Now let's come back to where we left the discussion on the subproblems of D-Set in Section 2; it has not been settled whether D-Set remains NP-complete when every subset S in the collection C satisfies $|S| \leq 2$. We now prove that this subproblem is NP-complete.

Here we look at a special case of this subproblem, namely, a "subsubproblem" of D-Set, subject it to two restrictions: (1) $\emptyset \in C$ and (2) $|S| = 2$ for all nonempty subsets $S \in C$. Let's call this special case *0-2 D-Set*. If 0-2 D-Set is proven to be NP-complete, so will be the original subproblem.

It's quite intuitive to understand a 0-2 D-Set instance in terms of a graph $G = (V, E)$, where $V = X$, the finite set of which every $S \in C$ is a subset, and

$$E = \Big\{ (u, v) \ \Big| \ \{u, v\} \in C \Big\}.$$

In other words, each element of the set X corresponds to a vertex in G while each subset, except $\emptyset$, corresponds to an edge. Clearly this correspondence from all 0-2 D-Set instances to all graphs is one-to-one. Since any discriminating set D for C has

$$D \cap S \neq D \cap \emptyset = \emptyset, \qquad \text{for all } S \in C \text{ and } S \neq \emptyset,$$

D must be a vertex cover for G. Let $d(u, v)$ be the *distance*, i.e., the length of the shortest path, between

vertices u, v in G (or ∞ if u and v are disconnected). A *3-independent set* in G is a subset $I \subseteq V$ such that $d(u, v) \geq 3$ for every pair $u, v \in I$. The following lemma captures the dual relationship between a discriminating set for C and a 3-independent set in G.

Lemma 3 *Let X be a finite set and C a collection of $\emptyset$ and two-element subsets of X. Let $G = (X, E)$ be a graph with $E = \{ (u, v) \mid \{u, v\} \in C \}$. Then a subset $D \subseteq X$ is a discriminating set for C if any only if $X \setminus D$ is a 3-independent set in G.*

Proof.   Let D be a discriminating set for C. Assume there exist two distinct elements (vertices) $u, v \in X \setminus D$ such that $d(u, v) < 3$. We immediately have $(u, v) \notin E$ since D must be a vertex cover in G; so $d(u, v) = 2$. Hence there is a third vertex, say w, that is connected to both u and v; furthermore, $w \in D$ holds since the edges (u, w) and (v, w) must be covered by D. But now we have $D \cap \{u, w\} = D \cap \{v, w\} = \{w\}$, a contradiction to the fact that D is a discriminating set.

Conversely, suppose $X \setminus D$ is a 3-independent set in G, for some subset $D \subseteq X$. Then D must be a vertex cover. Suppose it is not a discriminating set for C. Then there exist two distinct subsets $S_1, S_2 \in C$ such that $D \cap S_1 = D \cap S_2 = \{w\}$, for some $w \in S$. Writing $S_1 = \{u, w\}$ and $S_2 = \{v, w\}$, we have $d(u, v) = 2$; but in the meantime $u, v \in X \setminus D$. A contradiction again. $\square$

This lemma tells us that the NP-completeness of 0-2 D-Set, and therefore of our remaining open subproblem of D-Set, follows if we can show the NP-completeness of *3-Independent Set*. 3-Independent Set is among a family of problems defined, for all integers $k > 0$, as follows:

k-INDEPENDENT SET (k-IS)
Given a graph $G = (V, E)$ and an integer $0 < l \leq |V|$, is there a *k-independent set* of size at least l, that is, is there a subset $I \subseteq V$ with $|I| \geq l$ such that $d(u, v) \geq k$ for every pair $u, v \in I$?

Thus 2-IS is the familiar NP-complete Independent Set problem. We will see in Appendix A that every problem in this family for which $k > 3$ is also NP-complete. To avoid too much divergence from 0-2 D-Set, let's focus on 3-IS only here.

Lemma 4 *3-Independent Set is NP-complete.*

Proof. It is trivial that 3-IS $\in$ NP. To show NP-hardness, we reduce Independent Set (2-IS) to 3-IS. Let $G = (V, E)$ and $0 < l \leq |V|$ form an instance of Independent Set. A graph G' is constructed from G in two steps. In the first step, we introduce a "midvertex" $\alpha_{u,v}$ for each edge $(u, v) \in E$, and replace this edge with two edges $(u, \alpha_{u,v})$ and $(\alpha_{u,v}, v)$. In the second step, an edge is added between every two midvertices that are adjacent to the same original vertex. More formally, we have defined $G' = (V', E')$ where

$$
\begin{aligned}
V' &= V \cup \Big\{ \alpha_{u,v} \; \Big| \; (u, v) \in E \Big\}; \\
E' &= \Big\{ (\alpha_{u,v}, u) \; \Big| \; (u, v) \in E \Big\} \cup \\
&\quad \Big\{ (\alpha_{u,v}, \alpha_{u,w}) \; \Big| \; (u, v) \neq (u, w) \in E \Big\}.
\end{aligned}
$$

Two observations are made about this construction. First, it has the property that $d'(u, v) = d(u, v) + 1$ holds for any pair of vertices $u, v \in V$, where d and d' are the two distance functions in G and G' respectively. This equality can be verified by contradiction. Next, if $(u, v) \in E$, then any two midvertices $\alpha_{u,x}$ and $\alpha_{v,y}$ have

$$
\begin{aligned}
d'(\alpha_{u,x}, \alpha_{v,y}) &\leq d'(\alpha_{u,x}, \alpha_{u,v}) + d'(\alpha_{u,v}, \alpha_{v,y}) &\leq 2; \\
d'(\alpha_{u,x}, v) &= d'(\alpha_{u,x}, \alpha_{u,v}) + d'(\alpha_{u,v}, v) &\leq 2; \\
d'(\alpha_{v,y}, u) &= d'(\alpha_{v,y}, \alpha_{u,v}) + d'(\alpha_{u,v}, u) &\leq 2.
\end{aligned}
$$

Note strict '<'s appear in above three inequalities when $x = v$ or $y = u$, and in the first inequality when $x = y$. It is not difficult to see that the whole reduction can be done in time $O(|V|^3)$. Figure 4 illustrates an example of the reduction.

We claim that G has an independent set I of size at least l if and only if G' has a 3-independent set I' of the same size. Suppose I with $|I| \geq l$ is an independent set in G. Then I is also a 3-independent set in G'. This follows from our first observation. Conversely, suppose I' with $|I'| \geq l$ is a 3-independent set in G'. Then the set I, produced by replacing each midvertex $\alpha_{u,v} \in I'$ with either u or v, is an independent set in G. To see this, assume there exists two vertices $u, v \in I$ such that $d(u, v) = 1$. Thus $d'(u, v) = d(u, v) + 1 = 2 < 3$; so either u or v, or both, must have replaced some midvertices in I'. Let $s, t \in I'$ be the two vertices corresponding to u and v before the replacement respectively; that is, $s = u$ or $\alpha_{u,x}$ and $t = v$ or $\alpha_{v,y}$ for some $x, y \in V$. According to our second observation, we always have $d'(s, t) \leq 2$. Thus we have reached a contradiction, since $s, t \in I'$. That $|I'| = |I| \geq l$ is easy to verify in a similar way. $\square$

Combining Lemmas 3 and 4, we have the NP-completeness of 0-2 D-Set; this immediately resolves the complexity of our remaining subproblem of D-Set:

Theorem 4 *D-Set remains NP-complete even if $|S| \leq 2$ for all $S \in C$.*

6 Experiments

For geometric preprocessing, we implemented the plane sweep algorithm by Nievergelt and Preparata [26]. We modified the original algorithm so that the containing polygons of each swept region are maintained and propagated along during the sweeping.[7] The greedy approximation algorithm for Set Covering was implemented with a linked list to attain the running time $O(\sum_{S \in C} |S|)$. All code was written in Common Lisp and was run on a Sparcstation IPX.

We discuss simulation results on random polygons. These simulations empirically study how the number of sampling points varies with the "density" of polygons in the plane. The results suggest that the point sampling approach is most effective at sensing polygonal objects that have highly overlapping poses. Experiments on a Zebra robot are underway and the results will be presented in the near future.

6.1 Simulation Results

To generate random polygons, we precomputed an arrangement of a large number (such as 100) of random lines using Edelsbrunner and Guibas's topological sweeping algorithm [12]. A random polygon was extracted as the first "valid" cycle occurring in a random walk on this line arrangement, after being scaled to some random perimeter. By "valid" we mean that the number of vertices in the cycle was no less than some small random integer. This constraint was introduced merely to allow a proper distribution of polygons of

---

[7]This implementation has the same worst-case running time as a different version described in Section 1 which obtains the containment information by traversing the planar subdivision after the sweeping. But the implementation version is usually more efficient in practice.

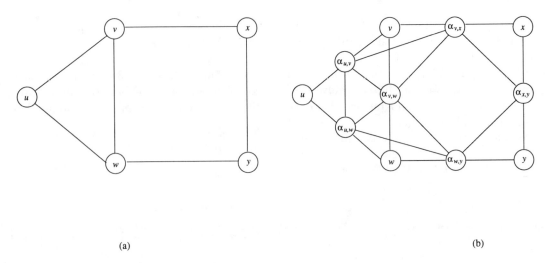

(a)   (b)

Figure 4: *An example of reduction from Independent Set to 3-IS. (a) An instance of Independent Set. (b) The constructed 3-IS instance: α vertices are added to increase the distances between the original vertices by exactly one.*

various sizes, for otherwise triangles and quadrilaterals would be generated with high probabilities according to our observations. In a sample run, a group of 1000 polygons generated (by this method) from an arrangement of 100 random lines had sizes in the range 3–30, with mean 5.545 and standard deviation 3.225.

All random polygons (or all random poses of a single polygon) in a test were bounded by some square, so that the "density", i.e., the degree of overlap, of these polygons mainly depended on their number as well as on the ratio between their average area and the size of the bounding square. Since polygons were generated randomly, the average area could be viewed as approximately proportional to the square of the average perimeter. The configuration of each polygon, say P, was assumed to obey a "uniform" distribution inside the square. More specifically, the orientation of P was first randomly chosen from $[0, 2\pi)$; the position of P was then randomly chosen from a rectangle inside the square consisting of all feasible positions at that orientation.[8]

To be robust against sensor noise, the sampling point of every region in the region basis was selected as the center of a maximum inscribed circle in that region. In other words, this sampling point had the maximum

distance to the polygon bounding that region. It is not difficult to see that such a point must occur at a vertex of the generalized Voronoi diagram inside the polygon, also called its *internal skeleton* or *medial axis function*.[9] Also for sensing robustness, regions with area less than some threshold were not considered at the stage of region basis computation.[10] Though this thresholding traded off the completeness of sampling, it almost never resulted in the failure of finding a region basis once the threshold was properly set.

The first two groups of six tests gave a sense of the number of sampling points required when polygons are sparsely distributed in the plane. The results are summarized in Table 1. Every test in group (a) was conducted on distinct, i.e., non-congruent, random polygons with perimeters between $\frac{1}{4}$ and $\frac{3}{4}$ of the width of the bounding square; every test in group (b) was conducted on distinct poses of a single polygon with

---

[8]If the diameter of P is greater than the width of the square, then not every orientation is necessarily feasible. However, this situation was avoided in our simulations.

[9]The construction of the internal skeleton of a polygon is a special case of the construction of the generalized Voronoi diagram for a set of line segments, for which $O(n \log n)$ algorithms are given in [15], [21], and [30]. Since the maximum region size for a region basis turned out to be very small in the simulations, we only implemented an $O(n^4)$ brute force algorithm.

[10]We thresholded on the region area rather than the radius of a maximum inscribed circle merely to avoid the inefficient computation on the latter for all the regions in the planar subdivision.

# polys	# regions	# sampling points
50	320	31
60	500	37
70	594	41
80	783	46
90	973	47
100	1422	51

(a)

# polys	# regions	# sampling points
50	362	36
60	609	34
70	741	34
80	1061	39
90	1125	49
100	1643	61

(b)

Table 1: *Tests on sampling sparsely distributed random polygons/poses. The twelve tests were divided into two groups: (a) All polygons in each test were distinct, with perimeters between $\frac{1}{4}$ and $\frac{3}{4}$ times the width of the bounding square. (b) All polygons in each test represented distinct random poses of a same polygon. The polygon perimeter was uniformly $\frac{5}{8}$ times the side length of the square for all six tests in the group.*

perimeter equal to $\frac{5}{8}$ of the width of the square. The scene of the last test from each group is displayed in Figure 5.

Without any surprise, the number of sampling points found were *around half* of the number of polygons, for all twelve tests in Table 1. This supports the fact that, for n sparsely distributed polygons in the plane, the minimum number of sampling points turns out to be $\Theta(n)$. As we can see from Figure 5, in such a situation every polygon intersects at most a few, or more precisely, a constant number of, other polygons. In other words, the number of polygon pairs distinguishable by any single region in the planar subdivision is $\Theta(n)$; but there are $\lfloor \frac{n^2}{4} \rfloor$ such pairs in total! Thus sensing by point sampling is inefficient in a situation with a large number of sparsely distributed polygons.

(a)

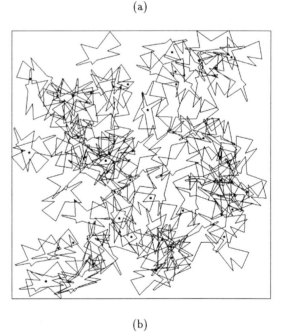

(b)

Figure 5: *Sampling 100 sparsely distributed random polygons/poses. (a) The scene of the last test from group (a) in Table 1: There are 1422 regions in the planar subdivision and 51 sampling points (drawn as dots) to discriminate the 100 polygons. (b) The scene of the last test from group (b) in Table 1: There are 1643 regions in the planar subdivision and 61 sampling points to discriminate the 100 poses.*

The next two groups of six tests were on polygons much more densely distributed in the plane, and the results are given in Table 2. In these two groups of

# polys	# regions	# sampling points
25	776	8
30	998	8
35	1270	12
40	1759	12
45	2153	11
50	2678	13

(c)

# polys	# regions	# sampling points
10	264	4
20	1121	7
25	1781	6
30	2796	7
35	3655	9
40	4995	9

(d)

Table 2: *Tests on sampling densely distributed random polygons/poses, divided into two groups (c) and (d). The width of the bounding square was reduced to $\frac{1}{4}$ times the width of the square used in groups (a) and (b) in Table 1. In group (c) all polygons in each test were distinct with perimeters in the range between $\frac{1}{2}$ and 2 times the width of the bounding square. In group (d) all polygons in each test were distinct poses of the same polygon as in Figure 5(b).*

tests, we used a bounding square with side length only $\frac{1}{4}$ of the width of the one used in test groups (a) and (b). Every test in group (c) was conducted on distinct polygons with perimeters in the range $\frac{1}{2}$–2 times the side length of the bounding square. All tests in group (d) were distinct poses of the same polygon used in the last test of group (b). Again the scene of the last test from each group is shown in Figure 6.

All twelve tests in groups (c) and (d) except the last one in group (c) found sampling points *at most twice* the lower bound $\lceil \log n \rceil$, while the first test in group (d) found exactly $\lceil \log n \rceil$ sampling points. The data in group (d) were more densely distributed than

the data in group (c) in that every two poses intersected each other. Since an extremely dense distribution of polygons may cause numerical instabilities in the plane sweep algorithm, smaller numbers of polygons were tested in these two groups than were tested in groups (a) and (b). The results of these two groups of tests show that the sampling strategy is very applicable to sensing densely distributed polygons.

Acknowledgments

Support for this research was provided in part by Carnegie Mellon University, and in part by the National Science Foundation through the following grants: NSF Research Initiation Award IRI-9010686, NSF Presidential Young Investigator award IRI-9157643, and NSF Grant IRI-9213993.

Many thanks to David S. Johnson for pointing to the results of [24] on the hardness of approximating Set Covering and for looking into the status of Discriminating Set and k-Independent Sets, which we found neither in the catalog [16] nor in the NP-Completeness Columns of *Journal of Algorithms* starting from [19]. Also thanks to Somesh Jha for his valuable suggestions on the proof of Lemma 4, and to Bruce Donald for his valuable comments and reading of this paper.

After presentation of this paper at the workshop we learned that similar results to our Theorem 2 and our approximation algorithm had appeared in [5], [27], [1], and [2]. We are very grateful to Jean-Daniel Boissonnat, Mark Overmars, Anil Rao, and Kathleen Romanik for pointing out these papers. We are particularly grateful to Kathleen Romanik for numerous interesting discussions on shape discrimination and shattered sets, and for her reading of this paper.

References

[1] Esther M. Arkin, Patrice Belleville, Joseph S. B. Mitchell, David Mount, Kathleen Romanik, Steven Salzberg, and Diane Souvaine. Testing Simple Polygons. In *Proceedings of the Fifth Canadian Conference on Computational Geometry*, August 5–9, 1993.

[2] Esther Arkin, Hank Meijer, Joseph S. B. Mitchell, David Rappaport, and Steven S. Skiena. Decision Trees for Geometric Models. In *Proceedings of the*

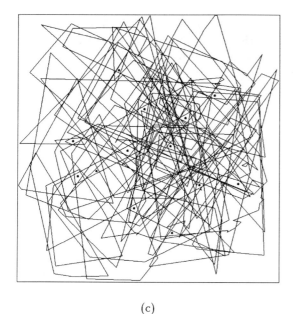

(c)

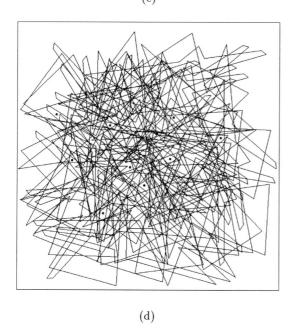

(d)

Figure 6: *Sampling densely distributed polygons. The bounding square has width $\frac{1}{4}$ times the width of the one shown in Figure 5. (c) The scene of the last test from group (c) in Table 2: There are 50 distinct polygons which form a planar subdivision with 2678 regions, and which can be discriminated by 13 sampling points. (d) The scene of the last test from group (d) in Table 2: There are 40 distinct poses of the polygon from Figure 5(b), which form a planar subdivision with 4955 regions, and which can be discriminated by 9 sampling points.*

Ninth Annual Symposium on Computational Geometry , pages 369–378, May 19–21, 1993.

[3] S. Arora, C. Lund, R. Motwani, M. Sudan, and M. Szegedy. Proof verification and intractability of approximation problems. In *Proceedings of the 33rd IEEE Symposium on Foundations of Computer Science*, pages 14–23, 1992.

[4] M. Bellare, S. Goldwasser, G. Lund, and A. Russell. Efficient probabilistically checkable proofs and applications to approximation. In *Proceedings of the 25th Annual ACM Symposium on Theory of Computing*, pages 294–304, 1993.

[5] Patrice Belleville and Tom C. Shermer. Probing Polygons Minimally is Hard. *Computational Geometry: Theory and Applications*, 2(5):255–265, 1993.

[6] R.C. Brost. Automatic grasp planning in the presence of uncertainty. *International Journal of Robotics Research*, 7(1):3–17, 1988.

[7] John F. Canny and Kenneth Y. Goldberg. A "RISC" paradigm for industrial robotics. Technical Report ESRC 93-4, University of California at Berkeley, 1993.

[8] Bernard Chazelle and Herbert Edelsbrunner. An optimal algorithm for intersecting line segments in the plane. *Journal of the ACM*, 39(1):1–54, 1992.

[9] V. Chvátal. A greedy heuristic for the set-covering problem. *Mathematics of Operations Research*, 4(3):233–235, 1979.

[10] Thomas H. Cormen, Charles E. Leiserson, and Ronald L. Rivest. *Introduction to Algorithms*. McGraw-Hill, 1990.

[11] Bruce R. Donald, Jim Jennings, and Daniela Rus. Towards a theory of information invariants for co-operating autonomous mobile robots. *Sixth International Symposium on Robotics Research*, October 1–5, 1993, Hidden Valley, Pennsylvania.

[12] Herbert Edelsbrunner and Leonidas J. Guibas. Topologically sweeping an arrangement. In *Proceedings of the 18th Annual ACM Symposium on Theory of Computing*, pages 389–403. ACM Press, 1986.

[13] Michael Erdmann. Understanding action and sensing by designing action-based sensors. *Sixth*

International Symposium on Robotics Research, October 1–5, 1993, Hidden Valley, Pennsylvania.

[14] Michael Erdmann and Matt Mason. An exploration of sensorless manipulation. *IEEE Journal of Robotics and Automation,* 4(4):369–379, 1988.

[15] Steven Fortune. A sweepline algorithm for Voronoi diagrams. *Algorithmica,* 2:153–174, 1987.

[16] Michael R. Garey and David S. Johnson. *Computers and Intractability: A Guide to the Theory of NP-Completeness.* W.H. Free and Company, 1979.

[17] Ronald L. Graham, Donald E. Knuth, and Oren Patashnik. *Concrete Mathematics: A Foundation for Computer Science.* Addison-Wesley, 1989.

[18] David S. Johnson. Approximation algorithms for combinatorial problems. *Journal of Computer and System Sciences,* 9:256–278, 1974.

[19] David S. Johnson. The NP-completeness column: an ongoing guide. *Journal of Algorithms,* 4:393–405, 1981.

[20] Richard M. Karp. Reducibility among combinatorial problems. In Raymond E. Miller and James W. Thatcher, editors, *Complexity of Computer Computations.* Plenum Press, 1972.

[21] David G. Kirkpatrick. Efficient computation of continuous skeletons. In *Proceedings of the 20th Annual Symposium on Foundations of Computer Science,* pages 18–27, 1979.

[22] Phokion G. Kolaitis and Madhukar N. Thakur. Approximation properties of NP minimization classes. In *Proceedings of the 6th Conference on Structure in Complexity Theory,* pages 353–366, 1991.

[23] Lászo Lovász. On the ratio of optimal integral and fractional covers. *Discrete Mathematics,* 13:383–390, 1975.

[24] Carsten Lund and Mihalis Yannakakis. On the hardness of approximation minimization problems. In *Proceedings of the 25th Annual ACM Symposium on Theory of Computing,* pages 286–293. ACM Press, 1993.

[25] B.K. Natarajan. On detecting the orientation of polygons and polyhedra. In *Proceedings of the 3rd Annual Symposium on Computational Geometry,* pages 146–152, 1987.

[26] J. Nievergelt and F. P. Preparata. Plane-sweep algorithms for intersecting geometric figures. *Communications of the ACM,* 25(10):739–747, 1982.

[27] Kathleen Romanik. Approximate Testing Theory. Ph.D. Thesis, Technical Report CS-TR-2947/UMIACS-TR-92-89, University of Maryland at College Park, August 1992.

[28] Kathleen Romanik and Steven Salzberg. Testing Orthogonal Shapes. *Proceedings of the Fourth Canadian Conference on Computational Geometry,* pages 216–222, August 10–14, 1992.

[29] Steven S. Skiena Problems in Geometric Probing. *Algorithmica,* 4:599–605, 1989.

[30] Chee K. Yap. An $O(n \log n)$ algorithm for the Voronoi diagram of the set of simple curve segments. *Discrete & Computational Geometry,* 2:365–393, 1987.

Appendix

A k-Independent Sets

We extend Lemma 4 to all k-IS with $k > 3$: They are NP-complete as well. The proof we will present is indeed a generalization of the proof of Lemma 4; it will again construct a k-IS instance with graph G' from an instance of Independent Set with graph G by local replacement. In the proof, each vertex v in G will be replaced by a simple path $\mathcal{P}_v$ of fixed length (depending only on k) that has v in the middle and an equal number of auxiliary vertices on each side; and each edge (u, v) will be replaced by four edges connecting the two end vertices on $\mathcal{P}_u$ with the two end vertices on $\mathcal{P}_v$, either directly or through a "midvertex". More intuitively speaking, all shortest paths between pairs of vertices in G, if they exist, get elongated in G' to such a degree that (1) (u, v) is an edge in G if and only if the distance between vertices u and v in G' is less than k; (2) any two vertices u' and v' in G' with a distance of at least k can be easily mapped to two *non*adjacent vertices in G. The first condition ensures that any given independent set in G will be a k-independent set in G', while the second condition ensures the construction of an independent set in G from any given k-independent set in G'.

Lemma 5 *k-Independent Set is NP-complete for all integers $k > 3$.*

Proof. Given an instance of Independent Set as a graph $G = (V, E)$ and a positive integer $l \le |V|$, a k-IS instance is constructed by two consecutive substitutions. A path

$$\mathcal{P}_v = \begin{cases} v_{k-3} \ldots v_1 v v_2 \ldots v_{k-2}, & \text{if } k \text{ even;} \\ v_{k-4} \ldots v_1 v v_2 \ldots v_{k-3}, & \text{if } k \text{ odd,} \end{cases}$$

first substitutes for vertex $v \in V$, where $v_1, \ldots, v_{k-3}$ (and v_{k-2} when k is even) are auxiliary vertices. And then a set of four edges

$$E_{u,v} = \begin{cases} \Big\{ (u_{k-3}, v_{k-3}), (u_{k-3}, v_{k-2}), \\ \quad (u_{k-2}, v_{k-3}), (u_{k-2}, v_{k-2}) \Big\}, & \text{if } k \text{ even;} \\ \Big\{ (u_{k-4}, \alpha_{u,v}), (u_{k-3}, \alpha_{u,v}), \\ \quad (v_{k-4}, \alpha_{u,v}), (v_{k-3}, \alpha_{u,v}) \Big\}, & \text{if } k \text{ odd,} \end{cases}$$

substitute for each edge $(u, v) \in E$, where $\alpha_{u,v}$ is an introduced midvertex. Figure 7 shows two subgraphs after applying the above substitutions on edge $(u, v) \in E$, for k even and odd respectively.

We can easily verify that, for any pair of vertices x on $\mathcal{P}_u$ and y on $\mathcal{P}_v$, both k even and odd, we have $d'(x, y) \le d'(u, v) = k - 1 < k$ if $(u, v) \in E$, where d' is the distance function defined on G'. On the other hand, if $(u, v) \notin E$, we have $d'(u, v) \ge k$ when k is even and $d'(u, v) \ge k + 1$ when k is odd. Thus an independent set I in G is also a k-independent set in G'. Conversely, suppose I' with $|I'| \ge l$ is a k-independent set in G. We substitute $u \in V$ for every auxiliary vertex $u_i \in I$ on path $\mathcal{P}_u$, and u or v for every midvertex $\alpha_{u,v} \in I$ when k is odd. Let I be the set after this substitution. It needs to be shown that I is an independent set in G and $|I| = |I'| \ge l$. This is obvious for the case that k is even. When k is odd, however, the situation is a bit more complicated due to the possible occurrences of those α vertices in I'. We observe, for any $\alpha_{u,v}$, $\alpha_{u',v'}$, and x on path $\mathcal{P}_w$ where $u, v, u', v', w \in V$,

$$d'(\alpha_{u,v}, x) \le \frac{k-3}{2} + 3 < k, \quad \begin{array}{l} \text{if } w = u \text{ or } v, \\ \text{or } (u, w) \in E, \\ \text{or } (v, w) \in E; \end{array}$$

$$d'(\alpha_{u,v}, \alpha_{u',v'}) \le 4 < k, \quad \text{if } (u, u') \in E.$$

In fact, these two conditions guarantee that I is an independent set in G and $|I| = |I'|$, which we leave for the reader to verify.

The reduction can be done in time $O(k|V| + |E|)$, which reduces to $O(|V| + |E|)$ if k is treated as a constant, in contrast to the time $O(|V|^3)$ required for the reduction from Independent Set to 3-IS. This time reduction is due to the fact that midvertices corresponding to the same vertex in V no longer have edges between each other. $\square$

Since 1-IS can be easily solved by comparing $|V|$ and l, we are now ready to sum up the complexity results on this family of problems in the following theorem.

Theorem 5 *k-Independent Set is in P if $k = 1$ and NP-complete for all $k \ge 2$.*

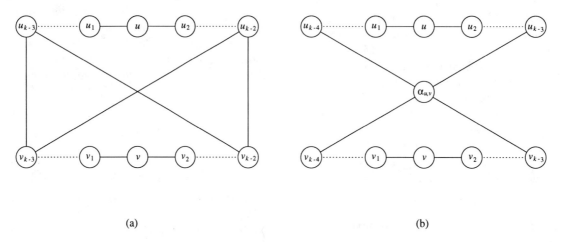

(a)   (b)

Figure 7: *Two subgraphs resulting from the described substitutions performed on edge $(u, v) \in E$ for the cases that (a) k is even; and (b) k is odd.*

Exploiting Visual Constraints in the Synthesis of Uncertainty-Tolerant Motion Plans

Armando Fox, *University of Illinois, Urbana, IL, USA*

Seth Hutchinson, *University of Illinois, Urbana, IL, USA*

We introduce visual constraint surfaces as a mechanism to effectively exploit visual constraints in the synthesis of uncertainty-tolerant robot motion plans. We first show how object features, together with their projections onto a camera image plane, define a set of visual constraint surfaces. These visual constraint surfaces can be used to effect visual guarded and visual compliant motions (which are analogous to guarded and compliant motion using force control). We then show how the backprojection approach to fine-motion planning can be extended to exploit visual constraints. Specifically, by deriving a configuration space representation of visual constraint surfaces, we are able to include visual constraint surfaces as boundaries of the directional backprojection. By examining the effect of visual constraints as a function of the direction of the commanded velocity, we are able to determine new criteria for critical velocity orientations (i.e. velocity orientations at which the topology of the directional backprojection might change).

1 Introduction

To perform effectively in real-world settings, robots must be able to plan and execute tasks in the presence of uncertainty. Typical sources of uncertainty in a robotic work cell include limited sensing accuracy, errors in robot control, and discrepancies between geometric object models and physical objects (including the parts to be manipulated and the robot itself). Because of this, the application of robotic technology to manufacturing problems has typically been restricted to situations in which uncertainty can be tightly controlled (for example, by using specialized fixturing devices).

The problems associated with uncertainty in robotic systems have long been the subject of research in the robotics community. A number of planning systems have been developed that characterize uncertainty by systems of constrained variables [3, 17, 26, 32]. These variables are propagated through transformations that represent the effects of actions on uncertainty in the planner's world description, providing the planner with worst case estimates at each step of the plan. There are two primary disadvantages to this approach: (i) the planner always assumes the worst, propagating upper bounds even when actual bounds are likely to be much tighter, and (ii) the planner must know a priori the effects of manipulation actions and sensing on the uncertainty in the description of the work cell. There is no mechanism for revising behavior at execution time to account for dynamically occurring uncertainties.

An alternative to a priori consideration of worst case error is to use sensory feedback to adapt task execution to the actual state of the world that the robot encounters. Force-based control can be used to effect guarded [35] and compliant motions [25, 34, 21, 28], making assembly actions robust by exploiting physical constraints imposed on motion by the geometry of the workspace. To date, the most successful planning systems that use this approach comprise the preimage family of planners [20, 10, 8, 19]. The key to the preimage planning paradigm is the transformation of physical constraints into geometric constraints, which can be expressed as C-surfaces (i.e. sets of points in the configuration space where the manipulator makes contact with a physical surface). Thus, equipped with a set of equations that govern motion and friction in the configuration space, preimage planners are capable of developing motion plans that are tolerant of uncertainties in the manipulator's position (represented by an error ball in the configuration space), its trajectory (represented by an error cone), and even in part dimensions (represented by added dimensions in the configuration space [8, 9]).

Informally stated, a preimage of a goal is the set of points from which a commanded motion is guaranteed to reach and terminate recognizably in the goal. Because preimages are often difficult to compute, Erdmann [10] introduces *backprojections* as a means of usefully approximating preimages by separating goal reachability from goal recognizability.

The primary limitation of force control is that it can only be used to constrain motion along directions normal to the C-surfaces. Position control must be used to control motions in directions tangent to C-surfaces. Therefore, hybrid position/force control is not sufficient when the exact manipulator and goal positions are not known in the dimensions of position control. As an example, suppose that the robot's task is to insert a peg into a hole in a block, and that the position of the hole is not precisely known. The insertion cannot proceed until the peg is directly over the hole; but this cannot be achieved using either position control (since the exact goal position is not known), or force control (since force control can only be used to constrain motion normal to the block surface).

One way to cope with the limitations of hybrid force/position control is to add vision sensing to the control servo loop. If the geometry of the imaging process is known, then the task geometry can be used to constrain the remaining degrees of freedom by using visual servo control. In recent years, the integration of computer vision with robot motion control has steadily progressed, from early look and move systems in which vision was used to recognize and locate an object prior to its manipulation [30, 29], to current systems in which visual feedback is incorporated directly into the control loop (e.g., [1, 33, 13, 31, 23, 24]). To date, however, the corresponding *motion planning* problem has not been addressed. Thus, even though visual servo control systems are now available, there is no motion planning system that is capable of exploiting such a control system.

In this paper, we present a geometric motion planner that exploits visual constraints in the synthesis of uncertainty-tolerant motion plans. Specifically, we extend the preimage formalism to exploit visual constraints. In Section 2, we introduce the concept of *visual constraint surfaces*, which are generated by projecting workspace features onto the image plane of a fixed camera [16]. A visual constraint surface can be

used to constrain visually controlled motions in the same way that physical surfaces can be used to constrain force controlled motions. In Section 3, we show how visual constraint surfaces in the workspace are mapped to configuration space constraint surfaces for the special case of $\mathcal{C} = \mathbf{R}^2$.

In Section 4, we review preimages [20], backprojections [10], and algorithms for the construction of backprojections [7]. In Section 5, we show how the directional backprojection (i.e. the backprojection with respect to a specified velocity) can be extended to exploit visual constraint surfaces. In particular, we discuss our implemented backprojection algorithm (which extends the plane-sweep algorithm presented in [7]), and evaluate the added computational complexity of considering visual constraint surfaces. By allowing visual constraint surfaces to be included as boundaries of the backprojection, we can often significantly increase the size of the backprojection, as illustrated in Figure 11. Following this, in Sections 6 and 7 we discuss the nondirectional backprojection, and show how it must be modified to exploit visual constraints.

2   A Geometric Specification for Visual Constraints

This section introduces *visual constraint surfaces, visual guarded motion* and *visual compliant motion*. Although the planning algorithms subsequently discussed will be implemented in configuration space (C-space), this section will show how visual constraint surfaces are developed in the workspace. Their C-space representation is deferred to Section 3.

2.1   Visual Constraint Surfaces

Consider a workspace containing a number of solid objects and a fixed camera. If the imaging process is modelled by perspective projection [15], projection rays from each point in the workspace converge on the camera focal center. In general, any one-dimensional object feature will project onto a planar curve on the camera image plane. We will refer to the projection of an object feature onto the camera image plane as an *image feature*. An object feature is said to be *unoccluded* if no projection ray emanating from that feature intersects the interior of any other object or the robot. Intuitively, this means that nothing is blocking the camera's view of the feature. In this paper, the only one-dimensional object features that will be considered are the 3D edges of objects. Note that for the special case of polyhedral objects, all image features are straight line segments.

Consider a 3D edge defined by the parametric space curve $\vec{X}(\tau)$ and its corresponding image feature defined by the planar parametric curve $\vec{x}(\tau)$. The projection equations relating $\vec{X}(\tau)$ and $\vec{x}(\tau)$ depend only on the position and orientation of the camera and a set of camera parameters (e.g. the focal length of the lens). Assume that the image plane is defined with origin at $\mathbf{r_0}$ and local coordinate system specified by the orthonormal vectors $\mathbf{I}$, $\mathbf{J}$. The normal to the image plane is specified by $\mathbf{K} = \mathbf{I} \times \mathbf{J}$. Then, for the case of perspective projection, the projection equations relating $\vec{X}(\tau)$ and $\vec{x}(\tau)$ are given by:

$$\begin{cases} x_I(\tau) = \dfrac{-d_f(\vec{X}(\tau) - \mathbf{r_0}) \cdot \mathbf{I}}{(\vec{X}(\tau) - \mathbf{r_0}) \cdot \mathbf{K} - d_f} \\[2mm] x_J(\tau) = \dfrac{-d_f(\vec{X}(\tau) - \mathbf{r_0}) \cdot \mathbf{J}}{(\vec{X}(\tau) - \mathbf{r_0}) \cdot \mathbf{K} - d_f} \end{cases} \quad (1)$$

where d_f is the distance from the image plane to the center of projection of the camera (i.e. the focal length of the lens). These parameters can be obtained by calibration procedures described in [5]. Once they are derived, if the camera remains fixed (as we assume), the same projection equations can be used to compute the image-plane coordinates of any image feature, given the world coordinates of an object feature.

A *visual constraint (VC) surface* is a ruled surface $\mathbf{S}(\tau, \lambda)$ bounded by $\vec{X}(\tau), \vec{x}(\tau)$, and the rays joining their respective endpoints, as in Figure 1. In the special case of polyhedral obstacles, all image features are straight line segments, so that the VC surfaces are polygons. A VC surface is defined by

$$\mathbf{S}(\tau, \lambda) = \vec{X}(\tau) + \lambda(\vec{X}(\tau) - \vec{x}(\tau)). \quad (2)$$

Recall that a ruled surface is generated by a family of lines [12]. For $\mathbf{S}(\tau, \lambda)$, a particular generating line is obtained for each valid value of τ. We will refer to such a line as a generating line of $\mathbf{S}(\tau, \lambda)$, or simply a generating line. For perspective projection, each generating line is a projection ray through the center of projection and the image plane point $\vec{x}(\tau)$.

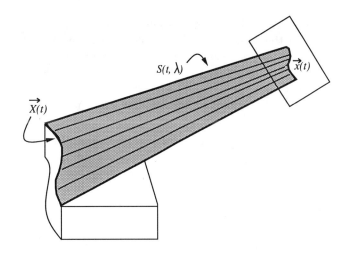

Figure 1: *Ruled VC surface generated by a curved 3D edge*

Note that a visual constraint surface does not intersect any obstacles. This is because a necessary condition for an edge to have a projection on the camera image plane is that the edge not be occluded. Similarly, if only a part of a particular 3D edge is visible (perhaps the rest of the edge is occluded by another object), only the visible part will have a projection on the image plane, so that the resulting surface will not intersect any obstacles.

Visual compliant motion. *Compliant motion* has been exploited in various motion planning and execution strategies [9, 11, 20, 21]. During compliant motion, a physical surface is used to constrain the motion of a robot along one or more degrees of freedom [25, 34, 21, 28]. For example, sliding motion along a surface might be achieved by ensuring that some constant force be maintained in the direction normal to the surface.

We define *visual compliance* as compliant motion along a (virtual) VC surface, such that the manipulator's motion is constrained to always remain in contact with a particular generating line of the the VC surface. Visual compliance can be achieved by means of a closed-loop visual servo system, as described in [5].

Visual guarded motion We define *visual guarded* motion analogously to guarded motion using physical surfaces, in which the robot moves until force feedback indicates contact with a physical surface [35]. We say that the force feedback provides a termination condition for the motion. VC surfaces can be used for visual guarded motion; that is, the manipulator can move along a trajectory that intersects a VC surface and be instructed to stop when this intersection occurs. This is possible because the intersection is a visually observable event.

Since VC surfaces are virtual rather than physical, the motion planner also has the option of ignoring them, in contrast to the force-based approach, which must explicitly consider all physical surfaces on which *sticking* may occur [10].

3 Visual Constraints in Two Dimensions

In this section, we describe the computation of visual constraint rays in the case of a two-dimensional workspace populated by polygonal obstacles. We begin by discussing the construction of visual constraint rays in the workspace, which is a special case of the formalism developed in Section 2. We then discuss the selection of robot features that will be used by the visual servo system. Following this, we describe how to map visual constraint rays into the C-space $\mathcal{C} = \mathbf{R}^2$. Finally, we discuss the time complexity of the algorithms presented in this section.

3.1 Workspace Visual Constraint Rays

In the case of a two-dimensional workspace, the camera is a one-dimensional sensor positioned in the plane. Using perspective projection, all projection rays converge on the camera projection center. We assume that if an object vertex is unoccluded (i.e. a projection ray from that vertex to the camera focal point intersects the interior of no workspace obstacle), then the projection of that vertex in the camera image can be located by the vision system. Workspace VC rays can be computed by extending rays from unoccluded workspace obstacle vertices to the camera projection center, as shown in Figure 2.

3.2 Selecting Robot Features for Visual Servo Control

In a two-dimensional workspace, visual compliant motion is effected by moving a particular robot vertex so

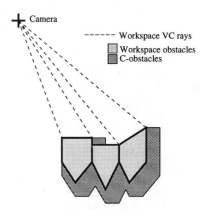

Figure 2: *Construction of workspace VC rays from unoccluded obstacle vertices*

that it remains in contact with some VC ray emanating from a workspace object vertex. This raises the question of which vertices of the robot should be used in visual compliant motions.

If the robot is a simple polygon, its projection on the camera image plane is a line segment whose endpoints represent the two furthest-apart robot vertices simultaneously visible to the camera. Note that there may be other robot vertices that project to points on the line segment; however, for the purposes of visual compliant motion, we assume that the vision system can only robustly distinguish in real time the two vertices whose projections are the endpoints of the image plane line segment (i.e. the silhouette of the robot). This restriction could be lifted if the vision system were capable of robustly distinguishing other unoccluded vertices in real time.

We will refer to the two robot vertices that project to the endpoints of the line segment as *CM vertices*, to indicate that they are the only robot vertices suitable for effecting **Compliant Motion** along a VC ray. Note that the particular robot vertices that are CM vertices can change with the position of the robot. Figure 3 shows the same robot in two different positions for which the CM vertices are different. In certain non-general configurations of the robot, two robot vertices may lie along the same projection ray. In such cases, we may arbitrarily select one of these as a CM vertex. Thus, for any specified position of the robot, we will obtain two CM vertices whose projections are the

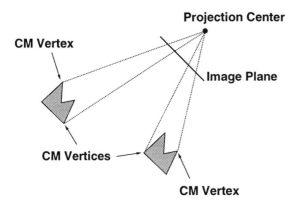

Figure 3: *Different positions of the polygonal robot give different CM vertices*

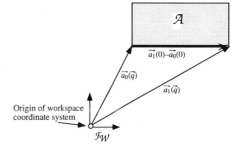

Figure 4: *The spatial relationship between the robot vertices*

endpoints of the image plane line segment representing the robot. There cannot be only one CM vertex unless the robot is itself a line segment and is collinear with a projection ray.

3.3 Configuration Space Representation of Visual Constraints

Visual constraint rays in the workspace give rise to C-space visual constraint rays (CVC rays). In mapping workspace VC rays to CVC rays, we must allow for either of the CM vertices to be moved compliantly along the workspace VC ray. Suppose that a particular vertex a_0 of the robot is taken as the origin of the robot's internal coordinate frame to compute a representation of the C-space $\mathcal{C}$. Since trajectories in $\mathcal{C}$ will specify the motion of vertex a_0 among the C-obstacles, we must find an appropriate representation for compliant motion of an arbitrary robot vertex a_j along a VC ray.

Since we are considering the case where $\mathcal{C} = \mathbf{R}^2$, the spatial relationship between a_0 and a_j is fixed. Specifically, if for some configuration $\vec{q}$ the world coordinates of a_0 are given by the vector $\vec{a_0}(\vec{q})$, then the world coordinates of a_j in the same configuration are given by $\vec{a_j}(\vec{q}) = \vec{a_0}(\vec{q}) + (\vec{a_j}(0) - \vec{a_0}(0))$. Figure 4 illustrates this relationship for vertices a_0 and a_1.

Let $e_{vc}^{\mathcal{W}}$ be a VC ray emanating from a workspace obstacle vertex b_i, and let a_j be a CM vertex of the robot when the robot is positioned such that a_j coincides with b_i. Then, as the robot moves compliantly, maintaining contact between a_j and $e_{vc}^{\mathcal{W}}$, vertex a_0 will

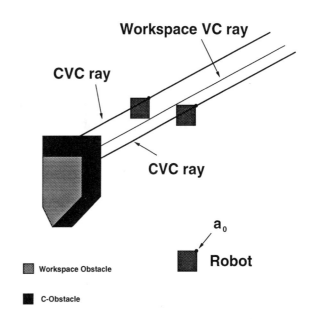

Figure 5: *Construction of two CVC rays from a single workspace VC ray*

move along a straight line trajectory parallel to $e_{vc}^{\mathcal{W}}$ but displaced from it by $\vec{a_0}(0) - \vec{a_j}(0)$. We construct a CVC ray e_{vc} in $\mathcal{C}$, whose endpoints are the endpoints of $e_{vc}^{\mathcal{W}}$ displaced by $\vec{a_0}(0) - \vec{a_j}(0)$. Motion of a_0 (the reference vertex) along e_{vc} corresponds to visual compliant motion of vertex a_j along $e_{vc}^{\mathcal{W}}$. Similarly, visual guarded motion of a_0 terminating on e_{vc} corresponds to visual guarded motion of a_j terminating on $e_{vc}^{\mathcal{W}}$, which is a visually observable event. The construction of CVC rays using this technique is illustrated in Figure 5.

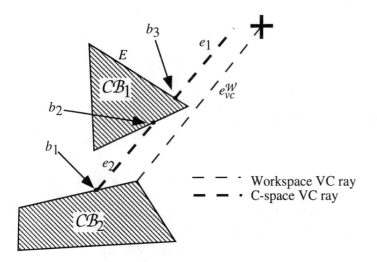

Figure 6: *A CVC ray may intersect the interior of* CB.

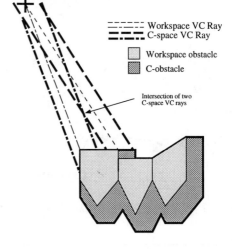

Figure 7: *Intersection of two CVC rays*

Intersection of CVC rays with C-obstacles.
When a CVC ray intersects a C-obstacle, visual compliant motion *cannot* be effected along the portion of the CVC ray that lies inside of the C-obstacle, since doing so would cause the robot to overlap a workspace obstacle. In this case, we must truncate the CVC ray at those points where it enters CB, retaining only those segments of the CVC ray that lie outside of CB (we will use the notation CB_i to indicate a particular C-obstacle, and CB to represent the union of all C-obstacles). In Figure 6, the CVC ray constructed from the workspace VC ray e_{vc}^{W} intersects the interior of C-obstacle CB_1. That part of the CVC ray that lies outside of CB includes two line segments: e_1 is the segment between the camera and artificial vertex b_3 on edge E, e_2 is the segment from artificial vertex b_2 to artificial vertex b_1. In this example, only the segments e_1 and e_2 are included in the set of CVC rays.

Intersection of multiple CVC rays. Although *workspace* VC rays intersect only at the camera projection center, two *C-space* VC rays may intersect at a point other than the camera projection center. Figure 7 shows one example of intersecting CVC rays. Since the CVC rays corresponding to a single workspace VC ray are parallel, the intersecting CVC rays cannot have originated from the same workspace VC ray. The physical interpretation of the intersection of two CVC rays is a change from executing compliant motion of vertex

a_i along workspace VC ray $e_{vc_k}^{W}$ to compliant motion of vertex $a_j, j \neq i$ along workspace VC ray $e_{vc_l}^{W}, k \neq l$. For example, such an intersection point might correspond to a change from compliant motion of the top right vertex of the square robot along VC ray $e_{vc_1}^{W}$, to compliant motion of its bottom left vertex along $e_{vc_2}^{W}$.

Algorithmic and Complexity Issues. The C-space representation of the VC rays can be computed by using two successive plane-sweep algorithms. The first constructs all of the CVC rays, and the second is used to truncate these CVC rays, as described in Section 3.3. Plane-sweep algorithms comprise a general class of algorithms that operate by stepping through a queue of geometrically interesting or critical events, and performing some local processing at each event in order to construct a global solution to some input problem. Conceptually, a line is "swept across" the plane, stopping at the critical events. At any time, the sweep-line divides the space into two half-spaces, such that the solution in one half-space has been computed and will not be affected by the computation of the solution in the other half-space. Further, between any two critical events, the solution that is being constructed does not change in a qualitative way.

The input to a plane-sweep algorithm is an arrangement of geometric objects and the output is some desired operation on the objects. Plane-sweep algorithms can be used, for example, to compute the intersection points of an arrangement of line segments or the union of an arrangement of polygons [27]. The advantage of a plane-sweep algorithm over a naive algorithm is usually reduced time complexity. As an example, a naive algorithm for computing the intersection points of n line segments runs in $O(n^2)$ time, but a plane-sweep algorithm can compute them in $O((n+c)\log n)$ time, where c is the number of intersection points. In the worst case $c = O(n^2)$, but in practice it is usually small and the plane-sweep algorithm is more efficient than the naive algorithm.

The initial set of CVC rays can be constructed by a variant of the traditional plane-sweep algorithm in which the sweep-line is a half-line that is rotated about the projection center of the camera. Let P represent the projection center of the camera. Then anchor a half line at P, and sweep this half line from θ_1 to θ_2, where θ_1 and θ_2 are the angles at which the sweep-line anchored at P intersects the two extreme points of the line segment that defines the camera image plane. The obstacle vertices define the set of events at which the sweep stops. By properly maintaining the status of the set of intersections of the sweep-line with obstacle edges, it is possible to determine whether each visited vertex is unoccluded. Making this determination and performing the processing necessary to update the status of the sweep-line requires $O(\log n)$ operations at each vertex. Since there are $O(n)$ vertices, determining the set of unoccluded vertices requires $O(n \log n)$ operations. This is a slight variation of the algorithm for computing visibility graph edges described in [18].

In addition to determining the set of unoccluded vertices, we must also determine which vertices of the robot are CM vertices for each unoccluded obstacle vertex. In general, for a robot with m vertices this can be done by a naive algorithm using $O(m)$ operations for each obstacle vertex. However, we can interleave the process of determining the CM vertices with that of determining the set of unoccluded vertices by making the following observation. The set of CM vertices changes only when the rotational sweep-line becomes parallel to an edge in the convex hull of the robot. In other words, as the sweep-line rotates, the same two robot vertices will be the CM vertices until the sweep-line becomes parallel to an edge in the convex hull of the robot. Therefore, by modifying the rotational sweep so that it also stops at orientations parallel to edges in the convex hull of the robot, we can simultaneously com-

pute the set of unoccluded obstacle vertices and the two CM vertices for each unoccluded vertex. The complexity of the resulting algorithm is $O(m \log m)$ to compute the convex hull of the robot, and $O((m + n) \log n)$ to perform the sweep. If we assume that $n > m$, the algorithm requires $O(n \log n)$ operations.

Truncating the CVC rays can also be accomplished by a plane-sweep algorithm. Here, a line is swept across the plane in the direction perpendicular to the camera image plane. The events are the vertices of C-obstacles and the intersections between C-obstacle edges and CVC rays. Such an algorithm requires $O((n + c) \log n)$ operations, where c is the number of intersections of CVC rays with C-obstacle edges.

4 Preimages and Backprojections

In this section, we provide a review of preimages and backprojections. We begin by reviewing the preimage formalism of Lozano-Pérez, Mason and Taylor [20]. Following this, we present a review of backprojections [10], including a discussion of issues related to goal recognizability. Finally we describe the algorithm of Donald and Canny [7] for computing a directional backprojection in $O(n \log n)$ time (where n is the number of vertices of the C-space obstacle region). Readers that are familiar with this work may wish to skip this section.

4.1 Preimage Planning

Lozano-Pérez, Mason and Taylor [20] present a formalism for the automatic synthesis of fine-motion strategies using *preimages*. Informally stated, a *preimage* for a specified goal region is a set of points from which a commanded motion is guaranteed to terminate recognizably in the goal region. The main advantage to the preimage formalism is that it allows the fine-motion planner to explicitly consider uncertainties in position and control.

In [20], position uncertainty is modelled by an error ball, $B_{ep}(p)$, in the C-space, centered on the actual position p. Velocity uncertainty is modelled by an *uncertainty cone*, whose vertex angle represents the maximum directional deviation between the commanded velocity and the actual velocity. If a position p_0 lies within the error ball centered on measured position p_0^*, then p_0^* is said to be *consistent with* p_0. Intuitively, this means that the sensor might "mistakenly" measure p_0 as p_0^*. A similar definition holds for measured *vs.* actual velocity vectors.

The velocity uncertainty cone plays a key role in the computation of preimages (and, as will be seen below, in the computation of backprojections). Specifically, both preimages and backprojections may include in their boundaries *free edges* (also called *free rays*). A free edge is a line segment that is parallel to an edge of the inverted velocity cone, erected at some C-obstacle vertex.

The formal definition of a *directional preimage* $P_\theta(G)$ is as follows. Let G be a goal region in $\mathcal{C}_{valid}$ (where $\mathcal{C}_{valid}$ is the set of valid configurations in the configuration space, $\mathcal{C}$). A motion command $M = (\vec{v}_\theta, TC)$, consists of a commanded velocity $\vec{v}_\theta$ (which is a considered to be a unit vector with orientation θ), and a termination predicate TC, which is used to determine when the motion has achieved the goal. The preimage of G for motion M is defined as a subset of points, $R \subseteq \mathcal{C}_{valid}$, such that if M commences from any point in R, TC will eventually return *true*, at which point the motion will terminate in G. A *maximal directional preimage* is the largest possible preimage relative to a given motion direction and goal region.[1]

A preimage planner works by backward-chaining from the goal region G to the C-space region I in which the initial configuration lies. If the backward-chaining process terminates successfully, the result is a sequence of directional preimages $P_1, P_2, \ldots, P_r$ such that: (a) P_i is the directional preimage of P_{i-1} relative to the commanded velocity $\vec{v}_{\theta_i}$ and termination condition TC_i, (b) P_1 is the directional preimage of G, and (c) $I \subset P_r$. The reverse sequence of motion commands $M_r, M_{r-1}, \ldots, M_1$, with $M_i = (\vec{v}_{\theta_i}, TC_i)$, is the generated r-step motion strategy guaranteed to recognizably reach the goal configuration from the initial configuration.

Mason has shown that the LMT approach of preimage backchaining is bounded complete (i.e., if a solution with bounded number of motions exists, the LMT preimage backchaining method will find it), and that it suffices to consider directional preimages as subgoals

---

[1]This explanation paraphrases Latombe's discussion of preimages [18], which also presents the relevant equations and formal definitions of the termination predicate.

in the recursive backchaining process [22]. These results, however, do not imply that preimages are computable. In fact, Erdmann has proven by a reduction from the halting problem, that, in arbitrary environments, preimages and backprojections are uncomputable [10]. That the recursively defined constraints in the proof do not generally occur in practice led Erdmann to conjecture without proof that in an environment with a known finite number of constraints preimages should be computable. Canny has shown that this is indeed the case when the set of possible robot trajectories has a finite parameterization, and the set of feasible trajectories is a semi-algebraic subset of the parameter space [4]. Canny's approach is to cast the fine motion planning problem as a decision problem in the theory of the real numbers, and to then use quantifier elimination algorithms (see, e.g., [6]) to derive parametric semi-algebraic sets that are preimages. Backprojections from recognizable goal regions also constitute valid preimages [10, 19]. Such backprojections are the topic of the next section.

4.2   Backprojections

A major problem in computing preimages is that there are many circumstances under which a real termination predicate TC may not be able to reliably detect entry into the goal region. There are two primary reasons for this: uncertainty in sensing, and limitations on the amount of information available to the termination predicate. Because of these difficulties, Erdmann introduces backprojections [10] as a means of approximating preimages. Essentially, a backprojection is a preimage without a termination predicate; that is, a backprojection is the set of all points from which an appropriate commanded velocity is guaranteed to enter the goal, regardless of whether entry into the goal is recognized. The lack of a termination predicate makes backprojections weaker than preimages, but backprojections are often easier to compute than preimages, and are appropriate for use in certain planning problems. The backprojection approach to motion planning can be characterized as recursively finding some subset of the goal region that is guaranteed to be recognized [19, 10], and constructing the backprojection for this region.

4.3 Computing Backprojections in $\mathcal{C} = \mathbf{R}^2$

Donald and Canny have implemented a plane-sweep algorithm for computing backprojections of polygonal goal regions for the case $\mathcal{C} = \mathbf{R}^2$ [7]. The algorithm works by sweeping a line across the plane in the direction opposite that of the commanded velocity. The sweep-line stops at events that are (1) vertices of C-obstacles, (2) vertices of the goal region, (3) the intersection of two free edges, (4) the intersection of a free edge with a boundary of the goal region, and (5) the intersection of a free edge with an edge of a C-obstacle. In each case, the backprojection is extended appropriately, using only local decision criteria.

Donald shows [7] that the algorithm is correct provided that the environment has a bounded number of vertices, and that the friction cone is larger than the velocity uncertainty cone. (This latter criterion is necessary because without it, the algorithm would not be able to rely only on local information to determine how to continue the backprojection.)

5 The Effect of Visual Constraints on the Directional Backprojection

In this section we show how the backprojection algorithm of Donald and Canny can be modified to exploit visual constraints. We will refer to a backprojection that includes CVC rays as a *VC-enlarged* backprojection, and we will denote a VC-enlarged backprojection by $B_{vc_\theta}(G)$. We begin by describing the new event types that must be considered by the plane-sweep algorithm. We then present two examples of backprojection continuation at such events. Following the examples, the formal decision criteria for determining whether to include a CVC ray in the backprojection boundary are presented. We then discuss the time complexity of the modified directional backprojection algorithm. Finally, we present examples of $B_{vc_\theta}(G)$ for $\mathcal{C} = \mathbf{R}^2$ computed by our implementation of the modified algorithm.

5.1 New Events for the Plane-Sweep Algorithm

The first step in modifying the Donald and Canny directional backprojection algorithm to exploit CVC rays is to determine the new events that must be considered

during the plane sweep. When CVC rays are included, there are three new types of events that must be considered:

1. the intersection of a CVC ray with a C-obstacle edge (or a C-obstacle vertex),

2. the intersection of a CVC ray with a free edge of the inverted velocity uncertainty cone,

3. the intersection of two CVC rays.

When a CVC ray intersects a C-obstacle edge, we create an artificial vertex at the intersection point. If a particular C-obstacle vertex has a CVC ray incident on it, that vertex is marked to indicate this fact, and the equation of the incident CVC ray is attached to it. Thus, only intersections of CVC rays with C-obstacle vertices will be considered in the remainder of the paper.

In the worst case, there will be $O(n)$ new artificial vertices for the CVC rays that intersect C-obstacle edges. There are $O(n^2)$ intersections of free edges with CVC rays, but during the construction of the backprojection, only the first intersection of a free edge with a CVC ray is considered. Therefore, the number of new events of this type that must be considered by the algorithm is $O(n)$.

Finally, there are, in the worst case, $O(n^2)$ pairwise intersections of CVC rays. To see this, consider that the intersection of two CVC rays occurs when the two CM vertices of the robot are simultaneously in contact with two distinct workspace VC rays, say $e_{vci}^{\mathcal{W}}$ and $e_{vcj}^{\mathcal{W}}$. Such an intersection point can be created by positioning one CM vertex of the robot on $e_{vci}^{\mathcal{W}}$, and then moving the robot compliantly along this ray until the remaining CM vertex contacts $e_{vcj}^{\mathcal{W}}$.

Thus the number of events considered by the modified plane-sweep algorithm is $O(n + c)$, where c is the number of intersections of pairs of CVC rays.

5.2 Example of Backprojection Continuation

Before presenting the formal decision criteria for the new events, we present the following two examples. These examples show how visual constraint surfaces can be used to bound the backprojection, which is made possible by exploiting visual compliance (which is analogous to physical compliance).

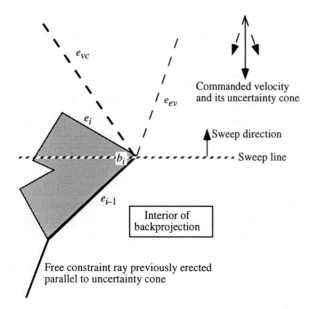

Figure 8: *Deciding how to continue the backprojection from a C-obstacle vertex*

Consider an obstacle vertex b_i with incident C-obstacle edges e_i and e_{i-1}, and incident CVC ray e_{vc}, as shown in Figure 8.[2] The commanded velocity is straight down. Edge e_{i-1} has already been added to the backprojection, and it forms the left edge of the current backprojection boundary. The algorithm must decide whether to continue the backprojection along e_i, e_{vc}, or the free edge of the inverted velocity uncertainty cone, e_{ev}.

Intuitively, the algorithm tries to make the backprojection as large as possible. Suppose e_i is a sliding edge. Then choosing it will always result in the maximal backprojection, because if either e_{vc} or e_{ev} made the backprojection larger, it would intersect the interior of the C-obstacle bounded by e_i. Conversely, suppose e_i is a sticking edge. In this case a simple comparison of the orientations of e_{vc} and e_{ev} suffices to determine which ray should be chosen, namely the one that forms the larger angle with the direction perpendicular to the commanded velocity.

---

[2]Although CVC rays are not necessarily incident on C-obstacle vertices, we may assume this without loss of generality since artificial vertices are introduced where CVC rays intersect C-obstacle edges.

When a free edge intersects a CVC ray, it is not always the case that the CVC ray should be added to the boundary of the backprojection. Figures 9 and 10 illustrate an example of such a case. Suppose that the motion begins at the point R with a commanded velocity straight down. When the robot intersects the CVC ray e_{vc}, the execution system equipped with visual feedback will begin visual compliant motion along the ray toward the goal. However, if visual compliant motion continues until the C-obstacle is contacted, the motion may not reach the goal. Instead, somewhere along e_{vc}, between points p and q, the execution system must resume motion in a downward direction. But the region of C-space in which this change of direction must occur is a free-space region, so that position sensing uncertainty becomes a problem. Specifically, since visual feedback cannot be used to determine where the robot is positioned along a projection ray, the robot will comply to e_{vc} and leave the backprojection region, continuing until contact is made with the C-obstacle upon which e_{vc} terminates. Therefore, we cannot guarantee that the single commanded motion would reach the goal, and we conclude that the CVC ray in this example should not be a backprojection boundary. We note that it would be possible to allow for the use of position sensing to detect when the robot has reached point p. However, determining that p has been reached is essentially equivalent to the goal recognizability problem, and therefore greatly complicates the computation of the backprojection (indeed, it was this difficulty that led Erdmann to separate goal recognizability and goal reachability in his original formulation of backprojections [10]).

5.3 Intersection of a CVC Ray and a C-Obstacle Edge

The decision criteria for an event corresponding to the intersection of a CVC ray and a (possibly artificial) C-obstacle vertex are as follows. As in the above example, the vertex event being processed is a (possibly artificial) C-obstacle vertex b, with incident obstacle edges e_i and e_{i-1} and incident CVC ray e_{vc}. We denote by e_{ev} the free edge of the velocity uncertainty cone erected at b. We assume e_{i-1} has already been added to the backprojection, as in Figure 8. We denote the *orientation* of the sweep-line by $\vec{y}$, i.e. the direction of the sweep itself is perpendicular to $\vec{y}$. We assume that $\vec{y}$ points to the *interior* of the backpro-

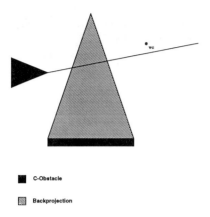

Figure 9: *The correct backprojection, when a CVC ray intersects a free edge, and the CVC ray does not terminate in the backprojection*

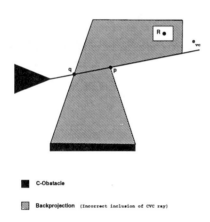

Figure 10: *An incorrect backprojection, when a CVC ray intersects a free edge, and the CVC ray does not terminate in the backprojection*

jection region that lies behind the sweep-line, so that it would point to the right in Figure 8. The decision criteria are:

1. If e_i is a sliding edge, continue the backprojection along e_i.

2. Otherwise, if the angle between e_{vc} and $\vec{y}$ is greater than the angle between e_{ev} and $\vec{y}$, continue the backprojection along e_{vc}.

3. Otherwise, add e_{ev} to the backprojection.

5.4   Intersection of a CVC Ray and a Free Edge

The CVC ray should be used to continue the backprojection at the intersection of a CVC ray and a free edge of the velocity uncertainty cone only when the termination point of the CVC ray on a C-obstacle edge is known to be in the backprojection. This becomes evident by enumerating the possible types of C-edges on which a CVC ray e_{vc} may terminate.

1. *CVC ray e_{vc} terminates on a nongoal sticking edge.* In this case, compliant motion along e_{vc} would result in contact with a sticking edge from which motion to the goal is not possible. Therefore e_{vc} should *not* be included in the backprojection.

2. *CVC ray e_{vc} terminates on a nongoal sliding edge e_j along which sliding motion* away *from the goal occurs.* In this case, a motion that brings the robot into contact with e_j will continue by sliding away from the goal. Therefore e_{vc} should *not* be included in the backprojection.

3. *CVC ray e_{vc} terminates on a nongoal sliding edge e_i along which sliding motion toward the goal occurs.* In this case, a motion that brings the robot into contact with e_i will continue by sliding towards the goal. Therefore e_{vc} can be included in the backprojection.

4. *CVC ray e_{vc} terminates on a goal edge.* Visual compliant motion along e_{vc} will bring the robot into contact with the goal, so e_{vc} can be included in the backprojection.

The cases enumerated above are exhaustive, and the cases in which e_{vc} should be included in the backprojection occur only when e_{vc} terminates on an edge already known to be in the backprojection.

The decision criteria at a vertex event, v, at which a free edge of the velocity uncertainty cone e_{ev} and a CVC ray e_{vc} intersect are as follows. As in Section 5.3, the vector $\vec{y}$ points along the sweep-line toward the interior of the backprojection.

1. If the two edges incident on v are already in the backprojection, no new edge is added to the backprojection boundary (this case is illustrated in Figure 13).

2. If e_{vc} is incident on a C-obstacle edge or vertex that is already included in the backprojection, *and*

the angle between e_{vc} and $\vec{y}$ is greater than the angle between e_{ev} and $\vec{y}$, continue the backprojection along e_{vc}.

3. Otherwise, continue the backprojection along e_{ev}.

5.5   Intersection of a Two CVC Rays

Since the intersection of two CVC rays is a visually observable event (i.e. the two CM vertices simultaneously contact two workspace VC rays), at such an intersection point, the backprojection algorithm should be continued along the CVC ray that maximizes the size of the enclosed backprojection. Let $\vec{y}$ be as defined above, and let the two intersecting CVC rays be e_{vc1} and e_{vc2}. Then:

1. If the angle between e_{vc1} and $\vec{y}$ is greater than the angle between e_{vc2} and $\vec{y}$, continue the backprojection along e_{vc1}.

2. Otherwise, continue the backprojection along e_{vc2}.

5.6   Asymptotic Time Bounds

Before beginning the plane-sweep to compute a directional backprojection, $O(n)$ free edges can be erected at sticking vertices and a separate plane-sweep can be used to intersect them with each other and obstacle edges in time $O(n \log n)$. Therefore we make the following proposition:

Proposition 1 *The time to compute the directional backprojection with visual constraint rays, $B_{vc_\theta}(G)$, is $O((n + c) \log n)$, where c is the number of pairwise intersections of CVC rays.*

Proof: The total number of events to be examined is $O(n + c)$, since there are $O(n)$ additional artificial vertices introduced by the CVC rays, and c vertices that correspond to the intersection of pairs of CVC rays. At each event, a constant number of local comparisons is required: (a) a vertex incident on a C-obstacle edge requires one test to determine whether e_i is a sliding or a sticking edge, and if the latter, one test to determine whether e_{vc} or e_{ev} should be used to continue the backprojection; (b) a vertex that corresponds to the intersection of a free edge with a CVC ray requires one test to determine whether e_{vc} or e_{ev} should be used to continue the backprojection; (c) a vertex that corresponds to the intersection of a pair of CVC rays requires one test to determine which CVC ray should

be used to continue the backprojection. Thus, the decision of how to continue the backprojection at any event is $O(1)$. Finally, as with all plane-sweep algorithms, at each event the algorithm must perform book-keeping operations that require time $O(\log n)$. Therefore the asymptotic running time of our modified version of the Donald and Canny algorithm for computing a directional backprojection becomes $O((n + c) \log n)$. $\square$

5.7   Two-Dimensional Examples

We now present some examples contrasting backprojections that contain CVC rays to those obtained without considering CVC rays. The two sets of examples illustrate the effect of considering *vs.* ignoring CVC rays. In each case, observe that the CVC rays never make the backprojection smaller, and frequently make it larger.

In all of the example figures, we use the following conventions. The directional backprojection is enclosed by a dashed line, with edges contributed by visual constraints highlighted in bold. Workspace obstacles are shaded; C-obstacles are outlined. Solid arrows denote the commanded velocity direction. The camera projection center (workspace coordinates) is indicated by a cross. The goal polygon is shaded black. The direction $\theta = 0$ corresponds to movement straight down the page.

Figures 11(a)–(d) compare the traditional directional backprojection with the VC-enlarged directional backprojection for a range of commanded velocities. In each frame, the traditional directional backprojection, $B_\theta(G)$, is shown on the left, and the VC-enlarged backprojection, $B_{vc_\theta}(G)$, is shown on the right.

Frame (a) corresponds to commanded velocity $\theta = 0$. In this case the backprojection is significantly enlarged by the CVC ray y. Note that free edge x, given by the inverted velocity uncertainty cone erected at vertex b in both backprojections, terminates on a workspace obstacle vertex, but x does not close the backprojection in the right-hand figure since y is nearly parallel to x. (The top horizontal edge of the backprojection is given by the environment's bounding box.) That the backprojection continues from b along x rather than along a CVC ray indicates that x results in a larger backprojection than any CVC ray incident on b.

Frame (b) corresponds to $\theta = -\pi/6$. In this case, even though a small part of a CVC ray contributes to

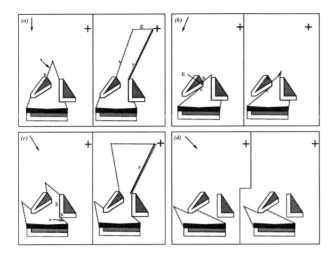

Figure 11: *Effect of considering CVC rays in computing the directional backprojection*

the backprojection, the backprojection is not significantly enlarged by considering it. This is because for the given position of the camera, there is no CVC ray incident on b that would enlarge the backprojection, since we have stipulated that visual compliant motion can only be robustly effected using CM vertices. This example suggests that the enlargement of backprojections due to visual constraints is sensitive to camera placement; which we discuss in [14].

Frame (c) corresponds to $\theta = +\pi/6$. This is similar to the case $\theta = 0$; the backprojection is significantly enlarged by the CVC ray y. It is interesting to note that the backprojection seems to be enlarged the most when the orientations of CVC rays are most nearly perpendicular to the commanded velocity direction (and therefore the free edges of the velocity uncertainty cone). When this occurs, the free edges and CVC rays "fan out" from C-obstacle vertices to enlarge the backprojection into a funnel-like region.

The commanded velocity of frame (d) is nearly identical to that of frame (c). However, the VC-enlarged backprojection is significantly different for these two cases. The reason for this, as will be shown in the next two sections, is that the commanded velocity direction in these two cases is very near a *critical* orientation, at which the topology of the VC-enlarged backprojection changes in a qualitative way.

6 The Nondirectional Backprojection

The backprojection algorithm presented in Section 5 computes a directional backprojection relative to a specific commanded velocity. A complete planner should consider all possible commanded velocities at each iteration of the backchaining algorithm. This can be achieved by considering the nondirectional backprojection, $B(G)$, which is defined as the union of all directional backprojections together with their respective velocity directions:

$$B(G) = \bigcup_{\theta} (B_\theta(G) \times \{\theta\}). \tag{3}$$

Donald has shown that the topology of the directional backprojection changes only at a finite set of critical velocity orientations, $\theta \in S^1$ [7]. Therefore, the nondirectional backprojection can be represented by a finite set of directional backprojections; one for each critical orientation, and one for each non-critical interval. In this section we review critical orientations, and the time complexity of computing the traditional nondirectional backprojection (i.e. the backprojection without visual constraints).

6.1 Critical Orientations

Critical orientations occur under the following three conditions [7]:

1. *A free edge becomes parallel to an edge in the obstacles' visibility graph.* To see this, notice that a free edge erected at some obstacle vertex b_0 will rotate with θ and may eventually rotate to an angle θ_1 at which it intersects another obstacle vertex b_1. When the ray rotates beyond θ_1, it will be truncated by the obstacle edge incident on b_1, and part of that obstacle edge may be included in the backprojection. Given this argument, note that the critical angle θ_1 occurs exactly when the free edge coincides with the visibility-graph edge connecting b_0 to b_1. Hence such orientations are called *vgraph-critical.*

2. *An obstacle edge changes from a sliding into a sticking edge or vice versa.* This occurs when a free edge of the velocity uncertainty cone is parallel or antiparallel to an edge of the friction cone. These orientations are called *sliding-critical.*

3. *The intersection point of two free edges of the back-projection intersects an obstacle edge.* Since the free edges rotate with θ, so do the backprojection vertices formed by their intersections. When any such vertex intersects an obstacle edge, one of the free edges incident on that vertex disappears, to be replaced by the obstacle edge. These are called *vertex-critical* orientations.

Donald presents an algorithm for computing these critical orientations. He then shows that the nondirectional backprojection may be represented by a finite set of directional backprojections: one directional backprojection for each critical orientation, and one representative directional backprojection for each noncritical interval (where the value of θ at which the backprojection is computed may be chosen arbitrarily).

Since the representative directional backprojection inside a noncritical interval may be computed for an arbitrary value of θ in that interval, it is possible that the algorithm will fail to compute a directional backprojection that entirely contains the polygonal start region R. To avoid this problem, Donald [9] suggests adding R to the arrangement of polygons, thus adding the following critical orientation criterion:

4. *An edge of R intersects a free edge of the backprojection.* These orientations are called *R-critical*.

For an input of n C-obstacle vertices, R has a constant number of edges and there are $O(n)$ free edges bounding the backprojection. Therefore there are $O(n)$ R-critical orientations. If the directional backprojection for some R-critical orientation θ_i contains all the vertices of R, then a commanded motion from R with velocity $\vec{v}_{\theta_i}$ will reach the goal.

6.2   Time complexity

Although Donald shows that there are $O(n^2)$ critical orientations of type (3), he proposes a naive $O(n^3)$ algorithm to compute them, as follows. There are $O(n)$ free edges, and therefore $O(n^2)$ possible intersections of free edges. These intersections are free-space vertices of the directional backprojection that trace out circles as the velocity orientation is changed. Each such circle may intersect $O(n)$ obstacle edges. Therefore the number of intersections of circles with obstacle edges is $O(n^3)$. The $O(n^2)$ critical orientations are contained in this set of size $O(n^3)$.

The motivation for this algorithm is that, of all possible $O(n)$ free-space backprojection vertices, the subset of these that will contribute to the critical orientations is not known in advance; however, if *all* intersections of possible free-space vertices with obstacle edges are computed in advance, this set is guaranteed to contain all of those that will contribute to critical orientations.

Donald's critical-slice algorithm recomputes the backprojection from scratch at each critical orientation and inside each noncritical interval. Recently, Briggs has presented an algorithm that incrementally computes the nondirectional backprojection, achieving an $O(n^2 \log n)$ time complexity [2]. The reduced complexity is due in part to an amortized analysis that shows there are at most $O(n^2)$ topological changes to the boundary of the backprojection over the entire range of θ. The algorithm also uses a dynamic data structure to keep track of the rotating free-space vertices, rather than computing all possible free-space vertices in advance.

7   The Effect of Visual Constraints on the Nondirectional Backprojection

In this section, we describe how the introduction of visual constraints affects the computation of the nondirectional backprojection. In particular, we discuss the new critical orientations that result from the introduction of visual constraints, and the time complexity of a modified nondirectional backprojection algorithm.

According to the procedure outlined in Section 5, the decision of whether to continue the backprojection along a CVC ray from a given vertex event depends, among other things, on whether the incident C-obstacle edge e_i is a sliding or a sticking edge. Sliding *vs.* sticking behavior changes only at sliding-critical orientations [7], so these orientations are also critical for VC-enlarged backprojections.

The introduction of visual constraints also adds two new criteria for critical orientations. The first is analogous to Donald's vertex-critical criterion, and the second to his vgraph-critical criterion.

7.1   Free-Edge-Critical orientations

Suppose f_i is a vertex of the backprojection formed by the intersection of two rays of the inverted velocity

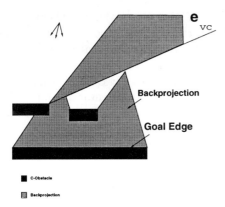

Figure 12: *A free-edge-critical orientation, just before an intersection of a free vertex with a CVC ray*

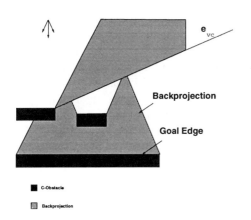

Figure 13: *A free-edge-critical orientation, just after the intersection of a free vertex with a CVC ray*

uncertainty cone. As the commanded velocity direction θ varies, f_i moves along a circular arc. A critical orientation occurs when this circular arc intersects a CVC ray, since the decision of which of the free edges or CVC ray should be used to continue the backprojection may change. This is illustrated in Figures 12 and 13. By analogy to Donald's vgraph-critical orientations, we express this new critical orientation as follows:

5. *A free-space vertex of the backprojection intersects a CVC ray.* We call such orientations *free-edge-critical.*

Proposition 2 *There are $O(n^2)$ free-edge-critical orientations.*

Proof: We showed that the $O(n)$ workspace obstacle vertices give rise to $O(n)$ CVC rays. Donald shows that there are $O(n^2)$ vertex-critical orientations resulting from the intersection of free vertices with $O(n)$ obstacle edges. The same argument applies by treating the $O(n)$ CVC rays as obstacle edges. □

Of course, it is not always the case that the backprojection topology changes at free-edge-critical orientations, as illustrated in Figures 9 and 10.

7.2 VC-Critical Orientations

Before describing the second critical orientation criterion added by CVC rays, we note the conditions from which it follows directly:

- The visibility of a vertex does not change with the commanded velocity direction θ, since the workspace obstacles and camera are fixed. Therefore the workspace VC rays do not change with θ.

- Consequently, the C-space representation of the VC rays does not change with θ, since CVC rays are constructed from workspace VC rays by considering only the vectors joining adjacent robot vertices. Since the robot cannot rotate, these vectors never change.

- With respect to the directional backprojection, CVC rays behave as if they were nonsticking obstacle edges terminating at the camera focal center, since they do not move and sticking can never occur on them.

With these statements in mind, we note the second new criterion for critical orientations added by CVC rays, constructed by analogy to Donald's condition (3):

6. *A free edge becomes parallel to a CVC ray.* At such an orientation, the decision of whether to add the free edge or the CVC ray to the backprojection may change. We will call such orientations *VC-critical.*

Figure 14 shows how the backprojection changes across such a critical orientation.

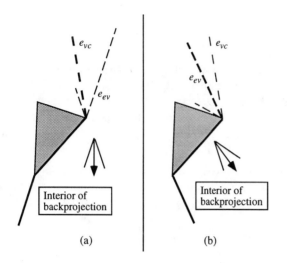

Figure 14: *How the backprojection changes across a VC-critical orientation*

Proposition 3 *There are $O(n)$ VC-critical orientations.*

Proof: As was shown in Section 7.1, there are $O(n)$ CVC rays in an environment that contains n C-obstacle vertices. Each of these introduces two critical orientations of the type described above, given by the two free edges of the velocity uncertainty cone. Hence there are $O(n)$ VC-critical orientations. □

7.3 New Asymptotic Time Bounds

If we denote by $B_{vc_\theta}(G)$ the directional backprojection with visual constraints, and by $B_{vc}(G)$ the nondirectional backprojection with visual constraints, we have

$$B_{vc}(G) = \bigcup_\theta (B_{vc_\theta}(G) \times \{\theta\})$$

Proposition 4 *A representation of the nondirectional backprojection with visual constraints, $B_{vc}(G)$, can be computed in time $O(n^3(n + c) \log n)$, where c is the number of pairwise intersections of CVC rays.*

Proof: Donald's critical-slice method [7] computes the nondirectional backprojection in time $O(n^4 \log n)$ when there are $O(n^3)$ critical orientations, by computing $O(n^3)$ slices each in time $O(n \log n)$. For the VC-enlarged backprojection, the complexity of computing

a slice, $B_{vc_\theta}(G)$, is $O((n+c) \log n)$. There are $O(n)$ additional VC-critical orientations, and $O(n^2)$ additional free-edge-critical orientations, but this does not asymptotically increase the total number of critical orientations since there are already $O(n^2)$ vgraph-critical orientations [7]. The critical orientations can be found using Donald's proposed naive algorithm in time $O(n^3)$. Hence the nondirectional backprojection with visual constraints can be computed in time $O(n^3(n+c) \log n)$. □

Conjecture 1 *A representation of the nondirectional backprojection with visual constraints, $B_{vc}(G)$, can be computed in time $O(n^2 \log n)$.*

Rationale: Using an amortization techniques, Briggs has shown that for the nondirectional backprojection (without visual constraints) there are at most $O(n^2)$ changes to the topology of the backprojection over all values of θ [2]. This same analysis should apply to backprojections that include visual constraints, since there are only $O(n^2)$ additional critical orientations due to visual constraints. Briggs's algorithm [2] computes the nondirectional backprojection in $O(n^2 \log n)$ time when there are $O(n^2)$ critical orientations. We believe that it should be possible to extend this algorithm to compute $B_{vc}(G)$ in time $O(n^2 \log n)$. It remains to provide a constructive proof of this conjecture. □

8 Conclusions

In this paper, we have introduced visual constraint surfaces as a mechanism to effectively exploit visual constraints in the synthesis of uncertainty-tolerant robot motion plans. Visual constraint surfaces can be used to effect visual guarded and visual compliant motions. By deriving a configuration space representation of visual constraint surfaces, we were able to include visual constraint surfaces as boundaries of the directional backprojection. We described an implemented backprojection planner for $\mathcal{C} = \mathbf{R}^2$ based on Donald and Canny's algorithm [7].

By examining the effects of visual constraints as a function of the direction of the commanded velocity, we were able to determine new criteria for critical orientations (i.e. orientations at which the topology of the directional backprojection, including visual constraint surfaces, might change). We presented an algorithm to compute the nondirectional backprojection modified to include visual constraint surfaces.

9 Acknowledgements

This work was supported by grant IRI-9110270 from the National Science Foundation. The authors are grateful to the anonymous reviewers, who provided many constructive comments that greatly improved the clarity of the presentation, and to Michael Barbehenn and Jean Ponce for their helpful comments on an earlier draft of this paper.

References

[1] P. Allen, B. Yoshimi, and A. Timcenko. Real-time visual servoing. In *IEEE International Conference on Robotics and Automation*, pages 851–856, April 1991.

[2] A. J. Briggs. An efficient algorithm for one-step planar compliant motion planning with uncertainty. *Algorithmica*, 8:195–208, 1992.

[3] R. A. Brooks. Symbolic error analysis and robot planning. *International Journal of Robotics Research*, 1(4), Winter 1982.

[4] J. F. Canny. On computability of fine motion plans. In *IEEE International Conference on Robotics and Automation*, pages 177–182, 1989.

[5] A. Castaño and S. A. Hutchinson. Hybrid vision/position servo control of a robotic manipulator. In *IEEE International Conference on Robotics and Automation*, pages 1264–1269, Nice, France, May 1992.

[6] G. E. Collins. Quantifier elimination for real closed fields by cylindrical algebraic decomposition. In *Lecture Notes in Computer Science, 33*, pages 135–183. Springer-Verlag, New York, NY, 1975.

[7] B. R. Donald. *Error Detection and Recovery for Robot Motion Planning with Uncertainty*. PhD thesis, Massachusetts Institute of Technology, 1987.

[8] B. R. Donald. A geometric approach to error detection and recovery for robot motion planning with uncertainty. *Artificial Intelligence*, 37(1-3):223–271, December 1988.

[9] B. R. Donald. Planning multi-step error detection and recovery strategies. *International Journal of Robotics Research*, 9(1):3–60, February 1990.

[10] M. Erdmann. On motion planning with uncertainty. Master's thesis, Massachusetts Institute of Technology, 1986.

[11] M. Erdmann. Using backprojections for fine motion planning with uncertainty. *International Journal of Robotics Research*, 5(1):19–45, Spring 1986.

[12] I.D. Faux and M. J. Pratt. *Computational Geometry for Design and Manufacture*. Ellis Horwood Ltd., Chichester, 1985.

[13] J. T. Feddema and O. R. Mitchell. Vision-guided servoing with feature-based trajectory generation. *IEEE Transactions on Robotics and Automation*, 5(5):691–700, October 1989.

[14] A. Fox and S. Hutchinson. Exploiting visual constraints in the synthesis of uncertainty-tolerant motion plans. Technical Report UIUC-BI-AI-RCV-92-05, The Beckman Institute, University of Illinois, 1992.

[15] Berthold Klaus Paul Horn. *Robot Vision*. MIT Press, Cambridge, 1986.

[16] S. A. Hutchinson. Exploiting visual constraints in robot motion planning. In *IEEE International Conference on Robotics and Automation*, Sacramento, CA, April 1991.

[17] S. A. Hutchinson and A. C. Kak. SPAR: A planner that satisfies operational and geometric goals in uncertain environments. *AI Magazine*, 2(1):30–61, Spring 1990.

[18] J. C. Latombe. *Robot Motion Planning*. Kluwer Academic Publishers, Boston, 1991.

[19] J.C. Latombe, A. Lazanas, and S. Shekhar. Robot motion planning with uncertainty in control and sensing. *Artificial Intelligence*, 52:1–47, 1991.

[20] T. Lozano-Pérez, M. T. Mason, and R. H. Taylor. Automatic synthesis of fine-motion strategies for robots. *International Journal of Robotics Research*, 3(1):3–24, Spring 1984.

[21] M. T. Mason. Compliance and force control for computer controlled manipulators. In B. Brady, J. M. Hollerbach, T. L. Johnson, T. Lozano-Pérez, and M. T. Mason, editors, *Robot Motion:*

Planning and Control, pages 373–404. MIT Press, Cambridge, Mass., 1982.

[22] M. T. Mason. Automatic planning of fine motions: Correctness and completeness. In *Proc. IEEE Int'l Conference on Robotics*, pages 492–503, 1984.

[23] N. Papanikolopoulos and P. K. Khosla. Adaptive robotic visual tracking: Theory and experiments. *IEEE Transactions on Automatic Control*, 38(3):429–445, March 1993.

[24] N. P. Papanikolopoulos, P. K. Khosla, and T. Kanade. Visual tracking of a moving target by a camera mounted on a robot: A combination of vision and control. *IEEE Transactions on Robotics and Automation*, 9(1):14–35, 1993.

[25] R. P. Paul and B. Shimano. Compliance and control. In *Proc. of the Joint American Automatic Control Conference*, pages 694–1699, 1976.

[26] J. Pertin-Troccaz and P. Puget. Dealing with uncertainties in robot planning using program proving techniques. In *Fourth Int'l Symposium on Robotic Research*, August 1987.

[27] F. P. Preparata and M. I. Shamos. *Computational Geometry: An Introduction*. Springer-Verlag, New York, 1985.

[28] M. H. Raibert and J. J. Craig. Hybrid position/force control of manipulators. *Journal of Dynamic Systems, Measurement and Control*, 102:126–133, June 1981.

[29] P. Saraga and B. M. Jones. Simple assembly under visual control. In Alan Pugh, editor, *Robot Vision*, pages 209–223. IFS Pub. Ltd., U. K., 1983.

[30] Y. Shirai and H. Inoue. Guiding a robot by visual feedback in assembling tasks. *Pattern Recognition*, 5:99–108, 1973.

[31] S. B. Skaar, W. H. Brockman, and R. Hanson. Camera-space manipulation. *International Journal of Robotics Research*, 6(4):20–32, Winter 1987.

[32] R. H. Taylor. The synthesis of manipulator control programs from task-level specifications. AIM-282, Stanford Artificial Intelligence Laboratory, 1976.

[33] L. E. Weiss, A. C. Sanderson, and C. P. Neuman. Dynamic sensor-based control of robots with visual feedback. *IEEE Journal of Robotics and Automation*, RA-3(5):404–417, October 1987.

[34] D. E. Whitney. Force feedback control of manipulator fine motions. *Journal of Dynamic Systems, Measurement and Control*, pages 91–97, June 1977.

[35] P. M. Will and D. D. Grossman. An experimental system for computer controlled mechanical assembly. *IEEE Trans. on Computers*, C-24(9):879–888, 1975.

A Machine Learning Perspective on Modeling Robot Actions*

Alan D. Christiansen

Department of Computer Science, Tulane University, New Orleans, LA 70118

This paper addresses the automated learning of action effects by an autonomous agent in the physical world. The learning agent's physical sensors and effectors provide ordered feature values for the description of actions. Imperfections in the agent's sensor-effector system and characteristics of physical actions combine to generate noise and non-determinism in observed results of commanded actions. Successful learning of action effects requires a noise-tolerant learning algorithm. Successful goal-seeking ability further requires that the agent reason about its uncertainty in predicting the effects of its actions.

Actions are represented by continuous operators called funnels. Funnels are computed by an empirical learning algorithm that is both noise-tolerant and biased away from making false positive prediction errors. This paper presents a learning algorithm for computing funnels and demonstrates the method's effectiveness with empirical tests of a robot on a physical manipulation task called tray-tilting.

1 Introduction

An autonomous agent interacts with its environment via its sensors and effectors. In order to accomplish its goals in the environment the agent uses its sensors to obtain information about the current task state and uses its effectors to accomplish changes in the task state. If the agent is to act effectively to accomplish its goals, the agent must possess an ability to predict the effects of its actions. Typically this predictive ability is programmed by a human before the agent interacts with its environment. An alternative approach is to allow the agent to learn the effects of its actions through interaction with the environment via its sensors and effectors [Vere, 1978, Rivest and Schapire, 1987, Shen and Simon, 1989]. This paper concentrates on two aspects of 'learning from the environment' that have received little attention in the literature: dealing with (1) continuous feature spaces and (2) noise and uncertainty. These two aspects of the problem are very important if the methodology is to be applied to physically embodied agents (robots).

From the point of view of a learning agent, its 'environment' is a black box. The agent provides *actions* as inputs to the environment and receives perceived states via its sensors. These outputs of the environment are called *states*, but in general these states only partially reflect the true state of the environment. In this paper, I assume that both states and actions are conjunctions of ordered feature values, corresponding to continuous measurable environmental characteristics. Thus states (and actions) are described by points in an ordered multi-dimensional feature space. The relevant spaces are called the *state space* $\mathcal{S}$ and the *action space* $\mathcal{A}$, respectively. It is also sometimes convenient to describe the current situation by means of a point in the *state-action space*. This point indicates the current sensed state and the action to be executed.

The primary problem addressed within this paper is to predict the result of executing an action from a sensed state. Assuming that the agent uses its sensors before and after the execution of every action, information about the effects of its actions comes to the agent as a stream of points $[s_0, a_0, s_1, a_1, s_2, \ldots]$. Another way to view this is as a sequence of state transitions $[s, a, r]$ (state, action, result state).

An interesting aspect of this learning problem is that the agent receives only positive examples of the concept of interest. The agent only sees what can happen as a result of its actions. It never sees what can *not* happen! One approach for inductive inference problems of this kind is to assume a restricted family of functional forms for the answer, and apply regression techniques to choose the best answer according to some measure of error. These regression techniques possess the ability to deal with noise and continuous feature values, and thus are relevant to the problems addressed in this paper. For these techniques to succeed requires that the family

---

*This paper was previously published in the Proceedings of the Ninth International Workshop on Machine Learning (ML92) as "Learning to Predict in Uncertain Continuous Tasks."

of acceptable answers is limited a priori. This limitation, which constitutes a form of inductive bias, is in essence knowledge possessed by the agent as to the form of the answer. For systems possessing large amounts of initial knowledge, such as explanation-based learning systems, the success of the learning system is highly dependent on the initial knowledge chosen by the human programmer. The research described in this paper adopts the stance of very weak prior knowledge, and is therefore at the empirical, rather than the analytical, end of the spectrum of learning methods. ([Bennett, 1990] reports research at the analytical end of the spectrum.)

2   Assumptions

In this paper, I assume that the agent initially has very weak knowledge about the environment. In particular, the agent does not initially possess any ability to predict the effects of its actions. The agent assumes that the observed features (corresponding to sensed state values) correspond to continuous measurable environmental characteristics, and it assumes that the controlled features (corresponding to the controlled parameters of its actions) represent continuous changes to the effects of its physical actions. The agent assumes that the environment changes state only by the influence of the agent's actions. Further, it assumes that the environment is not dynamic: as a result of each action, the environment settles into a stable state by the time that the agent uses its sensors to measure the resulting state. These assumptions obviate modeling the effects of external influences such as time or other agents. The assumed time-invariance allows the agent to generalize over observed state transitions, independent of the specific history of the state transitions.

3   Manipulation Tasks

Manipulation tasks are representative of the uncertain continuous tasks addressed in this paper. Manipulation tasks are primarily concerned with the purposeful movement of physical objects. Figure 1 illustrates some well-known manipulation tasks. In these problems, goals specify desired object configurations (position and orientation) which an agent must achieve. In robotic manipulation, a robot is the agent causing changes in task state. A fundamental problem for the robot is to choose a sequence of actions to achieve a desired configuration. The robot's ability to plan depends on its ability to predict the effects of single actions. In experimental results reported below, it is this predictive

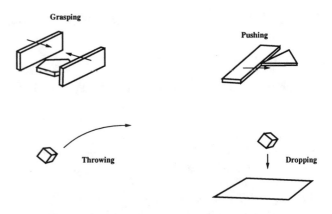

Figure 1: *Some familiar manipulation tasks.*

Figure 2: *The tray-tilting task. A robot holds a flat, walled tray from below. Applied tilting actions cause a planar object (in this case, a rectangle) to change its position and orientation.*

ability that the robot learns, and it is the improvement of its predictive abilities that are responsible for its improved proficiency at the task.

In this paper, we concentrate on a manipulation task called the *tray-tilting task* (illustrated in Figure 2). The tray-tilting task consists of a commercial robot arm that holds a flat walled tray from below. Within the tray is a planar rectangular object that slides in response to applied tilting actions by the robot. A camera is mounted above the tray, and a commercial vision system is used to extract the x-y position of the object as well as its orientation θ. The sensory information provided by the vision system is therefore three real-valued features. Actions are defined by the tilts performed by the robot. Although the actions undertaken by the robot are very complex (in general, the robot's motions are described by positions and velocities of six independent joints) the robot has been programmed to perform the tilting actions in response to a single real-valued input parameter defining the *tilt azimuth*. As the robot tilts the tray, there is a unique direction that gravity acts within the plane of the tray floor. That direction is called the tilt azimuth. Other parameters influencing the object, such as how steeply the tray is tilted and how fast the tilt is performed, are held constant for all tilts. In general, there is a tendency for the object to move in the direction of the tilt azimuth, but the actual behavior of the object is much more complex. The object can slide, rotate, bounce, or not move at all. The robot, which has no initial knowledge that allows it to predict how the object will move, builds a model based upon observed state transitions caused by selected tilt azimuths.

4   Noise and Uncertainty

The regression techniques seek to describe the effects of actions as a mathematical function: $r = result(a, s)$. The use of a function in describing the effect of an action embodies an assumption of determinism. That is, if the agent knows the current state s and the current action a, then it will be possible to predict the exact resulting state r. For robots, this assumption will often fail in practice. Due to noise characteristics of the robot's sensors and effectors, and due to the fact that the robot's sensors provide only partial information about the environmental state, the environment 'black box' appears to be non-deterministic. There will be cases where state transitions $[s, a, r_1]$ and $[s, a, r_2]$ will be observed, where $r_1 \neq r_2$. The approach followed in [Rivest and Schapire, 1987] and [Shen and Simon, 1989] was to assume that the environment was deterministic, and then to infer that perceived non-determinism was due to an improper description of state. Two main techniques, both examples of constructive induction, were used to disambiguate the state description. The first of these techniques involved the generation of expressions involving sensed features. The second technique used history expressions involving recent actions as well as

sensed features. Both techniques were demonstrated in simulated environments.

One very real possibility for physical environments is that no deterministic model of the actions can be found. It is possible that the physical environment itself is stochastic. Instead of viewing actions as deterministic (functions), we can choose to view actions as possibilistic (relations) or probabilistic (joint probability densities). In these views, it is not inconsistent to observe two state transitions with the same initial state and action, resulting in differing outcome states. We can treat *result* as a relation, or we can say that $result(a, s)$ is multivalued: $result(a, s) \in R$ where R is the set of possible outcomes. If the agent can describe the set R, it has some notion of the variability of the outcomes, and the size of R indicates the *specificity* with which the agent can make a prediction. The third alternative for modeling actions is to treat the actions as probabilistic: $p(s, a, r) = p(result(a, s) = r)$. In this case, to maintain consistent probabilities, we require that $0 \leq p(s, a, r) \leq 1$ for all s, a, and r, and that $\int_S p(s, a, r)\, dr = 1$, where S is the set of all possible result states r.

5   Generalization

For all of the possible forms (deterministic, possibilistic, and probabilistic) it is desirable for the agent to describe a subset of the state-action space where the results are equivalent. For a deterministic model we can describe such a subset β for every result state r:

$$\beta(r) = \{\; [s, a] \;\mid\; result(a, s) = r \;\}.$$

For the other model forms it may not make sense to insist that a state-action pair always produces exactly the same state. Therefore, for a possibilistic model we judge equivalent results as those that fall within a designated set:

$$\beta(R) = \{\; [s, a] \;\mid\; result(a, s) \in R \;\}.$$

Similarly, for a probabilistic model,

$$\beta(R) = \{\; [s, a] \;\mid\; \int_R p(s, a, r) dr \geq P_{min}. \;\}$$

The β stands for *backprojection* [Erdmann, 1986] (which should not be confused with 'backpropagation'). The terminology comes from projecting the effects of actions "forward" or "backward." A forward projection is what is usually thought of as a prediction—a description of what will result if an action is executed from a particular state (or some member of a set of states). A backprojection is the projection of a result state (or set of states) backward to give the set of states under which the action will achieve the desired result. This is the same notion as the 'goal regression' discussed in the planning literature.

The possibilistic action model is actually a special case of the probabilistic, where $P_{min} = 1$. This corresponds to what has been called the *strong backprojection*. Although there is a continuum of backprojections associated with every possible P_{min}, I will concentrate on strong backprojections within this paper.

6   Funnels

An alternative to the regression techniques is to treat the problem as one of learning *funnels* describing the state transitions. In this formulation we seek a description $[M, S]$ which generalizes the effect of many state transitions $[s, a, r]$. The set M is called the funnel *mouth* and S is the funnel *spout*. A funnel is a kind of operator [Fikes and Nilsson, 1971] for continuous state-action spaces that obeys the following semantics:

$$\forall \, [s, a] \in M \quad result(a, s) \in S,$$

where S specifies a desired outcome set, and M specifies the strong backprojection of S. The size of the funnel spout S is a direct indication of the *specificity* with which predictions can be made. Predictions indicate that the result will fall within the set S, but do not indicate anything further about the states within S. Similarly, the size of the funnel mouth M indicates the *scope* of a prediction. If we specify a particular action as well as a desired result set, then we can visualize the effect of an action as a mapping from a set of states to a set of result states. (See Figure 3a.)

Planning with Funnels

Once a set of funnels has been computed, the planning problem is very much like that of conventional operator-based planning systems. Because the funnels do not predict an exact result, but rather specify only a set of results, a simple form of constraint-based reasoning is required during planning. For simplicity of description, I assume that the funnel mouths do not generalize over actions (which means that every funnel is specific to some particular conjunction of controlled feature values). We seek a sequence of funnels that will transfer the initial state to a desired goal set. (See Figure 3b.) The computed plan is to be executed open-loop (without sensor-based execution monitoring), which places a large demand on the accuracy of computed funnels.

The planning process can be organized as a backward-chaining breadth-first search process to yield the short-

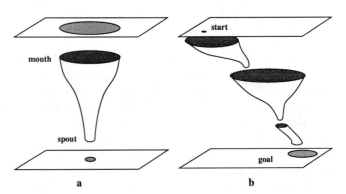

Figure 3: *(a) A funnel maps input states to result states. The funnel mouth indicates the set of allowed inputs, and the funnel spout indicates the possible results. The action represented in (a) reduces uncertainty, but in general actions can reduce uncertainty, increase it, or leave it more or less unaffected. The specificity of the prediction represented by the funnel is defined by the funnel spout and the scope of the prediction is defined by the funnel mouth. (b) Funnels can be combined into sequences to form plans.*

est plan. The global goal is defined as a region of the state space. If the current state (as given by the sensors) is within the goal, then a null plan is acceptable. Otherwise, the planner finds all known funnels that achieve the goal, and sets up the mouths of those funnels as subgoals. The search continues (in a breadth-first fashion) with each of those regions until some funnel mouth is found to contain the current observed state. At that time, the sequence of actions is extracted that leads from that funnel mouth to the global goal, and returned as the plan. The planning process is a simple search on a graph whose nodes are state space regions, and whose edges correspond to the agent's actions. When a search node is "expanded" the planner must compute whether or not the spout of each funnel is contained within the current subgoal region.

7   Learning Funnels

This paper proposes the following method for learning models of actions: First, choose a result region. Assuming that result region as a funnel spout, compute the (strong) funnel mouth based upon past observed state transitions. Combine the chosen spout and the computed mouth into a funnel, and store that funnel with other funnels thus computed, making it available to the planning system. The process of computing a funnel mouth from a given spout and observed state transitions is called *empirical backprojection* [Christiansen, 1991].

Representation of Mouths and Spouts

Funnel mouths and spouts correspond to geometric regions of the state-action space. We may choose from any of a number of representational primitives to describe these regions. For the tray-tilting task, it is convenient to describe regions with hyperrectangles, because features of the task (tray walls) are aligned with the x-y coordinate system for sensed object positions, and object positions tend to be near walls, because of the tilting actions. It is possible to describe any geometric region as a disjunction of hyperrectangles, but for certain feature spaces hyperrectangles may lead to unacceptably complex descriptions, and other representations may be preferred. As an example, suppose that in the tray-tilting task the robot's sensory system reported features in a frame of reference rotated 45 degrees from that of the tray. Since the object positions tend to be near walls, and wall-parallel rectangles are no longer within the descriptive language of the robot, the region descriptions become complex disjunctions of smaller hyperrectangles.

7.1 Choosing Spouts

Which results of actions should be of interest to the learning agent? There are many possible outcomes that might be of interest, because the state space is continuous and spouts are subsets of the state space. The agent can not afford to build funnels for every conceivable outcome set. A 'focus of attention' mechanism for choosing funnel spouts is required. There are at least two main strategies for this. First is the obvious strategy of clustering result states. From each observed state transition $[s, a, r]$, we extract the result state r. Taking the set of all observed result states, we look for regions with high density of result states. This is a reasonable strategy, as it embodies an assumption that since observed states are being repeated over and over, the agent would have an interest in achieving those results later on.

A second strategy requires knowledge of the intended goals for which the agent is to prepare. These goals may or may not correspond to the states that repeatedly occur during training. If the agent is given a set of goals to prepare itself for, it can also use these to determine subgoals of interest. The agent, during training, invokes its planner on one of the goals. Through the backchaining process, the planner can report to the learning system when a subgoal arises for which there is no funnel with a spout in the desired subgoal region. The learner can then attempt to identify funnels for the desired subgoal. (See [Porter and Kibler, 1986] for this

idea applied to a symbolic domain.) Thus, 'interesting' funnels are those which (1) achieve a known goal, or (2) achieve an intermediate result that is useful for attaining a known goal.

7.2 Computing Mouths

Once a spout S has been chosen, the computation of the required mouth is in many ways a conventional concept learning problem. In fact, now that a result set has been defined, the previously observed state transitions T can provide positive examples (that achieved some state within the spout) and negative examples (that did not):

$$pluses(S) = \{ [s, a] \mid [s, a, r] \in T \ \wedge \ r \in S \}$$
$$minuses(S) = \{ [s, a] \mid [s, a, r] \in T \ \wedge \ r \notin S \}$$

Previously Published Methods

Early in this research, a version space strategy [Mitchell, 1978] was tried to identify the funnel mouths. This did not work, and it is instructive to consider why the approach failed. The candidate elimination algorithm, as well as other more recent version space approaches, attempts to identify a necessary and sufficient condition for the concept. For the current problem of identifying a funnel mouth M for a given spout S, this would correspond to the condition

$$\forall \, [s, a] \in M \quad result(a, s) \in S \quad \wedge$$
$$\forall \, [s, a] \notin M \quad result(a, s) \notin S.$$

Figure 4 illustrates the approach. Figure 4a shows a concept region to identify. A version space strategy works by identifying two boundaries, one that includes all observed positive examples and another that excludes all negative examples (Figure 4b). As more examples are observed, either the boundaries converge to the correct boundary separating the classes, or the algorithm reports that no boundary consistent with the observed examples exists (this latter case is called version space collapse).

As the method was applied to the tray-tilting problem, it was observed that most computations of funnel mouths led to version space collapse. In retrospect, this is not surprising given the amount of noise and non-determinism present in the observed state transitions. Figure 4c illustrates the problem. Imagine that the agent observes examples of the true concept with a sensor with stochastic, but bounded error. When the sensor reports a particular sensed value, the true value can be anywhere within the bound indicated by the circle. The effect of this error characteristic is to "fuzz" the edges of the concept. The transition from 'the con-

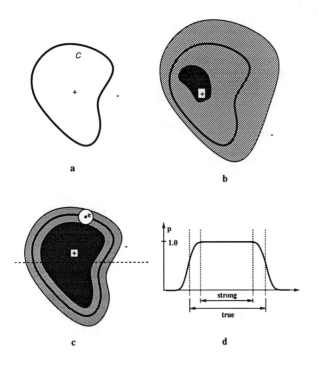

Figure 4: *A concept with observation error. The true concept boundary is shown in (a). Version space methods attempt to identify the boundary with two converging boundaries (b). With sensory error, the edges of the concept become fuzzy (c), leading to a probabilistic concept (d).*

cept' to 'not the concept' is no longer discrete in probability. If we consider a 'section' through the space of observations (indicated by the dashed line in Figure 4c), we see that the observed probability goes from a central kernel of true positive observations (the strong region) through a band of false negatives and false positives and finally to the region of true negative observations (Figure 4d). Inspection of the figure reveals that under the described noise characteristics *there can not exist any concept description that includes all positives and excludes all negatives*. Because of this fact, the funnel semantics previously described embody only a sufficient condition instead of the necessary and sufficient condition assumed by version space strategies. Identifying a sufficient condition is equivalent to determining the strong backprojection.

Some recent work [Hirsh, 1989] has produced learning algorithms that tolerate bounded error. But there are other cases, such as effectory errors associated with direction, where it is easy to construct examples of manipulation problems where the error is unbounded [Christiansen, 1992]. It is also easy to construct examples where, under the influence of the error, no strong region remains. (Imagine what will happen in Figure 4c

as the size of the sensory error is increased.) In these cases, the best concept description is probabilistic.

If we wish to identify the strong (100% probability) region associated with a spout, the learning algorithm must exhibit (1) a noise-tolerance of unbounded error, and (2) a bias of avoiding negative examples. The ID3 algorithm [Quinlan, 1986] is an example of a learning algorithm that is noise-tolerant, but not necessarily biased toward the strong region. ID3 uses an information-theoretic measure to decide on splits of the feature space. This measure is symmetric in the roles of the positive and negative examples—the only thing that matters is the relative number of positive and negative examples on either side of a split.

7.3 The Learning Algorithm

In many classification problems, there is no significant difference between a false positive and a false negative—an error is an error. But we can not treat positive and negative examples symmetrically in modeling actions. A computed funnel mouth that contains false positives will not be reliable. It will predict a result in cases where that result will not occur. Alternatively, a funnel mouth that makes false negative predictions does not err in correctness; it only errs in the scope of the prediction—the true funnel mouth is actually larger than what was computed.

We will assume that funnel mouths will be composed of disjunctions of hyperrectangles. Any such disjunction will be referred to as a *region*. If we are unconcerned with *description complexity* (i.e., the number of hyperrectangles in the region description), then there are very simple methods that suffice. One such method involves discretizing the feature space into rectangular cells without regard to where the example points lie. Once the discretization is done, positive and negative examples are associated with the cell in which they fall. For the region description, we take the disjunction of all cells that contain at least one positive example and no negative examples. Given a sufficiently fine discretization and a sufficiently large number of examples, the region computed will have the desired properties of avoiding all negative examples. This will yield a disjunction of hyperrectangles, but it will usually be a very complex description.

We desire a simple region description for two reasons. First, we should remember that the planner, in determining how to sequence actions, must compute whether the funnel spout (outcome) of an action falls within the funnel mouth of a subsequent action. Making region descriptions simple will thus help reduce the execution

```
FB(pluses,minuses) ≡
if unacceptable-support(pluses,nil)
    then return(empty-region)
bbp ← bounding-box(pluses)
relminuses ← {p ∈ minuses | p ∈ bbp}
if desirable-support(pluses,relminuses)
    then return(bbp)
bbm ← bounding-box(relminuses)
hpluses ← {p ∈ pluses | p ∈ bbm}
if ignorable(hpluses,pluses)
    then hole ← bbm
else hole ← choose-hole(relminuses,pluses,bbm,bbp)
ans ← empty-region
for each dimension i do
    ans ←
      ans
      ∪ FB({p ∈ pluses | pᵢ < minᵢ(hole)},
           {p ∈ relminuses | pᵢ < minᵢ(hole)})
      ∪ FB({p ∈ pluses | pᵢ > maxᵢ(hole)},
           {p ∈ relminuses | pᵢ > maxᵢ(hole)})
end do
return(ans).
```

Figure 5: *The FB algorithm.*

time of the planner. A second reason for having a simpler region description is that it will be easier for a human to comprehend, thus making it possible for us to judge whether or not the learning algorithm is computing a reasonable region description. Because of these considerations, the "discretize and threshold" approach described above was rejected, and a new algorithm was developed.

The FB algorithm is outlined in Figure 5. The algorithm takes as input a set of positive examples and a set of negative examples. It seeks to construct a description, using a disjunction of hyperrectangles, that excludes all negative examples and includes as many positive examples as it can. The algorithm proceeds generally as follows: First, if there are sufficient positive examples to warrant continuing,[1] the bounding box bbp of the positive examples is computed.[2] If there are sufficiently few negative examples within bbp,[3] then bbp

is returned as the concept description. Otherwise, some sub-box *hole* within bbp is identified.[4] The algorithm then subtracts out the *hole* from the box, and divides the positive and negative examples into sets inside and outside the *hole*. Points within the *hole* are ignored, and points not in the *hole* are divided up[5] and passed recursively to the FB algorithm. The recursion terminates whenever the number of positive examples falls below a minimum threshold (in which case the empty region is returned) or the number of negative examples is reduced to a satisfactory number (in which case the bounding box of the remaining positive examples is returned).

To identify the strong region, we want included boxes to contain only positive examples, and never any negative examples. To avoid including boxes containing only one example, or only a few examples, the bounding box is taken only if there is a sufficient number of positive examples (e.g., 10) in the set of positives.

The algorithm terminates quickly in cases where the negative examples can be eliminated quickly, such as when the bounding box of the relevant negative examples contains no positive examples. In such a case, the negative examples are eliminated all at once by subtracting out the box containing the negative examples. An 'L-shaped' concept region is such a case. If the true concept region is not easily described by hyperrectangles, or if there are areas of mixed positive and negative examples, then the algorithm requires more time to construct a description. For some pathological cases, the algorithm may require time exponential in the number of examples. When the bounding box bbm of relevant negative examples also includes a number of positive examples larger than should be ignored, the subroutine *choose-hole* is called to select (heuristically) a hole to be subtracted. This hole must contain at least one negative example (to insure that the recursion terminates) and it should include as many negatives and as few positives as possible.[6]

---

cept region—*desirable-support* implements this check.

[4] The bounding box of the relevant negative examples is computed, which becomes the first candidate for being the 'hole.' The number of positive examples in that box is counted, and the predicate *ignorable* is true when the number of such positives is low, say less than 5. If these positive examples are not ignorable, then the subroutine *choose-hole* is called to determine a more appropriate 'hole.'

[5] For each dimension of the state-action space, points are divided into those *below* the hole and those *above* the hole.

[6] For the experiments presented below, *choose-hole* was implemented as follows: The 'hole' bounding box was eroded (by removing the outermost negative example and

---

[1] The predicate *unacceptable-support* implements this check. *unacceptable-support(p,m)* is true when the number of points in p is greater than some threshold, say 10.

[2] The bounding box is the least hyperrectangle containing each of a set of points.

[3] zero negatives, in the case of identifying the strong con-

The FB algorithm is similar to some previously published approaches. EACH [Salzberg, 1991] also uses hyperrectangles as a representational primitive. However, the classification method is based on the *nearest* hyperrectangle to a point to be classified, not just whether a point falls within a hyperrectangle, as is the case for FB. In addition, Salzberg's method includes 'exceptions'—hyperrectangles explicitly representing holes in concepts. FB does not represent holes explicitly in the final region description, as those regions are subtracted out by FB during computation of the region. The Star Methodology [Michalski, 1983] also computes disjunctions of hyperrectangles. That algorithm proceeds bottom-up (from individual instances to regions) rather than top-down (from a starting region to a refined region). FB and ID3 are examples of algorithms that proceed top-down.

These previous methods could be modified for use on the present problem. With a suitable mechanism that makes the learning algorithm *conservative*, in the sense that false positive predictions are avoided, any such learning algorithm would be acceptable.

8   Experiments

The FB algorithm was applied to several artificially derived data sets. For concepts that can be described exactly as disjunctions of rectangular regions, FB finds an exact description of the concept. For cases where the true concept does not match the rectangle bias or where noise is present, FB computes an approximation to the true concept. The upper two examples of Figure 6 show the approximations computed for a triangular concept (left) and a circular concept (right). In both cases the algorithm avoids including negative examples in the computed representations. In the lower two examples the positive and negative examples are not cleanly separated. In the lower left, there are three vertical bands: on the left, all positives; in the middle, randomly mixed positives and negatives; and on the right, all negatives. FB approximates the all positives band. The example at the lower right of the figure is similar, except that the probability of a point being classified as positive follows a trapezoidal distribution with respect to the horizontal axis. Moving from left to right, we see first all negatives, second a narrow band where the probability of positives increases, third a band of all positives, fourth a band where the probability of positives decreases, and lastly another band of all negatives. The FB algorithm successfully finds the

recomputing the bounding box) until the number of remaining positive examples became 'ignorable.'

band of all positives (and a couple of regions where the random classification placed no negatives).

The Tray-tilting Task

The FB algorithm was also applied to some state transition data collected by a robot in the tray-tilting task. Figure 7 shows two computed funnels forming a plan to get the object into a horizontal orientation in the lower left corner of the tray. Each of the two funnels shows the position constraints for the funnel mouth (light gray) and the funnel spout (dark gray). Also shown are the orientation constraints—the long edge of the rectangle can have any orientation in the range shown by the 'bowtie' indicators. The prediction is: if the configuration of the rectangle is within the funnel mouth, then after executing the indicated action, the rectangle configuration will be within the funnel spout.[7]

Shown below the plan is a trace of one execution of the two step plan. The observed state transitions satisfy the predictions of the two funnels, and the goal is achieved. The robot was programmed to perform 3000 practice tilts, which required approximately 5 hours of real time. Based upon the recorded state transitions, the robot used knowledge of twelve anticipated goals[8] to build a set of funnels summarizing the state transitions. The robot was given a set of 100 randomly generated goals (chosen, with replacement, from the set of twelve anticipated goals), and asked to achieve each of the goals in sequence. In this test, the robot successfully achieved the requested goals in 96 of the 100 cases. Figure 8 shows a learning curve for the robot over its training history. In comparison, four human subjects observed example tray tilts and were then asked to generate tray-tilting plans for the same 100 goals, which the robot executed. The mean performance of the robot with the human-generated plans was 91%. Thus the robot-generated predictions were quite good.

9   Assumptions Revisited

The assumptions adopted in this paper are appropriate for manipulation tasks. States and actions are continuous, and noise is always present. Funnels are appropriate descriptions of manipulation actions because there

[7]The funnel mouths in Figure 7 show only the most significant single hyperrectangle (the hyperrectangle containing the most positive examples).

[8]The twelve goals were: each of the four corners of the tray, in both horizontal and vertical orientations, and four more goals in the middle of each wall, with orientations aligned with the wall. These goals were used as the basis of choosing funnel spouts (cf. Section 7.1).

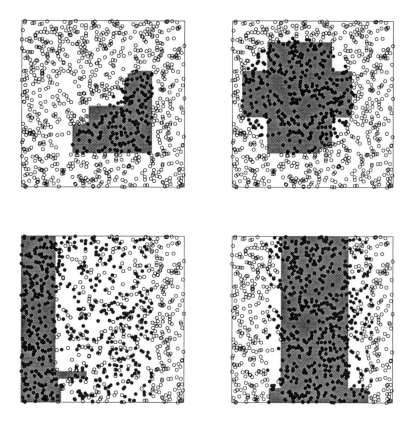

Figure 6: *The behavior of the FB algorithm on some artificial data sets.*

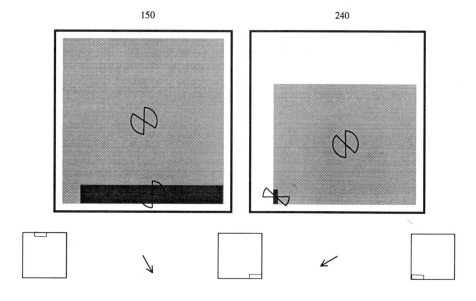

Figure 7: *A two step tray-tilt plan and an execution trace of the plan.*

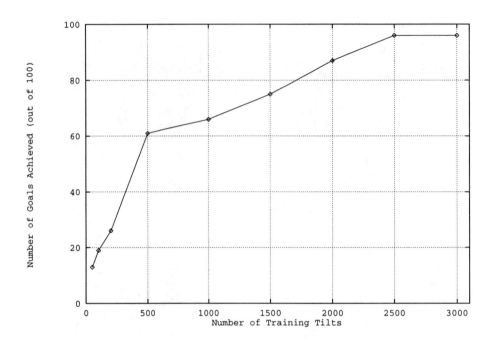

Figure 8: *A plot of robot proficiency (percentage of goals achieved) in the tray-tilting task, as a function of its experience with the task (number of tilts observed).*

often exist large, connected areas of the state-action space that lead to similar outcomes. This is particularly true of the tray-tilting task. The general approach outlined in this paper carries over to more complex continuous tasks, but more complex tasks will require a larger number of funnels, with each funnel covering a smaller volume of the state-action space. For example, imagine a tray-tilting task where, in addition to the bounding walls of the tray, there are walls internal to the tray. This additional physical structure increases the complexity of learning an adequate 'theory' (set of funnels) for the task, but the problem of learning individual funnels is unchanged.

The assumption of *continuity* of the underlying feature space is not strictly required. The computation of intervals of feature values and regions of the state-action space is predicated on the features being *ordered*. Without this ordering, the geometrical representation does not make sense, and perhaps a richer logic-based representation could be used. (Of course, hyperrectangles *are* logical expressions using conjunction ($\wedge$) and less than or equal ($\leq$).) The general approach of this paper does not require hyperrectangles, although the FB algorithm does. One can imagine funnels constructed from other geometric primitives, such as hyperspheres or arbitrary hyperpolygons.

The assumption of finding the strong (100% probability) funnel mouth can be relaxed. The FB algorithm was in fact designed with this in mind. In order to find a region corresponding to a lower probability, one only needs to change the predicates *unacceptable-support*, *desirable-support*, and *ignorable* so that they allow hyperrectangles with some negative examples to be included in the final result. In the limit where a high probability value is not required, one can always just take the bounding box of the positive examples as the single hyperrectangle region. The average probability over such a region must be greater than zero, because it contains some positive examples. We could also trade region complexity (number of disjuncts) against the probability associated with the region.

The assumption that the task changes state only as a result of the agent's actions is more difficult to relax. If a learning agent is to reason about environmental changes that it does not control, then it must have a vocabulary for the description of such changes. For example, if it is to reason about temporal regularities in the environment, it must be able to represent time in some way. For changes caused by other agents, it must at least be aware that other agents exist. This is not possible in the framework of this paper. *All* descriptions of the environment are based upon what the learning agent can sense and what it can do with its effectors.

Extension of the work of this paper to include reasoning about influences outside the learning agent's control would appear to require some a priori knowledge of the task. Similarly, cross-task learning (applying knowledge gained in one task to a new task) would appear to require prior knowledge of the two tasks. For example, suppose we have a robot that learns to push objects on a table. How could this knowledge be applied to a new task (e.g., grasping)?

This paper has said very little about which point along the empirical-analytical spectrum is most appropriate for developing predictive abilities. The paper has presented evidence that a point near the purely empirical end is feasible for some manipulation tasks. We know that the purely analytical end is also feasible, because programming of robots by humans has succeeded in some tasks. But we do not yet know which mix of the empirical and analytical is appropriate, and how the appropriate mix depends on the nature of the task.

10   Summary

This paper has demonstrated an approach to learning models of actions in uncertain continuous task domains. The approach identifies funnels, which are operators whose pre- and post-conditions are defined by geometric regions of the state-action space. Funnels are computed by choosing a desired result region (spout), backprojecting the region with respect to an action, and computing a region (the mouth) defining the preconditions under which the action will produce the desired result. The FB learning algorithm is noise-tolerant, and with the parameter settings described here, it is also biased toward finding strong (100% probability) backprojections. Using the derived funnels, planning is performed by a standard backward-chaining search strategy. The computed plans are then executed open-loop. The performance of a learning robot using this approach for the tray-tilting task was 96% after observing 3000 training actions.

In addition to demonstrating a methodology for learning action effects, this paper has discussed the problem of identifying concepts in the face of noise and non-determinism. By identifying strong backprojection regions, this approach seeks to find operators that, in a sense, embody deterministic semantics even though the underlying task may be non-deterministic. This is possible when funnel spouts are large enough that the result of an action is guaranteed to be within the predicted set. An alternative approach to creating ac-

tion models is to fix a funnel's mouth and spout a priori, and then identify the correct mouth-spout transition probability [Christiansen *et al.*, 1991]. It is also possible to identify the regions and the probabilities simultaneously[Christiansen, 1992].

Acknowledgements

This research was conducted at Carnegie Mellon University, where it was supported by NASA-Ames grant NCC 2-463 and by an AT&T Bell Laboratories Ph.D. Scholarship.

References

[Bennett, 1990] Bennett, Scott W. 1990. Reducing real-world failures of approximate explanation-based rules. In *International Conference on Machine Learning*. 226–234.

[Christiansen *et al.*, 1991] Christiansen, Alan D.; Mason, Matthew T.; and Mitchell, Tom M. 1991. Learning reliable manipulation strategies without initial physical models. *Robotics and Autonomous Systems* 8:7–18.

[Christiansen, 1991] Christiansen, Alan D. 1991. Manipulation planning from empirical backprojections. In *IEEE International Conference on Robotics and Automation*. 762–768.

[Christiansen, 1992] Christiansen, Alan D. 1992. *Automatic Acquisition of Task Theories for Robotic Manipulation*. Ph.D. Dissertation, Carnegie Mellon University, School of Computer Science.

[Erdmann, 1986] Erdmann, Michael A. 1986. Using backprojections for fine-motion planning with uncertainty. *International Journal of Robotics Research* 5(1).

[Fikes and Nilsson, 1971] Fikes, Richard E. and Nilsson, Nils J. 1971. Strips: A new approach to the application of theorem proving to problem solving. *Artificial Intelligence* 2:189–208.

[Hirsh, 1989] Hirsh, Haym 1989. *Incremental Version-Space Merging: A General Framework for Concept Learning*. Ph.D. Dissertation, Stanford University, Department of Computer Science.

[Michalski, 1983] Michalski, Ryszard S. 1983. A theory and methodology of inductive learning. In Michalski, R. S.; Carbonell, J. G.; and Mitchell, T. M., editors 1983, *Machine Learning: An Artificial Intelligence Approach*. Morgan Kaufmann. 83–134.

[Mitchell, 1978] Mitchell, Tom M. 1978. *Version Spaces: An Approach to Concept Learning*. Ph.D.

Dissertation, Stanford University, Dept. of Computer Science.

[Porter and Kibler, 1986] Porter, Bruce W. and Kibler, Dennis F. 1986. Experimental goal regression: A method for learning problem-solving heuristics. *Machine Learning* 1:249–286.

[Quinlan, 1986] Quinlan, J. R. 1986. Induction of decision trees. *Machine Learning* 1:81–106.

[Rivest and Schapire, 1987] Rivest, Ronald L. and Schapire, Robert E. 1987. A new approach to unsupervised learning in deterministic environments. In *International Workshop on Machine Learning*. 364–375.

[Salzberg, 1991] Salzberg, Steven 1991. A nearest hyperrectangle learning method. *Machine Learning* 6:251–276.

[Shen and Simon, 1989] Shen, Wei-Min and Simon, Herbert A. 1989. Rule creation and rule learning through environmental exploration. In *International Joint Conference on Artificial Intelligence*. 675–680.

[Vere, 1978] Vere, Steven A. 1978. Inductive learning of relational productions. In Waterman, and Hayes-Roth, , editors 1978, *Pattern-Directed Inference Systems*. Academic Press. 281–295.

Social Potential Fields: A Distributed Behavioral Control for Autonomous Robots

John H. Reif, *Duke University, Durham, NC, USA*

Hongyan Wang, *Duke University, Durham, NC, USA*

This paper is concerned with Very Large Scale Robotic (VLSR) systems consisting of from hundreds to perhaps tens of thousands or more autonomous robots. In the near future as the costs of robots are going down and the robots are getting more compact, more capable and more flexible, we expect to see industrial applications of VLSR systems, for example, many thousands of mobile robots perform tasks such as assembling, transporting, cleaning within the working space of factories. Traditional "global control" mechanism is not suitable for controlling VLSR systems because of its complexity, unreliability, and inflexibility.

Instead we propose a distributed control paradigm. We define simple artificial force laws between pairs of groups of robots and other components of the system. Each robot's motion is controlled by the resultant artificial force from the other robots and the other components of the system. Since the force calculations can be done in a distributed manner, the control is distributed. We show by simulations that such a simple control paradigm can yield interesting and useful cooperative behaviors among robots which achieve collision control, traffic control, and behavioral control for a VLSR system. We also develop methods for quantitatively defining behaviors of VLSR systems.

1 Introduction

1.1 Motivation: Very Large Scale Robotic (VLSR) Systems

Much of the earlier research in robotic motion planning and control has considered the case of only a single robot [10], or a small number of cooperative robots [5]. In this paper we are concerned with the motion planning and control for systems consisting of large numbers of autonomous robots. In the near future as the costs for individual robots are going down and the

robots are getting much more compact (Utilizing the emerging miniaturized manufacturing capabilities, the robots are likely to be made even at microscopic sizes for medical purposes), more capable and more flexible, with sensing systems that will handle outside information faster and more accurately, we expect to see industrial applications of systems of large numbers of robots, for example, robots can replace human beings to be the work force in factories. Note that in the 1980's electronic circuits went though a similar revolution in decreased physical size scale, cost and integration as the industry transitioned to VLSI.

We suggest the use of the term *Very Large Scale Robotics* (VLSR) for situations where we have robotic systems including hundreds to perhaps tens of thousands of autonomous robots.

As in the more traditional robotic systems, we need to worry about *collision control* (collision avoidance with obstacles and other moving objects), and *traffic control* (path or trajectory planning). In a VLSR system, collision control and traffic control are part of what we call *behavioral control*. In addition to collision avoidance and path planning, to make effective use of VLSR, we want the robots to exhibit some forms of useful cooperative behaviors. As example applications, we can use multiple robots to clean, harvest or patrol a large area instead of human beings. The robots may need at times to be clustering very tightly together, and at other times more loosely, scattering evenly in their working area. The behavior "scattering evenly" is important to make the work well partitioned among robots so the work can be done efficiently. Other industrial applications of VLSR systems include assembling, quality control sampling, and transportation. Military applications may include guarding, escorting, patrolling, strategic and aggressive behavior such as stalking and attacking, and so on.

In all the situations described above, the control of

the systems has the following properties:

- The control has to be efficient in order for the system to be practical.
- The control has to allow flexibility so that the robots can adapt to dynamically changing or unknown environment.
- The control does not have to be precise, i.e. we don't require that each robot take a precise path or position.

1.2 Drawbacks of Global Control Methods

A central control model assumes a control center which collects information of the entire VLSR system, then plans robot's motion and sends the control out to each individual robot. The drawbacks of a central control model are:

- Very high computational and communicational complexity (growing with the number of robots of the VLSR)
- Difficult to implement
- Lack of flexibility and robustness.

Another traditional way of robotic control is based on the computational paradigm that the global pattern (*global rules*), or the final goals are programmed. For example, each robot's path in an industrial assembly is scripted before hand explicitly for the whole assembly to display a global pattern. This method has the drawback of lacking flexibility too. It lacks adaptiveness to changing environment, thus the environment has to be precisely described for this method to work.

1.3 Distributed Control Paradigm

In a distributed control paradigm, each individual component of the system decides its behavior by applying a set of rules to the current state of the system. The rules are called *local rules* contrary to global control. In the study of Artificial Life, which is a field devoted to understanding and simulating the complexity of life-like behaviors, it is more and more realized that: Complicated behaviors do not necessarily have complicated descriptions. They are usually *emergent* behaviors as consequences of applying local rules by individuals [4].

This idea is certainly verified by studies in Sociobiology. Sociobiologist Wilson [11] gave many examples of intriguing complex behaviors shown by social animals (including humans) in preying, protecting, migrating, etc.. Yet there is no global control in an animal system. It is evident that complex behaviors of the animal groups are formed by applying local rules to information collected by perceptions by individual animals.

The distributed control paradigm has been applied to some other areas including Computer Graphics. For example C.W. Reynolds [8] has successfully synthesized some of the flocking behavior of birds, such as collision avoidance, velocity matching and flock centering. In the synthesis, each bird controls its speed and direction upon the *external states* of the nearby birds. The external states include speed, direction and position. The local rules are expressed as functions of variables describing the external states.

This distributed control paradigm can be found also in physical phenomena. It is well known that molecules, plasma gases, fluids, while obeying simple classical force laws, form interesting and complex structures. Moreover, they can exhibit a wide variety of dynamics. The local rules are classical force laws in this case. We have special interest in these physical force laws because they are very simple and they depend solely on distances among physical objects.

Our work differs from the work cited above in that:

- Research in Artificial Life and Computer Graphics is interested in specific rules to generate specific behaviors. The rules can be rather complicated and ad hoc. On the contrary, we are interested in finding a set of simple and general rules to generate interesting and practical robotic behaviors. We also want to quantitatively define behaviors of systems.

1.4 Our Proposed Solution: Social Potential Fields

We propose a simple distributed control mechanism, using artificial force laws (which are the local rules in our case), to achieve dynamic behavioral control for VLSR systems. A component of the VLSR system can be a robot, or an obstacle, or an objective. The global controller defines pair-wise force laws for pairs of components. Each robot then calculates the resultant force from other components in the system using the force laws and its motion is controlled in part by the resultant force. Once the force laws are defined, force calculation can be carried out by individual robots in a distributed manner, thus the control is completely distributed.

Our force laws are similar to those found in molecular dynamics. They are inverse power laws of distances incorporating both attraction and repulsion. Usually the attractive forces dominate for far away intervals and the repulsive forces dominate for close intervals. There are several reasons for choosing inverse power laws as the basic form of the force laws:

- The force laws can be made to guarantee collision avoidance between two components.

- The force gets smaller rapidly when two components get farther apart.

- The force laws are continuous in the domain of interest, thus enable us to perform some calculations.

- Studies in molecular dynamics, plasma gases, fluids showed that complicated dynamic behaviors can be generated by the inverse power laws.

Our force laws are more general than the usual molecular force laws; in fact the parameters of the force laws, i.e. the constants and the exponents, are chosen arbitrarily by the global controller. Indeed the parameters are chosen to reflect the relations of robots, e.g. they should stay close together or far part. Using the force laws, the resulting system displays "social" behaviors such as clustering, guarding, escorting, patrolling and so on. Thus we call our method *Social Potential Fields*.

Although the basic control is quite simple, the system can display quite complex and interesting behaviors (as will be shown in section 3). These systems are expected to have practical value in industrial and military applications in the near future.

1.5 Related Work in Robotic Motion Planning Using Artificial Potential Fields

The idea of using artificially defined potential fields in robotic motion planning is innovated by Khatib [3]. The idea is simple and elegant. Imagine in the configuration space of the robot (where the robot is a point object), we assign negative electrical charges to the robot and the configuration obstacles and positive charges to the goal configuration. Thus a potential field

is formed with very high potential close to the configuration obstacles and minimum potential at the goal configuration. The robot guided by the potential field will go from high potential configurations to low potential configurations and hopefully will eventually get to the the goal configuration. The potential functions can be functions other than electrical potentials.

One major weakness of the potential field methods is the Local Minima problem where the robot is caught in local rather than global minima (The local minima problem is also present in our VLSR systems). There has been a lot of research in trying to overcome or mitigate this problem (see Latombe's [6] for a review). An outstanding effort, for example, is made by Rimon and Koditschek ([9]). They proposed a potential function guaranteeing convergence to global minimum from almost all free configurations in restricted types of configuration space.

In the study of traditional potential field methods,

- The major concern is path planning for a single robot with collision free.

- In most situations, the force laws used in motion planning are either attractive force laws assigned to goals or repulsion force laws assigned to obstacles but not both in combination.

We are interested in the behavioral control of systems of large numbers of robots. Besides collision free, we want the system to display interesting and useful behaviors.

1.6 Organization of paper

The paper is organized as follows: Section 1 discusses the motivation for our work, introduces the basic idea of social potential fields, and surveys related works. Section 2 defines the general form of the force laws we use and describes our model of the dynamic system. Section 3 describes some examples of complicated and interested behaviors and some results of our computer simulations. Section 4 discusses how to define force laws given the desired behavioral pattern of a system. Section 5 discusses the use of another potential law, namely spring law. We draw conclusions in section 6.

2 Our Social Potential Fields Model

2.1 Inverse Power Force Laws

In our VLSR system, the robots are considered as point particles named $1, 2, \ldots, n$ with positions at $X_1, X_2, \ldots, X_n$ in fixed Euclidean space at fixed time. Let $r_{ij} = \|X_i - X_j\|_2$ be the Euclidean distance between robots i and j. The force defined from robot j to robot i has the form

$$F_{i,j}(r_{ij}) = \left(\sum_{k=1}^{L} \frac{c_{i,j}^{(k)}}{(r_{ij})^{\sigma_{i,j}^{(k)}}} \right) \left(\frac{X_j - X_i}{r_{ij}} \right).$$

Thus the force defined from j to i is the form of a sum of L inverse power laws where the kth inverse power law is of the form $\left(\frac{c_{i,j}^{(k)}}{(r_{ij})^{\sigma_{i,j}^{(k)}}} \right) \left(\frac{X_j - X_i}{r_{ij}} \right)$. Our force laws may differ from the molecular systems in that we allow the global controller to arbitrarily define distinct laws for separate pairs of robots. The coefficient $c_{i,j}^{(k)}$ and the inverse power $\sigma_{i,j}^{(k)}$ are constants depending on ordered pair $< i, j >$ and k. We generally require $\sigma_{i,j}^{(k)}$ to be positive but the constant coefficient $c_{i,j}^{(k)}$ can be either positive or negative, depending on whether an attractive or repulsive force, respectively, is required.

Note that if $\sigma_{i,j}^{(k)}$ is greater than $\sigma_{i,j}^{(k')}$ then the kth term of this sum dominates for small r_{ij} whereas the k'th term of this sum dominates for large r_{ij}. Typically (with the exception of pursuit applications), for fixed i, j, the kth term with the smallest $\sigma_{i,j}^{(k)}$ will have a negative (repulsive) coefficient $c_{i,j}^{(k)}$ to help insure that the robots do not collide. The force laws usually have the following property. As the distance between two robots becomes larger, the force between them gets smaller. But if two robots get too close to each other, the force might be very strong to avoid collision between these two robots.

Also, note that force laws need not be symmetric; force $F_{i,j}$ can be different from force $F_{j,i}$.

At a fixed time, the overall artificial force applied by the entire VLSR system upon a robot i is

$$F_i = \sum_{j \neq i} F_{i,j}(r_{ij}).$$

There are several ways that robot i's motion can be controlled by the force F_i. For examples, robot i can gain an acceleration, or it can move in the direction of F_i for a length proportional to the quantity of F_i, or it can move in the direction of F_i for a fixed length. The robot's motion is to reduce the quantity of F_i. With all the robots operate simultaneously, the system converges to an equilibrium state where all the F_i's are 0. For simplicity, in our system, robot i is controlled to move in the direction of F_i for a fixed length.

2.2 Succinct Definitions of Force Laws Between Groups of Robots

Robots are assigned to various (not necessarily disjoint) groups. Robots in the same group have certain common behaviors. For example, robots in a group called Cleaner are required to do cleaning and are required to have a even distribution pattern. Groups are not necessarily disjoint. A robot may belong to several groups and inherit all the behaviors from those groups.

We can define force laws between pairs of groups. Let $S_1, S_2, \ldots, S_m$ be fixed groups of robots $1, 2, \ldots, n$. Any robot in group S' imposes the same force on any robot in group S. Let $F_{S,S'}$ denote this force law. Suppose the robot in S' is at position X', the robot in S at position X, and the distance between them is r. Then the force law is

$$F_{S,S'}(r) = \left(\sum_{k=1}^{L} \frac{c_{S,S'}^{(k)}}{r^{\sigma_{S,S'}^{(k)}}} \right) \left(\frac{X - X'}{r} \right).$$

Thus the force law defined is again the form of a sum of L inverse power laws each of form $\left(\frac{c_{S,S'}^{(k)}}{r^{\sigma_{S,S'}^{(k)}}} \right) \left(\frac{X - X'}{r} \right)$, where the coefficient $c_{S,S'}^{(k)}$ and the inverse power $\sigma_{S,S'}^{(k)}$ are constants depending on ordered pair $< S, S' >$ and k (Again, we generally require $\sigma_{S,S'}^{(k)}$ to be positive). For $S = S'$, the force law defines the force for a pair of robots in the same group. $F_{S,S'}$ is 0 for the pairs of groups which we don't want to define force in between.

The force from robot j to robot i with distance r_{ij} apart, is defined to be

$$F_{i,j}(r_{ij}) = \sum_{\forall S, S' \ s.t. \ i \in S \wedge j \in S'} F_{S,S'}(r_{ij})$$

which is simply the sum of the forces $F_{S,S'}(r_{ij})$ induced by pairs of groups $< S, S' >$ where $i \in S$ and $j \in S'$.

The overall artificial force applied by the entire VLSR system upon a robot i at position X_i is again

$$F_i = \sum_{\forall j \neq i} F_{i,j}(r_{ij}).$$

We can see that once the robots are assigned to groups, the complexity of defining force laws is reduced. The number of force laws is reduced from $n(n-1)$ for all ordered pairs of individual robots to $m(m-1)$ for all ordered pairs of groups, where n is the number of robots and m the number of groups in the system. Usually, The number of robots is much larger than the number of groups in a VLSR system. Each robot can maintain a more succinct table of force laws defined for pairs of groups. It is easier to maintain or to modify the table.

2.3 Local Autonomous Control Forces

The resultant force calculated by each individual robot using force laws, i.e. F_i as defined in subsection 2.1, can be considered as global control force. The global control force mainly coordinates the robots and decides the distribution of the robots in the system. In order for a robot to work properly, we also need what we call *local control force*. The local control force may include force to avoid collision with unknown obstacles, force to approach an object, etc. This kind of local control can be accomplished by programs like "Generate a force to turn left if there is an obstacle in front at distance d". Consider a group of robots doing gold prospecting (or more realistically, garbage collecting). The global force laws are so defined as to allow robots to have a even distribution over the working area, neither crowding together nor getting too far away. At the same time, for a robot to work properly, it should move smoothly in an locally unknown and dynamic (unknown to the global controller) environment, avoiding unexpectedly discovered obstacles, not falling into holes and most importantly detect and gather the discovered gold (or garbage).

Thus a robot is controlled by the combination of a global and a local control force. There are various schemes in combining these forces, for example, one scheme may be assigning different weights to different kinds of forces and combining them by a weighted sum.

In the examples and simulations discussed in the next section, we focus on how the global control force controls the behaviors of the systems. The local control force is neglected.

2.4 Sensing

In order for a robot to calculate the potential forces, we need some methods for determining the distance between pairs of robots. Many techniques have been developed for range finding [2]. Among them are triangulation range finder, ultrasonic or laser time-of-flight range finder. While all these techniques have some drawbacks (e.g., computational complexity, limited surface orientation or limited spatial resolution.), our social potential technique is relatively robust and generally may not require exact measurements. Thus it seems likely that any of these sensing methods would suffice.

Also, for cooperative robots, information about the others can be accessed by communication via electromagnetic radiation, such as radio frequency, infra red, or ultra sonic.

2.5 Asynchronous Calculation of Forces by Distributed Robots

Even though the force laws are defined globally by a global controller, the actual control is carried out in a distributed manner. Each robot is equipped with sensors and also has a table of force laws, thus force calculation can be done simultaneously by individual robots. Each robot carries out a cycle of operations including sensing, calculating the potential force, and realizing the control to motion. The robots usually act asynchronously.

3 Examples and Simulations of Behaviors of VLSR Systems

In the following examples, the quantities of the force laws usually have the same form

$$F(r) = -\frac{c_1}{r^{\sigma_1}} + \frac{c_2}{r^{\sigma_2}} \quad c_1, c_2 \geq 0, \ \sigma_1 > \sigma_2 > 0 \quad (1)$$

with different $c_1, c_2, \sigma_1,$ and σ_2. Since $\sigma_1 > \sigma_2 > 0$, the repulsion, indicated by the term $-\frac{c_1}{r^{\sigma_1}}$, will dominate when two robots get close to each other, and the attraction, indicated by the term $\frac{c_2}{r^{\sigma_2}}$ will dominate when two robots separate. Collision avoidance is guaranteed since the force goes to infinity as two robots get closer and closer. The curve of the force law with a set of

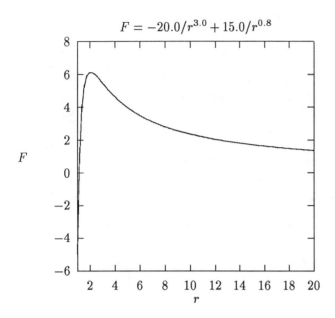

$$F = -20.0/r^{3.0} + 15.0/r^{0.8}$$

Figure 1: *The quantity of the force as a function of distance for a specified force law.*

specified parameters is shown in Figure 1. The *equilibrium distance* is defined as $d = (c_1/c_2)^{\frac{1}{\sigma_1 - \sigma_2}}$. If the force law defined above defines bilateral effect between two robots, then they reach equilibrium state if they are distance d apart. If the force law defines mutual effect among a group of robot, the group can either densely congregate if d is small or sparsely distribute if d is large. By defining a set of force laws for different pairs of groups, we can get more interesting behavior than clustering. If σ_1 is big, the repulsion will be strong at short range but decays rapidly with distance. If σ_2 is small, the attraction will have long distance effect. Our method is called "social potential fields" method, because the force laws reflect and enforce social relationships.

We will discuss some detailed concerns in the simulation. For simplicity in our computer simulations, all components, (e.g., robots and obstacles) are considered as point particles. An object with more complex shape can be approximated by putting many point particles along the boundary (e.g. obstacles as "walls" in the examples).

In Figure 2 and Figure 3, the pictures are dumped window images. The windows give visual presentation

of how the systems evolve. In the windows, the robots are represented by arrows pointing to their directions of motions.

In each iteration, a robot moves a fixed length to the direction pointed by the resultant force.

For simplicity, we use the very straightforward method to compute the forces, which takes $\Theta(n^2)$ time if we have n robots in the system. We are experimenting with about hundreds of robots and the efficiency is bearable. But keep in mind, there have been efficient algorithms for computing forces (developed for molecular simulations) with high precisions. The first one of them was given by Greengard [1]. Those efficient algorithms enable us to experiment with tens of thousands of robots in large systems.

In the following description of examples or simulations, we name separate groups and their members. The name of the group and the name of its members are the same except the name of the group starts with a capital letter.

3.1 A Single Cluster of Many Robots

A more or less evenly distributed cluster of robots is a useful pattern for jobs like cleaning, patrolling, or exploring an unknown area. Interpersonal distance is useful not only for robot collision avoidance, but also for getting work partitioned and distributed. For example, in the future we may send hundreds or thousands of robots to explore Mars before we send any human beings on it. To explore the planet efficiently, we want the robots to cover the surface of the planet evenly instead of crowding together. A central control mechanism is not applicable especially because of the communication complexity involved. A programmed script for each robot is impossible because of the lack of knowledge of Mars. And with a preprogrammed script, the robot will have no flexibility. Our simulation with a *single robot cluster* showed that in this case the potential fields method can help us to achieve the goal, i.e. let the robots form an evenly distributed cluster and allow flexibility for individual robots.

We considered a single group of robots with an identical pair-wise force law incorporating both attraction and repulsion. We need to fix the parameters in equation (1): $c_1, c_2, \sigma_1, \sigma_2$ and the initial conditions: the size of the cluster, and the initial distribution of the robots. We tried with many different sets of parameters and initial conditions and found that the equilibrium state depends very much on them. In some cases we achieved satisfactory results, i.e. after a moderate number of iterations, the group of robots form a stable even distribution. In the following paragraph, we will describe one of the simulations.

In the simulation, the size of the cluster is chosen to be 200, and the parameters are set to the following: $c_1 = 60.0$, $c_2 = 1.0$, $\sigma_1 = 2.0$ and $\sigma_2 = 1.0$. Thus the equilibrium distance $d = 60.0$. (The measure of all distances is pixel.) We also put "walls" surrounding the working area which a square area of size 400×400. The walls are the four edges of the square area. The walls impose a repulsive force to the robots. This repulsive force is very small when the robots are far away from the walls so the force won't affect too much the distribution of the robots. But the force will grow very strong once the robots are close to the walls so the robots won't get out of the working space.

Initially, the 200 robots are randomly distributed within a square area of size 400×400. Since at this time, for a robot at the periphery, there are many robots distant away (farther than 60.0), the robot feels an attraction toward the center of the cluster and the cluster starts to shrink. After a number of iterations, the cluster forms a disc with radius of about 280.0. At this time, each robot shows a periodic movement. Then we say that the system stabilizes. The initial and the equilibrium distributions are show in Figure 2

We have also tried with different initial distributions. For example, let the robots be initially in a small square area at the upper-left corner of the working area. The cluster expanded to a stable distribution.

3.1.1 A Cluster with an Arbitrary Shape

In the above simulations, the cluster usually forms discs. Actually we can form a cluster of an arbitrary shape. This is done by putting "walls" along the boundary of the shape, and having walls impose a repulsion on the robots, so the robots are confined with in the area. Then choose the parameters of the force laws to keep the equilibrium distance large. The robots are pushed to the walls and the area is filled.

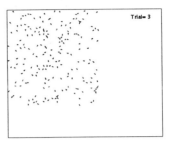

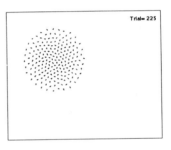

Figure 2: *To form a evenly distributed single cluster of robots using an identical force law. In the picture above is the initial distribution of robots. In the one below the equilibrium states forms.*

3.2 Moving Robot Clusters with Dynamic Collision Avoidance

A simple next step in increasing complexity of robotic behavior is the problem of avoiding obstacles while moving toward destination. The industrial application of this behavior might be, for example, using robots to do transportation through a region with filled with static as well as dynamic obstacles (including other robot vehicles).

Our computer simulation called *bypassing the walls* concerned the scenario where a group of robots bypass walls to approach a goal. To avoid collision with the walls, we let the walls impose repulsive forces on the robots; to approach the goal, we let the goal impose an attractive force on the robots. This is the same as in traditional motion planning using potential field methods. In addition, we have to coordinate robots since we have a group of them instead of just one. We want the robots to keep distance from each other yet to form a group. This can be done by imposing a force law combining both repulsive and attractive forces among the robots (as is done in example *a single cluster*). The details of the simulation and results are omitted here.

3.3 Guarding and Escorting Behaviors

Our computer simulation called *guarding a castle* showed how a group of robots can guard a fixed point object called *castle* by forming a ring structure around it and how they react to an invading robot. There are three groups in the simulation. The Guard consists of a number of identical point robots. The Castle has only one static point object representing the castle. The Invader has one (can be several) dynamic point robot (which may be either explicitly controlled by an adversary controller or simply robots which are attracted to the castle and which avoid the guards). The force defined from the Castle to the Guard combines both attraction and repulsion so the guards tend to stay around the castle (not too close and not too far away). Also because of the attraction, the guards will not wander too far away while chasing the invaders. The force defined from the Castle to the Invader is only an attraction. The Invader imposes both attraction and repulsion upon the Guard so the guards tend to move where the invaders are but to avoid contact. The Guard imposes only repulsion upon the Invader. There is also a force law defined among guards to prevent them from colliding with each other.

Our simulation effectively demonstrated the required guarding and escorting behavior. Initially the guards are scattered in a square area around the castle. The guards converge gradually to the castle and form a ring around the castle. Then an invader appears and tries to get into the castle. The guards will approach to where the invader is. They move from the castle to the invader but stop chasing when the invader is a certain distance away from the castle. The invader reacts by moving toward the weaker point of the guarding where there are fewer guards. Finally a dynamically balanced situation is formed, in which the guards are chasing the invader and the invader is moving around the castle to find the weaker point in guarding. See Figure 3 for the results.

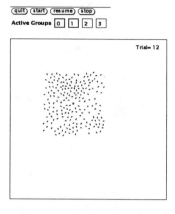

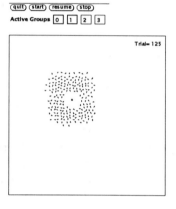

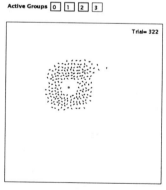

Figure 3: *Dynamic equilibrium between the guards and the invader. The top picture shows the initial distribution of guards around the castle. The middle picture shows an equilibrium state where the guards form a close ring around the castle. The bottom picture shows a dynamic equilibrium where the invader is trying to find weak points in the ring while the guards are chasing the invader.*

From the above simulation, we can see the advantages of "social potential fields" method. We avoid to define complicated rules as to where a particular robot guard should stay, which are the guards that should chase the invaders when they are at some positions, how many guards should chase, how far they should chase. By simply defining force laws, the above questions are answered. Furthermore the whole system showed more flexibility than one that is programmed.

More complex and interesting behavior can be simulated by modifying this pattern. The static castle can be replaced by some dynamic moving object, such as treasures to be transported. Thus the escorting behavior can be accomplished. The shape of the castle can be arbitrary instead of a point by putting point particles along the edge of the shape. There can be more than one invaders, then the guards should partition to groups to handle each invader. All these scenarios can be simulated by defining proper force laws.

In practice, we might want robots to guard places like important military bases, high priority labs, factories and so on. We might also want to use robots to escort transportation. We can also transport a large number of robots from place to place by simply defining a leading robot and let the other robots follow it as if they are escorting the leader.

3.4 Adapting Force Laws to Achieve Hierarchical Spatial Organization of Robots

Here we give an example of adapting force laws by the global controller to achieve hierarchical spatial organization of robots. The example is called *reorganizing after bivouacking* and imitates the following military behavior: Normally during maneuvers, an army will be spatially organized in a highly hierarchical manner (The army is divided into units. Different units occupy different areas; a unit can be further divided into subunits). While in camp, to efficiently utilize resources such as army canteen, the army may become disorganized (units cluster together in the same area). However, going back to maneuvers, the entire army must automatically spatially organize itself as before.

Figure 4 depicts a simplified version of the example where there are only two units (distinguished by the empty circles and the shaded circles). The organized army (shown in the leftmost picture) bivouacs (shown in the middle picture) and then reorganizes (shown in the rightmost picture).

In order to accomplish this *reorganizing after bivouacking* behavior, the global controller has to change the force laws at different stages. At the stage when the army is organized, there is no force law between the two units. The distribution of each unit is formed by (possibly differently) force laws among robots within each unit. To let the army bivouac, the global controller has to redefine the force laws. There should be an attractive force to the goal (camping area) imposed on both units. There should be a force law combing attraction and repulsion between individuals in both the units to achieve a clustering behavior. To reorganize, each unit has to be attracted to its own area while during the reorganization, there still should be a force law between the two units such that individuals do not collide with each other. Once the units get to their own areas, this force can be deactivated.

The spatial organization of robots occurs in industrial applications too. For example, at work in a factory, different groups of robots are dispatched to different areas. Within each group, the robots are further divided into subgroups for different tasks such as cleaning, transporting or assembling. At work, we want the groups of robots to be spatially distributed and there should be no interactions among them. At other times (say cleaning the factory), we want all the groups to be crowded together.

4 Defining the Behaviors of VLSR Systems

4.1 The Problem

In the previous section (section 3), we described some computer simulations of interesting behaviors of robotic systems. In those simulations, the force laws were derived from intuition and failure-and-trials. We noticed that the equilibrium behavior depends very much on the parameters, and the forms of the force laws. There are two interesting problems: t predict the equilibrium behavior of the system from the predefined force laws, or vise versa, to define force laws such that a desired equilibrium behavior can be achieved.

We propose a more systematic way of solving the above two problems than mere computer simulations.

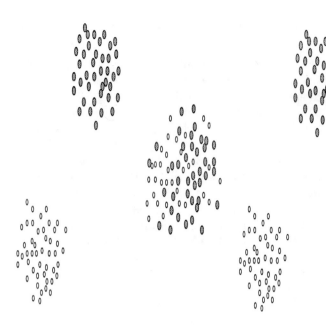

Figure 4: Reorganizing after bivouacking.

To do this, we use *density functions* to describe the distribution patterns (behaviors) of robotic systems. Then the two problems are re-phrased as: to compute density functions given predefined force laws; and to find force laws such that a desired density function can be achieved.

The methods we developed to solve the above two problems are iterative methods using approximations. They have some drawbacks: They only work for limited situations (e.g. a single cluster with an identical force law or its variations); we have to make assumptions about the forms of the density functions or force laws (e.g. polynomials of certain degrees); and we have to make assumptions about the structure of the distribution of robots. But the iterative methods have advantages over computer simulations.

- They are fast (constant time to compute density functions given force laws) compared to computer simulations which take linear time in one iteration and which take many iterations.

- Density functions can give us information of the overall distribution of the system, while the exact positions of individual robots can not.

The problem of computing density functions is dealt with in subsection 4.2, and the problem of finding force laws in subsection 4.3. Some details are omitted in this section.

4.2 Compute the Density Function given Force Laws

4.2.1 General Idea

Our method of computing the equilibrium density function is based on the assumption that the density function is smooth within a cluster of robots.

The density function (denoted by D) is computed in an iterative manner. The iterative equation is derived based on the fact that in equilibrium state, the overall force on any robot in the system should equal to 0. For a robot at position x within domain of interest, let $F_{nearby}(x)$ denote the total force from the nearby robots. Robots are nearby robots if they are within a distance μ, where μ is a parameter. Let $F_{faraway}(x)$ denote the total force from far away robots. Thus in equilibrium state, we hvae

$$F_{nearby}(x) + F_{faraway}(x) = 0.$$

(We will use the term particles instead of robots in the rest of this section, since our robots are actually point particles.) We will calculate $F_{nearby}(x)$ by summing up the forces from nearby particles discretely and $F_{faraway}(x)$ by integrating the forces from faraway particles. The force functions have the property that they have large quantities and large derivatives when the distance is small and have small quantities and small derivatives when the distance is large. Thus we decided to sum the forces from nearby particles discretely since an integration will cause large errors. On the other hand, the integration of forces from far away particles will be a good approximation and reduces the time complexity from proportional to the number of particles in the system to the time complexity of doing an integration which depends on the domain of interest and the precision required.

The above equation can be rewritten as

$$F_1(x) = -(F_{nearby-1}(x) + F_{faraway}(x)), \qquad (2)$$

where $F_1(x)$ denotes the force from one of the nearest neighbors and $F_{nearby-1}$ denotes the force from the rest of the nearby particles. In order to compute the forces from the nearby particles, we need to know the positions of them. The positions of nearby particles can be approximated by density functions. To do this, we make further assumptions about the structure of the particles. Specifically, the structure of particles in 2-D is hexagonal and in 3-D is cubic. $F_{faraway}(x)$ is an integration

$$\int_{\|x'-x\|\geq\mu,D(x')>\theta} f(x,x')D(x'),$$

where $f(x,x')$ is the force law from a particle at position x' to the particle at position x, μ is a threshold of distance to distinguish nearby robots from far away robots, and θ is a threshold of density.

In order to do the integration, the force law $f(x,x')$ has to be fixed within domain of interest. This imposes restrictions on the applicability of our method, namely the method can only be applied to computing density functions of a cluster of robots with an identical force law.

The density function appears on both sides of equation (2). By rearranging the equation, we can get a iterative equation for the density function. The next subsection shows an example of deriving the iterative equation for a specific form of force laws.

4.2.2 An Example of a Single Cluster of Working Robots

In this example the robots are point particles moving on a 2-D plane. The position of a particle is given by $z = (x,y)$. Assume that the density function is a polynomial of variables x and y, and is of degree k.

$$D(x,y) = \sum_{i+j=0}^{i+j=k} c_{ij} x^i y^j$$

To compute the density function is to determine the coefficients $c_{00}, c_{10}, c_{01}, c_{20}, c_{11}, c_{02}, \ldots, c_{0k}$, where c_{ij} is the coefficient of the term $x^i y^j$. Let C denote the vector $(c_{00}, c_{10}, c_{01}, c_{20}, c_{11}, c_{02}, \ldots, c_{0k})$. The length of C is $l = \frac{(k+2)(k+1)}{2}$. We need to determine the l coefficients.

Suppose the force law is the same as equation (1), with $\sigma_1 = 3.0$, $\sigma_2 = 2.0$, $c_1 = \alpha_1$ and $c_2 = \alpha_2$.

We assume that in 2-D, the structure of the equilibrium distribution of particles is hexagonal. If there is a particle at position $z = (x,y)$, and the density of

the equilibrium distribution is $D(x,y)$, then there is another particle at position $(x+1/D(x,y),y)$ which is one of the closest neighbours of the particle at $z = (x,y)$. Let F^x denote the projection of force F along x-axis. $F_1^x(z) = \frac{-\alpha_1}{(1/D(z))^3} + \frac{\alpha_2}{(1/D(z))^2}$ and equation (2) is transformed to

$$\frac{-\alpha_1}{(1/D(z))^3} + \frac{\alpha_2}{(1/D(z))^2} = -(F_{nearby-1}^x(z) + F_{faraway}^x(z)).$$

$F_{nearby-1}(z)$ is computed by approximating the positions of the nearby robots using density function and the hexagonal structure. $F_{faraway}(z)$ is approximated by computing the integration. Let the right hand side value be denoted by δ_z. Rearrange the equation, we get the following

$$-\alpha_1 D(z) + \alpha_2 = \delta_z / D(z)^2$$

$$D(z) = (\frac{\delta_z}{D(z)^2} - \alpha_2)/(-\alpha_1)$$

We consider the left hand side $D(z)$ as $D^{t+1}(z)$ and the $D(z)$ in right hand side $D^t(z)$, thus we get an iteration equation:

$$D^{t+1}(z) = (\frac{\delta_z}{D^t(z)^2} - \alpha_2)/(-\alpha_1)$$

To update $D(z)$ from this equation is to update the coefficient vector C.

Note that given the numerical values of a position $z_i = (x_i, y_i)$, the value $b_i = (\frac{\delta_{z_i}}{D(z_i)^2} - \alpha_2)/(-\alpha_1)$ can be computed numerically. Suppose we take m sample positions and compute the corresponding b_i's, we get a vector $B = (b_1, \ldots, b_m)$. Then we will have the following linear system

$$MC = B,$$

where M is a matrix of $m \times l$. The ith row of the matrix M is a vector

$$(1, x_i^1 y_i^0, x_i^0 y_i^1, x_i^2 y_i^0, x_i^1 y_i^1, x_i^0 y_i^2, \ldots, x_i^0 y_i^k),$$

with $(x_i, y_i) = z_i$. C can be solved using Least Square approximation. Thus we can compute $D^{t+1}(z)$ from $D^t(z)$. Iterating this procedure, we can find the density function.

4.2.3 A Single Cluster of Working Robots under Several Control Robots

Consider the scenario where a huge number of working robots form a single cluster, the shape of which is controlled by several control robots scattered around the cluster. The control robots are static. Their purpose is to impose forces on working robots so to achieve certain patterns. There needs only a slight change in the equation (2).

$$F_1(x) = -(F_{nearby-1}(x) + F_{faraway}(x) + F_{control}(x)),$$

where $F_{control}(x)$ is the total force from all the control robots. The method still applies here. Note that in this case, we can make no assumption about the shape of the cluster of working robots.

4.3 Define Force Laws for given Density Functions

Let $D^*(x)$ denote the given density function. Let $D_F(x)$ be the density function resulted from a set of force laws F (can be approximated using the method discussed in the last section). The problem of defining force laws to achieve the density function is to find the set F such that

$$\min_F \int (D_F(x) - D^*(x))^2$$

is achieved.

Assume that the forms of the force laws are known and only the parameters (coefficients and exponents) are to be determined. For example, all the force laws have the form as in equation 1, only the c_1, c_2, σ_1, and σ_2 are to be determined. Let A denote the vector of the parameters and D_A the density resulted from the parameters. Our method can help to determine the vector A.

Let $H = \int (D_A(x) - D^*(x))^2$. H is a function of A. Minimizing H is equivalent to solve the equation $\partial H/\partial A = 0$. Since H is not an explicit function of A, we can use Quasi Newton Method to solve it. The details are omitted here.

5 Spring Laws: Extensions to Molecular Bonding

5.1 Motivation

There are situations where we want a group of robots to form a certain structure, to keep the structure while moving around, and to change from one structure to another dynamically. For example, in military maneuvers or warfare, we often see that troops, battle flights or submarines, when attacking, withdrawing or moving, keep certain formations in order to protect or to attack more efficiently. These defensive or offensive phalanxes change from one form to another under different circumstances. When we want the robots to do some industrial jobs such as harvesting, cleaning, fishing, we want the robots to keep certain assemblies. For example, we may want the robots to form a harvesting line which sweeps from one end of a field to the other end. In this case, the robots should be equally spaced along the line so the work load is balanced. But when a robot is confronted with an obstacle, it is allowed the flexibility to move out of its way to get around the obstacle. After that, it should go back to its proper place.

For the same reasons as mentioned in section 1.2, we need a distributed control mechanism. The "social potential fields" as defined so far can only let the groups be well spaced but they are insufficient to form an exact structure. The control mechanism we desire is that by controlling directly a few robots, we can control the whole group to form a predefined structure, to keep the structure while moving around, and to change to other structures dynamically. Meanwhile the robots are allowed some flexibility.

There are other force laws from molecular dynamics that can be applied to dynamic robotic control. Another one is the spring law used in molecular bonding. We are interested in this force law because there are known theories about rigid graphs and spring laws. We define graphs, considered as vertices connected with (imaginary) ideal springs. If such a graph is rigid, it has a unique structure which carries the minimum potential energy. The minimum energy structure can be found by letting the vertices move according to physics laws and settle down to their equilibrium state. Furthermore the graph has the tendency to keep the minimum energy structure. After deformation, the structure will always be restored "automatically".

In section 5.2, we introduce the concept of rigid graphs. In section 5.3, we discuss the relation of graph connectivity and graph rigidity. Finally in section 5.4, we talk about the application of spring laws to robotic control. The computational aspect, i.e. how to define springs to get a rigid graph with a desired minimum energy structure, is omitted in this paper.

5.2 Rigid Graphs

Let G be a finite graph with a set of vertices $1, \ldots, n$ and with a non empty set E of undirected edges. Each element in E is designated as an idea spring. Let $L = \{< l_{ij}, k_{ij} >\}$ defines springs between i and j for all $(i, j) \in E$, i.e. the length of the spring l_{ij} without compression or extension and the force constant k_{ij}. Let G_L denote the graph with springs defined by L. If two vertices i and j connected by a spring $\{< l_{ij}, k_{ij} >\}$ are at distance r_{ij} apart. According to Hooke's Law, the force between the two vertices is:

$$F_{ij} = k_{ij} y_{ij}, \quad \text{where} \quad y_{ij} = r_{ij} - l_{ij}.$$

The potential energy stored is given by

$$P_{ij} = \frac{1}{2} k_{ij} (y_{ij})^2.$$

An *embedding* of G will be an assignment of the vertices into d-dimensional Euclidean space R^d. Let p_i be the position of vertex i in R^d and $p = (p_1, \ldots, p_n)$ be the embedding of G. The potential energy of a particular embedding p of G_L is

$$E_L(p) = \frac{1}{2} \sum_{ij} k_{ij} (\|p_i - p_j\| - l_{ij})^2.$$

The minimum of $E_L(p)$ is achieved when $\nabla E = 0$. This happens to be when all the vertices are at their equilibrium states. I.e. the minimum energy embedding p satisfies for all i,

$$F_i = \sum_{(i,j) \in E} F_{ij} = \sum_{(i,j) \in E} k_{ij} (r_{ij} - l_{ij}) \frac{P_i - P_j}{r_{ij}} = 0.$$

Thus the minimum energy embedding can be found by letting the vertices move according to physics laws and settle to their equilibrium state.

Definition 1 *A graph is* rigid *with respect to a given dimension d in Euclidean space if there is a unique embedding up to translation and rotation (i.e. with the relative positions of all vertices the same) that has minimum energy.*

Since a rigid graph has only one minimum energy embedding, the vertices will always settle to the same embedding (up to translation and rotation) after deformation.

5.3 Graph Connectivity and Graph Rigidity

Graph connectivity is a way of deciding if a graph is rigid.

G is *k-connected* if there is no subset of k vertices, which if deleted, disconnect G.

Definition 2 *Fix an arbitrary subset C of $d+1$ vertices with $C \subset V$ and fix an arbitrary embedding of the vertices of C in R^d. A C-embedding of G is an embedding $G(p)$ in R^d consistent with the already fixed embedding of C.*

Theorem 1 ([7]) *Let G be a graph on n vertices and $1 < k < n$. Then the following two conditions are equivalent:*

1. *G is $(d+1)$-connected.*

2. *For every $C \subset V$ with $|C| = d+1$, G has a convex C-embedding in R^d in general position.*

Theorem 1 reveals the connection between graph connectivity and graph rigidity. In particular, it shows that if a graph G is $(d+1)$-connected, and if $d+1$ vertices of G are fixed, G has a unique minimum energy embedding in R^d. It follows from the theorem that if G is 3-connected, G is rigid with any 3 vertices fixed in 2D. The positions of rest of the vertices are solely determined by the three fixed vertices. If G is 4-connected, then G is rigid with any 4 vertices fixed in 3D.

Assuming G is $(d+1)$-connected and that the edges of G are made of ideal springs, the convex C-embedding in R^d can be found by letting the remaining vertices move according to physics laws and settle into an equilibrium state. For a fixed embedding of C, the convex embedding of G is unique. If we deform the positions of the vertices in $V - C$, the embedding will be restored automatically after the deformation.

5.4 Apply Spring Laws to Robotic Control

To achieve a structure in d-dimension, we first design a $d+1$-connected graph G with edges as springs and with a desired minimum energy structure. Then we choose a subset of V as C. The vertices in C are assigned positions in d-dimensional space such that the unique C-embedding of G is the same as the desired structure.

If the embedding of C is fixed, we know that the structure is unique and can be restored automatically after deformation.

The vertices represent robots, with vertices in C the leading robots which will be controlled explicitly by a global controller. The other vertices are non-leading robots which will control their motions using force laws. The force laws are defined by springs. A non-leading robot i stores a table of $< l_{ij}, k_{ij} >$ for $(i, j) \in E$. The robot senses the distances r_{ij} for $(i, j) \in E$ and calculates a force using the following expression:

$$F_i = \sum_{(i,j)\in E} F_{ij} = \sum_{(i,j)\in E} k_{ij}(r_{ij} - l_{ij})\frac{P_i - P_j}{r_{ij}}$$

and its motion is controlled by the resultant F_i. The motion of robot i tries to reduce F_i to 0. With all the non-leading robots operate simultaneously in this way, the system reaches its minimum potential energy structure which is the structure desired.

An example of the application is to let the robots form a line on a 2-D plane, with robots equally spaced along the line. The spring force laws are defined among robots in such a way that the graph abstracted is 3-connected and has a minimum energy structure as a line, with all the vertices eqully spaced along the line. And 3 of the robots can be leading robots. If we control the leading robots to move forward (keeping the relative positions of them fixed), the original minimum energy structure is deformed. The remaining robots, controlled by the spring laws, will follow up, restoring the line structure. If we change the relative positions of the 3 leading robots, the structure can be changed to have some other forms.

Note that there are no physical springs between robots so the motion of the robots are not restricted by springs.

There are also applications where we do not desire entire rigid structures but instead flexible ensembles of rigid sub-components. The rigid sub-components can be accomplished by spring law control and the coordinations of the subcomponents can be accomplished by social potential fields control.

6 Conclusion

In this paper, we have proposed a so called "social potential fields" method for distributed autonomous multi-robot control. The force laws have simple forms, yet our simulations showed that these force laws can lead to a wide range of interesting and useful behaviors similar to basic behaviors in social animals. These behaviors may have practical utility in industry, military and other areas in the near future. The social potential fields method has the following advantages:

- The method is generic. For different systems, different tasks, or different environments, we only need to define different force laws. The robots carry out the same operations, just applying different force laws. They don't have to perform complicated and specifical algorithms.

- The method is very robust. Contrary to many of the motion planning algorithms, this method doesn't require precise sensors or precise actuators. Inaccuracies can be tolerated.

The future work is directed to the problems discussed in the following subsections.

6.1 Local Minima Problem

We found Local Minima problems in our computer simulations, for example, in the simulation of *guarding the castle*. In one simulation, the robots are initially distributed around the castle. The system stabilizes to a ring around the castle which is the equilibrium we desire. With the same set of parameters, but with the initial distribution of robots in a square area at the upper-left handside of the castle far away from it, the system fails to form a ring. Instead, it stabilizes to a disc on the way approaching the castle. It is very difficult to predict local minima in a dynamic system as such.

A partial solution to the problem resorts to the global controller which is responsible for defining the force laws. Once the global controller detects the local minimum situation, it can change force laws to help the system escape from the local minimum.

6.2 Loss of Information

In the distributed control paradigm, each individual applies *local rules* to the current state of the system. The description of the state can be very complicated (e.g. Reynolds' simulation of birds) and thus the local rules can be very complicated. In our VLSR system, the state of the system is simplified to only the distances. Thus information about the system is lost. Due to this, some behaviors can not be accomplished by social potential methods.

The force laws should be kept simple so they can be manipulated but they should be complex enough to generate behaviors we need.

6.3 Lack of Powerful Tools for Defining Force Law

As we have mentioned in subsection 4.1, our methods for defining force laws are quite restrictive. For example, the methods can not compute the density function given a cluster of robots with different force laws defined among them.

Furthermore, our iterative methods only deal with static equilibrium situations. This is because the iterative equation is derived from the fact that the resultant force on a single robot is equal to 0 in the equilibrium state. But for practical purpose, many equilibrium states are dynamic. For example, in the simulation *guarding a castle*, the invader and the guards form a dynamic chase-and-run behavior. How to define dynamic equilibrium behaviors and how to define force laws to achieve them need further study.

Acknowledgments

This research was supported by DARPA/ISTO Contracts N00014-88-K-0458, DARPA N00014-91-J-1985, N00014-91-C-0114, NASA subcontract 550-63 of prime contract NAS5-30428, US-Israel Binational NSF Grant 88-00282/2, and NSF Grant NSF-IRI-91-00681.

References

[1] L. F. Greengard. *The Rapid Evaluation of Potential Fields in Particle Systems.* MIT Press, 1987.

[2] R.A. Jarvis. A perspective on range finding techniques for computer vision. *IEEE Tran. on Pattern Analysis and Machine Intelligence*, PAMI-5(2):122–139, march 1983.

[3] O. Khatib. Real-time obstacle avoidance for manipulators and mobile robots. *Int. J. Robotics Research*, 5(1):90–99, 1986.

[4] C.G. Langton, editor. *Artificial Life, the Proceedings of an Interdisciplinary Workshop on the Syntheses and Simulation of Living Systems.* Addison-Wesley, Redwood City, CA, 1989.

[5] J. Latombe. *Robot Motion Planning*, chapter 8. Kluwer Academic Publishers, Norwell, MA, 1991.

[6] J. Latombe. *Robot Motion Planning*, chapter 7. Kluwer Academic Publishers, Norwell, MA, 1991.

[7] N. Linial, L. Lovasz, and A. Wigderson. Rubber bands, convex embeddings and graph connectivity. *Combinatorica*, 8(1):91–102, 1988.

[8] C.W. Reynolds. Flocks, herd, and schools: a distributed behavioral model. *Computer Graphics*, 21(4):25–34, July 1987.

[9] E. Rimon and D.E. Koditschek. Exact robot navigation using artificial potential functions. Technical report, Yale Univ., Dept. of Electrical Engineering, 1991.

[10] J. Schwartz, editor. *Planning, Geometry, and Complexity of Robot Motion.* Ablex Publishing Corp., Norwood, New Jersy, 1984.

[11] E.O. Wilson. *Sociobiology, the New Synthesis.* The Belknap Press of Harvard University Press, 1985.

Algorithms for Optimal Design of Robots in Complex Environments

Krasimir Kolarov *Interval Research Corp., 1801-C Page Mill Road, Palo Alto, CA 94304*

The goal of our work is to find the optimal design of a robot that can reach everywhere in an environment with obstacles without collisions. The main questions we are concerned with are: what is the most appropriate type for the links of the robot? what is the minimum number of links that are needed to cover every point in the environment? and what is the best placement for the robot?

We describe some algorithms for finding the set of points in the environment that can reach all other points with a minimum number of links. Initially the obstacles are modeled as convex polygons and subsequently we discuss extensively the modifications that those algorithms require to cover curvilinear, nonconvex and three-dimensional obstacles.

We derive several theorems that establish upper and lower bounds on the number of links for both planar and spatial cases. We describe some algorithms for minimizing the upper bounds to the optimal number of links for the environment.

We generalize the basic problem for the cases when both the robot and the environment are designed simultaneously, when we deal with moving robots and obstacles, and when we have multiple robots or robots with variable structure.

1 Introduction

Robots are used in industrial automation to perform repetitive tasks or to lift, move and manipulate objects that are heavy and dangerous to handle. They also work underwater, in space, and other environments hostile to human life. To achieve their purpose well, robots should have a high degree of autonomy, good reasoning and learning capabilities and appropriate construction for their tasks. However the vast majority of the existing industrial robots perform simple

tasks and have similar structures. Accordingly the majority of the research in the area of robotics has been concentrated on incorporating a high degree of intelligence, learning capabilities, control and mobility in the existing industrial robots. This is understandable, considering that the manufacturers and the users of robots systems would like their robots to have better capabilities.

Another way to improve performance is to design a robot for its specific workspace. That is why the main problem we are going to discuss in this paper, is how to design a robot that is most appropriate for a given workspace. In doing so we will introduce techniques similar to the ones used in "robot motion planning" and "computational geometry".

Researchers in the area of robot motion planning are aiming at providing robots with good reasoning capabilities (see [12]). Given a robot working in some given environment, they are trying to come up with some good algorithms and techniques that will allow the robot to grasp an object from one place and transfer it to another without colliding with the other moving or stationery objects in the environment. This operation is to be done in an optimal fashion (see [4]). Often this proves to be a difficult task, especially when the structure and the kinematics of the robot are not appropriate for the layout of the environment and the activities performed in it.

One can try to use a design optimization approach for the planning of a robot path which avoids the obstacles (see [21]), or to identify the kinematics of the robots with respect to their tasks [1]. However these approaches still do not guarantee the best fit between the environment and the robot operating in it, because they are both given a priori. In order to have an opportunity for optimization, we need to fix at most one of the components, and find the other one in such way

Algorithmic Foundations of Robotics
1-56881-045-8

that it best fits. We believe that the fixed component (if any) should be the environment, and the robot should be the one that varies in structure and placement to best fit the environment. Furthermore, we will also consider the problem when both the environment and the robot are allowed to vary in order to achieve more efficient operation of the robot.

The different aspects of optimal design of robots have been considered in the literature (see [3], [16] and [17]). We will concentrate in this paper on finding the best structure for a robot working in an environment with obstacles, and the best place for this robot in the environment. What we want to achieve is full autonomy, that is we want the robot to be able to reach every point in the environment, except of course the points interior to the obstacles. Thus the basic problem we will consider is:

Given an environment containing obstacles, what are the minimum number of links and the best placement of a manipulator operating in this environment that will allow it to reach any point in the environment that does not belong to the obstacles.

The set of points in the environment that do not belong to the obstacles is usually called "free space". We will use this term here and we will denote this space with "E". There are different types of joints that could be used between the links – revolute, prismatic, spherical, ball-and-socket, etc. The two most widely used joints are revolute and prismatic. We will combine those two joints in our robot in order to achieve a larger "reachable space" (the points that can be reached by the end-effector of the robot) and greater flexibility to reach around obstacles.

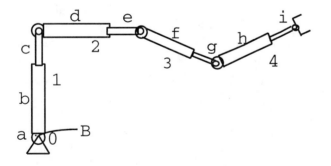

Figure 1.1: A robot with four telescoping links.

The primary joint type we will employ is a one degree

of freedom (dof) prismatic joint which folds into itself like a telescope to some minimal length. Our robot is formed by connecting each telescopically jointed link to a revolute joint at its other end. The revolute axes are perpendicular to the prismatic axes, as in Figure 1. We will call the links thus formed "telescoping links".

Each telescopic joint connects two links. So in Figure 1 we have 9 links (a, b, c, d, e, f, g, h, i) including the ground link but not the end-effector. However in this paper it will be more convenient to think of a link in a different way. Since the only function of the prismatic joint is to change the link's length, we consider the telescoping joint to be part of a link's internal structure. Hence we will say we have in Figure 1 only 4 moving links (*1*, *2*, *3*, *4*). In the plane, these types of links give more flexibility than a traditional fixed link-length, revolute-jointed manipulator. Such designs also make it easier to extend our analysis to other types of environments, such as ones where the obstacles can change their place and shape and the robot can move around rather than being on a fixed base. These structures can also be easily generalized to three dimensional spaces. In the limit, when the number of telescoping links goes to infinity, we can use these structures to consider problems similar to the ones discussed in [5].

Links with adjustable length are used in the "Adjustable Robotic Mechanisms" for low-cost automation, considered in [11] as well as in most of the walking robotic devices. Telescoping structures are especially popular in construction and building automation.

Once we have decided on the construction for the robot, the problem we want to solve becomes the one of finding the minimum number of telescoping links (we refer to them from here on simply as "links") that are needed to cover the whole free space.

In Section 2 of this paper we discuss this problem for a planar environment and obstacles represented as convex polygons. Using notions from computational geometry, we describe, in Section 2.1, an algorithm that finds the minimum number of links that cover E for any point of the free space. The complexity of this algorithm is in the worst case $O(m.n.log\ n)$, where m is the number of obstacles and n is the number of vertices of the environment and the obstacles. Section 2.2 describes algorithms for finding the set of points that are most appropriate for best placement of our robot in the environment.

Section 3 generalizes the discussions in Section 2 for more complex shapes of the obstacles and for higher dimensions of the environment. In this, an algorithm for general convex obstacles is described in Section 3.1, which is primarily based on a discussion of generalized convex obstacles. The discussion of general non-convex obstacles in Section 3.2 shows that in our problem sometimes the presence of reflex vertices simplifies the work of the algorithm. Finally a spatial (3-Dimensional) environment is considered in Section 3.3, in which it is shown that the algorithms developed in Section 2 can be easily extended to $2^{1/2}$D environments using horizontal slices of the environment.

Sections 2 and 3 give a complete algorithmic treatment of our problem, i.e. if we are given an environment with obstacles, applying those algorithms we can choose the best place for the robot operating in this environment and determine the minimum number of links that the robot needs.

Section 4 derives several estimates for this minimum number of links in terms of the number of obstacles and vertices in the environment, using a more theoretical approach. Several theorems estimating this number for the planar (Section 4.1) and the spatial (Section 4.2) case are derived. Some algorithms for optimizing these estimates are discussed in Section 4.3.

We also consider the case when we design both the robot and the environment simultaneously. In Section 5 it is shown that if certain conditions are satisfied, we can always place the obstacles in the environment in such a way that the best designs of the robots contain only three prismatic links (in the planar cases) or three telescoping links (in the spatial cases). A brief discussion of moving robots and obstacles, multiple robots and reconfigurable structures can be found in [7].

Finally Section 6 summarizes our results and makes some suggestions for future research in this area.

2 Convex Environment and Obstacles

In this section all obstacles will be represented by convex polygons in a two-dimensional environment. For simplicity of the figures illustrating the material, we will consider a rectangular outside boundary of the environment, although all algorithms are implemented for any convex outside boundary.

Our optimization problem is to find a set of points in the free space such that, if we place the base of the robot in any point of this set , we will need a minimum number of telescoping links to reach all the other points in the free space. In fact we have to solve two distinct problems:

I. How to find this set of points?

II. If we place the base of the robot at a point in the set above, how to find the minimum number of links needed to reach all points in the free space?

We will start with problem II which we will show to be equivalent to finding the visibility polygons for a given point in the environment.

2.1 Visibility From a Point

Let us denote with point B the base point of the robot and by P_{ij} the vertices of the polygons, where the first index $i = 0,1,2,...,m$ denotes the obstacle (0 for the environment EN) and $j = 1,2,...,m_i$ denotes the vertex of an obstacle. We will increase j clockwise for the obstacles and counterclockwise for EN. Let n be the total number of vertices in the environment.

We want to determine the minimum number of links that are needed to reach from point B any other point in the free space E. All points in E that can be seen from point B with one link (as mentioned before by 'link' we understand 'telescoping link') are depicted in Figure 2.1. Assuming no lower and upper bounds on the length of a link, those points constitute the shaded region in Figure 2.1 , which is sometimes referred to in the literature as the visibility polygon $V(B)$ for point B.

If we consider the obstacles in our environment as holes we can formulate our problem as a visibility problem for polygons with holes. Suri and O'Rourke [22] consider visibility polygons with holes, and show (theoretically) an algorithm that runs in $O(n.logn)$ time which they prove optimal by reduction from the problem of sorting n positive integers. Lee and Chen [14] present an algorithm for computing the visible edges of a set of m non-overlapping convex polygons that computes only the first visibility region from a given point. We want an algorithm that builds all the visibility polygons from a given point.

In our case all points that can be reached from B with two links are those that can be seen from the

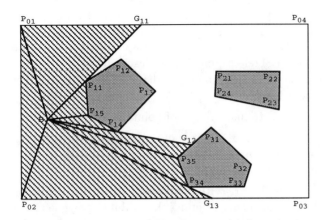

Figure 2.1: Region with Visibility 1 from point *B*.

points in *V(B)* with one link - we denote this set of points with $V_2(B)$. This set is depicted as the shaded area on Figure 2.2.

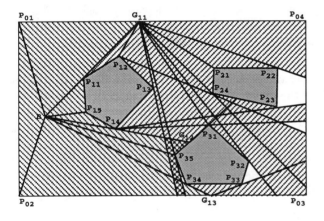

Figure 2.2: The visibility from point B is three.

In the same manner we can build the set $V_3(B)$ and so on. We continue this procedure until all the points in the free space *E* are included in one of the visibility polygons $V_i(B)$. We will denote the first number *i* for which all the points in *E* are covered as *VMIN*, which stands for "minimal visibility". In the computational geometry literature this is also known as the "link radius" of the environment, while the link distance between two points is the minimum number of straight segments (links) that connect the points without collisions with the obstacles.

The link center of a simple polygon P is the set of all points in P whose maximal link distance to any other point of P is the smallest possible. Our problem (#I) can be formulated to be the same as finding the link center for a polygon with holes. We do not know of any published research that deals with this problem.

In addition to the above notations, we can define the term "link diameter". This will be the maximal link distance between any two points in the polygon. In our notations we will also use *VMAX* (from maximal visibility) as a term for the link diameter in an environment with obstacles.

In Chapter 2 of [7] and in [8] we have described in detail an algorithm that builds all visibility polygons from a given fixed base point *B* in an environment with obstacles. The basic approach for building the visibility regions at a certain level is to draw tangents from certain points, called "generating points", towards the vertices of the environment and look for intersections of these tangents with the edges of the environment. In our example point B is the generating point for the visibility polygon on level 1, while points G_{11}, G_{12} and G_{13} are the generating points for the second level. The visibility polygons are sets of triangles that are formed by the tangents from the generating points and the edges of the obstacles and the environment. In addition, for completeness of the algorithm, we include generating points and triangles created by the mutual tangents between the obstacles.

The algorithm terminates when the union of the visibility regions $V_i(B)$ $(i=1,2,...,k)$, covers the whole free space with a minimal link distance *k*. Theorem 2.1 in [7] shows that this is equivalent to have the entire length of all the edges of the environment be reached from point *B* with *k* links.

The worst case complexity of our algorithm is *O(m.n.log n)* where *m* is the number of obstacles. In addition to the minimal number of links *k* for point *B*, we can automatically extract for <u>every</u> point in *E*: the visibility of this point from the base point; the visibility triangle it corresponds to and a possible link path from the base point to this point. We also find as a by-product of our program the minimum and the maximum lengths of the links for each level. This is useful information for planning the actual sizes of the links.

To be able to model the links as straight line segments we initially grow the obstacles by the width of the links. In fact we can also grow the obstacles with an additional safety margin if we do not want a hard contact between the links and the objects in the environment.

In the following section we consider the question of choosing a base point.

2.2 Link Center of the Environment

Once we know how to find the minimum number of links for a given point, we need to decide where to place the base point so that we have the smallest link radius overall (question I). As we can see from Figure 2.3 different points have different visibility of the environment. If we place the base of the robot at point A than we can see (reach) all points in the free space with two telescoping links. Point B however requires at least 3 links to reach the inside points in the shaded triangle.

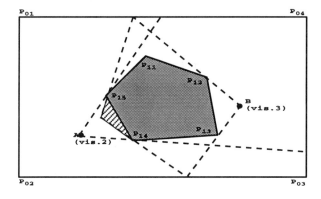

Figure 2.3: Points with different visibilities.

The algorithm for building the link center of the environment (the set of points that can reach all other points with minimal number of links) is based again on Theorem 2.1 from [7]

Theorem 2.1 *If the entire length of all the edges of the environment E can be reached from some point B with k revolute-jointed telescopic links, then all points in E can be reached from B with k revolute-jointed telescopic links.*

We can derive that he link center of E is the intersection, which is not empty, of the same lowest level k visibility regions for all points on the edges of E. This

is true because if a point A belongs to the link center with a link radius k, that means that this point can reach all the other point in E with maximum k links. Thus it can reach all points on the edges of E with k links, i.e. it can be reached from all those points with k links as well. Then this point A belongs to the intersection of the k-th visibility regions of all those points of the edges. As we show in Theorem 2.3 in [7] this k is necessarily the minimal one for which this event occurs. Thus the general algorithm for building the link center is as follows:

```
begin
    k:= 0;
    repeat
        k:= k+1;
        S_k:=E; {E is the whole free space}
        for i=1,...,m do
            for j=1,...,m[i] do
            begin
                Build visibility region V_k(e_ij);
                {e_ij denotes the edge P_ij P_ij+1}
                S_k = S_k ∩ V_k(e_ij);
            end;
    until S_k is non-empty ;
end;   {S_k is the exact link center}
```

As we can see from this algorithm we actually build the visibility polygons from all edges in the environment rather than using the infinite number of points along those edges. Such visibility polygons consist of the set of points in the environment that can see the entire edge, not just a point on the edge. As we can see in Figure 2.4 this set is the simply-connected component of the intersection of the visibility polygons of the two end-points of the edge.

This property is formally proved in [7] in Theorem 2.4. There, a detailed algorithm is explained of how to actually build such visibility region of an edge by going around clockwise along the visibility triangles defined by the two end points and merging them in a correct order. This algorithm builds the first visibility polygon $V_1(e)$ and has the same complexity as the one for visibility from a point.

One might think that we could find the next level visibility polygon by taking the generated points from the first level and use them to form the next level. However we can show that there exist points that cannot

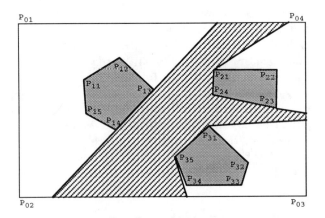

Figure 2.4: Visibility polygon of edge $P_{13}P_{14}$

see any of the vertices of the visibility polygon $V_1(e)$ of an edge e with k links, but can see the edge e with $k+1$ links.

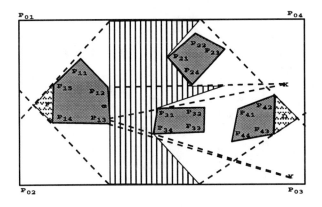

Figure 2.5: Partial visibility of higher order.

Consider the example in Figure 2.5. The visibility polygon $V(P_{12}P_{13})$ of the edge $P_{12}P_{13}$ is the vertically shaded area on the figure. We can draw the tangents from the vertices of this polygon, and determine the set of points that can see at least one point from $V(P_{12}P_{13})$ with one link. Those points can see the entire edge $P_{12}P_{13}$ with two links. There are two horizontally shaded triangles D and F on the figure that do not belong to any of the areas above, i.e. we might think that we need at least three links for the points inside those triangles to reach the entire edge $P_{12}P_{13}$. However this is not the case for the points in the right triangle D. They can see both points X and Y with only

one link, and every point on the edge $P_{12}P_{13}$ can be seen from either X or Y with one link. Consequently the points in D can see the whole edge $P_{12}P_{13}$ with two links. In other words when we go to higher levels of visibility the problem becomes much more complicated, because one point can see part of the edge from one side of the some obstacle, while the other part (or parts) of the edge are 'seen from another side of the obstacles. To deal with this theoretically we introduce a construction called "X-form" (see [7], [10]).

The terminology for the X-form is illustrated in Figure 2.6. The part of the edge that we want to see will be called the "base" of the X-form and the opposite side of the form will be called the "top". The tangents that outline the X-form are called "generating lines" of the form, and their intersection point V is the "vertex" of the X-form. The part inside the X-form between the vertex, the generating lines and the top is denoted as the "upper part", and the corresponding part for the base is called the "lower part" of the form. Any line l that intersects the lower part, passes through the vertex, or intersects the upper part but not the top of the X-form, is said to "cross" the form. Let U be a point in the free space E for which there exists a line l that crosses the X-form. We will denote the intersection points of l with the X-form with A and B (if l crosses through the vertex V only, then $A = B = V$ of the X-form).

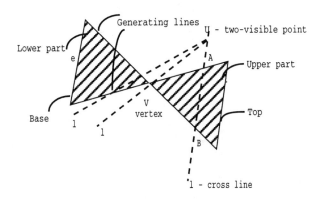

Figure 2.6: X-form of an edge.

We will say that U is a "two-visible" point for the X-form if:

- when l crosses the lower part, the segment UA is entirely in free space, i.e. does not intersect any of the obstacles and is inside the outside boundary;

- when l goes through the vertex V of the form, the segment UV is entirely in free space;

- when l crosses the upper part the segment UB is in free space.

From this definition follows that if U is two-visible for an X-form, then U can reach all points in the X-form with two telescoping links. Consequently a point U will be able to see the entire edge e of some obstacle with two links if there exist one or more X-forms that are two-visible from U and whose bases cover e completely.

If we want to check whether a point U can see an edge e with two links, we can apply the following hierarchical algorithm:

- Form all X-forms that are based on the whole edge e.

- Check if any of those forms is two-visible from the point U. If yes we are done and two links are enough for U to see e. Exit.

- If the above condition is not satisfied then draw all common tangents between the obstacles (both inside and outside), that intersect the edge e.

- Those tangents will form X-forms with bases on parts of the edge e. Find all forms that are crossed by the lines from U to the vertices of E.

- Form the union of the bases of all X-forms above that are two-visible from U.

- If this union covers the whole edge e then we are done, and point U can see the whole edge e with two links. Exit.

- If the union is part of e then e is not completely visible from U with two links. If the union is empty we cannot say anything. Exit.

- If however this union covers at least one point W of e then we can conclude that e can be reached completely from U with three links, because on the third level all points of e can be reached from W with one link. Exit.

The work of this algorithm is illustrated on Figure 2.7. Although there is no line through point U that crosses either of the X-forms $X1$, $X2$ or $X3$, using common tangents we can see that the two X-forms $X11$ and $X12$ are two-visible from U, and the union of their bases is the edge e. We conclude that U can see the entire edge e with two links.

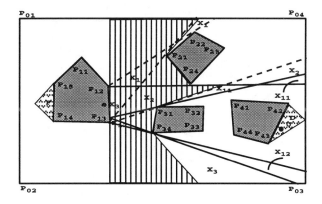

Figure 2.7: Three-visible polygons of an edge.

More details on the actual implementation of X-forms can be found in [7]. We can use the procedure above for finding third, fourth, etc. level visibility polygons of an edge (if necessary) and the criterion for termination is the same as for discrete points. In our example the maximum number of links necessary is 3.

The result of the application of the algorithm for finding the link center for the environment in Figure 2.1 is depicted in Figure 2.8. The link center is the set of points that belong to the unshaded area in the figure. Thus if we place the base of our robot in any point in this area, we are guaranteed to reach all other points in the environment with minimum 2 telescopic links. The choice of a particular point can be made using heuristics, knowledge of the environment, or by request.

As we can see on Figure 2.8, small part of the shaded area in the free space is darker than the rest of it. This is the difference between the exact link center in E and an approximate link center (ALC) built using the vertices and the midpoints of the edges in E.

As we mentioned before, the exact link center requires the use of all (infinite) number of points on the edges of the obstacles. However for some cases we can use approximations to this link center by using finite number of points along each edge. For the figure above we have used for each edge its end-points and its midpoint. The advantage of this approximation is that we don't have to go through the complicated process of building and intersecting X-forms. The complexity of this ALC algorithm is the same as the one for the exact link center, but ALC is faster and straightforward

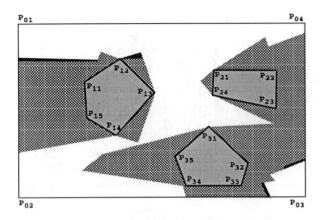

Figure 2.8: Link center for convex obstacles.

extention of the point visibility and the results are very close as illustrated in Figure 2.8 (the exact link center is a subset of any of the ALC's).

The algorithms for visibility from a point and building of an ALC are implemented in Think Pascal and running in real time on a Macintosh computer. Very detailed description of the procedures for building of the exact link center can be found in [7]. In the next section we will extend our analysis to general types of obstacles and environments.

3   General Shape and Dimension of the Obstacles

In this section we will use the algorithms for visibility and link properties above to generalize our discussion going from convex polygons to generalized polygons, to general convex obstacles, to general planar obstacles and finally to a 3D environment and obstacles.

3.1   Algorithms for Convex Obstacles

As a first step we will assume that the shape of the obstacles can include both straight edges and arcs of circles. A convex two-dimensional object, whose edges are either straight segments or arcs of circles is called a "generalized polygon" and is extensively discussed in [13].The significance of generalized polygons comes from the notion of "Configuration space" (or C-space) in motion planning (see [12]).

We can approximate the shape of a moving robot in a 2D environment with its minimal bounding circle.

Now instead of planning the motion of a circle around polygonal obstacles, we can grow the obstacles with the radius of the circle and plan the motion of a point around these grown obstacles (known as C-obstacles). In this case these C-obstacles are actually generalized polygons, i.e. there edges are either straight segments or arcs of circles.

We will also use generalized polygons as a stepping stone to more general obstacles.

The theory for generalized polygons is analogous to the treatment of convex polygons. To start with, the visibility region, from a point in an environment with obstacles represented as generalized polygons, is formed by drawing the tangents from the point to the obstacles. For example, the visibility polygon V(P) is the shaded area in Figure 3.1. The only difference between this area and the one in Figure 2.1 is that now we also have tangents from the point to an arc of circle. While the equation of the tangent in the case of convex polygons follows simply from the fact that the line passes through two known points, for generalized polygons this equation is a little more difficult to obtain. The corresponding equations, for this case, are derived in [7].

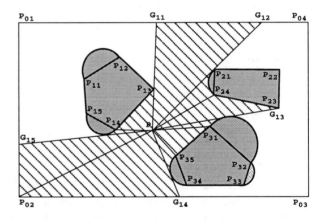

Figure 3.1: Visibility for generalized polygons.

The second visibility polygon $V_2(P)$ will be formed by the union of the visibility polygons $V(G_{11})$, $V(G_{12})$, $V(G_{13})$, $V(G_{14})$ and $V(G_{15})$ for all generating points. As in the case of convex polygons, we need to consider the mutual tangents between the obstacles for the generation of the higher level visibility polygons. We continue this way until we cover the entire environment.

In our example, the maximum number of links needed for P is 3.

Theorem 2.1 from Section 2 is valid for generalized polygons provided that now an edge of the free space E is either a straight segment or an arc of circle. Thus to complete the algorithm for visibility from a point, in this case we need to add an additional criteria for ending the algorithm. While the edges represented by straight segments can be treated analogously to the algorithm in Section 2, we need different considerations for the circular edges.

The complexity of the algorithm for visibility from a point is the same as the one for convex polygons because the only differences are in calculating the tangents, which takes constant time.

In building an ALC for convex polygons we used a property (Theorem 2.2 in [7]) that if we can reach all the vertices of the polygons with k links than we need maximum $k+1$ links to reach all points in the free space. However this is not true in this case because hat the points on an arc of circle cannot be seen with a straight line from the arc's vertices. In [7] we have outlined a procedure of dealing with this difficulty where using the tangents to the arc at its vertices, we search for an area in free space that contains a point or a set of points that can see the whole arc with one link only. In the worst case we might need to subdivide the arc into sub-arcs for analysis, but overall because of the connectedness of the environment, at the end we will get l points $D_{i},\ (i=1,...,l)$ forming a set, from which all points of the arc c are visible. Every point of the arc can be seen with one link from at least one of the points D_i. This procedure does not affect the overall complexity because it is done once in the beginning of the main algorithm.

As a criteria for "end", in the algorithm for the visibility from a point, for the edges that are arcs of circles we use Lemma 3.1 from [7]

Lemma 3.1 *If at level k all points D_{ij} for every arc have been reached, than the next level is necessarily the last, i.e. the visibility is maximum $k+1$.*

We can also use this criteria for building the visibility polygons for all vertices and all points D_i in the environment and then intersecting the corresponding levels bottom-up until the first non-empty intersection. This

will be an approximate link center for this environment. If we apply this procedure to our example, for all points P_{ij} and D_{ij} $(i=1,...,m;\ j=1,...,m_i)$ and intersect the corresponding triangles, the result is shown in Figure 3.2. The approximate link center on the figure is the unshaded portion of the free space. As we can see it is considerably smaller than the one in Figure 2.8.

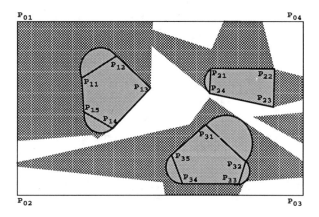

Figure 3.2: An ALC for generalized polygons.

To build the exact link center we need to build the visibility polygons from all edges in the free space. The visibility polygon of a circular arc is quite different however then the one for a line segment. In Figure 3.3 all points that can see the whole arc $Arc(P_{21}P_{24})$ with one link are shaded vertically.

The first characteristic of the shaded points is that they belong to the top part B in Figure 3.3, which we will call "hat" H_{ij} for every circular edge e_{ij}. In fact this hat is formed by intersection of the tangent lines t_1 and t_2 with the outside boundary and the vertices of the outside boundary between those intersections. In Figure 3.3 the hat is the polygon $D_{24}P\prime_{21}P_{12}P\prime\prime_{24}P\prime_{24}$. The hat is non-empty if the vertex D_{ij} is in free space and the circular triangle $P_{21}D_{24}P_{24}$ is also in free space. However if this hat is not completely in free space, we need to take just the part that is in free space and can see the whole edge with one link. Thus we can prove the following:

Theorem 3.1 *For every circular edge $c_i = Arc(P_iP_j)$, $V(c_i)$ is the simply-connected component of the intersection $V(P_i) \cap V(P_j)$ that is contained in the non-empty hat $H(c_i)$.*

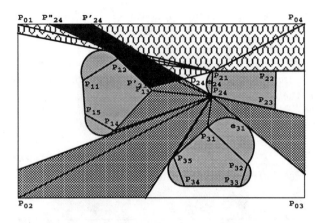

Figure 3.3: First Visibility region for an arc.

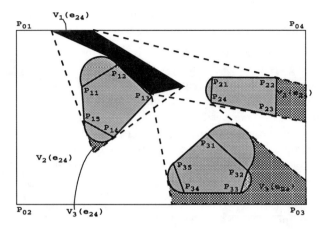

Figure 3.4: Visibility regions for circular edge.

We can build the shaded region on Figure 3.3 using the same algorithm as the one described in Section 2.2. The only difference is that now some of the edges of the resulting region are arcs of circles and instead of starting with the vertices P_i and P_j , we start and end with the point D_i. If an arc has an empty hat (see arc e_{31} in Figure 3.3), then there are no points that can see the whole arc with one link. In that case we can find the points that see the whole arc with two links by subdividing the arc into two (four, etc.) sub-arcs considering the intersection of the visibility polygons of the hats of the sub-arcs. This procedure is finite because there is no intersection between the obstacles and the outside environment, and every circular arc can be infinitely closely approximated with a finite polygonal chain.

For every circular edge after we find the first visibility region (in the case of non-empty hat) or the second visibility region when the first one is empty, we can build the rest of the visibility regions by using the X-forms introduced in Section 2.2. We have to add some conditions that reflect the specifics of the circular edges, explained in detail in [7].

Figure 3.4 illustrates the different visibility regions for edge e_{24}. In this the unshaded area represents $V_2(e_{24})$.

After we build the visibility polygons for all edges (straight and circular), we start intersecting them bottom-up from level two, until we obtain a non-empty intersection. For our example the link radius is again two, but this time the link center, shown as the unshaded portion of the free space in Figure 3.5, is sig-

nificantly smaller than both the one for straight edges in Figure 2.8 and the ALC for generalized polygons in Figure 3.3.

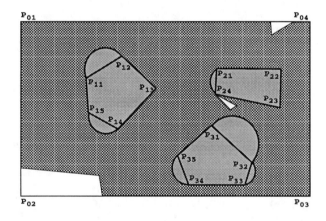

Figure 3.5: Exact link center for circular edges.

So far, in this Section, we have discussed visibility and link properties for generalized polygons. We emphasize that the procedure, described above, is valid for <u>any</u> set of convex obstacles. We never really actively used the fact that the edges are arcs of circles, as opposed to arcs of ellipses for example. All considerations about X-forms and visibility can be applied for any convex curved edges, because we can always draw and reason about the tangents. In addition for every convex obstacle we can use some reasonable heuristics to subdivide it in parts that are very closely approximated with straight lines and curved segments. Of course we need to trade-off between the closeness of the approximation and the number of edges that are introduced, because we want the program to run in reasonable time.

Thus using the considerations above and the detailed formulae in [7] we can conclude the description of our algorithm for general convex obstacles and environment. In the next section we will outline a method for dealing with non-convex two-dimensional objects.

3.2 Some Results for Non-Convex Obstacles

We start with non-convex polygons and we will point out the necessary modifications to the previous properties to accommodate this more general case. After that, using the results from Section 3.1, we will extend our treatment to general non-convex obstacles.

We will consider three general ways to deal with non-convex polygons:

- We can decompose them into convex components and use the properties of convex polygons to make conclusions for the non-convex ones.

- We can take the convex hulls of the polygons, solve the problem for them and then consider separately the concavities of the obstacles.

- We can try to deal directly with them, by modifying the actions of the algorithms for convex polygons, depending on whether we deal with convex or reflex vertices or edges.

The first method of decomposing the non-convex objects is widely used in the literature. We are looking for a subdivision where the polygon is partitioned into the minimum possible number of convex components.

There are two basic methods of minimum decomposition in convex polygons: 1) If we do not want to add new vertices to the partitioning, we can use the algorithm in [6] for decomposing a simple polygon with n sides into minimum number of convex pieces in time $O(n^3.log\ n)$. 2) If we do add additional vertices in the decomposition, we can sometimes achieve even a smaller number of convex components. Those additional vertices are called "Steiner" points, and if we allow them, we can use the algorithm in [18] for partitioning into the minimum number of convex pieces, which requires $O(n^3)$ time. After the partition we can

use all the algorithms described so far to build visibility polygons and link center for this convex pieces.

In the second method we deal with non-convex polygons by first building the convex hulls of the obstacles and applying our method for them directly. Then we apply any of the algorithms mentioned in Section 2.1 for visibility and link properties for the concavities in the original obstacles, i.e. the differences between the convex hulls and the obstacles.

Both methods above cause more computations than if we deal with the non-convex obstacles directly - the third method described in [7].

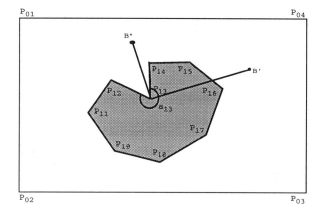

Figure 3.6: 'Inside visible' or 'not visible' reflex vertices.

In terms of the algorithm in Section 2.1 for visibility polygons from a point, the presence of reflex vertices does not introduce new generating points. If we look in Figure 3.6 we will see that for every point B in the free space, point P_{13} is either "not visible" (e.g. point B") or "inside visible" (e.g. point B ") from that point. Thus the reflex vertices just form visibility triangles on the current level and do not introduce new generating points on the next level.

Another important characteristic for non-convex polygons is that it is sufficient to be able to see the convex edges of the environment in order to see all points in E. The reasoning behind this is similar to the one in [15], where it is proven that the link center of a simple polygon P is built by considering the visibility regions only for the convex vertices of P. In fact for our case Theorem 2.1 can be reformulated by substituting "all convex" edges in E instead of "all" edges in E.

In that way the criteria for the end at level k is to be able to reach the lines of the convex edges in free space at level $k - 1$. The validity of these statements comes from the fact that a reflex edge lies in the halfplane of the neighboring edge that points to the outside of the polygon. That is why a reflex edge can be seen almost always from the first preceding convex vertex in the chain of vertices. If another obstacle is obstructing this view, the convex vertices of the obstructing obstacle will be able to see those reflex edges of the original obstacle.

These situations are illustrated and analyzed in detail in [7]. As it is shown there and from the discussion so far it is clear that the results in Section 2 for visibility polygons from a point are also valid for non-convex polygonal obstacles. The complexity of the algorithm here is the same as in Section 2.1, but in this case we can spend even less time because of the cutting-off of some of the checks for reflex vertices.

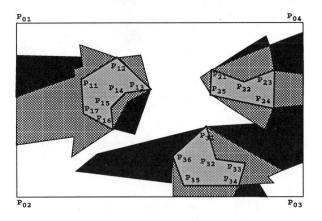

Figure 3.7: Link center for non-convex obstacles.

The approximate link center can be built analogously to the one in Section 2.2, but now again we need to consider only <u>convex</u> vertices, rather then all vertices in the environment. The same is true for the exact link center, where we will build only the visibility polygons of all convex edges in E. The X-forms and the higher level visibility polygons of an edge are treated the same way.

In Figure 3.7 the unshaded area is the exact link center for this environment, while the unshaded area and the darkly shaded one constitute the ALC for the vertices and midpoints of the edges. If we compare the two

we can see that additional reflex edges do not influence the exact link center. That is, if we have found the exact link center for the convex hulls of the obstacles and an ALC for the vertices of the non-convex obstacles, the exact link center for the non-convex obstacles is simply the intersection of the afore-mentioned link centers.

If we are dealing with general non-convex objects, not necessarily polygons, we can use the method, developed in Section 3.1. We approximate as close as possible the obstacles with "generalized non-convex polygons". (In fact the term "generalized polygons" refers to general simple polygons, not necessarily convex ones.) Now that we have generalized polygons, we can process each of the curved edges in the same manner as in Section 3.1. We point out that in this case we need to consider both the <u>convex</u> and the <u>reflex</u> <u>curved</u> edges. We illustrate the general algorithm for an approximate link center and the exact link center in an environment with non-convex obstacles in Figure 3.8. These obstacles are actually a combination of the obstacles in Figure 3.5 and Figure 3.7. Ss we can see the exact link center and the ALC here are very close to the link centers from Figure 3.5 and Figure 3.2.

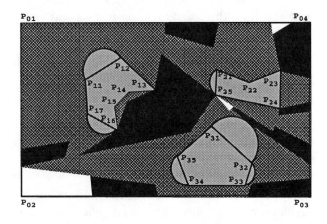

Figure 3.8: Link centers for general obstacles.

This completes our discussion of the two-dimensional case of an environment with obstacles. In the next section we will analyze the three-dimensional case.

3.3 3D Obstacles and Environment

As we have shown in this section so far, the transition from convex polygonal objects to general convex ones and then to general non-convex objects, requires only some additions and modifications to our program in Section 2. However the step from a two-dimensional to three-dimensional environment is not that obvious and easy to realize. In this Section we will discuss first an approximation to the real 3D case, which can be discussed conveniently using the 2D model we have developed. Next we will point out an approach for dealing with the real 3D case.

First of all , we need to determine the structure of our robot in 3D environments. It is clear that we cannot use only telescoping links because they are two-dimensional. One easy generalization of the existing structure is to add a prismatic joint between the manipulator's base and the ground in such a way that the translation action raises and lowers the entire planar manipulator structure normal to its original plane of action. In this way we will be able to reach any point in the free 3D environment by reaching first the horizontal plane it lies in and then reach the point itself with the telescoping links. This, of course, is not a general 3D solution, because we might be able to reach a point faster moving diagonally between horizontal planes. However this $2^{1/2}$D model of the environment is very convenient for direct implementation of our algorithm from Section 2, and we can use it to find an appropriate ALC for the 3D environment. That is why, we will concentrate first on this model.

Figure 3.9: Typical 3D environment

We will call each horizontal plane, on which the robot is working, a "horizontal layer" or "horizontal slice" of the environment. In a typical office or plant environment, the 3D objects usually have considerably uniform vertical structure. For example in Figure 3.9 we have

a sofa, a coffee table, an easy chair , a TV and a fireplace. We can describe the coffee table in $2^{1/2}$ D with three horizontal slices: one at the bottom of the legs, one at the transition from legs to surface, and one at the top of the surface of the table. Those slices are represented by the lines l_1, l_2 and l_3 respectively. Similarly we can find few horizontal slices that will define the other objects in the environment. If we take all slices for all the objects we will have a number of horizontal slices, say k , that will model the whole environment and its obstacles. Every slice at a height h is a 2D environment with obstacles, representing the horizontal section of the original obstacles at height h. For every slice we can build the link center for that environment. For example if we take the layer, corresponding to the height h_t of the top of the coffee table, the link center is depicted in Figure 3.10.

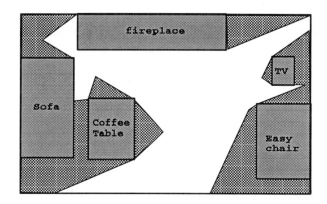

Figure 3.10: Link center for a horizontal slice.

We can apply the same procedure for every layer and find its corresponding link center. At the end we project all these link centers, that correspond to the same link radius, on one horizontal plane, and the intersection of the link centers for all layers will define a link center for the entire environment. The $2^{1/2}$ D link center will be the cylinder (or cylinders) with axial cross section the link center formed by the above intersection. The cylinder height is the same as the maximum height of the environment. This means that, if we put the fixed base of the first vertical prismatic link of the robot in any point of the base-plane link center, any point in the free space of the environment can be reached with our robot with a "sub-optimal" number of links. We say "sub-optimal", because we approximate the objects as being quite uniform vertically.

In [7] we discuss in detail this algorithm as well as the questions of: how do we choose the heights of the slices for each object, how many slices are going to be used for the overall environment with obstacles, how do we find the intersection of the link centers.

Sometimes, instead of using a large number of horizontal slices to closely model the shape of the obstacles, we can use a combination of horizontal and vertical slices. For every object we can take a few characteristic top-to-bottom, left-to-right and front-to-back slices. Of course the top-to-bottom slices are most important, because they reflect the structure of our robot. However, the vertical slices allow us to more easily discover holes or concavities in the objects, i.e. we can use them to determine <u>the shape</u> of the obstacles.

If we want to use the vertical slices for link center construction, we need to be able to have telescoping links in the vertical direction. This brings us to the problem of determining the best structure for our robot for real 3D obstacles. We need to generalize the telescoping structure in three dimensions. The reachable workspace for a telescoping link is a disk with center at the base of the link and radius equal to the maximum length of the link. In 3D we want the reachable space for the link to be a sphere. To achieve this in practice, we can define a RRP (revolute-revolute-prismatic) link, and call it a telescoping link in 3D. The joint's two revolute degrees of freedom are at the base joint, and the one prismatic stretches in and out like a telescope. This structure with the outbound of its reachable workspace is depicted in Figure 3.11, where point O is the base of the link and point P is its end-point. The telescoping link is stretched at distance r. The horizontal plane is the O_{yz} plane and the projection of the link OP on the horizontal plane is OP_1. The spherical coordinates a, b and r uniquely determine the position of point P with respect to point O.

Using those links we can generalize the procedure for building the visibility polygons described so far. We can draw tangents from a point to a 3D figure and they will correspond to reaching from the point with a 3D telescoping link. Of course now there can be an infinite number of tangents from a point to an obstacle. these are the generating lines of the cone with vertex the given point and base the element of the 3D obstacle. In fact the mutual tangents between two poly-

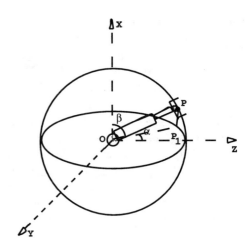

Figure 3.11: A 3D telescoping link.

hedral obstacles can also define a whole plane if the corresponding edges are intersecting lines (see [7]).

The actual implementation of the visibility algorithm in 3D requires a lot of work and depends strongly on the method of representation of the 3D objects in the algorithm. CAD software packages have been built for ray tracing of 3D objects. In [7] we have outlined few ideas for extending of the algorithms, described so far in this paper, to 3D environments. Those include determining visibility from a point using the normal vectors to the faces of the obstacles (with silhouette curves and sweeping lines), using breaks of continuity in the ray tracing for higher level visibility regions, and using visibility from a face for the link center. Some consideration is also given to non-convex and non-polyhedral bodies.

So far we have presented algorithms that build geometrically different sets of points related to the visibility and reachability of points in environments cluttered with obstacles. In the next Section we will introduce a complementary problem - the one of fast mathematical determination or estimation of the minimum necessary number of links to cover the whole free space of a given environment.

4 Best Estimates of the Number of Links

In this Section we want to derive formulas that estimate the minimum number of links for E in terms of the number of obstacles and the number of vertices for each obstacle. To do this we need to find characteristic points in the environment that can see large parts of it and then estimate the number of links needed to move between those points. This approach is closely related to the problem of guard placement in an art gallery.

4.1 The Planar Case

As described in [19], the art gallery is a simple polygon, with or without holes, and we want to find how many guards are needed and where they should be placed so that they can see the entire inside of the gallery. In Figure 4.1 the vertices of the gallery are $P_1, P_2, ..., P_{10}$.

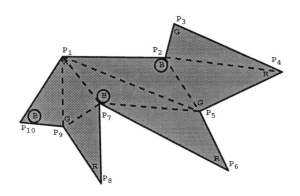

Figure 4.1: *Triangulation and three-coloring of an object.*

The basic approach taken in [19] is first to triangulate the free space E, i.e. to connect all vertices of E with straight edges, so that they form triangles that cover E. Then the vertices are three-colored in such a way that in every triangle the three different vertices have different colors. (It can be shown that there always exist triangulations for which three-coloring is possible.) Finally the guards are placed at those vertices, having the color that appears the least number of times.

If we apply this procedure to the example in Figure 4.1, where R, B and G denote red, blue and green respectively, we obtain three green, three blue and four red vertices. Consequently we can place the guards at

either the blue or the green vertices. In this case we choose the blue vertices. If we place three guards at vertices P_2, P_7 and P_{10}, they can see the entire inside of the polygon. The guard at P_2 sees triangles $P_2P_3P_4$, $P_2P_4P_5$ and $P_2P_5P_1$. The guard at P_7 sees triangles $P_7P_1P_5$, $P_7P_5P_6$, $P_7P_8P_9$ and $P_7P_9P_1$. Finally the guard at P_{10} sees triangle $P_{10}P_1P_3$. The union of all these triangles covers the polygon completely.

Using this procedure we can estimate the number of guards needed to cover the whole polygon. This follows from the basic theorem in [19], the Chvatal's Art Gallery Theorem, which states that *[n/3]* guards are occasionally necessary and always sufficient to see the entire interior of a polygon of n edges. Here the italic square brackets, *[?]* denote the largest integer less than or equal to x.

The term "polygon with holes" in [19] corresponds to our term "environment with obstacles". as we have shown in [9] and [7], a polygon with holes can be triangulated and the triangulation will have *n+2m-2* triangles.

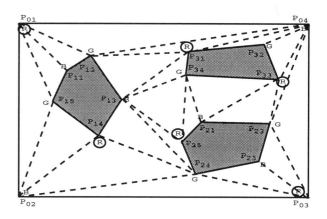

Figure 4.2: Guards placement in the environment.

In Figure 4.2 we illustrate an environment with 3 obstacles and 18 vertices overall. It is triangulated in 22 triangles. After three-coloring we have chosen the 6 red vertices and placed guards there.

The main result for polygons with holes, proved in [19], is Theorem 5.1 that states that *[(n+2h)/3]* vertex guards are sufficient to cover the whole free space E. However it is clear that this is not a necessary condition because for example everything that the guard at P_{25} can see, can also be seen by the guard at P_{14}, thus the guard at P_{25} is not needed. Conjecture 5.1 in [19] states that $[(n+h)/3]$ vertex guards will be actually sufficient, however this has not been proven in general.

In what follows, we will connect our visibility of the environment and the minimum number of links covering an environment with obstacles to the guard placement theory. We will recall from Section 1 that the minimal visibility, VMIN, denotes the minimum number of links needed to cover the whole free space E. *VMAX* (maximal visibility) is the minimum number of links that are needed so that every point in E can see every other point in E. We remember that all points in the link center have visibility of the environment equal to *VMIN* and naturally always $VMIN \leq VMAX$

In [9] and [7] we have proven a theorem that establish the connection we are looking for:

Theorem 4.1 *In the worst case, VMIN of the free space E is equal to the number of guards covering E.*

The proof of this theorem is based on a graph in which the vertices represent the guards in the gallery and an edge between two vertices indicates a relation of neighborhood. A guard is a neighbor with another one if it is a vertex in a triangle that has a side common to a triangle with a vertex at the other guard. The graph for Figure 4.2 is shown in Figure 4.3.

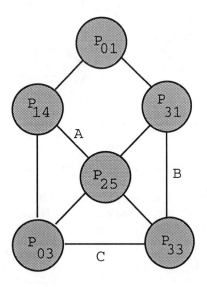

Figure 4.3: The graph of neighboring guards.

The longest possible chain of edges is one that connects all the guards. We can think of it as one big

obstacle. For example if the edges A, B and C in Figure 4.3 did not exist, we would have had one long chain visiting all the guards. In [7] we actually show an example, where the distance between two guards is *s-1*, where s is the number of guards. Thus we can show that there is always a guard Q in this conectivity graph that can reach all the rest of the guards with maximum $[(s-1)/2]$ edges. Finally every edge in the graph actually represents maximum two necessary links and thus all points in E can be seen from guard Q with a maximum of $2*[(s-1)/2] + 1 = s$ links.

We can also use similar arguments as above to prove that $VMAX = 2s$.

From Theorem 4.1 it follows that all results that are proven in [19] for polygons with holes can be directly transformed to results for *VMIN*. In particular we can prove:

Theorem 4.2 *For planar environments $VMAX \leq 2 * [(n + 2m)/3]$ and $VMIN \leq [(n + 2m)/3]$.*

It is clear that, for a given number of obstacles m, if we have smaller n we will need less links to cover E. Consequently if we can enclose the obstacles in polygons with less edges that can be placed in E without overlapping, we will get better results. This is the basis for yet another approach extensively discussed in [9] and [7]. In that we first enclose all obstacles in minimal bounding triangles. We use the fact that if we can reach all vertices of the triangles with k links, then we can reach all edges of the original convex obstacles with at most $k + 1$ links (Lemma 4.1 in [7]). Then by using graphical and algorithmic arguments, we show that:

- for a triangle in a convex environment we have $VMAX = 2$ and $VMIN = 1$;

- for two triangles $VMAX \leq 3$, $VMIN = 2$;

- for three triangles $VMAX \leq 4$, $VMIN = 2$;

By induction we can prove:

Theorem 4.3 *In an environment with m triangular obstacles we can reach all the vertices of the obstacles with $VMAX \leq m+1$ and $VMIN \leq [(m + 1)/2]+1$ links.*

If we want to include the case when the outside boundary is general convex or non-convex curve, we can formulate:

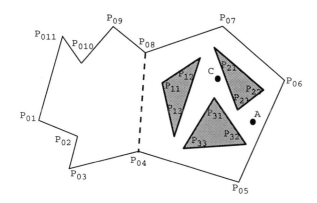

Figure 4.4: A non-convex outside boundary

Theorem 4.4 *For a general outside boundary shape we can reach all the vertices of the m triangular obstacles with:*

$$VMAX \leq m+1+2[(n-3m-1)/3] \quad and \quad VMIN \leq min(max(m+1,2[(n-3m-1)/3]),[(4n-9m+5)/6])$$

The detailed proof of this theorem can be found in [7]. It is based on analysis of the worst case situation where the environment can be subdivided into a big non-convex part S with almost no obstacles and another part with obstacles (see Figure 4.4). The part without obstacles can be treated as a general polygon with p vertices and no holes. Such polygon can be covered by $[p/3]$ guards (see [19]). For the part with holes we can use Theorem 4.3.

Sometimes triangular enclosures in 2D take a lot of space and it might happen that the resulting triangles intersect each other or the outside boundary. In that case we can try to use tighter enclosures with rectangles. Following exactly the same kind of reasoning as above we can show that Theorem 4.3 and Theorem 4.4 are also valid for rectangular obstacles.

If we cannot enclose the obstacles in rectangles, we can still apply this approach to derive upper bounds for *VMIN* and *VMAX* . If $m_i = 2k_i + 1$ or $m_i = 2k_i + 2$ for $i = 1,2,...,m$ (here m_i is the number of vertices of obstacle i) then we can use the following property: To go around a polygon with n vertices we need at most $[n/2]$ links. If we start with any vertex of the polygon, it can see two other vertices on the first level (that is with one link). These vertices can see two other vertices on the next level, and so on until we finish

all the vertices. Thus we can show that for a convex outside boundary:

Theorem 4.5 $VMAX \leq k_1 + k_2 + ... + k_m + 1$,

$VMIN \leq [(k_1 + k_2 + ... + k_m + 1)/2] + 1$ *to see all the vertices.*

In the general case we can always decompose the environment into sub convex parts with obstacles and non-convex parts with no obstacles by using the vertices that are the closest to the non-convex parts. Let us say there are s non-convex parts with no obstacles, and l separate sub convex parts with obstacles. In Figure 4, $s=1$ and $l=1$. Then for each sub convex part with obstacles we need $[n_j/3]$ links for $j = 1,2,...,l$ $(l \leq h)$ and for each non-convex part i we need a maximum of $[n_i/2]$ $(i = 1,2,...,s,$ $s \geq 0)$ links to cover it. Using Theorems 4.4 and 4.5 we can prove:

Theorem 4.6 *For a general shape of the outside boundary and the obstacles:*

$$VMAX \leq \sum_{j=1}^{l}[n_j/3] + \sum_{i=1}^{s}[n_i/2] + 1 \quad and \quad VMIN \leq [(\sum_{j=1}^{l}[n_j/3] + \sum_{i=1}^{s}[n_i/2] + 1)/2] + 1$$

We add one link at the end to cover the unreached edges between the last covered vertices. If we approximate above the expression for *VMIN* we get an estimate $[(n + 6)/4]$ which is better than $[(n + m)/3]$ for every $m \geq 2$.

In the next section we will discuss the extension of this analysis to 3D environments.

4.2   The Spatial Case

The 3D equivalent to triangulation is tetrahedralization. However as opposed to the 2D case, [19] shows that for the same polyhedron we can have tetrahedralizations with different numbers of tetrahedra. Moreover there exist non-tetrahedralizable polyhedra. In fact deciding whether or not a polyhedron can be tetrahedralized is NP-complete. Consequently we cannot readily use tetrahedralization to estimate the number of links that we need in three-dimensional environments.

However to obtain some estimates on the number of links needed to cover a 3D free space, we can use Theorem 10.1 in [19] which states that $[(2F-4)/3]$ vertex guards are sometimes necessary and always sufficient to see the exterior of a convex polyhedron of F faces for $F \geq 10$. The necessity holds for $F \geq 5$, and it is conjectured in [19] that the sufficiency also holds for that range.

Using the approach from Section 4.1 we get:

Theorem 4.7 *In three dimensional space*

VMAX $\leq 2[(2\sum_{i=1}^{m} F_i - 4m)/3]$ *and* *VMIN* $\leq [(2\sum_{i=1}^{m} F_i - 4m)/3]$ *, where F_i is the number of faces on the i -th obstacle.*

To derive another estimate of the number of links necessary to cover the free space in 3D, we can use the approach in [2]. There it is proven that using Steiner points as well as the vertices of the obstacles, we can construct an algorithm that decomposes a polytope (in general in 3D a polytope is a bounded polyhedra) with n vertices and r reflex edges into $O(n + r^2)$ tetrahedra in time $O(nr + r^2 log(r))$.

Let us denote with v_i and r_i, $(i=0,1,...,m)$ the number of vertices and reflex edges of obstacle i (0 stands for the outside boundary). Let us connect the outside boundary along a convex edge with the first obstacle (randomly chosen) along one of its convex edges. Then we connect in the same way the first obstacle to a neighboring one, etc. until we connect the one before the last obstacle to the last one. Then the connected figure will have $N = \sum_{i=0}^{m} v_i + 4m$ vertices and maximum $R = \sum_{i=0}^{m} r_i + 2m - 1$ reflex edges (if all the connections between the obstacles, except the one to the outside boundary, create reflex edges).

For example consider two cubes CB_1 and CB_2 inside a parallelepiped P, Figure 4.5. We connect the cubes with a plane through the edges $P_{13}P_{17}$ and P_{24} P_{28}, and cut them along these edges. In the same way we connect the cube CB_1 to the parallelepiped P along the edges P_{02} P_{06} and P_{11} P_{15}. The resulting figure is a polytope. Each of the cubes and the parallelepiped had eight vertices, twelve edges and six faces. The polytope has $8+8+8+4*2=32$ vertices, because at every cut we introduce 4 new vertices and we have $m=2$ cuts. Out of them there are $0 + 0 + 0 + 3 = 3$ reflex edges.

Using the consideration above and four-coloring of the vertices we can prove:

Theorem 4.8 *For a polytope with polygonal obstacles:*

VMAX $\leq 2[(O((\sum_{i=0}^{m} v_i + 4m) + (\sum_{i=0}^{m} r_i + 2m - 1)^2) + 3)/4]$, *VMIN* $\leq [(O((\sum_{i=0}^{m} v_i + 4m) + (\sum_{i=0}^{m} r_i +$

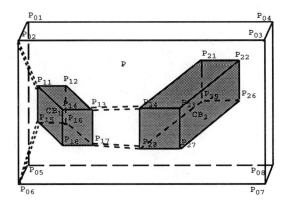

Figure 4.5: Decomposition of a 3D environment.

$2m - 1)^2) + 3)/4]$, *where v_i and r_i are the number of vertices and reflex edges respectively of the obstacle i.*

In analogy with our discussion of the 2D case, we can envision future analysis of general 3D obstacles with straight edges and faces by approximating them with triangular wedges and parallelepipeds. In the next section we propose some heuristic procedures to lower, even more, the estimates we derived in section 4.1 to the optimal values for *VMAX* and *VMIN*.

4.3 Optimizing the Number of Links

The estimates that we derived in section 4.1 are only upper bounds and in some cases they can be much higher than the actual maximums. For example for the environment in Figure 4.2, if we apply our results we obtain *VMAX* ≤ 7, *VMIN* ≤ 4. However using the algorithm for the link center from Section 2 we find that the actual values are twice smaller: VMAX= *3* and *VMIN = 2*. We can not always find the optimal numbers exactly, because that would be equivalent to finding the optimal number of guards in an art gallery, which was proven to be NP-hard in [20]. However we describe in [7] and [8] an algorithm that can find a sub optimal solution which in the majority of the cases is equal to the optimal one.

This algorithm uses triangulation of the environment to find the minimal visibility k for all edges. Every edge is a root of a tree with the vertices of depth 1 being the ones that are nodes of the triangles including this edge in all possible triangulations. For the example in Figure 4.2, edge $P_{25}P_{21}$ has in its tree vertices $P_{21}, P_{25}, P_{02}, P_{14}, P_{13}, P_{12}$ and P_{34}. The next level are the vertices that can see the vertices of the previous level with one link, and so on until we include all the vertices in the environment. In our example the only vertex on level 3 is P_{23}. However if we denote with A the intersection point of the lines containing edge $P_{21}P_{25}$ and edge $P_{23}P_{24}$, then A can see the root of our tree with one link, and it can see point P_{23} with one link. Thus the actual depth of point P_{23} is two, and the optimal number of links, p_1, that are needed to see the whole free space from edge $P_{21}P_{25}$ is two. After applying this relatively fast procedure for the other edges, we can find $p_2, p_3, ..., p_h$. Then it is clear that $VMIN = min(p_1, p_2, ..., p_h)$ and $VMAX = max(p_1, p_2, ..., p_h)$. In our example as we mentioned before $VMIN = 2$ and $VMAX = 3$.

Algorithmically this procedure can be realized using a module for determining whether an edge is visible from a given point with one link. We can optimize the running time by arranging all vertices and edges in E in a "visibility table" that can be processed in parallel to find the overall number of links. The method for building and analyzing this table is explained in detail in [7]. In addition the table allows for look-up estimates for the values of $VMIN$ and $VMAX$. Although this procedure is faster than the method in Section 2, it is not a constructive one. We don't know exactly where the link center is, we can just say which vertices lie in it. For an arbitrary point (different than the vertices) we cannot say what is its visibility and how many links are needed to reach some particular other point. Thus each of the methods has its advantages and serves its purpose.

The reason the approach in Section 4 does not give us optimal values is that we have restricted ourselves to vertex guards that are stationary points. With the telescoping links we are moving from region to region. Our visibility reflects the optimal solution for this environment, while the number of guards is just a solution for the art gallery problem. In the following algorithm we will remove the restriction on the guards to stay on the vertices of the environment, which will hopefully allow us to find the optimal number of guards for most cases. As before we first triangulate the environment, tricolor the vertices, choose the color with minimal number of occurrences and place the guards there.

In several instances the triangles that are completely visible by one vertex guard belong to another guard's patrol area. For example in Figure 4.2 triangles $P_{13}P_{14}P_{24}$ and $P_{13}P_{24}P_{25}$ are both visible to guards P_{14} and P_{25}. It is clear that if all the triangles visible to one guard are also visible to another guard, then this second guard is unnecessary. In our example everything visible to guard P_{25} is visible to P_{14}. Similarly, everything visible to guard P_{31} is visible to P_{01} or P_{33}. Consequently one of the guards P_{31} and P_{25} (say P_{25}) can be eliminated, and we still have complete cover of E, now with only five vertex guards.

Now the next step is to introduce non-vertex guards (Steiner points). Consider the guard at P_{33}. In the area covered by it there are two patches that cannot be seen by the rest of the guards. One of these patches is based on the edge $P_{21}P_{22}$, and is depicted on Figure 4.6 as the shaded area *1*.

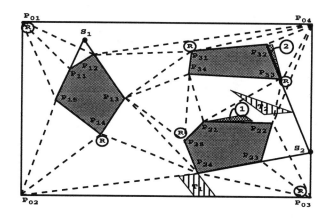

Figure 4.6: Optimal number of guards.

Fig.4.6 Optimal number of guards.

If we move the guard at P_{01} to a position from which it can see $P_{12}P_{13}$, the guards at P_{33} and P_{31} will be redundant from this side. The new guard should also be able to see $P_{11}P_{15}$, consequently we can place it at the intersection of the lines $P_{15}P_{11}$ and $P_{13}P_{12}$. If this point is in E, which in our case is conveniently true, this gives point S_1. Similarly, the other area that is seen by P_{33}, but not by the other guards, is the shaded area *2* based on the edge $P_{32}P_{33}$. Therefore we can move the guard in P_{03} to point S_2, the intersection of the lines $P_{24}P_{23}$ and $P_{32}P_{33}$, which is also inside E. Finally with guards at S_1, S_2 and $S_3 = P_{14}$ we can cover the whole environment E. This is the optimal number of guards because, by using arguments similar to those above we can show that it is not possible to reduce the number of guards by moving them around. Detailed description of this algorithm is given in [7].

We would like to point out that although we found the minimum possible number of guards for this environment, the visibility here is actually even smaller - it is two. In fact every point that can see all the guards with only one link will be able to see the entire free space with a maximum of two links. In Figure 4.6 those points form the horizontally shaded areas T_1 and T_2. These areas are actually part of the link center for this environment, as we can see in Figure 2.8 from Section 2.2. Of course they do not define the entire link center, because there might be other possible triples of guards covering the entire free space. This is still a fast way to get parts of the link center, provided we have managed to optimize the number of guards.

So far we have dealt with the situation when the environment is fixed and we are designing a stationary robot to operate in it. In the next chapter we present some extensions that look at our basic problem from a different point of view.

5 Extensions of the Basic Design Problem

There are several directions in which we can extend our analysis. Those including allowing the robot and/or obstacles to change their shape and/or position in the environment. The case of moving robots and obstacles as well as reconfigurable robots is discussed in [7] We will briefly mention here that all estimates in Section 4 are valid for both moving and stationary obstacles. We can look at the problem of moving obstacles with a fixed robot, as dual to the problem of fixed obstacles and a moving robot, because all that matters is the relative position of the obstacles and the robot. For a complicated environment with a large number of obstacles we may have a large link radius. That means that if we place the robot in a point of the link center, we will need a large number of telescoping links to reach all the points in the environment. From a practical point of view such a robot may be difficult to build because of the masses of the links, the actuator mechanisms, controls and so on. In that case we can introduce limited

mobility for more manageable structure of the robot. Alternatively we can use multiple robots operating in sub-spaces of the environment. Those robots can have reconfigurable structure (which is easy to achieve with our telescoping links) and can cooperate at their common boundaries.

In this paper we will concentrate on the situation that arises when we try to design both the robot and the environment simultaneously. Let us say that we have a set of objects (e.g. machines) that we want to place in an environment (e.g. a factory floor) that is going to be completely automated with robots. In this case we might want to take advantage of the fact that we know the basic structural elements of the robots (e.g. telescoping links) and place the objects accordingly.

To start with, the objects in the environment can have different shapes and dimensions. It is difficult to reason directly about the general case of arbitrary obstacles and in our formulation the exact initial position of the objects is not that important. That is why we will outline here an approach that uses approximations of the obstacles with rectangles, in 2D (or parallelepipeds in 3D), to find simultaneously the best design of the environment and the robot. The reason we are using rectangles is because they are easy to build and rectangular shapes are easier to place and analyze, especially if we think of the outside boundary of the environment as a rectangle. In fact we can use rotating calipers (see [23]) to find in linear time the smallest area rectangle enclosing given polygon.

If we assume that we have enclosed all the obstacles in rectangles and we have a rectangular bounding environment, the problem now is to place those rectangles in the environment in such a way that the robot design problem gives the best result overall. We will describe a method to place these rectangles, inside the boundary rectangle with sides parallel to each other.

We introduce the following notations: R_i denotes the enclosing rectangle for the obstacle OB_i, with $i = 1$, $2, ..., m$. For every rectangle R_i, a_i and b_i are the lengths of the sides, where $a_i \geq b_i$. We renumber the obstacles so that they are in decreasing order with respect to b_i. R_0 is the rectangular approximation of the outside boundary, a_0 and b_0 are the lengths of its sides, and $a_0 \geq b_0$. A simple condition that can be checked to verify that all enclosing rectangles can fit in the boundary rectangle, is $A : \sum_{i=1}^{m} a_i b_i < a_0 b_0$, i.e., the sum of the areas of the rectangular obstacles must be less than the area of the rectangular boundary. For the rest of this discussion we will refer to the enclosing rectangular obstacles as "obstacles" and to the rectangular boundary as "boundary". Even though condition A might be satisfied, we might not be able to place the obstacles inside the boundary without overlapping. We will assume that there is enough room for the obstacles. This assumption is a reasonable one, because in a realistic environment there is usually ample space for transportation and manipulation.

Let us first sort the sides b_i of the rectangles in decreasing order. We try to place the rectangles in the environment in horizontal layers where each layer has obstacles with similar height (see Figure 5.1). In other words we take the obstacle with the largest height and place it in the lower left corner so that there is distance e from the bottom and the left side of the boundary. This distance is such that the robot can fit and reach around the obstacle. We will always leave this distance between the obstacles and between the layers also.

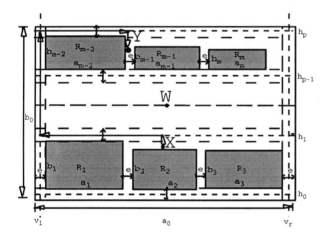

Figure 5.1: Placement of rectangular obstacles.

We place the next obstacle to the right of the first one on the same horizontal base at distance e from the first one. Similarly we place the next obstacles until we get to the right edge of the boundary. We make sure that there is at least e distance from the last obstacle in the layer to the right edge of the boundary. Then we continue with the next layer and so on, until we place all the obstacles. At the end if there is empty space

above the last layer we can translate half of the layers up in order to free the center or we can space the layers more generously.

For each layer we will have a condition:

$H : \sum_{s=l_{i-1}+1}^{l_i} a_s + (l_i - l_{i-1} + 1) * \epsilon \le a_0$ where l_i is the last index of the rectangles in layer i . The height of each layer is b_i and the vertical condition for existence of such placement is:

$U : \sum_{i=1}^{p} b_i + (p+1) * \epsilon \le b_0$ where p is the number of layers. If condition U is satisfied we can fit the obstacles using this algorithm.

We will label the lines along the middle of the horizontal passages between the layers "horizontal freeways" h_i , $(i=0,1,...,p)$. We define the lines v_l and v_r. as "vertical freeways". Here v_l is the mid-line between the left edge of the boundary and the line defined by the left edges of the first obstacles for all layers. In Figure 5.1 v_r is the mid-line between the right edge of the outside boundary and the line defined by the furthest to the right , right edge of the last obstacles for all layers. We will use the horizontal and the vertical freeways to reach from point to point in the free space.

The link distance between any two points in the free space is maximum five. If we take points X and Y in Figure 5.1 for example, we need one link to reach from X to the horizontal freeway above it, a second link to go to the extreme left (or right) on the freeway, a third link to go along the vertical freeway to the horizontal freeway above point Y, a fourth link to move along the freeway until we get directly above Y and a fifth link to actually reach Y.

Alternatively, we can always choose a point that can reach all other points in the free space with three links only. For example we can take any point Z along v_l. Then to reach point X for example we can move with one link along the v_l until we reach the horizontal freeway corresponding to X. Then we use a second link to move along this horizontal freeway until we get above X and the third link reaches X. Thus we have proven:

Theorem 5.1 *For a planar environment, if we can place the obstacles in such a way that condition U is satisfied then: VMIN ≤ 3 and VMAX ≤ 5.*

If we want to put the base of the robot in a more central location we can take a point at the middle of the horizontal freeway that is closest to the middle of the vertical sides of the outside boundary, for example point W in Figure 5.1. In that case we need a maximum of four links to reach all the points in the free space–one to reach the point of intersection of the horizontal freeway with v_l and then three more to reach all the other points. However, if this is an important condition for us, we can slightly modify our algorithm by introducing a vertical freeway v_c through the geometric center of the bounding rectangle, thus achieving maximum 3 links for the central point.

We note that once we have distributed the obstacles in the horizontal layers, we can reshuffle the obstacles in each layer so that they have more convenient placing within the corresponding layer. We can also incorporate some practical considerations in the initial numeration of the obstacles.

Sometimes it is also convenient to place the obstacles along the whole boundary of the environment. In that case we can use rectangular layers instead of horizontal ones. If we can place the obstacles in p rectangular layers, Theorem 5.2 in [7] shows that $VMIN \leq [(p+1)/2] + 3$and $VMAX \leq p + 3$.

There is a very important difference between the estimates in Chapter 4 and Theorem 5.1. In our discussion we have used the term "links" to denote "telescoping links", i.e. links with two degrees of freedom (one revolute and one prismatic). However the algorithms we discussed in this chapter only need links with one degree of freedom (prismatic links). We are moving along freeways that are perpendicular to each other and we do not really need the rotation of the links. So if we use a central highway and if condition U is satisfied, we can design the environment in such a way that no matter how many obstacles we have, a manipulator with three prismatic links is sufficient to cover the whole environment. This is a very strong result, even more because prismatic structures are easier to analyze and control than revolute.

This result can also be directly transferred to 3D environments. In 3D we will approximate the obstacles with parallelepipeds (or if possible with cubes). This approximation is also an open problem, and we do not know of substantial results and algorithms that achieve it. However, if we can approximate the original obstacles and the outside boundary with parallelepipeds, we can subdivide the environment into layers parallel to the X-Y plane, and arrange the obstacles according to

their height (which can be the smallest dimension of the parallelepipeds) and place them in order in layers. We can prove Theorems 5.1 and 5.2 for this case, the same way as in 2D.

6   Future Directions

The problem we discussed in this paper was: given an environment with obstacles, what is the optimal design of a robot that can reach everywhere in this environment without collision with the obstacles. There are several open problems which can be a topic for future research:

- the problem of 3D modeling of the obstacles for convenient application of the link center algorithm to obstacles with general shape;

- the problem of deriving estimates for the number of links in 3D using enclosures in simpler shaped obstacles;

- the problem of deriving a general condition, through optimization, for best fitting of rectangular obstacles in the environment;

In addition to those problems, it would be interesting to consider further generalizations of the problems discussed in this paper. Those include:

investigate the affect of the kinematic and dynamic properties of the telescoping links. One could also consider the affect of joint limits and singularities on the algorithms in Section 2.

incorporate uncertainty and imprecise knowledge of the shape and the position of the obstacles.

generalize to any n-dimensional space by using either topology of semi-algebraic sets.

The ideas for robot placement and design that we discussed can be used in space robotics, factory automation , construction industry and transportation.

7   References

[1] David Bennett and John Hollerbach, "Identifying the Kinematics of Robots and their Tasks," *Proceedings of the IEEE International Conference on Robotics and Automation*, 1989, pp.580-586.

[2] Bernard Chazelle and Leonidas Palios, "Triangulating a Nonconvex Polytope," *Proceedings 5th ACM*

Symposium on Computational Geometry, 1989, pp.393-400.

[3] P.Chedmail and Ph.Wenger, "Design and Positioning of a Robot in an Environment with Obstacles using Optimal Research," *Proceedings of the IEEE International Conference on Robotics and Automation*, 1989, pp.1069-1074.

[4] Yao-Chon Chen and Mathukumalli Vidyasagar, "Optimal Trajectory Planning for Planar n-Link Revolute Manipulators in the Presence of Obstacles," *Proceedings of the IEEE International Conference on Robotics and Automation*, 1988, pp.202-208.

[5] Gregory Chirikjian and Joel Burdick, "Parallel Formulation of the Inverse Kinematics of Modular Hyper-Redundant Manipulators," *Proceedings of the IEEE International Conference on Robotics and Automation*, 1991, pp.708-713.

[6] Marc Keil, "Decomposing a Polygon into Simpler Components," *SIAM Journal on Computing*, Vol.14, No.4, 1985, pp.789-817.

[7] Kolarov, K. *Optimal geometric Design of Robots for Environments with Obstacles*, PhD Thesis, Stanford University, 1992.

[8] Kolarov, K. and Roth, B. "On the Number of Links and Placement of Telescoping Manipulators in an Environment with Obstacles", *Proc. 5th International Conference on Advanced Robotics*, Pisa, Italy, 1991, pp.988-993.

[9] Kolarov, K. and Roth, B. "Best Estimates for the Construction of Robots in Environments with Obstacles", *IEEE International Conference on Robotics and Automation*, Nice, France, 1992.

[10] Kolarov, K. and Roth, B. "Best Placement of Telescoping Robots in Environments with Obstacles", Bangalore, India, 1993.

[11] Sridhar Kota and Tushchai Chuenchom, "Adjustable Robotic Mechanisms for Low-Cost Automation," *Cams, Gears, Robot and Mechanism Design*, ASME 1990, pp.297-306.

[12] Jean-Claude Latombe, *Robot Motion Planning*, Kluwer Academic Publishers, 1991.

[13] Jean-Paul Laumond, "Obstacle Growing in a Non-Polygonal World," *Information Processing Letters 25*, April 20, 1987, pp.41-50.

[14] D.Lee, I.Chen, "Display of Visible Edges of a Set of Convex Polygons," *Computational Geometry*, G.Toussaint (Ed.),1985, pp.249-265.

[15] W.Lenhart, R.Pollack, J.Sack, R.Seidel, M.Sharir, S.Suri, G.Toussaint, S.Whitesides and C.Yap, "Computing the Link Center of a Simple Polygon," *Proceedings of thc 63rd ACM Symposium on Computational Geometry*, 1987, pp.1-10.

[16] O.Ma and J.Angeles, "Optimum Architecture Design of Platform Manipulators," *Proceedings of the 4th International Conference on Advanced Robotics*, 1991, pp.1130-1135.

[17] J.A. Pamanes and Said Zeghloul, "Optimal Placement of Robotic Manipulators Using Multiple Kinematic Criteria," *Proceedings of the IEEE International Conference on Robotics and Automation*, 1991, pp.933-938.

[18] John Reif and James Storer, "Shortest Paths in Euclidean Space with Polyhedral Obstacles," *Technical Report CS-85-12*, Computer Science Department, Brandeis University, 1985.

[19] Joseph O'Rourke, *Art Gallery Theorems and Algorithms*, Oxford University Press, 1987.

[20] Joseph O'Rourke and Kenneth Supowit, "Some NP-Hard Polygon Decomposition Problems," *IEEE Transactions on Information Theory*, Vol. IT-29, No.2, March 1983, pp.181-190.

[21] E.Sandgren and S.Venkataraman, "Robot Path Planning and Obstacle Avoidance: A Design Optimization Approach," *15th Design Automation Conference* 1989, pp.169-175.

[22] Subhash Suri and Joseph O'Rourke, "Worst-Case Optimal Algorithms For Constructing Visibility Polygons with Holes," *Proceedings of the 2nd ACM Symposium on Computational Geometry*, 1986, pp.14-23.

[23] T.Toussaint, "Solving Geometric Problems with the 'Rotating Calipers'," *Proceedings IEEE MELECOM '83*, Greece.

Design of Modular Fault Tolerant Manipulators

Christiaan J. J. Paredis, *Carnegie Mellon University, Pittsburgh, PA, USA*
Pradeep K. Khosla, *Carnegie Mellon University, Pittsburgh, PA, USA*

In this paper, we deal with two important issues in relation to modular reconfigurable manipulators, namely, the determination of the modular assembly configuration optimally suited to perform a specific task and the synthesis of fault tolerant systems. We present a numerical approach yielding an assembly configuration that satisfies four kinematic task requirements: reachability, joint limits, obstacle avoidance and measure of isotropy. Further, because critical missions may involve high costs if the mission were to fail due to a failure in the manipulator system, we address the property of fault tolerance in more detail. We prove the existence of fault tolerant manipulators and develop an analysis tool to determine the fault tolerant work space. We also derive design templates for spatial fault tolerant manipulators. For general purpose manipulators two redundant degrees-of-freedom are needed for every order of fault tolerance. However, we show that only one degree of redundancy is sufficient for task specific fault tolerance.

1 Introduction

Conventional (serial or parallel link) manipulators are often considered to be general-purpose and flexible systems. Unfortunately, these systems are not general purpose. In order to understand this, consider a computer which is a general purpose computing engine if it can compute a computable function. Following a similar logic, a manipulator will be general purpose if it could do a 'doable' task. In defining a general purpose manipulator, one has, of course, to define a 'doable' task first. For the time being, let us avoid this open issue and consider two tasks that two different manipulators can perform, but that cannot be performed by either manipulator separately. If this is the case, then one may conclude that none of the above two manipulators are general purpose. So if one has to define a

general purpose manipulator, then one has first to define a criterion for 'doable' tasks (like 'doability'). Such a definition may lead to the development of models of 'doability' (like computability) and maybe to Turing-like machine models of manipulators. While such a development would certainly do a lot for advancing the state-of-the-art in manipulators, it is not our intention to address this general problem.

In order to make the problem tractable, let us define a set of tasks that we would like to perform with a manipulator. Let us also define a set of basic modules (consisting of joints and links) that we may combine to create various manipulators. Finally, let us assume the existence of a methodology that will accept a task (in the form of a program or as a set of requirements) as input and find a manipulator that can be created from the given set of modules to perform the task. This scenario is described in Fig. 1, and it allows us to put forth one possible definition of a general-purpose manipulator.

General-purpose Manipulator: If for every task in the set of tasks, it is possible to find a manipulator that can be created from the given set of modules to do the task, then we define the system of modules (or the system of all possible manipulators) as general purpose with respect to the set of tasks. We will call such a system a Reconfigurable Modular Manipulator System (RMMS).

Note that the above definition does not require us to define a set of all possible 'doable' tasks nor does it require us to define the criteria for determining 'doability' even though that is the ultimate goal of our research.

Our past work has addressed the development of the modules and the technology for the RMMS [21]. The RMMS has many potential applications in both hazardous and industrial environments. It puts forth the idea of designing a specific manipulator for a task and

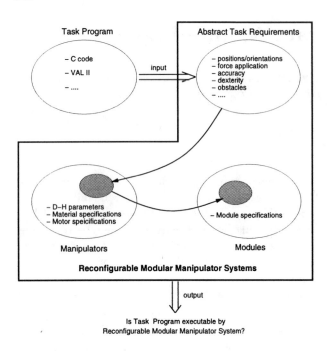

Figure 1: *Definition of a general purpose manipulator.*

also the notion of the user writing device (or manipulator) independent code. The RMMS raises several theoretical issues and it is our aim to address one of these in this paper. Specifically, we describe a design methodology that accepts a task specification as its input, determines a kinematic configuration of the desired manipulator and selects the modules to create this manipulator.

In order to support the current practice of picking the best configuration amongst available robots, several expert systems have been built to aid the user or the applications development engineer [15]. A straightforward extension of this selection process has been the inclusion of the design of new manipulators, optimally suited for a specific application [1, 16]. A totally different approach to the robot design problem finds its roots in simulation. A variety of commercial robot simulation packages are currently available [5, 22], providing designers with convenient tools to quickly check the implications of different design decisions. In general, however, these simulation packages still require a human to make the design decisions. Finally, a third way of dealing with the problem of robot design, has grown out of the field of mechanism design [13, 19]. Unlike the rule based expert systems, these programs are al-

gorithmic in nature. Commonly, the design process is subdivided in two stages: form synthesis and dimensional synthesis. The first stage is usually performed by searching over the set of feasible mechanism types, while the second stage consists of optimizing the set of dimensional parameters.

The approach we propose in this paper differs from the methods listed above, because we are specifically interested in modular manipulators. The interest in modular manipulators has grown steadily over the last decade [6, 24], and several related research issues have been addressed [2, 3, 7, 10, 14, 18]. However, the problem of determining the modular configuration optimally suited for one specific task, has never been addressed before to the best of our knowledge. In this paper, we investigate modular design from kinematic task requirements. These requirements affect only the kinematic structure of the manipulator, while dynamic requirements affect both its kinematic and dynamic structure. Examples of kinematic requirements are work space volume, maximum reach, and maximum positional error. Examples of dynamic requirements are maximum pay-load, maximum joint velocities, and maximum joint accelerations. Just as task requirements can be classified as kinematic or dynamic requirements, the design procedure can also be split into two parts: kinematic design and dynamic design [10]. Kinematic design determines the kinematic structure of the manipulator, while dynamic design determines the dynamic configuration. However, the dynamic design may require a change in kinematic structure, and thus a few iterations may be necessary to find a manipulator that satisfies both kinematic and dynamic requirements.

In the first part of this paper, we only consider reachability, joint limit, obstacle avoidance, and measure of isotropy requirements. A numerical procedure is proposed which determines a modular assembly configuration that meets all the requirements. In the second part, we focus our attention on one additional requirement, namely, fault tolerance. Recently, fault tolerant (or failure tolerant) robotics has been the subject of several publications [11, 23], in which different aspects of the problem are addressed. Visinsky et al. [23] propose a framework to include failure detection in fault tolerant robot systems. Lewis and Maciejewski [11], on the other hand, discuss the importance of the controller and the redundancy resolution. In this paper, we focus our attention on the design of fault tolerant manipulators. We define fault tolerance as the ability to continue the performance of a task even after immobilization of a joint due to failure. Several properties of fault tolerant manipulators are discussed and are illustrated with examples.

2 Kinematic Design: Preliminary Results

2.1 Problem Statement

The problem solved in this section is the determination of a modular assembly configuration, that satisfies all the kinematic task requirements. These requirements are that the manipulator must be able to reach a specified set of positions/orientations, p_j, (reachability requirement), without violating the motion constraints of the joint modules (joint limit requirement), and without colliding with any parallelepiped-shaped obstacles in the work space (obstacle avoidance requirement). Moreover, at the positions/orientations, p_j, the measure of isotropy, must be larger than a user specified minimum (measure of isotropy requirement).

In Section 2.2 and 2.3, we develop a numerical procedure to solve this design problem. To facilitate the implementation of our approach, we consider six types of modules, as shown in Fig. 2. The choice of these specific modules guarantees a simple conversion from the module dimensions and orientations into the Denavit-Hartenberg (D-H) parameters of the resulting manipulator (a set of 3 D-H parameters per degree-of-freedom, determines unambiguously the kinematic structure of any serial link manipulator). It has been shown by Kelmar and Khosla [7] that this conversion can be achieved for modules of arbitrary geometry. The actual number of different modules considered for the design can be far larger than six, due to variations in the parameterized dimensions, and our design method is general enough to allow for this.

We also require that the robot base be fixed and known, that the first joint module be of type 0 or 1 (i.e. the first joint axis is vertical), and that the last module be a wrist with three axes intersecting at a point. These restrictions result from our implementation of the inverse kinematics and can be relaxed by using iterative solutions to the inverse kinematics problem, as proposed in [2]. Also, the requirement that the robot

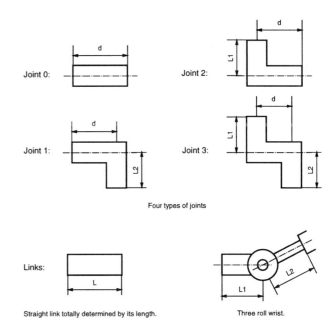

Figure 2: *The types of manipulator modules that are considered in the design procedure.*

base be fixed and known, can be relaxed as was shown by Kim [8], who addressed the problem of kinematic synthesis and base position synthesis simultaneously.

Finally, we would like to point out that this design problem can possibly have more than one solution. Consider the design of a 2-DOF planar manipulator, with link lengths L_1 and L_2, satisfying the task requirement that the manipulator should be able to reach a point located behind an obstacle without violating the joint limits, as is illustrated in Fig. 3. The region of the (L_1, L_2)-plane containing the solutions is bounded by the curves labeled c, d and e. All the manipulators inside this region satisfy all the design requirements and, therefore, are all equally good with respect to these requirements.

2.2 Solution Approach

In this section, we evaluate different approaches to the problem of determining the modular configuration, given some kinematic task specifications. The problem can be interpreted as a mapping from task specifications into constraints in the modular configuration space, as is shown in Fig. 1. This mapping is nontrivial due to the highly nonlinear character of the kinematic

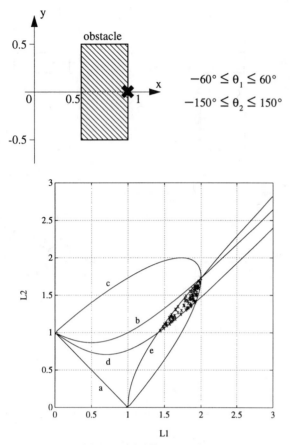

a: Bound due to reachability constraint.

b: Bound due to joint limits on θ_1, $|\theta_1| \leq \pi/3$

c: Bound due to joint limits on θ_2, $|\theta_2| \leq 5\pi/6$

d: Bound due to the obstacle, $|\theta_1| \geq \pi/4$

e: Bound due to the obstacle, $L_1^2 \geq L_2^2 + 1$

Figure 3: *A two-DOF planar design example with solution.*

relations and due to the complexity of the task specifications. Krishnan [10], therefore, suggested to solve the inverse problem first, namely, to analyze which task requirements are satisfied by a given modular configuration. This information is stored in lookup tables, which can then be used in a search procedure. One obvious disadvantage to this approach is the combinatorial explosion in the number of different configurations. Let the number of different modules available be N, and let R be the number of relative orientations in which one module can be mounted on the previous module. The total number of configurations that can be obtained from this set of modules is:

$$Num = \sum_{i=1}^{N} R^{(i-1)} \frac{N!}{(N-i)!} \tag{1}$$

This approach would therefore require a very large amount of memory storage for the lookup tables. Also, adding a new module requires that all the lookup tables be updated.

A different approach is to first design a manipulator defined by a set of continuously varying D-H parameters, as proposed in [17], and then transform this design into a modular configuration. The main problem here is the discretization of the continuous solution. As is known from integer programming, simply taking the discrete configuration nearest to the continuous solution might result in an infeasible solution. Therefore, we suggest working directly in the modular configuration space. Of course, an exhaustive search in this space suffers from combinatorial explosion in much the same way the look up table approach does. However, the efficiency of the search procedure can be improved drastically by 'guiding' the search to the most promising regions of the search space. Instead of answering the question whether a certain modular design meets all the task requirements with a simple 'yes' or 'no', we estimate the 'goodness' or 'badness' of the design, i.e., "How far are we away from a solution?" Guiding the search then means focusing the search effort on directions of decreasing 'badness'. This approach is usually referred to as a heuristic search technique [20], because in general, it is impossible to compute the 'badness', or the distance to the nearest solution, exactly. The heuristic function only estimates this distance so that it is possible that, locally, the heuristic decreases even though the actual distance to a solution increases. This corresponds to a local minimum in optimization terminology. To overcome this inadequacy, we have to employ a search method, such as simulated annealing, that allows for local hill climbing.

Simulated annealing was first proposed by Kirkpatrick [9] as a combinatorial optimization algorithm. The method is a random iterative improvement algorithm with the modification that, under certain conditions, an increase in the heuristic function is accepted (In order to be compatible with the standard terminology in discussions of simulated annealing, we use the term objective function instead of heuristic function, henceforth). A new trial configuration is generated randomly in the neighborhood of the current configuration. The condition for acceptance of this trial configuration is:

$$\begin{cases} \Delta F_{obj} \leq 0 & \Rightarrow \text{accept} \\ \exp(-\Delta F_{obj}/T) > \text{random}[0,1) & \Rightarrow \text{accept} \end{cases} \quad (2)$$

which depends on a control variable, T, the temperature. The algorithm is started at a high temperature for which most new configurations are accepted. After each iteration the temperature is decreased until no new acceptable configuration can be found. The search is then frozen. We adapted this basic algorithm to include the special properties of our objective function. In particular, the algorithm is stopped when a new trial configuration has an objective function value equal to zero, even if the search is not yet frozen. We know that a configuration with a 'badness' of zero satisfies all the design requirements.

2.3 Computation of the Objective Function

The goal in this section is to find an objective function which is zero when all the design specifications are satisfied and which is otherwise proportional to the amount of violation of these specifications. Minimizing the objective function by simulated annealing corresponds then to a search, guided towards the most promising regions of the search space. For a given modular configuration, the corresponding objective function can also be interpreted as a penalty for violating certain task specifications. The goal of the search is then to find a configuration with zero penalty.

We now propose a methodology for constructing a penalty function. Let us first define some terminology. A configuration is the set of D-H parameters which determines unambiguously the kinematic structure of a modular manipulator configuration. A posture is the position of a manipulator corresponding to a specific set of joint angles. A task point is a specified position/orientation of the end effector that the manipulator must be able to reach without violating the other task requirements.

By taking a closer look at the task requirements, one notices that all the requirements are defined for a specific configuration in a specific posture reaching for a specific task point. The penalty for such a posture should be defined such that, if any single requirement is not satisfied, the penalty for the posture is positive. This can be achieved by defining a nonnegative penalty for each task requirement, as described in [17], and summing these penalties for a posture:

$$P_{post} = \sum_{requirements} P_{req} \quad (3)$$

The task penalty is now defined as the minimum over all the posture penalties, so that it is zero when all the task requirements are satisfied for at least one posture:

$$P_{task} = \min_{post} P_{post} = \min_{post}(\sum_{req} P_{req}) \quad (4)$$

Finally, the total penalty of a manipulator configuration is given by the sum of all the task penalties:

$$P_{conf} = \sum_{tasks} P_{task} = \sum_{tasks} (\min_{post}(\sum_{req} P_{req})) \quad (5)$$

2.4 Example

The example solved in this section is the design of a seven degree-of-freedom manipulator that is able to reach eight different task points, located in an environment which includes five obstacles. The joint modules have a limited motion range and the task points must be reached in a posture with a measure of isotropy of at least 0.6. It is also required that the manipulator consists of a subset of the twenty different modules (4 joints, 1 wrist, 16 links: we also consider a link of length zero). It is assumed, at this point, that an unlimited number of each type of module is available, so that a design which includes the same module type several times is acceptable. All the task requirements are summarized in the input file in Fig. 4.

A quick calculation gives us an idea of the extent of the search space. A 7-DOF manipulator consists of five links, four joints and one wrist, and is further determined by five angles, specifying the relative orientations of the joint modules. Taking into account the restrictions, that the first joint module must be of type zero or one and the last module must be a wrist, the number of configurations in the search space equals:

$$(\#\text{links})^5(\#\text{joints})^3(\#\text{joints of type 0 or 1})$$
$$(\#\text{wrists})(\#\text{relative orientations})^5 \quad (6)$$
$$= 16^5 \cdot 4^3 \cdot 2 \cdot 1 \cdot 8^5 \approx 4.4 \times 10^{12}$$

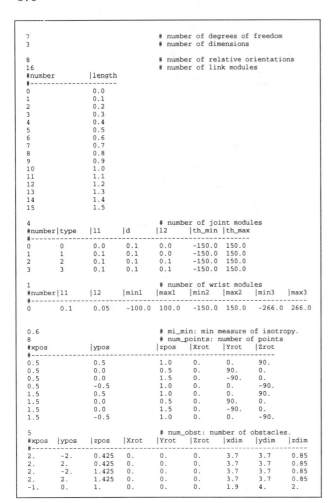

Figure 4: *The input file of the 7-DOF example.*

The content of Figure 4 (input file):

```
7                              # number of degrees of freedom
3                              # number of dimensions

8                              # number of relative orientations
16                             # number of link modules
#number       |length
#--------------------
0             0.0
1             0.1
2             0.2
3             0.3
4             0.4
5             0.5
6             0.6
7             0.7
8             0.8
9             0.9
10            1.0
11            1.1
12            1.2
13            1.3
14            1.4
15            1.5

4                              # number of joint modules
#number|type  |l1    |d     |l2     |th_min |th_max
#------------------------------------------------------
0      0      0.0    0.1    0.0    -150.0  150.0
1      1      0.1    0.1    0.0    -150.0  150.0
2      2      0.1    0.1    0.1    -150.0  150.0
3      3      0.1    0.1    0.1    -150.0  150.0

1                              # number of wrist modules
#number|l1    |l2    |min1   |max1  |min2   |max2  |min3    |max3
#----------------------------------------------------------------
0      0.1    0.05  -100.0  100.0  -150.0  150.0  -266.0   266.0

0.6                            # mi_min: min measure of isotropy.
8                              # num_points: number of points
#xpos         |ypos           |zpos   |Xrot   |Yrot   |Zrot
#---------------------------------------------------------------
0.5           0.5             1.0    0.     0.     90.
0.5           0.0             0.5    0.     90.    0.
0.5           0.0             1.5    0.     -90.   0.
0.5           -0.5            1.0    0.     0.     -90.
1.5           0.5             1.0    0.     0.     90.
1.5           0.0             0.5    0.     90.    0.
1.5           0.0             1.5    0.     -90.   0.
1.5           -0.5            1.0    0.     0.     -90.

5                              # num_obst: number of obstacles.
#xpos  |ypos  |zpos  |Xrot   |Yrot   |Zrot   |xdim   |ydim   |zdim
#----------------------------------------------------------------
2.     -2.    0.425  0.     0.     0.     3.7    3.7    0.85
2.     2.     0.425  0.     0.     0.     3.7    3.7    0.85
2.     -2.    1.425  0.     0.     0.     3.7    3.7    0.85
2.     2.     1.425  0.     0.     0.     3.7    3.7    0.85
-1.    0.     1.     0.     0.     0.     1.9    4.     2.
```

	link#	angle#	joint#
1	10	4	1
2	14	4	3
3	3	3	0
4	12	6	3
5	14	0	—

Table 1: *Module number of 7-DOF design.*

Starting from a random initial guess, the simulated annealing algorithm evaluated on the average only about 2700 configurations before finding a solution. One of these solutions is tabulated in Table 1 and Table 2. It is a SCARA-like manipulator with a nearly spherical joint at the end of the second link. The offset

DOF	d_i	a_i	α_i
1	1.1	1.6	180°
2	0.1	0.0	90°
3	1.8	0.0	-90°
4	0.1	0.0	90°
5	1.6	0.0	90°
6	0.0	0.0	-90°
7	0.05	0.0	—

Table 2: *D-H parameters of 7-DOF design.*

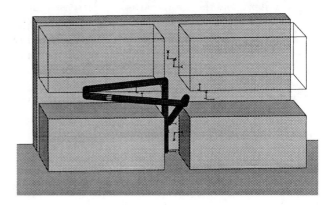

Figure 5: *The manipulator reaching point two while avoiding all the obstacles.*

along the first axis is 1 meter and the first twist angle is 180°, so that the first and second link move in a horizontal plane exactly between the four obstacles. Because of the spherical joint, link 3 can move either in a horizontal or a vertical plane, so that all the task points can be reached without hitting any obstacles, as shown in Fig. 5.

3 General Purpose Fault Tolerant Manipulators

In the rest of this paper, we focus our attention on one additional task requirement, namely, fault tolerance. To set the stage for our development, we define the following properties of fault tolerant manipulators [2]:

- **General Purpose Fault Tolerant Manipulator:** An n-DOF manipulator that will still be able to meet the task specifications, even if any one or more of its joints fail and are frozen at any arbitrary joint angles.

- **k-Reduced Order Derivative (k-ROD):** When k joints of an n-DOF manipulator fail, the effective number of joints is $(n-k)$. The resulting faulty manipulator is called a k-reduced order derivative.

- **Order of Fault Tolerance:** An n-DOF manipulator is fault tolerant of the k-th order, if and only if all k-reduced order derivatives can still perform the specified task. We call the manipulator k-fault-tolerant.

- **Fault Tolerant Work Space (FTWS):** The fault tolerant work space of a k-fault tolerant manipulator is the set of points reachable by all possible k-reduced order derivatives.

These definitions differ from the concept of fault tolerance as proposed by Maciejewski [12]. Instead of attributing the property of fault tolerance to a *manipulator*, he quantifies a measure of fault tolerance for a manipulator *posture* and describes a technique to determine the optimal fault tolerant posture, based on the singular value decomposition of the Jacobian matrix. If a joint fails in this optimal posture, the resulting reduced order derivative will have maximum possible dexterity. However, a failure at a different angle may make the execution of the task impossible.

In the rest of this section, if no specific task is mentioned, it is assumed that the task consists of reaching a nonzero volume of points in the task space, i.e., an m-dimensional manifold in the m-dimensional task space. A manipulator that can only reach a manifold of dimension lower than m in a fault tolerant way, is considered not to be fault tolerant.

4 Properties of General Purpose Fault Tolerant Manipulators

4.1 Existence

A general purpose manipulator has six DOFs which allow it to position its end effector in an arbitrary position and orientation anywhere in its work space. An obvious way to make this manipulator fault tolerant is to design every joint with a redundant actuator. If one of the actuators of the resulting $2n$-DOF fault tolerant manipulator were to fail, the redundant actuator could take over and the manipulator would still be functional.

Similarly, a k-fault tolerant manipulator can be constructed by duplicating every DOF k times, resulting in a $(k+1)n$-DOF manipulator.

4.2 Boundary of the Fault Tolerant Work Space

In this section, we show that a boundary point of the FTWS is a critical value. Consider a k-fault tolerant planar manipulator, $\mathcal{M}$. A boundary point, $\mathbf{p}_b$, of the FTWS has to be an element of the boundary of the work space of at least one ROD, $\mathcal{M}^*$, obtained by freezing k joints of $\mathcal{M}$. Indeed, if $\mathbf{p}_b$ were an interior point of the work spaces of all RODs, then it would by definition be an interior point of the FTWS and not a boundary point. The Jacobian of $\mathcal{M}^*$, J_{M^*}, can be obtained from the Jacobian of $\mathcal{M}$, J_M, by deleting the columns corresponding to the frozen DOFs. Because $\mathbf{p}_b$ is a boundary point of the work space of $\mathcal{M}^*$, the Jacobian of $\mathcal{M}^*$ at $\mathbf{p}_b$ is singular. We prove now that J_M is singular too. Suppose that J_M were non-singular, then at least one of the columns corresponding to a frozen DOF would be outside the column space of the singular matrix, J_{M^*}. Physically this means that a small change in the angle of that frozen DOF would cause the end effector of $\mathcal{M}$ to move in a direction with a component perpendicular to the boundary of the work space of the ROD, $\mathcal{M}^*$, as illustrated in Fig. 6. The ROD with this new frozen angle would be unable to reach the point, $\mathbf{p}_b$. As a result, $\mathbf{p}_b$ would be outside the FTWS, contradicting the fact that $\mathbf{p}_b$ is a boundary point of the FTWS. Thus, J_M is singular and $\mathbf{p}_b$ is a critical value.

Consequently, the FTWS is bounded by critical value manifolds. For planar positional manipulators, the critical value manifolds are concentric circles, and the FTWS is an annulus with inner radius R_{min}^{FTWS} and outer radius R_{max}^{FTWS}.

4.3 Required Degree of Redundancy

In Section 4.1, it is shown that, in general, kn redundant DOFs—i.e. $(k+1)n$ DOFs in total—are sufficient to achieve k-th order fault tolerance. For planar positional manipulators, however, we prove that $2k$ DOFs are also necessary for k-th order fault tolerance.

The proof shows that $(2k+1)$ DOFs (or $2k-1$ redundant DOFs) are insufficient, by finding a lower bound for R_{min}^{FTWS} and an upper bound for R_{max}^{FTWS} that are

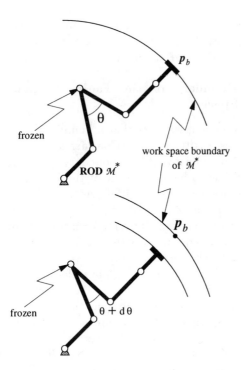

Figure 6: *A reduced order derivative that is unable to reach a point outside the FTWS.*

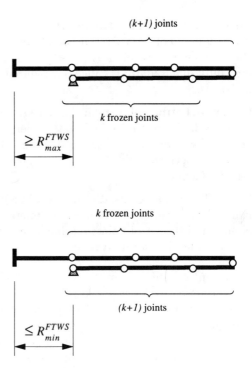

Figure 7: *An upper bound for R_{max}^{FTWS} and a lower bound for R_{min}^{FTWS}.*

equal to each other. First consider the ROD obtained by freezing the first k joints at 0 radians, as illustrated in Fig. 7. The maximum reach in the opposite direction is an upper bound for R_{max}^{FTWS}:

$$R_{max}^{FTWS} \leq -\sum_{i=1}^{k} l_i + l_{k+1} + \sum_{i=k+2}^{2k+1} l_i \qquad (7)$$

where l_i is the length of the i-th link. In order for R_{max}^{FTWS} to be positive, we must have that:

$$\sum_{i=1}^{k} l_i \leq l_{k+1} + \sum_{i=k+2}^{2k+1} l_i. \qquad (8)$$

Making this assumption, we find that R_{min}^{FTWS} is bounded below by the inner radius of the work space of the ROD obtained by freezing the k last joints at 0 radians, as illustrated in Fig. 7:

$$R_{min}^{FTWS} \geq \sum_{i=k+2}^{2k+1} l_i + l_{k+1} - \sum_{i=1}^{k} l_i \qquad (9)$$

From Equation (7) and Equation (9), it follows that at best

$$R_{max}^{FTWS} = R_{min}^{FTWS} \qquad (10)$$

resulting in a one-dimensional FTWS. Therefore, a $(2k + 1)$-DOF manipulator cannot be fault tolerant ∎

4.4 Including Orientation

Thus far, we have only considered planar positional manipulators. The results for positional manipulators can be easily extended to the case in which orientation is considered also, by converting the orientational problem into an equivalent positional problem:

An n-DOF manipulator, $\mathcal{M}$, is k-fault tolerant with respect to a set of points, $W = \{(x_i, y_i, \varphi_i)\}$, if and only if:

1. the positional manipulator, $\mathcal{M}'$, obtained from $\mathcal{M}$ by deleting its last link, l_n, is k-th order fault tolerant with respect to the set of points $W' = \{(x_i - l_n \cos \varphi_i, y_i - l_n \sin \varphi_i)\}$

DOF	d_i	a_i	α_i
1	0	1	90°
2	a	1	0°
3	−a	1	90°
4	b	1	0°
5	−b	1	—

Table 3: *D-H parameters of a 5-DOF first order fault tolerant spatial manipulator without orientation*

2. $\mathcal{M}'$ is $(k-1)$-fault tolerant while reaching the points in W' in any direction.

The positional manipulator, $\mathcal{M}'$, needs at least $(2k+2)$ DOFs to be k-fault tolerant with respect to W'; therefore, the manipulator $\mathcal{M}$ needs at least $(2k+3)$ DOFs. Now, consider a $(2k+3)$-DOF manipulator with the first links having length, l, and the last link having length zero. It is easy to verify that this manipulator's k-th order FTWS is:

$$W = \{(x, y, \varphi) \mid \sqrt{x^2 + y^2} \le 2l \text{ and } \varphi \in [0, 2\pi)\} \quad (11)$$

Thus, $(2k + 3)$ DOFs are necessary and sufficient for k-th order fault tolerance of planar manipulators when orientation is included

This result and the result obtained in Section 4.3 can be summarized in the following theorem:

Theorem:
For planar manipulators, $2k$ redundant DOFs are necessary and sufficient for k-th order fault tolerance.

4.5 Spatial Fault Tolerant Manipulators

For planar fault tolerant manipulators, we were able to prove that $2k$ is the required degree of redundancy. The proof was based on geometric work space analysis. However, the geometric analysis becomes too complex for spatial manipulators, especially since we are dealing with redundant manipulators. Therefore, we will demonstrate some properties of spatial fault tolerant manipulators using two examples.

As a first example, consider a 5-DOF spatial positional manipulator. Its D-H parameters are listed in Table 3. This manipulator is first order fault tolerant, and because of its simple kinematic structure, an analytic expression for the boundary of the FTWS can be

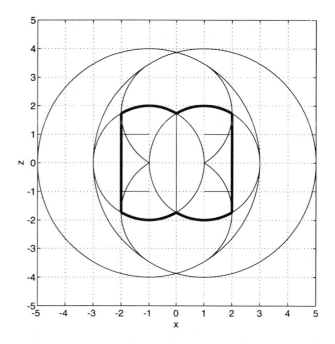

Figure 8: *A cross section of the boundary of the FTWS of a 5-DOF spatial manipulator (bold) as part of its critical value manifolds.*

derived. The FTWS is symmetric with respect to the first axis. A cross section (the X-Z plane), as shown in Fig. 8, can be described by two segments of a circle with radius 2 and center at $(x = 1, z = 0)$, and a straight line from $(x = 2, z = \sqrt{3})$ to $(x = 2, z = -\sqrt{3})$. An important property of this FTWS is that it does not have any holes or a central void, so that the FTWS of the same manipulator scaled by any factor, $\lambda > 1$, contains the original FTWS. As a result, this fault tolerant manipulator can be used as a design template. Any specified set of points can be reached in a first order fault tolerant way by a scaled version of the template.

In Section 4.2, it is shown that the boundary of the FTWS of a planar manipulator coincides with its critical value manifolds. Fig. 8 demonstrates that this property also holds for the 5-DOF spatial manipulator considered in this example. The critical value manifolds are computed using the algorithm described in [4] and are depicted in a solid line. The bold part of the critical value manifolds is the boundary of the FTWS.

As a second example, consider an 8-DOF manipulator, with D-H parameters listed in Table 4. It is

DOF	d_i	a_i	α_i
1	0	1	90°
2	a	1	0°
3	$-a$	1	90°
4	b	1	0°
5	$-b$	0	90°
6	1	0	90°
7	0	0	90°
8	0	0	—

Table 4: *D-H parameters of an 8-DOF first order fault tolerant spatial manipulator with orientation*

the same manipulator as in example one, with a zero-length 3-roll-wrist added at the end. Using a Monte-Carlo method, it has been determined that this manipulator is first order fault tolerant while reaching all the points in the FTWS of example one, *in any direction*. This property can be demonstrated with the following arguments. When one of the first five DOFs fails, the manipulator can still reach any position in the FTWS (because the 5-DOF positional manipulator is fault tolerant) and can take any orientation at this position using the intact 3-roll-wrist. When one of the DOFs in the wrist fails, we are left with a 7-DOF manipulator which has enough orientational capabilities to reach any point in the FTWS in any orientation. Consequently, one could call this the *dextrous* FTWS. Since there are again no holes or voids in the FTWS, this manipulator can also be used as a design template.

Finally, one should notice that both examples have only two redundant DOFs, which seems to indicate that the theorem in Section 4.4 is extendible to spatial manipulators.

5 Task Specific Fault Tolerant Manipulators

In the previous section, we considered the design of fault tolerant manipulators for general use. We proved that two redundant DOFs are necessary for first order fault tolerance. However, as we will show in this section, a simpler kinematic structure is often sufficient when one specific task is considered. This implies, of course, that a different kinematic structure might be needed for every task—a disadvantage that can be alleviated by the use of a reconfigurable modular manip-

ulator system.

We modify the definition of fault tolerance to include task specificity:

- **Task Specific Fault Tolerant Manipulator**: A manipulator is 1-fault-tolerant with respect to the task of following the Cartesian trajectory, $\mathbf{p}(t)$, if there exists a fault tolerant trajectory in joint space, $\theta(t)$, that maps into $\mathbf{p}(t)$, and which is such that when an arbitrary joint, j, were frozen at an instant, $^f t$, an alternate trajectory, $\theta(t, j, ^f t)$, could be followed to complete the task.

The difference between this definition and the one for general purpose fault tolerance, is that we no longer require that a point be reachable when a joint fails at an arbitrary angle, but only at an angle that occurred previously in the fault tolerant trajectory. Under this assumption, k-fault tolerance can be achieved with only k redundant DOFs.

Consider the task of reaching all the points in an ϵ-neighborhood, $B(\mathbf{p}, \epsilon)$, of the point $\mathbf{p} \in \Re^m$. Suppose that $\mathbf{p}$ can be reached by an n-DOF manipulator in a posture $\theta \in T^n$. If the posture, θ, is non-singular, then there exists an $\epsilon > 0$, such that the manipulator can reach any point in $B(\mathbf{p}, \epsilon)$. However, for k-fault-tolerance, any point in $B(\mathbf{p}, \epsilon)$ needs to be reachable even when k of the joints of the manipulator are frozen. This is possible if and only if the Jacobians of all k-ROD in the posture θ have at least rank m. We call such a posture, θ, locally fault tolerant. The Jacobian of a k-ROD can be obtained by deleting the columns of the fault-free Jacobian corresponding to the frozen DOFs; its dimensions are $m \times (n - k)$. In order for the rank to be at least m, n has to be larger than or equal to $(m + k)$, i.e., the manipulator needs to have at least k redundant DOFs. When the rank of a k-ROD Jacobian is less than m, the robot is in an internal singularity; otherwise, it is in a locally fault tolerant posture. The locus of internal singularities is a set of $(m + k - 1)$-dimensional surfaces in T^n; or $(n - 1)$-dimensional surfaces, when $n = m + k$. Thus, nearly all postures of a manipulator with k redundant DOFs are locally k-fault tolerant.

We now extend this result to larger trajectories for which a global condition has to be satisfied. This can best be illustrated with an example. Consider a 3-DOF planar manipulator with normalized link lengths

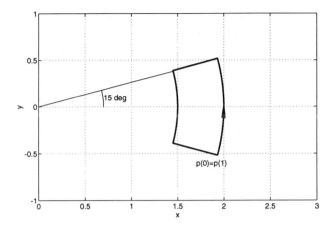

Figure 9: *The trajectory of the example of task specific fault tolerance.*

of 1. We want to determine whether this manipulator is able to execute the task of following the trajectory shown in Fig. 9, in a 1-fault tolerant way. The trajectory can be parameterized as $\mathbf{p}(\alpha)$ with $0 \leq \alpha \leq 1$. We assume for this example that $\mathbf{p}(0) = \mathbf{p}(1)$, and that the task is repeated from the beginning as soon as the end is reached. For every α, one can compute the preimage of $\mathbf{p}(\alpha)$. Since the manipulator has one degree of redundancy, the preimage of every $\mathbf{p}(\alpha)$ is a one-dimensional subset of T^n, and can be parameterized as $\theta = f(\mathbf{p}(\alpha), \beta)$ with $\beta \in T^1$. The continuous function, f, describes a 2-dimensional surface in T^3, as is shown in Fig. 10. Any joint trajectory that follows the specified Cartesian trajectory, $\mathbf{p}(\alpha)$, can be formulated as $\beta(\alpha)$, or $\theta(\alpha) = f(\mathbf{p}(\alpha), \beta(\alpha))$. According to the definition of task specific fault tolerance, the manipulator is fault tolerant if and only if a fault tolerant trajectory, $\theta(\alpha)$, can be found. It is clear that every posture of a fault tolerant trajectory, $\theta(\alpha)$, has to be locally fault tolerant. However, this requirement is not sufficient because a fault at a point, $\mathbf{p}(\alpha_1)$, might make another point, $\mathbf{p}(\alpha_2)$, unreachable, even when the posture $\theta(\alpha_1)$ is locally fault tolerant. Therefore, one should exclude as possible postures for a fault tolerant trajectory not only internally singular postures, but also postures that, in the case of failure, would cause an internal singularity elsewhere along the trajectory. For our example, the set of acceptable postures is shown in Fig. 11.

A fault tolerant trajectory exists when a continuous function, $\beta(\alpha)$ with $0 \leq \alpha \leq 1$, can be found for

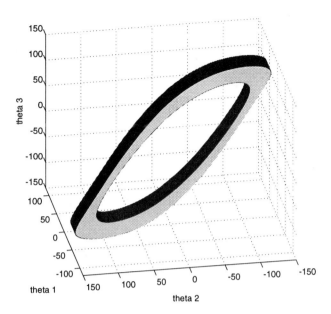

Figure 10: *The preimage of a trajectory.*

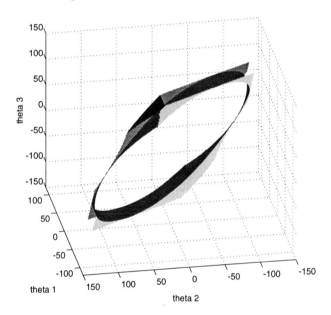

Figure 11: *The set of acceptable points for a fault tolerant trajectory.*

which all postures, $\theta = f(\mathbf{p}(\alpha), \beta(\alpha))$, are acceptable, i.e., satisfy the global fault tolerance condition. That such a trajectory exists for our example can be concluded from Fig. 11. The same conclusion follows more clearly from Fig. 12, in which only the postures of the 2-dimensional preimage of the trajectory are represented.

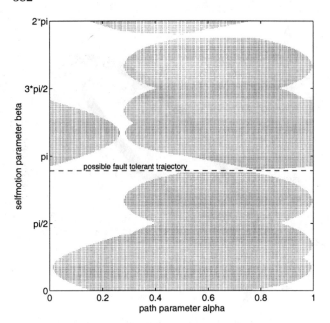

Figure 12: *A possible fault tolerant trajectory. Regions of unacceptable postures, (α, β), are marked in gray.*

The gray area is the set of postures that, in the case of a fault, would cause an internal singularity somewhere along the trajectory; these are the unacceptable postures. The manipulator is fault tolerant if a continuous trajectory, $\beta(\alpha)$, can be found that does not pass through any gray areas. One possible trajectory is shown in dashed line.

The conclusion of this section is that, if a fault-free trajectory is chosen carefully, one can possibly achieve first order fault tolerance with only one redundant DOF.

Our current research deals with the problem of translating the results of the analysis of task specific fault tolerance into specific design rules for kinematic structures of manipulators.

6 Summary

In this paper, we developed an approach for determining a configuration for a reconfigurable modular manipulator able to fulfill a specific task. We considered tasks that included four kinematic requirements: reachability, joint limits, obstacle avoidance and measure of isotropy. The attribution of a penalty to each manipulator configuration, enabled us to reduce the

search effort drastically, by guiding the search to the most promising regions of the assembly configuration space. Local minima in the penalty were avoided by using simulated annealing as a search algorithm. We also defined a property of a small class of redundant manipulators, called fault tolerance. We proved the existence of general purpose fault tolerant manipulators, obtained through joint duplication. When no joint limits are considered, we proved analytically that, $2k$ redundant DOFs are necessary and sufficient for general purpose fault tolerance of planar manipulators, and that the boundary of the FTWS consists of critical values. Also, 8- and 5-DOF design templates were introduced, for spatial general purpose fault tolerant manipulators with and without orientational capabilities, respectively. Finally, we demonstrated that a task specific 1-fault tolerant manipulator possibly only needs one degree of redundancy, versus the two needed for general purpose manipulators. This simplification of the kinematic structure can be achieved at the cost of having to reconfigure the manipulator for every task.

Acknowledgment

This research was funded in part by DOE under grant DE-F902-89ER14042, by Sandia under contract AC-3752-A, by the Department of Electrical and Computer Engineering, and by The Robotics Institute, Carnegie Mellon University.

References

[1] V. P. Agrawal, V. Kohli, and S. Gupta. Computer aided robot selection: the 'multiple attribute decision making' approach. *International Journal of Production Research*, 29(8):1629–1644, 1991.

[2] W. K. F. Au, C. J. J. Paredis, and P. K. Khosla. Kinematic design of fault tolerant manipulators. In *Proceedings of the Allerton Conference*, Urbana-Champagne, Illinois, October 2 1992.

[3] B. Benhabib, G. Zak, and M. G. Lipton. A generalized kinematic modeling method for modular robots. *Journal of Robotic Systems*, 6(5):545–571, 1989.

[4] J. W. Burdick. Kinematic analysis and design of redundant robot manipulators. Stanford Computer Science Report STAN-CS-88-1207, Stanford University, 1989.

[5] P. Fanghella, C. Gellatti, and E. Giannotti. Computer-aided modeling and simulation of mechanisms and manipulators. *Computer Aided Design*, 21(9):577–583, 1989.

[6] T. Fukuda, G. Xue, F. Arai, H. Asama, H. Omori, I. Endo, and H. Kaetsu. A study on dynamically reconfigurable robotic systems. assembling, disassembling and reconfiguration of cellular manipulator by cooperation of two robot manipulators. In *Proceedings of the IEEE/RSJ International Workshop on Intelligent Robots and Systems (IROS '91)*, pages 1184–1189, Osaka, Japan, November 3-5, 1991.

[7] L. Kelmar and Pradeep K. Khosla. Automatic generation of forward and inverse kinematics for a reconfigurable modular manipulator system. *Journal of Robotic Systems*, 7(4):599–619, 1990.

[8] J.-O. Kim. *Task Based Kinematic Design of Robot Manipulators*. PhD thesis, Carnegie Mellon University, The Robotics Institute, Pittsburgh, PA, August 1992.

[9] S. Kirkpatrick, C. D. Gelatt Jr., and M. P. Vecchi. Optimization by simulated annealing. *Science*, 220(4598):671–680, 1983.

[10] A. Krishnan and P. K. Khosla. A methodology for determining the dynamic configuration of a reconfigurable manipulator system. In *Proceedings of the 5th Annual Aerospace Applications of AI Conference*, Dayton, Ohio, October 23-27, 1989.

[11] C. L. Lewis and A. A. Maciejewski. Dexterity optimization of kinematically redundant manipulators in the presence of joint failures. *Computers and Electrical Engineering*, 20(3):273–288, 1994.

[12] A. A. Maciejewski. Fault tolerant properties of kinematically redundant manipulators. In *Proceedings of the 1990 IEEE International Conference on Robotics and Automation*, pages 638–642, Cincinnati, Ohio, May 1990.

[13] S. Manoochehri and A. A. Seireg. A computer-based methodology for the form synthesis and optimal design of robot manipulators. *Journal of Mechanical Design*, 112:501–508, December 1990.

[14] S. Murthy, P. K. Khosla, and S. Talukdar. Designing manipulators from task requirements: An asynchronous team approach. In *Proceedings of the 1st WWW Workshop on Multiple Distributed Robotic Systems*, Nagoya, Japan, July 1993.

[15] O. F. Offodile, B. K. Lambert, and R. A. Dudek. Development of a computer aided robot selection procedure (carsp). *International Journal of Production Research*, 25:1109–1121, 1987.

[16] O. F. Offodile, W. M. Marcy, and S. L. Johnson. Knowledge base design for flexible assembly robots. *International Journal of Production Research*, 29(2):317–328, 1991.

[17] C. J. J. Paredis. An approach for mapping kinematic task specifications into a manipulator design. Master's thesis, Carnegie Mellon University, Electrical and Computer Engineering Department, Pittsburgh, PA, September 1990.

[18] C. J. J. Paredis and P. K. Khosla. Kinematic design of serial link manipulators from task specifications. *The International Journal of Robotics Research*, 12(3):274–287, June 1993.

[19] V. Potkonjak and M. Vukobratovic. Computer-aided design of manipulation robots via multi-parameter optimization. *Mechanism and Machine Theory*, 18(6):431–438, 1983.

[20] E. Rich and K. Knight. *Artificial Intelligence*. series in artificial intelligence. Mc Graw-Hill Inc., New York, second edition edition, 1989.

[21] D. E. Schmitz, P. K. Khosla, and T. Kanade. The CMU reconfigurable modular manipulator system. In *Proceedings of the 19-th International Symposium and Exposition on Robots (ISIR)*, Australia, 1988.

[22] A. A. Tseng. Software for robotic simulation. *Advances in Engineering Software*, 11(1):26–36, January 1989.

[23] M. L. Visinsky, I. D. Walker, and J. R. Cavallaro. Layered dynamic fault detection and tolerance for robots. In *Proceedings of the 1993 IEEE International Conference on Robotics and Automation*, pages 180–187, Atlanta, GA, May 1993.

[24] R. H. Weston, R. Harrison, A. H. Booth, and P. R. Moore. Universal machine control system primitives for modular distributed manipulator systems. *International Journal of Production Research*, 27(3):395–410, 1989.

Computational Kinematics[*]

Leo Joskowicz, *IBM T.J. Watson Research Center, Yorktown Heights, NY, USA*
Elisha Sacks, *Princeton University, Princeton, NJ, USA*

We present a kinematic analysis algorithm for mechanisms built of rigid parts, such as door locks, gearboxes, and transmissions. The algorithm produces a concise and complete description of the kinematics of a mechanism. It optimizes the computation by decomposing complex mechanisms into subassemblies, deriving the kinematics of the subassemblies, and incrementally composing the results. We define a class of mechanisms for which kinematic analysis is feasible by restricting the shapes, motions, and interactions of parts. The feasible class contains linkages, mechanisms whose parts move along fixed spatial axes, and combinations of the two types. We show that the feasible class covers most mechanisms by surveying 2500 mechanisms from an engineering encyclopedia. We implement the kinematic analysis algorithm for fixed-axes mechanisms. The inputs are the shapes and initial configurations of the parts. The output is a region diagram, a partition of the mechanism configuration space into regions that characterize its operating modes. The program computes the region diagram by identifying motion axes and interacting pairs of parts, partitioning the pairwise configuration spaces, and composing them. Coupling the program with existing linkage analysis packages covers most feasible mechanisms. We identify classes of infeasible mechanisms and describe possible analysis strategies for them.

1 Introduction

This paper presents research on automating the kinematic analysis of mechanisms composed of rigid parts, such as door locks, gearboxes, and transmissions. The parts of a mechanism move and interact, giving rise to complex and useful behaviors. Kinematics specifies which behaviors are consistent with the shapes of the parts and the contacts among them. It abstracts away the forces acting on the parts, which determine the actual behavior. Kinematic analysis derives the kinematics from the shapes and initial positions of the parts. It is an essential component of mechanical engineering. It directly answers many questions about the workings of mechanisms and sets the stage for further analysis, such as tolerancing, dynamics, and stress analysis. Many tasks presuppose kinematic analysis, including mechanism design, evaluation, and classification.

We seek concise and complete descriptions of the kinematics of mechanisms. The descriptions should be both qualitative and quantitative. Qualitative descriptions characterize the working of mechanisms concisely and abstractly by identifying operating modes and mode transitions. They allow engineers to design, repair, and modify complex mechanisms without drowning in details. Quantitative descriptions provide the details for tasks where qualitative descriptions are overly general. Complete descriptions guarantee that no behavior is overlooked, allowing engineers to understand and evaluate mechanisms under all conditions. For example, here is a complete qualitative and quantitative kinematic description of a drill: "The drill has two operating modes, engaged and disengaged, and a switch that sets the mode. The drill rotates in the engaged mode and stands still in the disengaged mode. In the engaged mode, the transmission ratio between the motor and the drilling bit is 10:1."

Deriving the kinematics of a mechanism entails examining every potential interaction among its parts, an intractable task even for mechanisms with few parts. Engineers simplify the task by building mechanisms modularly and by restricting the shapes, motions, and interactions of parts. They derive the kinematics of a complex mechanism by decomposing it into manageable subassemblies, deriving the kinematics of the

---

[*]This paper first appeared in *Artificial Intelligence*, Vol. 51 No. 1-3, 1991, Elsevier, North-Holland. It is reproduced with permission from the publisher.

```
┌─────────────────────────────────────────┐
│   Input: mechanism structure            │
│       1. modeling                       │
│       2. subassembly analysis           │
│       3. composition                    │
│   Output: mechanism kinematics          │
└─────────────────────────────────────────┘
```

Figure 1: *Kinematic analysis algorithm.*

subassemblies, and composing the results. The most common subassemblies are sets of parts linked by permanent joints, called linkages, and sets of parts that move along fixed spatial axes, called fixed-axes mechanisms. Fixed-axes mechanisms decompose further into pairs of interacting parts, called kinematic pairs.

In this paper, we formalize the engineering strategy and systematize kinematic analysis. We identify the restrictions that make kinematic analysis feasible and define a feasible class of mechanisms. The feasible class contains linkages, fixed-axes mechanisms, and mechanisms with linkage and fixed-axes subassemblies. We demonstrate empirically that the feasible class covers most mechanisms by surveying some 2500 mechanisms from an engineering encyclopedia. We develop a modular kinematic analysis algorithm for feasible mechanisms (Fig. 1). The algorithm consists of modeling, subassembly analysis, and composition steps. The input is the mechanism structure: a description of the parts and their initial configuration. The output is the mechanism kinematics: a complete and concise description of its possible behaviors. The modeling step decomposes the mechanism into subassemblies and incorporates simplifying assumptions about their shapes and motions. The subassembly analysis step derives the kinematics of the subassemblies. The composition step derives the kinematics of the mechanism by composing the subassembly kinematics.

We formalize this strategy within the configuration space (CS) representation of mechanical engineering. Intuitively, the CS of a mechanism is the space of configurations of its parts. We show that partitioning the CS into regions yields a concise and complete description of the mechanism's kinematics. The regions represent the operating modes of the mechanism. Mode transitions occur when the mechanism configuration shifts between adjacent regions. We define feasible mechanisms in CS terminology and develop an analysis

algorithm that constructs a concise and meaningful CS partition.

We implement the kinematic analysis algorithm for fixed-axes mechanisms. The inputs are the shapes and initial configurations of the parts. The output is a partition of the mechanism CS. The modeling step decomposes the mechanism into interacting kinematic pairs and identifies the axes of motion. The subassembly analysis step partitions the CSs of the kinematic pairs. The composition step partitions the mechanism CS by composing the pairwise partitions. It optimizes the computation by incrementally enumerating and testing only the regions reachable from the initial configuration. It represents the partition as a region diagram whose nodes describe regions and whose links specify region adjacencies. (The region diagram corresponds to an attainable envisionment of the kinematics of the mechanism [29].)

The program extends the state of the art in automating the kinematic analysis of mechanisms. It is the first implementation of fixed-axes modeling. Previous research provides efficient analysis programs for linkages and for planar kinematic pairs, but provides only a limited composition algorithm that cannot handle many useful mechanisms, including gearshifts, differentials, and indexers. Our composition algorithm handles all mechanisms and is the first incremental algorithm. Our program handles most fixed-axes mechanisms.

The rest of the paper is organized as follows. We conclude this section with an informal illustration of the kinematic analysis program. In Section 2, we describe the CS representation and recast kinematic analysis as a CS partition problem. In Section 3, we define feasible mechanisms and demonstrate that they cover most mechanisms. In Section 4, we describe the kinematic analysis program for fixed-axes mechanisms. In Section 5, we identify classes of infeasible mechanisms and describe possible analysis strategies for them. We conclude with a review of previous work and a discussion of future work.

1.1 Kinematic analysis of a transmission

We illustrate the kinematic analysis program on a realistic engineering example: a two-speed transmission. Fig. 2 shows a side view of the transmission. The input shaft $S1$, the output shaft $S2$, and the gearshift P are mounted on the fixed frame F. Gears $G1$ and $G2$ are

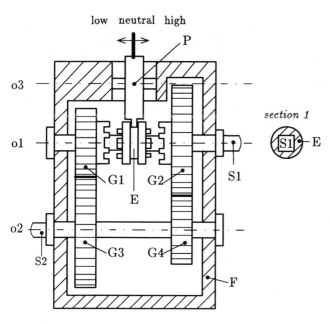

Figure 2: *Side view of the two-speed transmission.*

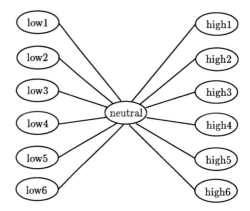

Figure 3: *Region diagram of the transmission.*

motion types	**motion relations**
$translation(P, o3, x_P)$	$1 \leq x_P \leq 2$
$translation(E, o1, x_E)$	$x_E = x_P$
$rotation(E, o1, \theta_E)$	
$rotation(S1, o1, \theta_{S1})$	$\theta_E = \theta_{S1}$
$rotation(G1, o1, \theta_{G1})$	$\theta_{G1} = -\frac{1}{2}\theta_{G3}$
$rotation(G2, o1, \theta_{G2})$	$\theta_{G2} = -2\theta_{G4}$
$rotation(S2, o2, \theta_{S2})$	
$rotation(G3, o2, \theta_{G3})$	$\theta_{S2} = \theta_{G3}$
$rotation(G4, o2, \theta_{G4})$	$\theta_{S2} = \theta_{G4}$

Figure 4: *Description of the neutral region.*

mounted on $S1$ and rotate freely around it. Engager E is mounted on a square section of $S1$ and translates along axis $o1$ (Fig. 6). Gears $G3$ and $G4$ are rigidly attached to $S2$. The engager E has six lateral teeth on each side that can engage with the six lateral teeth of $G1$ and $G2$. The three settings of the gearshift P define the three operating modes of the transmission. Input shaft $S1$ drives output shaft $S2$ via $G1$ and $G3$ when P is in low, rotates independently of $S2$ when P is in neutral, and drives $S2$ via $G2$ and $G4$ when P is in high. The gear ratios $G1/G3$ and $G2/G4$ define the transmission rates in the low and high modes.

The modeling step of the program determines the axes of motion and the motion types of the parts: $S1$, $G1$, and $G2$ rotate around axis $o1$, E rotates around and translates along $o1$, P translates along $o3$, and $S2$, $G3$, and $G4$ rotate around $o2$. It then decomposes the mechanism into kinematic pairs and eliminates the pairs that cannot interact. For example, $G1$ and $G3$ interact, but E and $S2$ do not.

The subassembly analysis step constructs region diagrams for the interacting pairs, which reveal the pairwise kinematics. The gearshift translates in unison with the engager. The engager rotates with $S1$ and translates along it. Gears $G1$ and $G2$ rotate freely around $S1$. The engager engages $G1$ and $G2$ at six

different angular positions. Gears $G3$ and $G4$ rotate together with $S2$. Gear pairs $G1/G3$ and $G2/G4$ mesh.

The composition step constructs the mechanism region diagram from the pairwise region diagrams (Fig. 3). The left, middle, and right regions represent the operating modes *low, neutral,* and *high*. The six *low* and six *high* regions represent the six different angular offsets in which $G1$ and $G2$ can mesh with the engager. The regions specify the motion axes and motion types of the parts and the algebraic relations among the part coordinates. Fig. 4 shows the description of the *neutral* region: $S1$ and $S2$ rotate independently because the rotation parameters θ_{S1} and θ_{S2} are independent.

2 Kinematics

Kinematics explores the ways in which assemblies of rigid parts move in space. An isolated part can be in any configuration (position and orientation) and is free to translate and rotate in any direction. But a part in

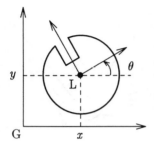

Figure 5: *The configuration (x, y, θ) of a 2D part with L its reference frame and G the global reference frame.*

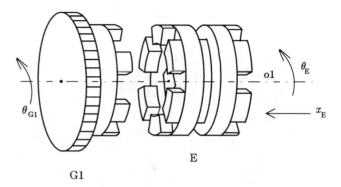

Figure 6: *The engager E and gear $G1$.*

an assembly can only be in configurations that do not overlap other parts, and can only move along or away from other parts. Parts moving together must satisfy the constraints imposed by their touching features (vertices, edges, and faces).

2.1 Configuration space

We study the kinematics of mechanisms within the standard configuration space representation of mechanical engineering. We assign a reference frame to each part. We express the configuration of the part by the position and orientation of its reference frame with respect to a global reference frame. A 2-dimensional (2D) part has a 3D CS specified by the two position coordinates of its reference frame and the angle between its reference frame and the horizontal axis of the global reference frame (Fig. 5). A 3D part has a 6D CS specified by the three position coordinates of its reference frame and the three angles between its reference axes and the global reference axes. (The CS is actually the product of a Euclidean translation space with a rotation group: $\Re^2 \times SO(1)$ in 2D and $\Re^3 \times SO(3)$ in 3D [28].) The dimension of the CS indicates the maximum number of independent motions, which we call the *potential degrees of freedom* of the part. Most potential degrees of freedom are unrealizable due to contacts with other parts. For example, the gearshift P in the transmission has one degree of freedom, horizontal translation along axis $o3$, out of a potential six because the frame prevents it from moving in any other direction.

The Cartesian product of the part CSs forms the mechanism CS. Its dimension equals the sum of the dimensions of the part CSs. The mechanism CS decomposes into *blocked space* where parts overlap and *free space* where no parts overlap. Only configurations

in free space are physically realizable. Free space is a closed set whose boundary consists of the configurations where parts touch. It decomposes into maximal, connected sets. The set containing the initial configuration of the mechanism is called *realizable free space*. A potential motion of the parts of a mechanism defines a curve in its CS. The mechanism can realize exactly those motions whose curves lie in realizable free space. Coupled part motions correspond to curves on the boundary of realizable free space.

We illustrate these concepts for the engager E and gear $G1$ from the transmission (Fig. 6). The engager rotates and translates along axis $o1$ with coordinates θ_E and x_E; $G1$ rotates around $o1$ with coordinate θ_{G1}. Fig. 7 shows a projection of their CS, which is 3D, into the 2D space $(\theta_E - \theta_{G1}, x_E)$. The shaded area is blocked space. The upper horizontal CS boundary contains the configurations where the tops of the engager's right teeth touch the gear's left face. The lower horizontal CS boundaries contain the configurations where the tops of the engager's left teeth touch the tops of the gear's teeth. The vertical CS boundaries contain the configurations where the engager teeth mesh with the gear teeth. Region r_0 contains the configurations where the engager is to the right of the gear and does not touch it. Region r_7 contains the configurations where the engager is to the left of the gear and does not touch it. The realizable free space is regions r_0–r_6, since the initial configuration p lies in r_0. The dashed lines denote the realizable motion where the engager rotates into alignment with the gear then translates until the parts engage.

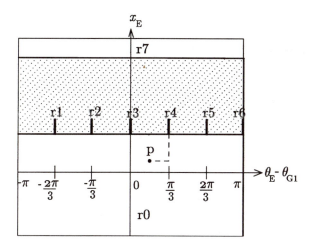

Figure 7: *The CS of engager E and gear G1.*

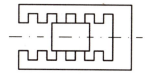

Figure 8: *(a) a ball in a circular tube, and (b) a block sliding on a jagged guiding surface.*

For example, Fig. 8 shows that the contact change criterion sometimes yields overly or underly specific descriptions. The ball has a single contact region, which conflates rising and falling motions, whereas the block has one contact region for each tooth, yet slides uniformly through the guide. The appropriate partition criterion is best determined by the task at hand [16].

2.2 Region diagrams

We describe the kinematics of a mechanism with a partition of its reachable free space into motion regions. We represent the partition with a *region diagram* whose nodes describe free space regions and whose links specify region adjacencies. The regions define the operating modes of the mechanism. A realizable motion of the mechanism translates into a path in the region diagram. The nodes in the path represent uniform portions of the motion and the links represent shifts in operating mode.

For example, we can partition the realizable free space of the engager E and gear $G1$ into regions r_0–r_6. The parts move independently in the 3D region r_0. In the 2D regions r_1–r_6, the parts rotate together ($\theta_E - \theta_{G1}$ is constant) and the engager translates independently. The parts must disengage in order to change angular offsets because every path between 2D regions goes through r_0.

We seek a partition that captures the relevant aspects of behavior as concisely as possible. The motion regions must be connected, pairwise disjoint, and maximal with respect to the remaining partition criteria. Different partition criteria capture different aspects of the kinematics. Common criteria are fixed part contacts, monotonic region boundaries (the signs of their partial derivatives are constant), or both. The partition of the engager/gear CS (Fig. 7) follows the monotonic boundary criterion. No single criterion can characterize all mechanisms kinematics appropriately.

2.3 Computing region diagrams

The CS representation enables us to recast kinematic analysis as an algebraic geometry problem. Algebraic geometry provides powerful algorithms for CS analysis and tight complexity bounds on their performance. We express the condition that no two parts overlap by multivariate polynomial inequalities in the CS coordinates. We must solve the equations to obtain the realizable free space and the partition into motion regions [28]. Although several algorithms are available, the problem is inherently intractable, since it requires time polynomial in the complexity of the parts and exponential in their degrees of freedom [3].

We construct the region diagram of a mechanism by the modular strategy pioneered by Reuleaux [24], used informally in mechanical engineering, and formalized by Joskowicz [15]. We decompose the mechanism into manageable subassemblies, construct subassembly region diagrams, and compose them into a mechanism region diagram. The modular strategy factors region diagram construction into simpler subtasks. We pick subassemblies for which region diagram construction is manageable, such as pairs of parts. We make composition manageable by exploiting the bounds on subassembly complexity and the limited subassembly interactions.

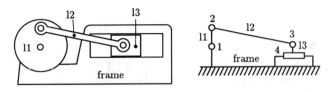

Figure 9: *A piston linkage and its schematic model. Joints 1–3 are revolute; joint 4 is prismatic. The frame is fixed.*

3 Feasible mechanisms

We define a class of mechanisms for which kinematic analysis is feasible. The definition restricts the shapes and motions of parts and the interactions among parts in ways that simplify region diagram construction. We show that the feasible class covers most practical mechanisms by surveying an encyclopedia of mechanisms.

3.1 Definition

The first type of feasible mechanism is the standard *linkage* of mechanical engineering. A linkage consists of parts permanently attached to each other by joints. Parts only interact with each other via joints. There are six types of joints: revolute, prismatic, helical, planar, cylindrical, and spherical. Each joint type constrains the attached parts to move in a simple manner and defines a joint relation. For example, two parts connected by a prismatic joint are constrained to translate relative to each other. The part shapes are irrelevant, as they only interact via joints; only the distances between joints matter. Linkages are modeled by one-dimensional rods linked by joint relations (Fig. 9).

The second type of feasible mechanism is a *fixed-axes mechanism*. Parts are 2.5D, meaning that each part consists of a finite number of *slices* over a reference plane. A slice is the Cartesian product of an interval orthogonal to the reference plane and one or more closed planar curves parallel to the reference plane. The closed curves consist of line segments and circular arcs. Fig. 10 shows a 2.5D part. Each part can translate along axes parallel or orthogonal to the reference plane and can rotate around an axis orthogonal to it. For example, all the transmission parts (Fig. 2) satisfy this condition. The subassemblies of the mechanism are 2.5D kinematic pairs, pairs of 2.5D parts with common reference planes (Fig. 6).

We require that the kinematic pairs in fixed-axes mechanisms have CSs of dimension two or lower. This

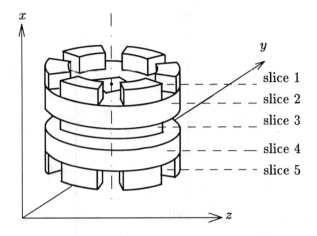

Figure 10: *The transmission engager E is 2.5D with five slices and reference plane yz.*

normally limits the parts to two joint degrees of freedom. However, we can compute a 2D CS for a pair with three or four degrees of freedom when the parts translate along parallel axes or rotate around the same axis. The motions are independent, so we can consider one of them fixed. We specify the relative coordinate of one part with respect to the other $(x_i - x_j)$ instead of both absolute coordinates (x_i and x_j), thus reducing the dimension by one. Examples are the engager/gear CS (Fig. 7), which is 2D instead of 3D, and the tile/tile CS (Fig. 19), which is 2D instead of 4D.

The third type of feasible mechanism is a collection of fixed-axes mechanisms connected by linkages. The linkages interact with the fixed-axes mechanisms solely via permanent connections to fixed-axes parts.

The modular analysis strategy makes region diagram construction manageable for feasible mechanisms. Linkages decompose into pairs of links connected by joints; fixed-axes mechanisms decompose into kinematic pairs; and heterogenous mechanisms decompose into linkages and fixed-axes mechanisms. In linkages, the standard joint types and permanent contacts make subassembly analysis unnecessary and simplify composition. Linkage packages exploit these restrictions, as described by Kramer [20]. In fixed-axes mechanisms, the restricted shapes, motions, and interactions among parts simplify subassembly analysis and composition. We describe the simplifications while presenting our analysis program for fixed-axes mechanisms. The results carry over to fixed-axes mechanisms

Volume	Total	CS dimension		Part dimension		Feasible
		≤ 2	> 2	2.5D	3D	
Lever mechanisms 1	159	130	29	102	57	94
Lever mechanisms 2	56	43	13	46	10	37
Gear mechanisms	180	174	6	140	40	136
Cam mechanisms	164	148	16	114	50	103
Total	559	495	64	402	157	370
Percentage	—	89%	11%	72%	28%	66%

Table 1: *Survey of kinematic pairs.*

Volume	Total	Motion Type				Varying Contacts	3D	Feasible
		Linkage	Fixed Axes	Fixed Axes Plus Linkages	Other			
Lever mechanisms 1	614	341	101	18	154	123	108	442
Lever mechanisms 2	677	323	105	66	183	110	70	448
Gear mechanisms	350	1	106	36	207	72	28	119
Cam mechanisms	271	4	101	47	119	40	72	99
Total	1912	669	413	167	663	340	278	1108
Percentage	—	35%	22%	9%	35%	18%	15%	58%

Table 2: *Survey of mechanisms.*

connected by linkages because of the restricted subassembly interactions.

3.2 Survey of mechanisms

We surveyed some 2500 common mechanisms to determine the percentage of practical mechanisms that are feasible and to identify significant exceptions. We chose Artobolevsky's four-volume *Mechanisms in Modern Engineering Design* [1] because of its size (more than 3500 mechanisms), uniform format, and comprehensiveness. The encyclopedia contains general-purpose, single-function mechanisms, such as couplers, indexers, and dwells. These constitute the functional components of larger, specialized mechanisms, such as printing presses, mills, motion-picture cameras, and cars. The other popular encyclopedias [4, 13] cover largely the same ground as does Artobolevsky. The survey shows that 66% of kinematic pairs and 58% of mechanisms are feasible. Readers uninterested in the details can skip to the next section.

The survey focuses on the shapes and motion types of rigid parts. We exclude mechanisms with flexible parts, such as belts, chains, and vibrators, because they fall outside the scope of standard kinematics.

The only exception is simple springs. We also exclude special-purpose mechanisms for which region diagrams are meaningless, such as curve generators and analog computers.

Table 1 summarizes the survey results for 559 stand-alone kinematic pairs. The number of features per part ranges from one (a circle) to 400 (an 80-tooth gear) with 15 the average. We classify the pairs in each volume by part shape and by CS dimension. A pair is feasible if it is 2.5D and has a CS of dimension two or lower. We assume that complex profiles of 2.5D parts, such as those in gear teeth and cams, are approximated with line and circular segments. The approximation is reasonable with the exception of a few delicate cams that require prohibitively many segments.

Feasible pairs comprise the majority within every category and 66% overall. A large majority (89%) of kinematic pairs have at most two degrees of freedom. Only 28% require 3D geometry.

Table 2 summarizes the survey results for 1912 mechanisms. The number of moving parts ranges from 3 to 25 and averages 7. We classify the mechanisms by motion type: linkages, fixed-axes, fixed-axes connected by linkages, and mechanisms with complex motions (other). We also count mechanisms with varying part contacts, such as brakes, clutches, and switching devices, mechanisms with 3D parts, and feasible mechanisms (Feasible). We consider standard infeasible kinematic pairs, such as spherical and screw joints and bevel and worm gears, as feasible, since our focus is on the motion properties of mechanisms. Even without this assumption, the coverage of kinematic pairs in mechanisms (80%) is higher than that of stand-alone kinematic pairs (66%). Except for linkages, most mechanisms contain at least one non-standard kinematic pair that engineers must analyze rather than look up.

Feasible mechanisms comprise 58% of all mechanisms. The mechanisms divide quite evenly between linkages (35%), fixed-axes mechanisms with or without linkages (31%), and others (35%). Some 15% of all mechanisms require 3D geometry. Part contacts vary in almost a third of the non-linkage mechanisms.

4 Analysis of feasible mechanisms

Now that we have defined feasible mechanisms and demonstrated that they include most known mechanisms, we turn to their analysis. We describe a modular region diagram construction program for fixed-axes mechanisms. The program extends the coverage of kinematic pairs from the planar pairs handled by existing programs (51%) to 2.5D pairs (66%). It covers for the first time mechanisms in which different parts interact in different configurations. The program performs well in practice, but cannot avoid the exponential worst-case running time intrinsic to CS construction.

Fig. 11 shows the program organization, which specializes the general kinematic analysis algorithm in the introduction (Fig. 1. The inputs are the part shapes and their initial configurations. The output is the mechanism region diagram. The modeling step finds the axes of motions of the parts, assigns a CS coordinate to each potential degree of freedom, and determines which pairs of parts interact. The pairwise analysis step constructs the region diagrams of the interacting pairs. The composition step constructs the mechanism region diagram from the pairwise region diagrams. Modeling greatly reduces the number and the complexity of the pairwise CSs, thus speeding up pairwise analysis and composition. Each pairwise CS can rule out potential degrees of freedom proposed by modeling, thus simplifying the construction of other

---

> **Input:** part shapes and initial configurations
> 1. modeling
> 2. pairwise analysis
> 3. composition
> **Output:** mechanism region diagram

Figure 11: *Region diagram construction algorithm for fixed axes mechanisms.*

pairwise CSs and of the mechanism CS. The program maximizes this benefit by constructing pairwise CSs in a breadth-first order starting from the frame, since interactions with the frame are typically the most restrictive.

The modular analysis strategy works for all feasible mechanisms. We can decompose the mechanism into linkage and fixed-axes subassemblies, construct the subassembly region diagrams, and compose them into a mechanism region diagram. We can construct the fixed-axes region diagrams with our program, analyze the linkages with a linkage package, and compose the outputs with the composition step of our program. We must manually set up the input for the linkage package and translate the output into a region diagram. Automating this interface would yield a complete CS construction program for feasible mechanisms.

In the following sections, we describe the three steps of the program and demonstrate it on the two-speed transmission.

4.1 Modeling

The modeling step finds axes of motions and assigns CS coordinates to the parts of a mechanism by analyzing the part shapes and the initial part contacts. It then determines the pairs of parts that potentially interact by intersecting motion envelopes. It never misses degrees of freedom or part interactions for fixed-axes mechanisms (under the guiding surface assumption stated below). It assumes that the input actually *is* a fixed-axes mechanism. It might miss degrees of freedom of other mechanisms.

The inputs are the shapes, initial configurations, and initial pairwise contacts of the parts. The outputs are the motion axes and CS coordinates of each part and the pairs of potentially interacting parts. Each part is

Stop

described by a boundary representation with respect to its local reference frame [12]. Parts are labeled as either mobile or part of the fixed frame. We represent the contacts by a graph whose nodes are the parts and whose links specify contacts between pairs of features. We could derives the contact graph from the shapes and initial configuration [30], but have not implemented this.

4.1.1 Axes of motion

The program identifies the rotational and translational axes of motion of the parts, the lines along or around which parts move without being completely blocked by immediate neighbors. The axes determine the potential degrees of freedom of the parts. (The pairwise analysis and composition steps determine the true degrees of freedom by eliminating potential degrees of freedom that are blocked by part interactions.) We assume that the axes of motion are determined by permanent surface to surface part contacts, called *guiding contacts*. Most mechanisms (and virtually all mechanisms in the survey) satisfy this condition by design because guiding contacts resist wear better than do line or point contacts or changing surface contacts. A guiding contact in a fixed-axes mechanism restricts the motion of each touching part to translation in a fixed plane or to motion along a fixed axis.

For 2.5D parts with planar and cylindrical faces, all surface to surface contacts occur between coaxial cylindrical faces and parallel planar faces. Contacts between coaxial cylindrical faces constrain the relative motions of the parts to rotation and translation along the axis of the cylindrical faces. Contacts between parallel planar faces constrain the relative motion of the part to translation along that plane. The combined contact constraints of a part determine its plane or axis of motion. In the transmission example, the cylindrical shaft $S1$ moves along axis $o1$ because of the contact with the cylindrical hole in the frame F (Fig. 2), and the gearshift P translates along axis $o3$ because of the vertical and horizontal contacts with the guiding slot in the frame (Fig. 12).

The program derives the axes of motion from the surface to surface contacts in the initial configuration. It collects the contact constraints by a breadth-first search of the contact graph starting from the frame. It replaces all intersecting planar constraints with their

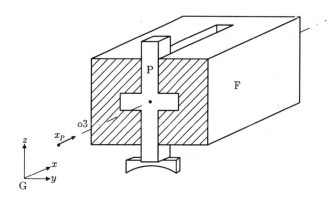

Figure 12: *Cross-section of gearshift P and the frame F. The global reference frame is G.*

lines of intersection. The resulting constraints define a set of potential axes and planes. The program derives a translational axis, a rotational axis, or a translational plane for each part from its potential axes and planes by the following rules. A single potential axis yields an axis of rotation and translation. Parallel potential axes yield an axis of translation. Two or more parallel planes yield a plane of translation. Every other set of potential axes and planes blocks the part.

For example, consider the contacts between the frame F and the gearshift P (Fig. 12). There are ten planar face contacts. The four horizontal contacts define four motion planes parallel to the xy plane. The six vertical contacts define six motion planes parallel to the xz plane. The program pairwise intersects the 10 planes, which yields 24 lines parallel to the x axis. It concludes that the gearshift translates along the x axis (or equivalently, along $o3$).

The program aligns the coordinates of the parts with their potential axes of motion. It assigns a CS coordinate to each potential degree of freedom along the axes and sets the other coordinates to constants. For example, the gearshift in Fig. 12 has translation coordinate x_P along axis $o3$. The other five coordinates, which specify the x orientation and the y and z position and orientation, are fixed.

The resulting coordinate system is compact and intuitive. It reduces the CS dimension by almost a factor of six because most mechanism parts have only one or two potential degree of freedom out of a possible six. Finding the axes of an n-part mechanism with m features per part takes at most $O(n^2m^2)$ time. The worst case happens when $O(m)$ faces of every part touch $O(m)$ faces of every other part.

4.1.2 Part interactions

The program determines potential interactions between parts by computing and intersecting motion envelopes. The motion envelope of a part is the volume that it sweeps as it moves along its axis or plane of motion. For two parts to interact, their motion envelopes must intersect. Although all $n(n-1)/2$ pairwise interactions can occur in principle, most mechanisms exhibit only $O(n)$ interactions because most parts interact only with a few neighboring parts. Motion envelope intersection rules out most of the non-interacting pairs without ever eliminating interacting pairs.

The program employs a rough test to detect potential interactions. It computes motion envelopes with respect to the bounding shapes of the parts. It bounds a fixed or translating part by an enclosing rectangular block whose dimensions are the maximum width, height, and length of the part. It bounds a rotating parts by an enclosing cylinder whose dimensions are the maximum diameter and height of the part along the axis. The motion envelopes are bounded or unbounded rectangular blocks or cylinders. The motion envelope of a fixed block is the block itself. The motion envelope of a block translating along an axis is a rectangular block that is unbounded in the direction of the axis. The motion envelope of a block translating on the plane is a block that is unbounded in the direction of the two axes that define the plane. The motion envelope of a rotating cylinder is the cylinder itself. The motion envelope of a rotating and translating cylinder is a cylinder that is unbounded in the direction of the axis. Testing if two motion envelopes intersect takes constant time.

Fig. 13 shows the engager E and the gearshift P, their bounding shapes, and their motion envelopes. The engager E rotates and translates along axis $o1$ and the gearshift P translates along axis $o3$. The bounding shape of the engager is a cylinder and the motion envelope is an unbounded cylinder along the x axis. The bounding shape of the gearshift is a rectangular block and the motion envelope is an unbounded block along the x axis. The parts interact, since their motion envelopes intersect.

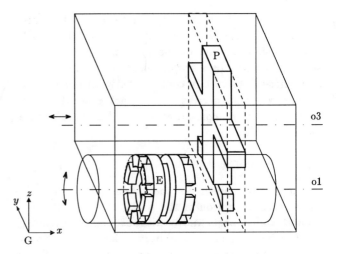

Figure 13: *The engager E and gearshift P (thick solid lines), their bounding shapes (dashed lines), and their motion envelopes (thin solid lines). Both motion envelopes are unbounded along the x axis. R is a global reference frame. Parts are shown disengaged for clarity.*

4.2 Pairwise analysis

The pairwise analysis step constructs a region diagram for the CS of a feasible kinematic pair (a 2.5D pair in which both parts move along fixed axes). The modeling step provides the axes and the motion types, which determine the CS topology (its dimensions and motion types). The CS describes all possible interactions between the parts. It depends solely on the shapes and degrees of freedom of the pair, not on the rest of the mechanism.

The CS consists of the joint configurations where no feature of the first part overlaps any feature of the second part. Every pair of features generates a *contact curve*, a 1D CS boundary consisting of the configurations where they touch. The contact curve separates the configurations where the features overlap from those where they do not touch. It may contain unrealizable segments where other contacts prevent its contact. The connected sequences of realizable segments form the free space boundary and define the CS regions. The program constructs the region diagram by deriving the contact curves and tracing the realizable segments, which form the region boundaries.

The motion types and shapes of a pair of features determine their contact curve. Parts in a feasible pair can

PLANAR

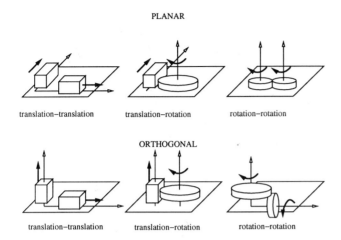

translation–translation   translation–rotation   rotation–rotation

ORTHOGONAL

translation–translation   translation–rotation   rotation–rotation

Figure 14: *The six feasible pairwise motion types with respect to a reference plane: planar and orthogonal pairs.*

translate along axes parallel or orthogonal to the reference plane and rotate around axes orthogonal to the reference plane. This yields six pairwise motion types (Fig. 14). Feasible 2.5D part features can be vertices, line segments and circular arcs parallel to the reference plane, and planar or cylindrical faces orthogonal to the reference frame. The 6 motion types and 25 feature interactions define 125 contact curves, which are specified by parameterized equations in the CS coordinates. For example, a rotating vertex A and a translating edge $B_1 B_2$ generate the contact curve:

$$x = R_A(\sin(\theta+\psi_A)-\cos(\theta+\psi_A)\tan\psi_B))+d\tan\psi_B-\delta_B$$

where θ and x are the CS coordinates and the other parameters are defined in Fig. 15. We reduce the 125 cases to 30 by exploiting symmetry and subsumption between contact types. For example, the contact curve of a rotating vertex touching a rotating edge is identical to the contact curve of a rotating edge touching a rotating vertex. Also, contacts between pairs of vertices are subsumed by contacts between vertices and the endpoints of line segments or circular arcs. We have compiled a contact curve table for all 30 cases indexed by feature and motion type.

As an example, consider the paiwise analysis of the reciprocating cam shown in Fig. 16, which consists of a cam and a follower mounted on a fixed base. Each time the cam rotates, the follower oscillates back and forth three times with short rest periods in between. The cam is pinned to the base and rotates around axis

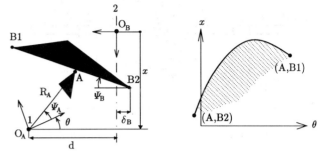

Figure 15: *A rotating vertex touching a translating line segment (left) and their contact curve (right). Vertex A rotates around axis 1 and is at a distance R_A and angle ψ_A from the origin of its reference frame O_A. Edge $B_1 B_2$ translates along axis 2 and is at distance δ_B and angle ψ_B from the origin of its reference frame O_B. The distance between the axes is d. x measures the offset of the origin of reference frame O_B from the axis of rotation. θ measures the angular offset of the axis of reference frame O_A from the horizontal axis passing through the axis of rotation. The endpoints of the contact curve, labeled (A, B_1) and (A, B_2), are where A touches the endpoints, B_1 and B_2, of the line. Shading indicates feature overlap.*

$o1$. The follower is mounted on the base and translates along axis $o2$. The angle θ and the displacement x are the CS coordinates.

The program first derives the contact curves of the pairs of features by looking up the defining equations in the table and filling in the feature parameters (such as length and endpoints positions). It eliminates the contact curves that are unrealizable because of local constraints. Examples of locally unrealizable contacts are features that are far apart, on different planes, or obstructed by neighbors. Each of the 30 contact curves has a constant-time feasibility test involving only the feature pair and the neighboring features. For example, for the rotating edge touching a translating line segment case in Fig. 15, we test that the edges incident to A form a convex corner and that $B_1 B_2$ is not too far to the left or to the right of A. (A similar test is used for the rotating circular arcs touching the translating line segments in Fig. 16.) Local constraints reduce the number of curves from $m_1 m_2$ to $O(m_1+m_2)$ on average with m_1 and m_2 the number of features per object. The worst case of $m_1 m_2$ occurs only if the parts

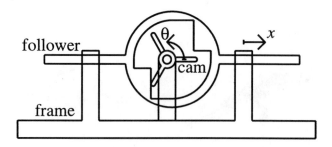

Figure 16: *Reciprocating cam mechanism.*

are close together, have many concave corners, and are not star-shaped.

The program partitions the CS into regions by tracing the locally realizable contact curves. It splits the contact curves at their intersection points by the Newton-Raphson method and sorts the resulting curve segments by the first coordinate of their endpoints. It traces the connected sequences of realizable segments by depth-first search. It repeatedly starts from the leftmost untraversed endpoint and traces through the segments until the traced path intersects itself or ends. At each step, a simple geometric test of the orientation and interior configuration of the current segment determines the next segment. If the sequence of segments intersects itself, the closed portion bounds a candidate region. The sequence is unrealizable otherwise. Fig. 17 shows the cam/follower's locally realizable contact curves and region diagram.

The program discards candidate regions that lie inside other candidate regions. It also discards regions that are unreachable from the initial configuration of the kinematic pair. This filtering greatly speeds up composition, as we illustrate in the discussion of the transmission below. For convenience in composition, the program subdivides the regions into convex regions using a line-sweep algorithm [23]. The resulting regions satisfy the monotonicity criterion of Section 2.2. (Other criteria can be substituted transparently by replacing the partition algorithm.) Fig. 17b shows a partition of the cam/follower's reachable free space into monotonic convex regions. The final number of regions is worst-case quadratic in the number of locally realizable constraints. Pairwise analysis of an n-part mechanism with m features per part takes $O(n^2 m^2 \log m)$ time in the worst case when all pairwise contacts and parts interactions are possible, and $O(nm \log m)$ on av-

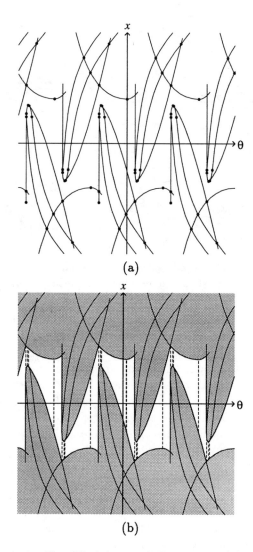

Figure 17: *The CS of the cam/follower pair: (a) CS contact curves and their intersection points; (b) region partition after line sweep. Dashed vertical lines indicate region boundaries (shading indicates part overlap).*

erage. [Note: See the Update Section.]

We have tested the pairwise analysis program on tens of examples, including the reciprocating cam and all the interacting pairs in the transmission. Running times range from less than a minute for the cam to ten minutes for a pair of 20-tooth gears.

4.3 Composition

The composition step composes the subassembly region diagrams of a mechanism into a mechanism region diagram. It handles subassemblies whose regions are polyhedral (defined by linear constraints) or of dimension two or lower. The legal inputs cover feasible kinematic pairs along with most linkages and other standard subassemblies. The program constructs the exact CS when all the subassembly regions are polyhedral and an approximate CS otherwise. The approximate CS usually is qualitatively correct and in good quantitative agreement with the true CS. Like other qualitative physics representations, it can contain unrealizable behaviors, but cannot miss true behaviors.

A mechanism configuration is realizable if no two parts of any subassembly overlap, that is if every subassembly configuration is realizable. Hence, the mechanism CS equals the intersection of the subassembly CSs. We obtain a mechanism region diagram by intersecting all combinations of subassembly regions, called *component sets*, and discarding the empty intersections. We guarantee that each component set yields at most one region by splitting the subassembly regions into convex regions. Two regions are adjacent if every pair of corresponding components is identical or adjacent in its subassembly CS.

The partition criteria of the subassemblies determine the mechanism partition criterion. In our case, monotonicity and convexity propagate from the kinematic pairs to the fixed-axes mechanism. The correctness of the composition algorithm does not depend on the partition method, but the computational cost grows with the number and complexity of the subassembly regions.

Enumerating and intersecting all the component sets is impractical for most mechanisms. The computation time equals the number of component sets times the intersection time, both of which are exponential in the number of parts. For example, a mechanism with ten parts has 45 kinematic pairs, which yield $2^{45} = 3.5 \times 10^{13}$ component sets when each pairwise CS has two regions. We develop an incremental algorithm that examines only the component sets reachable from the initial mechanism configuration. The algorithm performs well because of the design of mechanisms, although it cannot avoid the exponential worst-case time complexity of CS construction [3]. The tight coupling among parts makes most component sets unreachable

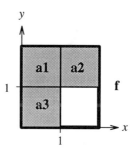

Figure 18: *The 3-puzzle.*

from the initial configuration. The restrictions on the shapes and motions of parts simplify component set intersection.

4.3.1 Component set enumeration

We implement component set enumeration for general mechanisms. The program initializes a search queue with the component set that contains the initial mechanism configuration. The components are the regions of the subassembly CSs that contain the initial configuration. (The program determines the components by projecting the initial configuration into the subassembly coordinates and finding the subassembly regions that contain the projections.) At each step, the program removes and intersects the first component set in the queue. If the intersection is nonempty, it records the new region, enumerates the reachable component sets, and adds the new ones to the queue. The reachable component sets are the adjacent component sets whose members are connected to the current region. Two component sets are adjacent if every pair of corresponding components is identical or adjacent. Two sets are connected if one contains a closure point of the other. For example, $(0, 1)$ connects to $[1, 2)$ but not to $(1, 2)$. The program can ignore unreachable component sets because the mechanism cannot enter the corresponding regions.

We illustrate the composition step on a 3-puzzle consisting of a fixed frame f containing square tiles a_1, a_2, and a_3 (Fig. 18). Tile a_i translates in the xy plane with coordinates (x_i, y_i) relative to a reference point at its bottom left corner. (We assume for simplicity that the tiles cannot rotate.) The initial configurations of a_1, a_2, and a_3 are $(0, 1)$, $(1, 1)$, and $(0, 0)$. Fig. 19 shows the pairwise CSs describing the interactions between

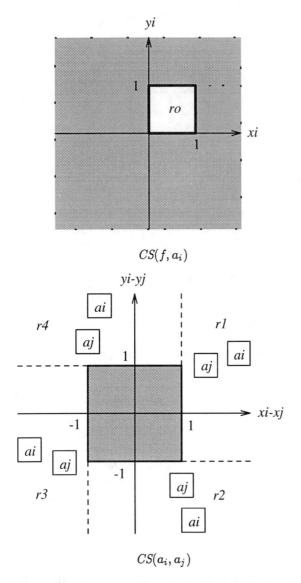

$$CS(f, a_i)$$

$$CS(a_i, a_j)$$

Figure 19: *Pairwise CSs. Dashed lines delimit regions, labeled with typical tile configurations. Shading indicates part overlap.*

CS pair	region	constraints	
$CS(f, a_1)$	r_0	$0 \leq x_1 \leq 1$	$0 \leq y_1 \leq 1$
$CS(f, a_2)$	r_0	$0 \leq x_2 \leq 1$	$0 \leq y_2 \leq 1$
$CS(f, a_3)$	r_0	$0 \leq x_3 \leq 1$	$0 \leq y_3 \leq 1$
$CS(a_1, a_2)$	r_3	$x_1 - x_2 \leq -1$	$y_1 - y_2 \leq 1$
$CS(a_1, a_3)$	r_4	$x_1 - x_3 \leq 1$	$y_1 - y_3 \geq 1$
$CS(a_2, a_3)$	r_1	$x_2 - x_3 \geq 1$	$y_2 - y_3 \geq -1$

Figure 20: *Components of the initial puzzle configuration.*

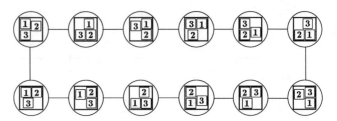

Figure 21: *Region diagram of the 3-puzzle CS. Regions are represented by typical tile configurations. The top left region is G_0.*

ing region G_0, a 1D submanifold of the 6D CS, defined by the constraints:

$$
\begin{aligned}
x_1 &= 0 & y_1 &= 1 \\
x_2 &= 1 & 0 &\leq y_2 \leq 1 \\
0 &\leq x_3 \leq 1 & y_3 &= 0. \quad (1)
\end{aligned}
$$

The first three components (interactions of the tiles with the frame) have no neighbors. The fourth has neighbors r_2 and r_4, but r_2 is unreachable because f and a_3 prevent a_2 from being above a_1. The fifth has no reachable neighbors because f and a_2 block a_1. The sixth has neighbors r_2 and r_4, but r_2 is unreachable because f and a_1 prevent a_3 from being above a_2. All told, G_0 has three neighboring component sets: $(r_0, r_0, r_0, r_4, r_4, r_1)$, $(r_0, r_0, r_0, r_3, r_4, r_4)$, and $(r_0, r_0, r_0, r_4, r_4, r_4)$. The program places them on the queue and searches them in turn. The first two yield the neighbors of G_0 and the third is empty. The full region diagram appears in Fig. 21.

The 3-puzzle demonstrates the effectiveness of the reachability criterion. The puzzle has 64 configurations because each tile has four possible configurations, but only 24 are reachable from the initial configuration. The 32 configurations where the tiles appear in counterclockwise order are not adjacent to any region. Another 8 configurations are not connected to any region

the frame and a tile and between two tiles. The first CS shows that the tiles stay inside the frame. The second CS shows that each tile in a pair can move around the other.

The program determines the initial component set by finding the regions in the pairwise diagrams that contain the initial configuration (Fig. 20). It intersects the components $(r_0, r_0, r_0, r_3, r_4, r_1)$ and records the result-

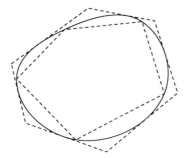

Figure 22: *Inner and outer CS boundary approximations of a 2D region.*

because parts cannot leave the frame.

4.3.2   Component set intersection

We implement component set intersection for mechanisms in which the subassembly CS regions are polyhedral or are of dimension two or lower. The program first tests whether the component set defines a region, that is whether the algebraic inequalities that define the components have a common solution. The worst-case time complexity of the test is exponential in the number of variables, making it impractical for most mechanisms [3]. Instead, the program approximates the nonlinear inequalities with a larger set of linear inequalities, which it tests in expected polynomial time with the BOUNDER inequality prover [25].

The program approximates every nonlinear component boundary from within and from without with piecewise linear segments (Fig. 22). It picks segments that preserve the topology of the subassembly CS and the monotonicity of its CS boundaries, and that differ from the original by a small tolerance [16]. The inner approximation specifies a subset of the true component and the outer approximation specifies a superset. If the intersection of the inner approximations is nonempty or the intersection of the outer approximations is empty, so is the region. Otherwise, the test is ambiguous. The program uses the outer result, which never misses regions, but can introduce spurious regions. The spurious regions correspond to part interactions beneath the granularity of the approximations. The program could resolve them by shrinking the granularity. Instead, we choose a reasonable tolerance based on standard machining assumptions.

After intersecting a component set, the program de-

motion types	motion relations
$fixed(a1, x, x_1)$	$x_1 = 0$
$fixed(a1, y, y_1)$	$y_1 = 1$
$fixed(a2, x, x_2)$	$x_2 = 1$
$translation(a2, y, y_2)$	$0 \le y_2 \le 1$
$translation(a3, x, x_3)$	$0 \le x_3 \le 1$
$fixed(a3, y, y_3)$	$y_3 = 0$

Figure 23: *Description of the initial 3-puzzle region.*

termines the true degrees of freedom of the parts in the intersection. Global interactions often rule out potential degrees of freedom by blocking axes of motion. In the 3-puzzle configuration of Fig. 18, each tile has two potential degrees of freedom, translation along both the x_i and y_i axes, but tile a_1 has zero true degrees of freedom and the other tiles have one apiece, as shown in Equation (1). The program bounds the coordinates of the parts with BOUNDER. It eliminates the degrees of freedom for which the lower and upper bounds coincide. For example, it eliminates translation along the x axis for a_1 because x_1 equals 0. Finding blocked degrees of freedom corresponds to finding implicit equalities in a set of linear inequalities [14]. The program uses the inner approximation for determining blocked degrees of freedom, so it ignores small motions caused by imperfectly fitting parts.

The algebraic relations that define a region succinctly characterize the precise kinematics of the mechanism. For fixed-axes mechanisms, the program annotates the region with symbolic motion predicates that describe the kinematics qualitatively. Each CS coordinate is associated with its part, motion axis, and motion type: *fixed, translation,* or *rotation.* Fig. 23 contains the description of the initial 3-puzzle region.

We have tested the composition program on a dozen examples, including the 3-puzzle, a tilted 6-puzzle, a door lock, and the transmission. Running times range from two minutes for the 3-puzzle to five minutes for the transmission.

4.4   The transmission revisited

The construction of the region diagram for the transmission example (Fig. 2) demonstrates the kinematic analysis program in a realistic setting. The transmission has 8 moving parts with an average of 35 faces per part. The program constructs an exact region diagram with 13 regions in about 40 minutes. Most of the time goes to computing pairwise CSs.

The modeling step determines the motion axes of the parts in contact with the frame: the gearshift P and cylindrical shafts $S1$ and $S2$. It deduces that the motion axes of $S1$ and $S2$ are $o1$ and $o2$ because the shafts are mounted on cylindrical holes in the frame. The shafts rotate around the axes because of the contacts of their shoulders with the frame. The gearshift has axis of motion $o3$ and translates along it, as described in the modeling section. The engager has axis of motion $o1$ because it is coaxially mounted on $S1$. Gears $G1$ and $G2$ rotate around $o1$. Gears $G3$ and $G4$ rotate around $o2$. The CS coordinates for each part are chosen along the axes of motion: x_P for P, x_E and θ_E for E, θ_{S1} for $S1$, θ_{G1} for $G1$, θ_{G2} for $G2$, θ_{S2} for $S2$, θ_{G3} for $G3$, and θ_{G4} for $G4$. The resulting CS is 9D, whereas the CS for 8 unconstrained parts is 48D.

The modeling step then computes and intersects the motion envelopes of the parts. The motion envelope of the frame F is a bounded block. The motion envelope of the gearshift P is an unbounded block. The motion envelopes of shafts $S1$ and $S2$ and gears $G1$ and $G2$ are bounded cylinders. The motion envelope of engager E is an unbounded cylinder. Motion envelope intersection retains 16 interacting pairs out of the possible 36. The frame interacts with all parts except the gears. Gearshift P interacts with $G1$, $G2$, and E. Shaft $S1$ interacts with $G1$, $G2$, and E. Gear $G1$ interacts with E and $G3$. Gear $G2$ interacts with E and $G4$. Shaft $S2$ interacts with $G3$ and $G4$.

The pairwise analysis step yields the following CSs for the 16 interacting pairs. The engager E and gear $G1$ have three joint degrees of freedom, x_E, θ_E, and θ_{G1}, but a 2D CS suffices because both parts rotate around the same axis (Fig. 7). In the region r_0, the engager and gear $G1$ turn independently and the engager is in *neutral* or in *high*. In the regions r_1–r_6 (one for each of the six lateral teeth of $G1$), the engager is in low and meshes with $G1$. Region transitions occur when the engager shifts between neutral and low. The engager and $G2$ have a dual CS in which $-x_E$ replaces x_E. The other pairwise CSs are 1D. The $G1/G3$ and $G2/G4$ CSs each consists of a single region, a line with negative slope. (These are really narrow 2D regions because of backlash. We abstract them into line segments as discussed in [17].) The engager/gearshift CS reduces

to an interval because the parts translate along parallel axes. The other CSs describe the interactions between the frame and the moving parts and between the shafts and the parts mounted on them. Each CS consists of one region.

The composition step constructs a region diagram with 13 regions (Fig. 3). It constructs the exact regions because all the constraints are linear. Gearshift P is in *neutral* and $S1$ and $S2$ turn independently in the 3D neutral region, which contains the initial mechanism configuration (Fig. 4). In the 2D regions low_1–low_6, the gearshift is in low and $S1$ drives $S2$ via $G1$ and $G3$ with six different angular offsets. In the 2D regions $high_1$–$high_6$, the gearshift is in high and $S1$ drives $S2$ via $G2$ and $G4$.

Pairwise analysis reduces the number of regions in the $G1/G3$ and $G2/G4$ CSs from m^2 to 1, with m the number of meshing teeth per gear, because the other regions are unreachable from the initial configuration. The composition step intersects 13 component sets out of 49 because the others are unreachable. Without this filtering, the program would have to intersect $49m^2$ component sets, an impossible task for realistic values of m. For example, $m = 20$ yields 19600 component sets.

4.5   Program extensions

Although the pairwise analysis step computes the exact pairwise CS constraints, the composition step linearizes them to test for feasibility. The approximation yields topologically correct and satisfactory CS approximations for most fixed-axes mechanisms, but the approximate CS might be either too coarse or incorrect for some linkages and cam mechanisms. We can obtain the exact CS by replacing the component set intersection module with Canny's decision procedure for polynomial inequalities [3], but at an exponential computational cost. A better approach is for the current module to identify the hard cases (where the inner and outer approximations disagree) and pass them to a nonlinear inequality reasoner, such as BOUNDER [25], Kramer's program [20], or Canny's algorithm. Piecewise linearization makes the extra computational cost for the hard cases affordable by handling most cases quickly.

5 Beyond feasible mechanisms

Now that we have developed a practical analysis algorithm for feasible mechanisms, we discuss infeasible mechanisms. We classify infeasible mechanisms according to the ways that they fail to be feasible based on our survey of 2500 mechanisms in Section 3.2. We treat kinematic pairs separately because they are the basic building blocks for other mechanisms. We estimate the sizes of the classes by surveying Artobolevsky's encyclopedia. We propose complete analysis strategies for some classes, propose partial strategies for others, and explain why the rest require new ideas. The extensions include standard component descriptions, special case algorithms, and abstraction. They cover 96% of kinematic pairs and 85% of mechanisms.

The extensions represent a trade-off between the assumptions we make about the input mechanism description and the capabilities of modeling and of subassembly analysis. For example, our program assumes that its input is a fixed-axes mechanism. If we remove this assumption, we must extend the modeling step to identify fixed-axes subassemblies. We aim to maximize the increase in coverage while minimizing input restrictions, running time, and programming effort.

5.1 Kinematic pairs

The survey of kinematic pairs in Table 1 indicates that 66% (370 out of 559) are feasible, that is they are 2.5D parts with a CS of dimension at most two. The infeasible pairs fall into three disjoint classes, depending on whether they violate the 2.5D part requirement, the 2D CS requirement, or both.

The first class, pairs with 3D parts and 2D CSs, contains 125 pairs (22% of all pairs). Of these, 15 are commonly used pairs, such as screws, worm gears, and bevel gears, and 32 contain a 3D cam. Fig. 24a shows an example of a 3D cam. The 3D conical cam rotates around the horizontal axis. The notch in the follower engages in the cam's helical groove. The follower reciprocates along the diagonal axis as the cam turns.

The second class, pairs with 2.5D parts and CSs of dimension three or higher, contains 32 pairs (6%). Ten pairs are planar gears. Fig. 24b shows a planar pair with a 4D CS: the operating claw mechanism of a motion picture camera. The pair consists of a driving cam, which rotates around an axis orthogonal to the

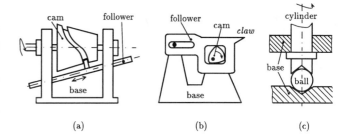

Figure 24: *Three examples of kinematic pairs not covered: (a) spatial conical cam (3032), (b) operating claw mechanism of a motion-picture camera (3107), and (c) ball-bearing turning pair (11). The numbers are indices from the encyclopedia.*

diagram, and a follower, which encloses the cam with a rectangular window and translates and rotates in the plane. The narrow slot in the follower is mounted on a pin attached to the base. As the cam rotates, the follower hooks the film with its claw (by grabbing it through a hole not shown in the figure), advances the film, and withdraws the claw. The cam has one degree of freedom and the follower has three.

The third class, pairs with 3D parts and CSs of dimension three or higher, contains 32 pairs (6%). Twelve pairs contain screws and 17 contain spherical parts. Fig. 24c shows a 3D ball-bearing pair with a 4D CS. The cylindrical link rotates around a vertical axis. The ball supports the link and has three independent rotational degrees of freedom.

The simplest way to extend the coverage of pairwise analysis is to hand-code the kinematics of common infeasible pairs. We can extend the coverage of kinematic pairs by 4% by hand-coding 25 standard pairs of meshed bevel gears, screw and nut pairs, and spherical joints. We can either extend the modeling step to recognize these pairs or require the user to identify them. The pairwise analysis step will then invoke the hand-coded modules with the parameters of the given parts. Hand-coding common feasible pairs also speeds up the analysis of feasible mechanisms.

After removing instances of the 25 standard pairs, 159 infeasible pairs remain. We can analyze most of them efficiently, thus covering 96% of kinematic pairs. We can analyze pairs in the first class (14%) with our pairwise analysis program because the contact curves and the CSs are 2D. We must augment the contact table with equations for all 3D feature contacts—a tedious task. We can analyze the important special case of cam pairs (6%) without tracing locally realizable CS curves, because of the single permanent contact between a cam and its follower.

For the second class, we must derive contact equations for high-dimensional feature contacts and must design a new boundary tracing algorithm for high dimensions. These are hard problems and have been solved only for planar pairs with 3D CSs (2%) [2]. We can analyze the important special case of gear pairs (2%) by abstracting away the details of the tooth interactions and reducing the CS dimension by symmetry. All told, the known methods cover 4% of the 6% of kinematic pairs that fall in the second class.

The third class is the hardest to automate because it combines the difficulties of the previous classes. However, only 2% of the pairs require a general algorithm. We can analyze pairs containing spherical parts (3%) by eliminating the rotational degrees of freedom, which do not affect the kinematics because of symmetry, thus lowering the CS dimension. For example, the ball-bearing pair in Fig. 24c has a 1D CS without the ball's rotational degrees of freedom.

5.2   Mechanisms

The survey of mechanisms in Table 2 shows that 58% (1108 out of 1912) are feasible. Feasible mechanisms are either linkages, fixed-axes mechanisms whose parts are 2.5D, or collections of fixed-axes subassemblies connected by linkages. The infeasible mechanisms fall into three disjoint classes depending on whether they violate the motion type requirements, the 2.5D part requirement, or both.

The first class, mechanisms whose parts are 2.5D but with infeasible subassemblies, contains 498 mechanisms (26% of all mechanisms). Of these, 209 (11%) have permanent contacts, including 94 (5%) planetary gears (gears that rotate about two different axes simultaneously) and 51 (3%) mechanisms in which a linkage subassembly interacts with a cam or a part with a guiding slot. The remaining 289 (15%) have varying part contacts, including brakes, clamps, and pawls. Fig. 25a shows a mechanism whose parts are in permanent contact: a slider-crank with a planetary gear. Driving

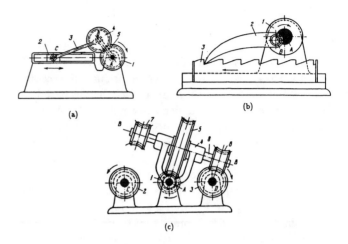

Figure 25: *Sample infeasible mechanisms: (a) planetary slider-crank (2421), (b) rack-type ratchet with complex pawl motion (2698), and (c) worm gears with alternately driven worms (2821). The numbers are indices from the encyclopedia.*

gear 1 rotates around axis A and meshes with planetary gear 4, which is rigidly attached to connecting rod 3. Crank 5 rotates around axis A and connects to rod 3 by revolute joint B. Rod 3 is connected by revolute joint C to slider 2, which translates along the fixed guides of the frame. As gear 1 rotates, gear 4 rotates simultaneously around axes A and B, causing slider 2 to reciprocate. Fig. 25b shows a mechanism whose part contacts vary: a ratchet-pawl mechanism. Driving wheel 1 rotates around axis A and connects to pawl 2 by revolute joint B. Rack 3 slides along a horizontal axis. As driving wheel 1 rotates, the tip of pawl 2 engages a tooth of rack 3, thereby moving it intermittently to the left. The contact between rack 3 and pawl 2 is broken when the pawl recedes and prepares to engage the next tooth. The pawl has 3 degrees of freedom and exhibits a complex motion.

The second class, mechanisms with 3D parts and fixed-axes or linkage subassemblies, contains 135 mechanisms (7%). Of these, 17 (1%) contain 3D cam pairs, such as the helical cam in Fig. 24a.

The third class, mechanisms with 3D parts and infeasible subassemblies, contains 171 mechanisms (9%). Of those, 70 have lower-pair constraints (surface to surface contacts) such as multiple motion joints, 48 have pairs with many degrees of freedom such as ball-bearings and screw adjustments, and 35 have varying part contacts. Fig. 25c shows an example of the later: an alternately driven worm-gearing mechanism. Driving worm 1 rotates about axis A and meshes with worm wheel 5. Output worms 2 and 3 rotate about fixed axes C and D. Worm gears 5, 6, and 7 are mounted on shaft 8 and rotate around axis B. Shaft 8 is mounted on yoke 4, which turns freely around axis A. When the yoke turns, worm wheel 6 engages with worm 3 on the right or worm wheel 7 meshes with wheel 2 on the left. The rotation of the driving wheel is transmitted to shaft C or D.

Coupling the fixed-axes analysis program with a linkage package covers 58% of all mechanisms. The simplest way to extend mechanism analysis is to hand-code the kinematic description of common subassemblies. We can extend the coverage to 62% of mechanisms by hand-coding some 70 subassemblies, including the 25 standard pairs, planetary gears, and lower-pair joints. To analyze the remaining infeasible mechanisms, we must treat the subassemblies that are neither standard nor feasible as collections of kinematic pairs. The extensions to pairwise analysis discussed above directly extend mechanism coverage. For example, handling 3D cams pairs adds 2% to the coverage. Although we have not counted the mechanisms handled by each extension, we estimate that 85% are covered. The remainder require general pairwise analysis and composition, which appear intractable. We believe that automating the approximation and abstraction of shape and behavior is the most promising strategy.

6   Related work

Previous research in automating the kinematic analysis of mechanisms involves linkages, geometrical simulation, CS construction, and qualitative physics.

Linkage analysis packages determine the response of a linkage to an input motion. The input specifies a mechanism schematically as one-dimensional rods connected by joints. The package sets up and solves a mixture of algebraic equations that constrain the configurations of the links and differential equations that relate the input motion to the configurations and velocities of the links. Haug [11] surveys the commercially available linkage packages, which solve linkage equations numerically. Kramer [20] describes a geometric solution technique. His program simulates a linkage by symbolically deriving its kinematic equations and calculating the linkage configurations for particular values of the driving input constraints.

Two recent research projects incorporate kinematics into dynamical simulators. Cremer [5] describes a simulator that uses a model-based 3D part representation and that handles contact changes, friction, and collisions. The program can only model a few shapes, such as spheres and polyhedra. It sets up the Newton-Euler equations for a set of objects in an initial configuration, simulates until two objects collide, modifies the equations to reflect the new configuration, and continues simulating. The output of the program is an animation of the objects and graphs of the time evolution of their configurations, velocities, and accelerations.

Gelsey [9] describes a kinematic analysis program for 3D mechanisms with standard kinematic pairs and permanent part contacts. The program roughly follows the modular algorithm shown in Fig. 1. It uses heuristics to identify kinematic pairs and retrieve their kinematic description. It composes the pairwise kinematic constraints with linear algebra and simple composition rules. The program handles five standard kinematic pairs and a few simple linkage mechanisms, such as cranks. Gelsey [10] augments the program with dynamical simulation. This allows him to analyze mechanisms with non-standard kinematic pairs and varying part contacts. The simulation algorithm resembles Cremer's.

Deriving the kinematics of a mechanism by simulating its dynamics can be expensive and unreliable. Each simulation run traces a curve through CS, which depends on the initial configuration and on the input motion. Mapping a high-dimensional CS with multiple regions requires many runs, yet can never guarantee complete coverage. We guarantee complete coverage and map the CS more efficiently by working directly with the kinematic equations. We can then use the CS to make the dynamical simulation faster and more reliable, as discussed in the conclusions.

CS construction has received much attention from the motion planning community. Schwartz and Sharir [28] formulate the problem algebraically. Canny [3] proves a tight bound on CS construction that is polynomial in the part complexity and exponential in the degrees of freedom. Motion planning research provides many CS construction programs for a single moving object amidst fixed obstacles. For example, Lozano-Pérez [21] and Brost [2] compute the CS of a planar polygonal object with 3 degrees of freedom. Donald [6] computes the CS of a polyhedral object with 6 degrees of freedom. These algorithms apply to pairwise analysis, but do not scale up to three or more moving parts.

Qualitative physics research in kinematics focuses on abstract analyses of fixed-axes mechanisms. Faltings and Joskowicz independently propose CS partitions as a representation for kinematics. Faltings [7, 8] analyzes 2D kinematic pairs whose parts have one degree of freedom and whose boundaries are line segments and circular arcs. His program partitions the 2D CS into regions where contacts do not change and contact relations are monotonic. This representation, which he calls a place vocabulary, is equivalent to a pairwise region diagram without the constant contact requirement (section 2.2) whose motion parameter relations are qualitative. Nielsen [22] provides a qualitative composition algorithm for the pairwise place vocabularies of a fixed-axes mechanism with one degree of freedom per part. The user must specify which pairs of parts interact. Composing qualitative descriptions produces potentially ambiguous kinematic descriptions. Our composition algorithm covers all mechanisms and avoids ambiguity. The current implementation covers most mechanisms and minimizes ambiguity by piecewise linearization. Joskowicz [15] presents an unimplemented algorithm for composing pairwise CSs using composition rules. The composition rules do not cover all cases. None of this work addresses modeling.

7   Conclusions

In this paper, we systematize the kinematic analysis of mechanisms and describe a partial automation. We build upon the configuration space representation, which provides a complete, but computationally intractable, first-principles theory of kinematics. We relate the complexity of analyzing a mechanism to engineering assumptions about the shapes and motion types of its parts. We identify an important class of mechanisms for which kinematic analysis is feasible. We empirically validate that most mechanisms fall in this class with a survey of 2500 mechanisms.

We develop an efficient kinematic analysis algorithm for feasible mechanisms and implement the algorithm for fixes-axes mechanisms. The program significantly extends the coverage of existing kinematic analysis programs. It is the first program to handle mechanisms in which different parts interact in different configurations. The modeling step identifies the axes along which parts move and the pairs of parts that interact. It is the first complete implementation of these capabilities. The pairwise analysis step constructs the region diagram of a pair of 2.5D parts with two (and sometimes more) joint degrees of freedom. It is the first program to handle 2.5D parts. It solves realistic problems in minutes on a workstation. The composition step composes the pairwise region diagrams into a mechanism region diagram. It is the first implementation of this capability.

The region diagram that the program constructs for a mechanism is a complete behavioral description of its kinematics that has the advantages of both qualitative and quantitative representations. The nodes and links describe the topology of the CS. The regions describe the operating modes of the mechanism. In each region, the algebraic relations among the part coordinates exactly specify the motions of the parts and relations among part configurations.

Our kinematic analysis algorithm provides the basis for many engineering applications, including kinematic simulation, design generation and validation, and on-line catalog construction. Kinematic simulations predict the behavior of a mechanism for a given set of input motions. They are commonly used for visualization and design validation. Existing design packages handle only linkages and cam pairs, leaving out many feasible mechanisms. On-line mechanism catalogs are useful for engineers and for intelligent CAD programs. Creating and maintaining such catalogs requires the ability to analyze, describe, compare, and classify mechanisms according to their function and behavior. Joskowicz builds upon our algorithm in recent work on kinematic simulation [16], kinematic pair design [18], and mechanism comparison and classification [17].

Our program sometimes produces overly detailed and complex kinematic descriptions. For example, the exact relationship between two meshed gears shows backlash and slight variations in the transmission ratio. In the transmission analysis, we can safely assume that the transmission ratio is linear and that there is no backlash. We can produce more focused descriptions by using the CS simplification and abstraction operators developed by Joskowicz [16, 17]. Simplification operators suppress irrelevant information by adding constraints and assumptions. The constraints restrict the type and range of input motions based on the operating context of the mechanism and on dynamical considerations. Abstraction operators suppress details by defining multiple levels of resolution. Simplification and abstraction are essential for many common tasks, such as comparing mechanisms or classifying mechanisms by behavior.

Integrating kinematics with dynamics is essential for a complete analysis of mechanisms. Dynamics determines the actual behavior of a mechanism, taking into consideration the laws of physics and the external forces acting on it. Many mechanisms exploit dynamical phenomena, such as friction, springs, gravity, and inertia, along with kinematics. Their analysis involves formulating and solving differential equations. The information in the CS helps us simplify the dynamical equations, solve them numerically, and interpret the solutions. The blocked degrees of freedom reduce the number of dynamical variables. The axes of motion provide a simple coordinate system for the dynamical equations. In the transmission example, we can relate the drive power to the configuration and velocity of the output shaft with a single second-order equation, as opposed to 48 equations for the 8 independent parts. We are working on systematizing and automating this kind of analysis.

8   Update

Four years after writing this paper, we are confident that configuration space computation is a powerful method for computer-aided mechanism design and analysis. We are currently working on practical applications and on basic research related to mechanism design.

First, we significantly improved the higher pair CS partition algorithm, which proved to be a key component for many tasks. We replaced it with a customized, robust line-sweep algorithm that exploits the structure of contact curves [27, 19]. The current implementation analyzes pairs with thousands of contacts in under a second, versus hours for the original implementation. It now covers parametric motion paths and parametric

part shape features such as splines, in addition to fixed-axes motion paths and point, line, or arc features. We also extended the composition algorithm to linkages.

Our subsequent work builds upon the improved configuration space computation algorithm. We developed a kinematic simulation program that derives the behavior of a mechanism for given driving motions by tracing the configuration space path that the mechanism traverses while constructing the configuration space regions along the path [26]. It simulates external forces and frictions using a simple model of dynamics that captures their steady-state effect without the conceptual and computational cost of dynamical simulation. It outputs a concise, symbolic interpretation of the simulation and a realistic, three-dimensional animation. We developed a parametric design program that helps designers pick parameter values that realize desired behaviors by locally inverting the mapping from part shapes to configuration spaces [19]. We have integrated these components into HIPAIR, a problem solving environment for the analysis and design of mechanisms [27] and tested it on numerous examples.

Acknowledgements

We thank Brian Williams and Rajan Ramaswami for helpful comments on drafts of this paper. We thank Sridhar Kota for referring us to Artobolevsky's encyclopedia and Brian Williams and Randy Davis for helping us finding it in the USA. Rajan Ramaswami collaborated in the development of the pairwise analysis program. Elisha Sacks was supported by the National Science Foundation under grant No. IRI–9008527 and by an IBM grant.

References

[1] Artobolevsky, I. *Mechanisms in Modern Engineering Design*, volume 1–4. (MIR Publishers, Moscow, 1979). English translation.

[2] Brost, R. C. Computing metric and topological properties of configuration-space obstacles. in: *Proceedings IEEE Conference on Robotics and Automation*, pages 170–176, 1989.

[3] Canny, J. *The Complexity of Robot Motion Planning*. (MIT Press, Cambridge, MA, 1988).

[4] Chironis, N. *Mechanisms, Linkages, and Mechanical Controls.* (Mc-Graw Hill, 1967).

[5] Cremer, J. F. and Stewart, A. J. The architecture of *newton,* a general-purpose dynamics simulator. in: *Proceedings of the 1989 IEEE International Conference on Robotics and Automation,* pages 1806–1811, 1989.

[6] Donald, B. R. A geometric approach to error detection and recovery for robot motion planning with uncertainty. *Artificial Intelligence* **37** (1988) 223–271.

[7] Faltings, B. Qualitative kinematics in mechanisms. in: *Proceedings of the Tenth International Joint Conference on Artificial Intelligence,* pages 436–442, 1987.

[8] Faltings, B. Qualitative kinematics in mechanisms. *Artificial Intelligence* **44** (1990) 89–120.

[9] Gelsey, A. Automated reasoning about machine geometry and kinematics. in: *Proc. of the Third IEEE Conference on Artificial Intelligence Applications,* Miami, Florida, 1987.

[10] Gelsey, A. Automated physical modeling. in: *Proceedings of the Eleventh International Joint Conference on Artificial Intelligence,* 1989.

[11] Haug, E. (Ed.). *Computer Aided Analysis and Optimization of Mechanical System Dynamics.* (Springer-Verlag, 1984).

[12] Hoffmann, C. M. *Geometric & Solid Modeling.* (Morgan Kaufmann, 1989).

[13] Horton, J. and Jones, F. (Eds.). *Ingenious Mechanisms for Designers and Inventors,* volume 1–4. (The Industrial Press, New York, 1951).

[14] Huynh, T., Joskowicz, L., Lassez, C., et al. Practical tools for reasoning about linear constraints. *Fundamenta Informaticae* **15** (1991).

[15] Joskowicz, L. Reasoning about the kinematics of mechanical devices. *International Journal of Artificial Intelligence in Engineering* 4 (1989).

[16] Joskowicz, L. Simplification and abstraction of kinematic behaviors. in: *Proceedings of the Eleventh International Joint Conference on Artificial Intelligence,* 1989. Reprinted in [29].

[17] Joskowicz, L. Mechanism comparison and classification for design. *Research in Engineering Design* **1** (1990).

[18] Joskowicz, L. and Addanki, S. From kinematics to shape: an approach to innovative design. in: *Proceedings of the National Conference on Artificial Intelligence,* 1988.

[19] Joskowicz, L. and Sacks, E. Configuration space computation for mechanism design. in: *Proceedings of the 1994 IEEE International Conference on Robotics and Automation.* IEEE Computer Society Press, 1994.

[20] Kramer, G. A. Solving geometric constraint systems. in: *Proceedings of the National Conference on Artificial Intelligence.* American Association for Artificial Intelligence, 1990.

[21] Lozano-Pérez, T. Spatial planning: A configuration space approach. in: *IEEE Transactions on Computers,* volume C-32. IEEE Press, 1983.

[22] Nielsen, P. E. *A Qualitative Approach to Rigid Body Mechanics.* PhD thesis, University of Illinois at Urbana-Champaign, 1988.

[23] Preparata, F. P. and Shamos, M. I. *Computational Geometry.* (Springer-Verlag, New York, 1985).

[24] Reuleaux, F. *The Kinematics of Machinery: Outline of a Theory of Machines.* (Dover Publications, 1963).

[25] Sacks, E. Hierarchical reasoning about inequalities. in: *Proceedings of the National Conference on Artificial Intelligence,* 1987. Reprinted in [29].

[26] Sacks, E. and Joskowicz, L. Automated modeling and kinematic simulation of mechanisms. *Computer-Aided Design* **25** (1993) 106–118.

[27] Sacks, E. and Joskowicz, L. Mechanism design and analysis using configuration spaces. submitted to CACM, 1994.

[28] Schwartz, J. T. and Sharir, M. On the piano movers II. general techniques for computing topological properties on real algebraic manifolds. *Advances in Applied Mathematics* **4** (1983).

[29] D. S. Weld and J de Kleer (Eds.), *Readings in Qualitative Reasoning about Physical Systems.* (Morgan Kaufman, San Mateo, Ca., 1990).

[30] Wilson, R. H. and Rit, J. F. Maintaining geometric dependencies in an assembly planner. in: *Proceedings of the IEEE International Conference on Robotics and Automation,* pages 890–895, 1990.

Impulse-based Dynamic Simulation

Brian Mirtich, *University of California, Berkeley, CA, USA*
John Canny, *University of California, Berkeley, CA, USA*

We introduce a promising new approach to dynamic simulation called impulse-based *simulation. The distinguishing feature of this method is the unification of all types of contact (colliding, rolling, sliding, and resting) under a single framework; non-colliding contacts are simulated as a series of tiny microcollisions. The approach is simpler and more robust than previous constraint-based methods. Simulation results agree with physical experiments, and the method is fast enough to make real time simulation possible. In the course of describing impulse-based simulation, we present an efficient collision detection scheduling scheme and a fully general treatment of frictional collisions. We conclude with some of the results generated by our simulator.*

1 Introduction

The goal of dynamic simulation is to make the computer into a tool which mimics the physical world; the applications of such a tool are countless. Electronic prototyping allows the engineer to interactively test and modify designs while they are still on the drawing board, before an actual prototype is ever constructed [8]. Experiments which are too costly or impractical to perform can be simulated, such as failure mode tests of bridges or dams. Even experiments which are performed today, such as automobile crash tests, can be done at much lower cost and under more varied conditions. Finally, dynamic simulation is an integral part of the expanding area of virtual reality. In everything from architectural walk-through programs to flight simulators, virtual environments need to behave as closely as possible to the actual physical world we inhabit.

The foremost requirement of dynamic simulation is physical accuracy. The goal of a simulation system is not simply to produce an animation sequence which "looks right" to a human; the sequence must *be* right.

The simulation is to take the place of a physical model, and hence its utility is directly related to how well it mimics this physical system. Assumptions such as frictionless collisions may be allowable for generating realistic looking graphics, but they have no place in a system designed to model reality.

The second major requirement is computational efficiency. Clearly in virtual reality applications, the simulation must run in real time. Furthermore, in engineering applications it is most beneficial when the user can make changes to a design, and see the results at fully interactive speeds. If the designer must wait hours or even days to analyze the behavior of a system, the electronic prototype loses its great advantage over an actual physical prototype.

This paper discusses a new approach to dynamic simulation called impulse-based simulation. We have focused on the twin goals of physical accuracy and computational efficiency. Our simulator can accurately model complex dynamic systems in real time. The organization of this paper is as follows. Section 2 gives an overview of the impulse-based method for dynamic simulation, highlighting its differences from and advantages over more traditional constraint-based methods. Section 3 describes collision check scheduling, and how this standard bottleneck in dynamic simulation can be streamlined. Section 4 discusses our method of resolving collisions between bodies. We treat collisions in a fully general manner, accounting for friction as well as non-perfectly elastic behavior. Correctly computing collision impulses is critical for achieving physically accurate simulations. Finally, section 5 describes some of the simulations we have performed with our simulator, illustrating the speed and accuracy of the approach, and mentions some future work.

Algorithmic Foundations of Robotics
1-56881-045-8

407

1.1 Related work

Moore and Wilhelms give one of the earliest treatments of two fundamental problems in dynamic simulation: collision detection and collision response [13]. Hahn also pioneered dynamic simulation, modeling sliding and rolling contacts using impact equations [7]. His work is the precursor of our method, although we extend the applicability of impulse dynamics to resting contacts, and give a more unified treatment of multiple objects in contact. Furthermore, these early approaches all suffered from inefficient collision detection and unrealistic assumptions concerning impact dynamics (e.g. infinite friction at the contact point). Our method combines fast collision detection with a fully general treatment of frictional collisions. Cremer and Stewart describe *Newton* [6], probably the most advanced general-purpose dynamic simulator in use today. Newton's forte is the formulation and simulation of constraint-based dynamics for linked rigid bodies. It has been used to simulate a high degree of freedom walking robot [16], although the contact modeling is fairly simplistic. Baraff has published a great deal on simulation of bodies in contact [1, 2]. His work focuses on the resolution of forces when bodies are in resting (non-colliding) contact. His earlier work is for frictionless collisions, and he later showed that computing contact forces in the presence of friction is NP-hard [3]. A summary of his work in this area can be found in [4]. There are few full treatments of frictional collisions. Routh [15] is still considered the authority on this subject, and is the source cited by most mechanics texts which address it. More recently, Keller gives a slightly different treatment of frictional collisions [9]; our approach is quite similar to his. Wang and Mason have studied impact dynamics for robotic applications; their approach is based on Routh's, but deals only with the two-dimensional case [17]. Finally, a number of researchers have investigated several problems and paradigms for dynamic simulation and physical-based modeling. We refer the reader to [5, 18, 19].

2 Constraint-based versus impulse-based simulation

One of the most difficult aspects of dynamic simulation is dealing with the interactions between bodies in contact. Most of the work which has been done in this area falls into the category of constraint-based methods [5, 18, 6, 4]. An example will illustrate the approach. Consider a ball rolling along a table top. The normal force which the table exerts on the ball is a constraint force that does not do work on the ball, but only enforces a non-penetration constraint. In a constraint-based system, this force is not modeled explicitly but is instead accounted for by a constraint on the configuration of the ball (in this case, the ball's z-coordinate is held constant). The problem with this method is that as a dynamic system evolves, the constraints may change many times, e.g. the ball may roll off the table, may hit an object on the table, etc. Determining the correct equations of motion for the ball means keeping track of these changing constraints, which can become complicated. Moreover, it is not even clear what type of constraint should be applied; there exist at least two models for rolling contact which in some cases predict different behaviors [10]. Finally, impacts are not easily incorporated into the constraint model, as they generally give rise to impulses, not constraint forces present over some interval. These collision impulses must be handled separately, as in [1].

In contrast, our system is based on a method we call impulse-based dynamic simulation. Unlike constraint-based methods, no constraints are imposed on the configurations of the moving objects; when the objects are not colliding, they are in ballistic trajectories. The key advantage of the impulse-based method is the unification of all types of contact under a single model. The model used for collisions between objects can also be used for continuous contact situations in which one object is resting, sliding, or rolling on another object. Consider for example a block resting on a table. Under impulse-based simulation, the block is actually experiencing many rapid, tiny collisions with the table, each of which can be resolved as any other collision. During this small, vibratory motion, different corners of the block will collide and rebound with the table. We call these small, frequent collisions between objects in continuous contact *microcollisions*.

Now consider the case of a ball bouncing along a terrain as shown in figure 1. Under constraint-based simulation, the constraints change as the ball begins traveling up the ramp, leaves the ramp, and settles into a roll along the ground. All these occurrences must be detected and processed. Impulse-based simulation avoids having to worry about such transitions. In fact,

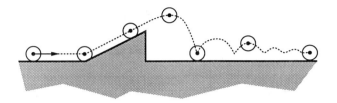

Figure 1: *A nightmare for constraint-based simulation.*

it is a more physically sound treatment since it does not establish an artificial boundary between, for example, bouncing and rolling, but instead handles the entire continuum of contact between these states.

Two obvious questions concerning impulse-based simulation are: (1) Does it work, i.e. does it result in physically accurate simulations?, and (2) Is it fast enough for simulation purposes? We defer more thorough answers to these questions to section 5, but for now state the following: impulse-based dynamic simulation *does* produce physically accurate results, even for cases which have been problematic for previous simulators. The microcollision contact model produces the correct macroscopic behavior; simulations agree with physical experiments. What is even more surprising is that the approach is extremely fast. Because of the simplicity of the model, the large number of collisions which must be resolved is not prohibitive. In fact, real time simulation is possible.

In summary, there are several advantages to impulse-based dynamic simulation:

1. All types of contact (colliding, rolling, sliding, sticking) are unified under a single approach; it is not necessary to classify the contact types and deal with them separately or in different manners.

2. It is not necessary to maintain a set of constraints, nor to determine when this set changes.

3. Simulating various types of contact with microcollisions gives rise to the correct macroscopic physical behavior, as verified experimentally.

4. The method is conceptually simpler and more robust then constraint-based dynamic simulation.

5. The method is fast. Despite the large numbers of collisions, real time simulation is possible.

3   Collision detection

Impulse-based dynamic simulation is obviously quite collision intensive—consider for example the high number of microcollisions which occur as a ball rolls across a table top. Furthermore, the naive approach to collision detection is inherently quadratic in the number of moving objects. Hahn and others have found collision detection to be the bottleneck in dynamic simulation [7], and certainly much effort should be put into streamlining it.

The heart of our collision detection system is the Lin-Canny closest features algorithm [12]. This algorithm returns the closest features (vertices, edges, or faces) between a pair of convex polyhedra. It is especially suited to applications like dynamic simulation, where a sequence of queries are made as objects move continuously in space. In these cases, geometric coherence can be exploited to achieve a constant expected query time. Non-convex objects can be handled by decomposition into convex pieces, and there is even an extension of the algorithm to curved surfaces [11]. Once the closest features are known, computing the distance between the two objects is a simple operation. The collision detection system reports a possible collision when this distance falls below some epsilon, ε_c. In our simulations using standard single-precision floating point arithmetic, ε_c is about three to four orders of magnitude smaller than the dimensions of the objects. For bowling simulations using a 60' alley and 18" inch pins, ε_c is one millimeter.

The basic simulation loop comprises three steps: dynamic state evolution, collision detection, and collision resolution. A naive approach for collision detection would test for possible collisions between all pairs of objects after each dynamic state evolution step. For a simulation involving n moving objects, this gives rise to an $O(n^2)$ collision detection step, despite the efficient constant time distance query. A second problem is how to choose the length of the integration time step, i.e. for how long should the dynamic state be evolved before the collision detection step is executed? Often, this integration step size is chosen to be "small enough so that no collisions are missed" (such as in [6]), but this is ad hoc and forces a small step size even when one is not necessary. We employ a strategy which reduces the number of collision checks from the naive approach, facilitates an adaptive step size for dynamic state evolution, and insures no collisions are missed.

3.1   Prioritizing collisions

The basic idea is to find a lower bound on the time of collision when two objects have not yet collided. Such a conservative bound is computed as a function of the distance between the closest points on the objects and the current dynamic state. It is also necessary to bound the linear and angular accelerations of the objects, since these also affect the minimum time to collision. By making the assumption that the objects will continue to travel in ballistic trajectories until impact, one can bound the linear acceleration of the center of mass to be g, the acceleration of gravity. Bounding the angular acceleration is a little trickier, but such a bound can be found as a function of the current angular velocity and the mass matrix of the object, again assuming a ballistic trajectory. We mention that the collision detection algorithm, which returns a soonest possible time of collision if the objects are not colliding, is in contrast to most other algorithms which simply compute a predicate indicating if the objects interpenetrate or not. For these latter algorithms, the exact time of collision is usually found by forward evolution and backtracking, using a binary search or other iterative method.

The information returned by the collision detection algorithm is stored in a heap; each heap item consists of a pair of objects and the soonest possible time at which these objects could collide. The heap is prioritized by collision time, so that at any moment, the object pair on the top of the heap is the nearest pair to collision. At each dynamic evolution step, the integration is performed up to the time of collision of the top item in the heap. At this point, collision detection is performed for this single pair of objects. If the objects have collided, the collision resolution step is performed, otherwise the time of impact is recomputed and the heap updated appropriately. When two objects do collide and experience collision impulses, the ballistic trajectory assumption is violated and all times of collision involving either of these objects must be updated in the heap. This strategy greatly reduces collision checks. If two objects are far apart or moving very slowly, collision checks between them will be very infrequent. As the objects approach one another, collision checks will increase as necessary.

3.2   Further reducing collision checks

Although the above strategy considerably lowers the asymptotic constant considerably, collision detection is still an $O(n^2)$ proposition. The problem is that collision checks between every pair of objects must still be performed at regular intervals, even if the pair of objects never come near one another. To alleviate this problem, some sort of spatial decomposition scheme must be built on top of the collision time priority queue. We now describe a method to eliminate these unneeded checks.

First, a basic time period t_f is chosen over which the swept volumes of moving objects will be bounded. It is convenient to choose t_f to be the frame period for the simulation, e.g. $t_f = \frac{1}{30}$ s. Let t denote the current time. For an object O, a rectangular bounding volume B_O with faces parallel to the global coordinate planes is computed. B_O bounds the volume occupied by O during the time interval $[t, t + t_f]$. Let r_O be the "radius" of object O, that is the greatest distance of any point on O from O's center of mass. B_O can be found by noting the position of O's center of mass at the current time t, at the time $t + t_f$, and possibly at the apex of its parabolic trajectory, should this occur during the interval $[t, t + t_f]$. The box which bounds these two or three points is then grown by r_O to give the final bounding volume, B_O (see figure 2).

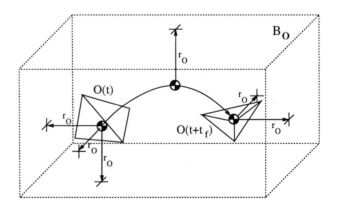

Figure 2: *The bounding box for an object O's swept volume.*

At the beginning of each frame, the bounding boxes for all objects are computed in linear time. The intersections between these n boxes are then found; for cubical boxes, Overmars has given an algorithm for doing this in $O(n(1 + \log R))$ time, where R is the ratio of largest to smallest box size [14]. If two bounding boxes do not intersect, then the corresponding objects will not collide during the next frame, and no distance or time of impact calculations need be performed for that pair of objects. If the boxes do intersect, these calculations are performed, and the upcoming collision is inserted into the heap. The simulation then proceeds as before, using an adaptive dynamic evolution step based on the pending time of collision at the top of the heap. Upon collision, the swept volume bounding boxes of the colliding objects are recomputed. These new bounding boxes are then intersected with the other bounding boxes, and object pairs are inserted or removed from the heap as appropriate. Note that distance and time of collision calculations are never performed for a pair of objects that never come near each other.

4 Computing collision impulses

When two bodies collide, an impulse $\mathbf{p}$ must be applied to one of the bodies to prevent interpenetration; an equal but opposite impulse $-\mathbf{p}$ is applied to the other. Once $\mathbf{p}$ and its point of application are known, it is a simple matter to compute the new linear center of mass velocity and the new angular velocity for each body. After updating these velocities, dynamic state evolution can continue, assuming ballistic trajectories for all moving objects. Thus, the central problem in collision resolution is to determine the collision impulse $\mathbf{p}$. Accurate computation of the impulses arising between colliding bodies or bodies in rolling or sliding contact is critical to the physical accuracy of the simulator.

4.1 Assumptions for collisions

For impulse-based simulation, it is not feasible to make gross simplifying assumptions such as frictionless contacts or perfectly elastic collisions. Our approach for analyzing general frictional impacts is similar to that of Routh [15], although we derive equations which are more amenable to numerical integration; Keller also gives an excellent treatment [9]. Before describing our method of computing $\mathbf{p}$, we state the assumptions.

Assumption 1 (Infinitesimal collision time) *The duration of a collision is negligible compared to the time*

over which simulated objects move appreciably. As a result, (1) the configurations of two colliding objects may be taken as constant during the entire collision, and (2) the effect of one object on the other can be described by an impulse, giving rise to instantaneous changes in the linear and angular velocities of the object.

This is a common assumption made in dynamic simulation [9]. The second part simply means that, unlike ordinary forces which can only affect accelerations instantaneously, the collision impulses can instantaneously affect velocities. Such behavior is necessary if we are to prevent objects from interpenetrating once they come into contact. What assumption 1 does *not* imply is that the collision can be treated as a discrete event. Even though the positions of the bodies are constant during the collision, the velocities are not. Since collision forces are functions of these velocities, it is necessary to examine the dynamics during the collision. One way to think of this is that the collision is a single point on the time line of the simulation, but to determine the collision impulses which are generated, we must use a magnifying glass to "blow up" this point, and examine what happens inside the collision.

In reality no body is completely rigid. When two bodies collide, there is a period of deformation in which elastic energy is stored in the bodies, and a restitution phase during which the bodies return to their original shapes (if the collision is non-destructive), rebounding off each other as the stored energy is released (see figure 3). One could use finite element analysis to study the stresses and strains occurring during a collision, but such a method is certainly not reasonable for real time simulation—furthermore, it is overkill. A simple empirical rule captures the essential behavior which occurs during this compression-restitution sequence.

Assumption 2 (Poisson's hypothesis) *For a collision between two objects, let p_{total} be the magnitude of the normal component of the impulse imparted by one object onto the other over the entire collision. Let p_{mc} be the magnitude of the normal component of the impulse imparted by one object onto the other up to the point of maximum compression. Then*

$$p_{total} = (1 + e)p_{mc}$$

where e is a constant between zero and one, dependent on the objects' materials, and called the coefficient of restitution.

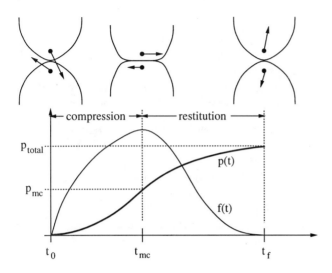

Figure 3: *A collision consists of a compression and a restitution phase. The boundary between these phases is the point of maximum compression, at which the relative contact velocity in the normal direction vanishes. $f(t)$ and $p(t) = \int f(t)dt$ are the force and total impulse delivered at time t in the collision.*

In other words, the normal impulses delivered during the compression and restitution phases are in the ratio $1 : e$. If $e = 1$, the collision is totally elastic; no energy is lost. If $e = 0$, the collision is totally plastic; in general not all the energy is lost, but the objects do not separate after collision. Poisson's hypothesis is useful for resolving collisions because it relates final impulse values to maximum compression impulse values. The point of maximum compression is easier to characterize than the point at which the collision ends. It is simply the point at which the normal component relative contact velocity vanishes.

The tangential component of the impulse has not yet been mentioned. Analyzing frictionless collisions is easy since this component vanishes, but in the presence of friction this component cannot be ignored.

Assumption 3 (Coulomb friction) *At a particular point during a collision between bodies 1 and 2, let* $\mathbf{u}$ *be the contact velocity of body 1 relative to body 2, let* $\mathbf{u}_t$ *be the tangential component of* $\mathbf{u}$, *and let* $\hat{\mathbf{u}}_t$ *be a unit vector in the direction of* $\mathbf{u}_t$. *Let* $\mathbf{f}_n$ *and* $\mathbf{f}_t$ *be the normal and tangential (frictional) component of force exerted by body 2 on body 1, respectively. Then*

$$\mathbf{u}_t \neq 0 \quad \Rightarrow \quad \mathbf{f}_t = -\mu \|\mathbf{f}_n\| \hat{\mathbf{u}}_t$$

$$\mathbf{u}_t = 0 \quad \Rightarrow \quad \|\mathbf{f}_t\| \leq \mu \|\mathbf{f}_n\|$$

where μ is the coefficient of friction.

While the bodies are sliding relative to one another, the frictional force is exactly opposed to the direction of sliding. If the objects are sticking (i.e. the tangential component of relative velocity is zero), all that is known is that the total force lies in the friction cone.

4.2 Initial collision analysis

A collision takes place between body 1 and body 2, as shown in figure 4. We introduce the following notation (the subscript i indicates the body number):

m_i	mass
J_i	mass matrix
$\mathbf{v}_i$	linear velocity of center of mass
$\mathbf{w}_i$	angular velocity
$\mathbf{u}_i$	absolute velocity of contact point
$\mathbf{r}_i$	contact point position relative to c.o.m.
$\mathbf{p}$	impulse imparted by body 2 on body 1

The vectors are expressed in the collision frame, which is some frame with z-axis perpendicular to the surfaces at the point of contact, and pointing from body 2 to body 1. When the colliding objects are polyhedra, the surfaces are not continuous, but reasonable surface normals can always be found. If one of the closest features is a face, the surface normal is the normal to this face; if the two closest features are edges, the normal is the vector mutually perpendicular to both edges; etc.

A possible collision is reported whenever the distance between two bodies falls below the collision epsilon, ε_c. The closest points on the objects are computed, and the collision frame is determined. A priori, the collision detection system only reports a *possible* collision, because the objects may be receding. If the closest points are moving away from each other, no collision impulse should be applied. (c.f. Baraff's constraint, $f\ddot{\chi}(\mathbf{f}) = 0$ [1].) The contact velocities are computed from

$$\mathbf{u}_i = \mathbf{v}_i + \mathbf{w}_i \times \mathbf{r}_i. \tag{1}$$

The relative contact velocity $\mathbf{u}$ is simply $\mathbf{u}_1 - \mathbf{u}_2$. If the z-component of $\mathbf{u}$ is non-negative, the objects are not colliding, and no action is taken.

When $\mathbf{u}$ has negative z-component, a collision impulse must be applied to prevent interpenetration; it is

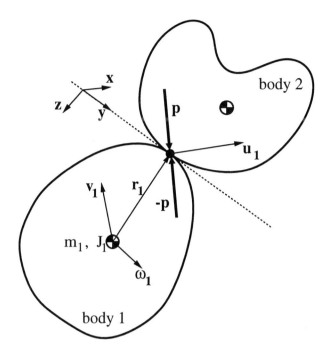

Figure 4: *Possible collision between two bodies.*

necessary to analyze the dynamics of the bodies during the collision to determine this impulse. We use γ to denote the collision parameter, that is γ is a variable which starts at zero, and continuously increases through the course of the collision until it reaches some final value, γ_f. All velocities are functions of γ, and $\mathbf{p}(\gamma)$ denotes the impulse delivered to body 1 up to point γ in the collision. The goal is to determine $\mathbf{p}(\gamma_f)$, the total impulse delivered.

Initially, one might choose γ to be "time since start of impact," but in fact this is not a very good choice. If the dynamics are studied with respect to time, the collision impulses are computed by integrating force. Unfortunately, the forces generated during a collision are not easily known; one can assume a Hooke's law behavior at the contact point, but this only leads to the question of how to choose the spring constants. Nonetheless, a variety of "penalty methods" do attempt to choose such spring constants. In addition to being chosen in a rather ad hoc way, these constants are very large, leading to stiff equations which are numerically intractable [18].

A way of avoiding all of these problems is to choose a different parameter for the collision, namely $\gamma = p_z$, the normal component of the impulse delivered to body 1. The scalar p_z is zero at the moment the collision begins, and increases during the entire course of the collision, so it is a valid parameter. In our analysis, we will continue to use γ to denote the collision parameter, for clarity. Consider the change in the contact point velocity of body 1 at a particular point during the collision

$$\Delta \mathbf{u}_1(\gamma) = \mathbf{u}_1(\gamma) - \mathbf{u}_1(0) = \Delta \mathbf{v}_1 + \Delta \mathbf{w}_1 \times \mathbf{r}_1. \quad (2)$$

Note that $\mathbf{r}_1$ is a constant throughout the collision, by assumption 1. Now $\Delta \mathbf{v}_1(\gamma)$ and $\Delta \mathbf{w}_1(\gamma)$ can easily be expressed in terms of the collision impulse:

$$\Delta \mathbf{v}_1(\gamma) = \frac{1}{m_1} \mathbf{p}(\gamma) \quad (3)$$

$$\Delta \mathbf{w}_1(\gamma) = J_1^{-1}[\mathbf{r}_1 \times \mathbf{p}(\gamma)]. \quad (4)$$

Substituting these into equation 2 and rearranging yields

$$\Delta \mathbf{u}_1(\gamma) = [\frac{1}{m_1} I - r_1^\times J_1^{-1} r_1^\times] \mathbf{p}(\gamma), \quad (5)$$

where I is the 3×3 identity matrix, and $r_1^\times$ is the canonical 3×3 skew-symmetric matrix corresponding to $\mathbf{r}_1$. Performing a similar analysis for $\Delta \mathbf{u}_2$ ($-\mathbf{p}$ must be used instead of $\mathbf{p}$), and using $\mathbf{u} = \mathbf{u}_1 - \mathbf{u}_2$ to compute *relative* contact velocity, we obtain

$$\Delta \mathbf{u}(\gamma) = M \mathbf{p}(\gamma), \quad (6)$$

where M is a 3×3 matrix dependent only upon the masses and mass matrices of the colliding bodies, and the location of the contact point relative to their centers of mass. By assumption 1, *M is constant over the entire collision.* We can differentiate equation 6 with respect to the collision parameter γ to obtain

$$\mathbf{u}'(\gamma) = M \mathbf{p}'(\gamma). \quad (7)$$

4.3 Sliding mode

While the tangential component of $\mathbf{u}$ is non-zero, the bodies are sliding relative to each other, and $\mathbf{p}'$ is completely constrained. Let $\theta(\gamma)$ be the relative direction of sliding during the collision, that is $\theta = \arg(u_x + iu_y)$.

Lemma 1 *If the collision parameter γ is chosen to be p_z, then while the bodies are sliding relative to one another,*

$$\mathbf{p}' = \begin{bmatrix} -\mu \cos \theta \\ -\mu \sin \theta \\ 1 \end{bmatrix}. \quad (8)$$

Proof: $p'_x = \frac{dp_x}{dp_z} = \frac{dp_x}{dt}\frac{dt}{dp_z} = f_x\frac{dt}{dp_z}$, where **f** is the instantaneous force exerted by body 2 on body 1. Under sliding conditions, assumption 3 implies $f_x = -(\mu\cos\theta)f_z = -(\mu\sin\theta)\frac{dp_z}{dt}$. Combining results gives $p'_x = -\mu\cos\theta$. The derivation for p'_y is similar. Finally, $p'_z = \frac{dp_z}{dp_z} \equiv 1$. $\square$

It is now clear why p_z is a good choice for the collision parameter. By applying the results of lemma 1 to equation 7, with θ expressed in terms of u_x and u_y, we obtain:

$$\begin{bmatrix} u'_x \\ u'_y \\ u'_z \end{bmatrix} = M \begin{bmatrix} -\mu\frac{u_x}{\sqrt{u_x^2+u_y^2}} \\ -\mu\frac{u_y}{\sqrt{u_x^2+u_y^2}} \\ 1 \end{bmatrix}. \qquad (9)$$

This nonlinear differential equation for **u** is valid as long as the bodies are sliding relative to each other. By integrating the equation with respect to the collision parameter γ (i.e. p_z), we can track **u** during the course of the collision. Because of the linear relationship between **p** and Δ**u** (equation 6), we can also track **p** throughout the collision. Projections of the trajectories into the u_x-u_y plane are shown in figure 5 for a particular matrix M; the crosses mark the initial sliding velocities. The trajectory plot is somewhat counterintuitive since for some initial conditions the sliding velocity *increases* although friction tends to resist sliding (for this plot $\mu = 0.55$). This is because the sliding velocity is also affected by the non-frictional (normal) component of the collision impulse, as shown in figure 6.

The basic impulse calculation algorithm proceeds as follows. After computing the initial **u** and verifying that u_z is negative, we numerically integrate **u** using equation 9. During this integration, u_z will increase[1]. When it reaches zero, the point of maximum compression has been attained. At this point, p_z is the total normal impulse which has been applied. Multiplying this value by $(1+e)$ gives the terminating value for the collision parameter, γ_f. The integration then continues to this point, and $\mathbf{p}(\gamma_f)$ is the desired total collision impulse.

---

[1]Baraff and others have noted that it is possible to construct cases for which u_z decreases as p_z increases [3]. However, this situation seems to be extremely rare; it has not occurred in any of our simulations.

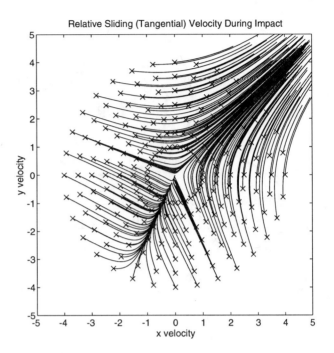

Figure 5: *Solution trajectories of equation 9 projected into the u_x-u_y plane.*

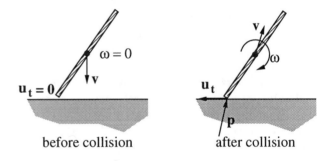

Figure 6: *A situation where the tangential relative contact velocity of the rod ($\mathbf{u}_t$) starts at zero and increases during the course of the collision, even though the frictional force resists this change in velocity.*

4.4 Sticking mode

Thus far we have not considered what happens if u_x and u_y both vanish, so that sliding motion ceases, and the objects are sticking. In this case, the direction of the frictional force is not known a priori, and lemma 1 no longer applies. The principle of virtual work implies that if the frictional force is strong enough to maintain the sticking condition, it will do so. To see if this is the case, we set $u'_x = u'_y = 0$ in equation 7, and solve for $\mathbf{p}'$. There is a unique solution for which $p'_z = 1$, say $\mathbf{p}' = (\alpha, \beta, 1)^T$. Now if

$$\alpha^2 + \beta^2 \le \mu^2, \tag{10}$$

the friction is sufficient to maintain sticking, and so $u_x = u_y = 0$ and $\mathbf{p}' = (\alpha, \beta, 1)^T$ for the duration of the collision.

If $\alpha^2 + \beta^2 > \mu^2$, the friction is not sufficient to maintain sticking, and sliding will immediately resume. Equation 9 is not valid when $u_x = u_y = 0$, and so is of no help in predicting the initial direction of sliding. In the case depicted in figure 5, there is a unique sliding direction leaving the origin; sliding must resume along this direction. It can be proven that the trajectories of equation 9 projected into the u_x-u_y plane never spiral around the origin, and we conjecture that in cases when the friction is not sufficient to maintain sliding there is always exactly one sliding direction away from the origin. Once u_x or u_y is nonzero, equation 7 again applies.

Our previous algorithm for computing collision impulses must be slightly modified to account for possible sticking. If at any point during the integration of $\mathbf{u}$, u_x and u_y both vanish, the integration is halted. If the criterion given by equation 10 is met, sticking is maintained for the duration of the collision and both $\mathbf{u}$ and $\mathbf{p}$ vary along a straight line. Otherwise, sliding resumes and the integration continues as before. Figure 7 illustrates some of the possible trajectories of $\mathbf{u}$ for different collisions. Path A represents a collision under low friction, in which the tangential component of relative contact velocity never vanishes, and the two objects slide on one another during the entire collision. Path C corresponds to a collision in which the frictional forces bring the sliding contact to a halt; as the object rebound off each other there is no relative sliding velocity. Finally, path B corresponds to a case in which sticking occurs momentarily, but the friction is insufficient to maintain this condition and sliding resumes.

4.5 Static friction under continuous contact

Consider a block sitting on a ramp, held at rest by friction. This continuous contact is modeled by a series of microcollisions. Under the model described thus far, the resulting behavior is exactly what would happen if

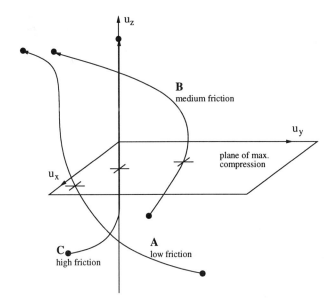

Figure 7: *Trajectories through relative contact velocity space for three different collisions.*

the ramp were experiencing very low amplitude, high frequency vibrations, namely, the block slowly creeps down the plane. The small collision impulses repeatedly bring the sliding to a stop, but during the ballistic phases gravity pulls the block slightly down the ramp. To eliminate this problem, a slight modification is needed to correctly model static friction under continuous contact.

A collision is characterized as a microcollision when the relative contact velocity in the normal direction is below some threshold. In this case, a check is first made to see if the impulse required to exactly reverse the contact velocity lies within the friction cone (using equation 6). If so, this impulse is applied, otherwise the full collision response computation is performed, as previously described.

There is physical basis for handling microcollisions in this manner. We are using impulses to model a contact force. In the case of one object statically resting on another, it is clear that this force does no work on the object. If we choose an impulse that accelerates the contact point velocity from $\mathbf{u}$ to $-\mathbf{u}$, and treat that impulse as a constant force acting for an infinitesimal time interval, than the impulse does no work. This is because the velocity of the contact point changes linearly from **u** to −**u** (equation 7, with γ denoting time), therefore the time integral of force times velocity, i.e. work done, vanishes. With this modification, the collision response algorithm correctly models static friction under continuous contact—the block does not creep down the ramp.

5   Results

We describe a few results produced by our simulator. All of these simulations were computed at close to real time speeds, and would be real time on a slightly faster platform (we are using a Silicon Graphics Iris Indigo XZ). For example, the colliding coins simulation, which involves eight moving objects and five fixed objects simulated for five seconds, takes less than 15 seconds to generate. These fast simulation times also reflect the efficiency of the collision detection algorithm: the bowling pins are 162-facet polyhedra; the marbles each have over 5000 facets.

5.1   Simulation descriptions

The colliding coins simulation involves eight coins being tossed or rolled into the center of a platform; complex interactions between the coins are followed by a segment dominated by microcollisions as the coins settle down or roll off the platform (figure 8).

In the bowling simulation, a bowling ball is thrown down an alley, with the same initial angular velocity that a bowler gives the ball upon release. With the low coefficient of friction between the alley and the ball, the ball initially slides down the alley, but it gradually accumulates a component of angular velocity in the forward rolling direction. The slow shift from sliding mode to rolling mode is complete as the ball hits the pins. This process gives the ball the familiar hook seen when good bowlers bowl. This portion of the simulation validates our collision model, and simulation of continuous contact by microcollisions. In the latter part of the simulation, the ball knocks down the pins, in a complex assortment of colliding and continuous contact (figure 9—it's a strike!).

We have run various marbles simulations which study the behavior of nearly elastic collisions between rolling balls. One of particular interest is a simulation in which one rolling black marble hits the end of a

line of four stationary marbles, causing the blue marble at the other end of the line to roll away, while the others remain basically at rest (figure 10). Constraint-based simulators often do not predict the correct response for these simultaneous or near-simultaneous collisions. However, under impulse-based dynamics this situation is treated no differently than any other series of collisions. Technically, the collisions are not simultaneous—they are transferred through the marble chain, just as they are in reality. It is not necessary to go through any contortions to get the proper response.

A final simulation worth mentioning is the ball on a spinning platter. What happens when a ball is placed on a spinning platter with a high coefficient of friction, such that the initial contact velocities match (i.e. the ball is rolling, not sliding)? The result is certainly non-intuitive, but has been verified by an actual experiment. The answer is that the ball rolls in circles, not centered at the axis of the platter, which gradually increase in radius until the ball rolls off of the platter [10]. When confronted with this problem, our simulator produced this very result, again verifying the collision model and the feasibility of generating correct macroscopic behavior through microcollisions (figure 11).

5.2   Conclusion

The impulse-based method is an excellent, new technique for dynamic simulation for two reasons: speed and accuracy. Simulations can be performed in real time, producing physically verifiable results. Most encouraging is the variety of systems and behaviors that can be modeled by our simulator; no ad hoc modifications or tweaking was necessary to produce results in agreement with the physical world. The impulse-based method is conceptually and algorithmically simpler than constraint-based methods, and perhaps the principle of Occam's razor applies here. After all, it seems most plausible that collisions forces in nature are based on local properties like contact velocity, and not on some global state configuration involving many bodies in contact.

There are situations in which constraint-based methods are more appropriate than the impulse-based approach—modeling an ideal hinge constraint by microcollisions between the hinge pin and its sheath would be slow and unnecessary (unless one were concerned with the actual vibration and slipping of the pin within the sheath). Future work should be done on integrating these two dynamic simulation methodologies into a hybrid system which can model linked bodies. We are also interested in enhancing our current simulator to make it more of an analysis tool. Statistics on contact forces and total impulses delivered would be useful, as well as a visualization capability for examining the forces at contact or collision points.

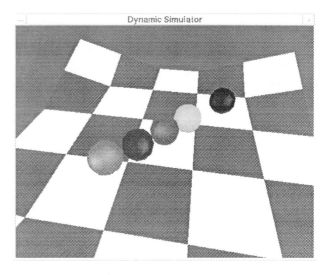

Figure 10: *Marbles. Snapshot taken just after the leftmost marble hit the marble chain.*

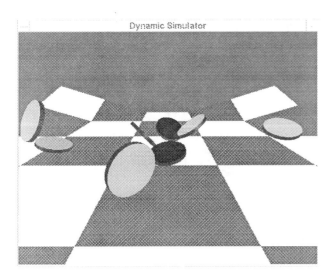

Figure 8: *Colliding coins.*

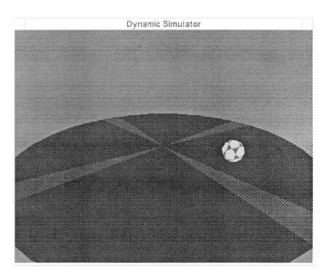

Figure 11: *Ball on a spinning platter.*

Acknowledgements

This research was supported in part by an NSF Graduate Fellowship, a David and Lucile Packard Fellowship, and a National Science Foundation Presidential Young Investigator Award (# IRI-8958577). We thank Jeff Wendlandt, Ming Lin, and numerous other inhabitants of the Berkeley Robotics Lab for many fruitful discussions.

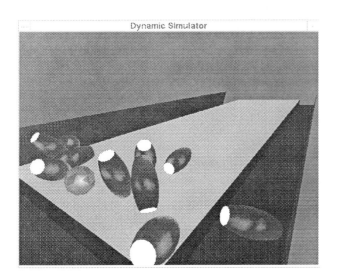

Figure 9: *Bowling.*

References

[1] David Baraff. Analytical methods for dynamic simulation of non-penetrating rigid bodies. *Computer Graphics*, 23(3):223–232, July 1989.

[2] David Baraff. Curved surfaces and coherence for non-penetrating rigid body simulation. *Computer Graphics*, 24(4):19–28, August 1990.

[3] David Baraff. Coping with friction for non-penetrating rigid body simulation. *Computer Graphics*, 25(4):31–40, August 1991.

[4] David Baraff. Issues in computing contact forces for non-penetrating rigid bodies. *Algorithmica*, 10:292–352, 1993.

[5] Ronen Barzel and Alan H. Barr. A modeling system based on dynamic constraints. *Computer Graphics*, 22(4):179–188, August 1988.

[6] James F. Cremer and A. James Stewart. The architecture of newton, a general-purpose dynamics simulator. In *International Conference on Robotics and Automation*, pages 1806–1811. IEEE, May 1989.

[7] James K. Hahn. Realistic animation of rigid bodies. *Computer Graphics*, 22(4):299–308, August 1988.

[8] John E Hopcroft. Electronic prototyping. *Computer*, pages 55–57, March 1989.

[9] J. B. Keller. Impact with friction. *Journal of Applied Mechanics*, 53, March 1986.

[10] A. Lewis, R. M'Closkey, and R. M. Murray. Modelling constraints and the dynamics of a rolling ball on a spinning table. Technical report, California Institute of Technology, 1993. Preprint.

[11] M.C. Lin and Dinesh Manocha. Interference detection between curved objects for computer animation. In *Models and Techniques in Computer Animation*, pages 431–57. Springer-Verlag, 1993.

[12] Ming C. Lin. *Efficient Collision Detection for Animation and Robotics*. PhD thesis, University of California, Berkeley, December 1993.

[13] Matthew Moore and Jane Wilhems. Collision detection and response for computer animation. *Computer Graphics*, 22(4):289–298, August 1988.

[14] M. Overmars. Point location in fat subdivisions. *Information Processing Letters*, 44:261–265, 1992.

[15] Edward J. Routh. *Elementary Rigid Dynamics*. 1905.

[16] A. James Stewart and James F. Cremer. Algorithmic control of walking. In *International Conference on Robotics and Automation*, pages 1598–1603. IEEE, May 1989.

[17] Yu Wang and Matthew T. Mason. Modeling impact dynamics for robotic operations. In *International Conference on Robotics and Automation*, pages 678–685. IEEE, May 1987.

[18] Andrew Witkin, Michael Gleicher, and William Welch. Interactive dynamics. *Computer Graphics*, 24(2):11–22, March 1990.

[19] Andrew Witkin and William Welch. Fast animation and control of nonrigid structures. *Computer Graphics*, 24(4):243–252, August 1990.

Completeness in Robot Motion Planning

Ken Goldberg, *University of Southern California, Los Angeles, CA, USA*

In robot motion planning, we say that an algorithm is complete *for a problem if it is guaranteed, for all instances of the problem, to find a solution when one exists and to return failure otherwise. Completeness is a desirable property. It provides a guarantee that the algorithm will work as expected for all inputs and thus can be dependably included in a larger system. This is especially important when algorithms are incorporated into industrial applications, where delays and failures can be extremely costly.*

In this paper we consider the completeness of several algorithms in robot motion planning. The two conditions for completeness, finding a solution if it exists, and terminating otherwise, suggest a distinction between algorithms that satisfy the former but not the latter: we refer to these as "exact" algorithms and give examples.

We then ask the stronger question: does a solution exist for all instances of a problem? We propose the term "solution-complete" to describe a property of problems and describe two recent results: a proof that a part feeding problem is solution-complete, and a proof that a fixturing problem is not solution-complete.

1 Introduction

Although robot motion planning is a general topic, we confine our discussion to the class of problems that can be formulated computationally, for example navigation, assembly, grasping, and the design of parts feeders. In all cases, we represent physical entities with mathematical models. Each problem defines a class of allowable inputs and a class of allowable solutions.

For example, the Piano Movers' problem asks if a robot can be moved from one configuration to another without colliding with obstacles. There are many versions of this problem depending on the class of allowable inputs (obstacles, robots, movements), and the

class of solutions (shortest or safest paths, etc.). We might consider a version of the Piano Movers' problem when inputs are restricted to 2D polygonal obstacles and a 2D polygonal robot, and solutions are restricted to shortest paths consisting of pure translations of the robot.

Latombe's text [17] gives a comprehensive review of developments in Robot Motion Planning. Latombe presents the concept of "configuration space" (C-space) as a unifying theme. In the Piano Movers' problem, the initial and desired final configurations of the robot can be specified as points in the continuous manifold of C-space. A problem instance defines surfaces in the C-space; solutions can be specified as free paths through this space. C-space can also be used to formulate problems involving contact mechanics and friction, for example the space of grasp configurations for three point contacts on a 2D object could be specified with a 3D C-space where friction defines subsets of the space that provide frictional form closure of the object.

As demonstrated in [20], problems that include uncertainty in the initial configuration can also be formulated in this framework, where all possible configurations of the robot is represented as a subset of C-space and commanded actions map between sets. Actions that allow compliant contact with a known boundary can be chained together to collapse a subset of C-space to a desired point.

In robot motion planning, a solution does not always exist. For example, it is easy to construct instances of the Piano Movers' problem such that there is no free path from initial to final configurations. Similarly with problems that include uncertainty in the initial configuration: it is easy to construct cases where there is no sequence of commanded actions that allow the initial subset to be collapsed to the desired point. For other problems, such as grasping, it may be difficult to construct instances where a solution does not exist;

Algorithmic Foundations of Robotics
1-56881-045-8

for some problems it may be possible to prove that a solution exists for all instances.

An *algorithm*, $\mathcal{A}$, for problem $\mathcal{P}$, accepts as input (a representation of) an instance of $\mathcal{P}$ and should return a solution. Some algorithms may fail to find a solution when one exists, or may fail to terminate. Examples will be given below. By convention, we say that $\mathcal{A}$ is *complete* for $\mathcal{P}$ if it is guaranteed to find a solution when one exists and to return failure otherwise [17, p.18][1].

Completeness is closely related to the way a problem is defined: algorithm $\mathcal{A}$ may not be complete for problem $\mathcal{P}$; however by adding additional assumptions limiting the class of inputs, it may be the case that $\mathcal{A}$ is complete for problem $\mathcal{P}'$.

Completeness is a desirable property. It provides a guarantee that the algorithm will work as expected for all inputs and thus can be dependably included in a larger system [8]. This is especially important when algorithms are incorporated into industrial applications, where a delay on the assembly line can cost thousands of dollars per minute. Completeness as specified above is related to the notion of *correctness* in program verification, where the behavior of a procedure is mathematically proved so that it can be reliably included in a large system.

As in program verification, the practice of proving completeness in robot motion planning is more often observed in the breach. Algorithms are often reported only with examples illustrating instances where they successfully find a solution. Completeness focuses attention on cases where algorithms may fail. There is a natural bias against thinking about such cases, and it can be extremely difficult to precisely characterize the class of inputs for which an algorithm is complete. Yet it is important to address this issue if robot motion planning algorithms are to be adopted outside of the laboratory.

In this paper we consider the completeness of several algorithms in robot motion planning. The two conditions for completeness, finding a solution if it exists,

and terminating otherwise, suggest a distinction between algorithms that satisfy the former but not the latter; we will use the term "exact" to describe the former property.

Closely related to the completeness of algorithms is the stronger question: does a solution exist for all instances of a problem? For example, can we specify conditions on inputs and outputs such that a solution exists for all instances of some version of Piano Movers' problem? In some cases it is possible to prove completeness; more commonly, it is sufficient to demonstrate a single counterexample. We introduce the term "solution-complete" to characterize this property of problems. We review two recent results: a proof that a part feeding problem is solution-complete, and a proof that a fixturing problem is not solution-complete.

2   Incomplete Algorithms

Consider a navigation algorithm that uses a naive potential field to trace a path from the initial pose toward a desired pose using gradient descent. It is well known that greedy algorithms such as this can lead to local minima which may cause the algorithm to terminate without finding a path when one exists. Thus this algorithm is incomplete: it is not guaranteed to find a solution even when one exists[2].

Algorithms that use a uniform grid (lattice) to discretely sample a continuous solution space generally sacrifice completeness since a solution may be hidden in the interstices. Consider the problem, introduced by Peshkin and Sanderson [26], of finding a sequence of fence angles that will orient a given polygonal part on a conveyor belt as it brushes past the sequence of fences. Consider an algorithm that samples the range of fence angles at $10°$ increments and enumerates all arrangements of length 1, 2, and so on, until a solution is found. Since this algorithm may overlook solutions that contain fences at odd angles, it is not guaranteed to find a solution when one exists.

Furthermore, this algorithm considers longer and longer sequences of fences until a predetermined length

---

[1]The term "complete" has other meanings. In logic, a theory (set of sentences closed under logical implication) is *complete* if every sentence or its negation can be proved. In complexity analysis, a problem to which all problems in a class can be reduced is *complete* for this class (*e. g.* NP-Complete).

[2]Rimon and Koditscheck recently gave conditions under which it is possible to construct a potential field with a unique minimum. These conditions define a class of problems that are solution-complete. For these problems the potential field algorithm is complete [28].

limit is used to terminate the algorithm. Thus there are two ways in which this algorithm may fail to find a solution: it may overlook a short solution that falls between lattice points, or it may terminate before considering a longer solution that lies on the lattice.

It is important to note that floating-point approximations can cause an algorithm to overlook solutions, as this is equivalent to imposing a uniform (yet fine!) grid on the set of available solutions. An otherwise complete algorithm may fail to find a solution if it is buried beneath the resolution of its arithmetic. In computational geometry, the *real-RAM* model of computation is generally assumed for the purposes of proof, although it is notoriously difficult to implement [16]. For computing with angular quantities, one approach is to use rational, or "exact", arithmetic [4, 23].

3 Exact Algorithms

Suppose an algorithm satisfies only the first the condition for completeness: it is guaranteed to find a solution when one exists. Such algorithms are thorough in that they will not overlook potential solutions; yet they are not necessarily complete. As suggested by John Reif, we use the term "exact" to describe such algorithms.

Consider a navigation algorithm that uses quad-tree decomposition to subdivide the configuration space. At each iteration the space is further divided in search of a collision-free path from start to goal. If a path exists, this algorithm will eventually find it. However, if a path does not exist, the algorithm may continue searching with finer and finer decompositions and may not terminate. This algorithm is exact but not complete[3].

One characteristic of exact robot motion planning algorithms is that they often partition a continuous C-space space into equivalence classes (cells) based on the geometry of the environment. For example, part vertices might define critical angles in a partition of C-space. Partitions induced by a collection of hyperplanes are called "arrangements" in Computational Geometry, which seeks precise bounds on the size of the partition [15].

---

[3]Suppose we terminate the decomposition at a pre-specified resolution. Latombe uses the term "resolution-complete" for the property that an algorithm is guaranteed to find a solution when a solution exists at that resolution.

4 Complete Algorithms

An algorithm is complete if it is exact and guaranteed to terminate with a negative report if a solution does not exist. A good example is the Visibility Graph algorithm for navigation, which searches a graph that contains a link between all pairs nodes that are "visible" in the configuration space. The algorithm is guaranteed to find the shortest path or to report that no path exists.

Other examples are Erdmann and Mason's tray tilting planner [10] and a search-based algorithm for planning sequences of grasp motions to orient polygonal parts [13]. Both of these algorithms partition the angular space of actions into a finite number of equivalence classes based on part shape and then enumerate all combinations of elements in this partition. Note that we can terminate a search path when we reach a state that has already been encountered since there is never any advantage to looping. Since there are a finite number of possible actions, and a finite number of possible state sets (the power set), these search based planners are guaranteed to find a plan if one exists and to eventually terminate otherwise.

Another example where the problem can be converted to a graph search is Erdmann, Mason, and Vanecek's algorithm for orienting three-dimensional parts [9]. Given an n-sided polyhedral part resting on a planar table, the objective is to find a sequence of tilting angles for the table that will bring a particular part face into contact with the table (thereby eliminating all but one degree of rotational freedom). The authors gave a graph-based algorithm that is guaranteed to find a plan if one exists and to report failure otherwise.

These algorithms test each cell in a finite partition of solution space. If a solution exists at any point in a cell, then all points in that cell are solutions. Thus the algorithm is exact and is guaranteed to terminate. Such partitions also characterize several algorithms based on what Donald calls a "non-directional backprojection". Donald originally proposed this partition for planning compliant motions in the plane [7]. It has since been applied to planning navigation with landmarks [19] and to Assembly Planning [30]. All of these algorithms are complete.

A related problem is finding grasps for polygonal parts. Here the solution can be characterized as 4 points along the continuous boundary of the part that achieves form closure (the contacts can resist any applied forces and torques). Nguyen [25] partitioned the part boundary into equivalence classes and gave an algorithm that finds a form closure grasp, if one exists, in time $O(n^4)$.

A variation of grasp planning restricts contacts to lie on a regular lattice, as is the case for modular fixtures. Modular fixtures are gaining wide use for flexible manufacturing and job shop machining. A modular fixture is an arrangement of fixture elements (fixels) that will locate and securely hold a given part. Recently, polynomial-time algorithms have been reported that are guaranteed to find a form closure fixture if one exists [3, 29].

5   Solution-Complete Problems

For algorithms that are complete, it is natural to ask: When does a solution exist?. In path planning, it is easy to construct cases where a solution does not exist. This is also true for motion planning with uncertainty by making the initial set of possible configurations sufficiently large. As stated above, it is sometimes possible to prove that a solution exists for all instances of a problem. In such cases we say that a problem is solution-complete. In other cases, we can construct a counterexample to prove the converse.

There are several examples in robot motion planning. One is Akella and Mason's proof that in the absence of obstacles, it is always possible to position and orient a polygonal part by pushing [1]. They prove this by defining a general linear program for the problem and using linear algebra to show that a positive solution must exist.

Another is Barraquand and Latombe's proof for a mobile robot, subject to (nonholonomic) rolling constraints, whose steering angle can take on only two distinct angles. The authors showed that in the absence of obstacles, a path always exists between any two planar configurations [2].

For a one-dimensional version of tray tilting, [9] showed that a plan always exists as long as the polygonal part has a unique angle in its transition diagram.

And although lower bounds on the number of contacts required to achieve form closure were known since the turn of the century, [24, 22] recently proved upper

bounds. In particular, that for any piecewise-smooth compact connected planar body, excluding surfaces of revolution, a form-closure grasp with 4 contacts always exists.

Solution completeness is similar to the notion of *controllability* in Control Theory: a linear system is controllable if, for any state of the system, there exists a solution (a control function) which will drive the system to the zero state in finite time [6, 18]. For example, it is known that in the absence of obstacles, a wheeled robot can reach any configuration in the plane. We might say that from any initial configuration we can always drive the robot to the zero configuration. Lynch and Mason recently used results from non-linear control theory to give conditions under which a plan exists to push a passive part to the zero configuration using a planar fence [21]. In such cases we say that the part is "controllable". It would be interesting to characterize a class of obstacles such that a plan always exists to reach the zero configuration. However it is not clear that results from Control Theory can be applied to this type of problem.

In the next section we describe a problem in Robot Motion Planning with Uncertainty and show that the problem is solution-complete.

6   Proving Solution-Completeness

Consider a planning algorithm that finds plans to orient a given part using a parallel-jaw gripper: given a list of n vertices describing a polygonal part whose initial orientation is unknown, find the shortest sequence of gripper actions that is guaranteed to orient the part up to symmetry in its convex hull. We prove that a plan exists for all polygonal parts and thus that the problem is solution-complete. The proof is based on an algorithm described in greater detail in [12].

When a polygonal part is grasped with the frictionless gripper, it assumes one of a finite number of "stable" configurations where at least one edge of the part's convex hull is in contact with a jaw. The outcome can be predicted with the **squeeze function**, $s : S^1 \to S^1$, such that if θ is the initial orientation of the part with respect to the gripper, $s(\theta)$ is the orientation of the part with respect to the gripper after the squeeze action is completed. For a polygonal part, the piecewise-constant squeeze function is derived as follows.

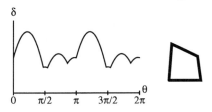

Figure 1: *A part and its width function.*

The **width function** for a $2D$ part is the distance between parallel supporting lines at angle θ as shown in Figure 1. All orientations that lie between a pair of adjacent local maxima in the width function will map to the orientation corresponding to the enclosed local minimum, *i. e.* the squeeze function is constant over this interval of orientations. The result is a step function.

We assume that all steps in the squeeze function are closed on the left and open on the right. Strictly speaking, the squeeze function has value $s(\theta) = \theta$ at its discontinuities, corresponding to an unstable equilibrium where the part is wedged between two exactly-aligned vertices. We could define a squeeze action at angle α to include closing and opening the gripper at angle α followed by closing the gripper at angle $\alpha + \epsilon$, rotating the gripper by $-\epsilon$, and then opening the gripper. In [14], we show how to find an appropriate ϵ for any polygonal part such that the combined action has a piecewise constant transfer function where each step is closed on the left and open on the right. In practice however, mechanical vibration in the gripping mechanism is sufficient to dislodge such wedged configurations, and after the first squeeze action causes the part to rotate into a stable configuration, the plan's margin for error (specified below) allows us to avoid actions that could produce a wedged configuration. For more on this issue, see [5].

In what follows, we use the term **interval** to refer to a connected subset of S^1. For an interval Θ, let $|\Theta|$ denote its Lebesgue measure. We define an **s-interval** to be a semiclosed interval of the form $[a, b)$ such that a, b are points of discontinuity in the domain of squeeze function s. For an s-interval Θ_X, let θ_X refer to its included bounding point. Since there are $O(n)$ discontinuities in the squeeze function, there are $O(n^2)$ unique s-intervals, each of which has non-zero measure. We define the **s-image** of a set, $s(\Theta)$, to be the smallest interval containing the following set: $\{s(\theta) | \theta \in \Theta\}$. Note that the s-image of any set will be a closed interval.

The algorithm begins with an s-interval whose image is a point. It continues, finding larger and larger s-intervals . When the algorithm terminates, the resulting sequence of s-intervals can be transformed into a sequence of squeeze actions that, in effect, "funnel" the largest s-interval into a unique final orientation. The algorithm is given below.

---

1. Compute the squeeze function.

2. Find the widest single step in the squeeze function and set Θ_1 equal to the corresponding s-interval . Let $i = 1$.

3. While there exists an s-interval Θ such that $|s(\Theta)| < |\Theta_i|$,

   - Set Θ_{i+1} equal to the widest such s-interval .
   - Increment i.

4. Return the list $(\Theta_1, \Theta_2, ..., \Theta_i)$.

We illustrate the algorithm using the squeeze function for the rectangular part as shown in Figure 2. Since this part has aspect ratio 1.5, let $a = \text{atan2}(3, 2)$.

In step 2 of the algorithm, the widest single step is found and Θ_1 is set to be the corresponding s-interval on the horizontal axis: $[\pi - a, \pi + a)$. Note that $s(\Theta_1)$ is the unique orientation at angle π.

In step 3 of the algorithm, we seek the widest s-interval whose s-image has smaller measure than Θ_1. As illustrated in Figure 3, this can be visualized by aligning the lower left corner of a box of dimension $|\Theta_1|$ with the leftmost point from each step in the squeeze function. If the squeeze function emerges from the right edge of the box, then the s-image of the corresponding s-interval has smaller measure than Θ_1. The largest such s-interval in this case is $\Theta_2 = [\pi - a, 2\pi - a)$. Note that $s(\Theta_2) = [\pi, 3\pi/2], |s(\Theta_2)| = \pi/2 < |\Theta_i| = 2a$,

Continuing in this manner, wider and wider s-intervals are found until the loop terminates. This will occur when $|\Theta_i| = T$, a period of symmetry in the squeeze function. For the rectangular part, the algorithm terminates with $i = 2$ since $|\Theta_2| = \pi$.

Theorem 1 *For any polygonal part, we can always find a plan to orient the part up to symmetry.*

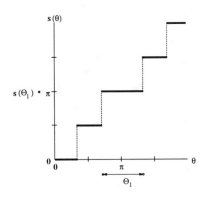

Figure 2: *In step 2, the widest single step in the squeeze function is identified. All the orientations in Θ_1 map into the single final orientation, $s(\Theta_1)$.*

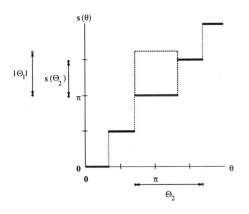

Figure 3: *Illustrating step 3 of the algorithm.*

Proof. Any polygonal part will generate a piecewise-constant squeeze function, $s : S^1 \to S^1$, where all s-intervals have non-zero measure and $s(\theta + T) = s(\theta) + T$, where T is a period of symmetry. To simplify the problem of wraparound at 0, in this section we extend s to a function on the real line, $s : \Re \to \Re$, that has exactly the same value as before on $[0, 2\pi)$. Elsewhere, it is specified as $s(\theta + T) = s(\theta) + T$.

We prove the claim by showing that for any such squeeze function, we can always find a sequence of s-intervals , $(\Theta_1, \Theta_2, ..., \Theta_i)$, of increasing measure with the condition that Θ_j has larger measure than the s-image of Θ_{j+1}. In other words, we must show that for any piecewise-constant monotonic squeeze function

and any s-interval , we can always find a larger s-interval unless it corresponds to a period of symmetry in the squeeze function.

Let h be the measure of some s-interval . Either we can find a larger s-interval whose s-image is smaller than h,

$$\exists \theta, s(\theta + h) - s(\theta) < h, \qquad (1)$$

Or h is a period of symmetry in the squeeze function:

$$\forall \theta, s(\theta + h) = s(\theta) + h, \qquad (2)$$

where the quantifiers range over the interval $[0, T]$.

To understand formula 1, consider that we've reached a point in the algorithm where the current s-interval is $\Theta_j = [\theta_j, \theta_j + h)$. Formula 1 says that there is some closed interval, $\Theta = [\theta, \theta + h]$, whose s-image is smaller than Θ_j. We can expand Θ (without increasing its s-image) by extending it to the right until we reach a discontinuity in the squeeze function. This yields an s-interval whose s-image is smaller that Θ_j. The difference between this s-interval and Θ is an open interval and hence has nonzero measure. Thus this s-interval will have larger measure than Θ_j.

We can also interpret formula 1 with reference to figure 3. The formula states that we can always find a position for the lower left hand corner of the box such that the squeeze function enters on the left edge of the box and exits on the right edge.

To show that for any such squeeze function and any h, either formula 1 or formula 2 must hold, consider the integral of the function $s(\theta + h) - s(\theta) - h$ over the interval $[0, T]$.

$$\int_0^T [s(\theta + h) - s(\theta) - h]d\theta$$

$$= \int_h^{T+h} s(\theta)d\theta - \int_0^T s(\theta)d\theta - hT \qquad (3)$$

$$= -\int_0^h s(\theta)d\theta + \int_T^{T+h} s(\theta)d\theta - hT \qquad (4)$$

$$= -\int_0^h s(\theta)d\theta + \int_0^h [s(\theta) + T]d\theta - hT \qquad (5)$$

$$= -\int_0^h s(\theta)d\theta + \int_0^h s(\theta)d\theta + hT - hT \qquad (6)$$

$$= 0. \qquad (7)$$

Since this integral is zero, either there is some point where the function is less than zero (formula 1 is true), or the function is uniformly zero (formula 2 is true, i.e. $h = T$).

Hence we can always continue to find larger s-intervals until we reach a period of symmetry in the squeeze function. We have shown earlier that we can transform this sequence of s-intervals into a plan to orient the part up to symmetry. Thus we've shown that a plan exists for all polygonal parts and hence that the problem is solution-complete. □

Rao and Goldberg recently extended this result to the class of algebraic parts [27]. We note that a proof of solution-completeness for problem $\mathcal{P}$ has consequences for its algorithms: if $\mathcal{A}$ is an exact algorithm for problem $\mathcal{P}$ and $\mathcal{P}$ is solution-complete, then $\mathcal{A}$ is complete. For example, the result given in this section implies that the exact algorithm in [13] is complete.

7 Disproving Solution-completeness

Brost and Goldberg [3] recently gave a complete algorithm that is guaranteed to find a form closure fixture for a given polygonal part in polynomial time and to terminate with a negative report otherwise. For all parts that we tested, the algorithm found dozens of solutions. This leads us to speculate that the problem is solution-complete. In this section we show that it is not.

The algorithm considers a class of fixtures using 4 frictionless point contacts: 3 locators and a clamp (3L/1C) such that each are attached to a square lattice of point holes. In particular, is any convex polygonal part fixturable in this fashion? Clearly if the part is very small with respect to the lattice spacing it may fall between the holes and thus not be fixturable. However, we might conjecture that for parts of sufficient "width" the method is complete: it is always possible to find a 3L/1C fixture to hold the part. Below we construct a counterexample; details can be found in [31].

The maximum and minimum values of the width function are well defined; we denote them with $\bar{d}(S)$ and $\underline{d}(S)$, respectively. The minimum width of a part is a useful way to characterize its size.

To construct a polygonal part of arbitrary size that is unfixturable, we show that for any given positive number M, we can construct a disk of size $> M$ that can make contact with at most 2 lattice sites. We then show how to transform this disk into a regular polygon

while preserving this property. Since this polygon can make contact with at most 2 locators, it cannot be fixtured under the 3L/1C model.

From geometry, we know that any three noncolinear points uniquely determine a circle. Thus, every non-trivial lattice site triplet, t, determines a disk which we denote by $s(t)$. If two triplets determine a disk of the same width, we say that these triplets are *equivalent*.

Let the *maximum width* of a triplet be the length of the longest side of the triangle it determines.

Lemma 1 *For any given width there exists a **disk** of greater width that can achieve contact with at most two lattice sites.*

Proof. For any given positive number M, let $S(M)$ be the set of disks with widths between M and $M + 1$ defined by triplets of lattice sites. $S(M)$ is finite since any disk $s(t)$ has a width no less than the maximum width of t and the number of non-trivial triplets with maximum width less than $M + 1$ is finite. Let us define the set of widths as

$$D(M) = \{d(s(t)) \mid s(t) \in S(M)\}$$

Then $D(M)$ is a finite set. Let c be any width in the interval $(M, M + 1]$ not in $D(M)$, the disk with c as its width can achieve contact with at most two fixel points. *(Figure 4)* □

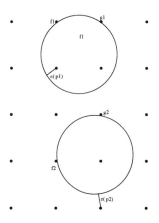

Figure 4: *The disk of indicated width can achieve contact with at most two lattice sites*

Based on this disk, we now construct a polygon that can achieve contact with at most two fixel contacts.

Theorem 2 *For any given width there exists a* **convex polygon** *of greater width that can achieve contact with at most two lattice sites.*

Proof. For any given positive number M, let S_c denote the disk with the width c, which is constructed in the lemma 1. By the proof of the lemma 1, S_c has at most two fixels on its boundary. Obviously for any fixel pair with a length less than or equal to c, we can always locate the disk S_c such that the two fixel points are on its boundary. Therefore we only need to consider the set of all non-identical pairs of sites with length less than or equal to c. Let us denote this set by Q. It is clear that Q is a finite set. Assume $|Q| = k, k > 0$,then we can represent Q as $Q = \{p_i \mid 1 \le i \le k\}$. We define $\epsilon(p)$ as the minimum distance from any fixels, other than the coordinates of p, to the boundary of the disk ∂S_c. Then we have $\epsilon(p_i) = \inf\{d_{\partial S_c}(f) \mid f \in F, f \notin p_i\} > 0$, for all $1 \le i \le k$, by the construction of the disk S_c. Let $\epsilon = \min\{\epsilon(p_i) \mid 1 \le i \le k\}$, then we have $\epsilon > 0$. There exists an inscribed regular polygon, P, of S_c, such that the length of its side is less than $\frac{1}{2}\epsilon$. *(Figure 5)* In order to achieve this, we only need to choose the number of sides of P, denoted by N, large enough, since the length of the side of P, denoted by L, satisfies the following

$$L = c \sin \frac{\pi}{N}.$$

Therefore we only need choose

$$N > \frac{\pi}{\sin^{-1} \frac{\epsilon}{2c}}.$$

Since $c > M$ (by the construction of S_c in the lemma 1), then we can select N to be large enough, such that $\underline{d}(P) > M$.

We denote the maximum distance between P and S_c by δ. Since $N \ge 3$, then

$$\delta = \frac{1}{2}L \tan \frac{\pi}{2N} < \frac{1}{2}L < L < \frac{1}{2}\epsilon.$$

Claim: Such a polygon, P, can achieve contact with at most two fixel points.

Proof of Claim: Let f, g and h be three fixel points, among which each pair has a length no greater than c. (If some pair has a length greater than c, then they cannot form a three-point contact, due to the fact that the maximum width of P is c.) Without loss of generality, we assume that P has f and g as its *two − point*

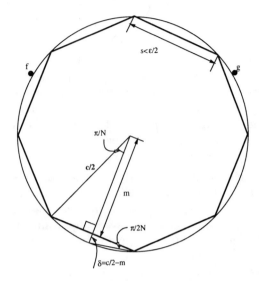

Figure 5: *The construction of the polygon that does not have a 3L/1C design.*

contact, and the contacting edges are e_f and e_g, respectively. Now we put P inside the disk S_c such that P is a inscribed polygon of S_c, then we position the disk (with P inscribed) on the lattice, such that f and g are on the arcs corresponding to e_f and e_g, respectively. (This can be always achieved.) By the construction of S_c, we know that $d_{\partial S_c}(h) \ge \epsilon$. Therefore, by the construction of the equilateral polygon P, the distance from h to the boundary of P is greater than $\frac{1}{2}\epsilon$, namely $d_{\partial P}(h) > \frac{1}{2}\epsilon$. In order to achieve *two − point* contact on the edges e_f and e_g, we can always rotate P inside the disk S_c first, then translate it to achieve the contact. After the rotation, f and g will be still on the corresponding arcs. Therefore the distance needed for translation is less than $\frac{1}{2}\epsilon$, because the distance of the translation is bounded by the length of the side of P, which is less than $\frac{1}{2}\epsilon$. But this translation won't be enough to make h the third point contact, since h is distant from the edge of P more than $\frac{1}{2}\epsilon$.

Hence P can achieve contact with at most two lattice sites.

□

Thus we have shown how to construct an infinite set of polygonal parts that cannot be fixtured using 3 point locators and a point clamp and so for any of these polygons the planning algorithm will terminate with no solutions. Thus the problem is not solution-complete.

Perhaps by suitably restricting the set of admissible parts we can show that a fixture always exists; this is currently an open problem.

8 Discussion

In this paper we focused on the issue of completeness in robot motion planning and specified three properties:

- An algorithm is *exact* if it is guaranteed to find a solution if one exists.

- An algorithm is *complete* if it is exact and guaranteed to terminate with a negative report if a solution does not exist.

- A problem is *solution-complete* if a solution exists for all instances.

There are many subtleties that are not addressed in this discussion, such as categories for probabilistic algorithms that terminate with probability one [11], and the relationship between completeness and complexity. Complexity analysis generally requires showing completeness: it is not clear how to define the asymptotic complexity of an algorithm that is not guaranteed to terminate. For solution-completeness, it is possible in some cases to prove that solutions always exist without providing a constructive algorithm for finding them. Similarly, the asymptotic complexity of the decision problem may be lower than the complexity of finding a solution.

As noted in the Introduction, algorithmic completeness bears some relation to the notion of correctness in program verification. In fact, Rimon and Koditschek use this language when they state that their results yield a "potential-function-based robot navigation algorithm that is provably correct" [28]. If these notions are identical, then why is "completeness" preferred in the motion planning literature? Etymology suggests that complete ("to make full") may be more appropriate than correct ("to make straight")!

We should note that all of our definitions are with respect to mathematical problems . The relationship between mathematical models and the physical world offers additional complications. For example, consider the navigation algorithm that uses the Visibility Graph. Although it is complete for the idealized mathematical model of the environment, this plan may perform extremely poorly when executed in the presence of disturbances in the physical environment.

Although the issue of completeness is familiar in Computational Geometry, it is often neglected in the Robotics literature. This may be due in part to the experimental nature of much robotics research, where empirical demonstrations are more common than mathematical proofs. Certainly it is often difficult to formally characterize complex problems in robotics; in many cases the abstractions used to simplify problems, such as assuming zero friction, may be of little relevance to practitioners. Yet empirical demonstrations of success without careful examination of failures leaves the impression that algorithms will work in all cases. This can lead to unpleasant surprises on the factory floor and hurt the credibility of future efforts.

As with asymptotic complexity, completeness focuses on worst-case scenarios. There is a natural bias against such "pessimism". Yet careful examination of such cases may lead to a proof of completeness or a counterexample, both of which are useful to practitioners. Completeness is also relevant to experimenters; although experiments cannot demonstrate completeness, they can be used to identify counterexamples. My hope is that focusing on these cases will ultimately lead to better algorithms.

Acknowledgements

This work was supported by the National Science Foundation under Award IRI-9123747 and by National Young Investigator Award IRI-9457523. I'm grateful to my students Yan Zhuang, Hadi Moradi, and Rick Wagner, and to the workshop participants, especially John Canny, Mike Erdmann, Danny Halperin, Dave Kriegman, Jean-Claude Latombe, Kevin Lynch, Anil Rao, John Reif, and Randy Wilson for their insightful feedback on early drafts of this paper.

References

[1] Srinivas Akella and Matthew T. Mason. An open-loop planner for posing polygonal objects in the plane by pushing. In *International Conference on Robotics and Automation*. IEEE, May 1992.

[2] Jerome Barraquand and Jean-Claude Latombe. Nonholonomic multibody mobile robots: Controllability and motion planning the the presence of obstacles. *Algorithmica*, 10(2):121–155, August 1993. Special Issue on Computational Robotics.

[3] Randy Brost and Ken Goldberg. A complete algorithm for synthesizing modular fixtures for polygonal parts. Technical Report SAND 93-2028, Sandia National Laboratories, Oct 1993.

[4] John Canny, Bruce Donald, and Gene Ressler. A rational rotation method for robust geometric algorithms. December 1991.

[5] Yui-Bin Chen and Doug Ierardi. The complexity of non-adaptive plans for orienting polygonal parts. Technical Report USC-CS-92-502, University of Southern California, December 1991.

[6] Bruce Donald. Guest editor's forward: The geometric theory of manipulation, planning and control. *Algorithmica*, 10(2):91–101, August 1993. Special Issue on Computational Robotics.

[7] Bruce R. Donald. The complexity of planar compliant motion planning under uncertainty. In *Fourth ACM Symposium on Computational Geometry*, 1988.

[8] Michael Erdmann. Personal Communication, 1990.

[9] Michael Erdmann, Matthew T. Mason, and George Vanecek Jr. Mechanical parts orienting: The case of a polyhedron on a table. *Algorithmica*, 10(2), August 1993. Special Issue on Computational Robotics.

[10] Michael A. Erdmann and Matthew T. Mason. An exploration of sensorless manipulation. In *IEEE International Conference on Robotics and Automation*, 1986. Also appears in IEEE Journal of Robotics and Automation, V. 4.4, August 1988.

[11] Mike Erdmann. Randomization in robot tasks. *IJRR*, 11(5), Oct 1992.

[12] Ken Goldberg. Orienting polygonal parts without sensors. *Algorithmica*, 10(2):201–225, August 1993. Special Issue on Computational Robotics.

[13] Ken Goldberg and Matthew T. Mason. Bayesian grasping. In *International Conference on Robotics and Automation*. IEEE, May 1990.

[14] Ken Goldberg and Anil Rao. Orienting planar parts. Technical Report IRIS-275, University of Southern California, September 1991.

[15] Dan Halperin and Micha Sharir. Arrangements and their applications in robotics: Recent developments. In *The First Workshop on the Algorithmic*

Foundations of Robotics. A. K. Peters, Boston, MA, 1994.

[16] C. M. Hoffman, J. E. Hopcroft, and M. S. Karasick. Towards implementing robust geometric computations. In *4th Symp. on Computational Geometry*. ACM, 1988.

[17] Jean-Claude Latombe. *Robot Motion Planning*. Kluwer Academic Press, 1991.

[18] Jean-Claude Latombe. Robot algorithms. In Takeo Kanade and Richard Paul, editors, *Robotics Research: The Sixth International Symposium*, 1993.

[19] Anthony Lazanas and Jean-Claude Latombe. Landmark-based robot navigation. Technical Report STAN-CS-92-1428, Stanford CS, May 1992.

[20] T. Lozano-Perez, M. T. Mason, and R. H. Taylor. Automatic synthesis of fine-motion strategies for robots. *International Journal of Robotics Research*, 3(1):3–24, Spring 1984.

[21] Kevin Lynch and Matt Mason. Stable pushing: Mechanics, controllability, and planning. In *The First Workshop on the Algorithmic Foundations of Robotics*. A. K. Peters, Boston, MA, 1994.

[22] Xanthippi Markenscoff, Luqun Ni, and Christos H. Papadimitriou. The geometry of grasping. *IJRR*, 9(1), February 1990.

[23] Matthew T. Mason. exact angle arithmetic package in commonlisp. (Ported to C++ by Jeff Wiegley (USC) in 1993)., August 1988.

[24] Bud Mishra, J. T. Schwartz, and M. Sharir. On the existence and synthesis of multifinger positive grips. *Algorithmica*, 2(4):641–558, 1987.

[25] Van-Duc Nguyen. Constructing force-closure grasps. *International Journal of Robotics Research*, 7(3), 1988.

[26] Michael A. Peshkin and Art C. Sanderson. Planning robotic manipulation strategies for workpieces that slide. *IEEE Journal of Robotics and Automation*, 4(5), October 1988.

[27] Anil Rao and Ken Goldberg. Manipulating algebraic parts in the plane. Technical Report 318, IRIS, December 1993. (To appear in the IEEE Transactions on Robotics and Automation).

[28] Elon Rimon and Dan Koditschek. Exact robot navigation using artificial potential functions. *IEEE Transactions on Robotics and Automation,* 8(5), October 1992.

[29] Aaron Wallack and John Canny. Planning for modular and hybrid fixtures. In *International Conference on Robotics and Automation.* IEEE, May 1994.

[30] Randy Wilson and Jean-Claude Latombe. One the qualitative structure of a mechanical assembly. In *National Conference on Artificial Intelligence,* 1992.

[31] Yan Zhuang, Yin-Chung Wong, and Ken Goldberg. On the existence of modular fixtures. In *International Conference on Robotics and Automation.* IEEE, May 1994. Also available as USC Techreport IRIS-93-314.

Information Invariants for Distributed Manipulation[1]

Bruce Randall Donald, James Jennings, and Daniela Rus

Robotics & Vision Laboratory
Computer Science Department
Cornell University
Ithaca, New York

In [Don4], we described a manipulation task for cooperating mobile robots that can push large, heavy objects. There, we asked whether explicit local and global communication between the agents can be removed from a family of pushing protocols. In this paper, we answer in the affirmative. We do so by using the general methods of [Don4] analyzing information invariants.

We discuss several measures for the information complexity of the task: (a) How much internal state should the robot retain? (b) How many cooperating agents are required, and how much communication between them is necessary? (c) How can the robot change (side-effect) the environment in order to record state or sensory information to perform a task? (d) How much information is provided by sensors? and (e) How much computation is required by the robot? To answer these questions, we develop a notion of information invariants. We develop a technique whereby one sensor can be constructed from others by adding, deleting, and reallocating (a) – (e) among collaborating autonomous agents. We add a resource to (a) – (e) and ask: (f) How much information is provided by the task mechanics? By answering this question, we hope to develop information invariants that explicitly trade-off resource (f) with resources (a) – (e). The protocols we describe here have been implemented in several differ-ent forms, and report on experiments to measure and analyze information invariants using a pair of coop-erating mobile robots for manipulation experiments in our laboratory.

1 Introduction

In this paper, we develop and analyze synchronous and asynchronous manipulation protocols for a small team of cooperating mobile robots than can push large boxes. The boxes are typically several robot diameters wide, and 1-2 times the mass of a single robot, although the robots have also pushed couches that are heavier (perhaps 2-4 times the mass, and 8×3 robot diameters in size). We build on the ground-breaking work of [Mason, EM] and others on planar sensorless manipulation. Our work differs from previous work on pushing in several ways. First, the robots and boxes are on a similar dynamic and spatial scale. Second, a sin-gle robot is not always strong enough to move the box by itself (specifically, its "strength" depends on the ef-fective lever arm). Third, we do not assume the robots are globally coordinated and controlled. (More pre-cisely, we first develop protocols based on the assump-tion that local communication is possible, and then we subsequently remove that communication via a series of source-to-source transformations on the protocols). Fourth, our protocols assume neither that the robot has a geometric model of the box, nor that the first moment of the friction distribution is known. Instead, the robot combines sensorimotor experiments and ma-nipulation strategies to infer the necessary information (the experiments have the flavor of [JR]). Finally, the pushing literature generally regards the "pushers" as moving kinematic constraints. In our case, because (i) there are at least two robot pushers and (ii) the robots are less massive than the box, the robots are really "force-appliers" in a system with significant friction.

[1] This paper describes research done in the Robotics and Vision Laboratory at Cornell University. Support for our robotics research is provided in part by the National Science Foundation under grants No. IRI-8802390, IRI-9000532, and by a Presidential Young Investi-gator award, and in part by the Air Force Office of Sponsored Re-search, the Mathematical Sciences Institute, Intel Corporation, and AT&T Bell laboratories. The third author has been supported by the Advanced Research Projects Agency of the Defense Department under ONR Contract N00014-92-J-1989, by ONR Contract N00014-92-J-39, and by NSF Contract IRI-9006137.
This paper is a revised version of a paper presented at the Interna-tional Symposium on Robotics Research in Hidden Valley, PA (1993).

431

In this sense, our task is in some ways closer in flavor to *dynamic manipulation* [ML], even though the box dynamics are essentially quasi-static.

Of course, our protocols rely on a number of assumptions in order to work. We develop a framework for analysis and synthesis, based on *information invariants* [Don4], to reveal these assumptions and expose the information structure of the task. We believe our theory has implications for the parallelization of manipulation tasks on spatially distributed teams of cooperating robots. To develop a parallel manipulation strategy, first we start with a perfectly synchronous protocol with global coordination and control. Next, in distributing it among cooperating, spatially separated agents, we relax it to an MPMD[2] protocol with local communication and partial synchrony. Finally, we remove all explicit communication. The final protocols are asynchronous, and essentially "uniform," or SPMD[2]—the same program runs on each robot. Ultimately, the robots must be viewed as communicating implicitly through the task dynamics, and this implicit communication confers a certain degree of synchrony on our protocols. Because it is both difficult and important to analyze the information content of this implicit communication and synchronization, we are wielding a fairly heavy hammer, namely the theory of information invariants.

1.1 The Big Picture

Our goal is to investigate the information requirements for robot tasks. This paper uses the theoretical framework introduced by Donald in [Don,Don4]. A central theme to previous work (see the survey article [Don1] for a detailed review) has been to determine what information is required to solve a task, and to direct a robot's actions to acquire that information to solve it. Key questions concern:

1. What information is needed by a particular robot to accomplish a particular task?

2. How may the robot acquire such information?

3. What properties of the world have a great effect on the fragility of a robot plan/program?

4. What are the capabilities of a given robot (in a given environment or class of environments)?

---

[2]SPMD (MPMD) = *Single (Multiple) Program, Multiple Data*.

These questions can be difficult. Structured environments, such as those found around industrial robots, contribute towards simplifying the robot's task because a great amount of information is encoded, often *implicitly*, into both the environment and the robot's control program. These encodings (and their effects) are difficult to measure. We wish to quantify the information encoded in the assumption that (say) the mechanics are quasi-static, or that the environment is not dynamic. In addition to determining how much "information" is encoded in the assumptions, we may ask the converse: how much "information" must the control system or planner compute? Successful manipulation strategies often exploit properties of the (external) physical world (eg, compliance) to reduce uncertainty and hence gain information. Often, such strategies exploit mechanical computation, in which the mechanics of the task circumscribes the possible outcomes of an action by dint of physical laws. Executing such strategies may require little or no computation; in contrast, planning or simulating these strategies may be computationally expensive. Since during execution we may witness very little "computation" in the sense of "algorithm," traditional techniques from computer science have been difficult to apply in obtaining meaningful upper and lower bounds on the true task complexity. We hope that a theory of information invariants can be used to measure the sensitivity of plans to particular assumptions about the world, and to minimize those assumptions where possible. We would like to develop a notion of information invariants for characterizing sensors, tasks, and the complexity of robotics operations. We may view information invariants as a mapping from tasks or sensors to some measure of information. The idea is that this measure characterizes the intrinsic information required to perform the task—if you will, a measure of complexity. For example, in computational geometry, a rather successful measure has been developed for characterizing input sizes and upper and lower bounds for geometric algorithms. Unfortunately, this measure seems less relevant in robotics, although it remains a useful tool. Its apparent diminished relevance in embedded systems reflects a change in the scientific culture. This change represents a paradigm shift from *offline* to *online* algorithms. Increasingly, robotics researchers doubt that we may reasonably assume a strictly offline paradigm. For example, in the offline model, we might assume that the robot, on

Figure 1: The Cornell mobile robot TOMMY. Note (mounted top to bottom on the cylindrical enclosure) the ring of sonars, the IR Modems, and the bump sensors. LILY is very similar.

booting, reads a geometric model of the world from a disk and proceeds to plan. As an alternative, we would also like to consider *online* paradigms where the robot investigates the world and incrementally builds data structures that in some sense represent the external environment. Typically, online agents are not assumed to have an *a priori* world model when the task begins. Instead, as time evolves, the task effectively forces the agent to move, sense, and (perhaps) build data structures to represent the world. From the online viewpoint, offline questions such as "what is the complexity of plan construction for a known environment, given an *a priori* world model?" often appear secondary, if not artificial. In this paper, we describe two working robots TOMMY and LILY, which may be viewed as online robots. We discuss their capabilities, and how they are programmed. These examples and the papers [JR,Don,Don4] link our work to the recent but intense interest in online paradigms for situated autonomous agents. In particular, in these papers, we discuss what kind of data structures robots can build to represent the environment. We also discuss the *externalization* of state, and the *distribution* of state through a system of spatially separated agents.

We believe it is profitable to explore online paradigms for autonomous agents and sensorimotor systems. However, the framework remains to be extended in certain crucial directions. In particular, sensing has never been carefully considered or modeled in the online paradigm. The chief *lacuna* in the armamentarium of devices for analyzing online strategies is a principled theory of sensori-computational systems. We attempt to fill this gap in [Don,Don4], where we provide a theory of *situated sensor systems*. We argue this framework is natural for answering certain kinds of important questions about sensors. Our theory is intended to reveal a system's information invariants. When a measure of intrinsic information invariants can be found, then it leads rather naturally to a measure of hardness or difficulty. If these notions are truly intrinsic, then these invariants could serve as "lower bounds" in robotics, in the same way that lower bounds have been developed in computer science.

In our quest for an intrinsic measure of the information requirements of a task, we are inspired by Erdmann's monograph on sensor design [Erd3], and the *information invariants* that Erdmann introduced to the robotics community in 1989 [Erd2]. We also observe that rigorous examples of information invariants can be found in the theoretical literature from as far back as 1978 (see, for example, [BK, Koz]). We note that many interesting lower bounds (in the complexity-theoretic sense) have been obtained for motion planning questions (see, eg, [Reif, HSS, Nat, CR]; see, eg, [Erd1, Don2, Can, Bri] for upper bounds). Rosenschein has developed a theory of synthetic automata which explore the world and build data-structures that are "faithful" to it [Ros]. His theory is set in a logical framework where sensors are logical predicates. Perhaps our theory could be viewed as a geometric attack on a similar problem. This work was motivated by the theoretical attack on perceptual equivalence begun by [DJ] and by the experimental studies of [JR]. Horswill [Hors] has developed a semantics for sensory systems that models and quantifies the kinds of assumptions a sensori-computational program makes about its environment. He also gives source-to-source transformations on sensori-computational "circuits." The paper [Don4] discusses the semantics of sensor systems. This formalism is used to explore some properties of what we call *situated sensor systems*. [Don4] describes a way to transform sensori-computational systems. When one

Figure 2: TOMMY and LILY pushing a couch in a straight line, using an asynchronous SPMD protocol requiring no explicit communication.

can be transformed into another, we say the latter can be "reduced" to the former, and we call the transformation a "reduction." We also derive algebraic algorithms for reducing one sensor to another. This machinery is only necessary if one wishes to automate the transformation process; it is quite easy to calculate reductions "by hand," using pencil and paper.

In addition to the work discussed here in sec. 1, for a detailed bibliographic essay on previous research on the geometric theory of planning under uncertainty, see, eg., [Don1] or [Don3].

The goals outlined here are ambitious and we have only taken a small step towards them. The questions above provide the setting for our inquiry, but we are far from answering them. We hope that information invariants can serve as a framework in which to measure the capabilities of robot systems, to quantify their power, and to reduce their fragility with respect to assumptions that are engineered into the control system or the environment. We believe that the equivalences that can be derived between communication, internal

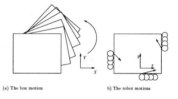

(a) The box motion   b) The robot motions

Figure 3: The box is being rotated by three cooperating autonomous agents. (a) The motion of the box viewed in world coordinates. (b) The relative motion of the pushing robots, viewed in a system of coordinates fixed on the box. The arrows illustrate the direction of the applied forces.

state, external state, computation, and sensors, can prove valuable in determining what information is required to solve a task, and how to direct a robot's actions to acquire that information to solve it. There are several things we have learned. We can determine a lot about the information structure of a task by (i) parallelizing it and (ii) attempting to replace explicit communication with communication "through the world" (through the task dynamics). Communication "through the world" takes place when a robot changes the environment and that change can be sensed by another robot. In this paper we give two different protocols (strategies) for a 2-robot pushing task: one protocol uses explicit communication and the other makes use of an encoding in the task mechanics of the same information. Our approach of quantifying the information complexity in the task mechanics involves viewing the world dynamics as a set of mechanically implemented "registers" and "data paths". This permits certain kinds of *de facto* communication between spatially separated robots. This "equivalence" of task mechanics and communication is operational in flavor, and we are still exploring its generality.

We believe that, by spatially distributing resources among collaborating agents, the information characteristics of a robot task are made explicit. That is, by asking, *How can this task be performed by a team of robots?* one may highlight the information structure. In robotics, the evidence for this is, so far, largely anecdotal. In computer science, one often learns a lot about the structure of an algorithmic problem by parallelizing it; we argue that a similar methodology is useful in robotics.

It is very difficult to analyze the interaction of sensing, computation, communication (a – e) and mechanics (f) in distributed manipulation tasks. The analyses of [Don4] focus on (a – e), and each analysis is "parameterized" by the task. This paper represents an attempt to integrate a measure of the "information content of the task mechanics" (f) into the theory. Nevertheless, the theory is still biased towards sensing, and it remains to develop a framework that treats action and sensing on an equal footing.

This paper draws extensively on the material reported in the draft monograph by Donald [Don4], and announced in an abbreviated, preliminary version in [Don]. We reported on our ideas on coordinated manipulation strategies in a preliminary form in [DJR].

1.1.1 Research Agenda

Robot builders make claims about robot performance and resource consumption. In general, it is hard to verify these claims and compare the systems. We really think the key issue is that two robot programs (or sensor systems) for similar (or even identical) tasks may look very different. We discuss why it is hard to compare the "power" of such systems. Our examples are distinguished in that they permit relatively crisp analytical comparisons: they represent the kinds of theorems about information trade-offs that we believe can be proved for sensorimotor systems. The analyses in sec. 2 reveal trade-offs in terms of resource consumption. We then ask, is there a general theory quantifying the power gained in such trade-offs? In sec. 3, we present a theory, which represents a systematic attempt to make such comparisons based on geometric and physical reasoning. In [Don4], we operationalize our analysis by making it computational; we give effective (albeit theoretical) procedures for computing our comparisons. See sec. 5 for a summary.

We wish to rigorously compare embedded sensori-computational systems. To do so, we define a "reduction" $\leq_1$ that attempts to quantify when we can "efficiently" build one sensor out of another (that is, build one sensor using the components of another).[3] Hence, we write $A \leq_1 B$ when we can build A out of B without "adding too much stuff." The last is analogous to "without adding much information complexity." Our measure of information complexity is *relativized* both to the information complexity of the sensori-computational components of B, and to the bandwidth of A. This relativization circumvents some tricky problems in measuring sensor complexity. In this sense, our "components" are analogous to *oracles* in the theory of computation. Hence, we write $A \leq_1 B$ if we can build a sensorimotor system that simulates A, using the components of B, plus "a little rewiring." A and B are modeled as *circuits*, with wires (data-paths) connecting their internal components. However, our sensori-computational systems differ from computation-theoretic (CT) "circuits," in that their spatial configuration—i.e., the spatial location of each component—is as important as their connectivity.

We develop some formal concepts to facilitate the analysis. *Permutation* models the permissible ways to reallocate and reuse resources in building another sensor. Intuitively, it captures the notion of repositioning resources such as the active and passive components of sensor systems (*e.g.*, infra-red emitters and detec-

---

[3] $\leq_1$ is also called $<_s$ in [Don].

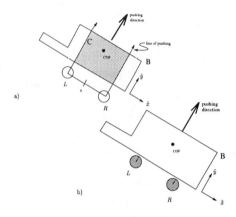

Figure 4: (a) the "two-finger" pushing task vs. (b) the two robot pushing task. The goal is to push the block B in a straight line.

Repeat :   measure(τ)
        case   ($\tau = 0$) $\Rightarrow$
                push(p)
            ($\tau < 0$) $\Rightarrow$
                move(R)
            ($\tau > 0$) $\Rightarrow$
                move(L)

Figure 5: Protocol O (for a two-fingered gripper).

tors). *Geometric codesignation constraints* further restrict the range of admissible permutations. *I.e.*, we do not allow arbitrary relocation; instead, we can constrain resources to be "installed at the same location", such as on a robot, or at a goal. *Output communication* formalizes our notion of "a little bit of rewiring." When resources are permuted, we often find they must be reconnected using "wires", or data-paths. If we separate previously colocated resources, we will often need to add a communication mechanism to connect the now spatially separate components. Like CT reductions, $A \leq_1 B$ defines an "efficient" transformation on sensors that takes B to A. However, we can give a generic algorithm for synthesizing our reductions (whereas no such algorithm can exist for CT.)[4] Whether such reductions are widely useful or whether there exist better reductions is open; however we try to demonstrate the potential usefulness both through examples and through general claims on algorithmic tractability. We also give a "hierarchy" of reductions, ordered on power, so that the strength of our transformations can be quantified.

Our ideas have the following applications:

1. *(Comparison).* Given two sensori-computational systems A and B, we can ask "which is more powerful?" (in the sense of $A \leq_1 B$, above).

[4]For example: no algorithm exists to decide the existence of a linear-space (or log-space, polynomial time, Turing-computable, etc.) reduction between two CT problems.

2. *(Transformation).* We can also ask "Can B be transformed into A?"

3. *(Design).* Suppose we are given a specification for A, and a "bag of parts" for B. The bag of parts consists of boxes and wires. Each box is a sensori-computational component ("black box") that computes a function of (i) its spatial location or pose and (ii) its inputs. The "wires" have different bandwidths, and they can hook the boxes together. Then, our algorithms decide, can we "embed" the components of B so as to satisfy the spec of A? The algorithms also give the "embedding" (that is, how the boxes should be placed in the world, and how they should be wired together). Hence, we can ask, can the spec of A be implemented using the bag of parts B?

4. *(Universal Reduction).* Consider application 3, above. Suppose that in addition to the spec for A, we are given an encoding of A as a bag of parts, and an "embedding" to implement that spec. Suppose further that $A \leq_1 B$. Since this reduction is relativized both to A and to B, it measures the "power" of the components of A relative to the components in B. By universally quantifying over the configuration of A, we can ask, "can the components of B always do the job of the components of A?"

More specifically: Let α be a "configuration" of

sensorimotor system A. Thus α encodes the spatial embedding of A as well as its wiring connectivity. Similarly, let β be a configuration of system B. Let $A(\alpha)$ and $B(\beta)$ denote systems A and B "installed" at these configurations. The gist of application 4 is that, we can decide whether or not $(\forall \alpha, \exists \beta): A(\alpha) \leq_1 B(\beta)$.

1.2 Outline

We discuss questions (a) – (f) from the Abstract for an experiment with communicating robots. We consider the task of coordinated manipulation of large objects (particularly, manipulation of a large box using a pair of communicating mobile robots). We foreground the task of pushing an object, using two communicating robots who need to infer the position of the first moment of the friction distribution with respect to their lines of pushing (see Figure 4a). In [Don4], we asked whether explicit communication could be removed from this protocol (by "explicit" we mean local communication, such as IR, or global communication, such as RF). In this paper we give a protocol with no explicit communication, and we analyze and compare the the protocols using the tools introduced in [Don4]. We believe our methods generalize to other manipulation tasks and to larger teams of robots, but work is still underway; for example, in [DJR], we considered the task of coordinated manipulation of large objects (particularly, rotations, or more accurately, *"reorientations"* of a large box using a team of communicating mobile robots). There, we examined an *offline* strategy, in which the robots have a geometric model of the object, and an *online* strategy, in which they do not. In sec. 7, we describe these strategies and sketch a general methodology for developing distributed manipulation protocols.

2 Pushing with Two Communicating Mobile Robots

To introduce our ideas we consider a task involving two autonomous mobile robots. The two robots must cooperate to push a box. Now, many issues related to information invariants can be investigated in the setting of a single agent. However, one of our ideas is that, by spatially distributing resources (a) – (f) among collaborating agents, the information characteristics of a task are made explicit. That is, by asking, *How can this task be performed by a team of robots?* one may highlight the information structure.

Here is a preview of how we will proceed. The goal of the *pushing task* is to move an object along a straight line (called the *pushing direction*) with two agents. We first describe a strategy for this task designed to be executed by a manipulator with two rigidly connected fingers and force feedback. We then propose variants of this algorithm that are suitable for execution by autonomous mobile robots. Finally, we compare these strategies with respect to questions (a) – (f), posed in the Abstract.

2.1 Three Pushing Protocols

Consider the task whose goal is to push a box B in a straight line. Figure 4a depicts one robot (the reader should picture a robot manipulator, or gripper) executing this task. The robot consists of two rigidly connected fingers L and R; for example, they could be the fingers of a parallel-jaw gripper. One complication involves the micro-mechanical variations in the slip of the box on the table [Mas]. This phenomenon is very hard to model, and hence it is difficult to predict the results of a one-fingered push; we will only obtain a straight-line trajectory when the center of friction (COF) lies on the line of pushing. However, with a two-fingered push, the box will translate in a straight line so long as the COF lies between the fingers. An advantage of the two-finger pushing strategy is that the COF can move some and the fingers can keep pushing, since we only need ensure the COF lies in some region C (see Figure 4a), instead of on a line. If the COF moves outside C, then the fingers can move sideways to "capture" it again. We have implemented the control loop described as Protocol O on our PUMA (see fig. 5). The basic idea is to sense the reaction torque τ about the point 0 in Figure 4a. If $\tau = 0$, push forward in direction $\hat{y}$. If $\tau < 0$ move the fingers in $\hat{x}$; else move the fingers in $-\hat{x}$.

From the mechanics perspective it might appear we are done. However, it is difficult to overstate how critically Protocol O above relies on global communication and control. Now, consider the analogous pushing task in Figure 4b. Each finger is replaced by an autonomous mobile robot such as those described in [RD]. The robots we have in mind are the Cornell mobile

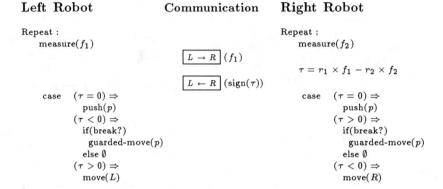

Left Robot   **Communication**   **Right Robot**

Repeat :   Repeat :
  measure(f_1)   measure(f_2)

               $\boxed{L \to R}$ (f_1)

               $\tau = r_1 \times f_1 - r_2 \times f_2$

               $\boxed{L \leftarrow R}$ (sign(τ))

  case ($\tau = 0$) $\Rightarrow$ case ($\tau = 0$) $\Rightarrow$
         push(p)   push(p)
        ($\tau < 0$) $\Rightarrow$ ($\tau > 0$) $\Rightarrow$
         if(break?)   if(break?)
           guarded-move(p)   guarded-move(p)
         else $\emptyset$ else $\emptyset$
        ($\tau > 0$) $\Rightarrow$ ($\tau < 0$) $\Rightarrow$
         move(L)   move(R)

Figure 6: Protocol I.

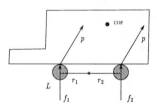

Figure 7: The mechanical observables for Protocol I

robots (see Figure 1), but the details of their construction are not important. The robots can move about by controlling motors attached to wheels. The robots are autonomous and equipped with a ring of 12 simple Polaroid ultrasonic sonar sensors. Each robot has onboard processors for control and programming. The description in [RD] is augmented as follows. (This description characterizes the robots in our lab). We equip each robot with 12 infra-red modems/sensors, arrayed in a ring about the robot body. Each modem consists of an emitter-detector pair. When transmitting or receiving, each modem essentially functions like the remote control for home appliances (*e.g.*, televisions). In addition, each robot has a ring of one-bit contact ("bump") sensors.

We assume the following: (1) robots can sense the relative orientation of objects with which they are in contact [JR]; (2) robots know that they are on the same flat face of the object; (3) both robots know the direction of pushing, p; and (4) robots can synchronize their velocities.[5] In addition, (5) by examining the servo-loop in [RD], it is clear that we can compute a measure of applied force by observing the applied power, the position and velocity of the robot, and the contact sensors.

The pushing strategy described as Protocol O can be

_____
[5] For the pushing task, it suffices to assume that the robots have identical control systems and can command the same speeds. More generally, velocity synchronization requires that the robots begin and end pushing at the same time; this can be necessary for more complicated manipulation tasks. A protocol for velocity synchronization is not hard to synthesize using our methods; see sec. 7.1.

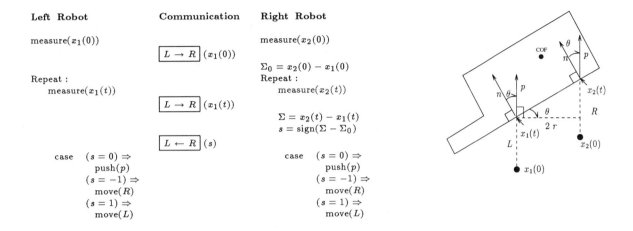

Left Robot

measure($x_1(0)$)

$\boxed{L \rightarrow R}\,(x_1(0))$

Repeat :
  measure($x_1(t)$)

$\boxed{L \rightarrow R}\,(x_1(t))$

$\boxed{L \leftarrow R}\,(s)$

case $(s = 0) \Rightarrow$
      push(p)
    $(s = -1) \Rightarrow$
      move(R)
    $(s = 1) \Rightarrow$
      move(L)

Communication

Right Robot

measure($x_2(0)$)

$\Sigma_0 = x_2(0) - x_1(0)$
Repeat :
  measure($x_2(t)$)

$\Sigma = x_2(t) - x_1(t)$
$s = \text{sign}(\Sigma - \Sigma_0)$

case $(s = 0) \Rightarrow$
      push(p)
    $(s = -1) \Rightarrow$
      move(R)
    $(s = 1) \Rightarrow$
      move(L)

Figure 8: Protocol I (Quasi-Static). The case for breaking contact is not shown; it can be handled as in fig. 6.

Figure 9: Protocol I(QS). The normal to the box face is denoted by n. The x axis is parallel to the direction of pushing p. The lines of pushing are distance $2r$ apart (perpendicular distance). θ is the angle between n and p.

approximated by observing the following (see Figs. 7, 6). Each robot can measure its applied force (f_1 or f_2) and communicate this data to the other. This allows the robots to compute the net torque τ about a point in between the two robots, and from the sign of this quantity, to infer the location of the first moment of the friction distribution of the box. If the first moment of the friction distribution is between the two lines of pushing, each robot continues pushing alone the line p. If there is a positive net torque, the instantaneous center of friction is to the left of both robots. In this situation, the left robot is in contact with the box, while the right robot may or may not be in contact. The left robot can "recapture" the center of friction between the two lines of pushing by moving left along the face of the box (move(L)). If the right robot is not in contact with the box (the predicate (break?) returns TRUE) it executes a guarded-move (motion until contact) in the direction p. Otherwise this robot takes the null action, $\emptyset$. The case when the net torque is negative is symmetric. We call this Protocol I (see Figures 7, 6).

A variant of this protocol can be derived for a quasi-static (QS) system. Here, relative displacements along the line of pushing p are measured instead of forces.

Figure 9 shows a configuration where the two robots originate at positions $x_1(0)$ and $x_2(0)$, respectively. Their locations at time t are $x_1(t)$ and $x_2(t)$. In this protocol (Figure 8), the initial locations of the robots are communicated to determine their offset Σ_0. Σ_0 "specifies" (or better, *parameterizes*) the pushing task: this offset determines the pushing direction p relative to the initial orientation of the box face. The robots exchange location information successively at each loop iteration; this information is used to infer the direction of motion of the box. We call this Protocol I(QS) (see Figures 9, 8).

We now derive a different version of Protocol I(QS) by observing that the information needed to determine the motion of the box (*i.e.* Σ_0 and Σ) is related to the angle θ between the normal to the face of the box n and the direction of pushing p as follows: $2r \tan \theta_0 = \Sigma_0$ and $2r \tan \theta = \Sigma$ (see Figures 9, 11, 10). Moreover, we observe that the tangent function is monotonic and sign preserving; this means we can adapt the control system to servo on θ instead of Σ, without knowing r. Specifically, the robots measure θ_0 (the initial angle between n and p; see [JR]), and compare this value to the angle $\theta(t)$ measured at time t in order to infer the direction of motion of the box. A negative change

Left Robot

$\theta_0 \leftarrow \theta(0)$
Repeat :
 case (break?) $\Rightarrow$
    guarded-move(p)
   ($\theta(t) \approx \theta_0$) $\Rightarrow$
    push(p)
   ($\theta(t) \gg \theta_0$) $\Rightarrow$
    move(L)
   ($\theta(t) \ll \theta_0$) $\Rightarrow$
 (**) $\emptyset$

Right Robot

$\theta_0 \leftarrow \theta(0)$
Repeat :
 case (break?) $\Rightarrow$
    guarded-move(p)
   ($\theta(t) \approx \theta_0$) $\Rightarrow$
    push(p)
   ($\theta(t) \gg \theta_0$) $\Rightarrow$
 (*) $\emptyset$
   ($\theta(t) \ll \theta_0$) $\Rightarrow$
    move(R)

Figure 10: Protocol P.II. This protocol is "almost uniform," and can be made uniform by changing the $\emptyset$ lines (*) to Move(L) and (**) to Move(R). Note that "uniform" does not quite imply SIMD, since the protocols run asynchronously.

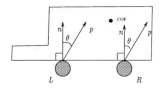

Figure 11: The mechanical observables for Protocol II.

in the value of this angle implies a clockwise rotation of the box. A large positive change implies a counterclockwise rotation. The robots adjust their pushing location on the face of the box accordingly. This is an example of how the robots can use the task dynamics instead of explicit communication to determine their next actions. We call this pushing strategy Protocol II (Figure 10).

2.2 Comparing the Protocols

Now, we ask, how do protocols I, I(QS), and II compare to one another with respect to the questions (a) – (f) posed above? We first note that the three protocols require different sensing capabilities. Protocol I relies on force sensing, Protocol I(QS) relies on position sensing, and Protocol II relies on orientation sensing. Next we observe that the robots must coordinate to find locations that result in a stable pushing along p. This coordination is accomplished differently in the three

protocols. In Protocol I and Protocol I(QS) the robots synchronize by exchanging their sensed values.

Robots executing Protocol I require communicating $\log \Bbbk(f_1)$ bits to transmit the value of force f_1, and 2 bits to transmit the sign of the torque τ ($\Bbbk(b)$ denotes how many values a variable b may take on). Robots executing Protocol I(QS) require $\log \Bbbk(x_1)$ bits to transmit the value of the distance x_1, and 2 bits to transmit the sign of s. In Protocol II there is no explicit communication and the synchronization is realized through the world, by monitoring the change in the angle θ between the normal to the face and the pushing direction. In other words, the robots infer the motion of the object by decoding changes in the task mechanics. Thus, protocols I and I(QS) rely on direct communication, while protocol II does not. The internal state requirements of the three strategies are also different. Protocol I requires no internal state. Protocol I(QS) requires a register to record the value Σ_0. Protocol II requires a register to record the value θ_0.

We can get a deeper understanding of the relationship between these protocols by attempting to "transform" or "reduce" one to the other. We do so below.

3 Reductions and Transformations

We now formalize our model of sensori-computational systems by viewing them as "circuits." The theory in secs. 3-6 is extracted from [Don4], but the particular examples and application (especially, claim 4.1) are new. We model these circuits as graphs. Vertices correspond to different sensori-computational components of the system (what we will call "resources" below). Edges correspond to "data paths" through which information passes. Different immersions of these graphs correspond to different spatial allocation of the "resources." Our idea involves asking: *What information is added (or lost) in a sensor system when we change its immersion?* and *What information is preserved under all immersions?* We also define an operator + as a way to "combine" sensori-computational systems. The operation + is like taking the union of two graphs. Below, we use the term *"sensor system"* to mean *"sensori-computational system"* where it is mellifluous.

3.1 Situated Sensor Systems

Definition 3.1 *A labelled graph $\mathcal{G}$ is a directed graph (V, E) with vertices V and edges E, together with a labelling function that assigns a label to each vertex and edge. Where there is no ambiguity, we denote the labelling function by ℓ.*

Definition 3.2 *A sensor system $\mathcal{S}$ is represented by a labelled graph (V, E). Each vertex is labelled with a* component. *Each edge is labelled with a* connection.

See figs. 12–13 for illustrations for the circuits—that is, the sensor systems for protocols I (QS) and II. We will use the terms *protocol* to refer to the computer programs in figs. 8–10, and *circuit* for the sensor systems in figs. 12–13. It is clear that each circuit implements its protocol. Now, in Protocol I(QS) (fig. 8), there are three communications operations. The first two $(L \rightarrow R)$ use the same datapath; they simply refer to the first use, and to subsequent use, of the same resource. In our circuits, we now restrict the *components* to be the resources such as those described in the figures; however, our definitions could, we feel, be generalized to other resources. In the figures, the components correspond to boxes: $\boxed{\text{odom}}$ is an odometer, the "signal"[6] coming out of this box is the odometry reading. The box $\boxed{-}$ performs subtraction, the boxes $\boxed{x_1(0)}$ and $\boxed{\Sigma_0}$ are registers, and they implement internal state. The part of the circuit labelled **case** interfaces to the control part of the circuits, which is the same for both protocols.

One interesting resource is $\theta(t)$ in fig. 13—we call this box a θ-source—it produces a signal indicating relative orientation. [JR] describe in detail how to implement a bounded-error θ-source. Here is the basic idea. A θ-source is an abstraction of relative normal sensing. It allows the normal of the box to be treated as an external "register" that both robots may read and write. We could implement an (approximate) θ-source as follows. The robot has a ring of 1-bit bump sensors. These are used to implement relative normal sensing [JR]. We specify the pushing direction p, by specifying θ_0 (see fig. 10), the direction of p relative to the box normal $n(0)$ at time 0. More specifically: at the beginning of the task, each robot does a guarded move along p until contact with the box. It then aligns normal to the box,

---

[6]We use *"signal"* as a metaphor; our circuits are strictly digital. Either *message* or *stream* would be better, but both have distracting religious connotations.

using the bounded-error algorithm of [JR]. Finally, the robot turns by angle θ_0 using pure position control.

INIT is one bit of state, and RUN= $\overline{\text{INIT}}$. The small crossed circles ($\oplus$) that these bits run into are gates; the $\downarrow$ input must be 1 for the $\leftrightarrow$ signal to pass. Thus in robot L in fig. 12, if INIT= 1, then the odometer writes into the register $x_1(0)$. When INIT= 0, then the $\boxed{-}$ component is fed the odometry input. We assume that numbers are represented as signed integers. The $\boxed{\text{sgn}}$ box just selects out the sign bit. We omit the logic to toggle INIT once the register $x_1(0)$ is written. Whereas L requires one bit of state INIT$_0$, R requires two, due to the asymmetry of the protocol P.I(QS). These are denoted INIT$_1$ and INIT$_2$.

Connections are like data-paths in that they carry information; a connection's label represents the information that will be sent along that path. Connections carry data between components. We adopt the convention that two components can communicate without an (explicit) connection when they are spatially colocated. Now, in figs. 12–13, many of the data paths are *internal*, i.e., $L \rightarrow L$ or $R \rightarrow R$. The most interesting datapaths are the *external* datapaths: the $L \rightarrow R$ edge (b) labelled COMM(Δx_1), and the $L \leftarrow R$ edge (b') labelled COMM(s). (b) has bandwidth $\log \Bbbk(\Delta x_1)$ bits, and (b') has bandwidth 2 bits.

Observe fig. 13. Here, there is a θ-source that is external to both robots. There are two interesting external data paths from the θ-source, one to L marked COMM($\cdot$), and one marked (b) to R. Both these datapaths have bandwidth $\log \Bbbk(\theta(t))$ bits.

Definition 3.3 *Let b be a variable that ranges over all possible values that a sensor system can compute. We call b the* output *of the system. Let $\Bbbk(b)$ be the number of values b can take on, and define $\log \Bbbk(b)$ to be both the* size *of b and the* output size *of the sensor. The output size is an upper bound on the bit-complexity of b. For example, if b takes on integer values in the range $[1, q]$, then $\Bbbk(b) = q$, and $\log \Bbbk(b) = \log q$. Now, suppose the information b is communicated over a datapath e. We will assume that the information is communicated repeatedly; without loss of generality, we take the unit of time to be the interval of the occasion to communicate the information. Thus we can take the size of the output b to be the* bandwidth *of e.*

So far, we have defined components and connections operationally. We now give a formal definition. Components and connections are defined by their *simulation functions*. Simulation functions describe the behavior of both components and connections.

Consider a component $\ell(v)$ associated with vertex v. To simulate a component, we need to know (i) its inputs and (ii) its configuration. Suppose a component has r inputs and s outputs, each of which lies in some space R. Let C be the configuration space of the component. A *simulation function* for a component $\ell(v)$ is a map[7] $\Omega_v : R^r \times C \to R^s$.

Now we connect the components together. Assume for a moment that all the components have the same input-output structure as Ω_v above (i.e., that r and s are fixed throughout the system, but that the components themselves may perform different functions). We model an edge e between vertices v and u by its label, $\ell(e) = b$, and by a pair of integers, (i, j). $\log \Bbbk(b)$ is the *bandwidth* of the edge and the index i (resp. j) specifies to which of the r outputs of $\ell(v)$ (resp., s inputs of $\ell(u)$) we attach e ($1 \leq i \leq r$ and $1 \leq j \leq s$).

Now, a simulation function for this edge e is taken to be a function $\Omega_e : R \to R$. We will usually restrict the edge functions to be the identify function (but they also check for bandwidth, i.e., that the transmitted data has size no greater than $\log \Bbbk(b)$).

We also define a resource called the "output device." Each sensor system must have exactly one vertex with this label, called the *ouput vertex*. The output vertex of the sensor system is where the output of the sensor is measured. The simulation function for the output device is the identity function, but the output value of this device defines the output value of the sensor system.

A simulation function $\Omega_{\mathcal{U}}$ for an entire sensor system $\mathcal{U}$, then, is a collection of component simulation functions such as Ω_v and edge simulation functions such as Ω_e. The function $\Omega_{\mathcal{U}}$ simulates all the component simulation functions in the correct configuration, and simulates routing the data between them using the edge simulation functions. We adopt the convention that two components can communicate without

an (explicit) connection when they are spatially colocated. When all these component and edge functions are semi-algebraic, then the sensor simulation function $\Omega_{\mathcal{U}}$ is also semi-algebraic (see Section 5).

Definition 3.4 *Consider a sensor system $\mathcal{U}$ with simulation function $\Omega_{\mathcal{U}}$. The* output value *of $\mathcal{U}$ at a particular configuration is the value $\Omega_{\mathcal{U}}$ computes for that configuration. Hence the output value of $\mathcal{U}$ is a function of $\mathcal{U}$'s configuration.*

The notions output value *and* output *(Definition 3.3) are related as follows. The output of $\mathcal{U}$ is a variable that ranges over all possible output values of $\mathcal{U}$. Given another sensor system $\mathcal{V}$, we say the* output *of $\mathcal{U}$ is the same as the output of $\mathcal{V}$ when $\Omega_{\mathcal{U}}$ and $\Omega_{\mathcal{V}}$ are identical.*

Under this model, we can simulate trees of embedded sensorimotor computation. It is also possible (in principle) to simulate more general graphs and systems with state, but in this case the value at the output vertex may vary over time (even for a fixed configuration). In this case we need some explicit notion of time and blocking to model the (a)synchronous arrival of data at a component. Such extensions are considered in [Jen94]; for now we restrict our attention to trees, which suffice to model our examples. In general our discussion is restricted to consider one clock-tick; however, generalizations are possible to consider the time-varying behavior of the system [Jen94]. We will treat the circuits P.I(QS) and P.II (below) as effectively being trees, and not graphs, even though there is data flow both from R to L and L to R. This is because to a first order approximation, data does not not feed back into the system.

To summarize: a *component* is a primitive device that computes a a function of (i) its inputs and (ii) its configuration $z \in C$. Each component is installed at a vertex of communication graph with d vertices, whose edges are the *connections* described above. The graph is immersed in a configuration space C^d, and the configuration z of a component is the configuration of its vertex. More generally, components can be *actuators*. An actuator is a component whose output forces the configuration of the graph to change or evolve through a *dynamics equation*. If the configuration of the entire graph is $\mathbf{z} = (z_1, \ldots, z, \ldots, z_d) \in C^d$, then the dynamics equation models a mapping from the actuator component $\ell(v)$'s output at z to the tangent space $T_{\mathbf{z}}C^d$ to

---

[7]Components that retain state can be modeled by a function $\Omega_v : R^r \times C \times S \to R^s \times S$, where S is a *store* that records the state. For example, a state element with k bits of state would be modeled with $S = \{0, 1\}^k$. Alternatively, S can be absorbed as a factor subspace in the configuration space of the component.

the configuration space. See [Jen94] for more discussion of actuators. The actuator systems of our circuits are identical and in this paper we do not consider them in detail. This actuator subsystem is represented by the elision CASE(s) $\cdots$ in figs. 12-13. The circuit for this system would look very much like a traditional plant diagram from control engineering.

Weaker forms of sensori-computational equivalence are possible. If we define the *state* of a sensor system $\mathcal{U}$ to be a pair $(\mathbf{z}, b)$ where $\mathbf{z}$ is the configuration of the system and b is the output value at $\mathbf{z}$, we can examine the equilibrium behavior of $\mathcal{U}$ as it evolves in state space. Consider the Definition 3.6; let us call this *strong simulation*. By analogy, let us say that a system $\mathcal{U}$ *weakly simulates* another system $\mathcal{V}$ when $\mathcal{U}$ and $\mathcal{V}$ have identical, forward-attracting compact limit sets in state space.[8] If we replace strong simulation ($\cong$ in Definition 3.6) with weak simulation, all of our structural results go through *mutatis mutandis*. The computational results also go through, if we can compute limit sets and their properties (a difficult problem in general). Failing this, if we can derive the properties of limit sets "by hand" then in principle, reductions using weak simulation instead of strong simulation ($\cong$) can also be calculated by hand.

In our sensor systems, there is no separate notion of "sensor inputs." Instead, the sensory inputs are encoded in the configuration space.

3.2 Transformations as Reductions

In sec. 4, we give a proof (claim 4.1) and an equation (3) relating the circuits P.I(QS) and P.II. Intuitively, eq. (3) indicates that, operationally speaking, one could transform a system which executes Protocol I(QS) into one which executes Protocol II by removing the odometry from both robots, and by adding a θ–source, which is a component that senses the orientation of the manipulated object. In describing Protocol II, we demonstrated that one implementation of such a sensor involves using some retained state (θ_0), and a relative orientation sensor such as the bumper system described in [JR]. In fact, equation (3) is a precise statement of this engineering fact. Below, we carefully define the operators $+$ and $-$, and formalize the no-

tions of *simulation* and *efficient reduction*, as well as *permutation, etc.*

Our reduction involves two concepts. The first is *permutation,* and it involves redistributing resources in a sensor system, without consuming new resources. Surprisingly, a redistribution of resources can add information to the system. In order for permutation to add information, it is necessary for the sensor system to be spatially distributed (as, for example, our circuits are). When permutation gains information, it may be viewed as a way of arranging resources in a configuration of lower entropy.

The second concept is *communication*. It measures resource (b) (from the abstract). We consider adding communication primitives of the form COMM($L \to R$, *info*), which indicates that L sends message *info* to R. Examples of this primitive are COMM(Δx_1) and COMM(s) in fig. 12. Like permutation, communication only makes sense in a spatially distributed sensor system. That is, because spatially colocated components can communicate "for free" in our model, only "external" datapaths add information complexity to the system. In effect, to transform system Protocol O into Protocol I(QS) (see figs. 5, 8), we view it as a system composed of autonomous collaborating agents L and R, each of which has certain resources. The COMM($\cdot$) primitive above we view as shared between L and R. We measure communication by counting the number of agents and the bits required to transmit *info*. This is the only kind of external communication we will consider here (i.e., $L \to R$ or $L \leftarrow R$); we will abbreviate it by COMM(*info*) when the direction is clear.

We can be sure of getting the semantics of COMM($\cdot$) correct by treating it as a sensor system in its own right (albeit, a small one). Now, COMM(b) defines the graph with vertices $\{u_1, u_o\}$ and a single edge $e = (u_1, u_o)$ with $\ell(e) = b$. The "head" vertex u_o of the edge $e = (u_1, u_o)$, is defined to be the *output vertex* of the sensor system COMM(b). The argument (parameter) b to COMM(b) determines the *bandwidth* of e. Thus, for example, COMM(b) specifies a graph with one edge e whose label is b. This specifies that the edge is a datapath that can carry information b; if b requires $k = \log \Bbbk(b)$ bits to encode then k is the *bandwidth* of e. Our model of communication is rather abstract. External communication is probably not possible without buffering by either the sender or the receiver. COMM($\cdot$)

---

[8]I am grateful to Dan Koditschek, who has suggested this formalism in his papers.

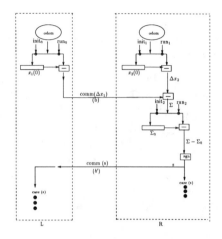

Figure 12: Sensor system P.I(QS). This is a circuit for Protocol I (QS). This circuit shows one possible implementation of the protocol. Figs. 12-13 do not show how to handle lost of contact (i.e., the (break?) case), but this circuitry is easily added, and is the same for both P.I(QS) and P.II.

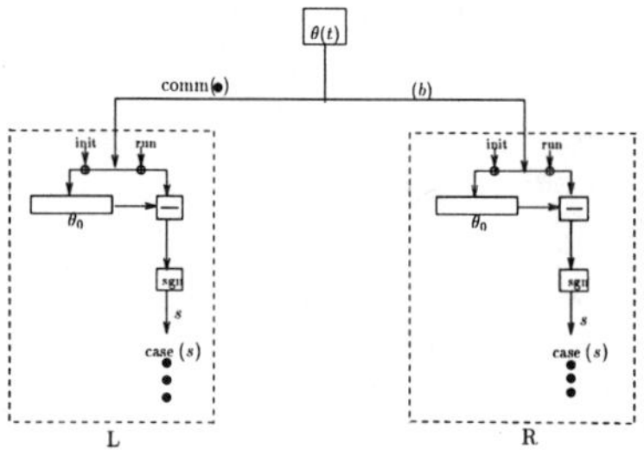

Figure 13: Sensor system P.II. This is a circuit for Protocol II.

should include this buffer to be more realistic about modeling internal state.

We now formalize the ideas above. Consider a sensor system with vertices V. For each vertex v in V, we assume there is a configuration space C. A point in this space C represents the configuration of the component.

Definition 3.5 *A* situated *(or* immersed*) sensor system* $\mathcal{S}$ *is a sensor system* $\mathcal{S} = (V, E)$, *together with an immersion* $\phi : V \to C$ *of the vertices. If* $v \in V$, *then we call* $\phi(v)$ *the* configuration *of the vertex* v. *When there is no ambiguity, we also call* $\phi(v)$ *the configuration of the component* $\ell(v)$.

A situated sensor system is essentially an immersed graph. If the map ϕ in def. 3.5 is injective, then we call ϕ an *embedding*. Immersions need not be injective. Moreover, in order to colocate vertices, it is necessary for immersions to be non-injective.

In def. 3.5, the immersion ϕ may be a partial (as opposed to total) function, indicating that we do not specify the spatial configuration of those components whose vertices are outside the domain of the immersion. We denote the *domain* of a (partial) immersion $\phi : V \to C$ by $\phi^{-1}C$. We denote its *image* by im ϕ.

We can now define what it means for two systems to be equivalent:

Definition 3.6 *Given two sensor systems* S *and* Q, *we say* Q simulates S *if the output of* Q *is the same as the output of* S. *In this case we write* $S \cong Q$. *More generally, suppose we write*

$$(\mathcal{S}, \phi) \cong (\mathcal{U}, \psi) \tag{1}$$

for two situated sensor systems. Eq. (1) is clearly well-defined when ϕ *and* ψ *are total. Now, suppose that* ϕ *and* ψ *are partial, leaving unspecified the configurations of components* $\ell(v)$ *of* $\mathcal{S}$ *and* $\ell(u)$ *of* $\mathcal{U}$. *Then eq. (1) is taken to mean that* $(\mathcal{U}, \psi)$ *simulates* $(\mathcal{S}, \phi)$ *for any configuration of* v *and* u.

For def. 3.6, in the case where (say) ϕ is partial, we operationalize eq. (1) by rewriting it as a statement about all *extensions* $\overline{\phi}$ of ϕ. That is, we define ex ϕ to be the set of all extensions of ϕ. Then, we write: "$\forall \overline{\phi} \in$ ex ϕ, eq. (1) holds (with bars placed over the immersions)." We treat ψ similarly, with an inner universal quantifier, although codesignation constraints (sec. 3.5) allow us to make the choice of extension $\overline{\psi}$ of ψ depend on the extension $\overline{\phi}$ that is bound by the outer quantifier. For example, def. 3.6 becomes, "for all configurations $x \in C$ of v, for all configurations $y \in D_{\mathcal{S}}(x)$ of u, eq. (1) holds." Here $D_{\mathcal{S}}(x)$ is a set in C that varies with x; the function $D_{\mathcal{S}}(\cdot)$ models the codesignation constraints. Def. 3.6 can be generalized to any number of "unbound" vertices; see eq. (8) and [Don4].

3.3 Permutation

We may consider two orthogonal kinds of permutation. In both models, the vertex and edge labels $\ell(v)$ and $\ell(e)$ never change. The first model is called *vertex permutation,* and is given in def. 3.7. In this model, the vertices can move, and they drag the components and wires with them. That is, the vertices move (under permutation), and as they move, the edges follow.

Vertex permutation of a situated sensor system corresponds to the choice of a different immersion with the same domain:

Definition 3.7 *Let* $\mathbb{S} = (\mathcal{S}, \phi)$ *be a situated sensor system. A permutation* $\mathbb{S}^*$ *of* $\mathbb{S}$ *is a situated sensor system* $(\mathcal{S}, \phi^*)$ *such that the domain* $\phi^{-1}C$ *of* ϕ *and the domain* $\phi^{*-1}C$ *of* ϕ^* *are the same.*

For technical reasons, we also permit a permutation to change which vertex has the "output device" label. Note that the definition of permutation (def. 3.7) also makes sense for partial immersions. However, see the appendix for a discussion of the semantics of permutation for unsituated sensor systems.

We can also consider an alternate model, called *edge permutation,* where the edge connectivity changes. An edge permutation can be modeled as follows. Consider a graph with vertices V and edges E. Start with any bijection $\sigma : V^2 \to V^2$. We call σ an *edge permutation,* since it induces the restriction map $\sigma_{|_E} : E \to \sigma(E)$ on the edge set E. An edge permutation says nothing about the immersion of a graph.

We can also compose the models. We define a *graph permutation* to be a vertex permutation followed by an edge permutation. In a graph permutation, the vertices and the edges move independently. That is, vertices may move, but in addition, the edge connectivity may change. To illustrate the different models, consider a sensor system $\mathcal{U}$ with three vertices $\{ v_1, v_2, v_3 \}$ with labels $\ell(v_i) = B_i$ ($i = 1, 2, 3$). $\mathcal{U}$ has one edge $e = (v_1, v_2)$ of bandwidth k that connects B_1 to B_2. So, the B_i are the components of the system, and e is a datapath. A *vertex* permutation $\mathcal{U}^*$ of $\mathcal{U}$ would move the vertices (and therefore the components) spatially, but in $\mathcal{U}^*$, e would still connect v_1 and v_2, (and therefore, B_1 and B_2). An *edge* permutation σ of $\mathcal{U}$ would change the edge connectivity. So, for example, an edge permutation $\sigma(\mathcal{U})$ could be a graph with one

edge $\sigma(e) = (v_2, v_3)$, connecting v_2 to v_3 (and hence B_2 to B_3). But in $\sigma(\mathcal{U})$ no edge would connect v_1 and v_2. Finally, consider a *graph* permutation $\mathcal{U}^*$ of $\mathcal{U}$. Suppose $\mathcal{U}^* = \sigma(U^*)$, that is, $\mathcal{U}^*$ is the vertex permutation $\mathcal{U}^*$ followed by the edge permutation σ above. $\mathcal{U}^*$ has the same edge connectivity as $\sigma(\mathcal{U})$. However, in $\mathcal{U}^*$, the vertices are immersed as in $\mathcal{U}^*$.

Let $(\mathcal{U}, \phi)$ be a situated sensor system. A graph permutation of $\mathcal{U}$ is given by $\mathcal{U}^* = (\mathcal{U}, \phi^*)$ where $\phi^* = (\phi^*, \sigma)$, ϕ^* is a vertex permutation, and σ is an edge permutation. In practice, we wish to impose some restrictions on edge and graph permutation. For example, suppose we have a sensor system $\mathcal{U}$ containing two cooperating and communicating mobile robots L and R. The sensori-computational systems for L and R are modeled as circuits. The datapaths in the system, in addition to bandwidth, have a *type,* of the form SOURCE→DESTINATION, where both SOURCE and DESTINATION $\in \{ L, R \}$. (Maintaining exactly two physical locations can be done using simple codesignation constraints). When permuting the edges of $\mathcal{U}$ to obtain $\mathcal{U}^*$, it makes sense to permute only edges of the same type. More generally, we may segregate the edge types into two *classes, internal* edges $L \to L$ and $R \to R$, and *external* edges $L \to R$ and $L \leftarrow R$. In constructing $\mathcal{U}^*$, we may reallocate an internal edge (of sufficient bandwidth) to connect any two components where SOURCE=DESTINATION. External edges (of sufficient bandwidth) can be (re)used when SOURCE≠DESTINATION. Hence, in *class* edge permutation, we permute edges within a type (or class). In this paper we will restrict our edge permutations to this kind of class edge permutation. Class edge permutation leaves unchanged the complexity bounds and the lemmas of [Don4].

To summarize: vertex permutation preserves the graph topology whereas edge permutation can move the edges around. Edge permutation permits arbitrary rewiring (using existing edges). It cannot add new edges, nor can it change their bandwidth. In sec. 6, we discuss further the consequences of allowing different kinds of permutation on our model. There we show that graph permutation can be permitted at no additional "cost" and without changing the power of our systems very much.

3.4 Combination

Definition 3.8 *Consider two graphs* $\mathcal{G} = (V, E)$ *and* $\mathcal{G}' = (V', E')$. *We define the* combination $\mathcal{G} + \mathcal{G}'$ *of* $\mathcal{G}$ *and* $\mathcal{G}'$ *as follows:*

$$\mathcal{G} + \mathcal{G}' = (V \cup V', E \cup E').$$

We may define $+$ on sensor systems (def. 3.2) by lifting the definition for graphs. We may define $+$ on two immersed graphs whenever the immersions are *compatible*. An immersion ϕ of $\mathcal{G}$ and an immersion ψ of $\mathcal{G}'$ are said to be *compatible* when the two immersions agree on the intersection $V \cap V'$ (for total immersions) or more generally, on $\phi^{-1}C \cap \psi^{-1}C$ (for partial functions). Similarly:

Definition 3.9 *Consider two graphs* $\mathcal{G} = (V, E)$ *and* $\mathcal{G}' = (V', E')$, *where* $\mathcal{G}'$ *is s subgraph of* $\mathcal{G}$. *We define the* difference $\mathcal{G} - \mathcal{G}'$ *of* $\mathcal{G}$ *and* $\mathcal{G}'$ *as follows:*

$$\mathcal{G} + \mathcal{G}' = (V - V', E - E').$$

Def. 3.9 may also be lifted to (partially) immersed graphs, and hence to situated sensor systems.

3.5 Reductions, Calibration, and Codesignation

As we observed in [Don], calibration exploits external state. We wish to order systems on how much information this external state (from calibration) yields, to obtain def. 3.10, below. *Calibration complexity* is defined formally in [Don4]. Here is the basic idea. Calibration complexity measures how much information we add to a sensor system when we install and calibrate it. Installing a sensor system may require physically establishing some spatial relation between two components of the system. In this case we say the two components *codesignate* by the spatial relation. More generally, we may have to establish a relation between a component and a reference frame in the world. Most generally, when we compare two sensor systems S and Q, we typically must install and calibrate them in some appropriate relative configuration—again, in a spatial relation. When all these relations are (in)equalities of configuration, we say the system is *simple*. When all the relations are semi-algebraic (s.a.), we say the system is *algebraically codesignated*.

Definition 3.10 *We write* $S \leq Q$ *when*

1. Q *simulates* S $(S \cong Q)$,

2. S *dominates* Q *in calibration complexity, and*

3. *The maximum bandwidth of* S *is at least that of* Q *(def. 3.12).*

Convention: We will now drop the notation $\star$, and use Q^* to denote any graph permutation of sensor system Q, as described above.

Definition 3.11 *We write* $S \leq^* Q$ *if there exists some (graph) permutation* Q^* *of sensor system* Q *such that* $S \leq Q^*$.

3.5.1 Relativized Information Complexity

Consider a sensor system with output z. The complexity of many sensors can essentially be characterized using the size $\log \Bbbk(z)$ of the *output* z. Let us now ask what is the "output" of our protocols, and, more important, what is its "size."

Suppose we suggest that the output of each system is at most 2 bits: the system's output chooses between three motor control states, the actions $\text{MOVE}(L)$, $\text{MOVE}(R)$, or $\text{PUSH}(p)$. In this case, we note that the sensor system has internal bandwidth that is much higher ($\log \Bbbk(\Delta\theta)$ or $\log \Bbbk(\Delta x)$ bits). The output in some sense encodes that information. That is, we may view the protocols as "recognizing" a "model" or a "signal" of size $O(\log \Bbbk(\Delta x))$ bits, and subsequently "hashing" that model to one of three equivalence classes. This argues that perhaps the intrinsic output complexity of the protocols should be more like $\log \Bbbk(\Delta x)$ bits.[9]

Another idea is to observe that the actuator output p in $\text{PUSH}(p)$ would be at a similar resolution to the orientation sensing $\theta(t)$ or odometry $x_i(t)$. This argues that the "output" of the protocol is something more like $O(\log \Bbbk(p))$ bits (since the $\text{MOVE}(L/R)$ decision is indeed binary).

This discussion reveals a more general issue with sensor systems. In particular, there are sensor systems whose complexity cannot be well-characterized by the number of bits of output.[10] For example: consider a

---

[9] To see that instrumenting $\Delta\theta$ and Δx require the same number of bits, requires an argument like the "decalibration" lemmas of [Hors]. For this paper, we can see this from the relation $2r \tan \theta(t) = \Sigma(t)$.

[10] This discussion devolves to a suggestion of Sundar Narasimhan [Personal communication], for which we are very grateful.

"grandmother" sensor. Such a sensor looks at a visual field and outputs one bit, returning #t if the visual field contains a grandmother and #f if it doesn't. Now, one view of the sensor interpretation problem is that of information reduction and identification (compare [DJ], which discusses hierarchies of sensor information). However consider a somewhat different perspective, that views sensors as *model matchers*. So, imagine a computational process that calculates the probability $P(G/V)$ of G (grandmother) given V (the visual field) — i.e., the probability that G is in the data (the visual field itself). The sensor in the former case is something specific only to detecting grandmothers, while the latter prefers to see a grandmother as the model that best explains the current data. The latter is a process that computes over model classes. For example, this sensor might output TIGER (when given a fuzzy picture that is best explained as a tiger).

In short, one may view a sensor system as storing prior distributions. These distributions bias it toward a fixed set of model classes. In principle, the stored distributions may be viewed either as calibration or internal state. To quantify the absolute information complexity of a sensor system, we need to measure the information complexity of model classes stored in the prior distribution of the sensor. This could be very difficult.

Instead, we propose to measure a quantity called the *maximum bandwidth* of a sensor system. Intuitively, this quantity is the maximum over all internal and external edge bandwidths (data-paths). That is:

Definition 3.12 *We define the* internal *(resp. external) bandwidth of a sensor system S to be the greatest bandwidth of any internal (resp. external) edge in S. The output size of S is given by Definition 3.3. We define the* maximum bandwidth *to be the greater of the internal bandwidth, external bandwidth, and the output size of S. We call a sensor system* monotonic *if its internal and external bandwidths are bounded above by its output size.*

The maximum bandwidth is an upper bound on the relative intrinsic output complexity (relativized to the information complexity of the components (secs. 3 and 3.5.1)). We explore this notion briefly below.

Maximum bandwidth is a measure of internal information complexity. The bandwidth is a measure

of information complexity only *relative* to the sensori-computational components of the system. For example, imagine that we had a sensor system with a single component that outputs one bit when it recognizes a complicated model (say, a grandmother). The only data path in the system has bandwidth one bit, because the single component in the system is very powerful. So, even though the maximum bandwidth is small, the absolute information complexity may be large.

So, some sensors are black boxes. We call a sensor system a *black box* if it is encoded as a single component. The only measure of bandwidth we have for a black box is its output size. For example, the odometry system ODOM and the θ-source system θ in sec. 4 are black boxes.

More generally, we call a sensor system *monotonic* if its internal bandwidth is bounded above by its output size. So, black box sensors are trivially monotonic. All the sensor systems in [Don4] are monotonic. If we believe that the output size of our protocols is $O(\log \Bbbk(p))$ bits, then our sensor systems are also monotonic. If we believe the output size is 2 bits, they are not. In any case, the bandwidths of Protocols I(QS) and II are $\log \Bbbk(\Delta x)$ and $\log \Bbbk(\Delta \theta)$ (resp.) Since $2r \tan \theta(t) = \Sigma(t)$, we argue that these two systems have the same maximum bandwidth.

3.5.2 Reductions using Communication

In light of this discussion, we now define the reduction $\leq_1$ from [Don], using relativized information complexity. Recall the construction of COMM($\cdot$) as a sensor system (sec. 3.2). First, let S be a monotonic sensor system with output z. In this case, we define COMM(S) to be COMM(z).

More generally, for (possibly) non-monotonic sensors, we will let COMM(S) be COMM(2^k) where k is the *relative intrinsic output complexity* of S. Measuring this (k) in general is difficult, but we will treat the *maximum bandwidth* (def. 3.12) of S as an upper bound on k.

Definition 3.13 *Consider two sensor systems S and Q. We say S is* efficiently reducible *to Q if*

$$S \leq^* Q + \text{COMM}(S). \tag{2}$$

In this case we write[11] *$S \leq_1 Q$.*

---

[11] $\leq_1$ is also called $<_s$ in [Don].

Now, permutation (the $*$ operation) and combination (the $+$ operation) "commute" for compatible partial immersions. This is formalized as a "distributive" property in [Don4]. So, for example, for any sensor system S, we have ensured that $S^* + \text{COMM}(\cdot) = (S + \text{COMM}(\cdot))^*$, i.e., we can do the permutation and combination in any order. Second we have ensured that the combination operation $+$ is commutative and associative. Third, in the reduction $\leq_1$ we have given the single edge e in $\text{COMM}(\cdot)$ enough bandwidth so that it still works when we switch it (e) around using permutation. Hence, the sensor system $(Q + \text{COMM}(S))^*$ from eq. (2) may be implemented as the sensor system Q permuted in an arbitrary way, plus one extra data path whose bandwidth is that of the largest flow in S.

Observe that even when $\leq^*$ is transitive, it appears that $\leq_1$ is not. To see this, suppose that $A \leq_1 B$ and $B \leq_1 C$. Then it appears that to reduce A to B we require one "extra wire" (namely, $\text{COMM}(A)$), and that to reduce B to C we could require (another) extra wire $\text{COMM}(B)$, and therefore, in the worst case, to reduce A to C we could require *two* extra wires. That is, it could be that A cannot reduce to C with fewer than two extra wires. We have yet to find a non-artificial example of this lower bound but it appears to indicate that $\leq_1$ is not transitive (even for simple sensor systems (sec. 3.5)).

Let us summarize. The reduction $\leq_1$ corresponds to a specific circuit transformation. This transformation can be understood as follows. Let S be a monotonic sensor system with output b. Let Q be another sensor system. We imagine Q as a "circuit" embedded in (say) the plane. Let $\text{COMM}(S)$ be a "sensor system" with one datapath e, that has bandwidth $\log \Bbbk(b)$. Then, adding output communication to Q can be viewed as the following transformation on sensor systems: $Q \mapsto Q + \text{COMM}(S)$. The transformation is parameterized by (the bandwidth of) S. The bounded-bandwidth datapath e can be spliced into Q anywhere. We note that this transformation can be composed with permutation (in either order):

$$\begin{array}{ccccc} Q & \mapsto & Q^* & \mapsto & Q^* + \text{COMM}(S) \\ \| & & & & \| \\ Q & \mapsto & Q + \text{COMM}(S) & \mapsto & (Q + \text{COMM}(S))^*. \end{array}$$

The reduction $\leq_1$ (def. 3.13) is a "1-wire" reduction. It does not appear to be transitive. The reduction $\leq^*$

(def. 3.11) is a *"0-wire"* reduction. It is transitive for simple sensor systems [Don4]. We could analogously define a 2-wire, or more generally, any *k-wire reduction* $\leq_k$ by modifying eq. (2) in def. 3.13 to

$$S \leq^* Q + k \cdot \text{COMM}(S), \tag{2'}$$

where $k \cdot \text{COMM}(S)$ denotes $\overbrace{\text{comm}(S) + \cdots + \text{comm}(S)}^{k \text{ times}}$.

Since $(\leq^*) = (\leq_0)$, this suggests a hierarchy of reductions, indexed by k. The k-wire reductions $\{ \leq_i \}_{i \in \mathbb{N}}$ form a *graded relation*. Even though we believe that $\leq_1$ is not transitive (in the elementary sense), the hierarchy has *graded transitivity* on simple sensor systems [Don4]. This means that for any simple sensor systems S, Q, and U, if $S \leq_i Q$ and $Q \leq_j U$, then $S \leq_{i+j} U$. This follows from a lemma that the 0-wire reduction $\leq_0$ (called $\leq^*$ in def. 3.11) is elementary transitive for *simple* sensor systems.

Consider the hierarchy of k-wire reductions $\{ \leq_i \}_{i \in \mathbb{N}}$. We say such a hierarchy *collapses* if it is isomorphic to an elementary relation. In particular, the hierarchy of k-wire reductions ($k > 0$) collapses if $\leq_1$ is elementary transitive [Don4].

4 Comparing Protocols Using Reductions

We now apply the ideas above to compare our protocols, P.I(QS) and P.II (the circuits in figs. 8–10). First, we define two black boxes (see sec. 3.5.1). Define the *odometry sensor system* ODOM to have one vertex, whose label is [odom]. It has a single component, an odometer. Similarly, define the *θ-source* sensor system θ to have a one vertex and a single component [$\theta(t)$], which is a θ-source. These systems can be installed (or better, spliced) "into" our circuits. Each black box comes with (simple) codesignation constraints. Vertex [odom] must be installed *on* a robot (either L or R). So, its vertex codesignates with L or R. Vertex [$\theta(t)$] must be installed at a location *not* on a robot. So, its vertex cannot codesignate with L or R. We now show:

Claim 4.1 *Let P.II, P.I(QS), ODOM, and θ be the sensor systems defined as above. Then,*

$$\text{P.II} \leq_k \text{P.I(QS)} - 2\text{ODOM} + \theta \tag{3}$$

for $k = 1$. Moreover, eq. (3) does not hold for $k = 0$.

Proof: Consider the sensor system $\mathcal{U}$ obtained by removing both odometers from circuit P.I(QS), and adding a θ-source:

$$\mathcal{U} = \text{P.I(QS)} - 2\text{ODOM} + \theta. \qquad (4)$$

Now, consider permutations of $\mathcal{U}$, and recall def. 3.11. We first ask whether P.II can be reduced to $\mathcal{U}$ using $\leq^*$. That is, P.II $\leq^* \mathcal{U}$? First, we note that we can move around the registers and $\boxed{-}$'s from $\mathcal{U}$ to situate all the components of P.II. We also have some leftover components (and wires). P.II requires two sign boxes; however, the $\boxed{\text{sgn}}$ box just selects out the sign bit. To build the extra sign box we can just ignore the other bits, or we can use the leftover hardware from $\mathcal{U}$ to build a small circuit to simulate $\boxed{\text{sgn}}$. We need to argue that a register big enough to hold $x_i(0)$ will also hold θ_0; this follows from $2r \tan \theta(t) = \Sigma(t)$, or from "decalibration" [Hors]. Next, we see that we can permute the internal edges of $\mathcal{U}$ to wire up the components of P.II *in situ*—internally. What about externally?

Permuting the external wiring almost works, but not quite. $\mathcal{U}$ has two external data paths, (b') and (b), with bandwidth 2 bits and $\log \Bbbk (\Delta x_1)$ bits (resp) (fig. 12). Now, since we only allow class edge permutation (as in sec. 3.3), we must permute external edges to external edges and internal edges to internal edges. Therefore, in fig. 13, the edge (b) from $\mathcal{U}$ will suffice as a datapath from $\theta(t)$ to R, since it has adequate bandwidth. However, the datapath (b') from $\mathcal{U}$ does not have adequate bandwidth to carry information from $\theta(t)$ to L. In order to build P.II from $\mathcal{U}$, we must add another external data path COMM($\cdot$) from $\theta(t)$ to L. Now, what is the argument to COMM($\cdot$)? This data path must have bandwidth of at least the relative intrinsic output complexity of P.II, or $\log \Bbbk(\Delta \theta)$ bits. Hence we may parameterize this new edge by writing COMM(P.II), following sec. 3.5.1. Hence, we see that

$$\text{P.II} \leq (\mathcal{U} + \text{COMM(P.II)})^*. \qquad (5)$$

Therefore by def. 3.11,

$$\text{P.II} \leq^* \mathcal{U} + \text{COMM(P.II)}, \qquad (6)$$

so using def. 3.13, we have P.II $\leq_1 \mathcal{U}$. Hence we conclude

$$\text{P.II} \leq_1 \text{P.I(QS)} - 2\text{ODOM} + \theta, \qquad (7)$$

which implies eq. (3) as desired. $\square$

This formalizes our intuition that, by removing odometry but adding relative normal sensing, we can accomplish the pushing task without explicit communication. More precisely, we show how to build one circuit P.II "efficiently" out of the other ($\mathcal{U}$). To transform P.I(QS) into P.II, the operators $+$ and $-$ quantify what resources we add and delete. Relative information complexity allows us to measure the effective communication "through the world." The permutation quantifies the redistribution of resources.

5 Computing Reductions

Claim 4.1 is a proof done "by hand." That is, we can in principle determine that eq. (3) holds (for $k > 0$) by showing—"by hand"—the existence of a suitable permutation. It is somewhat surprising that we can in fact automate this process: [Don4] gives algorithms for deciding the relation $\leq_1$. More precisely, given suitable encodings of two sensor systems $\mathcal{S}$ and $\mathcal{U}$, we can computationally decide whether $\mathcal{S} \leq_1 \mathcal{U}$ [Don4]. The algorithm is too complicated to describe here. We examine a special case to give a flavor for it; many details are omitted. The basic idea involves employing the theory of real closed fields with bounded quantification. Let us for a moment restrict our reductions to vertex permutation alone (def. 3.7). First, suppose that $\mathcal{S}$ and $\mathcal{U}$ each have d vertices. Then an immersion of $\mathcal{S}$ can be encoded as a point in C^d. More generally, a partial immersion ϕ whose domain contains $l \leq d$ vertices can be modeled as a point in C^l. We can "guess" a (vertex) permutation ϕ^* of ϕ by existentially quantifying over the configurations of the l vertices inside ϕ's domain. Hence, the space of permutations of ϕ, denoted $\Sigma(\phi)$, is isomorphic to C^l. Similarly, we can verify a Tarski sentence for all extensions $\bar{\phi}$ of ϕ, by universal quantification over the $d - l$ vertices *outside* the domain of ϕ. Hence, the space of all extensions of ϕ, $\text{ex}\,\phi$, is isomorphic to C^{d-l}. We will model (algebraic) codesignation constraints as a (possibly constant) semi-algebraic (s.a.) mapping $\bar{\phi} \mapsto D(\bar{\phi})$ taking an immersion $\bar{\phi}$ to a s.a. set $D(\bar{\phi}) \subset C^d$. All these methods generalize to graph permutation as well [Don4]. Now,

Definition 5.1 *A simulation function* $\Omega_{\mathcal{U}}$ *for* $\mathcal{U}$ *is a map* $\Omega_{\mathcal{U}} : C^d \to R$, *where* R *the space of outputs. We call the value* $\Omega_{\mathcal{U}}(\phi) \in R$ *of* $\Omega_{\mathcal{U}}$ *on a sensor configuration* ϕ *to be the* output value *or* sensor value *at* ϕ.

Definition 5.2 *We call a sensor system* $\mathcal{U}$ <u>algebraic</u> *if it is algebraically codesignated (sec. 3.5), has an algebraic configuration space* C, *and a semi-algebraic algebraic simulation function* $\Omega_{\mathcal{U}}$.

How do we construct and permute simulation functions? Recall the discussion of simulation functions for components and connections above. To decide $\leq^*$ means to deciding whether or not $(\mathcal{S}, \phi) \leq^* (\mathcal{U}, \psi)$. Hence we must decide whether there exists a permutation ψ^* of ψ such that $(\mathcal{S}, \phi) \cong (\mathcal{U}, \psi^*)$. Computationally, this requires deciding the Tarski sentence[12]

$$(\exists \psi^* \in \Sigma(\psi), \ \forall \overline{\phi} \in \mathrm{ex}\,\phi, \ \forall \overline{\psi^*} \in D_{\mathcal{S}}(\overline{\phi}) \cap \mathrm{ex}\,\psi^*):$$
$$\Omega_{\mathcal{S}}(\overline{\phi}) = \Omega_{\mathcal{U}}(\overline{\psi^*}). \tag{8}$$

Here, $D_{\mathcal{S}}(\cdot)$ models the codesignation constraints; they require the choice of extension $\overline{\psi^*}$ by the inner quantifier to depend on the extension $\overline{\phi}$ selected by the middle quantifier. When comparing two sensor systems $\mathcal{S}$ and $\mathcal{U}$, we typically must install and calibrate them in some appropriate relative configuration—i.e., in the spatial relation that the codesignation constraint $D_{\mathcal{S}}(\cdot)$ encodes.

If we can decide $\leq^*$, we can decide $\leq_1$. Here is why: to decide $\leq_1$, we must determine whether $(\mathcal{S}, \phi) \leq^* (\mathcal{U}, \psi) + \mathrm{COMM}(\mathcal{S})$, (def. 3.13). Recall the definition of *compatibility* for partial immersions (sec. 3.4). We first observe that permutation (the * operation) and combination (the + operation) "commute" for compatible partial immersions [Don4]. Our arguments above for guessing extensions and permutations can be generalized *mutatis mutandis* to compute the combination (def. 3.8) of two algebraic sensor systems. Since $\mathrm{COMM}(\mathcal{S})$ is a constant-sized sensor system (2 vertices, one edge) with only a constant number of codesignation constraints (at most 2), we may guess how to combine it with a permutation $(\mathcal{U}, \psi^*)$ of $(\mathcal{S}, \psi)$ within the same time bounds given below in lemma 5.3 and eqs. (9–10).

---

[12]We must also be able to compute dominance in calibration complexity (see def. 3.10). This can be done independently, and much faster than eq. (8) can be decided; see [Don4].

Let us suppose that $\mathcal{U}$ and $\mathcal{S}$ are algebraic. Let us define the *size* d of $\mathcal{U}$ to be the number of vertices in $\mathcal{U}$. Let the *simulation complexity* n_Ω be the size of the simulation functions $\Omega_{\mathcal{U}}$ and $\Omega_{\mathcal{S}}$. We let n_D measure the complexity of the codesignation constraints $D_{\mathcal{S}}(\cdot)$ in (8). Then, we can decide eq. (8) in the time bounds below:

Lemma 5.3 (Don4) *There is an algorithm for deciding the relations* $\leq^*$ *and* $\leq_1$ *for algebraic sensor systems. It runs in time polynomial in the simulation and codesignation complexity* $(n_\Omega + n_D)$, *and sub-doubly exponential in the size of the sensor systems. That is, if the system has size* d *the time complexity is*

$$(n_\Omega + n_D)^{(r_c d)^{O(1)}}, \tag{9}$$

where r_c *is the dimension of the configuration space* C *for a single component.* $\square$

Let us call $\mathcal{U}$ <u>small</u> if n_Ω and n_D are only polynomially large in d, i.e., $(n_\Omega + n_D) = d^{O(1)}$. Note that for "small" sensor systems, eq. (9) becomes

$$d^{(r_c d)^{O(1)}}. \tag{10}$$

Although complex, eq. (8) is simplified for presentation. The full Tarski sentence also contains codesignation constraints for the outer quantifiers, and is given in [Don4]. We must warn that in sec. 5 we have examined a special case, where $\mathcal{S}$ and $\mathcal{U}$ are partially situated (that is, the domains of ϕ and ψ are non-empty). A powerful generalization is given in app. A, where the sensor systems can be *unsituated*.

6 Sensitivity of the Model

We wish to explain whether or not our theory has revealed some universal truth about sensoricomputational information invariants, or whether the results are sensitive to the particular encodings (circuits) we chose to analyze. How sensitive is our model? We consider two ways to investigate this issue. First, we try changing our model of reduction (specifically, permutation) slightly, to see how that affects our results. Second, we ask, what if the input were reëncoded slightly differently? Would our results change a lot?

Specifically, we ask: how can we compare vertex permutation with graph permutation (sec. 3.7)? In particular: (i) if we permit graph permutation instead of vertex permutation, does it change our complexity bounds? and (ii) does graph permutation give us a more powerful reduction than vertex permutation? Question (i) gives us some insight into the sensitivity of our complexity bounds to the model of reduction we use. Question (ii) sheds light on whether we can cheaply and cleverly reëncode a sensor system so as to gain a lot of "power" (information complexity).

[Don4] first derives the complexity bounds in lemma 5.3 and eq. (9–10) for vertex permutation. Next, [Don4] asks: how expensive it is to compute the reductions $\leq^*$ and $\leq_1$ using graph permutation? By extending the configuration space C^d to include all possible edge permutations, we obtain an extended configuration space of sufficiently low dimension that we still obtain the same complexity bounds given in lemma 5.3 and eqs. (9–10), (so long as r and s are constants) [Don4].

We now address question (ii): does graph permutation give us a more powerful reduction? We show:

Lemma 6.1 (The Clone Lemma) *Graph permutation can be simulated using vertex permutation, preceded by a linear time and linear space transformation of the sensor system.*

Proof: Given a sensor system $\mathcal{U}$ we "clone" all its vertices, and attach the edges to the clones. The cloned system simulates the original when each vertex is colocated with its clone. Components remain associated with original vertices. We can move an edge independently of the components it originally connected, by moving its vertices (which are clones). Any graph permutation of $\mathcal{U}$ can be simulated by a vertex permutation of the cloned system.

More specifically: Given a graph $G = (V, E)$ with labelling function ℓ, we construct a new graph $G' = (V', E')$ with labelling function ℓ'. Let the cloning function $\mathrm{cl} : V \hookrightarrow \mathbb{V}$ be an injective map from V into a universe of vertices $\mathbb{V}$, such that $\mathrm{cl}(V) \cap V = \emptyset$. We lift cl to V^2 and then restrict it to E to obtain $\mathrm{cl} : E \to \mathrm{cl}(V)^2$ as follows: If $e = (u, v)$, then $\mathrm{cl}(e) = (\mathrm{cl}(u), \mathrm{cl}(v))$. Edge labels are defined as follows: $\ell'(\mathrm{cl}(e)) = \ell(e)$.

Finally we define $V' = V \cup \mathrm{cl}(V)$ and $E' = \mathrm{cl}(E)$. We define the labelling function ℓ' on V' as follows.

$\ell'(v) = \ell(v)$ when $v \in V$. Otherwise, $\ell'(v)$ returns the "identity" component, which can be simulated as the identity function.[13]

Suppose $\mathcal{U}$ has $d = |V|$ vertices and $|E|$ edges. This transformation adds only d vertices and can be computed in time and space $O(d + |E|)$. $\square$

Let us denote by $\mathrm{cl}(\mathcal{U})$ the linear-space clone transformation of $\mathcal{U}$ described in lemma 6.1. Now consider any k-wire reduction $\leq_k$ (sec. 3.5.2). We see that:

Corollary 6.2 *Let $k \in \mathbb{N}$. Suppose that for two sensor systems $\mathcal{U}$ and $\mathcal{V}$, we have $\mathcal{V} \leq_k \mathcal{U}$ (using graph permutation). Then $\mathcal{V} \leq_k \mathrm{cl}(\mathcal{U})$ (using only vertex permutation).* $\square$

7 Experiments

Using information invariants, we have presented a formal *post hoc* analysis of two manipulation protocols for a pushing task. However, we are also using these techniques in our laboratory for synthesis, to develop new manipulation protocols and analyze their robustness and information content. We believe that our techniques are useful both for transforming protocols so as to remove assumptions and thereby increase their generality and robustness, and also to develop new protocols for complex manipulation tasks. We give examples of these application below, in secs. 7.1 and 7.2. It is our belief that our methods can give valuable insights into the information structure of the task.

7.1 Removing Assumptions

The protocols we described depend critically on the assumption of velocity synchronization. For example, let v_1 and v_2 be the speeds of the robots. Let C_p be the region between the pushing rays p. Consider the situation shown in fig. 14. One explanation is the the COF is at location Y outside of C_p, and $v_1 = v_2$. But a second explanation is that the COF is at X, inside

---

[13]The proof can be strengthened as follows. Recall that two components can communicate without an (explicit) connection when they are spatially colocated. Therefore the proof goes through even if cloned vertices have no associated components, that is, $\ell'(v) = \emptyset$ for $v \notin V$. This version has the appeal of changing the encoding without adding additional physical resources.

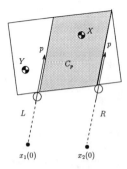

Figure 14: Two robots pushing a box. If their speeds are equal, we can infer that the COF is at Y, outside C_p. However, if the right robot is faster, it is possible that the COF lies between the pushing rays at X.

Left Robot	Communication	Right Robot
$x_0 \leftarrow \text{measure}(x_1(0))$		$x_0 \leftarrow \text{measure}(x_2(0))$
$v \leftarrow v_{\text{init}}$		$v \leftarrow v_{\text{init}}$
Repeat:		Repeat:
$\quad \text{measure}(\Delta x_1)$		$\quad \text{measure}(\Delta x_2)$
	$\boxed{L \rightarrow R}\ (\Delta x_1)$	
		$\Sigma \leftarrow \Delta x_1 - \Delta x_2$
		$s \leftarrow \text{sign}(\Sigma)$
	$\boxed{L \leftarrow R}\ (s)$	
$v \leftarrow v - s\Delta v$		$v \leftarrow v + s\Delta v$

Figure 15: A simplified version of protocol P.V. P.V performs velocity synchronization, using explicit local communication (it does not perform the pushing task!) The velocity of each robot is incremented or decremented by a fixed amount Δv, depending on which on which robot gets ahead. This simplified version assumes that $\Sigma_0 = 0$.

C_p, and $v_2 > v_1$. Velocity synchronization ($v_1 = v_2$) is key to ensuring that we can infer whether or not the COF is in C_p.

We have used methods similar to those in secs. 2-5 to develop protocols that do not require synchronization. Removing explicit synchronization from manipulation protocols is analogous to removing explicit communication. The ideal protocol we started with assumed global coordination and control—i.e., global velocity synchronization. Next, we developed a velocity synchronization protocol P.V with explicit local communication. Given our analysis above, it is very simple to develop such a protocol, and we give the basic idea in fig. 15. Next, we "composed" it with protocol I(QS) above–this corresponds to "splicing" the circuits for P.V and P.I(QS) together. This is slightly more difficult, but using the techniques outlined in this paper, it it not too hard to do a careful analysis and get all the special cases right. Finally, we removed all explicit communication; again the robots communicate "through the world" (through the task dynamics). We leave it as an exercise for the reader to develop the resulting, communication-free protocol for box pushing without explicit communication or synchronization.[14]

---

[14]Hint: the easiest way to compose the control loops, is to first add two bits of explicit communication. A one-bit $L \rightarrow R$ datapath tells R when L has broken contact. A symmetric $L \leftarrow R$ datapath tells L when R breaks contact. The need for this communication falls out of a synthesis similar

It was actually somewhat surprising to us that a uniform asynchronous protocol with no explicit communication can be developed for this task.

7.2   Reorienting Large Objects

Pushing Task	Reorientation Task	
Protocol O	⚠	global coordination and control
Protocol I(QS)	⚠	local IR comm, partial synchrony, MPMD
Protocol II	⚠	uniform, SPMD, asynchronous

Figure 16: Summary of parallel manipulation protocols.

We would like to show that our methods could also be useful in engineering new protocols for difficult, multi-robot manipulation tasks. In this direction, we have also considered three multi-mobot manipulation protocols for box reorientation (see fig. 3 and [DJR]). We denote them by ⚠, ⚠, etc. For these protocols, we started with the offline algorithm ⚠ of [Rus], which was designed for multi-fingered robot hands with global

---

to the one we presented. However, a careful analysis then shows that even these two bits of (explicit) communication can be removed.

coordination. Next, we developed a protocol ⚠ for three cooperating mobile robots with local IR communication. In this protocol, only one robot moves at a time, so, although the task has been "parallelized", it is not "load-balanced." Finally, ⚠ we removed all explicit communication between agents, and allowed the robots to perform simultaneous, asynchronous manipulation of the box. Protocol ⚠ has several advantages over protocol ⚠. Using protocol ⚠, two robots (instead of three) suffice to rotate the box. The protocol is "uniform" (SPMD) in that the same program (including the same termination predicate) runs on both robots. More interesting, in ⚠ it is no longer necessary for the robots to have an *a priori* geometric model of the box— whereas such a model is required for ⚠ and ⚠. Of course, various assumptions must hold for the task to succeed—the point of our analyses is to reveal these assumptions. We are currently completing such an analysis.[15]

In terms of program development, synchrony, and communication, we have the following approximate correspondence between these protocols:

We believe that a methodology for developing coordinated manipulation protocols is emerging, based on the tools described in this paper, [DJR], and [Don4]:

Developing Parallel Manipulation Protocols

1. Start with a sensorless [EM, EMV] or near-sensorless [Erd4, JR] manipulation protocol requiring global coordination of several "agents" (eg., fingers [Gol, Rus], or "fences" [PS]). Examples: ⚠ above [Rus] or protocol O (fig. 5).

2. Distribute the protocol over spatially separated agents. Synchronize and coordinate control using explicit local communication. Examples: Protocol I(QS), or ⚠ above.

3. Define a *virtual sensor*[16] that measures the key signal we wish to servo on. Example: Σ (or better, $s = \text{sign}(\Sigma - \Sigma_0)$) in P.I(QS) (fig. 8).

4. Find a way to implement this virtual sensor using concrete sensors on mechanical observables. Example: $n(t)$ in P.II (fig. 11).

5. Transform the communication between two agents L and R into shared data structures.

6. Implement the shared data structures as "mechanical registers." Example: $\theta(t)$ *is a "register" that L and R share.*

We believe that our methods are useful for developing parallel manipulation protocols. We think that information invariants can serve as a framework in which to measure the capabilities of robot systems, to quantify their power, and to reduce their fragility with respect to assumptions that are engineered into the control system or the environment. We believe that the equivalences that can be derived between communication, internal state, external state, computation, and sensors, can prove valuable in determining what information is required to solve a task, and how to direct a robot's actions to acquire that information to solve it.

8 Discussion

Our work raises a number of questions. For example, can robots "externalize," or record state in the world? The answer depends not only on the environment, but also upon the dynamics. A juggling robot probably cannot. On a conveyor belt, it may be possible (suppose "bad" parts are reoriented so that they may be removed later). However, it is certainly possible during quasi-static manipulation by a single agent. In moving towards multi-agent tasks and at least partially dynamic tasks, we are attempting to investigate this question in both an experimental and theoretical setting.

By analogy with CT reductions, we may define an equivalence relation $\equiv_k$, such that $A \equiv_k B$ when $A \leq_k B$ and $B \leq_k A$. We may also ask, does a given class of sensori-computational systems contain "complete" circuits, to which any member of the class may be reduced? Note that the relation $\equiv_k$ holds between any two complete circuits.

Finally, can we record "programs" in the world in the same way we may externalize state? Is there a "universal" manipulation circuit which can read these programs and perform the correct strategy to accomplish a task? Such a mechanism might lead to a robot which could infer the correct manipulation action by performing sensori-motor experiments.

---

[15] In particular, we have considerably improved the protocol ⚠ from [DJR].

[16] We use the term in the sense of [DJ]; others, particularly Henderson have used similar concepts.

Figure 17: TOMMY and LILY reorient a couch using an asynchronous SPMD protocol requiring no explicit communication.

Appendix

A   What is Permutation?

In sec. 5 we examined a special case, where $\mathcal{S}$ and $\mathcal{U}$ are partially situated (that is, the domains of ϕ and ψ are non-empty). A powerful generalization is given in [Don4], where the sensor systems can be *unsituated*. Using the ideas in sec. 5, we can give an "abstract" version of permutation that is applicable to partially immersed sensor systems with codesignation constraints. Each set of codesignation constraints defines a different arrangement in the space of all immersions. Each cell in the arrangement, in turn, corresponds to a region in C^d.

Permutation corresponds to selecting a different family of immersions, while respecting the codesignation constraints. Since this corresponds to choosing a different region of C^d, the picture of abstract permutation is really not that different from the computational model of situated permutations discussed in sec. 5. Suppose a simple sensor system $\mathcal{U}$ has d vertices, two of which are u and v. When there is a codesignation constraint for u and v, we write that the relation $u \sim v$ must hold. This relation induces a quotient structure on C^d, and the corresponding quotient map $\pi : C^d \to C^d/(u \sim v)$ "identifies" the two vertices u and v. Similarly, we can model a non-codesignation constraint as a "diagonal" $\Delta \subset C^d$ that must be avoided. Abstract permutation of $\mathcal{U}$ can be viewed as follows. Let $D_u = (C^d - \Delta)/(u \sim v)$. D_u is the quotient of $(C^d - \Delta)$ under π. For a partial immersion ψ^* to be chosen compatibly with the codesignation constraints, we view permutation as a bijective self-map of the disjoint equivalence classes

$$\left\{ \pi\left(\text{ex}\,\psi^* - \Delta\right)\right\}_{\psi^* \in \Sigma(\psi)}. \qquad (11)$$

Thus, in general, the group structure for the permutation must respect the quotient structure for codesignation; correspondingly, we call such permutations *valid*. Below, we define the "diagonal" Δ, precisely.

Now, an unsituated sensor system $\mathcal{U}$ could be modeled using a partial immersion ψ_0 with an empty domain. In this case $\text{ex}\,\psi_0 = C^d$ and eq. (11) specializes to the single equivalence class $\left\{D_\mathcal{u}\right\}$. In this "singular" case, we can take several different approaches to defining unsituated permutation. (i) We may define that $\psi_0^* = \psi_0$. Although consistent with situated permutation, (i) is not very useful. We choose a different definition. For unsituated permutation, we redefine $\Sigma(\psi_0)$ and $\text{ex}\,\psi_0$ in the special case where ψ_0 has an empty domain. (ii) When $\mathcal{U}$ is simple, we may define $\Sigma(\psi_0)$ to be the set of colocations of vertices of $\mathcal{U}$. That is, let $(x_1, \ldots, x_d)$ be a point in C^d, and define the ij^{th} diagonal $\Delta_{ij} = \left\{ (x_1, \ldots, x_d) \mid x_i = x_j \right\}$. Define permutation as a bijective self-map of the cells in the arrangement generated by all $\binom{d}{2}$ such diagonals $\left\{\Delta_{ij}\right\}_{i,j=1,\ldots,d}$. So, $\Sigma(\psi_0)$ is an arrangement in C^d of complexity $O(d^{2dr_c})$, $\text{ex}\,\psi_0^* \in \Sigma(\psi_0)$ is a cell in the arrangement, and $\psi_0^* \in \text{ex}\,\psi_0^*$ is a witness point in that cell. Hence ψ_0^* is a representative of the equivalence class $\text{ex}\,\psi_0^*$. As in situated permutation, unsituated permutation can be viewed as a self-map of the cells $\left\{\text{ex}\,\psi_0^*\right\}$ or (equivalently) as a self-map of the witnesses $\left\{\psi_0^*\right\}$. Perhaps the cleanest way to model our main examples (sec. 4) is to treat all the sensor systems as initially unsituated, yet respecting all the (non)codesignation constraints. This may be done by (1) "algebraically" specifying all the codesignation constraints, (2) letting the domain of each immersion be empty, (3) using (ii) above, choose unsituated permutations that respect the codesignation constraints. The methods of sec. 5 can be extended to guess unsituated permutations. In our examples (sec. 4), each guess (i.e., each unsituated permutation) corresponds to a choice of which vertices to colocate.[17]

---

[17] The codesignation relation $u \sim v$, the quotient map π, the non-codesignation relation Δ, and definition (ii) of unsituated permutation, can all be extended to algebraic sensor systems using the methods of sec. 5.

A.1 Example

Unsituated permutation is quite powerful. Consider deciding eq. (8) (in this example, we only consider vertex permutation of simple sensor systems). In particular, we want to see that (8) makes sense for unsituated permutation, when we replace ψ by ψ_0, ϕ by ϕ_0, etc., to obtain:

$$\left(\exists\psi_0^* \in \Sigma(\psi_0),\ \forall\overline{\phi_0} \in \text{ex}\,\phi_0,\ \forall\overline{\psi_0^*} \in D_\mathcal{S}(\overline{\phi_0}) \cap \text{ex}\,\psi_0^*\right):$$
$$\Omega_\mathcal{S}(\overline{\phi_0}) = \Omega_\mathcal{u}(\overline{\psi_0^*}). \qquad (8')$$

With situated permutation (8), we are restricted to first choosing the partial immersion ϕ, and thereby fixing a number of vertices of $\mathcal{S}$. Next, we can permute $\mathcal{U}$ to be "near" these vertices (this corresponds to the choice of ψ^*). This process gets the colocations right, but at the cost of generality; we would know that for any "topologically equivalent" choice of ϕ, we can choose a permutation ψ^* such that (8) holds. For simple sensor systems, "topologically equivalent" means, "with the same vertex colocations."

Unsituated permutation (8') allows us to do precisely what we want. In place of a partial immersion ϕ for $\mathcal{S}$, we begin with a witness point $\phi_0 \in C^d$. ϕ_0 represents an equivalence class $\text{ex}\,\phi_0$ of immersions, all of which colocate the same vertices as ϕ_0. So, ϕ_0 says *which* vertices should be colocated, but not *where*. Now, given ϕ_0, the outer existential quantifier in (8') chooses an unsituated permutation ψ_0^* of $\mathcal{U}$. ψ_0^* represents an equivalence class $\text{ex}\,\psi_0^*$ of immersions of $\mathcal{U}$, all of which colocate the same vertices of $\mathcal{U}$ as ψ_0^* does. The other, disjoint equivalence classes, are also subsets of C^d; each equivalence class colocates different vertices of $\mathcal{U}$, and the set of all such classes is $\Sigma(\psi_0)$ $(= \Sigma(\psi_0^*))$. Choice of ψ_0^* selects which vertices of $\mathcal{U}$ to colocate. The codesignation constraint $D_\mathcal{S}(\cdot)$ then enforces that, when measuring the outputs of $\mathcal{S}$ and $\mathcal{U}$, we install them in the same "place." More specifically: ϕ_0 (given as data) determines which vertices of $\mathcal{S}$ to colocate; choice of ψ_0^* determines which vertices of $\mathcal{U}$ are colocated; construction of $D_\mathcal{S}(\cdot)$ determines which vertices of $\mathcal{U}$ and $\mathcal{S}$ are colocated. Most specifically, given the configuration $\overline{\phi_0}$ of $\mathcal{S}$, $D_\mathcal{S}$ in turn defines a region $D_\mathcal{S}(\overline{\phi_0})$ in the configuration space C^d of $\mathcal{U}$. This region constraints the necessary coplacements $\overline{\psi_0^*}$ of $\mathcal{U}$ relative to $(\mathcal{S}, \overline{\phi_0})$.

Perhaps surprisingly, allowing unsituated permutation does not change the complexity bounds of sec. 5 [Don4].

Bibliography

[BK] **Blum, M. and Kozen, D.** *On the power of the compass (or, why mazes are easier to search than graphs)*, Proc. 19th Symp. Found. Computer Science, Ann Arbor, MI, pp. 132-42 (1978).

[Bri] **Briggs, Amy.** *An Efficient Algorithm for One-Step Compliant Motion Planning with Uncertainty*, Algorithmica, **8**, (2), 1992. pp. 195-208.

[Can] **John Canny** *On computability of fine motion plans*, IEEE ICRA, Scottsdale, AZ, (1989)

[CR] **Canny, J., and J. Reif,** "New Lower Bound Techniques for Robot Motion Planning Problems", *FOCS* (1987).

[Don] **Donald, B. R.** *Information Invariants in Robotics, Parts I and II*, IEEE International Conference on Robotics and Automation, Atlanta, GA. (1993).

[Don1] **Donald, B. R.** *Robot Motion Planning*, IEEE Trans. on Robotics and Automation, **(8)**, No. 2. (1992).

[Don2] **Donald, B. R.** *The Complexity of Planar Compliant Motion Planning with Uncertainty*, Algorithmica, **5** (3), pp. 353-382 (1990).

[Don3] **Donald, B. R.** *Error Detection and Recovery in Robotics*, Lecture Notes in Computer Science, Vol. 336, Springer-Verlag, New York (1989).

[Don4] **Donald, B. R.** *On Information Invariants in Robotics,* Cornell Computer Science Department Technical Report TR 93-1341. To appear in *Artificial Intelligence*. (1993).

[DJ] **Donald, B. R. and J. Jennings** *Constructive Recognizability for Task-Directed Robot Programming*, Jour. Robotics and Autonomous Systems, **(9)**, No. 1, Elsevier/North-Holland pp. 41-74. (1992).

[DJR] **Donald, B. R., J. Jennings, and D. Rus** *Experimental Information Invariants for Cooperating Autonomous Mobile Robots*, International Joint Conference on Artificial Intelligence (IJCAI) Workshop on Dynamically Interacting Robots. Chambery, France (Aug 28) (1993).

[Erd1] **Erdmann, M.** *Using Backprojections for Fine Motion Planning with Uncertainty*, IJRR Vol. 5 no. 1 (1986).

[Erd2] **Erdmann, M.** *On Probabilistic Strategies for Robot Tasks*, Ph.D. thesis, MIT Department of EECS, MIT A.I. Lab, Cambridge MIT-AI-TR 1155 (1989).

[Erd3] **Erdmann, M.** *Action Subservient Sensing and Design*, IEEE ICRA, Atlanta. See also the Carnegie-Mellon report CMU-CS-92-116. (1993).

[Erd4] **Erdmann, M.** *Randomization for Robot Tasks: Using Dynamic Programming in the Space of Knowledge States*, Algorithmica Vol. 10, Nos. 2/3/4, Aug/Sept/Oct. pp. 248-291 (1993).

[EM] **Erdmann, M., and M. Mason,** "An Exploration of Sensorless Manipulation", *IEEE International Conference on Robotics and Automation*, San Francisco, April, 1986.

[EMV] **Erdmann, M., M. Mason, and G. Vaněček** *Mechanical Parts Orienting: The Case of a Polyhedron on a Table*, Algorithmica Vol. 10, Nos. 2/3/4, Aug/Sept/Oct. pp. 266-247 (1993).

[Gol] **Goldberg, K. Y** *Orienting Parts without Sensors*, Algorithmica Vol. 10, Nos. 2/3/4, Aug/Sept/Oct. pp. 201-225 (1993).

[HSS] **Hopcroft, J. E., Schwartz, J. T., and Sharir, M.** 1984 On the Complexity of Motion Planning for Multiple Independent Objects; *PSPACE*-Hardness of the "Warehouseman's Problem." *International Journal of Robotics Research.* 3(4):76–88.

[Hors] **Horswill, I.** *Analysis of Adaptation and Environment,* Submitted to *Artificial Intelligence* (1992).

[JR] **Jennings, J. and Rus, D.** *Active Model Acquisition for Near-Sensorless Manipulation with Mobile Robots,* The IASTED International Conference on Robotics and Manufacturing, Oxford, UK (1993).

[Koz] **Kozen, D.** *Automata and Planar Graphs,* Fundamentals of Computing Theory, Proc. Conference on Algebraic, Arithmetic, and categorical methods in Computation Theory (Berlin) ed. L. Budach, Akademie Verlag (1979).

[Lat] **Latombe, J.-C.** *Robot Motion Planning,* Kluwer: New York (1991).

[LMT] **Lozano-Pérez, T., Mason, M. T., and Taylor, R. H.** *Automatic Synthesis of Fine-Motion Strategies for Robots,* Int. J. of Robotics Research, Vol 3, no. 1 (1984).

[Mas] **Mason, M. T.** 1986. Mechanics and Planning of Manipulator Pushing Operations. *International Journal of Robotics Research* **5**(3).

[ML] **Mason, M. and Lynch, K.** *Dynamic Manipulation,* Proc. of the 1993 IEEE/RSJ Int. Conf. on Intelligent Robots and Systems, Yokohama, Japan, July 26-30 (1993).

[Nat] **Natarajan, B. K.** *On Planning Assemblies,* Proc. of the 4th Annual Symposium on Computational Geometry, Urbana, Illinois, June. pp. 299-308. (1988).

[PS] **Peshkin, M.** *Planning Robotic Manipulation Strategies for Sliding Objects,* Ph.D. dissertation, Department of Physics, Carnegie-Mellon University (1986).

[RD] **Rees, J. and B. R. Donald** *Program Mobile Robots in Scheme,* Proc. IEEE International Conference on Robotics and Automation, Nice, France (1992).

[Reif] **Reif J.,** "Complexity of the Mover's Problem and Generalizations," Proc. 20th IEEE Symp. FOCS, (1979).

[Ros] **Rosenschein, S.J.** *Synthesizing Information-Tracking Automata from Environment Descriptions,* Teleos Research TR No. 2 (1989).

[Rus] **Rus, D.** "Fine Motion Planning for Dexterous Manipulation", PhD. Thesis available as CU-CS-TR 92-1323 (August) from Comp. Sci. Dept., Cornell University, 1992.

Acknowledgment and Historical Note. Many key ideas in this paper arose in discussions with Tomás Lozano-Pérez, Mike Erdmann, Matt Mason, and Ronitt Rubinfeld. Tomás suggested studying the two-finger pushing task in 1987, and laid out a framework for analysis. Many of the ideas in this paper develop suggestions of Mike Erdmann; in particular, he proposed the notion of calibration complexity. Any perspicuous reader will notice our indebtedness to Mike's *weltanschauung,* and to [Erd3]. The robots and experimental devices described herein were built in our lab by Jim Jennings, Russell Brown, Jonathan Rees, Craig Becker, Mark Battisti, and Kevin Newman; these ideas could never have come to light without their help. Thanks to Loretta Pompilio for drawing the illustration in figure 1. Debbie Lee Smith, Amy Briggs, and Karl-Freidrich Böhringer made suggestions and helped draw the other figures, and we are very grateful to them for their help.

Anticipating Computational Demands when Solving Time-Critical Decision-Making Problems

Lloyd Greenwald, *Brown University, Providence, RI, USA*
Thomas Dean, *Brown University, Providence, RI, USA*

An agent embedded in a dynamic environment may need to respond in a timely manner to sequences of events outside the agent's control. By anticipating computational demands and allocating processing time accordingly the agent can avoid costly delays often arising from trying to respond to a dynamic environment with high-complexity decision procedures. Deliberation scheduling, the process of allocating processing time among competing decision procedures to explicitly account for the costs and benefits of computational delays, may aid an agent that must solve time-critical decision-making problems in which the time spent in decision-making affects the quality of the responses generated. The more accurate the agent is in anticipating the computational demands of forthcoming problems the more successful it can be in allocating its decision-making time. We present an approach to solving time-critical decision-making problems by taking advantage of domain structure to expand the amount of time available for processing difficult combinatorial tasks. Our approach uses predictable variability in anticipated computational demands to allocate on-line deliberation time and exploits problem regularity and stochastic models of environmental dynamics to restrict attention to small subsets of the state space. This approach demonstrates how slow, high-level systems (e.g. for planning and scheduling) might interact with faster, more reactive systems (e.g. for real-time execution and monitoring) and enables us to generate timely solutions to difficult combinatorial planning and scheduling problems such as the traffic control of multiple robot vehicles.

1 Introduction

This paper is concerned with anticipating computational demands to assist in allocating processing time to decision-making procedures in time-critical applications. The mechanisms for anticipating computational demands are discussed in Section 2 and a particular approach to time-critical decision making called *response planning* is presented in Section 3. This approach focuses on solving time-critical decision-making problems in dynamic environments with large state spaces and resource limitations. Section 4 relates our response planning approach to the general idea of anticipating computational demands. In Section 5 we describe an example traffic-control problem for multiple robot vehicles that demonstrates the applicability of the response planning approach to solving time-critical problems. A general overview of this work motivated by an air traffic control problem is found in [10].

2 Anticipating Computational Demands

We are interested in the design of systems that make the best use of the time available for decision-making by explicitly accounting for the costs and benefits of computational delays. Applications such as multiple-vehicle traffic control involve solving high-complexity planning and scheduling problems in dynamic environments. These applications have time-critical properties that require timely responses to events outside the planner's control in order to satisfy hard or soft deadlines and to avoid costly delays.

The computational demand of a high-complexity planning and scheduling problem is the amount of processing time required to reach a given level of performance for a particular decision procedure applied to a particular problem instance. Each new state of the environment may be considered a unique problem instance. Alternatively, a group of states may be jointly considered as a problem instance. Dynamic events cause changes to the state of the environment and, thus, introduce new problem instances to be solved. An agent embedded in a dynamic environment may need to respond to a sequence of problem instances and must determine how to best allocate its decision-making time for varying decision procedures and problem instances. The more accurate the agent is in anticipating the computational demands of forthcoming problems the more successful it can be in allocating its scarce decision-making time.

The agent can anticipate computational demands by predicting the problem instances for a forthcoming time period and determining the expected performance of different decision procedures given varying processing allocations. These expectations are typically derived from past performance of the decision procedures applied to similar problem instances. Anticipation requires domain knowledge and the ability to make predictions about forthcoming problem instances based on prior environmental states. No anticipation is possible in a completely nondeterministic environment or without knowledge of the domain. In some cases precursor events may give advance notice of upcoming states or domain regularity may deterministically restrict the possible sequences of upcoming events or states.

By anticipating computational demands and allocating processing time accordingly, the agent can avoid costly delays often arising from trying to respond to a dynamic environment with high-complexity decision procedures. The agent does so by initiating decision procedures prior to the occurrence of a problem instance, based on prediction and anticipated computational demands. The process of allocating processing time among competing decision procedures to explicitly account for the costs and benefits of computational delays is referred to as *deliberation scheduling*. Deliberation scheduling is a resource allocation problem where the resource is limited processing time. Deliberation scheduling aids an agent that must solve time-critical decision-making problems in which the time spent in decision-making affects the quality of the responses generated.

The task of allocating decision-making time is simplified by the use of decision procedures that can be interrupted at any time to return a solution whose quality improves with more computation time. These interruptible algorithms are often referred to as *anytime algorithms* because they can output some result, of varying quality, at any time. The flexibility inherent in these decision procedures allows for a more flexi-

ble allocation of processing time to satisfy an overall performance criterion. Anytime algorithms are sometimes referred to as *flexible computations* [11] or *imprecise computations* [20]. Zilberstein [22] contains a nice comparison of related work in anytime algorithms.

An anytime algorithm may also take the form of a *contract algorithm*. A non-interruptible contract algorithm associates a specific processing time allocation (contract time) with any given performance level. As opposed to an interruptible algorithm, a contract algorithm does not guarantee results at any time and may, in fact, not return anything for processing times less than the allocated contract time and may not achieve improved performance for actual processing time greater than the allocated contract time.

The relationship of decision procedure performance (output quality) to allocated processing time is one special case of meta-level knowledge that can be represented by a *performance profile*. In general, these profiles capture the anticipated computational demands of an algorithm under differing problem parameters and inputs. One particularly relevant parameter is the desired level of performance (output quality). Additionally, the anticipated computational demands of an algorithm may be related to both the quality level of its input and the desired performance level of its output. This is represented by a *conditional performance profile*. Conditional performance profiles may be generalized to encompass other problem parameters such as resource limitations and time window of applicability of decision procedure output. Deliberation scheduling is a special case of the more general problem of composing anytime algorithms with conditional performance profiles to satisfy overall performance measures. This is the problem of *compiling* anytime algorithms introduced by Zilberstein [22].

2.1 Performance Profiles

Anticipated computational demands for a given problem and decision procedure are captured in the form of a performance profile. In the standard usage introduced by [4, 1] a performance profile is a mapping from processing time to expected performance (quality) for a decision procedure, derived from past experience. Performance profiles are used to allocate computational resources among anticipated problem instances and competing decision procedures. In this section we generalize the performance profile notion and extend some of the discussions of [22]. In Section 2.2 we expand upon methods for conditioning profiles based on differing problem parameters and inputs.

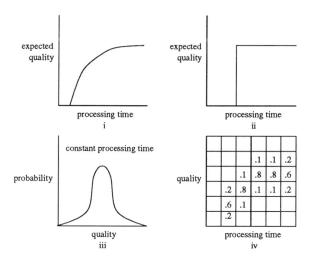

Figure 1: *Basic Performance Profiles*

A typical performance profile for an anytime algorithm is depicted in Figure 1.i. This figure indicates that the expected quality of the decision procedure summarized by this profile increases with diminishing returns (decreasing slope) as processing time increases. Another way to look at this profile is that the anticipated computational demands (processing time) increase as desired level of performance increases. Compare this to a typical performance profile for a traditional algorithm depicted in Figure 1.ii. The expected output quality of this decision procedure takes on exactly two values. It is zero until some threshold processing time and then a constant value thereafter.

By *expected quality* we are referring to a statistical measure of the output quality drawn from past experience with the decision procedure over varying problem instances. This corresponds to the actual observed quality only for deterministic algorithms in which quality is dependent only on processing time. However for nondeterministic algorithms and those for which quality may vary with the particular problem instance, actual observed quality may deviate from this expected value.

The variance in observed algorithm quality for varying problem instances can be interpreted in two ways.

For any constant desired quality there may be a probability distribution over processing times for varying problem instances. Similarly, for any constant allocation of processing time there may be a distribution over observed output quality for varying problem instances. A typical distribution is depicted in Figure 1.iii. We can capture both these interpretations in a three-dimensional performance profile in which probability is the third dimension. Zilberstein refers to this representation as a *performance distribution profile*. For discrete processing times and output qualities we can use a tabular representation like Figures 1.iv where the values in the table represent the probability that that particular quality and processing time will occur jointly. The granularity of the table is a tradeoff between accuracy and space requirements. Each column must sum to one.

The actual interpretation of the quantity *output quality* depends upon the particular decision procedure. An objective metric must be chosen for this parameter. It captures the difference between an approximation and the optimal solution for a problem instance. In ideal situations performance profiles should be machine independent. However, in general a performance profile is specific to the decision procedure and its implementation environment. This implies that a performance profile in a uniprocessor environment may not immediately generalize to a multiprocessor environment. Furthermore, resource limitations may affect the decision procedure and therefore become implicitly represented in the performance profile.

Performance profiles are generally derived from experience. In certain limited cases they can be derived from a structural analysis of the corresponding decision procedure. In addition to graphical and tabular representations, performance profiles can be represented in closed form. This becomes especially important when we consider multidimensional profiles for which graphical representation becomes difficult to handle. A closed form can be derived from statistical experience through curve fitting, from standard classes of distributions or from other approximations. A closed form for an expected performance profile may be a piecewise linear or negative exponential function whereas a closed form for a performance distribution profile may have a normal distribution about the expected output quality.

The previous discussion has focused on two parameters of a decision procedure; namely, output quality and processing time. There are many more parameters that may be of interest in deriving performance profiles. Some parameters are implicit. It is assumed that a performance profile and its decision procedure are only applicable to a specific domain of problems. As mentioned above a performance profile may be implicitly tied to a specific computation resource. For example the quality of plans produced by a decision procedure that has access to unlimited storage may be superior to one constrained by limited storage, given equal processing time. Likewise, a uniprocessor algorithm may not generalize to a multiprocessor architecture. Thus, any resource limitations may be integral to the applicability of profile.

Other parameters are explicit. Some work has been done in differentiating the *objective value* of the output of a decision procedure with the *comprehensive value* stemming from its use in a particular situation [11]. An example of such a differentiation would be a successive approximation integration procedure. The objective value of such a procedure is measured in terms of numerical error, but the contribution to comprehensive value differs depending on whether the procedure is used for medical diagnosis or stock trading. We may explicitly account for the target use of the decision procedure by constructing different profiles for different uses. Even within the same domain the target time of use of the decision procedure output and the time window of applicability of the output may be explicit features of a performance profile.

The decision whether a parameter is implicit or explicit is not always straight-forward. For example, it may be necessary to construct separate profiles for different levels of resource limitations. Alternatively we can add a new dimension to our performance profile to account for different resource limitations. The decision on whether a parameter need be explicitly represented as an additional dimension, explicitly represented as a set of conditional performance profiles or implicitly assumed as part of the single profile for a decision procedure depends upon how the profiles will be used in deliberation scheduling.

2.2   Conditional Performance Profiles

In the previous section we alluded to *conditional performance profiles*. In this section we clarify their im-

portance. Conditional performance profiles are a way to represent the dependence of performance of a deliberation procedure on various input parameters. For example, resource limitations and the time window of applicability of decision procedure output are two specific input parameters. Conditional performance profiles are an important tool in anticipating computational demands in that they allow the planner to reason about the context in which a decision procedure is used.

In general, a conditional performance profile conditions a performance profile on variations in the input parameters. One such parameter is the quality of the problem instance. This is a particularly important variation in that the quality of a problem instance of one decision procedure in a sequence is often directly related to the output quality of prior decision procedures. Other input parameters such as resource limitations may have no relation to the context in which a decision procedure is used and, therefore, have little direct impact on deliberation scheduling. Nevertheless, if resource limitations can be varied and reasoned about they may affect the output quality and, therefore, affect the deliberation scheduling problem. In deliberation scheduling, reducing time allocated to any decision procedure may affect any decision procedure that uses the results of the first and all subsequent procedures.

With conditional performance profiles, instead of deriving a single profile for all possible input parameters, we may instead have separate profiles that partition the set of possible values of input parameters. As mentioned, one particular input parameter of interest is the quality of the problem instances. The performance profile depicted in Figure 1.i is input independent if it is applicable to all possible input parameters (*e.g.* all problem instances, regardless of quality). A corresponding conditional expected performance profile is depicted in Figure 2.i. In this figure the problem instances are partitioned into three discrete classes of input quality, each with its own expected performance profile. Figure 2.ii depicts a profile for a decision procedure in which, for any constant desired output quality, the processing time to achieve that quality decreases as input quality increases. In Figure 2.iii the input quality is given its own dimension in a conditional probability distribution profile of the joint probability of input quality, output quality and processing time. If the problem inputs may be partitioned based on multiple

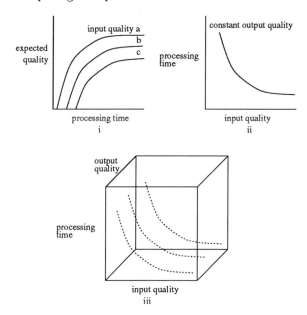

Figure 2: *Conditional Performance Profiles*

parameters (in addition to problem instance quality) we may either construct a new dimension for each parameter or combine the inputs into a single measure.

Zilberstein points out that a performance distribution profile in which the probability distribution is widely spread out may indicate that the decision procedure performs differently under differing inputs. It may be useful in this case to gather statistics independently for different classes of input parameter values in order to obtain more specific performance profiles. For probability distributions with small variance, an expected performance profile compactly represents the decision procedure applied to all relevant input parameter values including all varying problem instance qualities. Alternatively, variance in a performance distribution profile may be used as an indication of the *sensitivity* of a decision procedure to variations in input parameters.

An important use of conditional profiles is in applications in which a regular sequence of classes of problem instances can be expected. This is discussed in Section 4.

2.3 Compilation of Anytime Algorithms

Deliberation scheduling is a special case of the more general problem of composing anytime algorithms with conditional performance profiles to satisfy overall performance measures. Compiling anytime algorithms is the process of combining multiple algorithms with separate profiles into a single module with a new profile. For example we may compose anytime algorithms $g(x)$ and $h(y)$ into a new anytime algorithm $f(g(x), h(y))$ that combines the properties of both to optimize combined performance. A new profile describes the resulting behavior.

There are at least two aspects to consider in deliberation scheduling. One, how to determine the sequence of problem instances that need to be solved; and two, how to allocate processing time among decision procedures to solve those problem instances with satisfactory overall performance. These two problems are interrelated in that the allocation of time to solve earlier problem instances may help to determine the later problem instances (*i.e.* one measure of quality of a decision procedure may be its ability to improve the quality of subsequent problem instances). By anticipating the sequence of problem instances that will be realized we can anticipate the computational demands of decision procedures applied to these problem instances. We can then reason about the tradeoffs between allocating time among the procedures and the subsequent overall quality.

In some environments the sequence of problem instances may be predicted in advance of any deliberation. Alternately, the prediction of potential problem instances may be part of the decision procedure itself and reflected in its conditional performance profile. In [4, 1] the sequence of problem instances is indicated in advance by *precursor events*. In Sections 3 and 4 we discuss prediction by taking advantage of regularities in the domain.

Once we have a given set of problem instances or have narrowed the sequence of potential problem instances to classes of instances with similar conditional performance profiles, the deliberation scheduling problem can be considered to be a problem of compiling a sequence of anytime algorithms so that the overall performance is maximized. In general when performance profiles are represented in closed form the compilation problem involves solving differential equations; when represented in tabular format the compilation problem becomes a search problem. Compilation problems in which the problem instances are independent are considerably easier than those in which the problem instances are dependent.

An important special case of compilation of anytime algorithms is when we can guarantee that the decision procedures (with their associated conditional performance profiles) will be executed in a strict sequential order. Similar are predictably periodic sets of problem instances or classes of problem instances.

3 Response Planning Approach to Embedded Planning

In this section we present an approach to solving time-critical decision-making problems by taking advantage of domain structure to make better use of the time available for processing difficult combinatorial tasks. Our approach uses predictable variability in anticipated computational demands to allocate on-line deliberation time and exploits what we call problem regularity and stochastic models of environmental dynamics to restrict attention to small subsets of the state space. This approach demonstrates how slow, high-level systems (*e.g.* for planning and scheduling) might interact with faster, more reactive systems (*e.g.* for real-time execution and monitoring) and enables us to generate timely solutions to difficult combinatorial planning and scheduling problems such as traffic control for multiple robot vehicles.

We make use of domain structure in two principal ways. First, we use a stochastic model of the domain and on-line knowledge of the state of the environment in order to predict future states. In this context a set of possible states in some target time period corresponds to a problem instance. The ability to predict future states allows us to begin processing in advance of the actual states. Furthermore, careful prediction allows us to restrict our planning to a small subset of the state space. We may consider the size of the state space corresponding to a predicted problem instance a measure of input quality. In terms of conditional performance profiles, restricting the state space to smaller subsets of states reduces the anticipated computational demand of the corresponding problem instance. Prediction alone is insufficient to make progress with these types of problems. Second, we take advantage of large variabilities in anticipated computational demand across time. By removing the restriction of equal processing

allocation for each problem instance, we gain flexibility in allocating processing time. Additionally, some forms of regularity allow us to meta-reason off-line about the allocation of on-line deliberation time.

3.1 Embedded Planning

We define *embedded planning* to be the problem of determining actions for an agent embedded in an uncertain environment with dynamics outside the agent's control. A salient feature of this problem is that the agent must react to changes in the environment in *real-time*, *i.e.* the agent must sense the state of the environment and act appropriately in a timely manner. There is rarely the luxury of waiting for a deliberative planner to construct the optimal action in response to environmental dynamics. There are hard deadlines imposed by the environment that, if violated, can lead to disastrous conclusions for the agent.

As mentioned earlier, we may consider each new state or set of possible states of the environment during a specified time period to be a problem instance. We define reactive response to be the ability to sense a change in the state of the environment and respond before the environment changes state again. *Response time* for reacting to a problem instance is defined as the minimum time between sensing a state and the necessity of response. The minimum time between state changes (granularity) is referred to as a *time step* and is domain dependent. We assume for these purposes that response time is a fixed constant. We can then divide time into discrete steps where a state is sensed at the beginning of each time step (the state may be the same as in the previous time step) and a response is required by the end of the time step (in some situations a null response may be satisfactory).

Only in ideal situations may we determine in advance a sequence of individual states (one per time step) to be used as problem instances for deliberation scheduling. In general, the best we may do in advance is to restrict each time step to a set of reachable states. Each of these sets may then either individually or in temporal or similarity based partitions be considered problem instances for deliberation scheduling. The choice of partitioning determines the form of conditional performance profiles required for deliberation scheduling.

A solution that provides a response at each time step is called a *policy*. A policy π is a mapping from states to responses (actions). A *partial policy* provides responses to only a subset of possible states. A policy may be used as part of a real-time control system that responds to changes in the environment. In this model policies are constructed by applying decision procedures to problem instances that partition the phase space (product of state space and time steps). The focus of the response planning approach is to determine how to allocate on-line deliberation time to the decision procedures that jointly construct policies, such that particular performance criteria are satisfied. Additionally, response planning provides an interface between the combinatorial process of constructing the policy and the reactive execution of the policy.

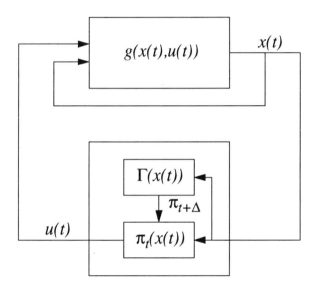

Figure 3: *Model for embedded planning and control*

Figure 3 captures the essential properties of embedded planning and control systems; $x(t)$ is the state of the system (including the embedded agent) at time t, $u(t)$ is the action taken at time t by the composite planning and control system, and the function $g(x(t), u(t))$ determines the dynamics of the system. In this work we assume that we are dealing with a discrete time system. The action is determined by a policy that can be executed by the real-time control system. The policy π_t has a temporal index to indicate that it may change under the control of the planning component Γ. Due to the combinatorics involved in planning there is typically a delay Δ between when a state triggers the planning process and when the resulting policy is avail-

able for execution. We distinguish between state and control in keeping with standard practice in the control literature. For a description of the motivation and use of this formalism for modeling dynamical systems see [5, 9, 13].

3.2 Response Planning

Response planning is an approach to embedded planning that makes specific assumptions about the underlying environment and computational resources. In [10] we outline the general motivations and options for designing composite planning and control systems.

The response planning approach combines the reactivity of off-line construction of *universal plans* [19] (complete policies that precompute actions for all world states) with the adaptability of on-line decision-making. Time-critical decision-making is accomplished despite the constraints of limited storage and computational resources in the face of large state spaces. We do so by noting that predictive knowledge of the underlying domain may be used to anticipate the reachable states in a problem instance before actually observing the environment at the corresponding time step(s). This knowledge is then used to compute partial policies for such restricted subsets in advance. Complete policies are produced by combining the partial policies with default responses for remaining states. Problem variability allows us to provide strategies in which difficult and uncertain problem instances consume more processing time than less computationally intensive problem instances. Domain-dependent regularity allows for off-line precomputing of deliberation scheduling (processing time allocation) strategies.

As defined in Section 3.1 the primary components of an embedded planning solution are the reactive component π and the deliberative component Γ. π must be able to execute a policy constrained by hard deadlines. Without loss of generality we consider the reactive process (π) to be comprised of a *reactive process table* of responses indexed by world states. In general π can be realized by any computation that is initiated in response to a state and completed before the arrival of the next state.

Restricting π (we abuse the notation by calling both the process and it's table π) to be a table simplifies the interaction between the deliberative process (Γ) and the reactive process (π). In particular, the deliberative process interacts on-line with the reactive process by populating the reactive process table with responses for upcoming time steps. To represent changes in the policy over time we index the reaction table by time steps, (*i.e.* π_j for time step t_j) and require a complete policy for each time step (though not necessarily unique).

One motivating assumption for the construction of π is that in typical embedded planning problems the state space is very large. In particular, the number of states that make up environment g is generally exponential in the number of state variables in the embedded planning problem. Given limited resources (storage and computation) large state spaces eliminate the possibility of computing and storing responses for every possible state in the environment. Therefore, we require that each π_j be bounded in size. In particular we bound the size k of π_j to be of low-order polynomial in the number of state variables. Define π_j^k to be a reactive process table of size k.

By introducing a bound on the size of each π_j we limit the possible policies that Γ can construct. In particular, for time steps in which greater than k states are reachable, Γ must determine the optimal set of states for which to plan, given some optimality criterion. Alternatively, we could redefine π_j^k as a table indexed by equivalence classes of states. In some embedded planning problems such as partially observable processes in which the actual state is difficult to measure with certainty, indexing by equivalence classes of similar states (similar in terms of policy) could be a more natural solution. On the other hand, doing so would sacrifice fine-grained control over the environment. In Figure 4 we depict π_j as a combination of both. A bounded number of responses triggered by equivalence classes of states and a bounded number of responses triggered by states. Equivalence classes may be used either to capture partial observability or to encompass states that are not included in a given partial policy π_j. In the latter case we can consider the responses triggered by equivalence classes to be default responses to be applied if none of the state-based responses are appropriate. Thus, the deliberation scheduler can reason about which states to include in π_j given the expected quality of executing default responses.

Bounding the size of π_j makes it even more important to provide accurate on-line prediction to reduce the size of the reachable state space. Since a con-

$$\pi_t(x(t))$$

Trigger	Response
$[x(t)]_{s_1}$	u_{s_1}
$[x(t)]_{s_2}$	u_{s_2}
...	...
$[x(t)]_{s_k}$	u_{s_k}
$s_1(t)$	u_1
$s_2(t)$	u_2
...	...
$s_k(t)$	u_k

Figure 4: *Tabular, bounded size representation of π_j^k partitioned into similarity-based equivalence class indexed responses and state-indexed responses; where s_k is a known state equal to $x(t)$ and $[x(t)]_{s_k}$ is the class of states determined by $x(t)$ and represented by s_k.*

ditional performance profile for a decision procedure takes into account both the limitations on the resources and the quality of the input, if we fix the resource limitations we are left with the quality of input as our most important degree of freedom in deliberation scheduling. In particular, the quality of input (*i.e.* the quality of the particular problem instance achieved for the time period) is dependent upon how far in advance of the time period the decision procedure is called and the ability to restrict the size of the reachable state space via prediction.

The response planning approach requires certain properties of the problem domain to hold. A first requirement is that computing high-quality responses for some states requires multiple time steps. If this requirement does not hold then prediction and deliberation in advance are unnecessary. Purely reactive approaches apply. In addition there must be states for which responses can be computed quickly (*i.e.* fractions of time steps) or for which responses, once computed, are relevant for many time steps. Otherwise, delay caused by multiple step processing would grow unbounded.

We must take limited resources into account when designing Γ. In general, Γ does not have the luxury of computing optimal responses for all reachable states for all future time steps. Γ is an on-line process that performs three separate functions. It performs domain specific processing to compute the responses to be taken for sensed states (*i.e.* constructs policies); it uses the underlying environment model g to predict reachable states in upcoming time periods; and, it performs deliberation scheduling by reasoning about its use of time based on conditional performance profiles, predictions and performance measures.

The ability to predict future reachable states depends upon Γ's knowledge of g. If g is deterministic then Γ can predict the exact state for any time step given some start state. If g is stochastic then the best Γ can do is predict a distribution over states for a given time step.

We divide Γ conceptually into two components, f and Γ. The on-line decision procedure, f, performs prediction of reachable states and construction of policies. f may be called with varying input parameters including the target time window for the policy and the time allocation for deliberation for policy construction. The deliberation scheduler, that maintains the name Γ, reasons with the aid of the conditional performance profile associated with f to determine the start time step and processing time to allocate to f in order to provide the best possible performance according to a given performance criterion.

It is important to note that, although it may be possible to compute stationary distributions of reachable states off-line and construct a single table π^k, there are advantages to performing prediction on-line. In particular, a stationary distribution is only applicable asymptotically. In the short-run it is beneficial to look at states that have recently occurred to prune the possible states reachable in the near future and calculate distributions over those states for upcoming time steps. In a sense, on-line processing allows us to compute *on demand* rather than process all possible states. If g can be modeled as a Markov chain then we need only look at the most recently occurred state in predicting future distributions. This can be further simplified by using dependencies among state variables to provide short-term predictions [3].

f is an on-line process that, given a trajectory of previously visited states, and a target time window uses the known model of g to predict and prioritize reachable states. It may then calculate responses for the

states so as the maximize the expected quality of the resulting policy. One particular formalization for f is as follows. Given a state trajectory $H = s_0, s_1, \ldots, s_i$ for times $t_0, t_1, \ldots, t_i$, a projected time t_n ($i \leq n$) and a *window size m*, f computes a series of reactive process tables $\pi_n^k, \pi_{n+1}^k, \ldots, \pi_{n+m-1}^k$ such that each respective table π_j^k contains a policy with state-based responses for k states reachable at time t_j. In addition, (for completeness) the policy may include class-based default responses as previously described. Given our resource limitations we must also restrict m to be low order polynomial in the size of the state space, just as we did for k. As a special case, if the underlying source g is a Markov chain then we need merely provide f with the most recent time and state (t_i, s_i) rather than the entire trajectory.

Decision procedure performance may be optimized by designing deliberation scheduling (Γ) such that the predicted set of reachable states is minimized for the target time window (thus, maximizing input quality). Short-term dependencies may be taken advantage of by designing Γ as a reactive process that schedules deliberation in response to states of the environment g. Specifically, at each time step Γ can either commence a new a call to f with a set of parameters H, t_n, m (that may cause an existing call to be exited) or do nothing. When completed, Γ populates $\pi_n^k, \ldots, \pi_{n+m-1}^k$ with the results of the call to $f(H, t_n, m)$. Given bounded resources we must limit the number of calls to f that are active at any given time. Since processing profiles are based on exclusive use of a uniprocessor, multiple calls to f at a time require a multiprocessor system. As mentioned previously, conditional performance profiles for uniprocessor systems may not generalize to multiprocessor use.

The above description of Γ as a reactive process overlooks some important points. First, the deliberation scheduling optimization problem that Γ must solve, namely determining the optimal call to f for a given time step, is combinatorial in nature. In other words, it may take multiple time steps to compute. Second, for the same reasons that π cannot be a universal plan, Γ cannot store actions (calls to f) computed off-line that are applicable to every state. In order to be effective Γ must be both reactive, to interrupt f during processing in response to actual state trajectories, and deliberative, to reason about f's processing requirements given the underlying process. As before, this dilemma can be solved two ways. Either limit the number of states that can be reacted to or perform deliberation in advance based on prediction. But, if we choose the latter approach to Γ we would still need to schedule calls to Γ which schedules calls to f, ad infinitum. We avoid the infinite regress in this case by the expedient of compiling Γ off-line into a reactive *strategy table*.

Γ is designed as a strategy table such that for each time step, based on the observed state, Γ defines a strategy to either call f with particular parameters (for one or a sequence of calls to f) and wait for the results (possibly terminating an existing call) or do nothing. Again when constructing a bounded on-line table we must decide between indexing by actual states and default computations or indexing by equivalence classes of states. The same arguments apply here as did for the construction of π, namely equivalence classes are a natural way to express some problems but they suffer from a loss of fine-grained control over the environment. We choose to construct Γ in general similarly to π as a bounded table partly indexed by environmental states and partly indexed by equivalence classes of states.

We have alluded to a particular domain structure we call *regularity*. This structure occurs naturally in problems such as traffic control for multiple robot vehicles. In such problems we can partition the state space into a sequence of equivalence classes of states corresponding to contiguous time periods. In general, this regularity is disrupted only by catastrophic (very low probability) events. Figure 5 demonstrates what we mean by a regular sequence of equivalence classes. As long as regularity is maintained the process will proceed in a fixed, possibly repeating sequence of equivalence classes of states. Catastrophic events force the process into a unique state that must be handled separately. The actual state trajectory may vary widely from one repetition to the next but, the sequence of classes is, for the most part, deterministic. Another way to look at regularity is as a property of inherent local dependencies among state variables. In this case the state at time t is only dependent on a subset of the state variables at times prior to t.

Regularity is applicable in the design of strategy tables in that it allows for the off-line construction of *rigid* strategies. A rigid strategy is computed by performing deliberation scheduling off-line using (temporally-

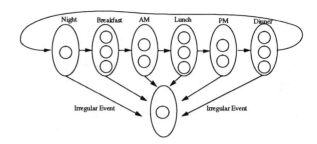

Figure 5: *Regularity demonstrated by deterministic sequence of equivalence classes; catastrophic events disrupt regularity.*

based) equivalence classes of states as the input to f rather than actual states. In other words, the conditional performance profiles for f have as input parameters the target class and the start class rather than specific states. Problems in the same equivalence class have similar anticipated computational demands. The argument here is that class-based partitioning of the problem provides the deliberative scheduler with as much information as state-based partitioning, given the additional structure of regularity. This is not true in all domains but is a nice property of some domains that we can exploit through rigid strategies.

Additionally, we provide a mechanism within Γ for state-based triggering of strategies. In domains in which regularity may be occasionally violated we can use state-based triggering to provide default strategies that attempt to regain regularity. This is similar to the principal of match-up scheduling used in operations research. Figure 6 depicts Γ as a strategy table with a rigid strategy indexed by time step (temporal equivalence class) and a bounded number of default state-based strategies. In Section 4 we further discuss the implications of regularity for deliberation scheduling. In other work we are considering indexing among a bounded number of rigid strategies as well. In nonregular domains we may provide for state-based deliberation scheduling in which a bounded number of states trigger strategies directly.

One formalization for Γ is to construct one strategy table per temporal equivalence class. For an environmental model with period d, Γ is constructed off-line into d strategy tables, $\Gamma_0, \ldots, \Gamma_{d-1}$, one for each temporal equivalence class. Each strategy table contains h specific strategies indexed on actual states and one

$$\Gamma_t(x(t))$$

Trigger	Strategy
$[x(t)]_t$	$(f(\sigma),\dots)$
$s_1(t)$	$(f(\sigma_1),\dots)$
$s_2(t)$	$(f(\sigma_2),\dots)$
$\dots$	$\dots$
$s_h(t)$	$(f(\sigma_h),\dots)$

Figure 6: *Tabular, bounded size representation of Γ_j partitioned into one temporal-based equivalence class corresponding to the domain regularity and state-indexed strategies to account for non-regularity.*

default strategy to take for this equivalence class if no specific strategies apply.

Strategies limited to a single call to f per time step will not generally lead to optimality, especially when the processing time for some states is considerably less than a single time step. Thus, we design Γ to reason about a sequence of calls to f at each time step. If preemption of decision procedures is desirable, the strategy table Γ may contain a sequence of *partial* calls that can be swapped out and resumed later. This requires small additional overhead in processing time and storage but does not violate any assumptions of our model. In this case the profile for a preempted job must correspond to the profile when initially started (no intermediate knowledge may be used).

We make one further restriction concerning the interaction of Γ, f and π given bounded resources. We have described Γ as populating π with policies generated by f, with one policy π_j per time step t_j. Although we have bounded the number of policies a single call to f can produce (by the time window m, which is polynomial in the number of state variables), we have not bounded the cumulative number of policies that can be produced by a sequence of calls to f and stored by the system before being executed. To do this we define a *time horizon l* that determines how far in the future for which f can plan. In particular for any current time step t_i, window start time t_n and window size m, $n + m - i - 1 \le l$. If we restrict l to be polynomial in the number of state variables then the total number of policies stored at any time is also polynomial in the number of state variables.

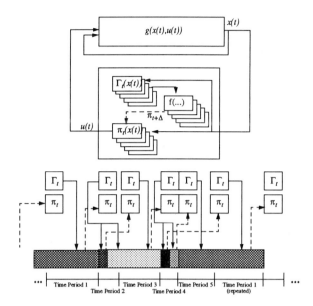

Figure 7: *Response Planning model and example*

Given the components of our response planning solution to embedded planning summarized in Figure 7 we have two algorithms to construct. The off-line construction of Γ taking f as input (deliberation scheduling) and the on-line execution of response planning. We can sketch the on-line algorithm that governs response planning in general as follows. At each time step t_j we perform the following functions:

1. determine state s_j of environment g;

2. determine temporal equivalence class equiv(s_j) for state s_j;

3. determine similarity-based equivalence class similar(s_j) for state s_j;

4. index into reaction table π_j to find response corresponding to state s_j;

5. if response found execute, else execute default response corresponding to similar(s_j);

6. index s_j into strategy table for time period t_j ($\Gamma_{\text{equiv}(t_j)}$) to find parameters for new (partial) call to f (or sequence of calls, subject to horizon limitation l) or no op;

7. if strategy found execute it, else execute rigid strategy corresponding to equiv(t_j);

8. if previous call to f has terminated successfully update π for that time window by storing the m

constructed policy tables on-line;

9. if executing new strategy requires previous call to f to be interrupted before completion then update π with most recent policy produced by anytime algorithm for f; and

10. discard any out-of-date policies.

Note that we make the assumption that the state, temporal equivalence class and similarity-based equivalence class can all be sensed immediately from the environment.

New calls to f include parameters such as target time window and processing time allocated. This procedure is exemplified at the bottom of Figure 7. In this figure we show a new call to Γ at the beginning of each time period for simplicity. In general, Γ can instigate a new call to f in response to any state. In the figure we see Γ calling f during first time period to construct policies for second time period. We further note that just prior to the second time period, f updates π with its constructed policy. This policy is then in effect until the end of the second time period at which point the policy for the third time period takes effect.

4   Exploiting Regularity in Anticipating Computational Demands

In this section we return to the notion of regularity and explore it in the context of deliberation scheduling and compilation of anytime algorithms. In doing so we tie together the particular approach to solving time-critical decision-making problems proposed by response planning with the general idea of anticipating computational demands. In Section 4.1 we compare the deliberation scheduling task of response planning to that introduced by Boddy and Dean [4, 1]. In Section 4.2 we show that the notion of compilation of anytime algorithms introduced by Zilberstein [22] encompasses the particular notions of deliberation scheduling used in response planning.

In response planning we introduce the notion of regularity so that deliberation scheduling may be performed off-line to construct rigid strategies. In particular, we assume that our environment g proceeds in a fixed, possibly recurrent pattern of temporally equivalent classes of states. Although at any given time step (or time

period) the class of states that can occur is deterministic, the actual state experienced from a given class is stochastically determined by the environment g. Given that there may be a very large (exponential in the number of state variables) number of states in any equivalence class, this regularity does not trivialize the response planning problem.

We use this deterministic knowledge of the sequence of equivalence classes in deliberation scheduling by reasoning about conditional performance profiles in which the input parameters are start class and target class rather than start state and target state. Inherent in this usage is the assumption that the performance of our decision procedure for a given allocation of processing time is determined by these two classes (as well as the window size). In [10] we use an air traffic control problem to motivate these assumptions. The problem becomes more involved if the responses constructed by f can alter the sequence of classes visited. We allow for this possibility in response planning by extending a rigid strategy to include some state-based strategies. More generally, we could extend Γ to include conditional or stochastic strategies but, that is beyond the scope of this treatment.

4.1   Deliberation Scheduling

Deliberation scheduling is a resource allocation problem in which the resource is processing time and it is allocated among competing decision procedures to be applied to varying problem instances. An embedded agent can anticipate computational demands by determining the problem instances for a forthcoming time period and accessing profiles indicating the expected performance for varying processing allocations for varying decision procedures applied to similar problem instances in the past.

In response planning the competing decision procedures involve computing policies for target time periods (problem instances). Regularity guarantees that the target time periods are known deterministically in a fixed sequence. Deliberation scheduling applied to response planning involves determining the start class, processing time allocation and time window for each target time period in order to maximize overall performance. In response planning, overall performance is determined by the expected performance of the policies executed by π.

The assumption of regularity simplifies the problem of determining the problem instances from which to base deliberation scheduling. Other researchers have used different techniques. In [4, 1] the sequence of problem instances is indicated by precursor events. A precursor event indicates to the deliberation scheduler a sequence of future events (problem instances) that need to be solved. However, determining the sequence of problem instances for deliberation scheduling based on precursor events requires that deliberation scheduling be performed on-line. By exploiting deterministic regularity response planning can move the deliberation scheduling problem off-line.

Once the sequence of problem instances is determined, the deliberation scheduling problem is to allocate processing time and, possibly, other parameters to the competing decision procedures. One important difference between response planning and the deliberation scheduling of Boddy and Dean [4, 1] is that our problem requires that the deliberation scheduler determine both time allocation and start time for processing a decision procedure for a particular problem instance. The treatment by Boddy and Dean [4, 1] does not require this because they assume that decision procedures are independent and only the relative processing allocation is important, not the start time. In the following we show that we can convert a class of response planning problems with start class dependence into start-class independent deliberation scheduling problems. Although this result is interesting it is specific to the properties of start-class dependent profiles and cannot be generalized to other conditional performance profiles.

Consider a deliberation scheduling problem for response planning that consists of a set of conditional performance profiles, one per equivalence class (corresponding to a fixed time window[1]). The dimensions of these performance profiles include start-class, processing time allocated and output quality. A typical start time independent performance profile is given in Figure 8.i and a corresponding set of start-time dependent conditional performance profiles are given in Figure 8.ii. Figure 8.iii shows that this dependent set of profiles demonstrates a situation in which there is benefit to starting later when there is more informa-

---

[1]We have argued that we require different profiles for differing size time windows so, we assume here that the size of the time window is fixed.

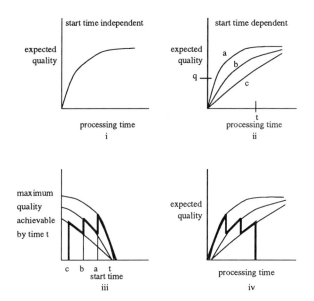

Figure 8: *Converting from start-time dependent conditional performance profiles to processing time versus quality performance profile.*

tion available (more accurate prediction). In particular, if we start processing before time b, performance is determined by the time c conditional performance profile which takes longer to achieve quality q than either the time b or time a conditional performance profiles. The variability in profiles based on start time could be related to the increased uncertainty in predicting reachable states further into the future.

In Figure 8.iv we see the resulting start-time independent profile. In a sense the profile is not start-time independent at all but, rather, the start-time dependence has been incorporated into the profile itself (and relies on the fact that processing takes place in time interval immediately prior to target time class). This profile shows that in the deliberation scheduling problems of interest to response planning, allocating additional processing time does not always improve quality because it requires starting processing sooner when less accurate prediction information is available.

While this transformation displays the relationship between the deliberation scheduling problems of response planning and those of Boddy and Dean it does not allow us to use the algorithms presented by Boddy and Dean because the resulting profiles do not display the required monotone increasing property. This clearly indicates that start-time dependent deliberation scheduling is in general more difficult than start-time independent deliberation scheduling.

4.2 Deliberation Scheduling as a Compilation of Anytime Algorithms

The deliberation scheduling problems of response planning and Boddy and Dean are both special cases of the general problem of compilation of anytime algorithms. We have already shown that deliberation scheduling with start-time independence is a special case of deliberation scheduling with start-time dependence. In this section we show that start-time dependence is a special case of input quality dependence.

The point to note in interpreting a response planning deliberation scheduling problem as a problem of compiling anytime algorithms is that conditioning profiles on target equivalence classes is equivalent to conditioning profiles on input quality. The input quality is a direct function of the target time period and start-time.

Zilberstein defines a conditional performance profile as corresponding to a decision procedure that may have several different performance profiles, each characterizing its performance when operating in a different environment. In response planning each problem instance corresponds to a different temporal equivalence class. The ability to restrict the set of reachable states within a class by prediction is equivalent to altering the input quality. Prior actions can affect the set of reachable states by forcing the state trajectory off the fixed sequence of equivalence classes defined by problem regularity.

Once we have a given set of problem instances or have narrowed the sequence of potential problem instances to classes of instances with similar conditional performance profiles, the deliberation scheduling problem can be considered to be a problem of compiling a sequence of anytime algorithms so that the overall performance is maximized. Algorithms for which problem instances are independent are easier to compile than dependent problem instances. In the deliberation scheduling problem of response planning we have a fixed sequence of problem instances for which to allocate processing time. In general, compilation of anytime algorithms deals with any partial order of problem instances.

One interesting point is that in response planning the input quality is not the output of a previous algorithm but rather a combination of the results of deliberation scheduling and the prediction component of f. This indirect dependence of input quality on deliberation scheduling is a subtlety that must be dealt with carefully. The anticipated computational demands of a decision procedure is based on the expectation that the results of a prior decision procedure will not cause the system to stray from the fixed regular sequence of classes. In response planning the algorithm that provides the expected input for a subsequent algorithm can actually be executed after that algorithm has finished deliberating as long as the result is ready before the first algorithm needs to execute the results of its deliberation. The first algorithm must use some expectation of the quality of the second algorithm to do its deliberation. If the second algorithm does not live up to that quality then it is too late and the deliberation provided by the first algorithm is severely diminished. This discussion points out the fact that response planning has three stages, deliberation scheduling, actual deliberation for policies and execution of policies. The execution of policies has a strict order used by the deliberation scheduling but the deliberation itself does not have a strict order and relies upon the a priori knowledge of the expectations of the other deliberations to provide accurate policies.

5 Response Planning Example: Traffic Planning and Control of Multiple Robot Vehicles

Consider the problem of employing multiple robot vehicles for delivering meals and pharmaceuticals and disposing medical waste in a large hospital. Robot vehicles are ideal for this task because of their low air contamination and ability to safely handle hazardous materials. In an artificial setting, such as a factory floor, tasks can be engineered so that path planning and routing problems are solved off-line and then implemented by a centralized controller. In the worst case there may exist delays and uncertainty that are handled in standardized ways. However, in a natural environment such as a hospital, tasks cannot be predicted with certainty ahead of time. Therefore, in order to provide effective use of resources, traffic planning and control must dynamically adjust to changing situations.

In this problem we assume a finite set of vehicles, a finite set of locations (source and/or destination), and a fixed network of limited-width undirected paths connecting the locations. The network serves as a pathway for the vehicles. The pathway may be augmented with infrared beacons to aid vehicle navigation. Additionally, vehicles can be controlled and tracked either through the network itself or through other devices such as radio ethernet or infrared transceivers. At any given time a subset of the vehicles are active on the network with specific destinations and, possibly, deadlines. The traffic planning and control problem is to schedule routes for these vehicles such that they reach their destinations with minimal tardiness. In order to do so the planner must avoid congestion, bottlenecks and contention within the limited-width network for currently active vehicles while taking into account the anticipated activation of inactive vehicles in the near future. It may be necessary to assume single-lane pathways in some problems (such as those of automatic guided vehicles), thus exacerbating the contention problems.

We can assume for simplicity that the traffic planner need only anticipate congestion caused by vehicles. The problem of avoiding obstacles and people can be handled by an on-board reflexive control system or may be engineered away by providing special conduits (such as in-ceiling crawl-space) for vehicle movement. This is consistent with traditional separation into central controller and on-board vehicle controls. Furthermore, we can assume that assignments of delivery and disposal tasks to specific vehicles are determined by an external source (or sources) and not under the control of the traffic planner. This includes obeying capacity constraints at the source and destination locations.

A combined planning and control system may be employed to mediate traffic congestion. The controller issues traffic control commands to the active vehicles in order to satisfy the routes determined by the planner. Prior to entering an intersection or approaching a source or destination location a vehicle is given a traffic control routing action indicating which of the alternative pathways to follow.

Formally, the following parameters describe the planning and control problem.

- a network including

  - vertices, N, corresponding to internal routing points,
  - locations, L, corresponding to source and destination points, and
  - edges, E, connecting vertices and locations;

- a set of vehicles, V; each vehicle $v \in V$ has

  - current location, $loc_v \in (N \cup L \cup E)$, ($loc_v \in L$ indicates that v is currently inactive),
  - current destination $dest_v \in L$,
  - optional current deadline $dead_v \in \mathbf{T}$ ($\mathbf{T}$ is the set of all time points, possibly infinite), and
  - current routing action, $act_v \in \mathbf{Z}$, indicating the pathway to follow at the next vertex or location; and

- a stochastic process g modeling environmental dynamics and future vehicle task assignments (described below).

The state of the system at any time step is the set of parameters for all vehicles. At each time step the controller issues a new vector of routing actions to the vehicles to satisfy routes determined by the planner. The subsequent state of the system is governed by a stochastic process g whose transitions model control actions and external, uncontrollable events (such as inactive vehicle assignment, active vehicle breakdown, velocity uncertainty, delay, etc.,).

The performance of the planner is measured by an objective function on the movement of vehicles. For delivery tasks with deadlines this may be to minimize tardiness whereas for other tasks we may try to minimize total, maximum or average completion times across all tasks. Tardiness and completion times are affected by congestion of vehicles in limited-width paths as well as uncontrollable delays modeled by the stochastic process.

Constructing routes to minimize the above objective functions is a difficult combinatorial problem that is made increasingly difficult by uncertainty in future parameters. For single-lane pathway problems we can consider the routing problem to be analogous to a space of $(N \cup L \cup E) \times \mathbf{T}$ points and the task to commit (assign) at most one vehicle to each point in this space, for some given horizon of time steps into the future. However, the dynamics of the problem require that any routing solution be modified regularly. Possibly as often as every time step. The planning algorithm determines a routing assignment based on a given set of problem parameters and conveys this information to the controller in the form of a policy. A policy describes a vector of routing actions for any given system state. A partial policy includes a vector of routing actions for some subset of system states and can be augmented by default control vectors (*e.g.* stop all vehicles) for the remaining states. Partial policies are necessary in large state spaces given limited on-line storage and computation resources.

In this paper we are not primarily concerned with the actual planning algorithm to solve the optimization problem. We are more interested in the ability to characterize planning algorithms by their processing time and quality measures given particular input parameters (*e.g.* planning window). The actual planning solution itself may be heuristic or exact.

Executing default control vectors is very costly and is to be avoided as often as possible. We focus on the task of scheduling the on-line deliberation consumed by the planning algorithm in order to minimize cost (maximize expected quality of policies).

The multiple robot vehicle problem of delivering meals and pharmaceuticals and disposing medical waste in a large hospital satisfies the regularity limitations that we require for our response planning approach. In particular, the pattern of delivery activity throughout the hospital varies widely throughout the day. Meal delivery vehicles are abundant during mealtimes and rarely active during other periods. Pharmaceutical delivery and waste disposal vehicles have some scheduled deliveries and some deliveries that are made on demand. The density of these on-demand deliveries varies throughout the day and week. Although the actual delivery tasks are not known ahead of time, the stochastic model g can be used to determine periods of relatively high activity versus periods of relatively low activity. The amount of time necessary to plan traffic control for periods of low activity is significantly less than those for high activity.

For example, consider a typical day. The night-time and early morning periods are characterized by low activity levels of some deterministically scheduled pharmaceutical and waste disposal deliveries and some deliveries due to emergency procedures. The traffic control problem during this period is relatively simple.

However the influx of deliveries necessary for the breakfast rush makes the traffic control problem much more difficult. Given the combinatorial nature of the route planning problem, deliberation time for this period is significant. We can take advantage of this situation by beginning to plan for the breakfast rush period during the night. While this involves considering many more reachable states than if we waited until morning, the additional time available for computation gained makes up for the increased computational difficulty. Furthermore, the pattern of peak and non-peak hours is fairly regular throughout the week. While the night-time and breakfast periods vary greatly in both problem complexity and problem parameters, the breakfast period on Tuesday is of the same difficulty level as the breakfast period Wednesday even if the actual combination of vehicles and delivery tasks varies. This regularity can be exploited in allocating on-line deliberation time between the night-time and breakfast tasks.

6 Discussion

We have focused on the specific issues involved in the construction of the response planning approach to embedded planning and the role of anticipating computational demands in providing time-critical responsiveness. There are many other issues and extensions of this work not included in this treatment. We briefly touch upon some here. In [10] we touch upon some others.

The response planning approach takes a very restricted view of possible strategies for Γ by enforcing domain regularity. As we have mentioned previously there are more flexible options available for the construction of Γ. In particular we may better handle opportunism and irregular events by composing Γ as a conditional or stochastic strategy. Certain events may trigger different strategy branches.

Given that processing time is dependent upon the actual problem instance, the time allocated to a decision procedure may not produce the desired quality. Zilberstein [22] introduces the concept of on-line monitoring to deal with these situations. Additionally, the concept of dynamically adjustable performance profiles and profiles conditioned on intervening states that pass during deliberation extend the dynamic properties of response planning.

It is interesting to think about multiprocess extensions of our two process, deliberative and reactive, architecture for response planning. In particular we can envision an architecture of multiple deliberative processes and multiple reactive tables. The multiple reactive tables may correspond to multiple agents, each table controlling the behavior of a unique agent. Multiple planners may compete for the control of each reactive table.

Finally the complexity of deliberation scheduling/compilation of anytime algorithms can be characterized for different classes of performance profiles. Zilberstein [22] and Boddy and Dean [4, 1] provide some results along these lines.

7 Related Work

Simon and Kadane [21] consider the case of allocating time to search in which deliberation time is quantized into chunks of fixed, though not necessarily constant, size. Etzioni [7] considers a similar model in which the deliberative chunks correspond to different problem solving methods. Russell and Wefald [18] describe a general approach in which deliberation is considered as a sequence of *inference steps* leading to the performance of an action. They explicitly compute the expected value of different sequences of inference steps using a technique similar to Howard's method for computing the expected value of information [12].

There are a number of approaches similar to the approach of using anytime decision procedures described in this paper. Anytime algorithms and various deliberation scheduling problems are described in [4, 1] and Chapter 8 of [5]. The advantages of the anytime algorithm approach include an ability to make use of any amount of time available, robust behavior in the presence of unexpected interruption, and simplifying the problem of optimal or near-optimal deliberation scheduling. Horvitz [11] uses what he calls flexible computations to allocate computational resources at run time. Lesser [16] uses a similar notion for solving time-critical problems. Liu *et al.* [20] uses the term *imprecise computation* in the context of real-time operating systems to refer to a computation that has both a necessary and an optional component. Zilberstein [22] discusses how anytime algorithms can be composed to perform more complicated computations.

Additionally, there have been a variety of planning systems that are related to the problem. Georgeff's *procedural reasoning system* [8] was designed for on-line use in evolving situations, but it simply executes user-supplied procedures rather than constructing plans of action on its own. Systems for synthesizing plans in stochastic domains, such as those by Drummond and Bresina [6], and Kushmerick, Hanks and Weld [14] do not directly address the problem of generating plans given time and quality constraints. Lansky [15] has developed planning systems for deterministic domains that exploit structural properties of the state space to expedite planning. Smith *et al.* [17] describe some initial efforts at building systems that modify plans incrementally. Neither Lansky or Smith *et al.*'s systems deal with uncertainty and Lansky's system cannot handle concurrent planning and execution.

Acknowledgements

This work was supported in part by a National Science Foundation Presidential Young Investigator Award IRI-8957601 and by the Air Force and the Advanced Research Projects Agency of the Department of Defense under Contract No. F30602-91-C-0041.

References

[1] Mark Boddy and Thomas Dean. Solving time-dependent planning problems. In *Proceedings IJCAI 11*, pages 979–984. IJCAII, 1989.

[2] Ronald J. Brachman, Hector J. Levesque, and Raymond Reiter, editors. *Proceedings of the First International Conference on Principles of Knowledge Representation and Reasoning*. Morgan-Kaufmann, Los Altos, California, 1989.

[3] Thomas Dean. Decision-theoretic planning and markov decision processes. In preparation, 1994.

[4] Thomas Dean and Mark Boddy. An analysis of time-dependent planning. In *Proceedings AAAI-88*, pages 49–54. AAAI, 1988.

[5] Thomas Dean and Michael Wellman. *Planning and Control*. Morgan Kaufmann, San Mateo, California, 1991.

[6] Mark Drummond and John Bresina. Anytime synthetic projection: Maximizing the probability of

goal satisfaction. In *Proceedings AAAI-90*, pages 138–144. AAAI, 1990.

[7] Oren Etzioni. Tractable decision-analytic control. In Brachman et al. [2], pages 114–125.

[8] Michael P. Georgeff and Amy L. Lansky. Reactive reasoning and planning. In *Proceedings AAAI-87*, pages 677–682. AAAI, 1987.

[9] M. Gopal. *Modern Control System Theory*. Halsted Press, New York, 1985.

[10] Lloyd Greenwald and Thomas Dean. Solving time-critical decision-making problems with predictable computational demands. In *Second International Conference on AI Planning Systems*, 1994.

[11] Eric J. Horvitz. Reasoning about beliefs and actions under computational resource constraints. In *Proceedings of the 1987 Workshop on Uncertainty in Artificial Intelligence*, 1987.

[12] Ronald A. Howard. Information value theory. *IEEE Transactions on Systems Science and Cybernetics*, 2(1):22–26, 1966.

[13] R. E. Kalman, P. L. Falb, and M. A. Arbib. *Topics in Mathematical System Theory*. McGraw-Hill, New York, 1969.

[14] Nicholas Kushmerick, Steve Hanks, and Daniel Weld. An algorithm for probabilistic planning. Unpublished Manuscript, 1993.

[15] Amy L. Lansky. Localized event-based reasoning for multiagent domains. *Computational Intelligence*, 4(4), 1988.

[16] Victor R. Lesser, Jasmina Pavlin, and Edmund Durfee. Approximate processing in real-time problem solving. *AI Magazine*, 9(1):49–61, 1988.

[17] P. S. Ow, S. F. Smith, and A. Thiriez. Reactive plan revision. In *Proceedings AAAI-88*. AAAI, 1988.

[18] Stuart J. Russell and Eric H. Wefald. Principles of metareasoning. In Brachman et al. [2], pages 400–411.

[19] Marcel J. Schoppers. Universal plans for reactive robots in unpredictable environments. In *Proceedings IJCAI 10*, pages 1039–1046. IJCAII, 1987.

[20] W-K. Shih, J. W. S. Liu, and J-Y. Chung. Fast algorithms for scheduling imprecise computations.

In *Proceedings of the Real-Time Systems Symposium*, pages 12–19. IEEE, 1989.

[21] Herbert A. Simon and Joseph B. Kadane. Optimal problem-solving search: All-or-none solutions. *Artificial Intelligence*, 6:235–247, 1975.

[22] Shlomo Zilberstein. *Operational Rationality through Compilation of Anytime Algorithms*. PhD thesis, University of California at Berkeley, 1993.

On the Algebraic Geometry of a Class of Contact Formation Cells

A.O. Farahat, *Dept of Aerospace Eng, Texas A&M University, College Station, Tx*

P.F. Stiller, *Dept of Mathematics, Texas A&M University, College Station, Tx*

J.C. Trinkle *Dept of Computer Science, Texas A&M University, College Station, Tx*

The efficient planning of contact tasks for intelligent robotic systems requires a thorough understanding of the kinematic constraints imposed on the system by the contacts. In this paper, we derive closed-form analytic solutions for the position and orientation of a passive polygon moving in sliding contact with two or three active polygons whose positions and orientations are independently controlled. This is accomplished by applying a simple elimination procedure to solve the appropriate system of contact constraint equations.

The benefits of having analytic solutions are numerous. For example, they eliminate the need for iterative nonlinear equation solving algorithms to determine the position and orientation of the passive polygon given the positions and orientations of the active ones. Also, because they contain the configuration variables of the active polygons and the relevant geometric parameters, models of geometric and control uncertainty can be readily incorporated into the solutions. This will facilitate the analysis of the effects of these uncertainties on the kinematic constraints.

We also prove that the set of solutions to the contact constraint equations is a smooth submanifold of the system's configuration space and we study its projection onto the configuration space of the active polygons (i.e., the lower-dimensional configuration space of the controllable parameters). By relating these results to the wrench matrices commonly used in grasp analysis, we discover a previously unknown and highly nonintuitive class of nongeneric contact situations. In these situations, for a specific fixed configuration of the active polygons, the passive polygon can maintain three contacts on three mutually nonparallel edges while retaining one degree of freedom of motion.

1 Introduction

Consider a planar system of rigid polygonal bodies in contact (see Figure 1) and assume that the positions and orientations of all but one of the polygons are actively controlled. The polygon that is not actively controlled, but rather is "grasped," is referred to as the workpiece or the passive polygon, and those that are actively controlled through joint actuation are collectively referred to as the manipulator or the active polygons. The workpiece and the manipulator taken together are referred to as the manipulation system, or just the system. The process of reorienting the workpiece by controlling the manipulator is known as dexterous manipulation. In contrast to the classical piano movers' problem [20] in which contact is avoided, every solution to a dexterous manipulation planning problem must include contact. If in addition, the level of uncertainty is a significant factor in the planning and execution of a dexterous manipulation task, then we refer to it as a fine motion or fine manipulation task.

Automatic fine manipulation planning is one of the most important unsolved problems in the field of robotics. Tasks in this class include mechanical assembly/disassembly and grasping operations. The development of a practical, reliable planner for this class of tasks would facilitate the automation of large portions of various manufacturing and service industries and would expand our ability to work in Space and other hazardous environments.

The worst case running times of general motion planning algorithms increase exponentially with the dimension of the system's configuration space (C-space). Nonetheless, nonpathological planning problems have been solved in several minutes indicating that the exponential worst-case complexity should not deter the future development of planning algorithms [8]. Since fine manipulation requires contacts, algorithms for fine mo-

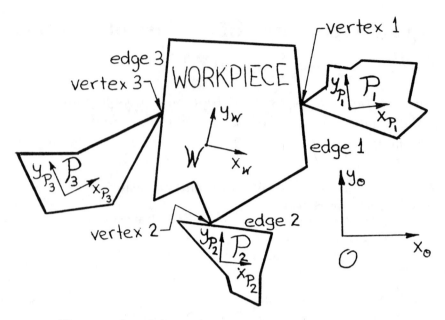

Figure 1: A workpiece and a three-polygon manipulator in contact.

tion planning can potentially be more efficient, because the kinematic constraint equation associated with each contact effectively reduces the dimension of C-space. However, fine manipulation planning algorithms can only fully benefit from this reduction if the relevant systems of kinematic constraints are thoroughly understood.

Besides the effective reduction in the dimension of C-space due to the contacts, there are two other good reasons to believe that practical fine manipulation planning algorithms can be developed: first, to find a plan, the entire C-space need not be decomposed (for example, see [8, 11, 22]); and second, cell decomposition techniques lend themselves to parallel and distributed computation.

In this paper, we present the geometric characterization and analytic representation of four fundamental types of kinematic constraint "surfaces" in C-space (called "contact formation cells" [24] [25]). These constraint "surfaces" are four of the most important ones that arise during the planar manipulation of a passive polygonal workpiece by a manipulator composed of up to three active polygons whose positions and orientations are independently controlled.

Using simple techniques from algebraic geometry, we derive analytic representations of these CF-cells which

yield a number of benefits. For example, they provide us with an in-depth understanding of the geometry of the cells and their projections onto the space of controllable variables. Also, they obviate the need for an iterative solution procedure to determine the position and orientation of the workpiece given the configuration of manipulator. Finally, they reveal configurations (analogous to the singular configurations of typical serial-link manipulators) in which controlling the motion of the system requires more careful planning. Kinematic constraints among the polygons due to linkage connections are not considered (see "Future Work" below). If present, those constraints could further restrict the subset of C-space, making planning even more efficient.

Each of the four types of contact formation cells (CF-cells) that we will be studying, corresponds to a distinct combination of the contact constraint equations associated with sliding, type A and/or type B contacts [13] (see definitions below). Each sliding contact yields one constraint equation. Under our assumption of pure position control of the manipulator, the number of contact constraints is generally three, because all degrees of freedom of the system except for the three associated with the workpiece are directly controlled. Contact combinations with four or more contact constraints could be maintained through compliant control, but those combinations are beyond the scope of this paper.

Even though we do not study three-dimensional systems, our analysis will be useful in planning the manipulation of solid objects with constant cross-sectional geometry. Also it will be possible to apply our results to manipulating extended solids which do not have constant cross-sectional geometries. For example, imagine a long slender object with a slightly varying polygonal cross-section and suppose that due to its length, one would prefer to pick it up at two locations (with two nominally identical dexterous manipulators) distributed along its length. Next, suppose that a manipulation plan is generated for the nominal cross-section. Applying techniques from deformation theory and differential topology, it will be possible to identify an envelope of geometric uncertainty and control error within which the system must remain for successful task execution. If both dexterous manipulators execute the plan "simultaneously" and remain in the envelope, then the task will succeed.

1.1 Previous Work

The results presented in this paper can be viewed as an extension of previous work on configuration obstacles (C-obstacles) by Brost. Specifically, in his thesis [4], he developed techniques for computing and representing the C-obstacle of two polygons in the plane. The C-obstacle is the three-dimensional subset of C-space (also three-dimensional) whose boundary is composed of curved surface patches, edges, and vertices corresponding to one, two, and three elemental contacts[1]. It is a complete description of all possible relative configurations of the two polygons. Points on the surface of the C-obstacle correspond to configurations in which the polygons are in contact (at one or more points).

Brost studied C-obstacles as a necessary first step in planning manipulation tasks involving contact between two polygons. He developed two planning algorithms: one based upon generalized energy arguments and the other based upon numerical integration. These algorithms can produce plans for generalized "dropping" tasks[2] and "pushing" tasks, respectively.

The ultimate goal of our work is the automatic planning of fine manipulation tasks, which can be viewed as a generalized "pushing" problem – our "pusher" is to be a manipulator of any desired kinematic structure. In the context of Brost's work, we are interested in the extended problem in which the geometry of the "pusher" is a function of a number of control parameters, in particular, the positions and orientations of the polygons composing the manipulator (*i.e.*, the active polygons) in contact with the workpiece (*i.e.*, the passive or "pushed" polygon). Like Brost, we begin by studying the kinematic constraints on the system.

We can relate the results of this paper to Brost C-obstacles from another perspective. Brost represented the facets and edges of the C-obstacle in parametric form and found the vertices by numerical solution. In the dexterous manipulation planning problem that we have been pursuing [22, 23, 24, 25], the vertices of the C-obstacle are extremely important, because during manipulation, the workpiece configuration commonly resides at a vertex when the coefficient of friction is small. Manipulation of the actual workpiece corresponds to the deliberate deformation of the C-obstacle (by varying the configuration of the manipulator) to cause desired motions of the C-obstacle vertex. Controlling two vertex points so as to come together and then separate corresponds to a discrete change in the contact topology of the actual system. Such changes typically take place during dexterous manipulation to effect significant reorientations the workpiece.

The contributions of this paper are closed-form, analytic solutions for the vertices of deformable C-obstacles. These solutions are written as functions of the configuration variables of the manipulator, thereby facilitating our future studies of the possible motions of the workpiece over finite, as opposed to infinitesimal, manipulation trajectories. We also show that the set of all possible workpiece configurations, the CF-cell, forms a submanifold of the system's C-space (the space of all possible configurations of the manipulator and the workpiece) and that the CF-cell forms a generically finite branched covering [3] (with up to four sheets)

---

[1] The term "elemental contact" [6] refers to either a type A or a type B contact.

[2] Generalized dropping tasks include compliant motion

tasks in which the moving polygon is firmly gripped by an end effector moving under generalized damper controlled.

[3] A branched covering of a space can be thought of as several sheets "above" the space. For example, the function $y^2 = x$ in Euclidean two-space forms a finite branched cover

of the C-space corresponding to just the configurations of the manipulator. Finally, by relating the results obtained through our algebro-geometric analysis to the existing research results on grasp analysis, we discover a new nonintuitive, nongeneric contact situation.

We would like to close this section with two remarks. First, the results in Brost's thesis could have been used to identify this class of nongeneric "grasps," but it's nonintuitive nature led Brost to overlook it [3]. Second, as we will discuss later, such a class of nonintuitive grasps could have been deduced from previous results on Cardan mechanisms [14, 16].

1.2   Layout of Paper

This paper is organized as follows. In Section 2, we define the class of systems that will be considered and state our assumptions. In Section 3, we derive the sets of contact constraints for the four relevant combinations of three sliding contacts and give a noniterative solution procedure for obtaining the workpiece's possible configurations given the configuration of the manipulator. This is done in detail mainly for the case of three type A contacts in Section 3.1, because the other three combinations of contact constraints can be analyzed in the same way. We also derive expressions for the relevant wrench matrices (commonly used in the analysis of grasps) as functions of the manipulator configuration and relate them to the multiplicity of the solution obtained. In Finally, in Section 4, we conclude and recommend avenues for further research.

2   System Model and Problem Statement

In studies of polygonal mobile robots operating in a plane among polygonal obstacles, two types of elemental contacts have been defined: type A, which is a contact between an edge of the robot and a vertex of an

---

of Euclidean one-space (under orthogonal projection onto the x-axis), because every point on the x-axis is "covered" by two, one, or zero points on the curve $y^2 = x$. Locally the sheets are diffeomorphic to the x-axis except at the branch point $(0, 0)$ where the sheets come together smoothly with infinite slope. Note that removing a branch point does not necessarily disconnect the sheets (*e.g.*, $z \rightarrow z^2$ mapping the complex plane to itself.)

obstacle; and type B, which is a contact between a vertex of the robot and an edge of an obstacle [12]. In the work presented here, we view the workpiece as a mobile robot and the manipulator as a deformable obstacle. Thus we are led to define type A and B elemental contacts as follows:

Type A Contact: an edge of the workpiece is in contact with a vertex of the manipulator.
Type B Contact: a vertex of the workpiece is in contact with an edge of the manipulator.

Figure 1 shows a workpiece and a three-polygon manipulator with three elemental contacts: edge 1 of the workpiece contacts vertex 1 of the first manipulator polygon (a type A contact), vertex 2 of the workpiece contacts edge 2 of the second manipulator polygon (a type B contact), and edge 3 of the workpiece contacts vertex 3 of the third manipulator polygon (a type A contact).

A set of elemental contacts constitutes a contact formation, CF [6]. Based on the numbers of type A and B contacts, we classify CF's into various types. For example, the CF shown in Figure 1 is of type 2AB. When the configurations of the manipulator polygons are independently controlled, a CF of this type will usually be maintained *only if* all three contacts slide as the manipulator moves. Only specially designed (compliant) trajectories could maintain the CF with one or more rolling contacts, but we have assumed that our manipulator is position controlled. Similarly, in maintaining a CF of type 2BA, 3A, or 3B, all the contacts will slide. We will be studying all four of these types in this paper.

In what follows, we will assume that as many as three rigid manipulator polygons are in contact with a rigid workpiece polygon; that the positions and orientations of the manipulator polygons can be controlled directly and independently; and that the position and orientation of the workpiece is controlled purely as a byproduct of maintaining (if possible) the specified CF. For each of the four types of CF's identified above, we will, among other things, answer the following:

1. Does the CF-cell form a submanifold of the system's C-space (the configuration space of the manipulator and workpiece taken together)?

2. For a given configuration of the manipulator, how many configurations of the workpiece achieve the

specified CF and what are the conditions under which there is exactly one such configuration, or perhaps an infinite number of such configurations?

3. How does the CF-cell project to the C-space of the manipulator? In particular, how does the CF-cell branch and when does it fail to be a finite covering? These are the places where controlling the manipulator requires special care.

3 CF-Cells with Three Contacts

Contact formation with three elemental contacts fall into the four basic types enumerated above. It turns out that the 3A and 3B cases and the 2AB and 2BA cases are dual in a sense that will be clarified later. For that reason, we will give complete derivations only for the 3A and 2AB cases and simply state our results for the other two.

3.1 3A CF-Cells

Referring again to Figure 1, let x_l, y_l, and θ_l denote the position and orientation of a frame $\mathcal{P}_l$, attached to the lth manipulator polygon, measured with respect to the world frame, $\mathcal{O}$. Similarly, let x, y, and θ, without any subscripts, denote the position and orientation of a workpiece-fixed frame $\mathcal{W}$ with respect to $\mathcal{O}$. Next define the workpiece configuration vector $\mathbf{q}$, the manipulator configuration vector $\mathbf{r}$, and the system configuration vector $\mathbf{p}$ as follows:

$$\mathbf{q} = [x, y, c, s] \qquad (1)$$
$$\mathbf{r} = [x_1, x_2, x_3, y_1, y_2, y_3, \theta_1, \theta_2, \theta_3] \qquad (2)$$
$$\mathbf{p} = [x, y, c, s, x_1, x_2, x_3, y_1, y_2, y_3, \theta_1, \theta_2, \theta_3] \qquad (3)$$

where c and s "represent" $cos(\theta)$ and $sin(\theta)$, respectively. The variables c and s are to be thought of as independent variables and are used to represent the orientation of the workpiece in place of θ, so that the contact constraints may be written as algebraic, rather than trigonometric, equations. This, however, will require the introduction of an additional algebraic constraint, namely, $c^2 + s^2 - 1 = 0$.

In this spirit, we define the system's *modified C-space*, $\mathcal{Z}$, to be the set of all possible vectors $\mathbf{p}$, and denote the three contact constraints by $C_l(\mathbf{p}) = 0$ for $l = 1, 2, 3$. We then define the resulting CF-cell, $\mathcal{CF}$, as follows:¡

$$\mathcal{CF} = \{\mathbf{p} \in \mathcal{Z} \mid c^2 + s^2 - 1 = 0 \text{ and}$$
$$C_l(\mathbf{p}) = 0 \text{ for every } l = 1, 2, 3\}. \qquad (4)$$

$\mathcal{Z}$ is topologically the product space $R^{10} \times T^3$, where R^{10} is ten-dimensional Euclidean space and T^3 is the three-torus (the product of three circles parametrized by the variables $\theta_1, \theta_2, \theta_3$). Note that $\mathcal{Z}$ is a manifold [12]. The usual C-space $\mathcal{X}$ for our four-polygon system would be the submanifold of $\mathcal{Z}$ cut out by the equation $c^2 + s^2 - 1 = 0$. It would be 12-dimensional, while $\mathcal{Z}$ is 13-dimensional.

Consider Figure 2. Let the pair (u_l, v_l) denote the position of vertex l, on polygon l of the manipulator, which is intended to contact edge l of the workpiece. Here u_l and v_l are measured with respect to the frame $\mathcal{P}_l$. Let the pair (w_l, z_l) denote the coordinates of an arbitrarily selected point on edge l of the workpiece measured with respect to $\mathcal{W}$. Also let ϕ_l denote the angle between the outward normal to edge l and the positive x-axis of $\mathcal{W}$. The signed distance between vertex l and the line supporting edge l, the so called C-function [12], is given by:

$$C_l(\mathbf{p}) = \mathbf{v}_l \cdot (\mathbf{g}_l - \mathbf{h}_l) \qquad (5)$$

where the vectors $\mathbf{g}_l$, $\mathbf{h}_l$, and $\mathbf{v}_l$ are:

$$\mathbf{g}_l = (x_l + cos(\theta_l)u_l - sin(\theta_l)v_l ,$$
$$y_l + sin(\theta_l)u_l + cos(\theta_l)v_l) \qquad (6)$$
$$\mathbf{h}_l = (x + cos(\theta)w_l - sin(\theta)z_l ,$$
$$y + sin(\theta)w_l + cos(\theta)z_l) \qquad (7)$$
$$\mathbf{v}_l = (cos(\theta + \phi_l), sin(\theta + \phi_l)). \qquad (8)$$

Here $\mathbf{g}_l$ is the position of vertex l, $\mathbf{h}_l$ is the position of the selected point on edge l, and $\mathbf{v}_l$ is the outward unit normal of edge l; all of these quantities are expressed with respect to the frame $\mathcal{O}$.

Notice that we have labeled the edges of the workpiece and the polygons so that contact l is between polygon l of the manipulator and the line supporting edge l of the workpiece. Note also that we handle cases in which an edge of the workpiece is to be contacted by more than one vertex of the manipulator by labeling that edge more than once.

The C-function corresponding to each type A contact can be written as a function of the configuration of the workpiece to yield the following system of contact constraint equations:

$$C_l(\mathbf{p}) = a_l + b_l c + d_l s - e_l xc + f_l xs$$
$$- f_l yc - e_l ys = 0; \quad l = 1, 2, 3 \qquad (9)$$

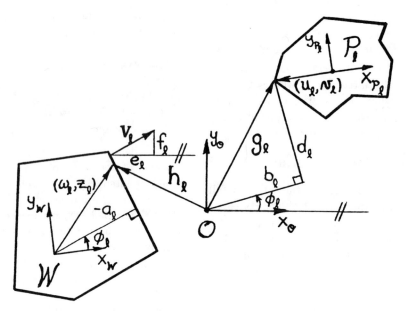

Figure 2: Illustration of parameters relevant to type A contacts.

$$c^2 + s^2 - 1 = 0 \qquad (10)$$

where the coefficients a_l, b_l, d_l, e_l, and f_l, illustrated in Figure 2, are functions of the geometry of the bodies in the system (including the workpiece) and the configuration of the manipulator:

$$a_l = -cos(\phi_l)w_l - sin(\phi_l)z_l \qquad (11)$$

$$b_l = cos(\phi_l - \theta_l)u_l + sin(\phi_l - \theta_l)v_l$$
$$+ sin(\phi_l)y_l + cos(\phi_l)x_l \qquad (12)$$

$$d_l = -sin(\phi_l - \theta_l)u_l + cos(\phi_l - \theta_l)v_l$$
$$+ cos(\phi_l)y_l - sin(\phi_l)x_l \qquad (13)$$

$$e_l = cos(\phi_l) \qquad (14)$$

$$f_l = sin(\phi_l). \qquad (15)$$

Note that the derivation of the equations in (9) uses equation (10) to eliminate terms involving c^2 and s^2 in $C_l(\mathbf{p})$.

Each C-function, $C_l(\mathbf{p})$, as well as the unit circle equation (equation (10)), defines a quadric hypersurface in C-space. The intersection of these four hypersurfaces defines the set of *geometrically admissible* 3A configurations of the workpiece as a function of the manipulator configuration vector, (*i.e.*, the intersection defines the CF-cell, $\mathcal{CF}$, in $\mathcal{Z}$). The reader should be cautioned that "geometrically admissible" means that

contact is maintained along the line supporting the edge, not the actual physical edge of the workpiece.

To obtain an analytic solution which describes the 3A CF-cell, notice that each equation in (9) is linear in x and y so that the system of equations (9-10) can be rewritten as:

$$D_l(\mathbf{r}, c, s) + E_l(\mathbf{r}, c, s)x + F_l(\mathbf{r}, c, s)y = 0;$$
$$\text{for all } l = 1, 2, 3 \quad (16)$$

$$c^2 + s^2 - 1 = 0 \qquad (17)$$

where

$$D_l = a_l + b_l c + d_l s \qquad (18)$$

$$E_l = -e_l c + f_l s \qquad (19)$$

$$F_l = -f_l c - e_l s. \qquad (20)$$

The first two of the three equations in (16) can be solved for x and y yielding:

$$x = \frac{(F_1 D_2 - F_2 D_1)}{E_1 F_2 - F_1 E_2} \qquad y = \frac{(E_2 D_1 - E_1 D_2)}{E_1 F_2 - F_1 E_2} \quad (21)$$

where the denominator $E_1 F_2 - F_1 E_2$ simplifies to $e_1 f_2 - e_2 f_1 = sin(\phi_2 - \phi_1)$. Note that if $sin(\phi_2 - \phi_1)$ is zero, a different pair of equations in (16) must be used. Thus solving for x and y will fail if and only if the rank

of $\begin{bmatrix} E_1 & F_1 \\ E_2 & F_2 \\ E_3 & F_3 \end{bmatrix} = \begin{bmatrix} e_1 & f_1 \\ e_2 & f_2 \\ e_3 & f_3 \end{bmatrix} \begin{bmatrix} -cos(\theta) & -sin(\theta) \\ sin(\theta) & -cos(\theta) \end{bmatrix}$

is less than two. In this event, we shall say that *elimination fails*.

Proposition 1: *For a 3A CF, elimination fails, i.e., the rank of* $\begin{bmatrix} e_1 & f_1 \\ e_2 & f_2 \\ e_3 & f_3 \end{bmatrix}$ *is less than two, if and only if the lines supporting the edges designated for contact are parallel. (This includes the possibility that two or three of the lines are coincident.)*

Proof: If the three contacted edges are parallel, then the angle between any two normals to the contacted edges is 0 or π radians. Consequently, $sin(\phi_l - \phi_m) = 0$ for any choice of l and m in the set $\{1, 2, 3\}$. This in turn implies that the determinant of every (2×2) minor of $\begin{bmatrix} e_1 & f_1 \\ e_2 & f_2 \\ e_3 & f_3 \end{bmatrix}$ is zero and that the rank of this matrix is less than two.

Conversely, the rank of $\begin{bmatrix} e_1 & f_1 \\ e_2 & f_2 \\ e_3 & f_3 \end{bmatrix}$ can only be less than 2 if the determinants of all (2×2) minors are zero. Thus elimination fails only if $sin(\phi_2 - \phi_1) = 0$, $sin(\phi_3 - \phi_2) = 0$, and $sin(\phi_1 - \phi_3) = 0$, that is, only if the three contacted edges are parallel. ....□

Apart from the case when elimination fails the contacts can always be relabeled so that $sin(\phi_2 - \phi_1)$ is not zero and equations (21) are valid. Substituting x and y into the third equation of the system (16) gives a polynomial in c and s of the form:

$$G(\mathbf{r})c + H(\mathbf{r})s + I(\mathbf{r}) = 0 \qquad (22)$$
$$s^2 + c^2 - 1 = 0 \qquad (23)$$

where G, H, and I are given by:

$$\begin{aligned} G &= b_1 e_2 f_3 - b_1 e_3 f_2 - b_2 e_1 f_3 + b_2 e_3 f_1 \\ &\quad + b_3 e_1 f_2 - b_3 e_2 f_1 \end{aligned} \qquad (24)$$

$$\begin{aligned} H &= d_1 e_2 f_3 - d_1 e_3 f_2 - d_2 e_1 f_3 + d_2 e_3 f_1 \\ &\quad + d_3 e_1 f_2 - d_3 e_2 f_1 \end{aligned} \qquad (25)$$

$$\begin{aligned} I &= a_1 e_2 f_3 - a_1 e_3 f_2 - a_2 e_1 f_3 + a_2 e_3 f_1 \\ &\quad + a_3 e_1 f_2 - a_3 e_2 f_1. \end{aligned} \qquad (26)$$

It is important to note that equation (22) can be rewritten as follows:

$$Det \begin{bmatrix} D_1 & E_1 & F_1 \\ D_2 & E_2 & F_2 \\ D_3 & E_3 & F_3 \end{bmatrix} = 0, \qquad (27)$$

which is a necessary and sufficient condition that the three equations (16), when homogenized (by multiplying each of the coefficients, $D_l(\mathbf{r}, c, s)$, by a homogenizing variable z) have a nontrivial solution. The additional condition, $rank \begin{bmatrix} E_1 & F_1 \\ E_2 & F_2 \\ E_3 & F_3 \end{bmatrix} = 2$, guarantees that the homogenizing variable z will not be zero. Linearly scaling x, y, and z to make z equal to one gives an x and y satisfying the original system of equations (16).

Expanding the determinant in equation (27) and recalling that $E_i F_j - E_j F_i = e_i f_j - e_j f_i$, we arrive at alternate forms for G, H, and I that are useful in the proofs which will follow:

$$G = Det \begin{bmatrix} b_1 & cos(\phi_1) & sin(\phi_1) \\ b_2 & cos(\phi_2) & sin(\phi_2) \\ b_3 & cos(\phi_3) & sin(\phi_3) \end{bmatrix} \qquad (28)$$

$$H = Det \begin{bmatrix} d_1 & cos(\phi_1) & sin(\phi_1) \\ d_2 & cos(\phi_2) & sin(\phi_2) \\ d_3 & cos(\phi_3) & sin(\phi_3) \end{bmatrix} \qquad (29)$$

$$I = Det \begin{bmatrix} a_1 & cos(\phi_1) & sin(\phi_1) \\ a_2 & cos(\phi_2) & sin(\phi_2) \\ a_3 & cos(\phi_3) & sin(\phi_3) \end{bmatrix}. \qquad (30)$$

To determine the geometrically admissible workpiece configurations $\mathbf{q}$, given a specific 3A CF and the manipulator configuration $\mathbf{r}$, we determine the intersection of the line (22) and the unit circle (23). When $G \neq 0$, we solve equation (22) for c and substitute into equation (23) to get a quadratic equation for s:

$$(H^2 + G^2)s^2 + 2HIs + (I^2 - G^2) = 0 \quad \text{when} \quad G \neq 0. \qquad (31)$$

When $H \neq 0$, we solve for s in equation (22) to find:

$$(H^2 + G^2)c^2 + 2GIc + (I^2 - H^2) = 0 \quad \text{when} \quad H \neq 0. \qquad (32)$$

The discriminants of these equations are $-4G^2(-G^2 - H^2 + I^2)$ when $G \neq 0$ and $-4H^2(-G^2 - H^2 + I^2)$ when $H \neq 0$. For geometrically admissible workpiece configurations to exist (*i.e.*, for the system of equations, (22) and (23) to have real solutions) when $G^2 + H^2 \neq 0$, the discriminant must be nonnegative. Thus we require that the following inequality be satisfied:

$$G^2 + H^2 \geq I^2. \qquad (33)$$

When $G = 0$ and $H = 0$, we get real solutions only if $I = 0$. In that case, there will be an infinite number of solutions, because any pair (c, s) satisfying equation (23) also satisfies equation (22) which is identically zero, and x and y can be found by back substitution into equations (21). When $G = 0$ and $H = 0$, but $I \neq 0$, equation (22) will be inconsistent and there will be no solutions. Note that even though inequality (33) is satisfied when elimination fails (because $G = H = I = 0$), it is *not* a necessary and sufficient condition for solutions to exist in that case (see Proposition 2).

Proposition 2: *If elimination fails, then $G = H = I = 0$ and for a fixed $\mathbf{r}$, if there is any workpiece configuration which attains the specified 3A CF, then there will be an infinite number.*

Proof: This proposition refers to the case of three parallel edges and its truth is obvious from the discussion above. . . . . □

When elimination fails, we will say that the particular manipulator configuration vector $\mathbf{r}$, is *nongeneric*. Brost identified this nongeneric situation and noted that either there are no geometrically admissible workpiece configurations or an infinite number [4]. In the latter case, the workpiece can be translated along the contacted edges while maintaining the specified CF. As we will see below, by relating the results of the above analysis to the wrench matrices used in grasp analysis, we discover a new nongeneric contact situation with three contacts on nonparallel edges for which there is an infinite number of admissible workpiece configurations.

Our use of the term "nongeneric" is similar to Brost's [4]: the system of contact constraint equations does not satisfy "general position." Loosely speaking, a loss of general position implies that either a system of n equations in m unknowns $(m > n)$ has more than the expected $m-n$ degrees of freedom, or that solutions occur with multiplicities, or that the equations are inconsistent. In the context of this work, a system of three contact constraint equations in general position (*i.e.*, the generic case) for a fixed $\mathbf{r}$ constrains the workpiece to a finite number of configurations.

In practice, nongeneric situations are rare, if not nonexistent, because it is impossible to manufacture the perfect arrangement of geometric features required

(*e.g.*, a perfectly straight edge, or three perfectly parallel and straight edges) and to precisely to control the position of the manipulator. Mathematically, this fact is related to standard results in algebraic geometry which say that every nongeneric system of polynomial equations can be made generic by a suitably small perturbation of the coefficients [5]. However, from a computational and modeling standpoint, nongeneric situations are common, and if they are not understood completely, they will compromise the robustness of any dexterous manipulation planning algorithm based on geometric models.

In the 3A case, we will find that the generic situations are those in which there are two distinct real solutions to the system of contact constraints (9) and (10) or two distinct complex solutions, implying that there are two or zero real geometrically admissible workpiece configurations, respectively. The nongeneric situations are those for which there is an infinite number of solutions or just one solution (of multiplicity two), or inconsistent equations leading to no solutions. The one exception to our definition of genericity (over the four CF-cells studied) occurs in the 3A case when elimination fails and there are no solutions. Strictly speaking, such a case is generic, *i.e.*, small perturbations of $\mathbf{r}$ will not produce solutions. However, the geometry itself is nongeneric because the three edges are parallel, and for that reason we do not consider this case as generic. (In future work when uncertainty is involved, we will be varying the geometry and the case of three parallel edges with certainly be nongeneric in that context.)

In contrast to the nongeneric situations, generic situations are those for which $G^2 + H^2$ is positive and not equal to I^2. In these cases, regardless of the particular values of G, H, and I, there will always be two distinct solutions to the system of equations, (22) and (23). If the solutions are real, then there will be two geometrically admissible configurations of the workpiece; if they are complex there will be none.

Theorem 1: *For a generic positioning, $\mathbf{r}$, of the manipulator, we will have either zero or two geometrically admissible workpiece configurations.*

3.2   The Wrench Matrices

The functions $G(\mathbf{r})$, $H(\mathbf{r})$, and $I(\mathbf{r})$ are closely related to the "wrench matrix" which arises in grasp analysis.

This matrix is particularly useful in determining the stability and mobility of the grasped workpiece [10]. If the system under consideration is planar with n_c contacts, then the wrench matrix $\mathbf{W}$, can be partitioned into normal and tangential components $\mathbf{W}_n$ and $\mathbf{W}_t$, which have size $(3 \times n_c)$. The partitions $\mathbf{W}_n$ and $\mathbf{W}_t$ appear in the kinematic relationships constraining the normal and tangential components of the relative velocities of the contact points and in the summations of the normal and tangential components of the contact forces [23].

The wrench matrices have the following form:

$$\mathbf{W}_n = \begin{bmatrix} \hat{\mathbf{n}}_1 & \hat{\mathbf{n}}_2 & \hat{\mathbf{n}}_3 \\ \mathbf{p}_1 \wedge \hat{\mathbf{n}}_1 & \mathbf{p}_2 \wedge \hat{\mathbf{n}}_2 & \mathbf{p}_3 \wedge \hat{\mathbf{n}}_3 \end{bmatrix} \quad (34)$$

$$\mathbf{W}_t = \begin{bmatrix} \hat{\mathbf{t}}_1 & \hat{\mathbf{t}}_2 & \hat{\mathbf{t}}_3 \\ \mathbf{p}_1 \wedge \hat{\mathbf{t}}_1 & \mathbf{p}_2 \wedge \hat{\mathbf{t}}_2 & \mathbf{p}_3 \wedge \hat{\mathbf{t}}_3 \end{bmatrix} \quad (35)$$

where $\hat{\mathbf{n}}_l$ is the inward pointing unit normal vector to edge l, $\hat{\mathbf{t}}_l$ is the tangential unit vector defined so that $\hat{\mathbf{n}}_l \times \hat{\mathbf{t}}_l$ points out of the page, $\mathbf{p}_l$ is the position of the contact point, and $\mathbf{p}_l \wedge \hat{\mathbf{n}}_l$ is given by $p_{l_x} n_{l_y} - p_{l_y} n_{l_x}$. Here p_{l_x} and p_{l_y} are the components of $\mathbf{p}_l$. The components of $\hat{\mathbf{n}}_l$ are defined analogously.

Given the geometric definitions of the coefficients shown in Figure 2, the wrench matrices can be rewritten as explicit functions of the system configuration as follows:

$$^{\mathcal{O}}\mathbf{W}_n = \begin{bmatrix} -\cos(\theta+\phi_1) & -\cos(\theta+\phi_2) & -\cos(\theta+\phi_3) \\ -\sin(\theta+\phi_1) & -\sin(\theta+\phi_2) & -\sin(\theta+\phi_3) \\ d_1 c - b_1 s & d_2 c - b_2 s & d_3 c - b_3 s \end{bmatrix} \quad (36)$$

$$^{\mathcal{O}}\mathbf{W}_{t,verts} = \begin{bmatrix} \sin(\theta+\phi_1) & \sin(\theta+\phi_2) & \sin(\theta+\phi_3) \\ -\cos(\theta+\phi_1) & -\cos(\theta+\phi_2) & -\cos(\theta+\phi_3) \\ -(b_1 c + d_1 s) & -(b_2 c + d_2 s) & -(b_3 c + d_3 s) \end{bmatrix} \quad (37)$$

where the superscript $^{\mathcal{O}}$ indicates that the matrices are expressed with respect to the frame $\mathcal{O}$. Note that in this definition of $^{\mathcal{O}}\mathbf{W}_{t,verts}$, $\mathbf{p}_i$ is the position of the vertex l of the manipulator polygon designated to contact the workpiece *even though achieving all three contacts may be impossible* for the manipulator configuration under consideration.

Finally, comparing these expressions with equations (28) and (29), we find that G and H are related to the determinants of $^{\mathcal{O}}\mathbf{W}_n$ and $^{\mathcal{O}}\mathbf{W}_{t,verts}$ by the following simple formulas:

$$Det(^{\mathcal{O}}\mathbf{W}_n) = cH - sG \quad (38)$$

$$Det(^{\mathcal{O}}\mathbf{W}_{t,verts}) = -(sH + cG). \quad (39)$$

Therefore, the important quantity, $G^2 + H^2$, that arose in the discriminant of equations (31) and (32), is equal to the sum of the squares of the determinants of the normal and tangential wrench matrices:

$$G^2 + H^2 = Det^2(^{\mathcal{O}}\mathbf{W}_n) + Det^2(^{\mathcal{O}}\mathbf{W}_{t,verts}). \quad (40)$$

Similarly, one can deduce that:

$$^{\mathcal{W}}\mathbf{W}_{t,edges} = \begin{bmatrix} \sin(\phi_1) & \sin(\phi_2) & \sin(\phi_3) \\ -\cos(\phi_1) & -\cos(\phi_2) & -\cos(\phi_3) \\ a_1 & a_2 & a_3 \end{bmatrix}, \quad (41)$$

where the subscript "*edges*" indicates that the contacts are assumed to be on the designated edges of the workpiece. (The matrix does not depend on the specific edge point chosen.) Comparing equations (30) and (41), we see that:

$$I = Det(^{\mathcal{W}}\mathbf{W}_{t,edges}). \quad (42)$$

The values of G, H, and I give us information about the existence and the number of geometrically admissible workpiece configurations for a given 3A CF and manipulator configuration $\mathbf{r}$. Clearly, the existence of geometrically admissible workpiece configurations (and the number of configurations, if any exist) is independent of the specific choices of coordinate frames. This motivates the following result.

Proposition 3: *The quantities I^2 and $G^2 + H^2$, and the quantities $Det(^{\mathcal{W}}\mathbf{W}_{t,edges})$, $Det(^{\mathcal{O}}\mathbf{W}_n)$, and $Det(^{\mathcal{O}}\mathbf{W}_{t,verts})$ are independent of the choice of all coordinate frames.*

Proof: None of the vectors, $\hat{\mathbf{n}}_l$, $\hat{\mathbf{t}}_l$, and $\mathbf{p}_l$, appearing in the wrench matrix definitions given in equations (34) and (35) depend on the choice of frames $\mathcal{P}_l$. Also, it is known that a wrench matrix, expressed in an arbitrary frame $\mathcal{A}$, can be expressed in an arbitrary frame $\mathcal{B}$ by premultiplying it by a force transformation matrix $^{\mathcal{B}}_{\mathcal{A}}\mathbf{T}_f$ [21]:

$$^{\mathcal{B}}\mathbf{W} = {}^{\mathcal{B}}_{\mathcal{A}}\mathbf{T}_f \, {}^{\mathcal{A}}\mathbf{W} \quad (43)$$

where in the planar case, $^{\mathcal{B}}_{\mathcal{A}}\mathbf{T}_f$ is $\begin{bmatrix} \cos(\psi) & \sin(\psi) & 0 \\ -\sin(\psi) & \cos(\psi) & 0 \\ \Delta y & -\Delta x & 1 \end{bmatrix}$. Here ψ is the angle of rotation of frame $\mathcal{B}$ with respect to frame $\mathcal{A}$ and Δx and Δy are the components of the displacement of the origin of frame $\mathcal{B}$ relative to the origin of frame $\mathcal{A}$ expressed in terms of frame $\mathcal{A}$. Since the determinant of $_A^B\mathbf{T}_f$ is one, the determinants of the (3×3) matrices $^{\mathcal{O}}\mathbf{W}_n$, $^{\mathcal{W}}\mathbf{W}_{t,edges}$, and $^{\mathcal{O}}\mathbf{W}_{t,verts}$ are all invariant under a change of $\mathcal{O}$ or $\mathcal{W}$ as appropriate. The frame invariance of I^2 and $G^2 + H^2$ now follow from formulas (41) and (42). We emphasize, however, that G, H, and the wrench matrices are not frame invariant. $\ldots\ldots \square$

The invariance of the determinants of the wrench matrices can also be seen geometrically. Each column of $^{\mathcal{O}}\mathbf{W}_n$ represents a line perpendicular to an edge of the workpiece and containing the vertex of the manipulator intended to contact that edge. The condition that $Det(^{\mathcal{O}}\mathbf{W}_n) = 0$ is exactly that these three lines meet at a point (possibly at infinity). Since these lines depend purely on the geometry of the bodies and their relative arrangement, it is clear that the determinant of $^{\mathcal{O}}\mathbf{W}_n$ is independent of any choice of frames.

The tangential wrench matrices, $^{\mathcal{W}}\mathbf{W}_{t,edges}$ and $^{\mathcal{O}}\mathbf{W}_{t,verts}$, each represent three lines parallel to the edges of the workpiece specified by the 3A CF. The lines corresponding to $^{\mathcal{W}}\mathbf{W}_{t,edges}$ are those supporting the actual edges of the workpiece, while the lines corresponding to $^{\mathcal{O}}\mathbf{W}_{t,verts}$ are those that contain the designated vertices (which, given the value of $\mathbf{r}$, may not lie on the edges specified by the 3A CF). Again, the condition that $Det(^{\mathcal{O}}\mathbf{W}_{t,verts}) = 0$ or $Det(^{\mathcal{W}}\mathbf{W}_{t,edges}) = 0$ is that the three lines involved intersect at a point (possibly at infinity).

3.3 A Nonintuitive Nongeneric Contact Situation

Recall that nongeneric situations are those which satisfy $G = 0$ and $H = 0$, or which satisfy inequality (33) by strict equality. Thus we can view the equation $G^2 + H^2 - I^2 = 0$ as a hypersurface in the manipulator configuration space where nongenericity occurs.

Proposition 4: *If elimination succeeds, then for a fixed* $\mathbf{r}$*, we will have an infinite number of solutions on the 3A CF-cell if and only if* $Det(\mathbf{W}_n) = Det(^{\mathcal{O}}\mathbf{W}_{t,verts}) = Det(^{\mathcal{W}}\mathbf{W}_{t,edges}) = 0$*, or equivalently,* $G = H = I = 0$*.*

Proof: Our elimination procedure reduced the search for geometrically admissible configurations of the workpiece for a given $\mathbf{r}$ to finding the intersections between a line and a circle. When elimination succeeds and $G = H = I = 0$, we are left with only the circle. Given

the relationships between G, H, and I and the wrench matrices, it is clear that the determinants of all three wrench matrices must be zero and conversely. $\ldots\ldots \square$

One can easily construct examples meeting the conditions of Proposition 4. First, construct a workpiece polygon that has three edges which intersect at a point. Second, place two manipulator vertices anywhere on two of those three edges. Third, find the intersection point of the two contact normals and project it perpendicular to the third edge to locate the third manipulator vertex. Figure 3 shows a workpiece (light gray) in several geometrically admissible configurations for a fixed configuration of the manipulator (dark gray). Note that every possible orientation of the workpiece corresponds to a point on the unit circle, equation (22), so that for every orientation, there is a workpiece position, (x, y), which achieves the specified CF. This position can be found by back substituting into equations (21). Recall that in geometrically admissible configurations, the contacts may be anywhere on the lines of support of the three edges.

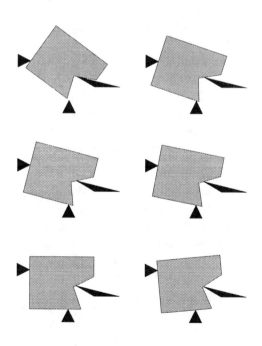

Figure 3: *A nongeneric situation for which elimination succeeds.*

To see the relationship of configurations satisfying Proposition 4 to Cardan mechanisms, recall that a Cardan mechanism is a mechanism which has two sliders traveling in linear slots (fixed to the ground) with a coupler (rigid link) connected to each sliders at a revolute joint [2]. A property of this type of mechanics is that there is a point on the coupler which moves on a line passing through the intersection point of the slots. Now view any two of the three edges of the workpiece as slider slots and the two corresponding contact points as the positions of the revolute joints' axes (which are perpendicular to both slots). Next imagine connecting the sliders by a rigid laminar link, shaped so that it contains the third contact point and mark that point on the coupler. This completes the construction of the Cardan mechanism. Finally, fix the workpiece in the plane and move the coupler. Because two "contacts" are connected to the sliders, they must move along their respective edges (slots), and it just so happens that the third contact point traces the line supporting the edge of the third contact.

The points, $\mathbf{r}$, in the manipulator configuration space, where Proposition 4 is satisfied will be referred to as *exceptional points*. These points "blow up" [19] to circles on the CF-cell. Each direction in manipulator configuration space in a neighborhood of an exceptional point potentially gives rise to a different limit point (in the CF-cell). Thus special care must be taken when planning a manipulation trajectory which pass near or through exceptional points to ensure that the workpiece follows its desired trajectory. Physically, the configuration of the workpiece when $\mathbf{r}$ is exceptional depends not only on the current $\mathbf{r}$, but on the direction from which $\mathbf{r}$ was reached.

We now turn to the CF-cell as a whole.

3.4 Global Properties of 3A CF-cells

Theorem 2: *Every 3A CF-cell is a nine-dimensional manifold.*

Proof: Let $\mathbf{f} = 0$ be the vector of constraint equations given by:

$$\mathbf{f}(\mathbf{p}) = [C_1(\mathbf{p}), C_2(\mathbf{p}), C_3(\mathbf{p}), c^2 + s^2 - 1]^T. \quad (44)$$

For $\mathcal{CF}$ to be a manifold, it is sufficient that at each point in $\mathcal{CF}$, the rank of the Jacobian matrix $\left(\frac{\partial \mathbf{f}}{\partial \mathbf{p}}\right)$ be four [17]. The following matrix is a four-by-four minor

of the Jacobian matrix always has nonzero determinant:

$$\begin{bmatrix} m_{1,\beta} & m_{1,i} & 0 & 0 \\ m_{2,\beta} & 0 & m_{2,j} & 0 \\ m_{3,\beta} & 0 & 0 & m_{3,k} \\ 2\alpha & 0 & 0 & 0 \end{bmatrix} \quad (45)$$

where $(\alpha, \beta) \in \{(c, 3), (s, 4)\}$, $i \in \{5, 8\}$, $j \in \{6, 9\}$, and $k \in \{7, 10\}$. The elements $m_{1,5}$ and $m_{1,8}$ are $cos(\theta + \phi_1)$ and $sin(\theta + \phi_1)$ respectively. Thus it is clear that both $m_{1,5}$ and $m_{1,8}$ cannot simultaneously be zero. An identical argument shows that at least one of $m_{2,6}$ and $m_{2,9}$ must be nonzero, as well as one of $m_{3,7}$ and $m_{3,10}$. Therefore, the Jacobian always has full rank and the $\mathcal{CF}$ is a manifold of dimension nine. $\square$

Asada and By [1] showed that the matrix of partial derivatives of the functions describing the surface of a part (taken with respect to the position and orientation variables of the workpiece) is a normal wrench matrix. In this paper, the orientation of the workpiece is represented by $cos(\theta)$ and $sin(\theta)$, which are treated as independent variables. Thus with our definitions, the four-by-thirteen Jacobian matrix contains a modified normal wrench matrix, $\mathbf{W}_n^*$, which has dimension (4×4). Its rows are the first four columns of that Jacobian matrix:

$$\mathbf{W}_n^* = \begin{bmatrix} m_{1,1} & m_{2,1} & m_{3,1} & 0 \\ m_{1,2} & m_{2,2} & m_{3,2} & 0 \\ m_{1,3} & m_{2,3} & m_{3,3} & 2c \\ m_{1,4} & m_{2,4} & m_{3,4} & 2s \end{bmatrix}. \quad (46)$$

The first two rows of the first three columns of $\mathbf{W}_n^*$ are the same as those in the traditional normal wrench matrix (expressed in frame $\mathcal{O}$). They represent the x- and y-components of the three contact normals. The last two rows of the first three columns are the moments (about the origin of frame $\mathcal{W}$) of the x- and y-components of the normal vectors at the contact points, respectively.

Using the chain rule relationship, $\frac{\partial f_i(c,s)}{\partial \theta} = -\frac{\partial f_i}{\partial c}s + \frac{\partial f_i}{\partial s}c$, we can derive the following simple relationship between $\mathbf{W}_n^*$ and the usual normal wrench matrix $\mathbf{W}_n$:

$$\begin{bmatrix} 1 & 0 & 0 & 0 \\ 0 & 1 & 0 & 0 \\ 0 & 0 & -s & c \\ 0 & 0 & -c & -s \end{bmatrix} \mathbf{W}_n^* = \begin{bmatrix} & & & 0 \\ & \mathbf{W}_n & & 0 \\ & & & 0 \\ (\cdot) & (\cdot) & (\cdot) & -2 \end{bmatrix} \quad (47)$$

¿From this relationship it is clear that the rank deficiency of $\mathbf{W}_n$ and $\mathbf{W}_n^*$ are the same. This fact will be important when planning manipulation tasks symbolically using only algebraic relationships.

Theorem 3: *For a 3A CF, a point $\mathbf{p} \in \mathcal{CF}$, mapping to a point $\mathbf{r}$ in manipulator configuration space, will be a branch point of the mapping of multiplicity two if and only if $G^2 + H^2 \neq 0$ and $\mathbf{W}_n$ is singular. In that case, the configuration will be the only geometrically admissible configuration. (This is equivalent to $G^2 + H^2 \neq 0$ and $G^2 + H^2 - I^2 = 0$.)*

Proof: We have shown that $Det(\mathbf{W}_n) = cH - sG$. But notice that $cH - cG = 0$ is the equation of a line through the origin in the c-s plane which is perpendicular to the line defined by $Gc + Hs + I = 0$. Since the set of C-functions reduces to the following system of equations:

$$Gc + Hs + I = 0 \qquad (48)$$
$$c^2 + s^2 - 1 = 0, \qquad (49)$$

the above system will have a unique solution if and only if the line represented by equation (48) is tangent to the circle represented by the equation (49). In that case, the line defined by $-Gs + Hc = 0$ is the one that contains the point of tangency. This implies that the determinant of the normal wrench matrix is zero.

Conversely, if the determinant of the normal wrench matrix is zero, then the line (48) is forced to be tangent to the unit circle. .... □

Situations that satisfy Theorem 3 are those for which the contact normals intersect at a point, but the edges designated for contact do not. For example, imagine placing three contacts at the midpoints of the three sides of an equilateral triangle. The contact normals intersect at the center of the triangle. This situation is nongeneric, because moving any one of the vertices straight through its edge and into the triangle even slightly results in no geometrically admissible workpiece configurations. Similarly, moving one vertex out from its edge results in two admissible configurations.

Corollary 1: *The branch locus of a 3A CF-cell under projection to the manipulator configuration space is precisely the locus where $G^2 + H^2 \neq 0$ and $\mathbf{W}_n$ is singular. By contrast, when $\mathbf{W}_n$ is nonsingular at $\mathbf{p} \in \mathcal{CF}$,*

we are forced to have $G^2 + H^2 \neq 0$, and the projection will be a local diffeomorphism.

Proof: The singularity of the normal wrench matrix at points where $G^2 + H^2 \neq 0$ indicates that the projection from the $\mathcal{CF}$-cell to the manipulator configuration space is not a local diffeomorphism. Since the CF-cell itself is a smooth manifold, the singularity of the wrench matrix indicates that the two-sheeted cover branches. .... □

In summary, every 3A CF-cell, $\mathcal{CF}$, is a smooth "surface" sitting over the space of manipulator configurations. For most values of $\mathbf{r}$, there are zero or two points on the $\mathcal{CF}$-cell corresponding to the geometrically admissible workpiece configurations. However, there are a "few" special manipulator configurations for which there is 1 or an infinite number of workpiece configurations. Despite these special configurations, every 3A CF-cell is smooth everywhere. Over the regions of manipulator configuration space where there is a finite number of workpiece configurations, the CF-cell can be viewed as two sheets which occasionally come together smoothly. When generating manipulation plans (*i.e.*, trajectories in $\mathbf{r}$) that pass near or through the points corresponding to an infinite number of workpiece configurations, extra care must be taken to ensure that the workpiece moves as desired.

3.5   3B CF-Cells

Suppose we are given a contact formation with three type B contacts. Consider Figure 4. Let (w_l, z_l) denote the position, with respect to the frame $\mathcal{W}$, of the vertex on the workpiece which is to be contacted by the designated edge of manipulator polygon l. Select a point on edge l, and let (u_l, v_l) denote its position with respect to the frame $\mathcal{P}_l$. Let $\mathbf{g}_l$, $\mathbf{h}_l$, and $\mathbf{v}_l$ be respectively, the position of vertex l, the position of the selected point (u_l, v_l) on edge l, and the unit outward normal of edge l, all expressed with respect to the world frame $\mathcal{O}$. The C-function for a single type B contact is then given as in equation (5):

$$C_l(\mathbf{p}) = \mathbf{v}_l \cdot (\mathbf{g}_l - \mathbf{h}_l). \qquad (50)$$

To make this explicit, let ζ_l denote the angle between $\mathbf{v}_l$ and the x-axis of frame $\mathcal{P}_l$. Then the C-functions

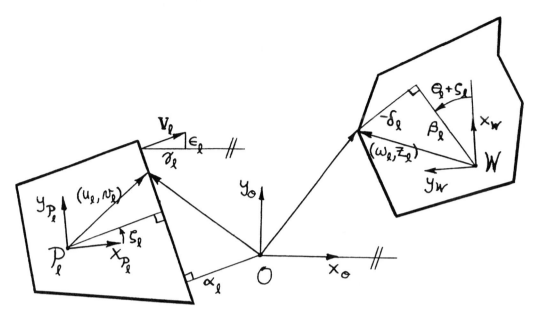

Figure 4: Illustration of parameters relevant to type B contacts.

for the three type B contacts can be expanded to yield:

$$C_l(\mathbf{p}) = \alpha_l + \beta_l c + \delta_l s + \gamma_l x + \epsilon_l y = 0$$
$$\text{for all } l = 1, 2, 3 \quad (51)$$
$$c^2 + s^2 - 1 = 0 \quad (52)$$

where the coefficients are given by:

$$
\begin{aligned}
\alpha_l &= -cos(\zeta_l + \theta_l)x_l - sin(\zeta_l + \theta_l)y_l \\
&\quad -cos(\zeta_l)u_l - sin(\zeta_l)v_l & (53) \\
\beta_l &= cos(\zeta_l + \theta_l)w_l + sin(\zeta_l + \theta_l)z_l & (54) \\
\delta_l &= sin(\zeta_l + \theta_l)w_l - cos(\zeta_l + \theta_l)z_l & (55) \\
\gamma_l &= cos(\theta_l + \zeta_l) & (56) \\
\epsilon_l &= sin(\theta_l + \zeta_l) & (57)
\end{aligned}
$$

Proceeding as we did in our study of 3A CF-cells, we again find (assuming elimination succeeds) that the feasible workpiece configurations correspond to the intersection of a line and the unit circle:

$$
\begin{aligned}
G(\mathbf{r})c + H(\mathbf{r})s + I(\mathbf{r}) &= 0 & (58) \\
c^2 + s^2 - 1 &= 0 & (59)
\end{aligned}
$$

where

$$
G = det \begin{bmatrix} \beta_1 & \gamma_1 & \epsilon_1 \\ \beta_2 & \gamma_2 & \epsilon_2 \\ \beta_3 & \gamma_3 & \epsilon_3 \end{bmatrix}, \quad
H = det \begin{bmatrix} \delta_1 & \gamma_1 & \epsilon_1 \\ \delta_2 & \gamma_2 & \epsilon_2 \\ \delta_3 & \gamma_3 & \epsilon_3 \end{bmatrix},
$$

and $I = det \begin{bmatrix} \alpha_1 & \gamma_1 & \epsilon_1 \\ \alpha_2 & \gamma_2 & \epsilon_2 \\ \alpha_3 & \gamma_3 & \epsilon_3 \end{bmatrix}.$

Not surprisingly, results similar to those obtained for 3A CF-cells can be proved for 3B CF-cells. It will become clear in the propositions and theorems below that the vertices and edges in the 3B case are simply the duals of the edges and vertices, respectively, in the 3A case. This duality can be explained in terms of the natural duality between lines and points in the projective plane [15]. Under this duality, the workpiece (viewed as a set of lines) gets mapped to a "dual" workpiece (given as a set of vertices) and a manipulator polygon (viewed as a set of vertices) gets mapped to a "dual" manipulator polygon (given as a set of lines).

Because the proofs for the propositions and theorems below exactly follow the logic of the proofs offered for the 3A CF-cell, they will not be given explicitly.

Proposition 5: *For a 3B CF, elimination fails, i.e., the rank of* $\begin{bmatrix} \gamma_1 & \epsilon_1 \\ \gamma_2 & \epsilon_2 \\ \gamma_3 & \epsilon_3 \end{bmatrix}$ *is less than two, if and only if the lines supporting the edges designated for contact are parallel. (This includes the possibility that two or three of the lines are coincident.)*

Proposition 6: *If elimination fails, then $G = H = I = 0$ and for a fixed $\mathbf{r}$, if there is any workpiece configuration which attains the 3B CF, then there will be an infinite number.*

Once again, the functions G, H, and I are related to the determinants of the wrench matrices and $G^2 + H^2$ and I are frame invariant, so Proposition 3 applies to the 3B CF-cell.

Proposition 7: *The quantities I^2 and $G^2 + H^2$, and the quantities $Det(^W\mathbf{W}_{t,edges})$, $Det(^O\mathbf{W}_n)$, and $Det(^O\mathbf{W}_{t,verts})$ are independent of the choice of all coordinate frames.*

The relationships among G, H, I, and the wrench matrices are:

$$Det(^O\mathbf{W}_n) = cH - sG \qquad (60)$$

$$Det(^O\mathbf{W}_{t,verts}) = sH + cG \qquad (61)$$

$$Det(^W\mathbf{W}_{t,edges}) = -I \qquad (62)$$

where

$$^O\mathbf{W}_n = \begin{bmatrix} \gamma_1 & \gamma_2 & \gamma_3 \\ \epsilon_1 & \epsilon_2 & \epsilon_3 \\ w_1 & w_2 & w_3 \end{bmatrix} \qquad (63)$$

$$^O\mathbf{W}_{t,verts} = \begin{bmatrix} -\epsilon_1 & -\epsilon_2 & -\epsilon_3 \\ \gamma_1 & \gamma_2 & \gamma_3 \\ w_4 & w_5 & w_6 \end{bmatrix} \qquad (64)$$

$$^W\mathbf{W}_{t,edges} = \begin{bmatrix} -\epsilon_1 & -\epsilon_2 & -\epsilon_3 \\ \gamma_1 & \gamma_2 & \gamma_3 \\ -\alpha_1 & -\alpha_2 & -\alpha_3 \end{bmatrix} \qquad (65)$$

with

$$w_1 = \delta_1 c - \beta_1 s + x\epsilon_1 - y\gamma_1 \qquad (66)$$

$$w_2 = \delta_2 c - \beta_2 s + x\epsilon_2 - y\gamma_2 \qquad (67)$$

$$w_3 = \delta_3 c - \beta_3 s + x\epsilon_3 - y\gamma_3 \qquad (68)$$

$$w_4 = \delta_1 s + \beta_1 c + y\epsilon_1 + x\gamma_1 \qquad (69)$$

$$w_5 = \delta_2 s + \beta_2 c + y\epsilon_2 + x\gamma_2 \qquad (70)$$

$$w_6 = \delta_3 s + \beta_3 c + y\epsilon_3 + x\gamma_3 \qquad (71)$$

Proposition 8: *If elimination succeeds, then for a fixed $\mathbf{r}$, we will have an infinite number of solutions on the 3B CF-cell if and only if $Det(^O\mathbf{W}_n) = Det(^O\mathbf{W}_{t,verts}) = Det(^W\mathbf{W}_{t,edges}) = 0$, or equivalently, $G = H = I = 0$.*

In cases when elimination succeeds, the nongeneric situations are dual to those for the 3A case. To highlight this fact, note that the nongeneric case shown in Figure 3 is nongeneric for the 3B type CF if one views the (dark) manipulator polygons as rigidly connected to become the workpiece and one cuts up the (light) workpiece polygon into disconnected pieces, each containing one of the contact edges, to become three manipulator polygons.

Theorem 4: *For a generic positioning, $\mathbf{r}$, of the manipulator, we will have either zero or two geometrically admissible workpiece configurations.*

Theorem 5: *Every 3B CF-cell is a nine-dimensional manifold.*

Theorem 6: *For a 3B CF, a point $\mathbf{p} \in \mathcal{CF}$, mapping to a point $\mathbf{r}$ in manipulator configuration space, will be a branch point of the mapping of multiplicity two if and only if $G^2 + H^2 \neq 0$ and $\mathbf{W}_n$ is singular. In that case, the configuration will be the only geometrically admissible configuration. (This is equivalent to $G^2 + H^2 \neq 0$ and $G^2 + H^2 - I^2 = 0$.)*

A nongeneric situation satisfying Theorem 6 is as follows. Let the workpiece be an equilateral triangle and let the edges of three manipulator polygons touch the three vertices such that the contact normals bisect the angles of the triangle.

Corollary 2: *The branch locus of the 3B CF-cell under projection to the manipulator configuration space is precisely the locus where $G^2 + H^2 \neq 0$ and $\mathbf{W}_n$ is singular.*

3.6 2AB CF-Cells

2AB CF-cells are characteristically different from 3A and 3B CF-cells, because finding the geometrically admissible configurations of the workpiece boils down to determining the intersections between a circle and a general quadric in c and s. Nonetheless, most of the steps followed in analyzing the 3A case apply directly to the 2AB case. Thus in this section, we will only include the steps that are different.

Building on the results of our analysis of the 3A and 3B CF-cells given above, the C-functions in the 2AB case are given by:

$$a_l + b_l c + d_l s - e_l xc + f_l xs$$
$$-f_l yc - e_l ys \;=\; 0 \qquad (72)$$
$$\alpha_l + \beta_3 c + \delta_3 s + \gamma_3 x + \epsilon_3 y \;=\; 0 \qquad (73)$$
$$c^2 + s^2 - 1 \;=\; 0 \qquad (74)$$

where l takes on the values 1 and 2. Eliminating x and y yields:

$$Jc + Ks + Lc^2 + Ms^2 + Ncs \;=\; 0 \qquad (75)$$
$$c^2 + s^2 - 1 \;=\; 0 \qquad (76)$$

Equation (75) can be rewritten as:

$$Det \begin{bmatrix} a_1 + b_1 c + d_1 s & -e_1 c + f_1 s & -f_1 c - e_1 s \\ a_2 + b_2 c + d_2 s & -e_2 c + f_2 s & -f_2 c - e_2 s \\ \alpha_3 + \beta_3 c + \delta_3 s & \gamma_3 & \epsilon_3 \end{bmatrix} = 0 \qquad (77)$$

and elimination will fail if and only if the rank of
$$\begin{bmatrix} -e_1 c + f_1 s & -f_1 c - e_1 s \\ -e_2 c + f_2 s & -f_2 c - e_2 s \\ \gamma_3 & \epsilon_3 \end{bmatrix}$$
is less than two. Note that the rows of this (2×3) matrix are unit normals to the edges written with respect to the frame $\mathcal{O}$.

Proposition 9: *For a 2AB CF, elimination fails if and only if the three designated edges are mutually parallel.*

Proposition 10: *If elimination fails for a fixed $\mathbf{r}$, then if there is any workpiece configuration which attains the 2AB CF, there will be an infinite number of workpiece configurations which attain the 2AB CF.*

Proposition 11: *If elimination succeeds, then we will have an infinite number of solutions for the 2AB CF if the two designated vertices of the manipulator contact a single edge of the workpiece at the same point. The designated edge of the manipulator can contact any vertex of the workpiece.*

Proof: If the two vertices of the manipulator are designated to coincide and contact the same edge, then one is redundant and may be removed. Thus we are left with two contacts. Using the two C-functions, we can solve for x and y as functions of c and s. Therefore, any c and s on the unit circle correspond to a valid (x, y) pair. ....□

The above proposition represents a sufficient condition for the existence of an infinite number of geometrically admissible configurations, but it *may* not be necessary. One might think that when considering all three contacts, the necessary and sufficient condition would be as before; the determinants of the normal and tangential wrench matrices are zero. This is not true. The 2AB configuration shown in Figure 5 has zero normal and tangential wrench matrix determinants, but this example has only two geometrically admissible workpiece configurations. The two admissible configurations are shown in solid lines. Clearly, for the orientation shown in dashed lines, the workpiece cannot satisfy the 2AB CF.

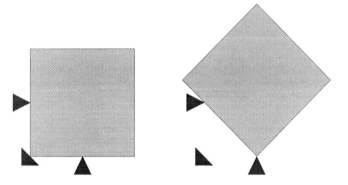

Figure 5: *Counter-Example that $Det(\mathbf{W}_n) = Det(\mathbf{W}_{t,edges}) = Det(\mathbf{W}_{t,edges}) = 0$ is a sufficient condition for an infinite number of geometrically admissible workpiece configurations.*

Theorem 7: *For a generic positioning $\mathbf{r}$ of the manipulator, we will have zero, two, or four geometrically admissible workpiece configurations.*

Proof: A valid configuration must satisfy equations (75) and (76) each of which has degree 2. By Bezout's Theorem [9], the number of intersections is 4 counting points at infinity, complex intersections, and multiplicities. For the case where all intersections are real, finite and of multiplicity one, the number of configurations is precisely 4. This is the maximum possible number of geometrically admissible configurations. The other cases occur when pairs of distinct solutions are complex. ....□

Theorem 8: *Every 2AB CF-cell is a nine-dimensional manifold.*

Theorem 9: *For a 2AB CF, a point* $\mathbf{p} \in \mathcal{CF}$, *mapping to a point* $\mathbf{r}$ *in manipulator configuration space, will be a branch point of multiplicity greater than one if and only if the normal wrench matrix* $\mathbf{W}_n$ *is singular.*

Proof: The condition that the normal wrench matrix is singular is given by:

$$-Js + Kc - Ns^2 + Nc^2 + 2(M - L)cs = 0. \quad (78)$$

We assume that elimination does not fail and that equation (75) is non-trivial.

Recall that the valid solutions are given by the intersection of the conic (75) and the unit circle (76). A point of intersection of these two conics will have multiplicity greater than one if and only if they are tangent at that point. Let U denote the equation for the unit circle and V denote the equation of the general conic, then the condition for tangency is then given by:

$$-\frac{\frac{\partial U}{\partial c}}{\frac{\partial U}{\partial s}} = -\frac{\frac{\partial V}{\partial c}}{\frac{\partial V}{\partial s}}. \quad (79)$$

which expands to yield:

$$-\frac{K + 2Nc + 2(M - L)s}{-J + 2Ns + 2(M - L)c} = -\frac{2c}{2s}, \quad (80)$$

Equation (80) which can be manipulated to give equation (78), which is precisely the condition that the wrench matrix is singular. .... □

Singularity of the normal wrench matrix does not indicate that we have a unique solution. However, it does indicate that we *don't* have four distinct solutions. Singularity is indicative of the fact that two or more of the sheets of the 2AB CF-cell over the manipulator configuration space come together (branch) and hence that the projection to the manipulator configuration space is not a local diffeomorphism.

3.7   2BA CF-Cells

The 2BA CF-cell is dual to the 2AB CF-cell with results carrying over in the obvious manner.

4   Conclusion

The results of this paper provide an in-depth understanding of the kinematic constraints imposed on a dexterous manipulation system by four fundamental systems of contact constraint equations. The CF-cells represented by these constraint systems are relevant to dexterous manipulation of a single passive workpiece by three position-controlled manipulator polygons in sliding contact with a workpiece polygon. The CF-cells have been found to be manifolds in the system's C-space. This result implies that one can predict the motion of the workpiece using well-established techniques for the integration of differential algebraic systems [18]. Either a dynamic or quasistatic rigid body model can be used.

When friction is sufficient to cause one contact to roll, then the results given here do not apply. However our approach has been applied and yielded similar results for the four fundamental CF-cells corresponding to one rolling and one sliding contact. Interestingly, in this case, elimination always succeeds and an infinite number of configurations achieve the given contact formation only if the rolling and sliding contact points coincide (see [7] for details).

The planning of dexterous manipulation tasks utilizing pure position control entails the generation of trajectories of the controllable variables of the system (*i.e.*, the positions and orientations of the manipulator polygons in contact with the workpiece) that cause the workpiece to follow a desired trajectory. Thus the mapping from the manipulator configuration space to the space of workpiece configurations must be well understood. The results of this paper include analytic formulas which will allow the efficient computation of the workpiece configuration(s) given the manipulator configuration. They also identify the points where a given manipulator configuration blows up to an infinite number of workpiece configurations. While further study of these blow-up points is needed, it is clear that special care must be taken when generating manipulation trajectories that pass near or through them. Another important benefit of our analytical formulas, is that they contain the relevant geometric model parameters and control settings of the manipulator polygons. Thus, models of geometric uncertainty and control error can be incorporated into the solutions to study the effects of uncertainty on the kinematic constraints.

Through our algebro-geometric analysis of the four CF-cells, we have discovered a previously unknown nonintuitive, nongeneric class of contact situations. In these situations, for a fixed configuration of the manipulator, (the lines of support of) three nonparallel edges of the workpiece can maintain contact with three distinct vertices of the manipulator while retaining one degree of freedom of motion. In fact, there is a unique position of the workpiece that achieve the three specified contacts for every orientation of the workpiece.

4.1 Future Work

Our results position us to study the connectivity of CF-cells and the properties of their intersections, so that we can plan dexterous manipulation actions that utilize several different CF's. We plan to apply techniques from deformation theory and differential topology to determine regions in the space of uncertain parameters for which transverse intersection and connectivity are maintained. Such results will allow us to modify nominal manipulation plans generated from a deterministic nominal model of the system to make them robust to variations in uncertain the parameters.

Finally, notice that if our manipulator is a linkage of prismatic and revolute joints, the forward kinematic map will go from the usual joint space to our manipulator configuration space. The usual workspace, which projects onto joint space, will be the fiber product (in the sense of algebraic geometry) of the forward kinematic map and the projection of the CF-cell to our manipulator configuration space. We plan to study this construction making use of the known behavior of the forward kinematic map. The results of this study will be expected to further enhance the efficiency of dexterous manipulation planning.

5 Acknowledgment

This research was partially supported by the National Science Foundation under grant IRI-9304734, NASA Johnson Space Center under the University's Space Automation and Robotics Consortium contract 28920-32525, subcontract 720-5-305-TAM, Amendment 2, and the Texas Advanced Technology Program under grant 999903-267. Any findings, conclusions, or recommendations expressed herein are those of the authors and do not necessarily reflect the views of the funding agencies. An expanded version of this paper has been submitted to the IEEE Transactions of Robotics and Automation for publication and is available as a technical report (TAMU-CS TR 93-048) of the Department of Computer Science, Texas A&M University.

References

[1] H. Asada and A. B. By. Kinematic analysis of workpart fixturing for flexible assembly with automatically reconfigurable fixtures. *IEEE Journal of Robotics and Automation*, RA-1(2):86–94, June 1985.

[2] O. Bottema and B. Roth. *Theoretical Kinematics.* North-Holland, 1979.

[3] R. Brost. personal communication, October 1993.

[4] R. C. Brost. *Analysis and Planning of Planar Manipulation Tasks.* PhD thesis, Carnegie Mellon University School of Computer Science, January 1991.

[5] J. F. Canny. *The Complexity of Robot Motion Planning.* PhD thesis, MIT Department of Electrical Engineering and Computer Science, May 1987.

[6] R. S. Desai and R. A. Volz. Identification and verification of termination conditions in fine motion in presence of sensor errors and geometric uncertainties. In *Proceedings, IEEE International Conference on Robotics and Automation*, pages 800–807, May 1989.

[7] A.O. Farahat, P.F. Stiller, and J.C. Trinkle. On the algebraic geometry of contact formation cells for systems of polygons. In *Proceeding, Workshop on the Algorithmic Foundations of Robotics*, February 1994. This paper is in essence equivalent to the one with the same title submitted to the IEEE Transactions on Robotics and Automation.

[8] K. Y. Goldberg, M. T. Mason, and A. Requisha. Geometric uncertainty in motion planning: Summary report and bibliography. Technical Report IRIS TR 297, Institute for Robotics and Intelligent Systems, University of Southern California, August 1992. NSF-sponsored workshop.

[9] J.H.Harris. *Algebraic Geometry : A First Course.* Springer Verlag, 1992.

[10] J. Kerr. *An Analysis of Multifingered Hands.* PhD thesis, Stanford University Department of Mechanical Engineering, 1984.

[11] K. Kondo. Motion planning with six degrees of freedom by multistrategic bidirectional hueristic free space enumeration. *IEEE Transactions on Robotics and Automation*, 7(3), June 1991.

[12] J.-C. Latombe. *Robot Motion Planning.* Kluwer Academic Publishers, 1991.

[13] T. Lozano-Pérez. Spatial planning: A configuration space approach. *IEEE Transactions on Computers*, C-32(2):108–119, February 1983.

[14] M. Mason. personal communication, October 1993.

[15] M.C.Berger. *Geometry Vols I & II.* Springer Verlag, 1992.

[16] M. McCarthy. personal communication, October 1993.

[17] Barret O'Neil. *Semi-Riemanian Geometry.* Academic Press, 1983.

[18] F. A. Potra and W. C. Rheinboldt. Differential-geometric techniques for solving differential algebraic equations. Technical Report ICMA-89-143, University of Pittsburgh, 1989.

[19] J.G. Semple & L. Roth. *Introduction to Algebraic Geometry.* Claredon Press, Oxford,Great Britian, 1985.

[20] J. Schwartz and M. Sharir. On the piano movers problem I: The case of a two-dimensional rigid body moving amidst polygonal barriers. *Communications of Pure and Applied Mathematics*, 36:345–398, 1983.

[21] J. C. Trinkle. *The Mechanics and Planning of Enveloping Grasps.* PhD thesis, University of Pennsylvania, Department of Systems Engineering, 1987.

[22] J. C. Trinkle and J. J. Hunter. A framework for planning dexterous manipulation. In *Proceedings, IEEE International Conference on Robotics and Automation*, pages 1245–1251, April 1991.

[23] J. C. Trinkle and R. P. Paul. Planning for dextrous manipulation with sliding contacts. *International Journal of Robotics Research*, 9(3):24–48, June 1990.

[24] J.C. Trinkle, A.O. Farahat, and P.F. Stiller. First-order stability cells for frictionless rigid body systems. *IEEE Transactions on Robotics and Automation*. conditionally accepted October 1993.

[25] J.C. Trinkle, A.O. Farahat, and P.F. Stiller. Second-order stability cells for frictionless rigid body systems. Technical Report TAMU-CS TR 93-020, Texas A&M University Department of Computer Science, April 1993.

Arrangements and their Applications in Robotics: Recent Developments

Dan Halperin, *Stanford University, Stanford, CA, USA*

Micha Sharir, *Tel Aviv University, Tel Aviv, Israel and New York University, NY, USA*

We survey a collection of recent combinatorial and algorithmic results in the study of arrangements of low-degree algebraic surface patches in three or higher dimensions. The new results extend known results involving 2-dimensional arrangements, and they almost settle several conjectures posed eight years ago. Arrangements play a central role in the design and analysis of geometric algorithms, and they arise in a variety of seemingly unrelated applications. In this survey we concentrate on the application of the new results to problems involving collision-free motion planning with three degrees of freedom, and visibility over polyhedral terrains.

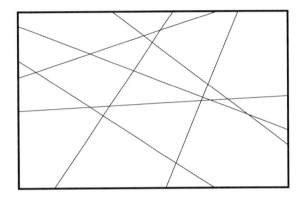

Figure 1: *An arrangement of lines (clipped within a window)*

1   Introduction

Let $\mathcal{L} = \{\ell_1, \ell_2 \ldots, \ell_n\}$ be a given collection of n lines in the plane. We denote by $\mathcal{A}(\mathcal{L})$ the *arrangement* of $\mathcal{L}$, i.e., the decomposition of the plane into vertices, edges, and faces, induced by the lines in $\mathcal{L}$. A vertex of $\mathcal{A}(\mathcal{L})$ is an intersection point of two lines, an edge is a maximal connected relatively open portion of a line that does not meet any vertex, and a face is a maximal connected open region of the plane not meeting any edge or vertex. See Figure 1. Similarly, an arrangement of hyperplanes in d-dimensional space is the subdivision of that space into cells of dimensions $0, 1, \ldots, d$, each being a maximal connected set contained in the intersection of a fixed subcollection of the hyperplanes and not meeting any other hyperplane.

. Arrangements of hyperplanes have been studied extensively in combinatorial and computational geometry [26]. In the preface to his book *Algorithms in Combinatorial Geometry* [26], Edelsbrunner writes: *"These [geometric] transforms led me to believe that arrangements of hyperplanes are at the very heart of computational geometry — and this is my belief now more than ever."*

However, when reasoning about geometric problems that arise in areas such as robotics or computer vision, hyperplanes are often not the most adequate objects to model these problems with. As will be seen below, it is often more appropriate to use objects that are (i) bounded, e.g., segments in the plane or polygons in 3-space; and/or (ii) non-linear, e.g., circular arcs or ellipses in the plane, ruled surfaces in 3-space, etc.

The definition of arrangements can naturally be extended to apply in more general situations, as follows. Let $\Sigma = \{\sigma_1, \ldots, \sigma_n\}$ be a given collection of n low-degree algebraic[1] surface patches in d-space (see Section 3 for a more precise statement of the properties that these surfaces are assumed to satisfy). The *arrangement* $\mathcal{A}(\Sigma)$ of Σ is the decomposition of d-space into (relatively open) cells of various dimensions, each

---

[1] We restrict ourselves to *algebraic* surface patches, of constant maximum degree, because the patterns of interaction among such surfaces are easier to analyze and to compute. Moreover, in practically all applications, the arrangements that arise do indeed involve low-degree algebraic surfaces.

Algorithmic Foundations of Robotics

1-56881-045-8

being a maximal connected set contained in the intersection of a fixed subcollection of Σ and not meeting any other surface of Σ.

Before continuing, let us give a simple example of the use of arrangements of surfaces in robotics. Consider the following simple case of *robot motion planning*. Let B be an arbitrary polygonal object with k sides, moving (translating and rotating) in some open planar polygonal region V bounded by m edges. Any placement of B can be represented by the triple $(x, y, \tan \frac{\theta}{2})$, where (x, y) are the coordinates of some fixed reference point on B, and θ is the orientation of B. We regard these triples as points in 3-space, referred to as the *configuration space* of B. In the motion planning problem, we want to compute the *free* portion of this space, denoted as FP, and consisting of those placements of B at which it is fully contained in V. We note that the boundary of FP consists of placements at which B makes contact with the boundary of V, but does not penetrate into the interior of the complement of V. We can therefore define, within the configuration space of B, a collection of 'contact surfaces', each being either the locus of all placements of B at which some specific corner of B touches some specific edge of V, or the locus of placements at which some side of B touches some vertex of V. Under the above parametrization, it is easily seen that each of the resulting $O(km)$ contact surfaces is a 2-dimensional algebraic surface patch of degree at most 4, and its relative boundary consists of a constant number of algebraic arcs, of constant maximum degree as well.

If B is placed at a free placement $Z \in FP$, and moves continuously from Z, then it remains free as long as the corresponding path traced in configuration space does not hit any contact surface. Moreover, once this path crosses a contact surface, B becomes non-free (assuming, as is customary, *general position* of B and V). It follows that the connected component of FP that contains Z is the cell that contains Z in the arrangement $\mathcal{A}(\Sigma)$ of the contact surfaces, and that the entire FP is the union of a collection of certain cells in this arrangement. Hence, bounding the combinatorial complexity of a single cell in such an arrangement, and designing efficient algorithms for computing such a cell, are natural and major problems in the study of motion planning. As we will see later, these notions can naturally be extended, in more general motion planning problems, to arrangements of other kinds of surfaces

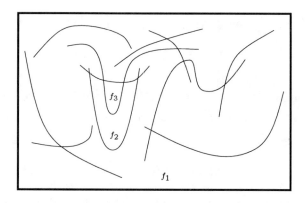

Figure 2: *An arrangement of curves*

and to higher dimensions.

To give the feeling of what makes the general setting more difficult to reason about (than arrangements of hyperplanes), let us consider the following problem. Let $\mathcal{A}$ be an arrangement of curves in the plane, and let π be a point in the plane, not lying on any curve. We wish to bound the *combinatorial complexity* of the face f_π of the arrangement that contains the point π, where this complexity is defined as the number of vertices and edges on the boundary of f_π. A face f in an arrangement of lines (see Figure 1) is clearly convex, as it is the intersection of halfplanes bounded by these lines; therefore, no line can contribute more than one edge to the boundary of f, and hence its complexity is at most $2n$. In contrast, a face in an arrangement of arbitrary curves can have a rather convoluted shape (see face f_1 in Figure 2, which makes up for most of the arrangement; the only other 2-dimensional faces are the small faces f_2 and f_3). Even simply-shaped faces in such an arrangement (such as face f_2 in Figure 2) need not be convex, and a curve may contribute more than one edge to the boundary of such a face. We are thus faced with an interesting and rather challenging problem of estimating the maximum complexity of a single face in general planar arrangements. (As just noted, this is a trivial problem for arrangements of lines; however, it is highly non-trivial even in the case of line segments.)

This and related questions are the topic of our survey. The new developments that we survey, however, concern arrangements in three or higher dimensions. The case of planar arrangements has been studied extensively, and rather satisfactorily, during the past decade, and we will mention the main known results in the background section 2.

The paper is organized as follows. In Section 2 we present the problems that this paper addresses and survey previous work on these problems. We state the basic new results in Section 3. We exemplify the usefulness of these results by applying them to problems involving robot motion planning (Section 4) and visibility and aspect graphs (Section 5). Section 6 deals with new results on Minkowski sums of convex polyhedra in three dimensions, which have applications in robot motion planning and in other related areas. The paper concludes in Section 7, with further applications of the new results and with some open problems.

2 Background and Previous Work

The problem posed at the introduction, which calls for estimating the complexity of a single face in a general planar arrangement, and a collection of related problems involving planar arrangements of curves, were settled, in a fairly complete and satisfactory fashion, more than five years ago. The main tool used in obtaining these results is *Davenport-Schinzel sequences*, a combinatorial tool that has found many applications in combinatorial and computational geometry. It is beyond the scope of this survey to give a detailed review of Davenport-Schinzel sequences, but we briefly present the basic relevant definitions and results, and refer the reader to the survey paper [60] or to the forthcoming book [63]; many of the other papers cited in our survey also make use of Davenport-Schinzel sequences.

Let n and s be positive integers. A *Davenport-Schinzel* sequence of order s composed of n distinct symbols (an (n,s)-DS sequence for short) is a sequence $U = (u_1, u_2, \ldots, u_m)$ that satisfies the following two conditions:

(i) $u_i \neq u_{i+1}$, for all $i < m$.

(ii) There do not exist $s+2$ indices, $1 \leq i_1 < i_2 < \cdots < i_{s+2} \leq m$, such that

$$\begin{aligned} u_{i_1} &= u_{i_3} = \cdots = a \\ u_{i_2} &= u_{i_4} = \cdots = b \,, \end{aligned}$$

and $a \neq b$.

Let $\lambda_s(n)$ denote the maximum length of an (n,s)-DS sequence. It is easy to see that $\lambda_1(n) = n$, and

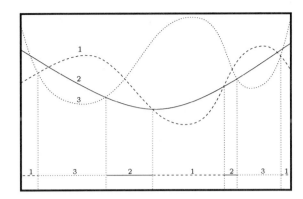

Figure 3: *A lower envelope and its associated sequence "1321231"*

that $\lambda_2(n) = 2n - 1$. Hart and Sharir [39] showed that $\lambda_3(n) = \Theta(n\alpha(n))$, where $\alpha(n)$ is the extremely slowly growing functional inverse of Ackermann's function, which, for all practical purposes, can be regarded as a constant. Still, from a theoretical point of view, $\lambda_3(n)$ grows faster than any linear function of n. The best bounds known to date for $\lambda_s(n)$, with $s > 3$, were obtained by Agarwal et al. [4], who showed that, for any fixed s, $\lambda_s(n)$ is an almost linear, slightly superlinear, function of n.

The connection of Davenport-Schinzel sequences to geometric problems lies in the following easy observation: Let $f_1, \ldots, f_n$ be n real-valued continuous functions defined on the real line, with the property that the graphs of any pair of these functions intersect in at most s points. Let $\psi(x) = \min_i f_i(x)$ denote the *lower envelope* of these functions, and let U be the sequence of function indices, in the order in which they appear along the graph of ψ from left to right; see Figure 3. Then U is an (n,s)-DS sequence. Conversely, any (n,s)-DS sequence U can be realized as the 'lower-envelope sequence' of a collection of n functions, with the above properties. Thus, Davenport-Schinzel sequences arise naturally in geometric (and other) problems involving minimization of a collection of univariate functions.

This observation can be extended to the case where the given functions are only partially defined, so that the graph of each function is a connected x-monotone arc. We state the result in the following theorem, and refer the reader to [39] for more details.

Theorem 1 *The complexity of the lower envelope of n x-monotone curves, having the property that no pair of them intersect in more than s points, is $O(\lambda_{s+2}(n))$. The envelope can be calculated, by a simple divide-and-conquer procedure, in time $O(\lambda_{s+2}(n) \log n)$.*

Remark: A slightly faster algorithm, with running time $O(\lambda_{s+1}(n) \log n)$, has been given by Hershberger [42].

However, the application of these sequences goes much further. Guibas et al. [32] obtained the following result:

Theorem 2 [32] *The complexity of a single face in an arrangement of n Jordan arcs, having the property that no pair of arcs intersect in more than s points, is $O(\lambda_{s+2}(n))$. Such a face can be calculated (deterministically) in time $O(\lambda_{s+2}(n) \log^2 n)$.*

This result, as well as most of the algorithmic results reported in this survey, assumes that the shape of the given arcs is sufficiently simple (e.g., that they are all algebraic of constant maximum degree), and assumes a standard model of computation in *computational geometry*, namely, the algorithms use infinite-precision real numbers and each arithmetic operation (or any other simple operation, such as extracting the real roots of any fixed-degree polynomial) has constant cost. See, e.g., [57, Section 1.4].

Remark: There are also several *randomized* algorithms for computing a single face in such an arrangement [16, 18, 23]. The expected running time of these algorithms, over the internal randomizations that they perform, is $O(\lambda_{s+2}(n) \log n)$.

The maximum possible combinatorial complexity of the entire arrangement of such a collection of arcs is $\Theta(n^2)$. Theorems 1 and 2 thus imply that the maximum complexity of the lower envelope and of any single face in an arrangement of 'well-behaved' curves in the plane are asymptotically equal, and they are both smaller by roughly a factor of n than the complexity of the entire arrangement.

In $d \geq 3$ dimensions, a prevailing conjecture (see, e.g., [53]) is that the maximum complexity of a single cell (or of the lower envelope) in an arrangement $\mathcal{A}(\Sigma)$ of n surface patches, as above, is close to $O(n^{d-1})$, which is again smaller by roughly a factor of n than the maximum complexity of the entire arrangement, which

can be $\Theta(n^d)$ [56].[2] (The complexity of the envelope, or of a cell, is the number of faces of the arrangement, of all dimensions, that appear along the envelope, or along the cell boundary.) A stronger version of the conjecture asserts that the maximum complexity of a single cell (or of the lower envelope) in such an arrangement is $O(n^{d-2}\lambda_s(n))$, where s is some constant that depends on the maximum degree of the given surfaces and of their boundaries.

For the single cell problem, there are only a few special cases in which the known bounds are better than the naive bound $O(n^d)$ (for the complexity of the entire arrangement). These include the cases of

(i) hyperplanes, where the complexity of a single cell, being a convex polyhedron bounded by at most n hyperplanes, is $O(n^{\lfloor d/2 \rfloor})$ (by the Upper Bound Theorem [50]);

(ii) spheres, where an $O(n^{\lceil d/2 \rceil})$ bound is easy to obtain by lifting the spheres into hyperplanes in $(d+1)$-space [26, 58];

(iii) $(d-1)$-simplices, where an $O(n^{d-1} \log n)$ bound has been established in [8]; and

(iv) several special types of surfaces in three dimensions, that arise in motion planning for various specific robot systems B with three degrees of freedom [34].

Similarly, improved bounds on the complexity of lower envelopes (better than the naive $O(n^d)$ bound) were previously obtained only for families of a few types of surfaces or surface patches, such as hyperplanes, balls, simplices, and, in three dimensions, also for a few other types of surfaces (see [53, 58]). The bounds for hyperplanes and for balls are the same as in the case of a single cell, and in fact are considerably smaller than the conjectured bounds. However, these surfaces are 'too simple'; it is easy to produce examples of a collection of n 'well-behaved' surfaces in d-space whose lower

---

[2]In these bounds, and in most of the other bounds mentioned in this paper, we are mainly concerned with the dependence of the quantities in question on the number n of curves or surfaces. In other words, we regard n as a large variable, and regard the dimension d and the maximum algebraic degree b of the surfaces as fixed constants. This is justified in most of the applications. The further dependence of the bounds on d and b is hidden in the constants of proportionality.

envelope has complexity slightly larger than $\Omega(n^{d-1})$. For example, the maximum complexity of the lower envelope of n $(d-1)$-simplices in d-space was shown in [27, 53] to be $\Theta(n^{d-1}\lambda_3(n)) = \Theta(n^{d-1}\alpha(n))$, so the conjecture is fully established for this special case.

In summary, these conjectures have been proved only for a few special cases of arrangements, and they were largely open in the general case stated above. In fact, no bounds better than $O(n^d)$ were known for the general case, even in three dimensions.

The situation has changed considerably over the past couple of years, when, in a series of papers [35, 36, 37, 62], the authors have obtained improved bounds for the complexity of lower envelopes and single cells in three and higher dimensions. These results have almost settled the above conjectures in the case of lower envelopes in any dimension, and in the case of single cells in three dimensions. In this survey, we summarize the recent developments, as reported in these papers, with an emphasis on their applications to robot motion planning and computer vision.

We also survey here another, different though related, problem, where sharp bounds were known in the two-dimensional case, and no non-trivial bounds were known in three (or higher) dimensions. The problem is that of bounding the complexity of the *Minkowski sum* of certain polygonal sets in the plane, or of certain polyhedral sets in 3-space.

Let A and B be two sets in $\mathbb{R}^2$ or in $\mathbb{R}^3$. The *Minkowski sum* (or vector sum) of A and B, denoted $A \oplus B$, is the set $\{a + b \mid a \in A, b \in B\}$. We will also use the notation $A \ominus B = A \oplus (-B) = \{a - b \mid a \in A, b \in B\}$.

Choose a reference point r rigidly attached to B, and suppose that the given placement of B is such that the reference point coincides with the origin. Suppose that B is allowed to translate (without rotating) in the plane or in 3-space, and that we represent each placement of B by the placement of the reference point r. In this case, $A \ominus B$ is the locus of all placements of the reference point for which the corresponding placement of B intersects A. Therefore, the Minkowski sum $A \ominus B$ is a useful construct in (translational) robot motion planning and in related problems [46, 49], and is often referred to as the *C-obstacle* (or *expanded obstacle*) induced by A.

We confine ourselves to Minkowski sums of polygonal sets in the plane, and of polyhedral sets in 3-space. As is well known, the Minkowski sum of polygonal sets is a polygonal set (see [31]), and, similarly, the Minkowski sum of polyhedral sets is a polyhedral set.

One of the main interesting special cases involving Minkowski sums, as above, is the case where B is a convex polygon in the plane, and the set A is the union of a collection of pairwise-disjoint convex polygons, $A_1, A_2, \ldots, A_m$ (it suffices to require that these polygons have pairwise disjoint interiors). We assume that the polygon B has a fixed number of sides, and that all the polygons A_i have a total of n sides. Kedem *et al.* proved the following result:

Theorem 3 [45] *The maximum combinatorial complexity of the boundary of the Minkowski sum $A \ominus B$, for A and B as defined above, is $O(n)$; moreover, this boundary contains at most $6m - 12$ points of intersection between the boundaries of pairs of different sums of the form $A_i \ominus B$. The sum $A \ominus B$ can be computed (deterministically) in $O(n \log^2 n)$ time.*

The three-dimensional variant of this problem has been open since 1986, when the paper [45] appeared. The goal is to bound the combinatorial complexity of the Minkowski sum $A \ominus B$, where B is a convex polyhedron (with a fixed number of faces), and A is the union of k pairwise-disjoint convex polyhedra with a total of n faces. It was conjectured that the complexity of this sum is roughly quadratic in n. A trivial cubic upper bound for this complexity can easily be deduced; however, no subcubic bounds were known, except for the special case where B is a line segment [44, 59]. Recently, a sharp bound for the case where B is a box was obtained by Halperin and Yap [38], and an almost tight bound for the general case was obtained by Aronov and Sharir [10]. The new results and their applications are described in Section 6.

3 Envelopes and Single Cells in Higher Dimensions: New Results

Let $\Sigma = \{\sigma_1, \ldots, \sigma_n\}$ be a given collection of n $(d-1)$-dimensional surfaces or surface patches in d-space. We assume that these surfaces satisfy the following conditions:

(i) Each σ_i is monotone in the $x_1 x_2 \cdots x_{d-1}$-direction (that is, every line parallel to the x_d-axis intersects σ_i in at most one point). Moreover, each σ_i is a portion of a ($(d-1)$-dimensional) algebraic surface of constant maximum degree b (i.e., a set of the form $\{(x_1, \ldots, x_d) \mid P(x_1, \ldots, x_d) = 0\}$, for some polynomial P of degree $\leq b$).

(ii) The projection in the x_d-direction of σ_i onto the hyperplane $x_d = 0$ is a *semi-algebraic set* (see, e.g., [12, 40]) defined in terms of a constant number of $(d-1)$-variate polynomials of constant maximum degree, say, b too.

(iii) The relative interiors of any d of the given surfaces intersect in at most s points, for some constant parameter s (by Bezout's theorem [40] and by Property (iv) below, we always have $s \leq b^d$).

(iv) The surface patches in Σ are in *general position*, meaning that the coefficients of the polynomials defining the surfaces and their boundaries are algebraically independent over the rationals. (This assumption excludes degenerate interactions among the surfaces; for example, in three dimensions, it excludes configurations where four surfaces meet at a point, the boundary of one surface meets a curve of intersection of two other surfaces, two boundary curves of distinct surfaces meet at a point, etc.)

We remark that the somewhat restrictive condition (iv) and the first part of condition (i) are not essential for the analysis. If condition (i) does not hold, we can decompose each surface into a constant number of pieces that are monotone in the $x_1 x_2 \cdots x_{d-1}$-direction by cutting it along the locus of points with x_d-vertical tangency. If condition (iv) does not hold, we can argue, by applying a small random perturbation of the given polynomials, that the complexity of the lower envelope (or of a single cell) in a degenerate arrangement of surfaces is at most proportional to the worst-case complexity of the lower envelope (or of a single cell) in arrangements of surfaces in general position (see [62]). Condition (iii) is stated explicitly, because the bounds given below depend more significantly on the parameter s than on any of the other parameters d and b.

3.1   Lower Envelopes in Higher Dimensions

We extend the definition of a lower envelope given in Section 2 to higher dimensions. The lower envelope E_Σ of Σ is the graph of the (partial) function $x_d = E_\Sigma(x_1, \ldots, x_{d-1})$ that maps each point $(x_1, \ldots, x_{d-1})$ to the smallest x_d-coordinate among those of the points of intersection between the x_d-parallel line through $(x_1, \ldots, x_{d-1})$ and the surfaces in Σ (if no such surface exists, E_Σ is undefined at $(x_1, \ldots, x_{d-1})$). If that point lies on the boundary of one or more surfaces, we take the maximal closed segment contained in the x_d-parallel line through $(x_1, \ldots, x_{d-1})$, whose bottom endpoint lies on the envelope and which does not cross the relative interior of any surface in Σ, and say that the envelope is *attained* at $(x_1, \ldots, x_{d-1})$ by all surfaces that touch that segment; if the envelope point does not lie on any surface boundary, we say that the envelope is attained by (the relative interior of) all surfaces incident to that point. If we project E_Σ onto the hyperplane $H : x_d = 0$, we obtain a decomposition $\mathcal{M} = \mathcal{M}_\Sigma$ of H into connected relatively-open semi-algebraic sets (which we call *cells*), such that each cell c of $\mathcal{M}$ is a maximal connected portion of H over which E_Σ is attained by a fixed combination of the relative interiors of some surfaces and/or the boundaries of other surfaces (by the general position assumption, the number of such surfaces is at most d). The *combinatorial complexity* of E_Σ is defined to be the number of cells (of all dimensions) of $\mathcal{M}$.

The main result for lower envelopes is:

Theorem 4 [35, 62] *The combinatorial complexity of the lower envelope of a collection of n $(d-1)$-dimensional surface patches in d-space that satisfy conditions (i)–(iv), is $O(n^{d-1+\varepsilon})$, for any $\varepsilon > 0$, where the constant of proportionality depends on ε, d, s, and on the maximum degree b and the shape of the given surfaces and of their boundaries.*

The main ingredients of the proof of this theorem appear in the paper [35], which addresses a special three-dimensional case, and exploits the probabilistic analysis technique of Clarkson and Shor (see [20, 61]) for obtaining generalized '($\leq k$)-level' bounds in arrangements. The proof of the general case is given in [62]; it uses induction on d, and extends the random-sampling technique of [35] in several ways, so as to make it apply to the more general situation.

The main idea of the proof is rather simple. We fix some threshold constant parameter k. For each vertex v of the envelope, we choose a curve γ of $\mathcal{A}(\Sigma)$ incident to v, and follow γ from v *away* from the envelope. If we encounter at least k vertices of $\mathcal{A}(\Sigma)$ along γ before reaching the envelope again (or before γ ends), we charge v to the block of the first k such vertices, and observe that the level in $\mathcal{A}(\Sigma)$ of each charged vertex is at most k. Hence the number of envelope vertices v of this kind is proportional to $1/k$ times the number of vertices of $\mathcal{A}(\Sigma)$ at level $\leq k$. The latter number, by the results of Clarkson and Shor [20, 61], is $O(k^d \phi(n/k))$, where $\phi(m)$ denotes the maximum number of vertices of the lower envelope of m surfaces, with the above properties. After bounding the number of envelope vertices for which the above charging scheme does not apply, we obtain a recurrence for $\phi(n)$, which has, roughly, the form $\phi(n) = O(k^{d-1}\phi(n/k))$ plus other 'overhead' terms. This recurrence solves to the asserted bound. See [35, 62] for more details.

In [62], Sharir also presents a randomized algorithm for computing lower envelopes in three dimensions. Alternative randomized algorithms have been given by Boissonnat and Dobrindt [13], and by de Berg, Dobrindt and Schwarzkopf [23]. Perhaps the simplest algorithm for computing lower envelopes in three dimensions is due to Agarwal, Schwarzkopf and Sharir [2]; this algorithm is deterministic and uses a straightforward divide-and-conquer approach. We summarize all these results in the following theorem.

Theorem 5 *The lower envelope of n surface patches in 3-space, satisfying conditions (i)–(iv) above, can be computed in deterministic or in randomized expected time $O(n^{2+\varepsilon})$, for any $\varepsilon > 0$, in an appropriate 'algebraic' model of computation.[3] The constant of proportionality depends on ε and on the maximum degree of the given surfaces.*

In $d > 3$ dimensions, efficient construction of lower envelopes, with time complexity close to $O(n^{d-1})$, is

---

[3]We assume here a model of computation in which primitive operations involving a constant number of the given surfaces can be performed in constant time. Since we have assumed that all the surfaces and their boundaries are algebraic of constant maximum degree, such a model is plausible, using standard machinery from computational real algebraic geometry for performing each operation of this kind in an exact manner, using rational arithmetic, and in constant time (see, e.g., [41]). Alternatively, we can always fall back to the more powerful model of infinite-precision real arithmetic, described earlier.

more problematic to design, for several technical reasons that we will not elaborate here. There have been two recent developments in this direction, both due to Agarwal, Aronov and Sharir [1]:

Theorem 6 *Let Σ be a collection of n surfaces or surface patches in d-space, satisfying conditions (i)–(iv).*

(a) All the vertices, edges, and 2-faces of the lower envelope of Σ can be computed in randomized expected time $O(n^{d-1+\varepsilon})$, for any $\varepsilon > 0$.

(b) In four dimensions, the lower envelope of Σ can be computed in randomized expected time $O(n^{3+\varepsilon})$, for any $\varepsilon > 0$, in the following strong sense: The algorithm produces a data structure, of size $O(n^{3+\varepsilon})$, so that, given any query point (x, y, z) in 3-space, the value of the envelope at (x, y, z), and the function(s) attaining the envelope at this point, can be computed in time $O(\log^2 n)$.

3.2 Single Cells and Zones in Three Dimensions

In this subsection, we consider the arrangement of a collection Σ of n low-degree algebraic surface patches, as above, in three-dimensional space. We are given a point Z not lying on any surface, and define $C = C_Z(\Sigma)$ to be the cell of the arrangement $\mathcal{A}(\Sigma)$ that contains Z.

The *combinatorial complexity* of a cell C is the number of lower-dimensional cells of $\mathcal{A}(\Sigma)$ appearing on the boundary of C. The authors have recently obtained in [36, 37] an almost-tight upper bound for the complexity of a single cell in such an arrangement:

Theorem 7 *The combinatorial complexity of a single cell in an arrangement of n algebraic surface patches in 3-space, satisfying the conditions (i)–(iv), is $O(n^{2+\varepsilon})$, for any $\varepsilon > 0$, where the constant of proportionality depends on ε, s and b.*

This bound is asymptotically the same as the bound for the complexity of lower envelopes in three dimensions, as just mentioned. This result almost establishes the conjecture mentioned above, in three dimensions.

The proof of Theorem 7 adapts the analysis technique of [36], used by the authors to tackle a special case that arises in motion planning (see Section 4 below), which in turn is based on the analysis technique of

[35, 62] for the case of lower envelopes. The lesson one can learn from the analysis in [36] is that, in studying the complexity of a single cell, one needs the following two preliminary results to 'bootstrap' the recurrences appearing in the analysis:

(a) a sharp bound on the number of 'x-extreme' vertices of the cell C (vertices whose x coordinate is smallest or largest in a small connected neighborhood of the vertex within C), and

(b) a sharp bound on the number of vertices bounding 'popular' faces of C (2-faces that are adjacent to C on both 'sides'; see [6, 8, 36], and below).

Bounds on these quantities were obtained in [36], using special properties of the surfaces that arise in the case studied there. The main technical contributions of the study of the general case [37] is a derivation of such bounds in the general algebraic setting assumed above. The bound (a) is obtained using considerations which are related to Morse theory (see e.g. [43]), but are simpler to derive in three dimensions. The bound (b) is obtained by applying the new probabilistic technique of [35, 36, 62] to counting only the vertices of popular faces (this idea is in the spirit of the methodology used in [6, 8]). Once these two bounds are available, the rest of the proof is rather similar to those used in [35, 36, 62], although certain non-trivial adjustments are required.

An interesting application of Theorem 7 is to bound the combinatorial complexity of the *zone* of a surface in an arrangement of other surfaces in 3-space. Specifically, let Σ be a collection of n algebraic surface patches in 3-space, and let σ be another such surface, so that the surfaces in $\Sigma \cup \{\sigma\}$ satisfy conditions (i)–(iv). The zone of σ in $\mathcal{A}(\Sigma)$ is the collection of all cells of $\mathcal{A}(\Sigma)$ that are crossed by σ. The complexity of the zone is the sum of the complexities of all its cells. The following theorem is a simple consequence of Theorem 7, based on an extension of a simple observation used in [28] for the analysis of zones in 2-dimensional arrangements:

Theorem 8 *The combinatorial complexity of the zone of σ in $\mathcal{A}(\Sigma)$ is $O(n^{2+\varepsilon})$, for any $\varepsilon > 0$, where the constant of proportionality depends on ε, s and b.*

4 Motion Planning with Three Degrees of Freedom

One of the main motivations for studying the 'single cell' problem is its applications to *robot motion planning*. Extending the discussion given in the introduction, let B be an arbitrary robot system with k degrees of freedom, moving in some environment V filled with obstacles. Any placement of B can be represented by a point in $\mathbb{R}^k$, whose coordinates are the k parameters controlling the degrees of freedom of B; as above, this space is called the *configuration space* of B. We want to compute the *free* portion of this space, denoted as FP, which consists of those placements of B at which it does not meet any obstacle. As above, the boundary of FP consists of placements at which B makes contact with some obstacle(s), but does not penetrate into any of them. Under reasonable assumptions concerning B and V, we can represent the subset of 'contact placements' of B (including those placements at which B makes contact with an obstacle but may also penetrate into other obstacles) as the union of a collection of a finite number of surface patches, referred to as *contact surfaces*, all algebraic of constant maximum degree, and whose relative boundaries are also algebraic of constant maximum degree.

As already noted, if B is placed at a free placement Z, and moves continuously from Z, then it remains free as long as the corresponding path traced in configuration space does not hit any contact surface. Moreover, once this path crosses a contact surface, B becomes non-free (under appropriate general position assumptions). It follows, as above, that the connected component of FP that contains Z is the cell that contains Z in the arrangement $\mathcal{A}(\Sigma)$ of the contact surfaces. (The entire FP is the union of a collection of certain cells in this arrangement.) Hence, the immediate problems that we face are to bound the combinatorial complexity of a single cell in such an arrangement, and to design an efficient algorithm for its construction.

In view of this discussion, Theorem 7 immediately implies the following corollary:

Theorem 9 *Let B be a given robot with three degrees of freedom, and with a fixed number of boundary features, and suppose that B is free to move in an environment filled with obstacles, whose boundaries have a total of n features. We assume that each contact*

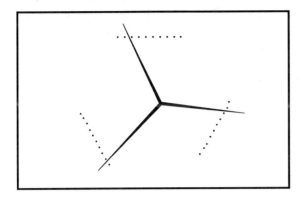

Figure 4: *A rigid polygonal robot moving among n point obstacles in the plane; the free configuration space has $\Theta(n^3)$ connected components.*

surface (representing contacts between some fixed robot feature and some fixed obstacle feature) is an algebraic surface patch of constant degree. Then the combinatorial complexity of a single connected component of the free configuration space of the moving robot is $O(n^{2+\varepsilon})$, where the constant of proportionality depends on ε and on the maximum degree of the contact surfaces and of their boundaries.

This combinatorial result implies that, in the set-up assumed above, the combinatorial complexity of the portion of the configuration space that the robot can reach from a fixed initial placement, is at most near-quadratic (in the complexity of the obstacles). To appreciate this result, we should note that the complexity of the entire free configuration space of such a robot can be $\Theta(n^3)$ (as is illustrated in Figure 4).

However, in order to solve the above motion planning problem, we still need to compute a reasonable representation of such a single cell. So far, we do not have an efficient algorithm (of near-quadratic complexity) for constructing a single cell in general 3-dimensional arrangements. Such efficient algorithms have been obtained only in a few special cases, one of them is that of the arrangement induced by moving (translating and rotating) an arbitrary polygon in a polygonal environment:

Theorem 10 [36] *A single connected component of the free configuration space of a polygon with a constant number of sides, moving in a polygonal region bounded by n edges, can be computed by a randomized*

algorithm with expected running time $O(n^{2+\varepsilon})$, for any $\varepsilon > 0$, with the constant of proportionality depending on ε.

This algorithm is based on random sampling, and is adapted from an algorithm for computing a single cell in an arrangement of triangles, due to Aronov and Sharir [7]. In fact, the same paradigm can be used to compute the single cell in the general setting of Theorem 7. As it turns out, the running time of the resulting algorithm depends on the number of subcells of 'constant description complexity', into which one can decompose the given cell. To obtain algorithms with good running time, using the above paradigm, one needs such a decomposition scheme that does not create too many subcells (in the above set-up, we need decompositions with only a near-quadratic number of subcells). Such decompositions were obtained for arrangements of triangles in 3-space [7, 24] (see also Theorem 18 below), and for the above case of a polygon moving in a polygonal environment [36]. Thus, the algorithmic open problem here—to extend the algorithm to the general case—can be reduced to the combinatorial problem of obtaining sharp upper bounds on the number of subcells of constant description complexity needed in a decomposition of any single cell in an arrangement of surfaces, as above. (Of course, there is always the possibility of obtaining an alternative efficient algorithm that does not depend on the availability of such an economical decomposition.)

5 Visibility Over a Polyhedral Terrain

In this section we study a special case of the so-called *aspect graph* problem, which has recently attracted much attention, especially in the context of three-dimensional scene analysis and object recognition in computer vision. The aspect graph of a scene represents all topologically-different views of the scene. For background and a survey of recent research on aspect graphs, see [14].

Here we will show how the new complexity bounds for lower envelopes (Theorem 4), with some additional machinery, can be used to derive near-tight bounds on the number of views of polyhedral terrains.

5.1   The Envelope of Rays over a Terrain

Let K be a *polyhedral terrain* in 3-space; that is, K is the graph of a continuous piecewise-linear bivariate function, so it intersects each vertical line in exactly one point. Let n denote the number of edges of K. A line ℓ is said to *lie over* K if every point on ℓ lies on or above K. Let $\mathcal{L}_K$ denote the space of all lines that lie over K. (Since lines in 3-space can be parametrized by four real parameters, we can regard $\mathcal{L}_K$ as a subset of 4-space.) The *lower envelope* of $\mathcal{L}_K$ consists of those lines in $\mathcal{L}_K$ that touch at least one edge of K. Assuming general position of the edges of K, a line in $\mathcal{L}_K$ (or any line, for that matter) can touch at most four edges of K. Our goal is to analyze the combinatorial complexity of the lower envelope, but to simplify matters (and with no loss of generality), we only count the number of its *vertices*, namely those lines in $\mathcal{L}_K$ that touch four distinct edges of K. We show:

Theorem 11 [35] *The number of vertices of $\mathcal{L}_K$, as defined above, is $O(n^3 \cdot 2^{c\sqrt{\log n}})$, for some absolute positive constant c.*

We give here a sketch of the proof (see [35] for a detailed proof). We fix an edge e_0 of K, and bound the number of lines of $\mathcal{L}_K$ that touch e_0 and three other edges of K, with the additional proviso that the three other contact points all lie on one fixed side of the vertical plane passing through e_0. We then multiply this bound by the number n of edges, to obtain a bound on the overall number of vertices of $\mathcal{L}_K$. We first rephrase this problem in terms of the lower envelope of a certain collection of surface patches in 3-space, one patch for each edge of K (other than e_0), and then exploit the results of Section 3.

The space $\mathcal{L}_{e_0}$ of oriented lines that touch e_0 is 3-dimensional: each such line ℓ can be specified by a triple (t, k, ζ), where t is the point of contact with e_0 (or, more precisely, the distance of that point from one designated endpoint of e_0), and $k = \tan\theta$, $\zeta = -\cot\phi$, where (θ, ϕ) are the spherical coordinates of the direction of ℓ, that is, θ is the orientation of the xy-projection of ℓ, and ϕ is the angle between ℓ and the positive z-axis.

For each edge $e \neq e_0$ of K, let σ_e be the surface patch in $\mathcal{L}_{e_0}$ consisting of all points (t, k, ζ) representing lines that touch e and are oriented from e_0 to e. Note that

if $(t, k, \zeta) \in \sigma_e$ then $\zeta' > \zeta$ iff the line (t, k, ζ') passes below e. It thus follows that a line ℓ in $\mathcal{L}_{e_0}$ is a vertex of the lower envelope of $\mathcal{L}_K$ if and only if ℓ is a vertex of the lower envelope of the surfaces σ_e in the $tk\zeta$-space, where the height of a point is its ζ-coordinate. It is easy to show that these surfaces satisfy conditions (i)–(iv) of Section 3 for $d = 3$; we omit here the rather straightforward details. Actually, it is easily seen that the number s of intersections of any triple of these surfaces is at most 2. The paper [35] studies the special case of lower envelopes of collections of surface patches in 3-space, which satisfy conditions (i)–(iv), with the extra assumption that $s = 2$. It is shown in [35] that the complexity of the lower envelope of such a collection is $O(n^2 \cdot 2^{c\sqrt{\log n}})$, for some absolute positive constant c, a bound that is slightly better than the general bound stated in Theorem 4. These arguments immediately imply Theorem 11.

Remarks: (1) The bound of Theorem 11 has been independently obtained by Pellegrini [54], using a different proof technique.

(2) Recently, de Berg [22] has given a lower bound construction, in which the lower envelope of $\mathcal{L}_K$ has complexity $\Omega(n^3)$, implying that the upper bound in Theorem 11 is almost tight in the worst case.

We can extend the result of Theorem 11, as follows. Let K be a polyhedral terrain, as above. Let $\mathcal{R}_K$ denote the space of all rays in 3-space with the property that each point on such a ray lies on or above K. We define the lower envelope of $\mathcal{R}_K$ and its vertices in complete analogy to the case of $\mathcal{L}_K$. By inspecting the proof of Theorem 11, one easily verifies that it applies equally well to rays instead of lines. This is because, after fixing an edge e_0, each 'ray-vertex' of $\mathcal{R}_K$ under consideration, when extended into a full line, becomes a 'line-vertex' of $\mathcal{L}_{K^*}$, where K^* is the portion of K cut off by a halfspace bounded by the vertical plane through e_0. Hence we obtain:

Corollary 1 *The number of vertices of $\mathcal{R}_K$, as defined above, is also $O(n^3 \cdot 2^{c\sqrt{\log n}})$.*

This corollary will be needed in the following subsection.

5.2 The Number of Orthographic Views of a Polyhedral Terrain

We next apply Theorem 11 to obtain a bound on the number of topologically-different orthographic views (i.e., views from infinity) of a polyhedral terrain K with n edges.

Following the analysis of [25], each orthographic view of K can be represented as a point on the sphere at infinity $\mathcal{S}^2$. For each triple (e_1, e_2, e_3) of edges of K we consider the locus $\gamma_{(e_1, e_2, e_3)}$ of views for which these three edges appear to be concurrent (that is, there exists a line parallel to the viewing direction which touches these three edges); each such locus is a curve along $\mathcal{S}^2$.

We next replace each curve $\gamma = \gamma_{(e_1, e_2, e_3)}$, as defined above, by its maximal *visible* portions; a point on γ is said to be visible if the corresponding line that touches the three edges e_1, e_2, e_3 either lies over K or else penetrates below K only at points that lie further away from its contacts with the edges e_1, e_2, e_3; in other words, we require the existence of a ray in the direction opposite to the viewing direction that touches e_1, e_2, and e_3, but otherwise lies fully above K. As is easily verified, each visible portion of γ is delimited either at an original endpoint of γ or at a point whose corresponding ray is a vertex of $\mathcal{R}_K$. Hence, by Corollary 1, the total number of the visible portions of all the loci γ is $O(n^3 \cdot 2^{c\sqrt{\log n}})$. We refer to these visible portions as *arcs of visible triple-contact views* (along $\mathcal{S}^2$).

Now, consider the arrangement, along $\mathcal{S}^2$, of the arcs of visible triple-contact views, and observe that the number of views that we seek to bound is proportional to the complexity of the arrangement of these arcs within $\mathcal{S}^2$. This is a consequence of the easy observation that a combinatorial change in the structure of an orthographic view can occur only in directions where either three edges of K appear to be concurrent (and this concurrency is visible), or a vertex of K and an edge of K appear to be coincident; both types of directions appear along arcs of the above arrangement. The view inside each face is combinatorially fixed. Thus we wish to count the number of faces in the arrangement, which is bounded by the complexity of the entire arrangement. We next apply a result of Cole and Sharir [21], which, rephrased in the context under discussion, states that each meridian of $\mathcal{S}^2$

crosses at most $k = O(n\lambda_4(n))$ arcs of visible triple-contact views. As shown in [25], this implies that the complexity of the arrangement of these arcs is $O(Nk)$, where N is the number of arcs. Hence we obtain:

Theorem 12 [25, 35] *The number of topologically-different orthographic views of a polyhedral terrain with n edges is*

$$O(n^4 \lambda_4(n) \cdot 2^{c\sqrt{\log n}}) = O(n^5 \cdot 2^{c'\sqrt{\log n}}),$$

for any constant c' slightly larger than c.

de Berg et al. [25] give lower bound constructions for the number of topologically-different views of a polyhedral terrain:

Theorem 13 [25] *The maximum number of topologically-different orthographic views of a polyhedral terrain, with a total of n edges, can be as large as $\Omega(n^5 \alpha(n))$. The number of topologically-different perspective views (views from any point in space) of such a terrain can be $\Omega(n^8 \alpha(n))$.*

Thus the upper bound in Theorem 12 is almost tight in the worst case. It is also instructive to note that, if K is an arbitrary polyhedral set in 3-space with n edges, then the maximum possible number of topologically-different orthographic views of K is $\Theta(n^6)$ [55].

5.3 The Number of Perspective Views of a Terrain

As it turns out, if we wish to extend the above analysis to the case of perspective views, where the viewpoint can be anywhere in 3-space, we require a bound on the number of vertices of the lower envelope of the space $\mathcal{E}_K$ of all line segments that lie over K, defined in analogy with the definition of the spaces $\mathcal{L}_K$ and $\mathcal{R}_K$. Unfortunately, we do not know how to extend our analysis to obtain non-trivial bounds for the complexity of $\mathcal{E}_K$. Nevertheless, Agarwal and Sharir [3] have recently obtained an almost tight bound on the number of topologically-different perspective views of a polyhedral terrain, using a different approach, where they apply Theorem 4 to the lower envelope of certain collections of surfaces in six-dimensional space. By Theorem 13, the following bound is almost tight in the worst case:

Theorem 14 [3] *The number of topologically-different perspective views of a polyhedral terrain with n edges is $O(n^{8+\varepsilon})$, for any $\varepsilon > 0$.*

Remarks: (1) Agarwal and Sharir [3] also analyze the number of topologically-different orthographic views of a polyhedral terrain, applying Theorem 4 to different collections of functions; they obtain the slightly inferior bound $O(n^{5+\varepsilon})$, for any $\varepsilon > 0$.

(2) If K is an arbitrary polyhedral set with n edges, the maximum possible number of topologically-different perspective views of K is $\Theta(n^9)$ [55].

6   Minkowski Sums of Polyhedral Sets

Minkowski sums of geometric figures, as defined in Section 2, play a central role in various applications of geometric computing, including robot motion planning, layout design, part machining, and assembly planning [48]. We have already explained in Section 2 the connection between Minkowski sums and motion planning applications in robotics. In this section we consider a special instance of this application, involving a convex polyhedral body translating among a collection of convex polyhedral obstacles in 3-space.

As mentioned in Section 2, the planar version of this problem, involving a convex polygon translating among convex polygonal obstacles, has been solved by Kedem et al. [45]. Their results, summarized in Theorem 3, imply that the complexity of the entire free configuration space of the translating polygon is only linear. (If the translating polygon is non-convex, the complexity of the free configuration space can be quadratic in the worst case [64].) This result is clearly stronger than the single-face bound of Theorem 2, since it applies to the complexity of the entire free space. See [59] for more background information on translational motion planning for convex objects.

In three dimensions, the complexity of the free configuration space of a non-convex polyhedron with a fixed number of vertices, translating among polyhedral obstacles with a total of n vertices, can be $\Omega(n^3)$ (see, e.g., [29]). It has long been conjectured that if the moving polyhedron is convex then the complexity of the free space is at most nearly quadratic. In [59], Sharir conjectures that the actual complexity is $O(n^2\alpha(n))$, and mentions that this conjectured bound is the best possible: using Davenport-Schinzel sequences, one can construct a polyhedral setting as above, where the free space has complexity $\Omega(n^2\alpha(n))$.

However, only the trivial upper bound of $O(n^3)$ was known for the complexity of the free space in this case. The only non-trivial result in support of the conjecture involves the case where B is a ladder (line segment). This result is described in [59], and was also independently observed by Ke and O'Rourke [44]. Their bound, $O(n^2)$, is slightly better than the bound in the general conjecture.

Six years after the conjecture had been proposed, Halperin and Yap [38] proved it for the case of a translating box:

Theorem 15 [38] *The combinatorial complexity of the entire free configuration space for a box translating among polyhedral obstacles, with a total of n vertices, is $O(n^2\alpha(n))$.*

The proof of the theorem is based on several rather simple observations that enable to reduce the original problem to that of translating a *triangle* in the same environment. To solve the latter problem, the authors adapt a technique devised by Leven and Sharir [47] for the case of a convex polygon translating and rotating among polygonal obstacles in the plane. This technique, in a tricky way, reduces the problem further to a problem involving lower envelopes of univariate functions.

It is not clear whether this bound for the box is tight. Indeed, there is a lower bound of $\Omega(n^2\alpha(n))$ for the complexity of the entire free configuration space in the case of a general translating convex polyhedron, but the best lower bound known for the case of a box is only $\Omega(n^2)$ [38]. Other works related to the translational motion planning problem in three dimensions are [7, 8].

Recently, Aronov and Sharir have (almost) settled the general case of a convex polyhedron translating among polyhedral obstacles. Let $A_1, \ldots, A_k$ be k convex polyhedra in 3 dimensions with pairwise disjoint interiors, and let B be another convex polyhedron, which, with no loss of generality, is assumed to contain the origin O. Suppose A_i has q_i faces and B has p faces, and put $q = \sum_{i=1}^{k} q_i$. Let $P_i = A_i \ominus B$ be the *Minkowski sum* of A_i and $-B$, for $i = 1, \ldots, k$, and let $U = \bigcup_{i=1}^{k} P_i$ be the union of these so-called *expanded obstacles*. As is well known (and noted in Section 2), the complement C of U represents the *free configuration space FP* of B, under the purely-translational motion allowed for B, in the sense that, for each point

$z \in C$, the placement of B, for which the reference point O lies at z, does not intersect any of the obstacles A_i, and all such free placements are represented in this manner.

As is well known, the complexity of each P_i is at most $O(pq_i)$, so the sum of the complexities of the expanded obstacles is $n = O(pq)$. In practice, though, n can be expected to be much smaller than pq, usually linear in $pk + q$.

The main result of this section is:

Theorem 16 [10] *The combinatorial complexity (i.e., the number of vertices, edges, and faces on the boundary) of the union U, and thus also of FP, is $O(nk \log^2 k) = O(pqk \log^2 k)$. This bound is almost tight, as there are constructions where the union complexity is $\Omega(nk\alpha(k))$. The union can be computed by a randomized algorithm whose expected running time is $O(nk \log^3 k)$.*

This should be compared to the recent bound by Aronov and Sharir [9] on the combinatorial complexity of the union of any k convex polyhedra in 3-space with a total of n faces. It is shown there that the maximum complexity of such a union is $O(k^3 + nk \log^2 k)$ and $\Omega(k^3 + nk\alpha(k))$ in the worst case. Thus Theorem 16 shows that the convex polyhedra P_i arising in translational motion planning have special properties that yield the above improved bound, without the cubic term of the general bound.

The proof of Theorem 16 is fairly involved. It is based on the recent inductive analysis technique of [30], but it also involves a careful analysis of the topological structure of the union.

The motion planning application of Theorem 16 is obvious:

Theorem 17 [10] *The combinatorial complexity of the entire free configuration space for a convex polyhedron with a fixed number of faces, translating among k convex polyhedral obstacles with a total of n faces, is $O(nk \log^2 k)$ (which is almost tight in the worst case), and it can be computed in randomized expected time $O(nk \log^3 k)$.*

We conclude the presentation of new results with the case where the moving polyhedron is non-convex. As mentioned earlier, the complexity of the entire free space in that case can be $\Omega(n^3)$ in the worst case. However, it might be the case that specific instances of the problem do not result in a free configuration space with such a high complexity. For example, if the environment is not too cluttered with obstacles, the complexity of the resulting *entire* arrangement of contact surfaces can be expected to be much smaller than cubic. To take advantage of such settings, we wish to design an *output sensitive* algorithm for computing arrangements of triangles in 3-space.

In the two-dimensional case, one can solve the corresponding problem of computing arrangements of segments (or of more general arcs) in an output-sensitive manner, by using a variant of the Bentley-Ottmann algorithm for detecting intersections between line segments (or arcs) in the plane [11]. Using this procedure, it is possible to compute the *vertical decomposition* of the collection of constraint segments or arcs, induced by the motion planning instance (see, e.g., [52] for this easy extension). Once the vertical (or trapezoidal) decomposition is computed, determining whether any two given placements of the robot can be reached from each other (by a collision-free translational motion) can be easily done in logarithmic time.

Recently, de Berg *et al.* [24] have extended this result to three-dimensions (see [19, 24] for the definition of vertical decomposition in 3-dimensional arrangements):

Theorem 18 *Given a collection T of n triangles in general position in three-dimensional space, one can compute the vertical decomposition of the arrangement $\mathcal{A}(T)$ in time $O(n^2 \log n + V \log n)$, where V is the combinatorial complexity of the vertical decomposition. Moreover, for any $\varepsilon > 0$, we have $V = O(n^{2+\varepsilon} + K)$, where K is the complexity of the arrangement $\mathcal{A}(T)$.*

It should be noted, however, that while the algorithm is sensitive to the size of the arrangement, it does not distinguish between free cells and "forbidden" cells. In other words, the complexity parameters K and V mentioned above might be much larger than the complexity of just the 'free' portions of $\mathcal{A}(T)$ (the same disclaimer applies to the planar version of the problem, mentioned above).

7 Conclusion

In this paper we have surveyed a collection of recent combinatorial and algorithmic results involving arrangements of surfaces in three and higher dimensions. These results extend previous results obtained more than 5 years ago for planar arrangements, and have numerous applications in robot motion planning, visibility problems in 3-space, and many other areas.

Additional applications of these new results are given in [1, 2, 51, 62] (see also the forthcoming book [63]). They include applications to Voronoi diagrams in higher dimensions and to dynamic Voronoi diagrams, which are useful tools in many application areas, and also to various problems involving lines in 3-space, including ray shooting among spheres, and computing the width of a point set in 3-space. The most significant of these developments, which is strongly related to the results reported in Section 6, due to Chew et al. [17], shows that the complexity of the Voronoi diagram of a set of n lines in 3-space, induced by a polyhedral convex distance function, is only $O(n^2\alpha(n)\log n)$.

While the results reported in this survey constitute a considerable advancement over previously known results, and they almost settle several long-standing open problems, a lot still remains to be studied. In particular, most of the new results yield improved combinatorial bounds, but fail to produce equally efficient algorithms. For example, it is still an open problem to design an efficient algorithm (of near-quadratic complexity) for constructing a single cell in an arrangement of algebraic surfaces in 3-space. As mentioned in section 4, this problem will be solved if we can show the existence of a decomposition of the single cell into a near-quadratic number of subcells of constant description complexity. This interesting combinatorial subproblem is also open. A few initial results on the algorithmic problem for a single cell are known in three dimensions [7, 24, 36], but the authors are not aware of any result of this kind in four or higher dimensions. Equally challenging is the problem of designing efficient algorithms for computing lower envelopes in five and higher dimensions.

Another open problem that the paper raises is to further tighten the complexity bounds reported here. For example, the bound for the complexity of lower envelopes in d-dimensional arrangements is conjectured

to be $O(n^{d-2}\lambda_s(n))$. Except for the bounds for hyperplanes and balls, such sharp bounds are known for lower envelopes of simplices in d-space (their complexity is $O(n^{d-1}\alpha(n))$) [27, 53]. For a single cell in 3-dimensional arrangements, a bound of $O(n^2\alpha(n))$ is known for arrangements arising in a certain special motion planning problem [33]. It would be interesting to improve the complexity bounds for lower envelopes, at least to $O(n^{d-1}\text{polylog}(n))$, and to improve the bound for a single cell in three dimensions, at least to $O(n^2\text{polylog}(n))$, as has been done in [8] for the case of simplices. Another open problem is to extend the combinatorial bound for the complexity of a single cell to arrangements in $d > 3$ dimensions. We believe that this is doable, and are currently exploring this problem.

Acknowledgements

Work on this paper by Dan Halperin has been supported by a grant from the Stanford Integrated Manufacturing Association (SIMA), by NSF/ARPA Grant IRI-9306544, and by NSF Grant CCR-9215219. Work on this paper by Micha Sharir has been supported by NSF Grant CCR-91-22103, and by grants from the U.S.-Israeli Binational Science Foundation, the G.I.F., the German-Israeli Foundation for Scientific Research and Development, and the Israel Science Fund administered by the Israeli Academy of Sciences.

References

[1] P.K. Agarwal, B. Aronov, and M. Sharir, Computing envelopes in four dimensions with applications, *Proc. 10th ACM Symposium on Computational Geometry*, 1994, pp. 348–358.

[2] P.K. Agarwal, O. Schwarzkopf and M. Sharir, The overlay of lower envelopes and its applications, Manuscript, 1993.

[3] P.K. Agarwal and M. Sharir, On the number of views of polyhedral terrains, *Proc. 5th Canadian Conf. on Computational Geometry*, 1993, pp. 55–60. (To appear in *Discrete Comput. Geom.*)

[4] P.K. Agarwal, M. Sharir and P. Shor, Sharp upper and lower bounds for the length of general Davenport Schinzel sequences, *J. Combin. Theory, Ser. A.* 52 (1989), 228–274.

[5] B. Aronov, J. Matoušek and M. Sharir, On the sum of squares of cell complexities in hyperplane arrangements, *Proc. 7th Symp. on Computational Geometry*, 1991, pp. 307–313. (To appear in *J. Comb. Theory Ser. A.*)

[6] B. Aronov, M. Pellegrini and M. Sharir, On the zone of a surface in a hyperplane arrangement, *Discrete Comput. Geom.* 9 (1993), 177–186.

[7] B. Aronov and M. Sharir, Triangles in space, or building (and analyzing) castles in the air, *Combinatorica* 10 (1990), 137–173.

[8] B. Aronov and M. Sharir, Castles in the air revisited, *Proc. 8th ACM Symp. on Computational Geometry*, 1992, pp. 146–156. (To appear in *Discrete Comput. Geom.*)

[9] B. Aronov and M. Sharir, The union of convex polyhedra in three dimensions, *Proc. 34th IEEE Symp. on Foundations of Computer Science*, 1993, pp. 518–527.

[10] B. Aronov and M. Sharir, On translational motion planning in 3-space, *Proc. 10th ACM Symposium on Computational Geometry*, 1994, pp. 21–30.

[11] J.L. Bentley and T.A. Ottmann, Algorithms for reporting and counting geometric intersections, *IEEE Transactions on Computers* 28 (1979), 643–647.

[12] J. Bochnak, M. Coste and M-F. Roy, *Géométrie Algébrique Réelle*, Springer-Verlag, Berlin 1987.

[13] J.-D. Boissonnat and K. Dobrindt, On-line randomized construction of the upper envelope of triangles and surface patches in $\mathbb{R}^3$, Manuscript, 1993.

[14] K. W. Bowyer and C. R. Dyer, Aspect graphs: An introduction and survey of recent results, *Int. J. of Imaging Systems and Technology* 2 (1990), 315–328.

[15] B. Chazelle, H. Edelsbrunner, L. Guibas and M. Sharir, A singly-exponential stratification scheme for real semi-algebraic varieties and its applications, *Proc. 16th International Colloquium on Automata, Languages and Programming*, 1989, pp. 179–192. Also in *Theoretical Computer Science* 84 (1991), 77–105.

[16] B. Chazelle, H. Edelsbrunner, L. Guibas, M. Sharir and J. Snoeyink, Computing a face in an arrangement of line segments and related problems, *SIAM J. Computing* 22 (1993), 1286–1302.

[17] L.P. Chew, K. Kedem, M. Sharir and E. Welzl, Voronoi diagrams of lines in 3-space, induced by a polyhedral convex distance function, Manuscript, 1994.

[18] K. Clarkson, Computing a single face in an arrangement of segments, Manuscript, 1990.

[19] K. Clarkson, H. Edelsbrunner, L. Guibas, M. Sharir and E. Welzl, Combinatorial complexity bounds for arrangements of curves and spheres, *Discrete Comput. Geom.* 5 (1990), 99–160.

[20] K. Clarkson and P. Shor, Applications of random sampling in computational geometry, II, *Discrete Comput. Geom.* 4 (1989), 387–421.

[21] R. Cole and M. Sharir, Visibility problems for polyhedral terrains, *J. Symbolic Computation* 7 (1989), 11–30.

[22] M. de Berg, Private communication, 1993.

[23] M. de Berg, K. Dobrindt, and O. Schwarzkopf, On lazy randomized incremental construction, *Proc. 26th ACM Symposium Theory Comput.*, 1994, pp. 105–114.

[24] M. de Berg, L.J. Guibas and D. Halperin, Vertical decompositions for triangles in 3-space, *Proc. 10th ACM Symposium on Computational Geometry*, 1994, pp. 1–10.

[25] M. de Berg, D. Halperin, M. Overmars and M. van Kreveld, Sparse arrangements and the number of views of polyhedral scenes, Manuscript, 1991.

[26] H. Edelsbrunner, *Algorithms in Combinatorial Geometry*, Springer-Verlag, Heidelberg, 1987.

[27] H. Edelsbrunner, The upper envelope of piecewise linear functions: Tight complexity bounds in higher dimensions, *Discrete Comput. Geom.* 4 (1989), 337–343.

[28] H. Edelsbrunner, L.J. Guibas, J. Pach, R. Pollack, R. Seidel and M. Sharir, Arrangements of curves in the plane – Topology, combinatorics, and algorithms, *Theoretical Computer Science* 92 (1992), 319–336.

[29] H. Edelsbrunner, L.J. Guibas and M. Sharir, The upper envelope of piecewise linear functions: algorithms and applications, *Discrete Comput. Geom.* 4 (1989), 311–336.

[30] H. Edelsbrunner, R. Seidel and M. Sharir, On the zone theorem in hyperplane arrangements, *SIAM J. Computing* 22 (1993), 418–429.

[31] L. J. Guibas, L. Ramshaw, and J. Stolfi, A kinetic framework for computational geometry, In *Proc. 24th Annu. IEEE Sympos. Found. Comput. Sci.*, 1983, pp. 100–111.

[32] L.J. Guibas, M. Sharir and S. Sifrony, On the general motion planning problem with two degrees of freedom, *Discrete Comput. Geom.* 4 (1989), 491–521.

[33] D. Halperin, On the complexity of a single cell in certain arrangements of surfaces related to motion planning, *Discrete Comput. Geom.* 11 (1994), 1–33.

[34] D. Halperin, *Algorithmic Motion Planning via Arrangements of Curves and of Surfaces,* Ph.D. Dissertation, Computer Science Department, Tel Aviv University, July 1992.

[35] D. Halperin and M. Sharir, New bounds for lower envelopes in three dimensions, with applications to visibility in terrains, *Proc. 9th ACM Symp. on Computational Geometry*, 1993, pp. 11–18. (To appear in *Discrete Comput. Geom.*)

[36] D. Halperin and M. Sharir, Near-quadratic bounds for the motion planning problem for a polygon in a polygonal environment, *Proc. 34th IEEE Symp. on Foundations of Computer Science*, 1993, pp. 382–391.

[37] D. Halperin and M. Sharir, Almost tight upper bounds for the single cell and zone problems in three dimensions, *Proc. 10th ACM Symposium on Computational Geometry*, 1994, pp. 11–20.

[38] D. Halperin and C.-K. Yap, Combinatorial complexity of translating a box in polyhedral 3-space, *Proc. 9th ACM Symposium on Computational Geometry*, 1993, pp. 29–37.

[39] S. Hart and M. Sharir, Nonlinearity of Davenport-Schinzel sequences and of generalized path compression schemes, *Combinatorica* 6 (1986), 151–177.

[40] R. Hartshorne, *Algebraic Geometry*, Springer-Verlag, New York 1977.

[41] J. Heintz, T. Recio and M.-F. Roy, Algorithms in real algebraic geometry and applications to computational geometry, in *Discrete and Computational Geometry: Papers from the DIMACS Special Year* (J.E. Goodman, R. Pollack and W. Steiger, Eds.), AMS Press, Providence, RI 1991, pp. 137–163.

[42] J. Hershberger, Finding the upper envelope of n line segments in $O(n \log n)$ time, *Information Processing Letters* 33 (1989), 169–174.

[43] M.W. Hirsch, *Differential Topology*, Springer-Verlag, New York 1976.

[44] Y. Ke and J. O'Rourke, An algorithm for moving a ladder in three dimensions, *Technical Report* JHU-87/17, Department of Computer Science, Johns Hopkins University, 1987.

[45] K. Kedem, R. Livne, J. Pach and M. Sharir, On the union of Jordan regions and collision-free translational motion amidst polygonal obstacles, *Discrete Comput. Geom.* 1 (1986), pp. 59–71.

[46] J.-C. Latombe, *Robot Motion Planning*, Kluwer Academic Publishers, Boston, 1991.

[47] D. Leven and M. Sharir, On the number of critical free contacts of a convex polygonal object moving in 2-D polygonal space, *Discrete Comput. Geom.* 2 (1987), 255–270.

[48] T. Lozano-Pérez, Spatial planning: A configuration space approach, *IEEE Trans. Comput.* C-32 (1983), pp. 108–120.

[49] T. Lozano-Perez and M. Wesley, An algorithm for planning collision-free paths among polyhedral obstacles, *Comm. ACM* 22 (1979), 560–570.

[50] P. McMullen, The maximum number of faces of a convex polytope, *Mathematika* 17 (1970), 179–184.

[51] S. Mohaban and M. Sharir, Ray shooting amidst spheres in 3 dimensions and related problems, Manuscript, 1993.

[52] K. Mulmuley, *Computational Geometry: An Introduction Through Randomized Algorithms*, Prentice Hall, New York, 1993.

[53] J. Pach and M. Sharir, The upper envelope of piecewise linear functions and the boundary of a region enclosed by convex plates: Combinatorial analysis, *Discrete Comput. Geom.* 4 (1989), 291–309.

[54] M. Pellegrini, On lines missing polyhedral sets in 3-space, *Proc. 9th ACM Symp. on Computational Geometry*, 1993, pp. 19–28.

[55] H. Plantinga and C. Dyer, Visibility, occlusion, and the aspect graph, *International J. Computer Vision*, 5 (1990), 137–160.

[56] R. Pollack and M.F. Roy, On the number of cells defined by a set of polynomials, *C.R. Acad. Sci. Paris*, t. 316, Série I (1993), 573–577.

[57] F.P. Preparata and M.I. Shamos, *Computational Geometry—An Introduction*, Springer Verlag, New York, 1985.

[58] J.T. Schwartz and M. Sharir, On the two-dimensional Davenport Schinzel problem, *J. Symbolic Computation* 10 (1990), 371–393.

[59] M. Sharir, Efficient algorithms for planning purely translational collision-free motion in two and three dimensions, *Proceedings of the IEEE International Conference on Robotics and Automation*, Raleigh, North Carolina, 1987, pp. 1326–1330.

[60] M. Sharir, Davenport-Schinzel sequences and their geometric applications, *Theoretical Foundations of Computer Graphics and CAD*, (R.A. Earnshaw, Ed.), NATO ASI Series, F40, Springer Verlag, Berlin, 1988, pp. 253–278.

[61] M. Sharir, On k-sets in arrangements of curves and surfaces, *Discrete Comput. Geom.* 6 (1991), 593–613.

[62] M. Sharir, Almost tight upper bounds for lower envelopes in higher dimensions, *Proc. 34th IEEE Symp. on Foundations of Computer Science*, 1993, pp. 498–507. (To appear in *Discrete Comput. Geom.*)

[63] M. Sharir and P.K. Agarwal, *Davenport Schinzel Sequences and Their Geometric Applications*, Cambridge University Press, to appear.

[64] S. Sifrony and M. Sharir, A new efficient motion planning algorithm for a rod moving in two-dimensional polygonal space, *Algorithmica* 2 (1987), 367–402.

A Toolkit for Non-linear Algebra

John Canny
University of California
Berkeley, CA, 94720

This paper describes an implementation, now in progress, of a toolkit for polynomial algebra. The toolkit is being written in C, and it can solve systems of equations over the complex numbers, and inequalities over the reals. There are many applications of such a toolkit, since problems from many branches of science and engineering can be formulated using systems of polynomial equations and inequalities. By several well-known theorems, in particular Tarski's on the decidability of the theory of the reals, it is possible to solve such problems in principle. But the worst case bounds are exponential or doubly-exponential in the number of variables, and no practical system has appeared that can deal with large problems. On the other hand, systems of polynomials may have a special structure, a kind of sparseness, that implies a complexity (measured by the algebraic degree) much lower than the worst case. Indeed, most of the applications we have studied are of this type. Very recently, algorithms have been developed by the author and others that can exploit this structure. This realization is a strong motivation for developing the toolkit at this time. In addition, there have been improvements in algorithms for sign determination and symbolic-numeric computation, and we feel that these methods are now advanced enough to warrant implementation.

1 Introduction

In this paper, we describe a collection of techniques which improve the efficiency of solving systems of polynomial equations and inequalities. Together, these techniques are being implemented in an algebraic-geometric toolkit, written in C. The system computes all the solutions, and provides either a symbolic description of them, or numerical approximations to real or complex solutions. The impetus for the develop-

ment of the toolkit was the realization that (a) Many practical algebraic problems that seem hard are in fact easy, i.e. have low effective degree (b) Algebraic algorithms have improved over the last few years to the point where we can achieve an overall complexity which is low-order polynomial in this effective degree.

The equation-solving problems we are most interested in come from robotics, computer vision, and geometric modeling. The equations are often polynomial because they arise from:

- Representing orientation of objects in 3D. Both quaternions and rotation matrix coefficients are polynomial descriptions.

- The most popular descriptions of smooth shapes are as tensor product surfaces, or as CSG models, both of which are algebraic surfaces.

- Geometric constraints, like contact, colinearity, distance, and lower pairs, are algebraic.

The theory we use to exploit the low effective degree of polynomial systems is called *sparse elimination theory*. In the late 1980's, Gel'fand and his colleagues began the study of discriminants and resultants of sparse polynomial systems. Sparseness leads to a lowering of effective degree, and the sparse theory provides a simple direct method for proving bounds on the number of solutions. Sparseness can also be exploited to speedup equation solving and elimination of variables. Algorithms to do this have appeared in the last year. An efficient homotopy algorithm for sparse systems was described in [HS92]. The first efficient algorithms for the sparse resultant were described by the author and a collaborator in [CE93], [EC93].

Development in symbolic computation on polynomials has a longer history. Since the work of Collins in 1975, and Schwartz and Sharir and Grigor'ev et al., in the early 1980s, there has been steadily increasing interest in algorithms for the first-order theory of the reals. Formulae in this language have real quantified variables, and predicates which are boolean functions of polynomial inequalities. Of particular interest to us are formulae with existential quantification only. Some readers may have seen the term *constraint satisfaction* applied to this family of problems. The theory of the reals is very powerful because it allows declarative description of an object via constraints, and leaves it to the decision algorithm to find an instance of the object.

Much recent work [HRS90], [Ren92] [GV92] has focussed on improving the asymptotic complexity bounds for the theory of the reals, but unfortunately, most of it ignores the complexity in practice. The work of the author [Can88a], [Can91b], [Can91a], [Can93b], [Can93a] on the other hand, has been specifically directed at practical algorithms. The collection of techniques developed in those papers permits symbolic calculation (i.e. exact calculation, even with singular inputs) whose complexity is polynomial in the effective degree mentioned earlier. The main ideas of those techniques will be described in this paper. In addition, we describe a new technique for performing arithmetic operations on very large integers with expected constant cost.

The paper is arranged as follows: The theory of sparse systems is at the heart of both the symbolic and numerical toolkits. So we begin in section 2 with a short introduction to this theory. Then in section 3 we give an overview of the numerical kit. The symbolic kit, which is considerably larger, occupies the bulk of the paper in section 4. Finally, in section 5 we describe some future projects.

2   Sparse Systems

In a moment we will be able to say precisely what we mean by a sparse polynomial system. But even before that, we should answer the question: Why study sparse systems? For us, the reason is that sparse systems are ubiquitous in robotics, vision and modeling. This is not a theorem that we can prove, but our experience shows that almost all the systems that arise there are sparse. We can exploit this sparseness in two ways: (i) To prove tight or tighter bounds for the number of solutions and (ii) To compute those solutions in a time that depends on the sparse bounds, not on the classical

degree bound (Bezout's bound) which is usually much larger.

Typical examples of sparse systems are those that describe the inverse kinematics for a 6R robot [MC92b], forward kinematics for the Stewart platform [Mer92], camera motion from point matches [FM90], and geometric constraints describing two- or three-dimensional objects [Owe91]. As the dimension of the problem increases, the difference between the sparse and nonsparse bounds increases dramatically. Very few algebraic problems with more than 4 variables can be solved with classical resultant methods, which ignore sparseness, but many practical problems which are much larger can be solved fast using a custom elimination formula.

As an example, we can take the 6R inverse kinematics problem from robotics. Given a robot with 6 rotational joints, and a placement of the gripper, this problem asks to find the six joint angles $\theta_1, \ldots, \theta_6$ that place the gripper in the wanted pose. The problem can be stated in terms of 4×4 matrices, which can represent any rigid transformation in 3D. We get

$$T_1(\theta_1) \cdots T_6(\theta_6) = T_e$$

where each T_i represents the transformation between links caused by rotation of joint i, and T_e is the transformation of the gripper. It is better to use a parametrization in terms of $t_i = \tan(\theta_i/2)$ rather than θ_i directly. Both sine and cosine are rational functions of t_i, and the matrix $T_i(t_i)$ then contains only rational functions of t_i. A little algebra shows that the inverse $T_i^{-1}(t_i)$ also contains only rational entries.

Now we have $T_1(t_1) \cdots T_6(t_6) = T_e$. The RHS matrix has 16 entries, so we have a system of 16 equations to be satisfied. But there are only 6 variables, and only 6 of the equations can be independent. Suppose we choose such an independent set. The matrix entries have degree 2 in the t_i's so the total degree of each equation is 12. We can reduce this degree by moving some of the joint transformations to the RHS:

$$T_1(t_1)T_2(t_2)T_3(t_3) = T_e T_6^{-1}(t_6) T_5^{-1}(t_5) T_4^{-1}(t_4) \quad (1)$$

From this we can choose 6 equations of degree 6. By Bezout's theorem, the number of solutions of such a system is $6^6 = 46656$. That is, we might have 46,000 tuples of angles $(\theta_1, \ldots, \theta_6)$ which satisfy the gripper pose constraint. This would be a very difficult problem to solve, if there really were this many solutions. But it has been shown that there are only 16, and [LL88] and [RR89] gave constructive proofs. A real-time solver based on [RR89] is described in [MC92b].

A large gap between the Bezout bound and the actual number of solutions is not unusual for geometric problems, although it is not always as dramatic as for inverse kinematics. Bezout's theorem gives an exact count of the number of solutions in projective space, so most of these solutions are at infinity. The problem with trying to solve a system like this is that most methods have a complexity that depends on the Bezout bound. The Bezout bound is exact if all the coefficients of a polynomial system of some given degree are *generic*. Genericity is the requirement that the coefficients do not satisfy a set of algebraic relations. For example, random coefficients would be generic with high probability. But systems that arise in inverse kinematics and other geometric problems are not generic, even if the robot design parameters were. For inverse kinematics, it is easy to see why. The long series of matrix multiplications leads to many common subexpressions. Each T_i matrix is determined by 4 parameters (called Denavit-Hartenberg parameters [SV89]), and with 6 joints, the whole robot is described by 24 parameters. The gripper has 6 degrees of freedom, so the T_e matrix depends on 6 parameters. Each inverse kinematics system is determined by $24 + 6 = 30$ parameters. But the 6×6 system of equations that we actually solve has hundreds of coefficients, all determined by those 30 parameters. Clearly the coefficients are strongly dependent.

The methods in this paper exploit the relation between solution count and *sparseness of the equations*. A polynomial is sparse if many of its coefficients, compared to a generic polynomial of that degree, are zero. A bound derived from the set of non-zero coefficients is called a Bernstein bound. Bernstein showed that his bound is exact if all the coefficients of the polynomial system are generic [Ber75]. In fact they are exact under much weaker assumptions: Only the coefficients of terms that lie at the vertices of the Newton polytopes (defined later in this section) need to be generic for the bounds to be exact [CR91]. So if we can write down a system in a form where these coefficients are independent, we know how to correctly count the solutions. Using sparse homotopy and resultant methods, we know also how to compute the solutions in a running time that depends on their number.

In contrast with sparse methods, which have appeared in the last year, most equation-solving approaches do not exploit the paucity of solutions, and instead have a complexity that depends on the Bezout bound. This has been true both of homotopy methods [TM85] and general homogenous resultants [Mac02]. Some progress has been made in homotopy methods in the last few years, [MS87] and [VC92]. These methods take advantage of some but not all forms of sparseness. Gröbner basis algorithms have a complexity that depends on the effective degree, and so they work well on systems with few roots. This is one reason they have been considered seriously as a practical equation-solving tool. But they also have high overhead, require arbitrary precision integer arithmetic to work over $\mathbb{R}$ or $\mathbb{C}$, and are difficult to parallelize. When their complexity is measured as a function of the number of solutions, their performance is poor. This has been clear for specific systems for which a sparse resultant was already known. An example is the Dixon resultant [Dix08] for tensor product surface implicitization [MC92a]. The solution using resultants can be computed in 1/100 the time of a Gröbner solution [Hof90].

The value of the special purpose resultants like Dixon's has been clear for some time. Certainly, one would like an analogue for polynomials with any given structure (the set of exponents appearing in the polynomials). These general resultants were defined first by Gel'fond et al., as a special case of an *A-discriminant* [GKZ90]. We will term them *sparse resultants*. A *Poisson formula* for the sparse resultant was given in [PS91]. The Poisson formula expresses the resultant in terms of symmetric functions, and makes it easy to find a sparse resultant's degree.

The most convenient description of a resultant is using a *Sylvester formula*. The Sylvester formula expresses the resultant as the determinant of a matrix whose elements are the polynomial coefficients or zero. Sylvester formulae were given for a class of multihomogeneous systems in [SZ93]. In general, sparse resultants cannot be computed via a Sylvester formula, but they all have a *determinantal formula*, which expresses them as a factor of a matrix determinant. The determinantal formula is important for several reasons. Firstly, it is efficient. The matrix size is small and under reasonable assumptions, polynomial in the sparse resultant degree. Secondly, as we describe in section 3, it allows us to solve non-linear equations using linear algebra tools, such as eigenvalue and characteristic polynomial routines. Thirdly, linear algebra algorithms are easily parallelizable, and we inherit this property when we transform from non-linear to linear with a determinantal formula. The first determinantal formulae for the sparse resultant were given by the author and a collaborator in [CE93] and [EC93]. We will refer to those later in this section.

Sparse equation solving is still a developing field, and sparse methods are not the full answer to exploiting low solution count. For systems with non-generic coefficients, the Bernstein bounds may still be poor. Returning to the inverse kinematics problem, recall that the matrix product gives us 6 equations of degree 6. These polynomials have only 53 non-zero coefficients, whereas a general polynomial of degree 6 in 6 variables would have 1716 coefficients. The Bernstein bound for the kinematics system written in this form is 2,304. Much less than Bezout at 46,000 but still excessive. But the Bernstein bound drops rapidly if we rewrite the equations in a form where there are fewer dependencies. For example, using some of the equations of [RR89] gives a Bernstein bound of 384. When the coefficient relations come from common subexpressions, there is a systematic way to remove them by introducing new variables. An example of this is given later in section 5.1. Since the kinematics equations do contain many common subexpressions, this method should give us a system with Bernstein bound close to 16. We expect to put a lot of future effort into dealing with coefficient dependencies, and some preliminary ideas are given in section 5.1.

2.0.1 Definitions

Suppose we are given m polynomials $f_1, \ldots, f_m$ in $x_1, \ldots, x_n$ with complex coefficients. We use x^e to denote the monomial $x_1^{e_1} \cdots x_n^{e_n}$, where $e = (e_1, \ldots, e_n) \in \mathbb{Z}^n$ is a multi-exponent. Let $\mathcal{A}_i = \{a_{i1}, \ldots, a_{im_i}\} \subseteq \mathbb{Z}^n$ denote the set of exponents occuring in f_i, then

$$f_i = \sum_{j=1}^{m_i} c_{ij} x^{a_{ij}} , \qquad \text{for } i = 1, \ldots, m , \quad (2)$$

and we suppose $c_{ij} \neq 0$ so that $\mathcal{A}_i$ is uniquely defined given f_i. We term the study of such systems *sparse elimination theory* because we consider the actual set of exponents $\mathcal{A}_i$ occuring in f_i rather than just the degree of f_i.

One unusual aspect of the theory of sparse systems is that we specifically discount solutions having a coordinate $x_i = 0$. That is, we count only solutions $x = \xi$ with $\xi \in (\mathbb{C}^*)^n$, where $\mathbb{C}^* = \mathbb{C} - \{0\}$. This point often causes confusion to readers seeing it for the first time. There are two natural questions to ask: (i) Why not count all the affine solutions rather than those in $(\mathbb{C}^*)^n$? (ii) Some genuine solutions having some $\xi_i = 0$ will be missed, how can they be recoved?

In answer to the first question: The most natural space to consider for the solutions of a polynomial system is the projective space $\mathbb{P}\mathbb{C}^n$, in which Bezout's theorem holds exactly. This space has coordinates $(x_0, \ldots, x_n)$ with scalar multiples identified, so $x \equiv \lambda x$ for all $\lambda \in \mathbb{C}^*$. The affine space $\mathbb{C}^n$, which is the one we are interested in in most applications of algebraic geometry, is obtained from $\mathbb{P}\mathbb{C}^n$ by removing the plane at infinity $x_0 = 0$. But this is not a "natural" space and Bezout's theorem appears only in a very weakened form. But if we remove all the coordinate planes $x_i = 0$ from $\mathbb{P}\mathbb{C}^n$, we obtain again a space with an exact degree theorem, this time Bernstein's theorem, which we state later in this section. Because of the removal of zero from the solution space, we can consider the more general case of f_i's which are polynomials in the x_i and their reciprocals, the *Laurent* polynomials $\mathbb{C}[x_1, x_1^{-1}, \ldots, x_n, x_n^{-1}]$.

The second question was how to recover the missing solutions having $x_i = 0$. This is a straightforward extension. We simply set $x_i = 0$ in the system of polynomials f, giving us a new system $f|_{x_i=0}$, and apply Bernstein's theorem to this system. We can do this for each $i = 1, \ldots, n$. Let the total number of roots found this way be N_1. There is a possibility of counting the same root twice, so we must also count roots where both $x_i = 0$ and $x_j = 0$ for each pair i, j. Let this number of roots be N_2. N_i is defined similarly by considering all i-tuples of polynomials. Finally, let N_0 be the number of roots of $f = 0$. Then applying the inclusion/exclusion principle, if N is the total number of roots of the system assuming generic coefficients,

$$N = N_0 + N_1 - N_2 + N_3 + \cdots$$

Note that the dimension of the Newton polytopes is at most $n - 1$ when we set $x_i = 0$. Because of this, unless one of the f_j's is zero after the specialization (meaning that it was divisible by x_i), the specialized system $f|_{x_i=0}$ will be overconstrained. So unless some f_j is divisible by x_i, the system generically will have no roots with $x_i = 0$.

In short, unless some of the polynomials in a system $f = 0$ are divisible by an x_i, all the affine roots will generically have non-zero coordinates. This is the more common situation in practice, and in this case the Bernstein bounds already count all the affine roots.

2.0.2 Newton Polytopes and Bernstein's Theorem

Definition 1 *The finite set $\mathcal{A}_i \subset \mathbb{Z}^n$ of all monomial exponents appearing in f_i is the* support *of f_i. The* Newton Polytope *of f_i is $Q_i = \text{Conv}(\mathcal{A}_i) \subset \mathbb{R}^n$, the convex hull of $\mathcal{A}_i$.*

Polynomials so defined are called *sparse* because we consider a general set of exponents $\mathcal{A}_i$ rather than all exponents of some degree.

Definition 2 *The* Minkowski Sum $A + B$ *of convex polytopes A and B in $\mathbb{R}^n$ is the set*

$$A + B = \{a + b | a \in A, b \in B\} \ .$$

$A + B$ is a convex polytope. Let $\text{Vol}(A)$ denote the usual n-dimensional volume of A:

Definition 3 *Given convex polytopes $A_1, \ldots, A_n \subseteq \mathbb{R}^n$, there is a unique real-valued function $\text{MV}(A_1, \ldots, A_n)$ called the* Mixed Volume *which is multilinear with respect to Minkowski sum, such that $\text{MV}(A_1, \ldots, A_1) = n! \, \text{Vol}(A_1)$. Equivalently, if $\lambda_1, \ldots, \lambda_n$ are scalars, then $\text{MV}(A_1, \ldots, A_n)$ is precisely the coefficient of $\lambda_1 \lambda_2 \cdots \lambda_n$ in $\text{Vol}(\lambda_1 A_1 + \cdots + \lambda_n A_n)$ expanded as a polynomial in $\lambda_1, \ldots, \lambda_n$.*

The Newton polytopes capture the combinatorial properties of the system in a remarkable way. We have the following bound on the number of roots of a system of $m = n$ polynomials in n variables, see [Ber75], [Kus76], [Kho78].

Theorem 1 *(Bernstein's Theorem) Let $f_1, \ldots, f_n \in \mathbb{C}[x_1, x_1^{-1}, \ldots, x_n, x_n^{-1}]$. The number of common zeros in $(\mathbb{C}^*)^n$ is either infinite, or does not exceed $\text{MV}(Q_1, \ldots, Q_n)$. For almost all specialization of the coefficients c_{ij} the number of solutions is exactly $\text{MV}(Q_1, \ldots, Q_n)$.*

2.0.3 Sparse Resultants

For systems of $m = n + 1$ polynomials in n unknowns, there are generically no solutions, and there is an algebraic condition on the coefficients for a solution to exist. That is, a solution exists whenever a certain polynomial in the coefficients of the system is zero. This polynomial is called the *resultant* of the system. We use the term *sparse resultant* to refer to the resultant of a system with particular supports, to distinguish from the term "resultant" which has traditionally meant either the Sylvester resultant or the resultant of a homogeneous system. The sparse resultant is the more general object, and includes the others as special cases.

The simplest and most efficient way to define the resultant is as the determinant of matrix whose entries are the coefficients of the polynomials, or zero. This is a generalization of the classical Sylvester formula for the resultant of two polynomials. In general, the sparse resultant cannot be expressed as the determinant of a single matrix, but as a factor of this determinant. But for our purposes, is just as good to define a matrix whose determinant is a multiple of the resultant, and to discard any extraneous factors.

Once a resultant matrix is defined, it can be used for numerical equation solving or for symbolic variable elimination. For the first use, we construct from the resultant matrix another matrix whose eigenvectors define the solutions of the system [MC93]. This method is based on the use of generalized companion matrices, and provides a particularly simple way to deal with non-linear systems. The resultant matrix is the same for all polynomial systems with a given set of exponents, and can be computed offline. Given this matrix in symbolic form, each particular system can be solved online with only an eigenvalue routine. See section 3.1.

To use a resultant for variable elimination, we repeatedly evaluate the resultant for specializations of the other variables (those not to be eliminated), and use Chinese remaindering and sparse interpolation to reconstruct the answer. This method is applied to surface implicitization in [MC92a].

2.1 The Sparse resultant matrix

For the sparse resultant, we assume we have $m = n + 1$ polynomials in n variables. We wish to define a square matrix M whose determinant is divisible by the sparse resultant of $f_1, \ldots, f_{n+1}$. Let Q_i be the Newton polytope of f_i. We need the notion of a mixed subdivision of $Q = Q_1 + \cdots Q_{n+1}$. Mixed subdivisions are defined fully in the appendix (section 7), but a short summary is helpful here:

Definition 4 *A mixed subdivision Δ of $Q = Q_1 + \cdots + Q_m$ is a polyhedral subdivision such that every element $F \in \Delta$ is of the form $F = F_1 + \cdots + F_m$, with F_i a face of Q_i. Furthermore $\dim(F) = \sum \dim(F_i)$.*

So a mixed subdivision may be thought of as a decomposition of the Minkowski sum $Q_1 + \cdots + Q_m$ into elementary Minkowski sums $F_1 + \cdots + F_m$. Certain faces of the subdivision play a special role:

Definition 5 *A face $F \in \Delta$ is called a* mixed facet *if $\dim(F) = n$ and every F_i has dimension ≤ 1.*

If $m = n$, then all the F_i in a mixed facet F have dimension exactly one. The mixed volume $MV(Q_1, \ldots, Q_n)$ is the sum of the volumes of the mixed facets of Δ in the mixed subdivision Δ of Q. The mixed subdivision, and hence mixed volume, can be computed effectively using a convex hull routine on a lifted polytope as described in the appendix. If only the mixed volume is needed, it can be computed by enumerating the mixed cells as described in section 3.2.1.

Returning to resultants, we assume from now on that $m = n + 1$. The rows and columns of M will be indexed by integer lattice points contained in Q, and the subdivision Δ will be used to select which coefficient c_{ij} appears in each matrix element.

For the selection to be well-defined, we must perturb the Minkowski sum slightly so that each integer lattice point lies in the *interior* of a facet of Δ. Thus we choose a generic vector $\delta \in \mathbb{Q}^n$, and the set of exponents that index rows and colums of M will be

$$\mathcal{E} = \mathbb{Z}^n \cap (\delta + Q)$$

If Δ_δ denotes the subdivision obtained by shifting all faces of Δ by δ, the choice of δ is satisfactory if every $p \in \mathcal{E}$ lies in the interior of a facet of Δ_δ.

We can now define our selection rule for elements of M. We define a function $RC : \mathcal{E} \to \mathbb{Z}^2$ (for row contents) as follows

Definition 6 (Row content function) Let $p \in \mathcal{E}$ be an exponent. It lies in the interior of a facet $\delta + F_1 + \cdots + F_{n+1}$ of Δ_δ. Let i be the largest integer such that F_i is a vertex, so $F_i = a_{ij}$ for some j. Then $RC(p) = (i, j)$.

The row of M indexed by $p \in \mathcal{E}$ contains the coefficients of f_i, and represents a multiple of f_i which is

$$x^{(p - a_{ij})} f_i \qquad (3)$$

where $(i, j) = RC(p)$. The coefficient of x^q in $x^{(p-a_{ij})} f_i$ appears in the qth column. More explicitly, the matrix M is constructed as

Definition 7 M is an $|\mathcal{E}| \times |\mathcal{E}|$ matrix whose rows and columns are indexed by elements of $\mathcal{E}$, and whose elements are

$$M_{pq} = \begin{cases} c_{ik} & \text{if } q - p + a_{ij} = a_{ik} \text{ where } (i,j) = RC(p) \\ 0 & \text{if } q - p + a_{ij} \notin \mathcal{A}_i \end{cases}$$

and therefore $M_{pp} = a_{ij}$ where $(i, j) = RC(p)$. For the matrix to be well-defined all exponent vectors of (3) must lie within $\mathcal{E}$, which we show in the next section.

2.2　A Sample Sparse Resultant

The construction is illustrated for a system of 3 polynomials in 2 unknowns;

$$\begin{aligned} f_1 &= c_{11} + c_{12}xy + c_{13}x^2y + c_{14}x \\ f_2 &= c_{21}y + c_{22}x^2y^2 + c_{23}x^2y + c_{24}x \\ f_3 &= c_{31} + c_{32}y + c_{33}xy + c_{34}x \end{aligned}$$

Pick generic functions

$$\begin{aligned} l_1(x, y) &= L^5 x + L^4 y \\ l_2(x, y) &= L^3 x + L^2 y \\ l_3(x, y) &= Lx + y \end{aligned}$$

where L is a sufficiently large integer. These functions define a mixed subdivision of Q as described in the appendix. The input Newton polytopes are shown in figure 1 and the mixed subdivision $\Delta + \delta$ is shown in figure 2.

To illustrate the construction, we will generate one row of the matrix. Choose any point in the subdivision of figure 2, say the point $(1, 2)$, which represents the monomial xy^2. We will fill the matrix row indexed by

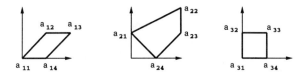

Figure 1: The Newton polytopes and the exponents a_{ij}.

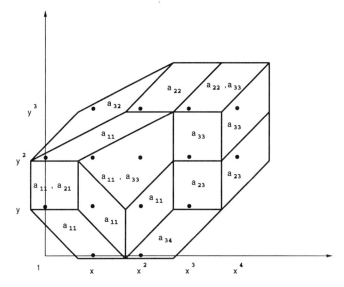

Figure 2: *The mixed subdivision Δ_δ of $Q + \delta$. Each facet is labeled with the vertices which contribute to optimal sums within that facet.*

$(1, 2)$. This point lies in a face $F + \delta$ labelled "a_{11}, a_{33}". This means that F is the Minkowski sum $F_1 + F_2 + F_3$ with $F_1 = a_{11}$ and $F_3 = a_{33}$ both vertices, and F_2 a non-vertex. In this case $F_2 = Q_2$. Either of the points a_{11} or a_{33} would define a suitable row, but the row contents function is defined to choose the larger i value, so $RC(1, 2) = (3, 3)$.

Denote $RC(1, 2) = (3, 3) = (i, j)$ where i is the number of the polynomial whose multiple fills row $(1, 2)$. The multiplier is chosen so that the coefficient c_{33} lies on the leading diagonal. The exponent of the multiplier is $p - a_{33} = (0, 1)$. So row $(1, 2)$ is filled with $y f_3$ etc.

The matrix M is given in [CE93]. It has some useful computational properties, namely (i) it is very sparse and (ii) its leading diagonal is always non-zero. This second property is important in proving that the determinant of M is non-zero in [CE93].

2.2.1 The Size of M

M is an $|\mathcal{E}| \times |\mathcal{E}|$ matrix. The cardinality of $|\mathcal{E}|$ to a good approximation equals the volume of Q. Ideally, we would like the size of M to be the total degree of the resultant, which can be shown to be $MV(Q_1, \ldots, Q_{n+1})$ or $MV(Q)$ by a slight abuse of notation. So our construction, while it does depend on the Newton polytopes, is suboptimal by a factor of

$$\frac{\text{Vol}(Q)}{MV(Q)}$$

In general, this ratio is difficult to determine. Worse than that, there are collections of polytopes whose mixed volume is zero, but whose Minkowski sum volume is finite. For these systems, the suboptimality ratio is infinite.

For reasonable cases, the ratio is better. As an example, suppose that all the Newton polytopes are identical:

$$Q_1 = \cdots = Q_{n+1} .$$

Then the total degree of the resultant $MV(Q)$, is the sum of all $n+1$ n-fold Mixed Volumes, each being equal to $n! \text{Vol}(Q_1)$. Hence

$$\deg R = (n + 1)! \ \text{Vol}(Q_1) .$$

The Minkowski Sum has volume $\text{Vol}(Q) = (n + 1)^n \text{Vol}(Q_1)$ and the number of lattice points in it is asymptotically the same [Kan92]. Then $|\mathcal{E}| = O\left(\frac{n^n \deg R}{(n+1)!}\right)$. Using Sterling's approximation and letting e be the base of natural logarithms, we arrive at

Lemma 1 *For unmixed systems*

$$|\mathcal{E}| = O(e^n \deg R) .$$

This is exponential in n, but of course, the mixed volume itself typically grows exponentially with dimension. e.g. if we posit a family of systems with Newton polytopes all equal to dQ_1 for integer d, then the resultant degree is $O(d^n)$. Our resultant matrix size is $O((de)^n)$, and so is polynomial in the resultant degree considered either as a function of d or a function of n. But there is still room for considerable improvement.

2.2.2 A smaller sparse resultant matrix

An improved construction is described in [EC93]. That paper defines a matrix M' whose determinant is a multiple of the resultant, and whose size is at most the size of M. In general, it is known to be impossible to construct a matrix whose determinant is exactly the sparse resultant. So M' has to be larger than optimal. But we claim the construction of M' leads to close-to-optimal size. We cannot make this statement quantitative yet, but have found some strong supporting evidence. Namely, in all the cases where optimal-size matrices are known to exist (enumerated in [SZ93]), the matrix M' has optimal size. This is the algorithm we plan to use in the toolkit for the elimination task. Even if we cannot prove better bounds on its size, we will gather plenty of empirical data on its effectiveness.

3 Toolkit Overview: Numerical Kit

Figure 3 shows the breakdown of the numerical part of the toolkit. It is actually two independent systems. One is based on resultants and eigenvalues, and the other on homotopy methods. These two methods are clear front-runners for numerical equation-solving, both in terms of simplicity and sheer speed, over other current techniques. Between the two, there is no clear winner, and their virtues are complementary, so both are included in the toolkit. Homotopy methods are faster in practice for very large problems. Their complexity for well-conditioned inputs is linear in the number of solutions N of $f = 0$. Their weakness is on singular problems, where they may diverge or run intolerably slowly. Resultant-eigenvalue methods work very well on small to medium sized problems, and they are simple to implement. Because of extensive study of singular eigenvalue problems, a resultant-eigenvalue method can provide more information about singular solutions, and can compute them accurately if their multiplicity is known. For now, the complexity of the resultant-eigenvalue method is $O(N^3)$, assuming an $O(N \times N)$ resultant matrix is available. But because this matrix is very sparse, it may be possible to use more efficient eigenvalue algorithms whose complexity is quadratic or even pseudo-linear in N. This is a topic that warrants further study.

3.1 Equation-Solving with Resultants and Eigenvalues

The idea of solving a non-linear polynomial system with resultants and eigenvalues is both general and simple. We believe it will find very wide application in the

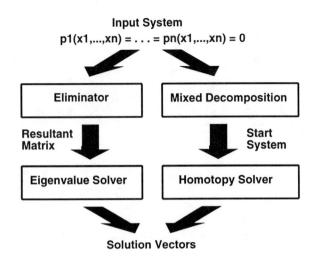

Figure 3: Numerical subsystem components

future, and along with homotopy methods will become the methods of choice for polynomial equation solving. It is described more fully in [MC91b] and [MC93], and applied to a robotics problem in [MC92b] and to geometric modeling in [MC91a].

In the last section, we summarized the theory of sparse systems and defined the resultant of a system of $n + 1$ equations in n unknowns. Before this general formula was known, there were many resultant formulae for specific systems. Almost all these formulae also express the resultant via matrix determinants. e.g. For a pair of polynomials $f(x)$ and $g(x)$ in a single variable, the resultant is the determinant of the Sylvester matrix:

$$\begin{bmatrix} f_n & \cdots & \cdots & f_0 & \cdot & 0 \\ \cdot & \ddots & & & \ddots & \cdot \\ 0 & \cdot & f_n & \cdots & \cdots & f_0 \\ g_m & \cdots & \cdots & g_0 & \cdot & 0 \\ \cdot & \ddots & & & \ddots & \cdot \\ 0 & \cdot & g_m & \cdots & \cdots & g_0 \end{bmatrix}$$

The form of the resultant matrix in the sparse case [CE93] is analogous, although in the multivariate case, the structure is more varied, as can be seen in the example resultant from section 2.2. But one important common feature is that each row of the matrix represents a multiple of one of the polynomials. That is, all the entries in that row are coefficients of that polynomial (or zero). This form is particularly convenient because it allows us to reduce the non-linear equation-solving problem directly to an eigenvalue problem.

3.1.1 Method 1: Adding a linear polynomial

The simplest way to perform the reduction is to add a generic linear polynomial. Suppose we are given $f_1, \ldots, f_n$ as polynomials in $x_1, \ldots, x_n$. We add the linear (in x) polynomial

$$r(s, x) = s - l(x)$$

where $l(x)$ is linear in x, and s is a new variable. We now have $n+1$ equations in n variables, so the resultant is well-defined. It will be a polynomial $R(s)$ in the new variable s.

$R(s)$ must vanish when all the equations have a common solution, and this is precisely when $s = l(\xi)$, where $x = \xi$ is a solution of $f = 0$. By ordering the rows of the resultant matrix A so rows containing $r(s, x)$ appear after all the others, A can be written in block form as:

$$\begin{bmatrix} A_{11} & A_{12} \\ A_{21}(s) & A_{22}(s) \end{bmatrix}$$

where the elements of A_{21} and A_{22} are linear in s. After some elementary row operations, we obtain:

$$\begin{bmatrix} A_{11} & A_{12} \\ 0 & A_{22}(s) - A_{21}(s)A_{11}^{-1}A_{12} \end{bmatrix}$$

whose determinant differs by only a constant factor $(\det(A_{11}))$ from the determinant of $B(s) = A_{22}(s) - A_{21}(s)A_{11}^{-1}A_{12}$. Now $B(s)$ is again a matrix whose elements are linear in s. So it can be written $B(s) = sB_1 + B_0$. Multiplying through by B_1^{-1}, we obtain

$$B'(s) = sI + B_0 B_1^{-1}$$

and considered as a polynomial in s, the determinant of $B'(s)$ has the same roots as $\det(B(s))$ and $\det(A(s))$ which both differ from it by constant factors. The roots of $B'(s)$ are the eigenvalues of $-B_0 B_1^{-1}$.

By the construction, these eigenvalues are the values of $l(\xi)$ for various roots $x = \xi$ of $f = 0$. So, for example, we could set $l(x) = x_1$, and the eigenvalues of $-B_0 B_1^{-1}$ will be the x_1 coordinates of the solutions ξ. But a better approach is to choose a generic linear polynomial for $l(x)$, say by specifying $l(x)$ with n random numbers. Then, with high probability, the values of $l(x)$ will be distinct for distinct solutions ξ and ξ'. So long as the solution $x = \xi$ has multiplicity one, $l(\xi)$

will be a simple eigenvalue. In this case, all the coordinates of the solution ξ can be recovered from the eigenvector corresponding to the eigenvalue $l(\xi)$.

This method is even simpler if the sparse resultant matrix of [CE93] is used. It is possible to define the matrix M of section 2.1 in such a way that the constant coefficient of $g(s, x)$, which is the only coefficient depending on s, falls always on the leading diagonal. Consequently, the matrix B_1 is diagonal, in fact the identity matrix, so no inversion of it is needed.

3.1.2 Method II: Generalized Companion Matrices

Suppose we have the same system of polynomials $f_1, \ldots, f_n$ in $x_1, \ldots, x_n$ and we would like to solve $f = 0$. Rather than adding another polynomial, we can reduce the number of variables by hiding one in the base field. So we rewrite $f_1, \ldots, f_n$ as polynomials in $x_2, \ldots, x_n$ whose coefficients are now polynomials in x_1. This time, we have n equations in $n - 1$ variables, so a resultant is well-defined. Again, let A be the matrix whose determinant is the resultant. The elements of A are coefficients of f and so are polynomials in x_1. We can arrange A in powers of x_1 as

$$A(x_1) = x_1^d A_d + \cdots + A_0$$

Assuming A_d is non-singular, we can define

$$A'(x_1) = A_d^{-1}A(x_1)$$

whose determinant has the same roots as $\det(A(x_1))$. Then we use theorem 1.1 [GLR82] to construct a *generalized companion matrix* of the form

$$C = \begin{bmatrix} 0 & I_n & 0 & \cdots & 0 \\ 0 & 0 & I_n & \cdots & 0 \\ \vdots & \vdots & \vdots & \ddots & \vdots \\ 0 & 0 & 0 & \cdots & I_n \\ -A'_0 & -A'_1 & -A'_2 & \cdots & -A'_{d-1} \end{bmatrix}, \quad (4)$$

such that the eigenvalues of C correspond exactly to the roots of $\det(A(x_1)) = 0$. C is a numeric matrix of order dN, where N was the original size of A.

In this case, the eigenvalues will be the x_1-coordinates of the root vectors ξ. This is true because we have hidden x_1 in the coefficient field in order that the system $f = 0$ be overconstrained and have a resultant. But this doesnt change the solutions of the system. So there can only be a sequence of values

$(x_2, \ldots, x_n)$ satisfying $f = 0$ when x_1 is specialized to the first coordinate of a solution. The existence of a solution $(x_2, \ldots, x_n)$ for some value of x_1 implies that the resultant must vanish for that value of x_1, and by the construction above, we have shown that the roots of the resultant are the eigenvalues of C.

3.2   Sparse Homotopies

The second subsystem of the numerical kit is based on homotopies. For those not familiar with the homotopy method, a very short explanation follows. Suppose we have a system $f = 0$ that we would like to solve. Suppose also, that we have another system $g = 0$, and suppose both systems comprise n polynomials in n variables. Now consider

$$(1 - \lambda)g + \lambda f = 0$$

For $\lambda = 0$, this reduces to the system $g = 0$. As λ increases from 0 to 1, the solutions vary continously as functions of λ, and when $\lambda = 1$ they must have either diverged to infinity or converged to solutions of $f = 0$. The tracking of solutions is typically done with Newton iterations, taking small discrete steps in λ, from $\lambda_0 = 0$ to $\lambda_N = 1$. Each solution of the system $(1 - \lambda_{i-1})g + \lambda_{i-1}f = 0$ is used as a seed for solution of $(1 - \lambda_i)g + \lambda_i f = 0$.

Under appropriate conditions, we can choose a system $g = 0$ whose solutions converge to all of the solutions of $f = 0$. For example, if f_i and g_i both have the same set of exponents for each i, g has distinct roots, and if we track solutions in complex projective space, we will obtain all the solutions of $f = 0$, provided we dont run into a singularity along the way. This is very useful for sparse systems, because both $f = 0$ and $g = 0$ have a number of solutions given by Bernstein bounds. And the complexity of homotopy methods is very good, in practice almost linear in the number of solutions. So it would seem that homotopy methods immediately offer a way to fully exploit sparseness.

The story is not so simple, because there is no way known in general to design a system $g = 0$ which has given exponents and known, distinct roots. On the other hand if one is always solving systems with the same structure, i.e. set of exponents, one can perform a computation offline to find the roots of one such system $f = 0$ from a somewhat larger start system $g = 0$. Then the roots of $f = 0$ can be saved, and subsequent

calculations can use these roots and the system $f = 0$ as the start system. For example, a suitable start system is $g_i = x_i^{d_i} - 1$, for $i = 1, \ldots, n$, where d_i is a bound on the degree of f_i. The number of solutions of this system is given by Bezout, rather than Bernstein bounds so the first calculation may be very expensive. Subsequent calculations will only need to track the Bernstein number of solutions.

But this is a poor solution when one has to deal with polynomial systems that do not all have the same structure. Fortunately, the problem of finding start systems was recently solved in [HS92]. The solution is not in the form of a single start system $g = 0$, but a family of start systems $f|_{\mathcal{F}} = 0$, where $\mathcal{F}$ ranges over the mixed facets of a mixed subdivision. Specifically, in the notation of section 6, let Q_i be the Newton polytope of f_i, and let Δ denote a mixed subdivision of $Q_1 + \cdots + Q_n$. Let $\mathcal{F} = F_1 + \cdots + F_n$ be an n-dimensional face of Δ. Then each F_i is a face of Q_i, and the sum of the dimensions of the F_i will be n. $\mathcal{F}$ will be a mixed cell if and only if all the faces F_i are one-dimensional.

The start system $f|_{\mathcal{F}}$ consists of the equations

$$f_i|_{F_i} = 0, \qquad \text{for } i = 1, \ldots, n$$

where $f_i|_{F_i}$ is the polynomial f_i with all terms set to zero except those whose exponents lie in F_i. Thus the Newton polytope of $f_i|_{F_i}$ is F_i. While it has a one-dimensional Newton polytope, $f_i|_{F_i}$ is not a univariate polynomial, and finding the roots of $f|_{\mathcal{F}} = 0$ is still non-trivial. But after a monoidal change of variables $x_i \mapsto y^{b_i}$, where $y = (y_1, \ldots, y_n)$, $b_i \in \mathbb{Z}^n$, we obtain polynomials $f_i'|_{F_i}(y_i)$ which are univariate, and which are easily solved. The number of roots of $f|_{\mathcal{F}} = 0$ either before or after the monoidal transformation is exactly the volume of the polytope $\mathcal{F}$. Applying the monoidal transformation to the y-solutions obtained by solving the univariate system gives us the x-solutions of $f|_{\mathcal{F}}(x) = 0$. We track these solutions from $f|_{\mathcal{F}}(x) = 0$ to $f = 0$, using a homotopy method.

There is a very satisfying proof in [HS92] which shows that every root of $f = 0$ is the endpoint of a path begining at a root $f|_{\mathcal{F}} = 0$ for some mixed facet $\mathcal{F}$. The number of roots of $f = 0$ is therefore the total of the numbers of roots of the $f|_{\mathcal{F}} = 0$. This is exactly the total volume of all the faces $\mathcal{F}$, which is exactly the mixed volume of $Q_1, \ldots, Q_n$. In this way, [HS92] provides a constructive proof of Bernstein's theorem.

3.2.1 Computing Mixed Cells and Volumes

The idea of a mixed subdivision of a Minkowski sum of polytopes is described in an appendix (section 6). We hinted in the previous section how the mixed subdivision furnishes information about a family of homotopy start systems, one for each mixed cell of the subdivision. To complete the implementation of a homotopy solver for sparse systems then, we need an efficient algorithm for computing either mixed subdivisions, or enumerating directly the mixed cells of a subdivision.

Because of the complexity of a mixed subdivision in high dimensions, we have found it much more effective to enumerate only the mixed cells. Let $\mathcal{F} = F_1 + \cdots + F_n$ be a mixed cell of a subdivision Δ of the Minkowski sum $Q_1 + \cdots + Q_n$. Then every F_i is an edge of Q_i. We could enumerate all possible mixed cells by enumerating all n-tuples of edges, one from each polytope, and checking them, but this method soon runs into a computational brick wall. Imagine for example, that $n = 6$ and each polytope has 20 edges - there would be 65 million possibilities to check. Fortunately, there is a much better search strategy.

Define the Minkowski sum of the first i faces contributing to $\mathcal{F}$:

$$\mathcal{F}_{\leq i} = F_1 + \cdots + F_i$$

Similarly, in the space of lifted polytopes, $\mathbb{R}^{n+1}$, we have

$$\hat{\mathcal{F}}_{\leq i} = \hat{F}_1 + \cdots + \hat{F}_i$$

A necessary condition for $\mathcal{F}$ to be a mixed facet is that every $\hat{\mathcal{F}}_{\leq i}$ is a facet of the partial Minkowksi sum $\hat{Q}_1 + \cdots + \hat{Q}_i$, for $i = 1, \ldots, n$. This immediately suggests a search strategy:

Our mixed volume algorithm constructs a tree of candidate $\hat{\mathcal{F}}_{\leq i}$'s. The root of the tree is level zero. The first level of the tree contains all lifted edges of $\hat{Q}_1$. The kth level of the tree contains candidate k-fold sums of edges. Let $\hat{\mathcal{F}}_{\leq k}$ be the k-fold sum corresponding to tree node v. We explore v by checking whether $\hat{\mathcal{F}}_{\leq k}$ lies on the boundary of $\hat{Q}_1 + \cdots + \hat{Q}_k$ using linear programming. If not, then no further expansion of v occurs. If $\hat{\mathcal{F}}_{\leq k}$ is on the boundary, then a new child u of v is created for each edge $\hat{e}$ of $\hat{Q}_{k+1}$. The facet at u is the Minkowksi sum $\hat{\mathcal{F}}_{\leq k} + \hat{e}$. These nodes are explored in the same manner. The search continues down to level n, and those leaves whose facets lie on the boundary of $\hat{Q}$ define the mixed cells.

n	Bernstein bound	Computing time
7	924	1m 2s
8	2560	9m 37s
9	11016	1h 42m 50s
10	36650	20h 22m 1s

Figure 4: *Bernstein bounds for the cyclic n-roots problem*

This search strategy works well in practice, and has allowed us to solve some very large mixed volume calculations, in up to 28 dimensions. As a benchmark, we give its running times for bounding the number of *cyclic n-roots* for various n. The cyclic n-roots problem is of some independent interest, but it has primarily served as a computational benchmark for algebraic algorithms. More details and references are given in [BF91].

The equations themselves arise in Fourier analysis. There are n equations in the variables $x_1, \ldots, x_n$, and they have the form:

$$
\begin{aligned}
x_1 + \cdots + x_n &= 0 \\
x_1 x_2 + x_2 x_3 + \cdots + x_n x_1 &= 0 \\
x_1 x_2 x_3 + x_2 x_3 x_4 + \cdots + x_n x_1 x_2 &= 0 \\
&\vdots \\
x_1 x_2 \cdots x_n &= 1
\end{aligned}
\tag{5}
$$

Using our mixed volume algorithm, we computed the Bernstein bounds for cyclic systems for various values of n, and these are shown in figure 3.2.1. The mixed volume algorithm is written in $ANSI-C$, and the test data is from a SUN Sparc-10 workstation.

The last equation $x_1 \cdots x_n = 1$ forces all the solutions to lie in $(\mathbb{C}^*)^n$. So Bernstein's theorem provides an exact count allowing for multiplicities. In the one case where the number of solutions is known, $n = 7$ [BF91], the multiplicities are all 1, and the Bernstein bound is exact. To our knowledge, the bounds for $n = 9$ and $n = 10$ have not been computed before, and the value for $n = 7$ tooks many hours to compute using a Gröbner algorithm called Bergmann in [BF91]. This is an encouraging sign for the effectiveness of the mixed volume algorithm.

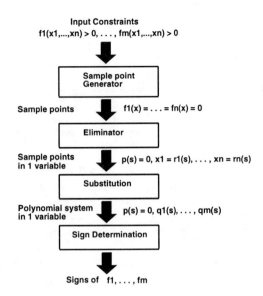

Figure 5: Symbolic subsystem modules

4   Toolkit Overview: Symbolic Kit

As shown in figure 5, the primary function of the symbolic toolkit is to determine solvability of systems of polynomial inequalities over the real numbers. A numerical approximation to the solution, if any, can be easily extracted. The symbolic toolkit can also compute optimal solutions to systems of inequalities, given a polynomial objective function.

We plan also to use various modules of the toolkit for other purposes. For example, computing connected components of algebraic curves, and ordering points along those curves. This leads ultimately to an algorithm for computing connected components [Can88a]. We also plan to implement some fast query algorithms for point location in arrangements of polynomial surfaces. The algebraic algorithms needed in these applications are all contained in the toolkit. This explains the toolkit moniker, and the module-by-module description in this paper. But for reasons of space, the only calculation described here will be solution of systems of inequalities.

The input is a predicate that specifies a *semi-algebraic set*. The predicate can be a general boolean combination, rather than simply a conjunction, of polynomial inequalities:

Definition 8 *A semi-algebraic set* S *is the set of*

points in $\mathbb{R}^n$ satisfying a predicate of the form $B(A_1, \ldots, A_k)$ where $B : \{0,1\}^k \to \{0,1\}$ is a boolean function and each A_i is an atomic formula of one of the following types:

$$(f_i = 0), (f_i \neq 0), (f_i > 0), (f_i < 0), (f_i \geq 0), (f_i \leq 0)$$
$$(6)$$

with each f_i a polynomial in $x_1, \ldots, x_n$ with rational (for our computational purposes) coefficients.

Semi-algebraic (SA) sets are a versatile class that includes the forms of most familar objects in $\mathbb{R}^3$, such as cones, cylinders, spheres, and combinations of these. SA sets in 3D are slightly more general than CSG (constructive solid geometry) models, for example, SA sets include sets of mixed dimension, and sets which are not topologically closed. Both of these are forbidden in CSG.

SA sets are defined in any dimension, and can also be used to represent the set of legal (e.g. obstacle-avoiding) configurations of a mechanical system, such as a robot, or the set of camera transformations consistent with some geometric constraints, or the set of positions at a given distance from a surface to guide a milling machine. The defining formulae for SA sets are the real analogue of SAT formulae, and they provide the ability to search over a space of real values for a solution satisfying some property.

The goal of the inequality solver then, is to decide if a semi-algebraic set S is non-empty, and to produce a sample point if it is. We describe each module of the solver in the following sections. The next section, on the sample point enumerator, gives also an overview of the whole system.

4.1   Sample point enumerator

In order to decide if the predicate is satisfiable, that is, if the semi-algebraic set S it defines is non-empty, the solver tries to find a witness point $v \in S$. The existence of such a point is a proof of satisfiability. It is also possible to prove unsatisfiability by enumerating sufficiently many potential witness points, one of which must lie in S if S is non-empty.

For simplicity of explanation, we will assume that the set S is closed and bounded, and that the surfaces $f_i = 0$ are in *general position*. These assumptions are not necessary, and methods for dealing with other cases are given in [Can93b].

Since we know the polynomials f_i are in general position, the intersection of any $n+1$ of them in n dimensions will be null. Any $j \leq n$ of them will intersect in a manifold of dimension $n - j$. Let P be an extremal point of π in the set S. Then P is also an extremal of π in a manifold M which is the set of zeros of some polynomials $f_{i_1}, \ldots, f_{i_j}$. These polynomials are precisely the f_i which are zero at P.

So to enumerate all potential witness points, we enumerate the critical points of π on the set of common zeros of $f_{i_1}, \ldots, f_{i_j}$, for every set of $j \leq n$ polynomials. First we define a polynomial

$$g = \sum_{l=1}^{j} f_{i_l}^2$$

and solve the system

$$g = \mu \qquad \frac{\partial g}{\partial x_2'} = \cdots = \frac{\partial g}{\partial x_n'} = 0 \qquad (7)$$

"in the limit" as $\mu \to 0$. The coordinates $x_2', \ldots, x_n'$ are a basis for the $n-1$ dimensional linear space which is the kernel of π. The process of solving this system in the limit is described in [Can88b] and [Can90], and involves computing the resultant of the system, arranging it in powers of μ, and retaining the lowest degree coefficient. The result is a polynomial $p(s)$ and rational functions $(r_1(s), \ldots, r_n(s))$ such that the solutions to the system (7) are all the tuples $(r_1(\alpha_i), \ldots, r_n(\alpha_i))$ where α_i are the roots of $p(s) = 0$.

To compute the signs of the other polynomials at these critical points, we substitute $x_i \mapsto r_i(s)$ for $i = 1, \ldots, n$, giving $q_i(s) = f_i(r(s))$ and the set of signs we are looking for is precisely the sign sequences of

$$(q_1(s), \ldots, q_k(s)) \text{ at roots of } p(s) = 0$$

and to find these we simply apply the sign determination algorithm of the last section (to numerator and denominator, since the q_i here are rational functions).

4.1.1 Summary of the inequality solver

We first choose a generic linear map $\pi : \mathbb{R}^n \to \mathbb{R}$ by selecting n random integers $\pi_1, \ldots, \pi_n$. Then the algorithm proceeds as follows:

- **Sample point enumeration:** Enumerate all subsets of $j \leq n$ polynomials $\{f_{i_1}, \ldots, f_{i_j}\}$. Do this in order of increasing j, so that "easy" witness points will be found early.

- **Elimination:** For each subset, construct a representation of the critical points of π as a polynomial $p(s)$ and rational functions $r_1(s), \ldots, r_n(s)$.

- **Substitution:** Substitute $r_i(s)$ for x_i in the other polynomials, giving $q_i(s) = f_i(r(s))$ for $i = 1, \ldots, k$.

- **Sign Determination:** Determine the signs of the $q_i(s)$ at roots of $p(s) = 0$ using the algorithm of section 4.3. Substitute these signs into the formula B to check if the corresponding critical point lies in S. If yes, S is non-empty, so return "true".

- Continue until no more sample points, and then return "false".

4.2 Reduction to a univariate problem

The function of this module is to take a system of n polynomials in n variables

$$f_1(x_1, \ldots, x_n) = \cdots = f_n(x_1, \ldots, x_n) = 0$$

which has solutions $\xi^{(i)} \in \mathbb{C}^n$, for $i = 1, \ldots, N$, and produce from it a univariate polynomial $p(s)$ and rational functions $r_1(s), \ldots, r_n(s)$. The pair (p, r) is a symbolic description of the roots $\{\xi^{(i)}\}$ in the following sense: If the roots of $p(s) = 0$ are $\alpha^{(i)} \in \mathbb{C}$, for $i = 1, \ldots, N$, then

$$\xi^{(i)} = r(\alpha^{(i)})$$

Put another way, $r(s)$ is a parametric curve in $\mathbb{C}^n$ which passes through all the roots $\xi^{(i)}$, $i = 1, \ldots, N$. The values of s at which it arrives at a root are precisely the roots of $p(s) = 0$. The system $f = 0$ will usually be derived from equation 7 of the last section.

There are several ways to construct (p, r) from f. One method uses polynomial GCDs and was described in [Can88b]. A more efficient method based on differentiation was given by Renegar [Ren89], and that is the one we use in the toolkit. Compared to the cost of computing the polynomial $p(s)$, Renegar's method computes all the $r_i(s)$'s in only n times as many operations.

To start, we add to the system $f = 0$, a linear polynomial

$$s - l(x) = 0, \qquad \text{where } l(x) = l_0 + l_1 x_1 + \cdots l_n x_n$$

and compute the resultant $R(s, l)$. Suppose that l is specialized to a random linear polynomial L, then with probability one, the values $l(\xi^{(i)})$ will be distinct for distinct $\xi^{(i)}$. The roots of the resultant $R(s, l = L)$ will be $s = L(\xi^{(i)})$, and will also be distinct. We set $p(s) = R(s, l = L)$ and we are halfway to our goal.

We obtain the functions r_i by differentiating $R(s, l)$. We define r_i as

$$r_i(s) = \left. \frac{\left(\frac{dR(s,l)}{dl_i}\right)}{\left(\frac{dR(s,l)}{dl_0}\right)} \right|_{l=L}$$

The reader can verify that the r_i's have the correct values at $s = \alpha^{(j)}$ by computing the derivatives of $R(s, l)$ in its factored form $R(s, l) = \prod_{j=1}^{N}(s - l(\xi^{(j)}))$.

In our implementation, it is particularly easy to compute the derivatives. All the algorithms in the symbolic toolkit use an SLP (straight-line program) representation of arithmetic. See section SLP-arithmetic below. The polynomial $p(s) = R(s, l = L)$ is computed explicitly. That is, we compute SLPs for each of the coefficients $p_0, \ldots, p_m$ of $p(s)$. Because they depend on the coefficients of l, they can be written $p_i(l)$. Then we compute the derivative of the SLP for $p_i(l)$ with respect to some l_j. For each node v of $p_i(l)$, we add a new node representing the derivative of v with respect to l_j. This roughly doubles the size of the SLP (see section 4.5.1). Computing all the derivatives of $R(s, l)$ increases the size of the SLP by a factor of $n + 2$.

This method works even if the system $f = 0$ has roots with multiplicity > 1. When this happens, $p(s)$ will have repeated factors for all choices of l. The partial derivatives above will all be zero. But we can remove the repeated factors by first choosing a random l, setting $p(s) = R(s, l)$ and then defining

$$\hat{p}(s) = p(s)/GCD(p(s), \frac{dp}{ds})$$

This time, we define $r_i = \frac{d\hat{p}}{dl_i} / \frac{d\hat{p}}{dl_0}$. The partial derivatives are again computed using SLPs, and they will be non-zero for almost all choices of l.

4.3   Sign Determination

Most of the recent work on real algebraic algorithms makes use of a sign-determination lemma due to Ben-Or, Kozen and Reif [BOKR86]. This lemma, henceforth called "BKR", takes a univariate polynomial $p(s)$, and polynomials $q_1(s), \ldots, q_k(s)$, and returns m sign sequences $\sigma \in \{-, 0, +\}^k$, where m is the number of

real roots of $p(s)$. Each sign sequence σ corresponds to a particular root α_j of $p(s)$, in such a way that $\sigma_i = \text{sign}(q_i(\alpha_j))$.

Given the importance of BKR, it is natural to look for improvements and simplifications. In [Can91b] and [Can93b], we described a faster and simpler version of BKR, which allows purely symbolic elimination. It eliminates the matrix rank tests of BKR, replacing them with a simple recursive algorithm.

4.3.1   Sign-Determination Algorithm

The input is the polynomial $p(s)$, and polynomials $q_1(s), \ldots, q_k(s)$ with rational coefficients, and of degree at most d. The output is the set of sign sequences $\{(q_1(\alpha_j), \ldots, q_k(\alpha_j))\}$ at the $m \leq d$ real roots α_j of $p(s)$. Note that the set of sign sequences is an unordered set, and the algorithm does not output the sign sequences in the order of the α_j as real numbers.

1. We assume at the ith step that the algorithm knows for each sign sequence of $q_1, \ldots, q_{i-1}$, how many roots of $p(s)$ produce this sign sequence. Most of these are zero, and the algorithm only stores the sign sequences with at least one root, and the number of these is $m_{i-1} \leq d$.

2. There are $3m_{i-1}$ possible sign sequences for $q_1, \ldots, q_i$, and each of these defines a column of the matrix $K_{i,3m_{i-1}}$. These columns are linearly independent, so there are $3m_{i-1}$ rows which together with the specified columns, define a square submatrix J_i of $K_{i,3m_{i-1}}$.

3. We solve the $3m_{i-1} \times 3m_{i-1}$ system corresponding to this matrix, to find the actual sign sequences of $q_1, \ldots, q_i$, and repeat the above steps for $i = 1, \ldots, k$.

The first task then, is to give a procedure that accepts a list of m columns of the matrix $K_{n,m}$, and returns a list of m rows, such that the resulting $m \times m$ matrix is non-singular. This procedure needs to run in polynomial time in n and m and not the size of $K_{n,m}$.

The second task, which is very easy, is to determine the entries of this submatrix of $K_{n,m}$. Again this must be in polynomial time in m, so we cannot aford to construct all of $K_{n,m}$. This task reduces to determining the value of a single element of $K_{n,m}$ given its row and column indices. Both steps can be solved with simple recursive procedures whose running time is $O(nm)$ and $O(n)$ respectively. Refer to [Can91b] and [Can93b] for details.

Let m denote the number of real roots of $p(s)$, then m is not more than d, the degree of $p(s)$. The overall running time is:

$$O(n(md^2 \log^2 m + m^3)) \qquad (8)$$

assuming naive algorithms for polynomial arithmetic. The improvement over the original BKR is that we have reduced the maximum number of polynomials in a Sturm query from n to $\log m$. Since the actual number of real roots of a polynomial is small compared to its degree, the complexity of BKR without this improvement would typically be nm^3d^2 times some log factors. With the improvement, we typically get nmd^2 times some logs.

4.4 Efficient Arithmetic

There are traditionally two routes to take when writing code to solve algebraic problems (i) Using floating point, and (ii) using exact arithmetic and arbitrary precision integers. Neither approach is satisfactory for large problems.

If a single floating point number is used to approximate a real number, then approximate tests for equality, based on an "epsilon" parameter, have to be used. Two numbers that are nearly equal are delared to be equal, and there is no way to tell if this is a false assumption. The author experimented with this approach in the mid 1980s and found that even with double precision arithmetic, Sturm sequence calculations were unreliable for polynomials of degree > 10. Conversations with other researchers since have supported this conclusion. A refinement of this approach is to use interval arithmetic. A real number r is represented as a pair of floats (ϕ_1, ϕ_2), such that $\phi_1 < r < \phi_2$, and $|\phi_1 - f_2|$ is as small as possible. This eliminates some types of error, but if two intervals overlap, it is still impossible to tell if the real numbers they represent are equal or not.

Finally, it is possible to construct "arbitrary precision" floating point, which works at fixed precision until an operation is performed which is a test for zero (or equality). Then the precision is automatically increased until the result of the test is known for certain. An arbitrary precision floating point system is described in [Pri91]. This approach is as good as a symbolic approach and guarantees correct results. The problem is that if the two quantities to be tested really are equal, the precision must be extended all the way to the equivalent of exact integer calculation. In the applications we have in mind, this may involve integers of hundreds of words in length. Arithmetic on such integers takes quadratic time in practice, so the calculation of the two values to be compared, and all those they depend on, slows down by a factor of 10^4 to 10^5 compared to double-precision floating point.

Exact calculation requires arbitrary precision integer arithmetic, and to be efficient, it also requires finite field arithmetic and chinese remaindering. Modular integer arithmetic using chinese remaindering has complexity $O(cp + ap^2)$ with naive algorithms, where c is the number of intermediate arithmetic steps, a is the number of integers in the result, and p is the precision in bits. Often cp dominates ap^2. For example, for solving an $n \times n$ system of linear equations, c is $O(n^3)$, a is n, and p for an exact result is $O(n)$. In these cases chinese remaindering offers linear complexity in p compared to quadratic complexity if arbitrary precision arithmetic is done at the intermediate steps. The only problem is that p may not be known exactly, and using an a priori bound on p often leads to unnecessarily high precision. To get around this problem, in [MC93] we used a probabilistic scheme that does chinese remainder lifting incrementally, and stops when it has sufficient precision. At this point the lifted integers stop changing with from one iteration to the next.

Even though incremental chinese remaindering method brings down the complexity of large integer arithmetic, it suffers from the same problems as arbitrary precision floating point. For our problems of interest, the integers may be hundreds of words long. Even with chinese remaindering, this slows the algorithm down by a factor of several hundred compared to fixed-precision floating point. This is still too large a penalty to pay for exact computation.

4.4.1 A Mixed Approach

We will take a new approach, based on the use of both finite field and floating point arithmetic, to achieve constant cost per arithmetic step. The idea is to use finite fields to test for equality, and then floating point to compute the approximate values. For example, in Sturm sequence or polynomial GCD calculation, it is critical to be able to tell if the leading coefficient of a remainder is zero. But to use the sequence, all that is required is the signs of the actual leading coefficients, and these are usually far from zero. In what follows, we assume the numbers we are trying to represent are (large) integers. It is possible to extend the method, with some effort, to rational numbers, but all the calculations we forsee in the toolkit require integers only.

In our method, we represent a (possibly very large) integer k as a structure with two values. One is the fixed precision integer $k \bmod p$, and the other is a floating point approximation ϕ_k, to k. An arithmetic operation on arguments of this type consists of separate operations on the mod p and float fields. To check two mixed numbers for equality, we check if their mod p values are equal. If so, we declare them to be equal. The probability of an incorrect answer is very small, and depends on the size of p. If we require the relative order of two mixed arguments, we compare first the mod p values to check for equality, and then their float values, and order them accordingly.

In the algorithms we use, equality comparisons are much more frequent than ordering (e.g. every addition or subtraction of polynomials includes a check for cancellation of the leading coefficient). It is still possible using mixed arithmetic that we try to compare two unequal numbers, and find that their floating point descriptions are too close for us to be able to order them. But because the ordering comparisons are much less common, the chances of this happening are much lower than if we tried to use floating point alone for equality testing. Nonetheless, ordering tests are a critical part of algorithms such as the sign determination algorithm, and they cannot be done with the mod p representation alone. For slightly greater cost than either, a mixed representation give us correct sign information, and very low probability of failure.

For simplicity, we have described the mixed representation using a mod p integer and a floating point number. But we also plan to try a mixed representation with an integer mod p and an interval of two floats. This is a safer representation since the ordering check can determine if the float descriptions are too close to order, and flag failure rather than producing an incorrect ordering.

4.5 Computing with Infinitesimals

In the algorithms of [HRS90], [Ren92], [GV92] and [Can91a], various singularities are dealt with by perturbing the input polynomials with infinitesimals. This moves the problem away from the singularity, and when done carefully preserves the important properties (like connectivity or non-emptiness) of the input. Computations with an infinitesimal ϵ are done in the rational field $\mathbb{Q}(\epsilon)$. That is, each number a or b in this extension field is a rational function (a quotient of polynomials) in ϵ. To perform arithmetic, we use the usual rules for arithmetic on rational functions. To determine the sign of such an element, we exclusive-or the signs of its numerator and denominator, which are polynomials in ϵ. To determine the sign of a polynomial in ϵ, we use the sign of the lowest degree non-zero coefficient.

But it is very expensive to compute with explicit rational functions. For example, in the extension $\mathbb{R}(\mu, \epsilon, \delta, \rho)$ that we have been using, an element of degree 10 would have several hundred coefficients. But the sign of the element, which is all we need for the sign-determination algorithm, is determined by just one of these coefficients. This element is the lowest degree element under the lexicographic ordering $\mu \prec \epsilon \prec \delta \prec \rho$

If we knew that this element was say $\mu^4 \epsilon \delta^2 \rho$, we could find it by computing modulo the ideal $(\mu^5, \epsilon^2, \delta^3, \rho^2)$, which effectively discards higher-degree terms. Since we dont know the degree, we would have to do some search, gradually increasing degree until we obtain a non-zero term. Rather than doing this repeatedly, we can obtain the lowest degree term by differentiating a straight-line program.

4.5.1 Straight-Line Program Arithmetic

In our approach, we use arithmetic straight-line programs to represent the intermediate values in a calculation, as advocated by Kaltofen et al. in [Kal88, Kal89, FIKY88]. An SLP may be represented as a directed acyclic graph. Each node of the graph holds an operation type, such as "+" and a value. If the operation is a binary operation, the node will have indegree two, with two incoming edges from the nodes which are arguments to the operation.

Computation with SLPs is split into two phases (i) creation and (ii) evaluation. During creation, unevaluated SLP nodes are created. e.g. If a and b are numbers represented as SLPs, computing $c = a + b$ means adding an SLP node with undefined value, operation type "+" and two edges from the nodes representing a and b. During evaluation of a node, the value field of that node is computed by applying the operation to the value fields of the two argument nodes. Evaluation involves a depth-first search of the SLP graph.

In our implementation, the creation and evaluation phases are transparent, and only normal arithmetic operations are visible. We do this by implementing arithmetic operations with creations, and comparisons with evaluations. This is a form of lazy evaluation, although the SLP sub-graph on which a node depends is always retained, even after the node is evaluated.

The data structure for an SLP node has several essential fields: two argument pointers, which point to a and b in the example above, the operation field (e.g. "+"), and a value field, which holds the value of the node during evaluation. There will usually also be a flag field to indicate that a node has already been evaluated, so that the SLP can be evaluated with depth-first search. And in the toolkit, we will need the values of various derivatives and anti-derivatives of SLP nodes, which adds two additional pointers to the data structure.

Suppose we have computed an element $a \in \mathbb{R}$ from some other real values $b_1, \ldots, b_m$ via a series of arithmetic operations. For example, such a could be a coefficient of $p(s)$ described in section 4.2, or a coefficient of one of the Sturm query polynomials. We can represent a as an SLP rooted at the values $b_1, \ldots, b_m$. Now suppose that b_1 is specialized to the infinitesimal value ϵ, and that the other b_i take on integer values. We would like to know the sign of a. For simplicity we assume that $a = a(\epsilon)$ is a polynomial in ϵ. This is all we need in our applications.

We could substitute $\epsilon = 0$ and evaluate the SLP over the rationals. If we are lucky, $a(0)$ will have a non-zero value, and this gives the sign of $a(\epsilon)$. If not, we can construct an SLP for the derivative $\frac{da(\epsilon)}{d\epsilon}$. This has roughly double the size of the original SLP. Now evaluating this program at $\epsilon = 0$ gives us a_1, the coefficient of ϵ in $a(\epsilon)$. If this is non-zero, it gives us the sign of $a(\epsilon)$, otherwise we compute the second derivative, and continue. The extra program for the kth derivative is about $k + 1$ times the size of the original program, and it uses nodes from the first $k - 1$ derivatives. The *total* program size to compute the kth derivative is $\binom{k+2}{2}$

times the original.

This process generalizes easily to multivariate elements, using randomization. For example, to find the sign of $a(\mu, \epsilon, \delta)$ with $\mu \ll \epsilon \ll \delta$, we first substitute random integer values for ϵ and δ. With high probability, this doesnt change the degree of the lowest degree term in μ. Then we apply the procedure above to obtain an SLP for the first non-zero derivative wrt μ at $\mu = 0$. Let $aa(\mu, \epsilon, \delta)$ denote this derivative. Then $aa(0, \epsilon, \delta)$ is the lowest-degree coefficient of a in μ, times the constant $k_1!$, where k_1 is the order of the derivative.

We iterate the process, and set δ to a random integer, μ to zero, and run the univariate procedure on the SLP for aa as a polynomial in ϵ. This gives us an SLP for the first non-zero derivative wrt ϵ at $\epsilon = 0$, which we denote $aaa(\mu, \epsilon, \delta)$.

Finally, we run the univariate routine on the SLP aaa with μ and ϵ both set to zero. Evaluating the resulting program at $\mu = \epsilon = \delta = 0$ gives the sign of the lexicographically first term, which is what we need.

Some simple analysis shows that the SLP for computing the sign when the lowest degree term is $\mu^{k_1} \epsilon^{k_2} \delta^{k_3}$ is $k_1^2 k_2^2 k_3^2$ times the original. More generally we have

Proposition 1 *Let $P(\epsilon_1, \ldots, \epsilon_m)$ be a polynomial, represented as an SLP with L vertices. If $\epsilon_1 \prec \cdots \prec \epsilon_m$ are infinitesimals, and if the lexicographically first term in P is $c\epsilon^{k_1} \cdots \epsilon^{k_m}$, then an SLP for this term can be constructed having size $\leq L k_1^2 \cdots k_m^2$, in the same number of steps.*

We claim this method is useful in practice because the k_is are typically small constants, independent of the degree of a in μ, ϵ, δ. Each infinitesimal is used to perturb away from a possibly singular input, and the degree in that infinitesimal is a measure of the multiplicity of the singularity. Where the input is not singular at all, the degree in that infinitesimal will be zero. Most of the time, we expect small multiplicities, and the cost of working over the infinitesimal extension should be only a small constant factor more than integer arithmetic, this factor being the increase in the SLP size.

4.6 Complexity

We measure the complexity of a predicate with four quantities, the number of polynomials k, the number of variables n, the maximum degree of the polynomials d, and the maximum coefficient length c of the coefficients of the polynomials.

The time complexity of the inequality solver is

$$k^{n+1} d^{(O(n))} c^2$$

arithmetic steps. This is an improvement over the previously published algorithms for the existential theory [Ren92], [HRS90] and [GV92], and is within a factor of k of optimal. More importantly, the algorithm is simple, and there are no large constants hidden in the exponents.

The $d^{O(n)}$ bound can be made more precise. The use of sparse elimination, mixed arithmetic and lazy limit-taking gives a *typical* complexity of $O(d^{3n})$. This assumes that the input is non-singular or has only low-order singularities. Higher order singularities may require many derivatives to be taken, pushing the complexity up as a polynomial in the number of derivatives.

This is a dramatic improvement over other published algorithms [Ren92], [HRS90] and [GV92], all of which give only a "big-O" estimate of the exponents of both k and d, that is, their bounds are of the form $(kd)^{O(n)}$. Based on the number of infinitesimals used, it is unlikely that any of these algorithms has typical complexity less than $O((kd)^{7n})$.

5 Future Work

5.1 Coefficient relations

We have now a good understanding of Newton polytope volumes and their effect on the number of solutions. Given a system with general coefficients, we can compute the sparse resultant, and that is the smallest possible eliminant. But often in practice, the coefficients of the system we are given will satisfy some relations, and the sparse bounds are no longer the tightest possible.

As an example, consider the system $f_1 = 1 + x + y + xy$, and $f_2 = 1 + 2x + 3y + 4xy$, whose Newton polytopes are the same unit square. The mixed volume of the system is 2, indicating two solutions. Now suppose we make a change of coordinates, say $x = u + v$, $y = u - v$. Then the equations become $f_1' = 1 + 2u + u^2 - v^2$, and $f_2' = 1 + 5u - v + 4u^2 - 4v^2$. Both these equations have Newton polytopes which are simplices of side length

2, and the mixed volume is 4. The number of affine solutions has not changed, but the coefficients are now dependent, and the leading terms (terms of degree two) of the equations now have roots in common. In this situation, Bernstein's theorem is an overestimate. If we used a sparse homotopy method to track the roots, we would find that two of them diverged to infinity. If we compute the sparse resultant of a system of polynomials whose coefficients are dependent, we may find that it is indentically zero. This indicates that there some roots always present, usually at infinity.

This situation is quite common, and perhaps the most striking example is the inverse kinematics of a rotary-joint robot. The simplest way of writing down these equations gives rise to 6 equations of degree 6. The Bernstein bound is about 2,000, while the Bezout bound is 46,000. The number of affine solutions is only 16 [RR89]. The kinematic equations of [RR89] for the same problem give a Bernstein bound of 384.

The large fluctuation in the bounds is caused by the heavy dependencies between the coefficients of the equations. For kinematics, the polynomial system consists of 6 of the elements of the matrix product $A_1(t_1) \cdots A_6(t_6) = B$. Each matrix $A_i(t_i)$ has elements which are quadratic functions of t_i. This would give degree 12 overall, but we can take instead $A_1(t_1)A_2(t_2)A_3(t_3) = BA_6^{-1}(t_6)A_5^{-1}(t_5)A_4^{-1}(t_4)$, because the inverses are also quadratic functions of t_i's. Clearly, the matrix products lead to many common subexpressions in the final polynomials. The dependencies that result reduce the number of solutions, but this is not manifest in the Newton polytopes.

Surprisingly, we can make the structure manifest by *introducing new variables*. We assign these variables to the common subexpressions and replace each occurence of the expression with the corresponding variable. This increases the dimension of the problem, and may or may not reduce the degree. The Bezout bounds will usually increase significantly as a result, but the Bernstein bounds decrease. This runs against the conventional wisdom for dealing with multivariate polynomials, which is to keep the number of variables as low as possible. But by reducing the Bernstein bound, we can use methods based on sparse homotopies or sparse resultants to solve the problem with much better time bounds.

We have applied this idea to a problem in computer

vision. The problem, discussed in [FM90], is to determine the camera displacement given the x, y coordinates of a set of points from two views. The equations given there are:

$$(u_2 - u_0)(v_2 - v_1) = (v_2 - v_0)(u_2 - u_1)$$

$$(u_2 e_2 - u_0 e_0)(v_2 e_2' - v_1 e_1') = (v_2 e_2' - v_0 e_0')(u_2 e_2 - u_1 e_1)$$

$$(\delta_{01} u_2{}^2 + \delta_{02} u_1{}^2 + 2\delta_0 u_2 u_1)(\delta_2' v_0 v_1 - \delta_{01}' v_2{}^2 - \delta_0' v_1 v_2 - \delta_1' v_0 v_2) =$$
$$(\delta_{01}' v_2{}^2 + \delta_{02}' v_1{}^2 + 2\delta_0' v_2 v_1)(\delta_2 u_0 u_1 - \delta_{01} u_2{}^2 - \delta_0 u_1 u_2 - \delta_1 u_0 u_2)$$

$$(\delta_{01} u_2{}^2 + \delta_{12} u_0{}^2 + 2\delta_1 u_2 u_0)(\delta_2' v_0 v_1 - \delta_{01}' v_2{}^2 - \delta_0' v_1 v_2 - \delta_1' v_0 v_2) =$$
$$(\delta_{01}' v_2{}^2 + \delta_{12} p v_0{}^2 + 2\delta_0' v_2 v_0)(\delta_2 u_0 u_1 - \delta_{01} u_2{}^2 - \delta_0 u_1 u_2 - \delta_1 u_0 u_2)$$

The system is bi-homogeneous in the two sets of variables (u_0, u_1, u_2) and (v_0, v_1, v_2). The main result of the paper [FM90] is that this system has 10 solutions, meaning that there are 10 possiblities for the camera displacement. (In fact this result was proved earlier by Demazure, but the proof was non-elementary, and the proof in [FM90] is straightforward) If we evaluate the Bernstein bound for the system as given, we find that there should be 18 solutions.

But inspection of the equations shows that there are common subexpressions in the last two equations. We can introduce two new variables, x_1 and x_2 to describe these, leading to the apparently equivalent system:

$$(u_2 - u_0)(v_2 - v_1) = (v_2 - v_0)(u_2 - u_1)$$
$$(u_2 e_2 - u_0 e_0)(v_2 e_2' - v_1 e_1') = (v_2 e_2' - v_0 e_0')(u_2 e_2 - u_1 e_1)$$
$$x_1 = (\delta_2' v_0 v_1 - \delta_{01}' v_2{}^2 - \delta_0' v_1 v_2 - \delta_1' v_0 v_2)$$
$$x_2 = (\delta_2 u_0 u_1 - \delta_{01} u_2{}^2 - \delta_0 u_1 u_2 - \delta_1 u_0 u_2)$$
$$(\delta_{01} u_2{}^2 + \delta_{02} u_1{}^2 + 2\delta_0 u_2 u_1) x_1 = (\delta_{01}' v_2{}^2 + \delta_{02}' v_1{}^2 + 2\delta_0' v_2 v_1) x_2$$
$$(\delta_{01} u_2{}^2 + \delta_{12} u_0{}^2 + 2\delta_1 u_2 u_0) x_1 = (\delta_{01}' v_2{}^2 + \delta_{12}' v_0{}^2 + 2\delta_0' v_2 v_0) x_2$$

which has a sparse bound of 12, almost optimal. It seems strange that the second system has fewer solutions, but remember that Bernstein's theorem counts roots in $(\mathbb{C}^*)^n$. These solutions have all their coordinates non-zero. By introducing x_1 and x_2, we have forced the corresponding subexpressions to be non-zero for a valid solution. It was the vanishing of these expressions, together with the first two equations, that led to 6 spurious roots in the first system.

This kind of substitution is a useful heuristic for improving bounds, but we would like a better understanding of the phenomenon underlying it. Specifically, we would like to extend the theory of sparse systems from systems $f_1 = \cdots = f_n = 0$ to *compositions*

of polynomial maps $F_i : \mathbb{C}^{n_{i-1}} \to \mathbb{C}^{n_i}$, of the form $F_1(F_2(\cdots)) = 0$.

In the example above, the original system can be thought of as a composition of two maps $F_1(F_2())$. The map $F_2 : K^6 \to K^8$ has 8 output values which are $(u_0, u_1, u_2, v_0, v_1, v_2, x_1, x_2)$. The map $F_1 : K^8 \to K^4$ gives, say, the differences between LHS and RHS of the 4 equations. But the equations in F_1 are the reduced equations 5.1 which use x_1 and x_2 instead of the expressions they replace. By performing this decomposition, we have produced polynomial maps F_1 and F_2 with coefficients that are generic, or at least, more likely to be generic. Bounds for this system should be tighter than for the original equations because the dependencies have been removed.

Tight bounds for the number of roots of compositions of maps would have many implications. For example, the kinematics of any mechanism with rotary or sliding joints can be written in that form. The number of solutions for the mechanism could be determined algorithmically. As we have seen, an understanding of sparse bounds often leads to good algorithms for solving the system. If this too were possible for compositions of maps, efficient solvers could be constructed automatically. This is particularly important for new types of complex mechanisms, such as silicon microstructures, which have large numbers of interacting links.

5.2 Pfaffian equations

One of the most important extensions of the theory of the real numbers is to Pfaffian systems. These are systems which satisfy ordinary differential equations in a single dependent variable, time, of the form

$$\dot{X} = A(X)u$$

where X is a vector of states, A is a matrix whose entries are polynomials in X, and u is a vector of control inputs. If $B(X)$ is a matrix whose null rowspace is the column span of A, the system can be written in equivalent form as

$$B(X)\dot{X} = 0$$

Pfaffian systems include most of the systems studied by control theorists and many dynamic problems in other branches of engineering and physics. The complexity of deciding properties of such systems seems hopeless at first because of the possibility of infinite oscillation, limits cycles etc and chaotic behaviour. But if one places some reasonable constraints on the domain of the state variables, some remarkable properties emerge. This line of work began with the remarkable book by Khovanski [Kho91]. He showed, among other things, that Bezout-like bounds apply to the number of intersections of trajectories ("p-curves") of Pfaffian systems.

Recently there has been progress on extension of the theory of the reals with exponential functions [Ric92], which are the simplest examples of p-curves.

We plan to work on further extensions of the theory of the reals to include Pfaffian constraints, and eventually incorporate this capability into the toolkit.

Acknowledgements

This research was supported in part by a Packard Foundation Fellowship and in part by NSF PYI award IRI-8958577.

References

[Ber75] D. N. Bernstein. The number of roots of a system of equations. *(Translated from) Funktsional'nyi Analiz i Ego Prilozheniya*, 9(3):1–4, Jul-Sep 1975.

[Bet92] U. Betke. Mixed volumes of polytopes. *Arch. der Math.*, 58:388–391, 1992.

[BF91] J. Backelin and R. Fröberg. How we proved that there are exactly 924 cyclic 7-roots. In *International Symposium on Symbolic and Algebraic Computation*, pages 103–111, 1991. Bonn.

[BOKR86] M. Ben-Or, D. Kozen, and J. Reif. The complexity of elementary algebra and geometry. *J. Comp. and Sys. Sci.*, 32:251–264, 1986.

[Can88a] J.F. Canny. *The Complexity of Robot Motion Planning*. M.I.T. Press, Cambridge, 1988.

[Can88b] J.F. Canny. Some algebraic and geometric computations in PSPACE. In *ACM Symposium on Theory of Computing*, pages 460–467, 1988.

[Can90] J.F. Canny. Generalized characteristic polynomials. *Journal of Symbolic Computation*, 9(3), 1990.

[Can91a] J.F. Canny. Computing roadmaps of general semi-algebraic sets. In *AAECC-91*, 1991. New Orleans.

[Can91b] J.F. Canny. An improved sign determination algorithm. In *AAECC-91*, 1991. New Orleans.

[Can93a] John Canny. Computing roadmaps of general semi-algebraic sets. *Computer Journal*, 1993. Special Issue on Quantifier Elimination.

[Can93b] John Canny. Improved algorithms for sign-determination and existential quantifier elimination. *Computer Journal*, 1993. Special Issue on Quantifier Elimination.

[CE93] John Canny and Ioannis. Emiris. An efficient algorithm for the sparse mixed resultant. In G. Cohen, T. Mora, and O. Moreno, editors, *Proc. 10th Intern. Symp. on Applied Algebra, Algebraic Algorithms and Error-Correcting Codes*, pages 89–104. Springer Verlag, May 1993. Lect. Notes in Comp. Science 263.

[CR91] J.F. Canny and J.M. Rojas. An optimal condition for determining the exact number of roots of a polynomial system. In *International Symposium on Symbolic and Algebraic Computation*, 1991. Bonn, Germany.

[Dix08] A.L. Dixon. The eliminant of three quantics in two independent variables. *Proceedings of London Mathematical Society*, 6:49–69, 209–236, 1908.

[EC93] Ioannis Emiris and John Canny. A practical method for the sparse resultant. In M. Bronstein, editor, *Proc. of ACM Int. Symp. Symbolic Algebr. Computation*, pages 183–192, Kiev, July 1993.

[FIKY88] T. S. Freeman, G. Imirzian, E. Kaltofen, and L. Yagati. Dagwood: A system for manipulating polynomials given by straight-line programs. *ACM Trans. Math. Software*, 14(3):218–240, 1988.

[FM90] O. D. Faugeras and S. Maybank. Motion from point matches: Multiplicity of solutions. *International Journal of Computer Vision*, 4(3):225–246, 1990.

[GKZ90] I. M. Gel'fand, M. M. Kapranov, and A. V. Zelevinsky. Discriminants of polynomials in several variables and triangulations of Newton polytopes. *(Translated from) Algebara i Analiz*, 2:1–62, 1990. (English Translation in) *Leningrad Math. J.* 2 (1991).

[GV92] D.Y. Grigor'ev and N.N. Vorobjov. Counting connected components of a semialgebraic set in subexponential time. *Computational Complexity*, 2:133–186, 1992.

[Hof90] C.M. Hoffmann. Algebraic and numeric techniques for offsets and blends. In W. Dahmen, M. Gasca, and C. Micchelli, editors, *Computations of Curves and Surfaces*, pages 499–528. Kluwer Academic Publishers, 1990.

[HRS90] J. Heintz, M.F. Roy, and P. Solerno. Complexité du principe de Tarski-Seidenberg. *Bull. Soc. Math. France*, 118:101–126, 1990.

[HS92] Birkett Huber and Bernd Sturmfels. A polyhedral method for solving sparse polynomial systems. Cornell University, manuscript, 1992.

[Kal88] Erich Kaltofen. Greatest common divisors of polynomials given by straight-line programs. *Journal of the ACM*, 35(1):231–264, 1988.

[Kal89] Erich Kaltofen. Factorization of polynomials given by straight-line programs. In S. Micali, editor, *Randomness and Computation*, volume 5 of *Advances in Computing Research*, pages 375–412. JAI Press, Greenwhich, Connecticut, 1989.

[Kan92] J-M. Kantor. Sur le polynôme associé à un polytope à sommets entiers dans r^n. *C. R. Acad. Sci. Paris*, 314:669–672, 1992.

[Kho78] A.G. Khovanskii. Newton polyhedra and the genus of complete intersections. *(Translated from) Funktsional'nyi Analiz i Ego Prilozheniya*, 12(1):51–61, Jan-Mar 1978.

[Kho91] A. G. Khovanskii. *Fewnomials*. AMS Press, Providence, Rhode Island, 1991.

[Kus76] A.G. Kushnirenko. Newton polytopes and the Bezout theorem. *(Translated from) Funktsional'nyi Analiz i Ego Prilozheniya*, 10(3), Jul-Sep 1976.

[LL88] H.Y. Lee and C.G. Liang. A new vector theory for the analysis of spatial mechanisms. *Mechanisms and Machine Theory*, 23(3):209–217, 1988.

[Mac02] F.S. Macaulay. On some formula in elimination. *Proceedings of London Mathematical Society*, pages 3–27, May 1902.

[MC91a] D. Manocha and J.F. Canny. A new approach for surface intersection. *International Journal of Computational Geometry and Applications*, 1(4):491–516, 1991. Special issue on Solid Modeling.

[MC91b] Dinesh Manocha and John Canny. Efficient techniques for multipolynomial resultant algorithms. In *ISSAC-91*, 1991. Bonn, Germany.

[MC92a] D. Manocha and J.F. Canny. The implicit representation of rational parametric surfaces. *Journal of Symbolic Computation*, 13:485–510, 1992.

[MC92b] D. Manocha and J.F. Canny. Real time inverse kinematics of general 6R manipulators. In *IEEE Conference on Robotics and Automation*, pages 383–389, 1992.

[MC93] Dinesh Manocha and John Canny. Multipolynomial resultant algorithms. *Journal of Symbolic Computation*, 15(2):99–122, 1993.

[Mer92] J-P. Merlet. Direct kinematics and assembly modes of parallel manipulators. *International Journal of Robotics Research*, 11(2):150–162, 1992.

[MS87] A.P. Morgan and A.J. Sommese. A homotopy for solving general polynomial systems that respects m-homogeneous structures. *Applied Mathematics and Computations*, 24:101–113, 1987.

[Owe91] J. C. Owen. Algebraic solution for geometry from dimensional constraints.

In J. Rossignac and J. Turner, editors, *Symp. on Solid Modeling Foundations and CAD/CAM Applications*, pages 397–407. ACM Press, 1991.

[Pri91]   Doug Priest. Algorithms for arbitrary precision floating point arithmetic. In *Proc. 10th Symp. on Computer Arithmetic*, 1991. edited by K. Kornerup and D. Matula.

[PS91]   Paul Pedersen and Bernd Sturmfels. Product formulas for sparse resultants and chow forms. Manuscript, 1991.

[Ren89]   J. Renegar. On the computational complexity and geometry of the first-order theory of the reals, parts I, II and III. Technical Report 852,855,856, Cornell University, Operations Research Dept., 1989.

[Ren92]   J. Renegar. On the computational complexity of the first-order theory of the reals, parts I, II, III. *Journal of Symbolic Computation*, 13(3):255–352, 1992.

[Ric92]   Dan Richardson. Computing the topology of a bounded non-algebraic curve in the plane. *Journal of Symbolic Computation*, 14(6), 1992.

[RR89]   M. Raghavan and B. Roth. Kinematic analysis of the 6r manipulator of general geometry. In *International Symposium on Robotics Research*, pages 314–320, Tokyo, 1989.

[Stu92]   Bernd Sturmfels. Sparse systems of polynomial equations. Lecture Notes, NSF Regional Geometry Institute, Amherst College, Amherst, Mass., July 1992.

[SV89]   M.W. Spong and M. Vidyasagar. *Robot Dynamics and Control*. John Wiley and Sons, 1989.

[SZ93]   Bernd Sturmfels and André Zelevinsky. Multigraded resultants of sylvester type. *Journal of Algebra*, 1993. to appear.

[TM85]   L.W. Tsai and A.P. Morgan. Solving the kinematics of the most general six and five-degree-of-freedom manipulators by continuation methods. *Transactions of the ASME, Journal of Mechanisms, Transmissions and Automation in Design*, 107:189–200, 1985.

[VC92]   J. Verschelde and R. Cools. Symbolic homotopy construction. manuscript, 1992.

6   Appendix: Mixed Subdivisions

Let Q denote the Minkowski Sum of all the Q_i:

$$Q = Q_1 + Q_2 + \cdots + Q_{n+1} \subset \mathbb{R}^n$$

Define an $(n+1)$-argument vector sum $\oplus : (\mathbb{R}^n)^{(n+1)} \to \mathbb{R}^n$ as $(p_1, \ldots, p_{n+1}) \mapsto p_1 + \cdots + p_{n+1}$, where $p_i \in \mathbb{R}^n$. Q may be thought of as the image of $Q_1 \times \cdots \times Q_{n+1}$ under $\oplus$. This is clearly a many-to-one mapping, but it is desirable to define a unique inverse by regularization. That is, for each $q \in Q$ we choose a unique $(p_1, \ldots, p_{n+1})$ in $\oplus^{-1}(q) \cap Q_1 \times \cdots \times Q_{n+1}$. To achieve this, the method outlined in [Stu92] is employed.

Choose $n + 1$ generic linear forms $l_1, \ldots, l_{n+1} \in \mathbb{Z}[x_1, \ldots, x_n]$. Then the regularized inverse under $\oplus$ of of $q \in Q$ is the point $p = (p_1, \ldots, p_{n+1}) \in Q_1 \times \cdots \times Q_{n+1}$ minimizes

$$l(p) = \sum_{i=1}^{n+1} l_i(p_i)$$

There is also a geometric interpretation of the inverse. Define, for $1 \le i \le n + 1$, *lifted* Newton polytopes

$$\hat{Q}_i \triangleq \{(p_i, l_i(p_i)) : p_i \in Q_i\} \subset \mathbb{R}^{n+1}.$$

Let the Minkowski Sum of the lifted Newton polytopes be

$$\hat{Q} = \hat{Q}_1 + \cdots + \hat{Q}_{n+1} \subset \mathbb{R}^{n+1}.$$

The $(n + 1)$-st coordinate of a point in $\hat{Q}_i$ is to be interpreted as the cost of using that particular point in the vector sum, that is, as the value of $l(p)$. To minimize $l(p)$, we simply choose the lowest point in $\hat{Q}$ which lies over q.

Let $\pi : \mathbb{R}^{n+1} \to \mathbb{R}^n$ denote projection on the first n coordinates, and $h : \mathbb{R}^{n+1} \to \mathbb{R}$ denote projection on the $(n + 1)$st. Let $s : \mathbb{R}^n \to \mathbb{R}^{n+1}$ map points in Q to points on the lower envelope of $\hat{Q}$ above them:

$$s(q) = \hat{q} \in \pi^{-1}(q) \cap \hat{Q} \text{ such that } h(\hat{q}) \text{ is minimized}$$

The lower envelope of $\hat{Q}$ is then $s(Q)$. For generic choices of l_i's every point $\hat{q}$ on the lower envelope can be *uniquely* expressed as a sum of points $\hat{q}_1 + \cdots + \hat{q}_{n+1}$ with $\hat{q}_i \in \hat{Q}_i$. See [Stu92] or [Bet92] for an explanation.

Let $\hat{\Delta}$ denote the natural (coarsest) polyhedral subdivision of the lower envelope of $\hat{Q}$. Each facet (n-dimensional face) of $\hat{\Delta}$ is a Minkowski sum $\hat{F}_1 + \cdots + \hat{F}_{n+1}$ with $\hat{F}_i$ a face of $\hat{Q}_i$, and because lower envelope points have unique expressions as sums,

$$\sum_{i=1}^{n+1} \dim(\hat{F}_i) = n$$

The image of $\hat{\Delta}$ under π induces a polyhedral subdivision Δ of Q.

Definition 9 *The subdivision Δ is a* mixed subdivision *of the Minkowski sum Q.*

The facets of Δ are of the form $F_1 + \cdots + F_{n+1}$ with the same dimension property as $\hat{\Delta}$, a corollary of which is the following:

Observation For every facet $F = F_1 + \cdots + F_{n+1}$ in Δ, F_i a face of Q_i, at least one of the F_i is zero-dimensional, i.e. a vertex.

Definition 10 A *mixed facet* of the induced subdivision is a facet which is a sum $F_1 + \cdots + F_{n+1}$ where *exactly one F_i is a vertex*. Thus the remaining F_j for $j \neq i$ are edges.

Computing a Set of Points meeting every Cell Defined by a Family of Polynomials on a Variety

Saugata Basu *CIMS, New York University, New York, NY USA*

Richard Pollack, *CIMS, New York University, New York, NY USA*

Marie-Françoise Roy, *IRMAR, Université de Rennes, Campus de Beaulieu, Rennes ,FRANCE*

We consider a family of s polynomials $\mathcal{P} = \{P_1, \ldots, P_s\}$ in k variables with coefficients in an ordered domain D contained in a real closed field R, each of degree at most d, and a variety $\mathcal{V}$ of real dimension k', defined as the zero set of a family of polynomials $\mathcal{Q}$, where each polynomial in $\mathcal{Q}$ has coefficients in D and has degree at most d. We prove that the number of semi-algebraically connected components of each non- empty sign condition on $\{P_1, \ldots, P_s\}$, over $\mathcal{V}$ is $s^{k'}(O(d))^k$. We also present three different algorithms for computing a set of points, of size $s^{k'}(O(d))^k$, intersecting each non- empty semi-algebraically connected component of $\mathcal{V}$ intersected with each sign condition of $P_1, \ldots, P_s$. The bound on the number of points produced by the last of these algorithms provides the claimed bound on the number of these connected components. The output is the list of points together with the sign condition at each point. This interpolates a sequence of theorems between the algorithm of Ben-Or, Kozen and Reif [3] which is the case $k' = 0$, in one variable, and the algorithm of Basu- Pollack-Roy ([2]) which is the case $k' = k$.

1 Introduction:

A sign condition for a set of s polynomials $\mathcal{P} = \{P_1, \ldots, P_s\}$ is specified by a sign vector $\sigma \in \{-1, 0, +1\}^s$ and the sign condition σ is called *non-empty over a variety $\mathcal{V}$* (with respect to $\mathcal{P}$) if there is a point $x \in \mathcal{V}$ such that

$$\sigma = (\text{sign}(P_1(x)), \ldots, \text{sign}(P_s(x))) = \sigma_{\mathcal{P}}(x).$$

For a sign condition σ over $\mathcal{V}$, the *realization space* of σ over $\mathcal{V}$ is the set $R(\sigma) = \{x \mid x \in \mathcal{V}, \sigma = \sigma_{\mathcal{P}}(x)\}$. If $R(\sigma)$ is not empty then each of these non-empty semi-algebraically connected components is called a *cell* of $\mathcal{P}$ on $\mathcal{V}$.

We say that a family $\mathcal{P} = \{P_1, \ldots, P_s\}$ of polynomials in k variables is in *general position* with respect to a variety $\mathcal{V}$ of real dimension k', if no $k' + 1$ polynomials of $\mathcal{P}$ have a common zero in $\mathcal{V}$. This is a weak notion of general position as we assume neither smoothness nor transversality.

For a finite set of polynomials $\mathcal{Q}$ we write $Z(\mathcal{Q})$ for the set of common zeros in R^k of the polynomials in $\mathcal{Q}$.

We say that a family $\mathcal{Q} = \{Q_1, \ldots, Q_{k-k'}\}$ is *transversal* if at each point of $Z(\mathcal{Q})$ the gradients of $Q_1, \ldots, Q_{k-k'}$ are linearly independent.

We will prove the following theorems:

Theorem 1 *Let $\mathcal{Q}$ and $\mathcal{P}$ be finite sets of polynomials in k variables of degrees at most $d \geq 2$, with coefficients in a real closed field R such that $\mathcal{V} = Z(\mathcal{Q})$ is a variety of real dimension k'. Then,*

$$\sum_{|\sigma| \geq 1} |\sigma| = s^{k'}(O(d))^k,$$

where $|\sigma|$ denotes the number of semi-algebraically connected components in the realization space of the sign condition σ over $\mathcal{V}$ with respect to $\mathcal{P}$ (these are also called the cells *of $\mathcal{P}$ on $\mathcal{V}$).*

The bound in theorem 1 on the number of cells of $\mathcal{P}$ on the variety $\mathcal{V}$ is separated into a combinatorial part (the dependence on s) which depends only on the real dimension of the variety and an algebraic part (the dependence on d) which depends on the dimension k of the ambient space.

Theorem 1 is a corollary of the following algorithmic theorem.

Theorem 2 *Let D be a subring of the real closed field R and $\mathcal{V} = Z(\mathcal{Q})$ be a variety of real dimension k' where $\mathcal{Q}$ is a finite set of m polynomials in*

Algorithmic Foundations of Robotics
1-56881-045-8

$D[X_1, \ldots, X_k]$, each of degree at most d and such that the polynomial $Q = \sum_{Q_i \in \mathcal{Q}} Q_i^2$ is already computed. Let $\mathcal{P} = \{P_1, \ldots, P_s\} \subset D[X_1, \ldots, X_k]$ with each $P_i \in \mathcal{P}$ of degree at most d.

There is an algorithm $\mathcal{A}$, which determines whether the family $\mathcal{P}$ is in general position with respect to $\mathcal{V}$ and if it is in general position outputs a set of cardinality $s^{k'}O(d)^k$, and which contains at least one point algebraic over D in each non-empty cell of $\mathcal{P}$ on $\mathcal{V}$. The algorithm also provides the signs of all the polynomials of $\mathcal{P}$ at each of these points. Algorithm $\mathcal{A}$ uses at most $\binom{O(s)}{k'}sd^{O(k)} = s^{k'}sd^{O(k)}$ arithmetic operations in D.

If $\mathcal{Q}$ has $k - k'$ elements and $\mathcal{Q}$ is transversal, independent of whether or not the polynomials in $\mathcal{P}$ are in general position with respect to $\mathcal{V}$, we use an algorithm, $\mathcal{B}$, which has the same output as algorithm $\mathcal{A}$ and also uses at most $\binom{O(s)}{k'}sd^{O(k)} = s^{k'}sd^{O(k)}$ arithmetic operations in D. Note that if $k = k'$, then the conditions for applying algorithm $\mathcal{B}$ are vacuously satisfied.

In the general case, with no assumptions on the polynomials of $\mathcal{Q}$ or of $\mathcal{P}$, there is an algorithm $\mathcal{C}$, which outputs a set X of cardinality $s^{k'}O(d)^k$ together with the sign vector $\sigma_{\mathcal{P}}(x)$ at each $x \in X$. For every non-empty cell C of $\mathcal{P}$ on $\mathcal{V}$ there is at least one point x in X which is

1. algebraic over a ring D' which contains D,

2. the extension (see item 9 of section 2) of C to a real closed field R' which contains $D' \cup R$, contains x.

This algorithm uses at most $s^{k'+1}d^{O(kk')}$ arithmetic operations in D.

It is worth noting, that using less than $((k-k')d)^{O(k)}$ arithmetic operations in D, we can check whether the conditions for applying algorithm $\mathcal{B}$ are satisfied.

A Remark about our measure of complexity:

We say that an algorithm is *well-behaved* if it satisfies the following conditions.

- If $D = \mathbf{Z}$, the bit-size of the output and of any intermediate computation is obtained by multiplying the number of arithmetic operations by the bit-size of the inputs.

- It is well parallelizable i.e. it can be described by an arithmetic network whose depth is a polynomial in the log of the sequential complexity (see [14]).

All our algorithms are all well-behaved, so we define the complexity of our algorithms to be the number of arithmetic operations in the ring D, keeping in mind that the bit size of the integer coefficients and consequently the number of bit-operations as well, can be easily evaluated.

Algorithms $\mathcal{B}$ and $\mathcal{C}$ are obtained by making suitable perturbations and reducing the problem to one in which Algorithm $\mathcal{A}$ applies.

An important aspect of our algorithms is that its combinatorial complexity depends on the real dimension of the variety rather than on the dimension of the ambient space. It is this dependence on k' which makes possible the bound in theorem 1. In the case $k = k'$ we recover the result of Basu-Pollack-Roy ([2]) using Algorithm $\mathcal{B}$, noting that the requirements on the variety $\mathcal{V}$ are vacuously satisfied in this case (see also [7]). In the case $k' = 0$ we have a better version of this result, theorem 3, which improves [18].

We gratefully acknowledge Michel Coste for many clarifying conversations.

2 Mathematical Preliminaries

We shall need the terminology and properties of infinitesimals, semi-algebraically connected components and paths in non-archimedean extensions. A full discussion of these can be found in [4] but we offer a brief summary below.

The order relation on a real closed field R defines, as usual, the euclidean topology on R^k. A semi-algebraic set is the finite union of sets defined by a finite number of polynomial equalities and inequalities, and a semi-algebraic map is one whose graph is a semi-algebraic set.

In particular we have the following elementary properties of semi- algebraic sets, over a real closed field R (see also [4]).

1. A semi-algebraic set S is *semi-algebraically connected* if it is not the disjoint union of two non-empty closed semi-algebraic sets in S.

2. A *semi-algebraically connected component* of a semi-algebraic set S is a maximal semi-algebraically connected subset of S.

3. A semi-algebraic set has a finite number of semi-algebraically connected components, each of which is a semi- algebraic set.

4. Let D be a subring of the real closed field R. A semi-algebraic set S is *defined over* D if it can be described by polynomials with coefficients in D.

5. A *semi- algebraic path* between x and x' in R^k is a semi-algebraic subset γ, which is the image of a semi-algebraic continuous map f_γ, which maps the unit interval of R to γ, with $f_\gamma(0) = x$ and $f_\gamma(1) = x'$. A semi-algebraic path γ is *defined over* D if the graph of f_γ is defined over D.

6. A semi-algebraic set is semi-algebraically connected if and only if it is semi-algebraically path connected.

7. In the case of real numbers, semi-algebraically connected components of semi- algebraic sets are ordinary connected components.

8. The *curve selection lemma* holds for semi-algebraic sets. Given a semi-algebraic set S and a point x in the closure $\bar{S}$ of S there exists a semi-algebraic continuous function f defined on $[0, 1]$ with $f(0) = x$ and $f((0, 1]) \subset S$.

9. Let S be a semi-algebraic set in R^k , and R' a real closed field containing R, we write $S_{R'}$ for the *extension of S to R'* i.e. the subset of R'^k defined by the same boolean combination of inequalities that define S.

10. Let S be defined over $D \subset R$ and let R' be a real closed field extension of R. The semi-algebraically connected components of $S_{R'}$ are the extensions to R' of the semi-algebraically connected components of S.

We also need the following definitions and properties of ordered rings and Puiseux series. Again, fuller details can be found in [4]. We suggest that in order to understand the following definitions it is helpful to think of ϵ as positive and very small.

11. For a subring $D \subset R$ we define the order on $D[\epsilon]$ making ϵ infinitesimal and positive by saying that $P \in D[\epsilon]$ is positive if and only if the tail coefficient of P, i.e. the coefficient of the least power of ϵ in P, is positive. Similarly for $Q \in R(\epsilon)$, the order making ϵ infinitesimal and positive is defined by saying that Q is positive if and only if the tail

coefficients of the numerator and denominator of Q have the same sign. It follows that $P(\epsilon) > 0$ if and only if for $t \in R$ sufficiently small and positive, $P(t) > 0$.

12. Let $R\langle\epsilon\rangle$ (resp. $\bar{R}\langle\epsilon\rangle$) be the field of Puiseux series in ϵ with coefficients in R (resp. in $\bar{R} = R[i]$). Its elements are 0 and series of the form

$$\sum_{i \geq i_0, i \in \mathbf{Z}} a_i \epsilon^{\frac{i}{q}}$$

with $i_0 \in \mathbf{Z}$, $a_i \in R$ (resp. $\bar{R}$), $a_{i_0} \neq 0$ and $q \in \mathbf{N}$. The field $R\langle\epsilon\rangle$ is real closed ([4]) and the field $\bar{R}\langle\epsilon\rangle$ is algebraically closed ([22]). An element of $R\langle\epsilon\rangle$ is *infinitesimal* (with respect to R) if and only if its absolute value is strictly smaller than any positive element in R. The element ϵ of $R\langle\epsilon\rangle$ is infinitesimal positive. The elements of $R\langle\epsilon\rangle$ bounded over R form a *valuation ring* denoted $V(\epsilon)$; the elements of $V(\epsilon)$ are 0 or Puiseux series

$$\sum_{i \in \mathbf{N}} a_i \epsilon^{\frac{i}{q}}.$$

We denote by eval_ϵ the ring homomorphism from $V(\epsilon)$ to R which maps $\sum_{i \in \mathbf{N}} a_i \epsilon^{\frac{i}{q}}$ to a_0. A Puiseux series is infinitesimal if and only if it is mapped by eval_ϵ to 0. We can think of eval_ϵ as the evaluation of the Puiseux series at 0. Whenever we write $\mathrm{eval}_\epsilon(S)$ we understand it to be the map eval_ϵ restricted to points of S where eval_ϵ is defined.

13. An element $\alpha \in R$, which is a root of a polynomial $f(t) \in D[t]$, is uniquely specified by the polynomial f, and the sign vector

$$(\mathrm{sign}(f(\alpha)), \mathrm{sign}(f'(\alpha)), \ldots, \mathrm{sign}(f^{(deg(f))}(\alpha))),$$

known as the Thom encoding ([10]) of the root α.

14. If α belongs to the valuation ring of $R\langle\epsilon\rangle$ and $f(\alpha) = 0$, where $f \in D[\epsilon, 1/\epsilon][t]$, and $f(t) = \sum_{\nu(f) \leq i} f_i(t) \epsilon^i$, for some integer $\nu(f) \geq 0$, then $\mathrm{eval}_\epsilon(\alpha) \in R$, and is a root of the polynomial $f_{\nu(f)}(t)$.

15. Suppose a Puiseux series $a(\epsilon) \in R\langle\epsilon\rangle$ is algebraic over $D[\epsilon]$. Then it is a root of a polynomial $P_\epsilon(X)$ with coefficients in $D[\epsilon]$. The root $a(\epsilon)$ is distinguished among the other roots of P_ϵ, by its Thom encoding (see [10]). For t sufficiently small and positive, we may replace ϵ by t in the polynomials

which are the coefficients of $P_\epsilon(X)$ to obtain the polynomial $P_t(X)$ and (since inequalities in $D[\epsilon]$ will agree with the corresponding inequalities in $D[t]$ for t positive and small enough) $P_t(X)$ will have the same number of roots in R as $P_\epsilon(X)$ has in $R\langle\epsilon\rangle$, with the corresponding Thom encodings and the algebraic Puiseux series $a(\epsilon)$ thus defines a semi-algebraic function $a(t)$ for t small enough. Conversely a semi-algebraic function from $(0,t)$ into R^k (where $t > 0$ is in R) defined over D gives a point in $R\langle\epsilon\rangle^k$.

16. Let $S(\epsilon)$ be a semi-algebraic set in $(R\langle\epsilon\rangle)^k$ defined over $D[\epsilon]$ and for $t \in R$ let $S(t)$ be the semi-algebraic set in R^k obtained by substituting t for ϵ. Let $P(S(\epsilon))$ be a property of the semi-algebraic set $S(\epsilon)$ in $R^k\langle\epsilon\rangle$ defined over $D[\epsilon]$ which is expressible by a first order formula $\Phi(\epsilon)(X_1,\dots,X_k)$ with parameters in $D[\epsilon]$. Then for t positive and small enough $P(S(t))$ is satisfied. This is an easy consequence of quantifier elimination and item 15.

We will use this property to replace ϵ by t in a path defined over $D[\epsilon]$ to obtain a "neighboring" path defined over $D[t]$.

We now develop some mathematical tools that will be useful for the proofs of our theorems.

The first proposition (see also [15]) gives information about the eval map.

Proposition 1 *If $S' \subset R\langle\epsilon\rangle^k$ is a semi-algebraic set defined over $D[\epsilon]$ and $S = \text{eval}_\epsilon(S')$, then S is a semi-algebraic set. Moreover, if S' is connected then S is connected.*

Proof: Suppose that $S' \subset (R\langle\epsilon\rangle)^k$ is described by a quantifier-free formula $\Phi(\epsilon)(X_1,\dots,X_k)$. Introduce a new variable X_{k+1} and denote by $\Phi(X_1,\dots,X_k,X_{k+1})$ the result of the substitution of X_{k+1} instead of ϵ in $\Phi(\epsilon)(X_1,\dots,X_k)$.

Embed R^k in R^{k+1} by sending $(X_1,\dots,X_k)$ to $(X_1,\dots,X_k,0)$ so that S is a subset of $Z(X_{k+1})$. We prove that $S = \overline{T} \cap Z(X_{k+1})$ where

$$T = \{ \ (x_1,\dots,x_k,x_{k+1}) \in R^{k+1}|$$
$$\Phi((x_1,\dots,x_k,x_{k+1}) \text{ and } x_{k+1} > 0\}$$

and $\overline{T}$ is the closure of T in the euclidean topology.

If $x \in S$ then there exists $z \in S'$ such that $\text{eval}_\epsilon(z) = x$. Let $B_x(r)$ denote the open ball of radius r centered

at x. Since (z, ϵ) belongs to the extension of $B_x(r) \cap T$ to $R\langle\epsilon\rangle$ it follows that $B_x(r) \cap T$ is non-empty, and hence that $x \in \overline{T}$.

Conversely, let x be in $\overline{T} \cap Z(X_{k+1})$. Using the semi-algebraic curve selection lemma (see item 8 or[4]), there exists a semi-algebraic function f from $[0,1]$ to $\overline{T}$ with $f(0) = x$ and $f((0,1]) \subset T$. This semi-algebraic function defines a point z whose coordinates lie in $R\langle\epsilon\rangle$ (see item 15 above) and belongs to S' and moreover $\text{eval}_\epsilon(z) = x$.

If S' is connected then there exists a positive t in R such that $T \cap (R^k \times [0,t])$ is semi-algebraically connected. It follows easily that $S = \overline{T} \cap Z(X_{k+1})$ is connected. □

The following Proposition makes it possible to reduce the problem of constructing points on connected components of semi-algebraic sets to the problem of constructing points on connected components of certain algebraic sets. It appears in [2].

Proposition 2 *Let C be a connected component of a non-empty sign condition of the form $P_1 = \dots = P_\ell = 0, P_{\ell+1} > 0,\dots,P_s > 0$, then we can find an algebraic set V in $R\langle\epsilon\rangle^k$ defined by equations $P_1 = \dots = P_\ell = P_{i_1} - \epsilon = \dots P_{i_m} - \epsilon = 0$, such that a connected component of V is contained in $C_{R\langle\epsilon\rangle}$.*

Proof: If C is closed, it is a semi-algebraically connected component of the algebraic set defined by $P_1 = \dots = P_\ell = 0$. If not, we consider Γ, the set of all semi-algebraic paths γ in R going from some point $x(\gamma)$ in C to a $y(\gamma)$ in $\overline{C} \setminus C$ such that $\gamma \setminus \{y(\gamma)\}$ is entirely contained in C. For any $\gamma \in \Gamma$, there exists an $i > \ell$ such that P_i vanishes at $y(\gamma)$. Then on $\gamma_{R\langle\epsilon\rangle}$ there exists a point $z(\gamma,\epsilon)$ such that one of the $P_i - \epsilon$ vanishes at $z(\gamma,\epsilon)$ and that on the portion of the path between x and $z(\gamma,\epsilon)$ no such $P_i - \epsilon$ vanishes. We write I_γ for the set of indices between $\ell+1$ and s such that $i \in I_\gamma$ if and only if $P_i(z(\gamma,\epsilon)) - \epsilon = 0$. Now choose a path $\gamma \in \Gamma$ so that the set $I_\gamma = \{i_1,\dots,i_m\}$ is maximal under set inclusion and let V be defined by $P_1 = \dots = P_\ell = P_{i_1} - \epsilon = \dots = P_{i_m} - \epsilon = 0$.

It is clear that at $z(\gamma,\epsilon)$, defined above, we have $P_{\ell+1} > 0,\dots,P_s > 0$ and $P_j - \epsilon > 0$ for every $j \notin I_\gamma$. Let C' be the semi-algebraically connected component of V containing $z(\gamma,\epsilon)$. We shall prove that no polynomial $P_{\ell+1},\dots,P_s$ vanishes on this semi-algebraically connected component, and thus that C' is contained in $C_{R\langle \epsilon \rangle}$.

Let us suppose on the contrary that some new P_i ($i > l, i \notin I_\gamma$) vanishes on C', say at y_ϵ. We can suppose without loss of generality that y_ϵ is defined over $D[\epsilon]$. Take a semi-algebraic path γ_ϵ defined over $D[\epsilon]$ connecting $z(\gamma, \epsilon)$ to y_ϵ with $\gamma_\epsilon \subset C'$. Denote by $z(\gamma_\epsilon, \epsilon)$ the first point of γ_ϵ with $P_1 = \ldots = P_\ell = P_{i_1} - \epsilon = \ldots = P_{i_m} - \epsilon = P_j - \epsilon = 0$ for some new j not in I_γ.

For t in R small enough, the set γ_t (obtained by replacing ϵ by t in γ_ϵ) defines a semi-algebraic path from $z(\gamma, t)$ to $z(\gamma_\epsilon, t)$ contained in C. Replacing ϵ by t in the Puiseux series which give the coordinates of $z(\gamma_\epsilon, \epsilon)$ defines a path γ' containing $z(\gamma_\epsilon, \epsilon)$ from $z(\gamma_\epsilon, t)$ to $y = \text{eval}(z(\gamma_\epsilon, \epsilon))$ (which is a point of $\bar{C} \setminus C$). Let us consider the new path γ^* consisting of the beginning of γ (up to the point z_t for which $P_{i_1} = \ldots, P_{i_m} = t$), followed by γ_t and then followed by γ'. Now the first point in γ^* such that there exists a new j with $P_j - \epsilon = 0$ is $z(\gamma_\epsilon, \epsilon)$ and thus $\gamma^* \in \Gamma$ with I_{γ^*} strictly larger than I_γ. This is impossible by the maximality of I_γ. $\qquad \square$

When the polynomials $\mathcal{P}$ are not in general position with respect to $\mathcal{V}$ we shall make various perturbations of the polynomials of $\mathcal{P}$ or $\mathcal{V}$ so that the perturbed polynomials satisfy this condition.

For Algorithm $\mathcal{B}$, we will use the following propositions which involve the adjunction of three additional infinitesimals.

First we introduce some notation. Let $\delta, \delta', \delta''$ be new variables.

We order $D[\delta]$, (resp. $D[\delta, \delta']$, $D[\delta, \delta', \delta'']$), so that δ is positive and smaller than any positive element of D (resp. δ' positive and smaller than any positive element of $D[\delta]$, δ'' positive and smaller than any positive element of $D[\delta, \delta']$). In short, the ordering is $\delta \gg \delta' \gg \delta''$ where $a \gg b$ stands for , " b is positive and infinitesimal with respect to a".

Let $R\langle \delta \rangle$ be the field of Puiseux series in δ with coefficients in R, $R\langle \delta, \delta' \rangle$ the field of real Puiseux series in δ' with coefficients in $R\langle \delta \rangle$, $R\langle \delta, \delta', \delta'' \rangle$ the field of real Puiseux series in δ'' with coefficients in $R\langle \delta, \delta' \rangle$.

Let $V(\delta, \delta', \delta'')$ be the valuation ring of elements of $R\langle \delta, \delta', \delta'' \rangle$ bounded by elements of $R\langle \delta \rangle$, and $\text{eval}_{\delta'}$ the

corresponding map from $V(\delta, \delta', \delta'')$ to $R\langle \delta \rangle$ sending δ' and δ'' to 0.

In Algorithm $\mathcal{B}$, we follow Renegar [19] and perturb the polynomials in $\mathcal{P}$, and $\mathcal{Q}$, using certain special polynomials $H_i = 1 + \sum_{1 \le j \le k} i^j X_j^{d'}$, for $1 \le i \le s + k - k'$, where d' is the least even integer greater than d, the maximum degree of the polynomials in $\mathcal{P}$ and $\mathcal{Q}$.

Below we prove some useful properties of these polynomials. Henceforth, for any polynomial $P(X_1, \ldots, X_k)$, we let $P^h(X_0, X_1, \ldots, X_k)$ denote the homogenization of P with respect to an additional variable X_0.

Lemma 1 The $H_i^h = (X_0^{d'} + \sum_{1 \le j \le k} i^j X_j^{d'})$ are in general position in projective (over $\bar{R} = R[i]$, the algebraic closure of R) k-space.

Proof: Take $H = (X_0^{d'} + \sum_{1 \le j \le k} t^j X_j^{d'})$. If $k + 1$ of the H_i^h (say $H_{j_1}^h, \ldots, H_{j_{k+1}}^h$) had a common zero $\bar{x}$, in projective k-space (over $\bar{R}$), substituting this root in H would give a non-zero univariate polynomial of degree at most k with $k+1$ distinct roots $(j_1, \ldots, j_{k+1})$, which is impossible. $\qquad \square$

Since a common zero of a set of H_i would certainly produce a common zero of the corresponding set of H_i^h we have:

Corollary 1 The polynomials H_i are in general position.

Lemma 2 The polynomials $(1-\delta)P_i + \delta H_i$ are in general position.

Proof: Let $Q_{i,t} = (1-t)P_i + tH_i$. Consider the set T, of those t such that the $Q_{i,t}$ are in general position in projective k-space over $\bar{R}$, which means that no $k + 1$ subset of the $Q_{i,t}^h$ have a common zero in projective k-space over $\bar{R}$. The set T contains 1, according to the preceding lemma. Hence it contains an open interval containing 1, since being in general position in projective k-space over $\bar{R}$ is a stable condition. It is also Zariski constructible, since it can be defined by a first order formula of the language of algebraically closed fields. So the transcendental element δ belongs to the extension of T to $R\langle \delta \rangle$, which proves the lemma. $\qquad \square$

In Algorithm $\mathcal{B}$, we perturb $\mathcal{P}$, and $\mathcal{Q}$ so that the perturbed $\mathcal{P}$ is in general position with respect to the perturbed $\mathcal{Q}$. The following proposition ensures that after finding points in every cell on the perturbed variety, we can go back to the original variety using an eval map and still get a point in every cell.

Proposition 3 *Let us suppose that the variety $\mathcal{V}$ is described by a set of transversal polynomials $\mathcal{Q} = \{Q_1, \ldots, Q_{k-k'}\}$. Let $\mathcal{P}$ be any set of s polynomials, $P_1, \ldots, P_s$ and let C be a connected component of the sign condition,*

$$Q_1 = Q_2 = \ldots, Q_{k-k'} = 0,$$

$$P_1 = P_2 = \ldots = P_l = 0,$$

$$P_{l+1} > 0, \ldots, P_s > 0.$$

Then there is a connected component C' in $R\langle\delta, \delta', \delta''\rangle^k$, of the semi-algebraic set defined by the following equalities and inequalities,

$$(1 - \delta)Q_i - \delta''\delta'\delta H_{i+s} = 0, \ 1 \le i \le k - k',$$

$$-\delta'\delta H_i < P_i < \delta'\delta H_i, \ 1 \le i \le l,$$

$$P_i > \delta H_i, \ l + 1 \le i \le s,$$

such that $\mathrm{eval}_{\delta'}(C')$ is contained in the extension of C to $R\langle\delta\rangle$.

Proof: The proof is a consequence of the following lemma and of Proposition 1. □

Lemma 3 *Let $\mathcal{P}$ and $\mathcal{Q}$ be the same as in the above proposition. Given a point $x \in R^k$, satisfying the following equalities and inequalities:*

$$Q_1 = Q_2 = \ldots, Q_{k-k'} = 0,$$

$$P_1 = P_2 = \ldots = P_l = 0,$$

$$P_{l+1} > 0, \ldots, P_s > 0,$$

there exists a point $z \in R\langle\delta, \delta', \delta''\rangle^k$, satisfying the inequalities,

$$(1 - \delta)Q_i - \delta''\delta'\delta H_{i+s} = 0, \ 1 \le i \le k - k',$$

$$-\delta'\delta H_i < P_i < \delta'\delta H_i, \ 1 \le i \le l,$$

$$P_i > \delta H_i, \ l + 1 \le i \le s,$$

such that $\mathrm{eval}_{\delta'}(z) = x$.

Proof: The given point x satisfies the inequalities,

$$-\delta'\delta H_i < P_i < \delta'\delta H_i, \ 1 \le i \le l,$$

$$P_i > \delta H_i, \ l + 1 \le i \le s,$$

in the extension $R' = R\langle\delta, \delta'\rangle$. Since, these inequalities define an open set in R'^k, there exists a neighborhood of x in R'^k, where all these inequalities hold. Hence, we can choose an open ball B, centered at x, of radius $r < \delta'$, such that all points in B satisfy the above inequalities.

Let $Z(Q_i)$ denote the zero set of the polynomial Q_i, and let $Z' = Z(Q_2) \cap Z(Q_3) \cap \ldots Z(Q_{k-k'})$.

We can choose the ball B sufficiently small so that the gradients of $Q_1, \ldots, Q_{k-k'}$ are linearly independent throughout B.

The transversality hypothesis implies that there exists a point $y \in Z' \cap B$ such that $Q_1(y) > 0$. Choose a semi-algebraic path, γ from x to y in $Z' \cap B$. By the intermediate value theorem, there exists a point x' on the semi-algebraic path $\gamma_{R'\langle\delta''\rangle}$, such that $(1 - \delta)Q_1(x') = \delta''\delta'\delta H_{s+1}(x')$. Thus, x' satisfies $(1 - \delta)Q_1 - \delta''\delta'\delta H_{s+1} = Q_2 = \ldots = Q_{k-k'} = 0$, since δ'' is smaller than any positive element of R'.

Now replace Q_1 by Q_2, and Q_2 by $(1 - \delta)Q_1 - \delta''\delta'\delta H_{s+1}$, and we continue by working inside $R'\langle\delta''\rangle$. Since the gradients of $Q_1, \ldots Q_{k-k'}$ are linearly independent and $Q_2(x')$ is not infinitesimal with respect to δ'' we can find a point y' in $Z'' = Z((1 - \delta)Q_1 - \delta''\delta'\delta H_{s+1}) \cap Z(Q_3), \ldots \cap Z(Q_{k-k'}) \cap B$ such that $Q_2(y') > \delta''$. Choose a semi-algebraic path, γ' from x' to y' in $Z'' \cap B$. By the intermediate value theorem, there exists a point x'' on the semi-algebraic path γ', such that $(1-\delta)Q_2(x'') = \delta''\delta'\delta H_{s+2}(x'')$. Now, x'' satisfies $(1-\delta)Q_1 - \delta''\delta'\delta H_{s+1} = (1-\delta)Q_2 - \delta''\delta'\delta H_{s+2} = \ldots = Q_{k-k'} = 0$. Repeating this argument $k - k' - 1$ times we obtain a point $z' \in B$, which satisfies,

$$(1 - \delta)Q_i - \delta''\delta'\delta H_{i+s} = 0, \ 1 \le i \le k - k'.$$

Now let z be the closest point to x satisfying

$$(1 - \delta)Q_i - \delta''\delta'\delta H_{i+s} = 0, \ 1 \le i \le k - k'.$$

it is clear that $\mathrm{eval}_{\delta'}(z) = x$, which proves the lemma. □

For Algorithm $\mathcal{C}$, we shall perturb the polynomials by infinitesimals $\bar{\delta} = \{\delta, \delta_1, \ldots, \delta_s\}$ with $\delta_1 \gg \ldots \gg \delta_s \gg \delta$. The next proposition shows that this new set of polynomials, $\mathcal{P}_{\bar{\delta}} = \cup_{1 \leq i \leq s} \{P_i - \delta_i, P_i + \delta_i, P_i - \delta\delta_i, P_i + \delta\delta_i\}$, are in general position with respect to $\mathcal{V}$ so that we can invoke algorithm $\mathcal{A}$ for this perturbed set of polynomials. Proposition 5 will then be used to show that any cell C of the family $\mathcal{P}$, contains $\mathrm{eval}_\delta(C')$, where C' is a cell of the family $\mathcal{P}_{\bar{\delta}}$ and eval_δ sends δ to 0; the correctness of algorithm $\mathcal{C}$ will follow.

Proposition 4 *Given a family $\{P_1, \ldots, P_s\}$, of polynomials in $R[X_1, \ldots, X_k]$ and a variety $\mathcal{V}$ of real dimension k' and algebraically independent transcendentals $\delta, \delta_1, \ldots, \delta_s$, the perturbed family $\cup_{1 \leq i \leq s} \{P_i - \delta_i, P_i + \delta_i, P_i - \delta\delta_i, P_i + \delta\delta_i\}$, is in general position with respect to the variety $\mathcal{V}$.*

Proof: The argument depends on the following simple observations which are classical:

- If $\mathcal{V}$ has real dimension k' then $\mathcal{V}$ is the union of a finite number of connected semi-algebraic sets of real dimension less than or equal to k' whose Zariski closure is irreducible (see [4]).

- If C is a connected semi- algebraic set whose Zariski closure is irreducible then any polynomial is either constant on C or the zero set of the polynomial meets C in a semi-algebraic set of real dimension less than the dimension of C. This is immediate from the definition of irreducibility.

As a consequence, we see that any of the perturbed polynomials meets the variety $\mathcal{V}$ in a variety of lower real dimension. The proposition is proved by repeating this argument at most k' times. □

Let $R\langle \delta_1, \ldots, \delta_s \rangle$ be a real closed field containing the ordered ring $D[\delta_1, \ldots, \delta_s]$, $R\langle \delta_1, \ldots, \delta_s, \delta \rangle$ the field of real Puiseux series in δ with coefficients in $R\langle \delta_1, \ldots, \delta_s \rangle$, V the corresponding valuation ring and eval_δ the map from V to $R\langle \delta_1, \ldots, \delta_s \rangle$ sending δ to 0.

Proposition 5 *Let C be a nonempty connected component in $\mathcal{V} = Z(Q)$, of the sign condition $P_1 = \ldots = P_\ell = 0, P_{\ell+1} > 0, \ldots, P_s > 0$, and let C' be the extension of C to $R\langle \delta_1, \ldots, \delta_s \rangle$. Then C' contains some $\mathrm{eval}_\delta(C'')$, where C'' is a connected component of the semi-algebraic set defined by the sign conditions $Q = 0, -\delta\delta_1 < P_1 < \delta\delta_1, \ldots, -\delta\delta_\ell < P_\ell < \delta\delta_\ell, P_{\ell+1} > \delta_{\ell+1}, \ldots, P_s > \delta_s$ over $R\langle \delta_1, \ldots, \delta_s, \delta \rangle$.*

Proof: If $x \in C'$ then x satisfies the equalities and inequalities, $Q = 0, -\delta\delta_1 < P_1 < \delta\delta_1, \ldots, -\delta\delta_\ell < P_\ell < \delta\delta_\ell, P_{\ell+1} > \delta_{\ell+1}, \ldots, P_s > \delta_s$, in $R\langle \delta_1, \ldots, \delta_s, \delta \rangle$. Let C'' be the semi-algebraically connected component of the semi-algebraic set in $(R\langle \delta_1, \ldots, \delta_s, \delta \rangle)^k$ defined by the above equalities and inequalities, which contains x.

Using Proposition 1 it is easy to see that $\mathrm{eval}_\delta(C'')$ is contained in C', which proves the lemma. □

Remark: The reason for introducing the additional infinitesimal δ is that, otherwise, two connected components of a sign condition defined by $\mathcal{P}$ could be contained in the same connected component of a sign condition defined by the perturbed polynomials. To illustrate this situation, consider the family of two bivariate polynomials in the plane defined by, $P_1 = X_2, P_2 = X_1^2 + X_2^2$, then the two cells C and C' of $P_1 = 0, P_2 > 0$ consisting of two half lines on the X_1-axis are contained in the same semi-algebraically connected component defined by the family $P_1 - \delta_1, P_1 + \delta_1, P_2 - \delta_2, P_2 + \delta_2$. However, if we consider $P_1 - \delta_1, P_1 + \delta_1, P_1 - \delta\delta_1, P_1 + \delta\delta_1, P_2 - \delta_2, P_2 + \delta_2, P_2 - \delta\delta_2, P_2 + \delta\delta_2$, and use the fact that $\delta_1 \gg \delta_2 \gg \delta$, we find that the two cells, C and C', are contained in two different cells after perturbation.

In all our algorithms, we will make use of a subroutine that computes a set of points which meets every connected component of the zero set of a polynomial. In the next section we describe an algorithm for doing this, which first perturbs the polynomial so that the zero set becomes smooth and bounded, and the projection map onto the first coordinates over this set has a finite number of critical points. The following Lemma and Proposition ensures that the perturbation has the required properties.

Lemma 4 *For $Q \in R[X_1, \ldots, X_k]$ of degree d, we let*
$$Q_1 = Q^2 + (X_1^2 + \cdots + X_{k+1}^2 - \Omega^2)^2,$$
with Ω an infinitely big positive variable. Then the algebraic set $Z(Q_1) \subset (R\langle \frac{1}{\Omega} \rangle)^{k+1}$ is contained in the open ball of center 0 and radius $\Omega + 1$. Moreover, the extension of every semi-algebraically connected component of $Z(Q)$ to $R\langle \frac{1}{\Omega} \rangle$, contains the projection onto $(R\langle \frac{1}{\Omega} \rangle)^k$ of a semi-algebraically connected component of $Z(Q_1) \subset (R\langle \frac{1}{\Omega} \rangle)^{k+1}$.

Proof: The first part of the lemma is obvious from the definition of Q_1. The second part follows from the fact, that the extension of a connected component, C of $Z(Q)$, to $R\langle\frac{1}{\Omega}\rangle$, will contain at least one point, say x, that is bounded over R. Then the projection of the component of $Z(Q_1)$ containing the point, $(x, (\Omega^2 - |x|^2)^{\frac{1}{2}})$ is contained in $C_{R\langle\frac{1}{\Omega}\rangle}$. □

Given a polynomial $Q \in R[X_1, \ldots, X_k]$ define the *total degree of Q in X_i* as the maximal total degree of the monomials in Q containing the variable X_i.

Proposition 6 *For a real closed field R, let $Q \in R[X_1, \ldots, X_k]$ be a non-zero polynomial, non-negative over R^k, whose degree is bounded by d, such that $Z(Q) \subset R^k$ is contained in the open ball with center 0 and radius r, for some $r \in R$. Let $d_1, d_2 \ldots, d_k$ be the total degrees of Q in $X_1, X_2, \ldots, X_k$, respectively (and, without loss of generality, $d_1 \geq d_2 \ldots \geq d_k$). Let*

$$Q_1 = (1-\zeta)Q + \zeta(X_1^{2(d_1+1)} + \ldots + X_k^{2(d_k+1)} - k(r^{2(d_1+1)}))$$

where ζ is an infinitesimal. Then the following holds:

1. *The algebraic set $Z(Q_1) \subset (R\langle\zeta\rangle)^k$ is bounded and smooth.*

2. *The set $K \subset Z(Q_1)$ of critical points of $Z(Q_1)$ (over $\bar{R}\langle\zeta\rangle$) of the projection map onto the X_1 co-ordinate, is finite (where $\bar{R} = R[i]$, is the algebraic closure of R).*

   The polynomials,

   $$Q_1, \frac{\partial Q_1}{\partial X_2}, \ldots, \frac{\partial Q_1}{\partial X_k}$$

   whose zero set defines K, form a Gröbner basis for the degree lexicographical ordering with $X_1 > \ldots > X_k$.

3. *Let K' be the set of real critical points of this projection map (with coordinates in $R\langle\zeta\rangle$). For every connected component, C_R of the algebraic set $Z(Q)$ there exists a point $p \in K'$, such that $eval_\zeta(p)$ belongs to C_R.*

Proof: Since Q is assumed to be non-negative over R^k, any zero of Q_1 satisfies the inequality,

$$X_1^{2(d_1+1)} + \ldots + X_k^{2(d_k+1)} - k(r^{2(d_1+1)}) \leq 0,$$

which shows that the zeros of Q_1 lie inside a bounded ball with center 0.

To prove that $Z(Q_1)$ is smooth, consider the family of hypersurfaces

$$Q_{1,t} = (1-t)Q + t(X_1^{2(d_1+1)} + \ldots + X_k^{2(d_k+1)} - k(r^{2(d_1+1)})).$$

The variety $Z(Q_{1,t})$ is smooth if and only if the set of solutions to the system of equations,

$$Q_{1,t} = \frac{\partial Q_{1,t}}{\partial X_1} = \ldots = \frac{\partial Q_{1,t}}{\partial X_k} = 0$$

is empty. The set of t's for which this system has no solutions is constructible in the Zariski topology, open, and contains $t = 1$. Hence, it contains ζ which is transcendental and thus $Z(Q_2)$ is smooth.

To prove that the set of critical points (over $\bar{R}$) for the X_1 function over Z is finite, we consider the ideal defined by the polynomials

$$Q_1, \frac{\partial Q_1}{\partial X_2}, \ldots, \frac{\partial Q_1}{\partial X_k}.$$

The quotient ring for this ideal is a vector space, which, since these polynomials form a Gröbner basis of the ideal with respect to the degree lexicographical ordering with $X_1 > \ldots > X_k$ (see the next paragraph), is spanned by the power products not occurring in the ideal generated by the terms of highest degrees of the polynomials. Hence, the quotient ring is spanned by power products $X_1^{e_1} X_2^{e_2} \ldots X_k^{e_k}$, $e_1 < 2(d_1 + 1), e_i < 2d_i + 1$, $2 \leq i \leq k$. Thus, the quotient ring is a finite dimensional vector space, and hence the number of zeros of the system (over $\bar{R}\langle\zeta\rangle$) is finite.

The set of polynomials,

$$Q_1, \frac{\partial Q_1}{\partial X_2}, \ldots, \frac{\partial Q_1}{\partial X_k}.$$

form a Gröbner basis of the ideal with respect to the degree lexicographical ordering with $X_1 > \ldots > X_k$. This follows immediately from Buchberger's algorithm ([5]), and the following lemma.

We first need some notation. We are given an admissible total ordering on monomials (a total order compatible with multiplication) such that $X_1 > \ldots > X_k$. Given a polynomial P we write $l(P)$ for its leading monomial with respect to this order, $c(m, P)$ for the coefficient of the monomial m in the polynomial P. Thus $c(l(P), P)$, the leading coefficient of P, is the coefficient of the leading monomial in P. Given two polynomials P_1, P_2 with leading coefficients 1, the *S- Polynomial* of P_1, P_2 is defined as,

$$S(P_1, P_2) = m \times P_1 - n \times P_2,$$

where $m = l(P_2)/gcd(l(P_1), l(P_2))$, $n = l(P_1)/gcd(l(P_1), l(P_2))$. Given a finite set of polynomials, G, with leading coefficients 1, we say that P

is reduced to P_1 modulo G if there is a $Q \in G$ and a monomial m occurring in P such that, $l(Q)|m$, and $P_1 = P - c(m, P)\frac{m}{l(Q)}Q$. Moreover, we say that P is *reducible* to P_1 modulo G if there is a finite sequence of reductions modulo G going from P to P_1. According to Buchberger's algorithm ([5]) an ideal basis, G, is a Gröbner basis if all the S-polynomials for all pairs of polynomials in the basis G, can be reduced to zero modulo G.

Lemma 5 *If* $c(l(Q_1), Q_1) = c(l(Q_2), Q_2) = 1$ *and* $gcd(l(Q_1), l(Q_2)) = 1$ *then the S-polynomial of* Q_1 *and* Q_2, $S(Q_1, Q_2) = l(Q_2) \times Q_1 - l(Q_1) \times Q_2$ *is reducible to* 0 *modulo* Q_1 *and* Q_2.

Proof: Let $R_1 = Q_1 - l(Q_1), R_2 = Q_2 - l(Q_2)$ Then $S(Q_1, Q_2) = l(Q_2)R_1 - l(Q_1)R_2$ and there is no monomial of $l(Q_2)R_1$ appearing in $l(Q_1)R_2$ (and vice versa). This is because all monomials in R_1 and R_2 are smaller (in the given ordering) than $l(Q_1)$ and $l(Q_2)$ respectively, and $gcd(l(Q_1), l(Q_2)) = 1$..

We consecutively reduce, every monomial of $S(Q_1, Q_2)$ coming from $l(Q_2)R_1$ using Q_2 and the result is $-l(Q_1)R_2 - R_1R_2$ and the monomials in $-l(Q_1)R_2$ are distinct from the monomials in R_1R_2. Then we reduce consecutively every monomial of $-l(Q_1)R_2 - R_1R_2$ coming from $-l(Q_1)R_2$ using Q_1 and the result is 0.

Thus $S(Q_1, Q_2)$ is reducible to 0 modulo Q_1 and Q_2. □

We now prove Part 3 of the Proposition. Since the image of a bounded connected semi-algebraic set under eval_ζ is again semi-algebraically connected, it is enough to prove that every point y in $Z(Q)$ belongs to the image of $\text{eval}_\zeta(Z(Q_1))$.

Let $y \in Z(Q)$. Then $Q_1(y)$ is a strictly negative element of $R\langle\zeta\rangle$. Now since Q is everywhere non-negative, and 0 on a subset of codimension at least 1, fixing a ball B, centered at y, of radius r_1 (r_1 in R) there exists a point y' in B with $y' \in R^k$ and $Q(y') > 0$ in R. Hence $Q_1(y') > 0$ in $R\langle\zeta\rangle$ (because ζ is infinitesimal). This implies that the sign of Q_1 changes inside the ball $B_{R\langle\zeta\rangle}$ and so there is a z with coordinates in $R\langle\zeta\rangle$ such that $Q_1(z) = 0$ in every ball $B_{R\langle\zeta\rangle}$. Hence, the point of $Z(Q_1)$ with minimal distance to y is infinitesimally close to y and is sent by eval_ζ to y. □

In order to solve a zero-dimensional system given by a system of k polynomials in k variables and give its solutions as univariate rational functions we make use of the notion of a Chow polynomial. We give a brief description of a Chow polynomial of a system of k polynomials in k variables, and an algorithm to compute it.

The notion of the u-resultant of a system of k polynomials in k variables can be found in Van der Waerden ([21]), where the u-resultant is defined using the more general multivariate resultants and has been used by Canny ([6]) and Renegar ([19]) for reducing the solving of multivariate zero-dimensional systems to the solving of univariate polynomials. Here we use a similar but more direct approach with the Chow polynomial computation, following Alonso, Becker, Roy and Wormann ([1]).

In what follows, R and $\bar{R}$, will denote any real closed field and its algebraic closure respectively.

Let I be a zero-dimensional ideal i.e. an ideal of the polynomial ring, $R[X_1, \ldots, X_k]$, such that the affine variety $Z(I) = \{\alpha \in \bar{R}^k | Q(\alpha) = 0, \forall Q \in I\}$, is a finite set. Let $A = R[X_1, \ldots, X_k]/I$. A is then a finite dimensional algebra over R.

In order, to understand the structure of A, we extend the scalars, and define $\bar{A} = A \otimes_R \bar{R}$. For $\alpha = (\alpha_1, \ldots, \alpha_k) \in Z(I)$, let M_α, be the maximal ideal $(X_1 - \alpha_1, \ldots, X_k - \alpha_k)$ in $\bar{R}[X_1, \ldots, X_k]$. Let, $\bar{M}_\alpha$ be the image of M_α in $\bar{A}$ under the canonical homomorphism, $\bar{R}[X_1, \ldots, X_k] \to \bar{A}$. Then, there exists an integer l such that, $\bar{A} \cong \prod_{\alpha \in Z(I)} \bar{A}/(\bar{M}_\alpha)^l$.

Defining $J = \cap_{\alpha \in Z(I)} M_\alpha$, and using the Chinese Remainder Theorem we have,

$$\bar{R}[X_1, \ldots, X_k]/J^l = \prod_{\alpha \in Z(I)} \bar{R}[X_1, \ldots, X_k]/M_\alpha^l.$$

Noting, that $J^l \subset I$, and each $M_\alpha \supset I$, we can take quotients of both sides to obtain,

$$\bar{A} \cong \prod_{\alpha \in Z(I)} \bar{A}/(\bar{M}_\alpha)^l.$$

We denote by $\bar{A}_\alpha$ the subspace $\bar{A}/(\bar{M}_\alpha)^l$. Thus, the above isomorphism can be expressed as,

$$\bar{A} \cong \prod_{\alpha \in Z(I)} \bar{A}_\alpha.$$

We denote by μ_α the dimension of $\bar{A}_\alpha$ as a $\bar{R}$-vector space. We call μ_α, the multiplicity of the zero $\alpha \in Z(I)$.

For any element $u \in A$, we can define a linear map, $L_u : \bar{A} \to \bar{A}$, given by $L_u(v) = uv$, $\forall v \in \bar{A}$.

Let $\chi(u,t)$ be the characteristic polynomial for the linear transformation, L_u. Then,

$$\chi(u,t) = \prod_{\alpha \in Z(I)} (t - u(\alpha))^{\mu_\alpha}.$$

(cf [18] or [1]). The univariate polynomial $\chi(u,t)$ is called a *Chow polynomial*.

Let us assume that u has the property that for any two distinct zeros, $\alpha, \beta \in Z(I)$, $u(\alpha) \neq u(\beta)$. Such an element of A is called a *separating element*.

Let $Z_\mu = \{\alpha \in Z(I) | \mu_\alpha = \mu\}$. We next show how to express the zeros $\alpha \in Z_\mu$, as rational functions of the roots of $\chi(u,t)$ provided u is separating. In particular, the real points in the variety $Z(I)$, can be expressed as rational functions of the real roots of $\chi(u,t)$.

We introduce a new variable s, and consider the polynomial $\chi(u+sv,t) \in R[s,t]$, for some $v \in A$. Now, if u is a separating element in A, then so is $u + sv$ for almost all values of s.

Let $g(u,v,t) = \dfrac{\partial \chi(u+vs,t)}{\partial s}\Big|_{s=0}.$

Let us define $g(u,t) = \chi'(u,t)$. Substituting , $t = u(\alpha)$ in the expression for $g^{(\mu-1)}(u,t)$, we have,

$$g^{(\mu-1)}(u,u(\alpha)) = \mu! \prod_{\beta \in Z(I), \beta \neq \alpha} (u(\alpha) - u(\beta))^{\mu_\beta}.$$

Similarly, computing $g^{(\mu-1)}(u,v,t)$ and substituting $t = u(\alpha)$ we have,

$$g^{(\mu-1)}(u,v,u(\alpha)) = -v(\alpha)\mu! \prod_{\beta \in Z(I), \beta \neq \alpha} (u(\alpha) - u(\beta))^{e_\beta},$$

(here $g^{(i)}(u,v,t)$, $g^{(i)}(u,t)$ are the ith derivatives with respect to t of $g(u,v,t)$ and $g(u,t)$ resp.).

Thus we have proved:

Proposition 7 *Restricted to the points α in Z_μ, the following equality holds :*

$$v(\alpha) = -\frac{g^{(\mu-1)}(u,v,u(\alpha))}{g^{(\mu-1)}(u,u(\alpha))}.$$

Thus, taking $v = X_i$, for $i = 1, \ldots, k$, one can express the coordinates of the points $\alpha \in Z_\mu$, in terms of rational functions of $u(\alpha)$, which in turn are roots of the polynomial $\chi(u,t)$.

In some of our algorithms, we perturb the input polynomials, introducing infinitely large and small elements. As a result, when we construct points, their co-ordinates lie in some non-archimedean extension of R.

Now we outline a method to get back points whose co-ordinates are in R, by replacing the infinitely large (resp. small) elements by suitable sufficiently large (resp. small) elements from R.

We first give a brief review of the definition of the subresultant coefficients of two polynomials, and their properties that we utilize. All this material is classical, and can be found in any standard text on the subject (see [16]).

Given two polynomials $P = a_0 X^d + \cdots + a_d$ and $Q = b_0 X^e + \cdots + b_e$ of degrees d and e respectively, with $a_i, b_j \in R$, and a_0 and b_0 are both not zero. The resultant of P and Q is the determinant of the *Sylvester matrix* of P and Q, which is the following square matrix of size $d + e$:

$$\begin{pmatrix}
a_0 & a_1 & a_2 & \cdots & \cdot & \cdot & \cdot & a_d & 0 & \cdot & \cdots & 0 \\
0 & a_0 & a_1 & \cdots & \cdot & \cdot & \cdot & a_{d-1} & a_d & 0 & \cdots & 0 \\
\vdots & \vdots & \vdots & \vdots & \vdots & \vdots & \vdots & \vdots & \vdots & \vdots & \vdots & \vdots \\
0 & \cdot & \cdot & \cdots & 0 & a_0 & a_1 & a_2 & \cdot & \cdot & \cdots & a_d \\
b_0 & b_1 & \cdot & \cdots & \cdot & \cdot & b_e & 0 & \cdot & \cdot & \cdots & 0 \\
0 & b_0 & \cdot & \cdots & \cdot & \cdot & b_{e-1} & b_e & 0 & \cdot & \cdots & 0 \\
\vdots & \vdots & \vdots & \vdots & \vdots & \vdots & \vdots & \vdots & \vdots & \vdots & \vdots & \vdots \\
0 & \cdot & \cdot & \cdots & 0 & b_0 & b_1 & \cdot & \cdot & \cdot & \cdots & b_e
\end{pmatrix}$$

It is well known that the resultant of P and Q is zero if and only if P and Q have a common factor.

The properties of the subresultant coefficients generalize those of the resultant.

For $0 \leq j < \min(d,e)$, the jth-*subresultant coefficient* of P and Q, denoted by $\mathrm{sres}_j(P,Q))$ is the minor of size $d + e - 2j$ of the Sylvester matrix of P and Q, obtained by removing the j last lines of coefficients of P, the j last lines of coefficients of Q, and the last $2j$ columns. The resultant of P and Q is $\mathrm{sres}_0(P,Q)$.

Proposition 8 *Let l be an integer, $0 \leq l < \min(d,e)$. The GCD of the polynomials P and Q is of degree strictly greater than l if and only if $\mathrm{sres}_0(P,Q) = \cdots = \mathrm{sres}_l(P,Q) = 0$.*

Proof: We first prove that if there exist non-zero polynomials U and V, with $\deg(U) < e - l$ and $\deg(V) < d - l$, such that $\deg(UP + VQ) < l$, then $\mathrm{sres}_l(P, Q) = 0$, and conversely. The proof follows easily once we note that the existence of the polynomials U, V with the stated properties, is equivalent to the existence of non-trivial solutions of a homogenous system of equations, whose determinant is $\mathrm{sres}_l(P, Q)$.

Next note that, the fact that P and Q have a GCD of degree strictly greater than l is equivalent to the existence of polynomials U and V, with $\deg(U) < e - l$ and $\deg(V) < d - l$, such that $UP + VQ = 0$.

The lemma is true when $l = 0$, because $\mathrm{sres}_0(P, Q)$ is just the Sylvester resultant of P and Q.

Suppose $l > 0$, and, that by the induction hypothesis, it is known that P and Q have a GCD of degree greater than *or equal* to l if and only if $\mathrm{sres}_0(P, Q) = \ldots = \mathrm{sres}_{l-1}(P, Q) = 0$.

If the degree of the GCD of P and Q is strictly greater than l, then there exists non-zero polynomials U and V, with $\deg(U) < e - l$ and $\deg(V) < d - l$, such that $UP + VQ = 0$ and hence $\deg(UP + VQ) < l$, and thus $\mathrm{sres}_l(P, Q) = 0$.

Conversely, if $\mathrm{sres}_0(P, Q) = \ldots = \mathrm{sres}_l(P, Q) = 0$, then there exists non-zero U and V, with $\deg(U) < e - l$, $\deg(V) < d - l$ and $\deg(UP + VQ) < l$. Since according to the induction hypothesis the degree of the GCD of P and Q is greater than or equal to l, and since this GCD must divide $UP + VQ$, we have $UP + VQ = 0$. Hence P and Q have a GCD of degree strictly greater than l. □

Corollary 2 *The number of distinct real roots of P is determined by the signs (> 0, < 0 or $= 0$) of the subresultant coefficients of P and P'.*

Definition 1 *Given a polyomial $Q = c_q X^q + \cdots + c_p X^p, q > p, c_q, c_p \neq 0$ with coefficients in an ordered domain D, define*

$$C(Q) = 1 + \sum_{p \leq i < q} (\tfrac{c_i}{c_q})^2 \ and,$$
$$c(Q) = (1 + \sum_{p < i \leq q} (\tfrac{c_i}{c_p})^2)^{-1}.$$

Given a list of polynomials $\mathcal{L}$, define $C(\mathcal{L})$ as the maximum of $C(Q)$ for for $Q \in \mathcal{L}$ and $c(\mathcal{L})$ as the minimum of $c(Q)$ for $Q \in \mathcal{L}$.

Lemma 6 *Given a polyomial Q with coefficients in an ordered domain D, contained in a real closed field R,*

the greatest absolute value of the roots of Q is smaller than $C(Q)$, while the smallest absolute value of the non-zero roots of Q, is greater than $c(Q)$.

Proof : Let α be a root of $Q = c_q X^q = \ldots + c_p X^p$ in R. Then

$$c_q \alpha = - \sum_{p \leq i < q} c_i \alpha^{-(q-1-i)}.$$

If the absolute value of α is greater than 1, this gives

$$c_q^2 \alpha^2 \leq (\sum_{p \leq i < q} (c_i)^2).$$

Otherwise we can bound the absolute value of α by 1. In both cases the absolute value of α is bounded by $C(Q)$. The bound on $c(Q)$ is obtained by considering the polynomial, $X^q Q(\frac{1}{X})$. □

Notation 1 *Let $T_i(g)(t)$ be the polynomial obtained by dropping all the terms with degree greater than i from a polynomial $g(t)$.*

Given a list $\mathcal{L} = \{f(t), g_1(t), \ldots, g_m(t)\} \subset D[t]$, of polynomials with maximum degree d, define $\mathcal{SL}$ as the collection of all the subresultant coefficients for the following pairs of polynomials : $(T_i(f), T_j(f^{(l)})), (T_i(f), T_j(g_w)), 0 \leq i, j, l \leq d, 1 \leq w \leq m$.

Proposition 9 *Let $\mathcal{L}$ be a list of polynomials, $\{f(\Omega, t), g_1(\Omega, t), \ldots, g_m(\Omega, t) \in D[\Omega, t]\}$, with maximum degree d and $\sigma = \{\sigma_1, \ldots, \sigma_m\}$ be a sign condition such that f has a root $\bar{t} \in R\langle \frac{1}{\Omega} \rangle$ with*

$$\mathrm{sign}(g_1(\Omega, \bar{t})) = \sigma_1, \ldots, \mathrm{sign}(g_m(\Omega, \bar{t})) = \sigma_m.$$

Then, using the notation of the previous paragraphs, for any v in R, $v > C(\mathcal{SL})$,

$$\mathrm{sign}(g_1(v, \bar{t}')) = \sigma_1, \ldots, \mathrm{sign}(g_m(v, \bar{t}')) = \sigma_m,$$

where $\bar{t}'$ is the root of $f(v, t)$, having the same Thom encoding as $\bar{t}$.

Proof : If $v > C(\mathcal{SL})$, then v is greater than the absolute value of all roots of every Q in $\mathcal{LS}$. Hence, by construction, the number of roots of the polynomial $f(v, t)$, as well as the number of its common roots with the polynomials $g_1(v, t), \ldots, g_m(v, t)$ and the Thom's encoding of its roots, remain invariant for all v satisfying $v > C(\mathcal{SL})$.

Now it is clear from property 15 (section 2) of the field of Puiseux series, that $(sign(g_1(v, \bar{t}')), \ldots, sign(g_m(v, \bar{t}'))) = (sign(g_1(\Omega, \bar{t})), \ldots, sign(g_m(\Omega, \bar{t})))$. □

The following proposition is analogous to the previous one, with the difference that we deal with an infintesimal, rather than an infinitely large variable. The proof is completely analogous and hence omitted.

Proposition 10 *Let $\mathcal{L}$ be a list of polynomials, $f(\epsilon, t), g_1(\epsilon, t), \ldots, g_m(\epsilon, t) \in D[\epsilon, t]$, with maximum degree d and $\sigma = \{\sigma_1, \ldots, \sigma_m\}$ be a sign condition such that f has a root $\bar{t} \in R\langle\epsilon\rangle$ with*

$$\mathrm{sign}(g_1(\epsilon, \bar{t})) = \sigma_1, \ldots, \mathrm{sign}(g_m(\epsilon, \bar{t})) = \sigma_m.$$

Then for any v in R, $v < c(\mathcal{SL})$,

$$\mathrm{sign}(g_1(v, \bar{t}')) = \sigma_1, \ldots, \mathrm{sign}(g_m(v, \bar{t}')) = \sigma_m,$$

where $\bar{t}'$ is the root of $f(v, t)$, having the same Thom encoding as $\bar{t}$.

It is clear from the above propositions, that given a list of polynomials, $\mathcal{L} = \{f(\epsilon, t), g_1(\epsilon, t), \ldots, g_m(\epsilon, t)\}$ (resp. $\{f(\Omega, t), g_1(\Omega, t), \ldots, g_m(\Omega, t)\}$), whose degrees are bounded by d, we can compute $c(\mathcal{SL})$, (resp. $C(\mathcal{SL})$) which belong to the field of fractions of D, using no more than $md^{O(1)}$ arithmetic operations in D.

3 Algorithmic Preliminaries

Throughout the paper, a root in R, say α, of a polynomial h with coefficients in D is characterized by h and the sign vector

$$(\mathrm{sign}(h(\alpha)), \mathrm{sign}(h'(\alpha)), \ldots, \mathrm{sign}(h^{\{deg h\}-1}(\alpha))).$$

This sign vector is called the *Thom encoding* of α after Thom's lemma (see item 12 in section 2 and [10]).

Whenever we output a point $x = (x_1, \ldots, x_k)$ algebraic over a ring D contained in a real closed field R what we actually output is:

1. A polynomial f, with coefficients in D,

2. A root α in R of f, characterised by its Thom encoding,

3. $k+1$ polynomials $g_0, \ldots, g_k$ with coefficients in D such that $x_i = \dfrac{g_i(\alpha)}{g_0(\alpha)}$.

3.1 Univariate representation subroutine

The following algorithm constructs univariate representations of the zeros of a zero dimensional variety, using the theory and notations for Chow polynomials given in section 2.

We describe how to compute the polynomial $\chi(u, t)$ and how to find a separating element u .

Assume for the moment that we have a separating element u. In order to compute the characteristic polynomial $\chi(u, t)$ of the linear transformation L_u, it is enough to know a basis of the finite-dimensional vector space A and the multiplication table in A in this basis. This information is known as soon as we know a Gröbner basis for the ideal I. With a multiplication table for the finite dimensional algebra A, we can compute the characteristic polynomial $\chi(u, t)$, of the linear transformation L_u, corresponding to multiplication by u.

In order to find a separating element u, we make use of the following strategy as in [19]. We consider u to be of the following form:

$$u = u_1 X_1 + \ldots + u_k X_k.$$

We need to find the proper values for the u_i's. Let N be the dimension of the vector space A. By a simple counting argument we see that the vector $U = (u_1, \ldots, u_k)$ can be chosen from the set, $\{(1, i, i^2, \ldots, i^{k-1}) | 1 \leq i \leq k\binom{N}{2} + 1\}$ so that the polynomial u, will be separating. To see this notice that any k of the above set of vectors are linearly independent. Let u_i be the polynomial $X_1 + iX_2^2 + \ldots + i^{k-1} X_k$. For any two distinct zeros α and β, and $1 \leq i_1 < i_2 < \ldots < i_k \leq k\binom{N}{2} + 1$, it cannot be the case that,

$$u_{i_l}(\alpha) = u_{i_l}(\beta), \quad \forall l, 1 \leq l \leq k.$$

Since, the number of zeros is bounded by N, by the pigeon-hole principle, it follows that a separating u can be found from the above set.

Input A Gröbner basis of a zero- dimensional ideal, I, of polynomials in k variables.

Output A set consisting of $(k+2)$-tuples of univariate polynomials, $(f, g_0, \ldots, g_k)$ such that the complex zeroes of I are among the points obtained by evaluating the rational functions $(\dfrac{g_1}{g_0}, \ldots, \dfrac{g_k}{g_0})$ at the roots of the univariate polynomial f.

Description Let N be the dimension of the vector space A (following the above notation). The output is obtained by taking all Chow polynomials $f(u,t) = \chi(u,t)$ for all $u \in \{X_1 + iX_2 + i^2X_3 + \ldots + i^{k-1}X_k | 1 \leq i \leq k\binom{N}{2} + 1\}$ and constructing the $(k+2)$-tuples of polynomials

$$
\begin{aligned}
(f(u,t), & \\
g_0(u,t) &= g^{(\mu-1)}(u,t), \\
g_1(u,t) &= -g^{(\mu-1)}(u,X_1,t) \\
&\vdots \\
g_k(u,t) &= -g^{(\mu-1)}(u,X_k,t)),
\end{aligned}
$$

for $1 \leq \mu \leq degree(f)$.

Complexity For $O(kN^2)$ choices of the polynomial u, we produce $degree(f) = O(N)$ tuples of polynomials. Thus, the set of tuples is of size $O(kN^3)$. Moreover, the degrees of all the polynomials in the output is bounded by $O(N)$.

3.2 The cell representatives subroutine

Our algorithms rely on the *cell representatives subroutine* for constructing points in every connected component of a given algebraic set. Given a polynomial Q of total degrees $d_1,\ldots,d_k$ in $X_1,\ldots,X_k$ (with $d_1 \geq \ldots \geq d_k$), with coefficients in an ordered domain D, the subroutine computes a set of size $(d_1\ldots d_k)^{O(1)}$ having as elements, $k+2$-tuples of univariate polynomials, $(f(t),g_0(t),\ldots,g_k(t))$, with coefficients in D, where each of the polynomials f and g_i have degrees bounded by $(O(d_1)\ldots O(d_k))$. The subroutine uses $(d_1\ldots d_k)^{O(1)}$ arithmetic operations in D. The points obtained by substituting the real roots of f in $(\frac{g_1}{g_0},\ldots,\frac{g_k}{g_0})$ give at least one point in every semi-algebraically connected component of the real algebraic set defined by Q. This result is already implicit in Canny [6] and Renegar [19].

In the algorithm we will introduce two new variables Ω and ζ. In order to prove the correctness of our algorithm using Proposition 6, we will interpret Ω to be infinitely big, and ζ to be an infinitesimal with the ordering $\frac{1}{\Omega} \gg \zeta$. However, the *Cell Representatives Subroutine* itself, treats Ω and ζ as just extra variables. Since we do not perform any sign determination in this subroutine, we do not make use of the ordering in the field of coefficients.

Input A polynomial $Q \in D[X_1,\ldots,X_k]$, of total degrees $d_1,\ldots,d_k$ in $X_1,\ldots,X_k$ (with $d_1 \geq \ldots \geq d_k$) and $d \geq d_i$.

Output A set whose elements are $(k+2)$-tuples composed of a univariate polynomial f and $k+1$ polynomials $(g_0,\ldots,g_k)$. The set of points obtained by evaluating the rational functions $\frac{g_i}{g_0}$ at the real roots of the corresponding polynomial f, intersect every semi-algebraically connected component of the set $Z(Q)$

Step 0 Introduce two new variables X_{k+1} and Ω, and replace Q by

$$Q^2 + (X_1^2 + \ldots + X_k^2 + X_{k+1}^2 - \Omega^2)^2.$$

Step 1 Introduce a new variable ζ and define

$$
\begin{aligned}
Q_1 &= (1-\zeta)Q + \zeta(X_1^{2(d_1+1)} + \ldots + \\
&X_k^{2(d_k+1)} + X_{k+1}^6 - (k+1)(\Omega+1)^{2(d+1)})
\end{aligned}
$$

Step 2: Apply the univariate representation subroutine to

$$Q_1, \frac{\partial Q_1}{\partial X_2}, \ldots, \frac{\partial Q_1}{\partial X_{k+1}}$$

to get a set of $(k+3)$-tuples $(f,g_0,\ldots,g_{k+2})$. Notice that according to Proposition 6, the above set of polynomials form a Gröbner basis, G, for the ideal they generate, with the ordering being the degree lexicographical ordering. Moreover, its quotient ring is spanned by the power-products $\notin (LT(G))$, where $(LT(G))$ is the ideal generated by the leading terms (in the degree lexicographical ordering) of the polynomials in G. This is the set of all power products $X_1^{e_1}\ldots X_{k+1}^{e_{k+1}}$, with $e_1 < 2(d_1+1), e_i < 2d_i+1, 2 \leq i \leq k, e_{k+1} < 5$. Thus, the quotient ring is a finite dimensional vector space of dimension $5(2(d_1+1))(2d_2+1)\ldots(2d_k+1)$ The multiplication table for this basis is of size $(d_1\ldots d_k)^{O(1)}$, and the Chow polynomial $f(t) = \chi(u,t)$ is a polynomial in t of degree at most $(O(d_1)\ldots O(d_k))$.

Notice, that the construction of the multiplication table for the algebra A, involves reductions by polynomials in the Gröbner basis G. Moreover, the leading terms of the polynomials in G are, $\zeta X_1^{2(d_1+1)}$, $\zeta X_i^{2d_i+1}$, for $2 \leq i \leq k$, and

ζX_{k+1}^5. Thus, the product of two basis monomials in the multiplication table, will be a polynomial $\in D[\Omega, \zeta, \frac{1}{\zeta}, X_1, \ldots, X_{k+1}]$. Hence, the polynomials in the tuples produced by the univariate representation subroutine, $f, g_0, \ldots, g_k$ are in $D[\Omega, \zeta, \frac{1}{\zeta}][t]$.

Step 3: In this step, we let ζ go to 0 and retain only those points which do not go to infinity in the process. We do this purely algebraically. However, if we interpret ζ to be an infinitesimal, then this has the same effect as applying the eval_ζ map on the points (which are bounded) represented by the tuples $(f, g_0, \ldots, g_{k+1})$, produced in the previous step.

Let us describe this more precisely.

Given a non-zero polynomial $h \in D[\Omega, \zeta, \frac{1}{\zeta}, t]$ we write it as

$$h(\Omega, \zeta, \frac{1}{\zeta}, t) = \sum_{i \geq \nu(h)} h(i)(\Omega, t)\zeta^i$$

($h(\nu(h))$ being non zero), and define $h(i) = 0$ for $i < \nu(h)$. We call $\nu(h)$ the order of h with respect to ζ. Note that $\nu(f) = \nu(g_0)$. In the case where for some $i = 1, \ldots, k$, $\nu(g_i) < \nu(f)$, $\text{eval}_\zeta(f, g_0, \ldots, g_k)$ is not defined. These correspond to the points which go to infinity when $\zeta \to 0$. Otherwise we define $\text{eval}_\zeta(f, g_0, \ldots, g_{k+1})$ as

$$(f(\nu(f)), g_0(\nu(f)), \ldots, g_{k+1}(\nu(f))) .$$

Step 4: Project onto the first k-coordinates.

Step 5 For each tuple,

$$(f(\Omega, t), g_0(\Omega, t), \ldots, g_k(\Omega, t)),$$

obtained above, substitute $g_i(\Omega, t)$ for X_i in the polynomial

$$X_0^d Q(\frac{X_1}{X_0}, \ldots, \frac{X_k}{X_0}),$$

to obtain a new polynomial $\bar{Q}(\Omega, t)$. Let $\mathcal{L}$ be the list of polynomials $f(\Omega, t), \bar{Q}(\Omega, t)$, and compute $C(\mathcal{SL})$ using Proposition 9. Substituting, $C(\mathcal{SL}) + 1$ for Ω in the polynomials,

$$(f(\Omega, t), g_0(\Omega, t), \ldots, g_k(\Omega, t)),$$

and clearing denominators we obtain polynomials with coefficients in D.

3.2.1   Proof of correctness

Notice that the polynomial Q after Step 0, satisfies the conditions of Proposition 6, with $r = \Omega + 1$. Thus it follows from Proposition 6, and the correctness of the univariate representative subroutine, that the points obtained by evaluating the associated rational functions, $\frac{g_i}{g_0}$, at the real roots of the corresponding polynomials f, give at least one point in every semi-algebraically connected component of the real algebraic set defined by Q over $(R\langle\frac{1}{\Omega}\rangle)^k$. Moreover, it follows from Proposition 9, that after eliminating Ω we get a set of points lying in R^k whose co-ordinates are algebraic over D, and which intersects every connected component of $Z(Q)$. It is also clear that the algorithm described above achieves the stated bounds on the number of arithmetic operations. $\qquad\square$

3.3   Reality checking, real root characterization and sign determination

We describe below a few more subroutines which we use in our algorithms.

The input of the reality checking subroutine is a univariate polynomial f together with a list of $k+1$ univariate polynomials $g_0, \ldots g_k$ and a homogeneous polynomial $P(X_0, \ldots, X_k)$. Its output is yes if f has real roots which satisfy $P(g_0, \ldots, g_k) = 0$ and no if f has no real roots satisfying $P(g_0, \ldots, g_k) = 0$.

The input of the real root characterization subroutine is a univariate polynomial f. Its output is a list of the Thom encodings of the real roots of f.

The input of the sign determination subroutine is a univariate polynomial f together with a list of $k+1$ polynomials $g_0, \ldots g_k$ and a list of homogeneous polynomials $P_1, \ldots, P_s$ in $k+1$ variables. Its output is the Thom encoding of the real roots of f and the sign vector

$$(\text{sign}(P_1(g_0, \ldots, g_k)), \ldots, \text{sign}(P_s(g_0, \ldots, g_k)))$$

at each of these roots of f.

These three subroutines are based on Sturm-Habicht sequences ([11] or [12]) and the variation of the sign determination algorithm of [3] in [20] (see also [8] and [9]). For the reality checking subroutine one computes the number of real roots of f with $P(g_0, \ldots, g_k) = 0$. For the real root characterization subroutine we apply the sign determination algorithm to the family $\{f', \ldots, f^{\{\deg f\}-1}\}$ while for the sign determination subroutine we apply it to the family $\{f', \ldots, f^{\{\deg f\}-1}\} \cup_{1 \leq i \leq s} \{P_i(g_0, \ldots, g_k)\}$.

The complexity of the root characterization subroutine is $N^{O(1)}$ while the complexity of the sign determination subroutine is $sN^{O(1)}$ (see [20]) where N is a bound on the degrees of all the univariate polynomials in $\{f\} \cup_{1 \leq i \leq s} \{P_i(g_0, \ldots, g_k)\}$.

In all calls to these subroutines the polynomials will have degree $N = (O(d))^k$ (see Step 2 of the cell representatives subroutine).

3.4 The zero-dimensional case

We now prove the following theorem improving the complexity results in [18] and theorem 2 when $k' = 0$. Note that we are able to check if the dimension is 0 within the same complexity and that the only thing we need is that the real dimension of $\mathcal{V}$ is 0 (the dimension of the complex part could be greater).

Theorem 3 *Let $\mathcal{V} = Z(\mathcal{Q})$ be a variety where $\mathcal{Q}$ is a finite set of polynomials in $D[X_1, \ldots X_k]$ of degree at most d such that the polynomial $Q = \sum_{Q_i \in \mathcal{Q}} Q_i^2$ is already computed. There is an algorithm which checks if the real dimension of the variety $\mathcal{V}$ is 0. If so, the algorithm also outputs a univariate representation of the real points in $\mathcal{V}$ in $d^{O(k)}$ arithmetic operations in D.*

Let $\mathcal{P} = \{P_1, \ldots, P_s\}$ be a family of polynomials in k variables each of degree at most d and each with coefficients in an ordered domain D contained in a real closed field R. The signs of all the polynomials at the points of $\mathcal{V}$ can be computed in $sd^{O(k)}$ arithmetic operations in D.

Proof: In order to check whether the real variety $Z(Q)$ is zero-dimensional, we apply the *Cell Representatives Subroutine* to Q. We then use the real root characterization and the sign determination subroutines, applied to the homogeneization of Q, to retain from amongst the finite set of points obtained by the *Cell Representatives Subroutine*, only those satisfying $Q = 0$. Call this finite set of points which intersects every connected component of the real variety $Z(Q)$, K. Now, $Z(Q)$ is zero-dimensional, if and only if every point in K has a sufficiently small sphere centered around

it, which does not intersect $Z(Q)$. For every tuple, $(f(t), g_0(t), \ldots, g_k(t))$, and for every root of f that is retained above, we introduce a new polynomial,

$$
\begin{aligned}
P(X_1, \ldots, X_k, t) = \\
Q(X_1, \ldots, X_k) + f^2(t) + \\
((g_0(t)X_1 - g_1(t))^2 + \ldots + (g_0(t)X_k - g_k(t))^2 - g_0^2(t)\epsilon)^2
\end{aligned}
$$

where ϵ is a new infinitesimal. We apply the cell representative subroutine to this $(k+1)$-variate polynomial, and again using the sign determination subroutine, retain only those points which satisfy $P(X_1, \ldots, X_k, t) = 0$. Notice, that the last co-ordinate of the points so obtained, are also roots of $f(t)$. Moreover, if any of these roots correspond to a point on $Z(Q)$, then $Z(Q)$ is not zero-dimensional. Use the univariate sign determination subroutine to construct the Thom encodings of these roots, and check whether any of these roots correspond to a real point on the variety $Z(Q)$.

For the second part we just apply the sign determination subroutine. The first part uses only the cell representatives and the sign determination subroutine, and $d^{O(k)}$ calls to the sign determination subroutine. Hence, the complexity is still $d^{O(k)}$. The second part of the algorithm has complexity $sd^{O(k)}$. □

4 Proof of theorem 2:

4.1 Algorithm $\mathcal{A}$

Recall that that the variety $\mathcal{V}$, is given as the zero set of a finite family $\mathcal{Q}$ of m polynomials, whose sum of squares is Q.

The algorithm will first check whether the polynomials in $\mathcal{P}$ are in general position with respect to $\mathcal{V}$.

Step 1: For every $k' + 1$-tuple of polynomials $P_{i_1}, \ldots, P_{i_{k+1}} \in \mathcal{P}$, consider the polynomial

$$
\bar{Q} = P_{i_1}^2 + \ldots + P_{i_{k'+1}}^2 + Q + (X_1^2 + \ldots + X_k^2 + X_{k+1}^2 - \Omega^2)^2
$$

introducing a new variable X_{k+1}, and an infinitely large variable Ω in the process, and determine whether $Z(\bar{Q})$ is the empty set or not, using Steps 1, 2 and 3 of the cell representatives and the reality checking subroutine. If every such $k' + 1$-tuple defines the empty set then we know that the set of s polynomials are in general position with respect to $\mathcal{V}$. If any of these polynomials has a zero, stop.

Step 2: Enlarge the set of s polynomials $\mathcal{P}$, to the following set of $3s$ polynomials.

$$\{P_{i,e}\}_{1 \le i \le s, e \in \{0,1,-1\}} = \cup_{1 \le i \le s} \{P_i, P_i - \epsilon, P_i + \epsilon\},$$

where $P_{i,e} = P_i + e\epsilon$, $e \in \{0, 1, -1\}$ and ϵ is a new infintesimal.

Step 3: For every $\ell \le k'$-tuple of polynomials $P_{i_1,e_1}, \ldots, P_{i_\ell,e_\ell}$ let

$$\bar{Q} = P_{i_1,e_1}^2 + \ldots + P_{i_\ell,e_\ell}^2 + Q + (X_1^2 + \ldots + X_{k+1}^2 - \Omega^2)^2,$$

where X_{k+1} is a new variable, and call Steps 1 to 5 of the cell representatives subroutine with $\bar{Q}$ as input and obtain a set of $(k+2)$-tuples of univariate polynomials denoted by $(f, g_0, \ldots, g_k)$. The polynomials obtained are defined over $D[\epsilon]$.

Step 4: Use the reality checking subroutine on each tuple $(f, g_0, \ldots, g_k)$ to decide if the set $Z(\bar{Q})$ is empty. If it is not empty use the sign determinations subroutine to characterize the real roots of the univariate polynomials f and determine the signs of $\mathcal{P}^h(g_0, \ldots, g_k)$ at these roots (abusing notations and denoting by $\mathcal{P}^h$ the set of polynomials obtained by homogenizing the polynomials in $\mathcal{P}$ with respect to X_0, and multiplying by X_0 if the degree of the polynomial is odd.)

Step 5: For each univariate representation $(f, g_0, \ldots, g_k)$ retained after step 4, use proposition 10 on $\{f\} \cup_{P \in \mathcal{P}} \{\mathcal{P}^h(g_0, \ldots, g_k)\}$ to replace ϵ with an appropriately small number from the field of fractions of D, and then clear denominators to obtain polynomials with coefficients in D.

Since there is a non-trivial output only if the polynomials are in general position with respect to the variety $\mathcal{V}$, the correctness of this algorithm is immediate from Proposition 2.

4.2 Complexity analysis of Algorithm $\mathcal{A}$:

Step 1: The number of $k'+1$ tuples to be examined is $\binom{s}{k'+1} = (O(s))^{k'+1}$ The cell representatives and reality checking subroutine tests whether each of the corresponding polynomials has a root in $d^{O(k)}$ arithmetic operations.

Step 3: The total number of $\ell \le k'$-tuples to be examined is $\sum_{\ell \le k'} \binom{3s}{\ell} = (O(s))^{k'}$.

Step 4: For each of these ℓ-tuples, we test whether the corresponding polynomial Q defines the empty set

or not and if not compute a point in each of its connected components. For each subset of size ℓ for which the corresponding polynomial has real zeros, we produce at most $(O(d))^k$ points. So the total number of points constructed is $s^{k'}(O(d))^k$. Each use of the sign determination subroutine is done with complexity $sd^{O(k)}$.

Step 5: For each of the $s^{k'}d^{O(k)}$ univariate representations generated, replacing ϵ by an appropriately small number requires $sd^{O(k)}$ arithmetic operations in D by proposition 10.

Thus the complexity of algorithm $\mathcal{A}$ is $s^{k'}sd^{O(k)}$. Note that the number of points constructed is $s^{k'}O(d)^k$.

4.3 Algorithm $\mathcal{B}$

We now describe algorithm $\mathcal{B}$, for the case when the variety $\mathcal{V}$ is given as the common zeros of $k - k'$ transversal polynomials, $Q_1, \ldots, Q_{k-k'}$.

We remark that though it is not part of the algorithm, this condition can be checked in time $((k - k')d)^{O(k)}$ as follows. Let M be the jacobian matrix of the polynomial map $R^k \to R^{k-k'}$ given by, $x \mapsto (Q_1(x), \ldots, Q_{k-k'}(x))$. Let,

$$Q = det(MM^T) + Q_1^2 + \ldots + Q_{k-k'}^2 + (X_1^2 + \ldots + X_{k+1}^2 - \Omega^2)^2,$$

where X_{k+1} is a new variable, and determine whether $Z(Q)$ is the empty set or not, using Steps 1, 2 and 3 of the cell representatives and reality checking subroutines. If $Z(Q)$ is non-empty, then the polynomials $Q_1, \ldots, Q_{k-k'}$ are not transversal.

Step 1: Introduce two new infinitesimals δ and δ' and replace the set $\mathcal{P}$ by the set $\bar{\mathcal{P}}$ of $4s$ polynomials $\cup_{i=1,\ldots,s}\{(1-\delta)P_i - \delta H_i, (1-\delta)P_i + \delta H_i, (1-\delta)P_i - \delta'\delta H_i, (1-\delta)P_i + \delta'\delta H_i\}$ where $H_i = (1 + \sum_{1 \le j \le k} i^j X_j^{d'})$ where d' is the least even number greater than d.

Step 2: Introduce a new infinitesimal δ'' and replace the set Q by the set $\bar{Q}$ of polynomials $\cup_{1 \le i \le k-k'}\{(1-\delta)(Q_i)^2 - \delta''\delta'\delta H_{i+s}\}$ where H_i and d' are defined as in Step 1.

Step 3: Apply steps 2 and 3 of algorithm $\mathcal{A}$ and apply the map $eval_{\delta'}$ sending $\delta', \delta'', \epsilon$ to 0, and keeping δ to obtain a set of tuples $eval_{\delta'}(f, g_0, \ldots, g_k)$ defined over $D[\delta]$. We describe the map eval$_{\delta'}$ more precisely. Given a non zero polynomial $h \in D[\delta, \delta', \delta'', \epsilon, t]$ we write it as $\sum_{i \geq \nu(h)} h(i)(\delta, t)\delta'^{i_1}\delta''^{i_2}\epsilon^{i_3}$ where the multi-index $i = (i_1, i_2, i_3)$ is ordered lexicographically and $h(i) \in D[\delta, t]$, $h(\nu(h))$ being non-zero, and define $h(i) = 0$ for $i < \nu(h)$. We call $\nu(h)$ the order of h with respect to $\delta', \delta'', \epsilon$. Note that $\nu(f) = \nu(g_0)$. In the case where for some $i = 1, \ldots, k$, $\nu(g_i) < \nu(f)$, eval$_{\delta'}(f, g_0, \ldots, g_{k+1})$ is not defined. Otherwise we define

$$\text{eval}_{\delta'} \quad (f, g_0, \ldots, g_{k+1}) = $$
$$(f(\nu(f)), g_0(\nu(f)), \ldots, g_{k+1}(\nu(f))).$$

Step 4: Use the sign determination subroutine to characterize the real roots of the univariate polynomials f using Thom's lemma and determine the signs of $\mathcal{P}^h(g_0, \ldots, g_k)$ at these roots , where $\mathcal{P}^h$ is the same as in Algorithm $\mathcal{A}$.

Step 5: For each univariate representation $(f, g_0, \ldots, g_k)$ retained after step 3, use proposition 10 on $\{f\} \cup_{P \in \mathcal{P}} \{\mathcal{P}^h(g_0, \ldots, g_k)\}$ to replace δ with an appropriately small number from the field of fractions of D, and then clear denominators to obtain polynomials with coefficients in D.

4.4 Proof of the correctness of Algorithm $\mathcal{B}$

We first note that the perturbed polynomials are in general position with respect to each other. This is an easy consequence of lemma 2.

The correctness of the algorithm is now a consequence of Propositions 2 and 3.

Notice that when $k = k'$, . the variety $\mathcal{V}$ is the whole affine space R^k. In this case, the algorithm is simpler, since we do not need to introduce the additional infinitesimal δ''.

$\square$

4.5 Complexity analysis of Algorithm $\mathcal{B}$:

Step 3: Since, we apply steps 2 and 3 of algorithm $\mathcal{A}$, with $4s$ polynomials in the input (other than the ones defining the variety), this step uses no more than $s^{k'}d^{O(k)}$ arithmetic operations. Moreover, the number of points constructed is bounded by $s^{k'}(O(d))^k$.

Step 4: Each call to the sign determination subroutine costs $sd^{O(k)}$ and hence the total cost for this step is bounded by $s^{k'}sd^{O(k)}$.

Thus the total complexity of algorithm $\mathcal{B}$ is again $s^{k'}sd^{O(k)}$. Note that the number of points constructed is again $s^{k'}O(d)^k$.

Since it is easy to produce $(sd)^{k'}$ cells with s polynomials of degree d, it follows that our algorithms $\mathcal{A}$ and $\mathcal{B}$ are nearly optimal in the sense that we are using not many more steps than the potential size of our output.

4.6 Algorithm $\mathcal{C}$:

Now we describe Algorithm $\mathcal{C}$ which works in the most general case, with no assumptions on the polynomials in $\mathcal{P}$ or $\mathcal{Q}$.

In this algorithm we perturb the polynomials of $\mathcal{P}$ so that the perturbed polynomials are in general position and then we use algorithm $\mathcal{A}$. To accomplish this we add new variables, with the ordering $\frac{1}{\Omega} \gg \delta_1 \gg \ldots \delta_s \gg \epsilon \gg \delta \gg \epsilon \gg \zeta$ and consider a real closed field R' containing the ordered ring $D[\delta_1, \ldots, \delta_s]$. In this algorithm, unlike algorithms $\mathcal{A}$ and $\mathcal{B}$, the coordinates of the points that we output belong to R'. A formal description of algorithm $\mathcal{C}$ follows.

Step 1: Replace each polynomial P_i by the four polynomials $P_i - \delta_i, P_i + \delta_i, P_i - \delta\delta_i, P_i + \delta\delta_i$.

Step 2: Apply steps 2 and 3 of algorithm $\mathcal{A}$ to the $4s$ polynomials produced by Step 1, and obtain $s^{k'}d^{O(k)}$ univariate representations defined over $D[\delta_1, \ldots, \delta_s, \delta, \epsilon]$. Then use the eval$_\delta$ map sending δ, ϵ to 0 and keeping $\delta_1, \ldots, \delta_s$, on the set of univariate representations constructed above. The details of applying the eval$_\delta$ map are similar to the ones described previously.

Step 3: Use the sign determination subroutine to characterize the the real roots of the univariate polynomials f using Thom's lemma and determine the signs of $\mathcal{P}(g_0, \ldots, g_k)$ at these roots (abusing notation and denoting by $\mathcal{P}$ the homogenization of $\mathcal{P}$ with respect to the smallest possible even power of X_0).

4.7 Proof of the correctness of algorithm $\mathcal{C}$:

It is clear that the δ_i and δ are algebraically independent transcendentals over R and thus by Proposition 3, these polynomials are in general position with respect to the variety $\mathcal{V}$. Once again the correctness of the algorithm follows from Propositions 2, 4 and 5. $\square$

4.8   Complexity analysis of algorithm $\mathcal{C}$

Since in algorithm $\mathcal{A}$ we never deal simultaneously with more than k' polynomials from the family of polynomials $\mathcal{P}$, and we are perturbing only the polynomials in $\mathcal{P}$, we never deal with more than $k' + 4$ infinitesimals ($\frac{1}{\Omega}, \delta, \epsilon, \zeta$ and at most k' among $\delta_1, \ldots, \delta_s$) in any call to the various subroutines.

Hence, the complexity of the algebraic part will be $d^{O(k)}$ arithmetic operations in a polynomial extension of D with $k' + 4$ variables and because of degree bounds in these variables coming from the fact that all our subroutines are based on linear algebra, $d^{O(kk')}$ arithmetic operations in D. Hence the total complexity up to Step 3 is $s^{k'} s d^{O(kk')}$.

Moreover, it is clear from the algorithm that the number of points it produces is bounded by $s^{k'}(O(d))^k$.

The points that we obtain have co-ordinates in a non-archimedean extension of R. We remark that it is possible to eliminate the infinitesimals $\delta_1, \ldots, \delta_s$ using refinements of the techniques used in proposition 10 at an extra cost of $(s + d^k)^{k'}(d^k)^{O(k')}$ for each tuple generated by the algorithm.

4.9   Proof of theorem 1:

Each call to the cell representatives subroutine produces at most $(O(d))^k$ points. So the total number of points constructed in Algorithm $\mathcal{C}$ will be $s^{k'}(O(d))^k$, and hence this quantity also bounds the number of cells defined by s polynomials of degree d in k variables on a variety of real dimension k' (see [17,23]) which proves Theorem 1. $\square$

References

[1] M. E. ALONSO, E. BECKER, M.-F. ROY. T. WORMANN *Zeros, Multiplicities; and Idempotents for Zero Dimensional Systems.* to appear in MEGA 94.

[2] S. BASU, R. POLLACK, M.-F. ROY *A New Algorithm to Find a Point in Every Cell Defined by a Family of Polynomials* in "Quantifier Elimination and Cylindrical Algebraic Decomposition", B. Caviness and J. Johnson Eds., Springer- -Verlag, to appear.

[3] M. BEN-OR, D. KOZEN , J. REIF The complexity of elementary algebra and geometry. *J. of Computer and Systems Sciences*, 32:251– 264, (1986).

[4] J. BOCHNAK, M. COSTE, M.-F. ROY *Géométrie algébrique réelle.* Springer-Verlag (1987).

[5] B. BUCHBERGER *Gröbner bases: an algorithmic method in polynomial ideal theory.* Recent trends in multidimensional systems theroy, Reider ed . Bose, (1985).

[6] J. CANNY *Some Algebraic and Geometric Computations in PSPACE*, Proc. Twentieth ACM Symp. on Theory of Computing, 460-467, (1988).

[7] J. CANNY *Some Practical Tools for Algebraic Geometry*, Technical report in Spring school on robot motion planning, PROMOTION ESPRIT, (1993).

[8] J. CANNY *An improved Sign Determination Algorithm*, AAECC (1991), New Orleans.

[9] J. CANNY, *Improved algorithms for sign determination and existential quantifier elimination*, the Computer Journal, 36:409–418, (1993).

[10] M. COSTE, M.-F. ROY *Thom's lemma, the coding of real algebraic numbers and the topology of semi-algebraic sets.* J. of Symbolic Computation 5 121-129 (1988).

[11] L. GONZALEZ, H. LOMBARDI, T. RECIO, M.-F. ROY *Sturm-Habicht sequence.* ISSAC'89 Portland, ACM Press136-145 (1989) .

[12] L. GONZALEZ, H. LOMBARDI, T. RECIO, M.-F. ROY *Sturm-Habicht,determinants and real roots of univariate polynomials.* Quantifier Elimination and Cylrindical Algebraic Decomposition, Texts and Mongraphs in Symbolic Computation, B. Caviness

[13] D. GRIGOR'EV, N. VOROBJOV *Solving Systems of Polynomial Inequalities in Subexponential Time*, J. Symbolic Comput., 5:37–64, (1988).

[14] J. HEINTZ, M.-F. ROY, P. SOLERNÒ *Sur la complexité du principe de Tarski-Seidenberg.* Bull. Soc. Math. France 118 101-126 (1990).

[15] J. HEINTZ, M.-F. ROY, AND P. SOLERNÓ *Single Exponential Path Finding in Semi-Algebraic sets: Part II: The general case*, .

[16] M. MIGNOTTE Mathematics for Computer Algebra Springer Verlag, New York, 1992.

[17] R. POLLACK, M.-F. ROY *On the number of cells defined by a set of polynomials*, C. R. Acad. Sci. Paris, 316:573–577, (1993).

[18] P. PEDERSEN, M.-F. ROY, A. SZPIRGLAS *Counting real zeroes in the multivariate case.* , Computational algebraic geometry, Eyssette et Galligo ed. Progress in Mathematics 109 , 203-224, Birkhauser (1993).

[19] J. RENEGAR *On the computational complexity and geometry of the first order theory of the reals*, J. of Symbolic Comput., (1992).

[20] M.-F. ROY, A. SZPIRGLAS *Complexity of computations with real algebraic numbers*, J. of Symbolic Comput., 10:39–51, (1990).

[21] B. L. VAN DER WAERDEN *Modern Algebra, Volume II*, F. Ungar Publishing Co. (1950).

[22] WALKER *Algebraic curves* Princeton University Press (1950).

[23] H. E. WARREN *Lower bounds for approximation of nonlinear manifolds.* Trans. Amer. Math. Soc., 133:167–178, (1968).

Author Index